Microbial
Food Safety
AND
Preservation
Techniques

Microbial Food Safety
AND
Preservation Techniques

EDITED BY
V. Ravishankar Rai
Jamuna A. Bai

CRC Press
Taylor & Francis Group
Boca Raton   London   New York

CRC Press is an imprint of the
Taylor & Francis Group, an **informa** business

CRC Press
Taylor & Francis Group
6000 Broken Sound Parkway NW, Suite 300
Boca Raton, FL 33487-2742

First issued in paperback 2016

Version Date: 20140728

ISBN 13: 978-1-138-03380-1 (pbk)
ISBN 13: 978-1-4665-9306-0 (hbk)

Library of Congress Cataloging-in-Publication Data

Microbial food safety and preservation techniques / editors, V. Ravishankar Rai, Jamuna A. Bai.
  p. ; cm.
  Includes bibliographical references and index.
  ISBN 978-1-4665-9306-0 (hardcover : alk. paper)
  I. Rai, V. Ravishankar, editor. II. Bai, Jamuna A. (Jamuna Aswathanarayn), editor.
  [DNLM: 1. Food Microbiology. 2. Food Safety. 3. Food Handling. 4. Food Preservation--methods.
WA 695]

  RA601
  363.19'26--dc23   2014027997

Contents

Section I   Microbial Food Safety and Hygiene

Section II   Detection of Food-Borne Pathogens

Section III Food Preservation and Intervention Techniques

Section IV Modeling Microbial Growth in Food

Preface

This book provides a comprehensive coverage of the fundamental and applied aspects of food safety, describes the control measures employed, and explores the advances in microbial food safety. It is divided into four sections. Section I, Microbial Food Safety and Hygiene, covers the hazards caused by food-borne pathogens and assesses the microbiological risk of raw, fresh produce and ready-to-eat (RTE), minimally processed and processed foods. Section II, Detection of Food-Borne Pathogens, deals with the detection of pathogens using advanced molecular techniques, biosensors, and nanotechnology. Section III, Food Preservation and Intervention Techniques, provides a detailed discussion on the various intervention and preservative techniques that are used to ensure high-quality and safe foods. The topics covered include smart/intelligent and active packaging techniques, hurdle technology, plasma technology, nanotechnology, use of natural flora belonging to lactic acid bacteria, and antimicrobials such as phytochemicals and essential oils. Novel food preservatives based on quorum sensing inhibitors are also addressed. Section IV, Modeling Microbial Growth in Food, comprises chapters on modeling microbial growth in food for enhancing the safety and quality of foods.

In recent years, rapid strides have been made in the fields of microbiological aspects of food safety and quality, predictive microbiology and microbial risk assessment, microbiological aspects of food preservation, and novel preservation techniques. Hence, this book comes as a timely guide and summarizes the latest advances and developments in these fields. All the contributing authors are international experts in their research fields. Therefore, the book will be an invaluable resource for graduate students, researchers, and professionals involved in food safety, hygiene, and quality control.

We acknowledge all the contributors for sharing their knowledge and expertise. We also thank the publisher for encouragement and technical support in publishing this book.

V. Ravishankar Rai
Jamuna A. Bai

MATLAB® is a registered trademark of The MathWorks, Inc. For product information, please contact:

The MathWorks, Inc.
3 Apple Hill Drive
Natick, MA 01760-2098 USA
Tel: 508-647-7000
Fax: 508-647-7001
E-mail: info@mathworks.com
Web: www.mathworks.com

Editors

V. Ravishankar Rai earned his MSc and PhD from the University of Mysore, India. He is currently a professor in the Department of Studies in Microbiology, University of Mysore, Mysore, India. He has been awarded a fellowship from the UNECSO Biotechnology Action Council, Paris (1996); an Indo-Israel Cultural Exchange Fellowship (1998); Biotechnology Overseas Fellowship, Government of India (2008); and Indo-Hungarian Exchange Fellowship (2011). He was invited by Academia Sinica, Taiwan, as a visiting fellow in 2010. Dr. Rai has recently edited a book entitled *Biotechnology: Concepts and Applications* published by Alpha Science International, Oxford, UK. His research interests include food biotechnology, bacterial quorum sensing, microbiological corrosion, bioprospecting of medicinal plants, and forest diseases and management. He is currently the coordinator for the innovative Program on Food Quality and Safety, funded by the University Grants Commission, New Delhi.

Jamuna A. Bai is an Indian Council of Medical Research senior research fellow with the Department of Studies in Microbiology, University of Mysore. She has authored four research papers, a review article, and four book chapters. She is currently conducting research on the role of quorum sensing in food-borne bacteria for regulating the expression of spoilage phenotypes and the production of virulence factors. Her research interests include studying quorum sensing and biofilms in food-related bacteria, developing quorum-sensing inhibitors, and investigating antimicrobial and anti-quorum-sensing activities of phytochemicals.

Contributors

María J. Andrade
Faculty of Veterinary Science
University of Extremadura
Cáceres, Spain

Miguel A. Asensio
Faculty of Veterinary Sciences
University of Extremadura
Cáceres, Spain

Jamuna A. Bai
Department of Studies in Microbiology
University of Mysore
Mysore, India

Augusto Bellomio
Facultad de Bioquímica, Química y Farmacia
Instituto Superior de Investigaciones Biológicas
and
Instituto de Química Biológica
    "Dr. Bernabé Bloj"
Universidad Nacional de Tucumán
San Miguel de Tucumán, Argentina

Elena Bermúdez
Faculty of Veterinary Sciences
University of Extremadura
Cáceres, Spain

Debabrata Biswas
Department of Animal and Avian Sciences
and
Department of Molecular and Cellular Biology
and
Center for Food Safety and Security Systems
University of Maryland
College Park, Maryland

Vasiliki A. Blana
Laboratory of Microbiology and Biotechnology
    of Foods
Department of Food Science and Human
    Nutrition
Agricultural University of Athens
Athens, Greece

Barry Byrne
School of Biotechnology
and
National Centre for Sensor Research
and
Biomedical Diagnostics Institute
Dublin City University
Dublin, Ireland

M.D. Calzada
Responsable del Laboratorio de Innovación en
    Plasmas
Departamento de Física
Universidad de Córdoba
Córdoba, Spain

Carmen A. Campos
Industry Department
Faculty of Exact and Natural Sciences
University of Buenos Aires
and
Scientific and Technical Research Council of
    Argentina
Buenos Aires, Argentina

Marcela P. Castro
Scientific and Technical National Research
    Council of Argentina
Austral Chaco University
Buenos Aires, Argentina

Juan J. Córdoba
Faculty of Veterinary Science
University of Extremadura
Cáceres, Spain

Philip G. Crandall
Department of Food Science
and
Center for Food Safety
University of Arkansas
Fayetteville, Arkansas

Manuela Curcio
Department of Pharmacy, Health and Nutrional
    Sciences
University of Calabria
Rende, Italy

Paulo Ricardo Santos da Silva
Laboratory of Technology and Food Processing
    Engineering
Department of Chemical Engineering
Federal University of Rio Grande do Sul
Porto Alegre, Brazil

Josué Delgado
Faculty of Veterinary Sciences
University of Extremadura
Cáceres, Spain

Fernando Dupuy
Facultad de Bioquímica, Química y Farmacia
Instituto de Química Biológica
    "Dr. Bernabé Bloj"
Universidad Nacional de Tucumán
San Miguel de Tucumán, Argentina

Gary A. Dykes
School of Science
Monash University
Selangor, Malaysia

Steve H. Flint
Institute of Food, Nutrition and Human Health
Massey University
Palmerston North, New Zealand

Alonzo A. Gabriel
Department of Food Science and Nutrition
College of Home Economics
University of the Philippines Diliman
Quezon City, Philippines

Niamh Gilmartin
School of Biotechnology
and
National Centre for Sensor Research
and
Biomedical Diagnostics Institute
Dublin City University
Dublin, Ireland

María F. Gliemmo
Industry Department
Faculty of Exact and Natural Sciences
University of Buenos Aires
and
Scientific and Technical National Research
    Council of Argentina
Buenos Aires, Argentina

Michael G. Johnson
Department of Food Science
and
Center for Food Safety
University of Arkansas
Fayetteville, Arkansas

Anastasia E. Kapetanakou
Laboratory of Microbiology and Biotechnology
    of Foods
Department of Food Science and Human
    Nutrition
Agricultural University of Athens
Athens, Greece

Salih Karasu
Faculty of Chemical and Metallurgical
    Engineering
Department of Food Engineering
Yıldız Technical University
Istanbul, Turkey

Haider Khan
National Agricultural Research Centre
Animal Sciences Institute
Islamabad, Pakistan

Robert E. Levin
Department of Food Science
University of Massachusetts
Amherst, Massachusetts

Stavros G. Manios
Laboratory of Microbiology and Biotechnology
    of Foods
Department of Food Science and Human
    Nutrition
Agricultural University of Athens
Athens, Greece

Sara R. Milillo
Department of Food Science
Pennsylvania State University
University Park, Pennsylvania

Carlos Javier Minahk
Facultad de Bioquímica, Química y Farmacia
Instituto Superior de Investigaciones Biológicas
and
Instituto de Química Biológica
   "Dr. Bernabé Bloj"
Universidad Nacional de Tucumán
San Miguel de Tucumán, Argentina

Lourdes Moyano
Grupo de Viticultura y Enología
Departamento de Química Agrícola y Edafología
Universidad de Córdoba
Córdoba, Spain

Azlin Mustapha
Food Science Program
Division of Food Systems and Bioengineering
University of Missouri
Columbia, Missouri

Félix Núñez
Faculty of Veterinary Sciences
University of Extremadura
Cáceres, Spain

George-John E. Nychas
Laboratory of Microbiology and Biotechnology
   of Foods
Department of Food Science and Human
   Nutrition
Agricultural University of Athens
Athens, Greece

Corliss A. O'Bryan
Department of Food Science
and
Center for Food Safety
University of Arkansas
Fayetteville, Arkansas

Semih Otles
Department of Food Engineering
Ege University of Izmir
Izmir, Turkey

İsmet Öztürk
Engineering Faculty
Department of Food Engineering
Erciyes University
Kayseri, Turkey

Ortensia Ilaria Parisi
Department of Pharmacy, Health and Nutrional
   Sciences
University of Calabria
Rende, Italy

Vandana B. Patravale
Department of Pharmaceutical Sciences and
   Technology
Institute of Chemical Technology
Mumbai, India

Mengfei Peng
Department of Animal and Avian Sciences
and
Department of Molecular and Cellular Biology
University of Maryland
College Park, Maryland

Fernando Pérez-Rodríguez
Department of Food Science and Technology
University of Cordoba
Córdoba, Spain

Nevio Picci
Department of Pharmacy, Health and Nutrional
   Sciences
University of Calabria
Rende, Italy

Sofia Poimenidou
Laboratory of Microbiology and Biotechnology
   of Foods
Department of Food Science and Human
   Nutrition
Agricultural University of Athens
Athens, Greece

G.D. Posada-Izquierdo
Department of Food Science and Technology
University of Cordoba
Córdoba, Spain

Francesco Puoci
Department of Pharmacy and Health and
   Nutrional Sciences
University of Calabria
Rende, Italy

V. Ravishankar Rai
Department of Studies in Microbiology
University of Mysore
Mysore, India

Andreja Rajkovic
Faculty of Agriculture
Department of Food Safety and Food Quality
Institute for Food Technology and Biochemistry
University of Belgrade
Belgrade, Serbia

Donatella Restuccia
Department of Pharmacy, Health and Nutrional
   Sciences
University of Calabria
Rende, Italy

Steven C. Ricke
Department of Food Science
and
Center for Food Safety
University of Arkansas
Fayetteville, Arkansas

Rocío Rincón
Laboratorio de Innovación en Plasmas
Departamento de Física
Universidad de Córdoba
Córdoba, Spain

Alicia Rodríguez
Faculty of Veterinary Science
University of Extremadura
Cáceres, Spain

Mar Rodríguez
Faculty of Veterinary Science
University of Extremadura
Cáceres, Spain

Osman Sagdic
Faculty of Chemical and Metallurgical
   Engineering
Department of Food Engineering
Yıldız Technical University
Istanbul, Turkey

Serajus Salaheen
Department of Animal and Avian Sciences
University of Maryland
College Park, Maryland

Prashant Singh
Food Science Program
Division of Food Systems and Bioengineering
University of Missouri
Columbia, Missouri

Panagiotis N. Skandamis
Laboratory of Microbiology and Biotechnology
   of Foods
Department of Food Science and Human
   Nutrition
Agricultural University of Athens
Athens, Greece

Nada Smigic
Faculty of Agriculture
Department of Food Safety and Food Quality
Institute for Food Technology and Biochemistry
University of Belgrade
Belgrade, Serbia

Umile Gianfranco Spizzirri
Department of Pharmacy, Health and Nutrional
   Sciences
University of Calabria
Rende, Italy

Yasmina Sultanbawa
Queensland Alliance for Agriculture and Food
   Innovation
Centre for Nutrition and Food Sciences
University of Queensland
Brisbane, Queensland, Australia

Fatih Tornuk
Faculty of Chemical and Metallurgical
   Engineering
Department of Food Engineering
Yıldız Technical University
Istanbul, Turkey

Subramanian Viswanathan
Department of Industrial Chemistry
School of Chemical Sciences
Alagappa University
Karaikudi, India

Swati Vyas
Department of Pharmaceutical Sciences and
    Technology
Institute of Chemical Technology
Mumbai, India

Yi Wang
School of Science
Monash University
Ehsan, Malaysia

Erin S. Whaley
Danisco USA, Inc.
New Century, Kansas

Buket Yalcin
Department of Food Engineering
Ege University of Izmir
Izmir, Turkey

Mustafa T. Yilmaz
Faculty of Chemical and Metallurgical
    Engineering
Department of Food Engineering
Yıldız Technical University
Istanbul, Turkey

Pak-Lam Yu
School of Engineering and Advanced Technology
College of Sciences
Massey University
Palmerston North, New Zealand

Cristina Yubero
Departamento de Física
Universidad de Córdoba
Córdoba, Spain

Luis Zea
Grupo de Viticultura y Enología
Departamento de Química Agrícola y Edafología
Universidad de Córdoba
Córdoba, Spain

Evangelia Zilelidou
Laboratory of Microbiology and Biotechnology
    of Foods
Department of Food Science and Human
    Nutrition
Agricultural University of Athens
Athens, Greece

Marija Zunabovic
Department of Food Science and Technology
Institute of Food Science
University of Natural Resources and Life
    Sciences
Vienna, Austria

G. Zurera
Department of Food Science and Technology
University of Cordoba
Córdoba, Spain

Section I

Microbial Food Safety and Hygiene

1

Microbiological Risk Assessment of Raw, Fresh Produce

Vasiliki A. Blana and George-John E. Nychas

CONTENTS

1.1 Introduction

The production, sale, and consumption of fresh fruits and vegetables have increased rapidly in the last two decades. Raw, fresh produce, especially vegetables, available as pre-cut and ready-to-eat salads are recognized as important sources of nutrients, vitamins, and fiber for humans. Their beneficial health effects as well as the changes in peoples' lifestyle have increased the demand for fresh produce, which in turn has increased imports of such products from various producing countries and regions to those of increased consumption. These procedures encourage introduction and dissemination of hazards from regions of production into other wider geographical areas (Nguyen-the and Carlin, 1994; De Roever, 1998). The potential for foodborne pathogen contamination is apparent in the different aspects of fruit and vegetable production practices. Contamination of raw, fresh produce with either spoilage or pathogenic microorganisms can occur by cultivation, handling, and processing procedures (Tournas, 2005). Indeed, these products have been implicated in numerous foodborne disease outbreaks and identified as sources of foodborne pathogens (Sewell and Farber, 2001; Sivapalasingam et al., 2004). The most well-documented pathogens involved in human disease associated with the consumption of fresh produce are the psychrotrophic *Listeria monocytogenes* and the mesophilic *Salmonella* and *Escherichia coli* O157:H7. In 2011 in the European Union, vegetables were implicated in 37 outbreaks according to the scientific report of EFSA and ECDC (EFSA, 2013a). Among the major causative agents were *Salmonella* (21.6%) and pathogenic *E. coli* (18.9%). Five outbreaks were related to sprouts, including the enterohemorrhagic *E. coli* (EHEC) outbreak that occurred in Germany and the linked outbreaks in Denmark, France, and the Netherlands due to fenugreek sprouts or seeds, and a S. Newport outbreak related to bean sprouts (EFSA, 2013a). Thus, the implementation of procedures that will highlight the hazards and identify the control measures during food production, processing, and distribution at the retail level and through commercial/domestic preparation should be considered in detail. Special emphasis should be placed on the detection methods used to identify the pathogenic hazards and on the preservative interventions required for pathogenic control.

The nature of fresh fruit and vegetable production and the consumption practices lead to a wide range of potential hazards. Risk assessments for fresh fruits and vegetables should therefore ideally cover multiple hazards/multiple commodities; however, because of the complexity of such a task and possibly the visibility of the disease burden, there is a lack of published risk assessments in the public domain to date, for these commodities.

1.2 Risk Assessment

Risk assessment for bacterial pathogens in raw, fresh produce may consist of the following elements:

- *Hazard identification*: The purpose of hazard identification is to identify all the pathogenic agents capable of causing adverse health effects in humans through their presence in fruits and/or vegetables. The hazards can be identified from relevant data sources e.g., scientific literature, government agencies, and international organizations.

- *Exposure assessment*: This refers to the qualitative and/or quantitative evaluation of the likely intake of pathogenic agents through fruits and/or vegetables, as well as exposures to other related sources.

- *Hazard characterization*: This refers to the qualitative and/or quantitative evaluation of the nature of the adverse health effects associated with the above identified hazards. This step requires a dose–response assessment, which may be feasible for chemical agents but is more difficult for biological as well as physical agents.

- *Risk characterization*: This is the process of determining the qualitative and/or quantitative estimation of the probability of occurrence and the severity of illness or adverse health effects in a given population following exposure to toxic doses of identified hazards. Risk characterization involves the combination of the hazard identification, exposure assessment, and hazard characterization to develop a risk estimate (output) (Codex Alimentarius Commission, 1999).

1.3 Hazard Identification

Raw, fresh produce can be contaminated with pathogenic organisms in a wide variety of ways. Contamination can occur during production, harvest, processing, at the retail outlets, in foodservice establishments, and in the home. The animal manure used as natural fertilizer in agriculture is a major source of contamination for fresh fruits and vegetables (FAO, 2004). Preharvest sources of contamination also include soil, compost, irrigation water, water used for pesticide application, insects, wild and domestic animals, and air (Neetoo et al., 2012). Most of these sources supply nutrients enhancing growth and crop yields, while all of them contain a diverse microbial flora, including human pathogens. Enteric pathogens can be spread in soil and growing crops through the use of animal waste of fecal origin as fertilizer in the field, a common agricultural practice that can contaminate fruits and vegetables and subsequently carry enteric pathogens through production to reach consumers. Many cases of direct or indirect contamination of food with *E. coli* O157:H7, *Salmonella*, *L. monocytogenes*, and other pathogens through manure, slurries, composts, soil, and irrigation water have been reported in the literature (Solomon et al., 2002; Vernozy-Rozand et al., 2002; Duffy, 2003; Johnson et al., 2003; Islam et al., 2004).

Special attention should also be paid to human hygiene, including proper and frequent hand washing, to reduce or eliminate contamination by fecal pathogens along the food chain.

1.4 Exposure Assessment

Treatments during harvest and postharvest management are considered the main stages during which pathogens gain access to the internal tissues of fruits and vegetables. The last few years foodborne outbreaks linked to fresh fruits and vegetables have been reported extensively by the European Food Safety Authority (EFSA), the US Centers for Disease Control and Prevention (CDC) and others. The CDC after an update on the reported cases of yearly foodborne illness concluded that approximately 48 million persons acquire foodborne illnesses with 128,000 hospitalizations and 3,000 deaths each year (Scallan et al., 2011). Furthermore, Scharff (2010) estimated that fresh produce causes 22 million illnesses every year in the United States. Outbreaks of *E. coli* O157 infections linked to consumption of leafy green vegetables have been increasingly identified (Hilborn et al., 1999; Friesema et al., 2008; Grant et al., 2008; Wendel et al., 2009). In particular, in 2011, a total of 9485 vero toxin-producing *E. coli* infections were reported in the European Union as associated with sprouts consumption, resulting in the largest *E. coli* outbreak to date (EFSA, 2013a). It is evident, therefore, that a quantitative microbial risk assessment should be made to assist industry and regulators in preventing these illnesses. This is an important issue considering that the very young and the elderly who are significant consumers of fresh fruits and vegetables, are the most susceptible groups (Crandall et al., 2001). It is estimated that almost 7%–8% of the population is under age 5 while the elderly constitute about 12%–14% of the population. In 20 years, the elderly will constitute almost 20%, and when these figures are combined with the increasing numbers of persons who are immunocompromised, it becomes important that we understand the risks associated with consumption of fruits and vegetables (Crandall et al., 2001).

The probability of contamination, the likely consumption levels, and the amount of contaminated food consumed are all factors in the level of exposure to foodborne pathogens. The Economic Research Service (ERS) of the US Department of Agriculture has data on per capita consumption of fresh fruits and vegetables based on production and sales, and the Continuing Survey of Food Intakes by Individuals (CSFII) has data based on food recall (Lin, 2004) as a result of complaints. The top five fruits consumed include citrus, melons, bananas, apples, and grapes, while the top five vegetables are potatoes, lettuce, onions, tomatoes, and carrots. Assumptions are required to compensate for limitations in the consumption data, the lack of consumption data for certain foods, and the extrapolation of servings from short times to an annual basis (Lin, 2004).

Salmonella spp., enterohemorrhagic *E. coli* (EHEC), *L. monocytogenes* and *S. aureus* are the dominant pathogens present and capable of surviving storage of fruit and vegetables (for review see Nychas and Sofos, 2009). It is notable that *E. coli* O157:H7 and *Salmonella* cause the largest outbreaks of foodborne illnesses linked to fresh produce (Sivapalasingam et al., 2004; Warriner et al., 2009). To ensure the safety of fresh fruits and vegetables, and consequently to generate reliable exposure assessment reports, it is essential to know the survival and proliferation mechanisms of these pathogens in the produce. This knowledge may be derived from challenge tests of pre- and postprocess inoculation of pathogens in the produce and final products, respectively. The effects of intrinsic, extrinsic, and processing factors on the growth, survival, or inactivation of pathogens are elucidated through these approaches. In addition, using the obtained data predictive models describing the growth, survival, or inactivation of pathogens in produce can be developed (Halder et al., 2010). A review of the reports on the growth/survival of the major pathogens in a variety of vegetables and fruits is presented in the following paragraphs. It needs to be noted that literature reports have focused considerably on Enterohemorrhagic *E. coli* (EHEC) and *L. monocytogenes* due to the severe pathogenicity of these organisms and their ability to survive under stressful environmental conditions. Cutting, slicing, skinning, and shredding remove or damage the protective surfaces of the vegetable or fruit. There are different consequences for food safety associated with such processes. For example, during washing some microorganisms will be removed from the product, nutrients will become available, and pathogens can be spread from contaminated parts to uncontaminated parts.

The growth/survival of *L. monocytogenes*, *Salmonella* spp., pathogenic *E. coli*, and *S. aureus* under different storage conditions with or without pretreatments has been reported (for review, see Nychas and Sofos, 2009). Chinese cabbage, spinach, cucumber, endive leaves, iceberg, lettuce, tomato, orange peels, radish, asparagus puree, carrot, and chive were inoculated with different inoculum sizes of the pathogenic bacteria mentioned earlier and the pathogen responses have also been reported. Although *L. monocytogenes* is among the most frequently detected pathogens in vegetables/fruits, reaching relatively high prevalence levels (up to 4.8), critical listeriosis cases have yet to be linked to consumption of vegetables/fruits contaminated with *L. monocytogenes* (EFSA, 2006). Several studies have reported data finding *L. monocytogenes* in a variety of ready-to-eat products. A recent report by EFSA provides the results of testing samples of fruits, vegetables, and bakery products (EFSA, 2006). The proportion of positive findings was generally low (<4.5%). Only Latvia reported a higher occurrence (6.9%) in sprouted seeds. The United Kingdom carried out investigations on prepackaged mixed raw vegetable salads containing either meat or fishery products. Out of the 2686 samples 130 (4.8%) tested positive for *L. monocytogenes* and in 2 samples the concentration of this pathogen exceeded the limit of 100 cfu/g.

In 2005, 12 Member States (MS) in the European Union reported data from investigation of fruits and vegetables. In total, 5798 samples were analyzed in these MS. The highest occurrence of positive samples was found in Sweden (3 of 564, 0.5%) and in Ireland (1 of 3365, 0.03%). The numbers of tested samples varied between MS, ranging from 2 to 3079. Juices from fruits and vegetables were investigated by 4 MS (46 samples) with no positive findings. Very few positive findings of *Salmonella* were made from fruit and vegetables (EFSA, 2006). However, quite a substantial proportion of positive samples were reported in spices and herbs (3%–7%) (EFSA, 2006). Germany, Ireland, and Slovenia analyzed 56, 22, and 45 samples of sprouted seeds, respectively, but Ireland detected only 1 sample (4.5%) positive for *Salmonella* Fresno and *S.* Fanti.

Enterohemorrhagic *E. coli* is considered as acid (low pH) tolerant or resistant (Bacon and Sofos, 2003; Samelis and Sofos, 2003). In the United States 350 outbreaks of *E. coli* O157:H7 infections have been reported for the period 1983–2002 (Rangel et al., 2005), and 38 of them were attributed to produce. Quantitative risk assessment for fresh produce, as for most other foods, is still hampered by a lack of data and information, particularly in relation to the sources of contamination during growth. Outbreak information sometimes points to environmental sources of pathogens (e.g., from irrigation water contaminated by run-off from livestock; Ackers et al., 1998), wastewater discharge or fecal contamination from farm and wild animals (Sagoo et al., 2003), but there is often little hard evidence to identify the levels or prevalence of contamination and the exact source or sources. In a recent investigation of irrigation water quality in the United Kingdom, Tyrrel et al. (2006) emphasized the need for risk assessment of sources, credible monitoring practices, and benchmarking of monitoring results against a reference point or standard. It is not unusual to find an emphasis on microbial testing as a method for control of pathogen contamination in harvested produce but this may only provide a false sense of security because of the likelihood that contamination will not be detected, particularly for low level contamination and low infectious-dose pathogens. For these reasons, it is important that the emphasis is placed on the practices and measures for control and that testing is used to validate and verify that these interventions are as effective as expected. Food producers should use good agricultural practices/good manufacturing practices (GAP/GMP) and hazard analysis critical control point (HACCP) concepts to assure safety of fresh produce based on a risk assessment approach. Application of such measures will vary, depending on local conditions and supply chains used. Validation of identified interventions can be *cumulative*, as for example, when multiple hurdles are applied as sequential or simultaneous interventions in meat decontamination (Stopforth and Sofos, 2006), to achieve a defined performance objective. Additional measures may be required for preservation and maintenance of quality aspects (e.g., the control of spoilage).

1.5 Hazard Characterization

Factors affecting the microbial ecology of fruits and vegetables can be intrinsic or extrinsic factors or interventions or preservative methods.

1.5.1 Intrinsic Factors

Moisture content, pH, acidity, nutrient content, biological structure, naturally occurring and added antimicrobials (e.g., sulfur dioxide and organic acids), as well as the indigenous microbial association, comprise the intrinsic factors that are intrinsic to the food itself. The pH is considered to be the most important determinant of bacterial growth on fresh fruit (De Roever, 1998). Many acidic fruits may not support the growth of pathogenic bacteria and may even inactivate them, while fruits with significant higher pH values may allow the microbial growth. Although outbreaks have been reported from low acid vegetables and fruits, illnesses have more commonly been associated with higher pH fruits. Epidemiological data may suggest that there is a relationship between pathogen presence at consumption and pH of the fruit/vegetable. The type of organic acids that contribute to lowering of the pH may also be important. Indeed some organic acids are more effective antimicrobials than others (Brul and Coote, 1999). Dividing fruits/vegetables into groups that include those that have a pH value below (or equal to) 4.0, and those with a value above 4.0 is based on the ability of the latter to allow better and faster growth of more microbes. This, however, may prove to be the wrong decision because growth is also affected by additional factors, such as temperature, a_w, and competing flora, and their interactions. Indeed this issue is usually overlooked by those dealing with risk analysis/assessment.

Raw fruits and vegetables have biological structures that may prevent the entry and growth of pathogenic microorganisms, despite their rich nutrient content by way of carbohydrates, amino acids, vitamins, and minerals providing suitable substrates for the growth of many microorganisms. The intact biological structures are important in preventing the entry and the subsequent growth of microbes, which, therefore, may be influenced by several factors like maturity, physical damage due to handling, insect invasion, and food preparation processes (FDA, 2001). The ability of pathogens to survive in a less favorable environment (e.g., low pH) has been reported extensively in the literature (Lin et al., 1996; Waterman and Small, 1998; Booth, 2002).

It is to be noted that the effect of structure and the physicochemical characteristics (buffering capacity, local pH gradients, nutrient availability, or diffusion, etc.) of vegetables/fruits have not been taken into account in microbial safety risk assessments of these products. On the other hand, it is well established that damage to the fruit integument by bruising, penetration, or cutting leads to faster spoilage (Tassou, 1993; Jay, 1996; Zagory, 1999). Although the effect of damage on pathogens has not been quantified, *E. coli* and *L. monocytogenes* have been shown to grow well on bruised or wounded apples (Janisiewicz et al., 1999; Conway et al., 2000; Dingman, 2000) as *Salmonella* on cut surfaces of tomatoes (Lund, 1992). Factors such as growth habitat (ground or tree, although it may affect the extent of exposure to microbial sources such as soil, manure, and water), organic acid type, a_w, nutrients, competing microflora, and naturally occurring antimicrobial substances may also affect pathogen survival and growth on fruits and vegetables, but their contribution is difficult to estimate because of inadequate published data.

Some fruits and vegetables intrinsically contain naturally occurring antimicrobial compounds that provide some level of microbiological stability to them. There are some antimicrobial compounds of plant origin such as essential oils, tannins, glycosides, and resins, which can be found in a number of fruit and vegetable products. These compounds, in combination with other factors including pH, a_w, presence of other preservatives, types of food constituents, presence of certain enzymes, processing temperature, storage atmosphere, and partition coefficients, may produce greater stability against spoilage organisms or pathogens (FDA, 2001).

Most of the bacteria present on freshly harvested produce are also present on the produce in the field. Part of the epiphytic microflora, comprising of bacteria, yeasts, and fungi, is present at the time of fruit and vegetable consumption. The occurrence of major pathogens such as *L. monocytogenes*, *Salmonella* spp., *E. coli*, and *Staphylococcus aureus* has been reported in the literature (Nychas and Sofos, 2009). The *Pseudomonas* group and the Enterobacteriaceae are the major parts of the natural microbial flora found on the surface of plants and are normally nonpathogenic for humans (Lund, 1992). The season and the climatic conditions affect the number of these bacterial organisms, which range between 10^4 and 10^8 cfu/g. Bacteria can be found in the inner tissues of fruits in low numbers as a result of the uptake of water through certain irrigation or washing procedures. If these waters are contaminated with human pathogens, these may also be introduced into the tissues. It is known that moulds are the spoilage agents of fruits and vegetables, with members of the genera *Penicillium, Aspergillus, Sclerotinia, Botrytis,* and *Rhizopus* being the main causative agents involved in this process (ICMSF, 2005). The spoilage is usually associated with pectinolytic or cellulolytic activity, which causes softening and weakening of plant structures. These structures are important barriers that prevent invasion and growth of contaminating microbes in the products.

1.5.2 Extrinsic Factors

Surface decontamination and cutting are important processing steps to control indigenous microflora and the undesirable growth of pathogenic bacteria on raw, fresh produce. Numerous methodologies/techniques have been developed in this field and are discussed extensively in the following text. Apart from these processing steps, extrinsic factors such as relative humidity, the storage and holding temperature, and the type of packaging and atmosphere surrounding the product are all crucial in controlling microbial growth. In addition, specific information on the growth kinetics and spatial distribution of pathogenic and nonpathogenic bacteria on fruits and vegetables could be important to manage effectively the growth and metabolic activities of the microbial association (Fleet, 2001). The latter is important due to dynamic changes of pH and a_w as a consequence of the evolution of diverse microbial associations dominated by Gram-negative (pseudomonads, Enterobacteriaceae) and lactic acid bacteria (Nguyen-the and Carlin, 1994) depending on the storage conditions. The development of mathematical models predicting the growth of these bacteria on specific fruit and vegetable products should be also considered as a useful key in microbial control (Halder et al., 2010; Liu et al., 2013).

1.5.3 Interventions or Preservative Techniques

Intervention methods are necessary to ensure food safety and maintain nutritional and sensory quality (Sela and Fallik, 2009). Interventions can be divided into three major categories, namely physical, chemical, and biological as shown in Figure 1.1. A combined intervention, characterized as hurdle technology, focused on the inhibition or inactivation of several pathogens and factors responsible for food spoilage can also be applied (Ramos et al., 2013). Furthermore, alternative strategies could make use of quorum sensing inhibitors (QSIs) that have been recently suggested as biopreservatives and have not yet been investigated (Skandamis and Nychas, 2012). These studies have been focused on food safety through the

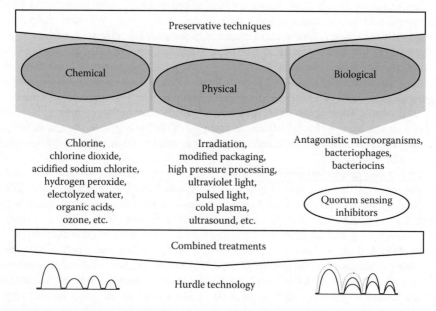

FIGURE 1.1   The interventions can be divided into three major categories, namely, physical, chemical, and biological.

blockage of various QS-regulated phenotypes such as virulence, toxin production, bacterial resistance, and biofilm formation (Smith et al., 2004). These are

1. *Chemical interventions*: To reduce/eliminate the undesirable growth of pathogenic bacteria on raw, fresh produce a number of sanitizing agents can be used such as chlorine, electrolyzed water, and ozone.
2. *Physical interventions*: High pressure processing, ionizing irradiation, ultraviolet-C (UV-C) irradiation, and modified atmosphere packaging (MAP) with an edible coating and refrigeration have been reported to effectively inactivate common foodborne pathogens on the aforementioned products (Ramos et al., 2013).
3. *Biological interventions*: These methods improve food safety using antagonistic microorganisms and/or their metabolites such as organic acids and bacteriocins, while bacteriophages as well as quorum sensing inhibitors have also been used to control pathogens (Skandamis and Nychas, 2012).

1.5.3.1 *Listeria monocytogenes*

Members of the genus *Listeria* are ubiquitous and widely distributed in the environment. This genus consists of six species and only one is pathogenic. *L. monocytogenes* is the causative agent of listeriosis and one of the most virulent foodborne pathogens. Listeriosis is a life-threatening disease for humans with compromised immune systems and most often affects the uterus during pregnancy, the central nervous system, and the bloodstream. Symptoms vary, ranging from mild flu-like symptoms and diarrhea to highly dangerous infectious diseases characterized by septicemia and meningoencephalitis. In pregnant women, the infection spreads to the fetus, which is either born severely ill or dies in the uterus resulting in miscarriage or stillbirth. The high morbidity and mortality rates in vulnerable populations make listeriosis an important disease worldwide. In United States, 1651 cases of listeriosis were reported from 2009 through 2011 of which 292 were fatal (CDC, 2013), whereas in Europe, it remains a relatively rare disease (EFSA, 2013b). The occurrence of this pathogen in fruits and vegetables has also been reported (Nychas and Sofos, 2009). Fresh fruits and vegetables that are consumed without any further thermal treatment, and are contaminated with more than 100 *L. monocytogenes* cells per gram are considered to be a direct risk to human health. Ready-to-eat food products, cooked and exposed to contamination after cooking or prepared but not cooked such as vegetable and fruit salads, have been typically identified as risk products for contamination with *Listeria*. The European Union MS have been requested to report data on *Listeria* in food. Thus far data comparison between countries is difficult because of the different sample sizes and protocols used. Although, it is generally considered that concentration levels of *L. monocytogenes* exceeding 100 cfu/g are necessary to cause disease in healthy humans, the qualitative positive presence of the pathogen does not necessarily constitute a risk indicator (EU Regulation 2073/2005).

1.5.3.2 *Salmonella*

Salmonella enterica is a leading cause of foodborne illness worldwide, commonly found in the gastrointestinal tract of numerous animals and humans. Salmonellosis is the disease caused by the *Salmonella* bacteria, the second most commonly reported gastrointestinal infection and an important cause of foodborne outbreaks in the European Union (ECDC, 2012). Most cases originate from meat, poultry, and their products (Tauxe et al., 1997). However, an increasing number of salmonellosis outbreaks are occurring as a result of contaminated produce (Hanning et al., 2009). The factors that possibly make fruits and vegetables more likely to be sources of *Salmonella* contamination are their ability to attach or enter into produce items. Most persons infected with *Salmonella* develop diarrhea, fever, and abdominal cramps 12–72 h after infection, and the illness usually lasts 4–7 days. Persons with impaired immune system, the elderly, and infants might have more severe illness and their hospitalization might be necessary. The high prevalence of *Salmonella* in a wide range of fruits and vegetables has been demonstrated (Nychas and Sofos, 2009). Environmental factors including contaminated water used to irrigate and wash produce crops have also been implicated in a large number of outbreaks (Sivapalasingam et al., 2003; Greene et al., 2005, 2008; FDA, 2008). *Salmonella* is carried by a wide range of domestic and wild animals and can contaminate freshwater by direct or indirect contact. In some cases, direct contact of produce or seeds with contaminated manure or animal wastes can lead to contaminated crops. Transmission often occurs when organisms are introduced in food preparation areas and are allowed to multiply in food (e.g., due to improper storage temperatures, inadequate cooking, or cross contamination of cooked food). The most common *Salmonella* serotypes found in 2010 in the EU countries were *S. enteriditis* and *S. typhimurium*, accounting for 45% and 22% of all reported serotypes, respectively. In 2011, four multinational *Salmonella* outbreaks were reported in the European Union; mung bean sprouts were the infection vehicle in one of them (ECDC, 2012). Recently, an outbreak of *Salmonella* Saintpaul infections linked to imported cucumbers was reported from 18 states in the United States. A total of 84 persons were infected, 28% of whom were hospitalized, but no deaths were reported (CDC, 2013).

1.5.3.3 Escherichia coli

Escherichia coli is a bacterium that is commonly found in the gut of humans and other warm-blooded animals; some *E. coli* strains can cause severe foodborne disease. Infection with vero/shiga toxin-producing *E. coli* (VTEC/STEC) is characterized by an acute onset of diarrhea, which may be bloody, and is often accompanied by mild fever and sometimes vomiting. Most patients recover within 10 days, although in a few cases the disease may become life-threatening. The infection may lead to the potentially fatal hemolytic uremic syndrome (HUS), characterized by acute renal failure, anemia, and lowered platelet counts requiring hospital care. EHEC are a subset of the VTEC harboring additional pathogenic factors. More than 150 different serotypes of VTEC have been recognized; however, the majority of reported outbreaks and sporadic cases of VTEC infections have been attributed to serotype O157. *E. coli* infection is usually transmitted through consumption of contaminated food, such as undercooked contaminated beef or contaminated vegetables, or water; person to person and direct transmissions from animals to humans may also occur. Although the occurrence of pathogenic *E. coli* strains have been reported in produce (Nychas and Sofos, 2009), in recent years, a great number of outbreaks have been recorded. For example, in 2012 in the United States, 33 persons were infected by STEC O157:H7 present in organic spinach and spring mix blend, 13 of them were hospitalized and 2 developed HUS (CDC, 2012). One year earlier (in 2011), a total of 9485 confirmed VTEC infections were reported in the European Union; an increase of 159.4% compared with 2010, the result of a large VTEC/STEC outbreak that occurred primarily in Germany (EFSA, 2013a). The German outbreak was the largest HUS outbreak ever reported with 845 cases. It affected adults (88%) predominantly and the causative agent appeared to be a rare and novel pathotype, most likely the result of horizontal gene transfer between two different types of pathogenic *E. coli* strains with distinct reservoirs (animals and humans). Sprouts were identified as a source of the outbreak, highlighting the risks related to raw and ready-to-eat food products, some of which have become popular health food items in Europe (ECDC, 2012).

1.5.3.4 Staphylococcus aureus

This pathogen has been detected in vegetables and is transmitted by person–to-person contact and through food as a result of improper food handling and inadequate hygiene of food handlers (for review see Nychas and Sofos, 2009). Some strains of *S. aureus* produce enterotoxins, which can cause illness in humans when consumed at even a low dose level of 100 ng (Kauffman and Roberts, 2006). Symptoms like nausea, vomiting, abdominal cramps, and diarrhea characterize staphylococcal food poisoning and usually appear 1–6 h after enterotoxin ingestion (Asao et al., 2003). Seven types of toxins have been distinguished (based on their antigenic properties) among the enterotoxigenic strains, of which types A and D are the most common involved in food poisoning. They are formed during the exponential and stationary phases of growth (ICMSF, 1996). In the United States during the period 1975–1979, 540 food poisoning outbreaks were reported, with *S. aureus* responsible for 28% (153 outbreaks) (Smith et al., 1983). During a survey that took place among street vendors located in India, out of 120 samples comprising different types of raw vegetables, fruits, and sprouts, *S. aureus* presented a total recovery rate of 58.3% (Viswanathan and Kaur, 2001).

1.5.3.5 Other Bacterial Pathogens

Several other pathogenic bacteria have been linked to outbreaks of produce-associated illness, including *Shigella* spp. The *Shigella* species, transmitted through raw vegetables mainly because of the fecal contamination of water and unsanitary food handling practices, can cause bacillary dysentery also known as shigellosis and the infection dose is 10 cells (Kothary and Babu, 2001; Warren et al., 2006). *Shigella sonnei* was implicated in outbreaks related to melon and raw lettuce contaminated consumption (Fredlund et al., 1987; Kapperud et al., 1995). In 2010, the confirmed shigellosis cases in 28 EU countries were 1.64 per 100,000 population, suggesting that it is a relatively uncommon disease (ECDC, 2012). *Clostridium botulinum* is also capable of causing illness by toxin production in foods. Its toxins and spores were found in coleslaw dressing which contained cabbage and carrot pieces (Solomon et al., 1990). In addition, *C. perfringens* and *Bacillus cereus* are found in soil and may be naturally present on some fruits and vegetables (Beuchat and Ryu, 1997).

1.5.3.6 Viruses

Hepatitis A virus, calicivirus, and norovirus have been associated with outbreaks related to the consumption of produce (Seymour and Appleton, 2001). These outbreaks were linked to frozen raspberries or strawberries, lettuce, melons, salads, watercress, diced tomatoes, and fresh-cut fruit consumption. Most of these outbreaks were the outcome of virus transmission to food by infected food handlers during final preparation. Hepatitis A and Norovirus are the most commonly documented viral food contaminants. Viruses have been isolated from sewage and untreated wastewater used for agricultural crop irrigation, while infected people can also excrete large numbers of viruses. Despite the fact that they are unable to grow in and on foods, their presence on fresh produce, which may serve as vehicles of infection, is of interest (FDA, 2001). Hedberg and Osterholm (1993) reported 8 out of 14 viral gastroenteritis outbreaks, which were attributed to a food handler who was ill before or while handling the implicated food. Salads were the implicated vehicle in five outbreaks (36%), and cold food items or ice was implicated in all but one outbreak. Outbreaks of hepatitis A in 2010 and 2011, associated with the consumption of semi-dried tomatoes, underline the threat of foodborne transmission and the need of tracing the origin of contamination in food. However, the outbreaks in Estonia in 2011 and among an orthodox Jewish community in London, demonstrate that person to person transmission is also a likely source of infection, particularly in populations with low or decreasing immune system (ECDC, 2012).

1.6 Risk Management and Control Measures

Contamination intervention strategies like HACCP that were implemented for the first time in meat products are being designed for produce production. The application of HACCP has been reviewed for minimally processed crops, with the aim to highlight the hazards and identify control measures in key risk areas of processing (Sperber, 1992; Hurst, 2006; Leifert et al., 2008). According to the Codex Alimentarius Commission (1997), Good Hygienic Practices (GHP) in combination with HACCP can be the basis for safe food, including vegetable and fruit products. The application of HACCP principles in the efforts to enhance the safety of fruits and vegetables is not easy. Indeed it is considered to be problematical, in particular, the identification of robust (because of harvesting of produce from the field) Critical Control Points (CCP) for the destruction of pathogens and for maintaining a record of measures taken at the CCP. Within the area of fruit and vegetable safety, the development and application of measures applicable in GAP can be an alternative good tool. Only few retailers appear to demonstrate or to accept responsibility for the safety of fruits and vegetables. This is reflected in specifications to be met by producers with respect to factors affecting the safety of these commodities. In guidelines for the U.S. industry (FDA, 2007), the FDA defines practices to be followed with respect to microbial food safety hazards and good agricultural and management practices common to growing, packaging, and transport of fruits and vegetables. The guideline focuses on water quality for different purposes, proper handling practices for manure and municipal biosolids, health and hygiene of workers, field sanitation facilities, packaging facility sanitation, proper transportation, and product trace-back.

The ability of various vegetable-borne pathogens to survive in many high acid vegetable fruit products makes it unlikely that complete suppression of growth can be achieved by the application of control measures at a single source. Thus, effective control strategies must consider the multiple points at which pathogens can gain access to the human food chain. The persistence, and the ability of very small numbers of these organisms to cause life-threatening infections with serious long term clinical consequences, particularly among the *at risk* sections of the human population, mean that many elements of our food safety strategies will have to be improved.

Measures to control pathogens during food production, processing, and distribution, at the retail level and during commercial/domestic preparation should be considered in detail. Therefore, the best approach to control pathogens is to implement HACCP principles as the food safety management system at all stages of food production and distribution. However, because of the unusual tolerance of some pathogens (e.g., *Listeria*, *E. coli*, and *Salmonella*) to low pH, especially those of low infective doses, these should be considered as of higher risk than other pathogens (e.g., *S. aureus*) in vegetables/fruits which have a low pH, and are minimally processed or are not cooked before consumption. The basic control measures should be based on:

- A quantitative microbial hazard analysis or quantitative microbial risk assessment of current practices carried out for products in various categories.
- Training of personnel working in the food preparation, processing, and food service industries coupled with consumer education on hygienic handling and adequate cleaning or cooking of food.

REFERENCES

Ackers M.-L., Mahon B.E., Leahy E., Goode B., Damrow T., Hayes P.S., Bibb W.F. et al. 1998. An outbreak of *Escherichia coli* O157:H7 infections associated with leaf lettuce consumption. *Journal of Infectious Diseases* 177:1588–1593.

Asao T., Kumeda Y., Kawai T., Shibata T., Oda H., Haruki K., Nakazawa H., and Kozaki S. 2003. An extensive outbreak of staphylococcal food poisoning due to low-fat milk in Japan: Estimation of enterotoxin A in the incriminated milk and powdered skim milk. *Epidemiology and Infection* 130:33–40.

Bacon R.T. and Sofos J.N. 2003. Food hazards: Biological food; characteristics of biological hazards in foods. In *Food Safety Handbook*, R.H. Schmidt and G. Rodrick, eds. Wiley Interscience, New York, pp. 157–195. ISBN 0-471-21064-1.

Beuchat L.R. and Ryu J.H. 1997. Produce handling and processing practices. *Emerging Infectious Diseases* 3:459–465.

Booth I.R. 2002. Stress and the single cell: Intrapopulation diversity is a mechanism to ensure survival upon exposure to stress. *International Journal of Food Microbiology* 78:19–30.

Brul S. and Coote P. 1999. Mode of action and microbial resistance mechanisms. *International Journal of Food Microbiology* 50:1–17.

Centers for Disease Control and Prevention (CDC). 2012. Multistate outbreak of shiga toxin-producing *Escherichia coli* O157:H7 infections linked to organic spinach and spring mix blend. http://www.cdc.gov/ecoli/2012/O157H7-11-12/index.html. Accessed date 05/07/2013.

Centers for Disease Control and Prevention (CDC). 2013. Outbreak-associated *Salmonella enterica* serotypes and food commodities, United States, 1998–2008. http://wwwnc.cdc.gov/eid/article/19/8/12-1511_article.htm. Accessed date 05/07/2013.

Codex Alimentarius Commission. 1997. Report of the twenty-second session of the Codex Alimentarius Commission. Ref. No. ALINORM 97/37. Geneva, Switzerland, June 23–28, 1997.

Codex Alimentarius Commission. 1999. Principles and guidelines for the conduct of microbial risk assessments. FAO/WHO, Rome, Italy.

Conway W.S., Leverentz B., Saftner R.A., Janisiewicz W.J., Sams C.E., and Leblanc E. 2000. Survival and growth of *Listeria monocytogenes* on fresh-cut apple slices and its interaction with *Glomerella cingulata* and *Penicillium expansum. Plant Disease* 84:177–181.

Crandall P.G., O'Bryan C., Li Y., and Zakariadze I. 2001. Hazard identification and exposure assessment for consumers of fruits and vegetables. www.riskworld.com/Abstract/2001/SRAam01/ab01aa064.htm. Accessed date 03/09/2013.

De Roever C. 1998. Microbiological safety evaluations and recommendations on fresh produce. *Food Control* 9:321–347.

Dingman D.W. 2000. Growth of *Escherichia coli* O157:H7 in bruised apple (*Malus domestica*) tissue as influenced by cultivar, date of harvest, and source. *Applied Environmental Microbiology* 66:1077–1083.

Duffy G. 2003. Verocytoxigenic *Escherichia coli* in animal faeces, manures and slurries. *Journal of Applied Microbiology* 94:94S–103S.

European Centre for Disease Prevention and Control (ECDC). 2012. Annual epidemiological report—Reporting on 2010 surveillance data and 2011 epidemic intelligence data. http://www.ecdc.europa.eu/en/publications/publications/annual-epidemiological-report-2012.pdf. Accessed date 23/07/2013.

European Food Safety Authority (EFSA). 2006. The community summary report on trends and sources of zoonoses, zoonotic agents, antimicrobial resistance and foodborne outbreaks in the European Union in 2005. *The EFSA Journal* 94:3–288.

European Food Safety Authority (EFSA). 2013a. The European Union summary report on trends and sources of zoonoses, zoonotic agents and food-borne outbreaks in 2011. *EFSA Journal* 11:3129. http://www.efsa.europa.eu/en/efsajournal/pub/3129.htm.

European Food Safety Authority (EFSA). 2013b. Analysis of the baseline survey on the prevalence of *Listeria monocytogenes* in certain ready-to-eat foods in the EU, 2010–2011 Part A: *Listeria monocytogenes* prevalence estimates. http://www.efsa.europa.eu/en/efsajournal/pub/3241.htm. Accessed date 23/07/2013.

Fleet G.H. 2001. Biodiversity and ecology of Australasian yeasts (Fungi). *Australian Systematic Botany* 14:501–511.

Food and Agriculture Organization of the United Nations (FAO). 2004. Manual for the preparation and sale of fruits and vegetables: From field to market. http://www.fao.org/docrep/008/y4893e/y4893e00.HTM. Accessed date 03/07/2013.

Food and Drug Administration (FDA). 2001. Analysis and evaluation of prevention control measures for the control and reduction/elimination of microbial hazards on fresh and fresh-cut produce. Center for Food Safety and Applied Nutrition, Washington, DC. http://www.fda.gov/Food/FoodScienceResearch/SafePracticesforFoodProcesses/ucm090977.htm. Accessed date 10/07/2013.

Food and Drug Administration (FDA). 2007. Guidance for industry guide to minimize microbial food safety hazards of fresh-cut fruits and vegetables. www.cfsan.fda.gov/~dms/prodgui3.html. Accessed date 10/07/2013.

Food and Drug Administration (FDA). 2008. FDA warns of *Salmonella* risk with cantaloupes from Agropecuaria Montelibano: The agency detains products from the Honduran manufacturer. http://www.fda.gov/bbs/topics/NEWS/2008/NEW01808.html. Accessed date 10/07/2013.

Fredlund H., Back E., Sjoberg L., and Tornquist E. 1987. Watermelon as a vehicle of transmission of Shigellosis. *Scandinavian Journal of Infectious Diseases* 19:219–221.

Friesema I., Sigmundsdottir G., van der Zwaluw K., Heuvelink A., Schimmer B., de Jager C., Rump B. et al. 2008. An international outbreak of Shigatoxin-producing *Escherichia coli* O157 infection due to lettuce, September–October 2007. *Eurosurveillance* 13:1–5.

Grant J., Wendelboe A.M., Wendel A., Jepson B., Torres P., Smelser C., and Rolfs R.T. 2008. Spinach-associated *Escherichia coli* O157:H7 outbreak, Utah and New Mexico, 2006. *Emerging Infectious Diseases* 14:1633–1636.

Greene S.K., Daly E.R., Talbot E.A., Demma L.J., Holzbauer S., Patel N.J., Hill T.A. et al. 2005. Recurrent multistate outbreak of *Salmonella* Newport associated with tomatoes from contaminated fields. *Emerging Infectious Diseases* 136:157–165.

Greene S.K., Daly E.R., Talbot E.A., Demma L.J., Holzbauer S., Patel N.J., Hill T.A. et al. 2008. Recurrent multistate outbreak of *Salmonella* Newport associated with tomatoes from contaminated fields, 2005. *Epidemiology and Infection* 136:157–165.

Halder A., Black D.G., Davidson P.M., and Datta A. 2010. Development of associations and kinetic models for microbiological data to be used in comprehensive food safety prediction software. *Journal of Food Science* 75:R107–R120.

Hanning I.B., Nutt J.D., and Ricke S.C. 2009. Salmonellosis outbreaks in the United States due to fresh produce: Sources and potential intervention measures. *Foodborne Pathogens and Disease* 6:635–648.

Hedberg C.W. and Osterholm M.T. 1993. Outbreaks of food-borne and waterborne viral gastroenteritis. *Clinical Microbiological Review* 6:199–210.

Hilborn E.D., Mermin J.H., Mshar P.A., Hadler J.L., Voetsch A., Wojtkunski C., Swartz M. et al. 1999. A multistate outbreak of *Escherichia coli* O157, H7 infections associated with consumption of mesclun lettuce. *Archives of Internal Medicine* 159:1758–1764.

Hurst W. 2006. HACCP: A process control approach for fruit and vegetable safety. In *Microbiology of Fruits and Vegetables*, G. Sapers, J. Gorny, and A. Yousef, eds. CRC Press, Boca Raton, FL, pp. 339–364.

International Commission on Microbiological Specifications for Foods (ICMSF). 1996. *Microorganisms in Food 5: Microbiological Specifications of Food Pathogens*. Blackie Academic & Professional, London, U.K., pp. 299–333.

International Commission on Microbiological Specifications for Foods (ICMSF). 2005. *Microorganisms in Foods 6: Microbial Ecology of Food Commodities* (2nd ed.), T.A. Roberts, J.-L. Cordier, L. Gram, R.B. Tompkin, J.I. Pitt, L.G.M. Gorris, and K.M.J. Swanson, eds. Kluwer Academic & Plenum Publishers, New York, pp. 277–353.

Islam M., Morgan J., Doyle M.P., Phatak S.C., Millner P., and Jiang X. 2004. Persistence of *Salmonella enterica* serovar Typhimurium on lettuce and parsley and in soils on which they were grown in fields treated with contaminated manure composts or irrigation water. *Foodborne Pathogens and Disease* 1:27–35.

Janisiewicz W.J., Conway W.S., and Leverentz B. 1999. Biological control of postharvest decays of apple can prevent growth of *Escherichia coli* O157:H7 in apple wounds. *Journal of Food Protection* 62: 1372–1375.

Jay J.M. 1996. Intrinsic and extrinsic parameters of foods that affect microbial growth. In *Modern Food Microbiology* (5th ed.), J.M. Jay, ed. Chapman & Hall, New York, pp. 38–66.

Johnson J.Y.M., Thomas J.E., Graham T.A., Townshend I., Byrne J., Selinger L.B., and Gannon, V.P.J. 2003. Prevalence of *Escherichia coli* O157:H7 and *Salmonella* spp. in surface waters of southern Alberta and its relation to manure sources. *Canadian Journal of Microbiology* 49:326–335.

Kapperud G., Rørvik L.M., Hasseltvedt V., Høiby E.A., Iversen B.G., Staveland K., Johnsen G., Leitao J., Herikstad H., and Andersson Y. 1995. Outbreak of *Shigella sonnei* infection traced to imported iceberg lettuce. *Journal of Clinical Microbiology* 33:609–614.

Kauffman N.M. and Roberts R.F. 2006. Staphylococcal enterotoxin D production by *Staphylococcus aureus* FRI 100. *Journal of Food Protection* 69:1448–1451.

Kothary M.H. and Babu U.S. 2001. Infective dose of foodborne pathogens in volunteers: A review. *Journal of Food Safety* 21:49–73.

Leifert C., Ball K., Volakakis N., and Cooper J.M. 2008. Control of enteric pathogens in ready-to-eat vegetable crops in organic and 'low input' production systems: A HACCP-based approach. *Journal of Applied Microbiology* 105:931–950.

Lin B.-H. 2004. Fruit and vegetable consumption looking ahead to 2020. *Economic Research Service Agriculture Information Bulletin*, No. (AIB-792-7). U.S. Department of Agriculture, Economic Research Service, Washington, DC, 4pp.

Lin C.-M., Fernando S.Y., and Wei C.I. 1996. Occurrence of *Listeria monocytogenes*, *Salmonella* spp., *Escherichia coli* and *E. coli* O157:H7 in vegetable salads. *Food Control* 7:135–140.

Liu C., Hofstra N., and Franz E. 2013. Impacts of climate change on the microbial safety of pre-harvest leafy green vegetables as indicated by *Escherichia coli* O157 and *Salmonella* spp. *International Journal of Food Microbiology* 163:119–128.

Lund B.M. 1992. Ecosystem in vegetable foods. *Journal of Applied Bacteriology* (*Symposium Supplement*) 73:115S–126S.

Neetoo H., Lu Y., Wu C., and Chen H. 2012. Use of high hydrostatic pressure to inactivate *Escherichia coli* O157:H7 and *Salmonella enterica* internalized within and adhered to preharvest contaminated green onions. *Applied and Environmental Microbiology* 78:2063–2065.

Nguyen-the C. and Carlin F. 1994. The microbiology of minimally processed fresh fruits and vegetables. *Critical Review Food Science and Nutrition* 34:371–401.

Nychas G.-J.E. and Sofos J. 2009. Pathogens in vegetables and fruits: Risks and control, Chapter 4. In *Survival and Control of Pathogens in Fresh Cut Vegetables*, G. Spano and G. Colelli, eds., Research Signpost, Kerala, India, pp. 77–110.

Ramos B., Miller F.A., Brandão T.R.S., Teixeira P., and Silva C.L.M. 2013. Fresh fruits and vegetables— An overview on applied methodologies to improve its quality and safety. *Innovative Food Science and Emerging Technologies* 20:1–15.

Rangel J.M., Sparling P.H., Crowe C., Griffin P.M., and Swerdlow D.L. 2005. Epidemiology of *Escherichia coli* O157:H7 outbreaks United States 1982–2002. *Emerging Infectious Diseases* 11:603–609.

Sagoo S.K., Little C.L., Ward L., Gillespie I.A., and Mitchell R.T. 2003. Microbiological study of ready-to-eat salad vegetables from retail establishments uncovers a national outbreak of salmonellosis. *Journal of Food Protection* 66:403–409.

Samelis J. and Sofos J.N. 2003. Strategies to control stress-adapted pathogens and provide safe foods. In *Microbial Adaptation to Stress and Safety of New-Generation Foods*, A.E. Yousef and V.K. Juneja, eds. CRC Press, Inc., Boca Raton, FL, pp. 303–351. ISBN 1-56676-912-4.

Scallan E., Griffin P.M., Angulo F.J., Tauxe R.V., and Hoekstra R.M. 2011. Foodborne illness acquired in the United States—Unspecified agents. *Emerging Infectious Diseases* 17:16–22.

Scharff R.L. 2010. Health-related costs from foodborne illness in the United States. The Produce Safety Project at Georgetown University, Washington, DC. http://www.producesafetyproject.org/admin/assets/files/Health-Related-Foodborne-Illness-Costs-Report.pdf-1.pdf. Accessed date 03/09/2013.

Sela S. and Fallik E. 2009. Microbial quality and safety of fresh produce, Chapter 13. In *Postharvest Handling: A Systems Approach* (2nd ed.), W.J. Florkowski, R.L. Shewfelt, B. Brueckner and S.E. Prussia, eds. Elsevier, New York, pp. 351–398.

Sewell A.M. and Farber J.M. 2001. Foodborne outbreaks in Canada linked to produce. *Journal of Food Protection* 64:1863–1877.

Seymour I.J. and Appleton H. 2001. Foodborne viruses and fresh produce. *Journal of Applied Microbiology* 91:759–773.

Sivapalasingam S., Barrett E., Kimura A., VanDuyne S., DeWitt W., Ying M., Frisch A. et al. 2003. A multistate outbreak of *Salmonella enterica* serotype Newport infection linked to mango consumption: Impact of water-dip disinfestations technology. *Clinical Infectious Diseases* 37:1585–1590.

Sivapalasingam S., Friedman C.R., Cohen L., and Tauxe R.V. 2004. Fresh produce: A growing cause of outbreaks of foodborne illness in the United States, 1973 through 1997. *Journal of Food Protection* 67:2342–2353.

Skandamis P.N. and Nychas G.-J.E. 2012. Quorum sensing in the context of food microbiology. *Applied and Environmental Microbiology* 78:5473–5482.

Smith J.L., Buchanan R.L., and Palumbo S.A. 1983. Effect of food environment on staphylococcal enterotoxin synthesis: A review. *Journal of Food Protection* 46:545–555.

Smith J.L., Fratamico P.M., and Novak J.S. 2004. Quorum sensing: A primer for food microbiologists. *Journal of Food Protection* 67:1053–1070.

Solomon E.B., Yaron S., and Mathews K.R. 2002. Transmission of *Escherichia coli* O157:H7 from contaminated manure and irrigation water to lettuce plant tissue and its subsequent internalization. *Applied and Environmental Microbiology* 68:397–400.

Solomon H.M., Kautter D.A., Lilly T., and Rhodehamel E.J. 1990. Outgrowth of *Clostridium botulinum* in shredded cabbage at room temperature under modified atmosphere. *Journal of Food Protection* 53:831–833.

Sperber W. 1992. Determining critical control points. In *HACCP Principles and Applications*, M. Pierson and D. Corlett, eds. Van Nostrand Reinhold, New York, pp. 39–49.

Stopforth J.D. and Sofos J.N. 2006. Recent advances in pre- and post-slaughter intervention strategies for control of meat contamination. In *Advances in Microbial Food Safety, Recent Advances in Intervention Strategies to Improve Food Safety*, V.J. Juneja, J.P. Cherry, and M.H. Tunick, eds. ACS Symposium Series, Vol. 931. American Chemical Society, Washington, DC, pp. 66–86.

Tassou C.C. 1993. Microbiology of olives with emphasis in antimicrobial compounds. PhD thesis, University of Bath, Bath, U.K.

Tauxe R., Kruse H., Hedberg C., Potter M., Madden J., and Wachsmuth K. 1997. Microbial hazards and emerging issues associated with produce: A preliminary report to the National Advisory Committee on Microbiologic Criteria for Foods. *Journal of Food Protection* 60:1400–1408.

Tournas V.H. 2005. Spoilage of vegetable crops by bacteria and fungi and related health hazards. *Critical Reviews in Microbiology* 31:33–44.

Tyrrel S.F., Knox J.W., and Weatherhead E.K. 2006. Microbiological water quality requirements for salad irrigation in the United Kingdom. *Journal of Food Protection* 69:2029–2035.

Vernozy-Rozand C., Montet M.P., Lequerrec F., Serillon E., Tilly B., Bavai C., Ray-Gueniot S., and Richard Y. 2002. Prevalence of verotoxin-producing *Escherichia coli* (VTEC) in slurry, farmyard manure and sewage sludge in France. *Journal of Applied Microbiology* 93:473–478.

Viswanathan P. and Kaur R. 2001. Prevalence and growth of pathogens on salad vegetables, fruits and sprouts. *International Journal of Hygiene and Environmental Health* 203:205–213.

Warren B.R., Parish M.E., and Schneider K.R. 2006. *Shigella* as a foodborne pathogen and current methods for detection in food. *Critical Reviews in Food Science and Nutrition* 46:551–567.

Warriner K., Huber A., Namvar A., Fan W., and Dunfield K. 2009. Recent advances in the microbial safety of fresh fruits and vegetables. *Advances in Food and Nutrition Research* 57:155–208.

Waterman S.R. and Small P.L.C. 1998. Acid-sensitive enteric pathogens are protected from killing under extremely acidic conditions of pH 2.5 when they are inoculated onto certain solid food sources. *Applied and Environmental Microbiology* 64:3882–3886.

Wendel A.M., Johnson D.H., Sharapov U., Grant J., Archer J.R., Monson T., Koschmann C., and Davis J.P. 2009. Multistate outbreak of *Escherichia coli* O157:H7 infection associated with consumption of packaged spinach, August–September 2006: The Wisconsin investigation. *Clinical Infectious Diseases* 48:1079–1086.

Zagory D. 1999. Effects of post-processing handling and packaging on microbial populations. *Postharvest Biology and Technology* 15:313–321.

2

Listeria monocytogenes *in Seafood with a Special Emphasis on RTE Seafood*

Marija Zunabovic

CONTENTS

2.1 Introduction

Ready-to-eat (RTE) foods are usually processed food products that can be consumed without any further bactericidal treatment (e.g., heating) or preparation step, except for thawing (Jaroni et al. 2010). The EC Regulation No. 2073/2005 (Amd. No. 1441/2007) on "microbiological criteria for foodstuffs" defines "ready-to-eat foods" as "food for direct human consumption without the need for cooking or other processing effective to eliminate or reduce to an acceptable level of microorganisms of concern." Certain types of RTE food such as minced meat served as *steak tartare* or foods needing a simple warm-up step (Dufour 2010) are not included in this definition. This regulation concerns all food business operators' (FBOs') processing, manufacturing, handling, and distribution of food, including retail and catering.

Different refrigerated RTE foods can be named as examples, such as meats (e.g., deli meats, sausages, hot dogs), dairy products (e.g., pasteurized milk, yoghurt, cheeses), seafood and fish (e.g., cold smoked salmon and tuna, sushi, fish salad), fruits and vegetables (e.g., salads, leafy greens), and poultry (e.g., chicken salad and wraps). Even certain frozen foods, such as ice cream, frozen fruits, and frozen yoghurt, can be included (Jaroni et al. 2010).

Listeria monocytogenes is able to grow in various (refrigerated) RTE products, which, from a practical point of view, necessitates a categorization into products that allow or inhibit the growth of this foodborne pathogen during a certain shelf life. This theoretically simple classification posed a challenge for FBOs and research groups when performing challenge tests to estimate the growth boundaries of *Listeria* spp. The aim of such tests is to artificially inoculate the target bacterium into a certain food matrix with specified intrinsic properties (pH, a_w-value, redox potential) and to monitor its growth potential (δ) during storage under defined conditions. There are several factors influencing the individual growth performance: product characteristics (physicochemical properties), choice of strains (source, growth behavior), inoculum preparation, packaging, and storage conditions (Beaufort 2011). Another issue is the product-dependent applicability of challenge testing. As soon as the processing conditions or the recipes are changed, the

TABLE 2.1

Overview of Inoculation Trials with *L. monocytogenes* in RTE Fishery Products

Matrix	Strain Designation	Inoculum Concentration (CFU[a]/mL)	Study Focus	Reference
Vacuum-packed smoked herring; surimi salad	*L. monocytogenes* serotype 4b and 1/2a	10^6–10^7	Variability of growth parameters	Augustin et al. (2011)
Smoked salmon products (raw, salted, salt-cold smoked, salt–sugar–pepper–dill)	*L. monocytogenes* CIP 78.38 serotype 4b; field strain *L. monocytogenes* 1/2a-3a	3×10^8	Comparison between new salmon preparations and conventional products	Midelet-Bourdin et al. (2010)
Cooked meat, smoked fish, and mayonnaise based deli-salads	*L. monocytogenes* strain cocktail	5×10^3	Different packaging conditions and varying physicochemical parameters	Uyttendaele et al. (2009)

[a] CFU, Colony forming unit.

test is no more valid for the tested product. Table 2.1 summarizes some formats of challenge tests applied to *L. monocytogenes* in different product matrices. Therefore, a technical guidance document intended for conducting shelf life tests for *L. monocytogenes* was prepared by the EU Community Reference Laboratory (CRL) (Beaufort et al. 2008). This document aids in overcoming the aforementioned pitfalls and hence has been rapidly accepted throughout the international reference laboratories. According to the Regulation (EC) No. 2073/2005, RTE foods with pH ≤ 4.4 or $a_w \leq 0.92$ and products with pH ≤ 5.0 or $a_w \leq 0.94$ are categorized as RTE foods unable to support the growth of *L. monocytogenes* (Beaufort 2011).

FBOs producing RTE foods that pose a potential risk to public health because of the presence or growth of *L. monocytogenes* should include additional monitoring schemes in the processing areas and equipments (Food Standards Agency 2009). As *L. monocytogenes* is capable of growing at refrigeration temperatures, special attention should be given to these RTE product types as well as to products with extended shelf life that allow *L. monocytogenes* enough time to multiply in the matrix. Next to ensuring appropriate manufacturing temperatures, one of the most critical factors is the in-home food handling practice contributing to elevated risk of listeriosis in humans (Liu 2008). However, refrigerated storage in domestic refrigerators is estimated to range from 5.2°C to 7.0°C but temperatures above 8°C were even measured in 20%–35% of the cases (EFSA 2007; Vermeulen et al. 2011).

L. monocytogenes growth in RTE food can be controlled through the reduction of a_w and pH, via improved packaging conditions and the addition of antimicrobials (e.g., nisin and pediocin). In general, a combination of treatments is more effective (Jaroni et al. 2010). Research initiatives have been undertaken in the field of decontamination strategies applied to various RTE foods (Jaroni et al. 2010; Tirpanalan et al. 2011). However, ongoing trends toward mildly preserved food (nonthermal treatment, using natural preservatives) need to be properly controlled by FBOs because weak processing steps could lead to accidents. Further, consumers should be aware of this type of perishable food category by carefully reading labels and following the preparation requirements (Havelaar et al. 2010). Labeling, such as *use within × days of opening* and *store below*, may be instructive as well as effective for controlling *L. monocytogenes* in RTE foods (Havelaar et al. 2010).

2.2 Relevance of *L. monocytogenes* in Seafood Products

2.2.1 *L. monocytogenes*—A Remarkable Foodborne Pathogen

L. monocytogenes is a Gram-positive, nonspore-forming bacterium causing severe diseases in humans, namely, listeriosis. Its link to foodborne transmission was recognized long after its first description in 1926 in Cambridge, United Kingdom. *L. monocytogenes* emerged as a new pathogen due to the high mortality rate (typically about 20%), the high economic impact in the food industry, the official control required in case of outbreak situations, and the accompanying market withdrawals (Wagner and McLauchlin 2008). The genus *Listeria* belongs to the *Clostridium* subbranch together with *Staphylococcus, Streptococcus, Lactobacillus,* and *Brochothrix* characterized by an average low GC content (38%) in this phylogenetic position. *L. monocytogenes* is one of ten phenotypically similar species—*Listeria ivanovii, Listeria seeligeri, Listeria marthii, Listeria rocourtiae, Listeria fleischmannii, Listeria innocua, Listeria welshimeri, Listeria grayi,* and *Listeria weihenstephanensis* (Lang-Halter et al. 2013; Bertsch et al. 2013; Graves et al. 2010; Khelef et al. 2006; Leclerq et al. 2010; McLauchlin and Rees 2009).

The formation of mono- and mixed culture biofilms on different surfaces is a well-described property of *L. monocytogenes* that persists in food processing environments, for example, in drains, pipes, and other frequently used equipment (Carpentier and Cerf 2011; Gandhi and Chikindas 2007). Biofilms physically protect the bacterium against disinfection treatments and desiccation. Further, the tolerance to antibiotics is increased because of slow growth rates (McMeekin et al. 2013). The formation of biofilms is affected by several factors, including material surface properties (texture, surface charge, and hydrophobicity), physiological state of cells, strain variation, nutrient levels, and temperature (Srey et al. 2013). Also a lineage-specific formation of biofilms by *L. monocytogenes* is evident but controversial among research outcomes (lineage I serotype 4b vs. lineage II serotype 1/2a strains) (Kadam et al. 2013). *L. monocytogenes* can produce strong biofilms on plastic surfaces (e.g., conveyor belts), but strains that produce weak biofilms have also been found. The application of huge amounts of water in seafood processing (e.g., hatcheries) is a relevant factor contributing to biofilm formation because during the processing the water quality is impaired regardless of the type of water treatment (Srey et al. 2013).

Various publications have reported that *Listeria* spp. are susceptible to antibiotics (Carpentier and Courvalin 1999; Ruiz-Bolivar et al. 2011; Troxler et al. 2000; Walsh et al. 2001; Yücel et al. 2005). First reports on (multi-) resistant strains of *L. monocytogenes* were published in 1988 (resistance to >10 µg of tetracycline per mL). Even if the resistance levels were low at that time, several authors claimed that there was a risk of altering resistance patterns due to changing food production practices. The resistance acquisition in *Listeria* spp., such as enterococcal and streptococcal plasmids, has also been described (Walsh et al. 2001). *L. monocytogenes* is naturally resistant to cephalosporins, aztreonam, fosfomycin, pipemidic acid, dalfopristin, and sulfamethoxazole (Carpentier and Courvalin 1999; Ruiz-Bolivar et al. 2011; Troxler et al. 2000). *Listeria* infection is associated with a high mortality rate and therefore, ampicillin and penicillin G combined with an aminoglycoside, such as gentamicin, are first choice antibiotics. As an alternative therapy, the association of trimethoprim with sulfonamide, such as sulfamethoxazole in co-trimoxazole, is recommended (Carpentier and Courvalin 1999). Other therapy antibiotics such as linezolid and rifampin have also been applied (Ruiz-Bolivar et al. 2011). Fallah et al. (2013) examined the prevalence and resistance patterns of 278 *L. monocytogenes* strains, isolated from RTE-seafood products and associated environments, against 12 antibiotics. Results indicated resistance to first choice antibiotics (ampicillin, penicillin) and a serotype specific (1/2a: 52%; 1/2c: 40%; 4b: 37%) multi-resistance pattern. It could also be concluded that there is a remarkable regional difference of determined antibiotic resistance patterns among *L. monocytogenes* serotypes (1/2a, 1/2b, 1/2c, and 4b).

Because of their tolerance to extreme pH (some cultures grow at pH 9.6), temperature (<0°C–45°C), and salt conditions (tolerance up to 20% (w/v) NaCl), *Listeria* species can be found in a variety of environments, foods, and clinical samples (Farber and Peterkin 1991; Gandhi and Chikindas 2007; Hof 2003). The presumably ubiquitous presence of *L. monocytogenes* requires that its metabolism go through a series of adaptations (Adrião et al. 2008; Hof 2003; Sergelidis and Abrahim 2009) necessary to establish it in difficult niches, resulting in an extensive regulatory repertoire (7% of *Listeria* genes are dedicated to regulatory proteins) (Buchrieser et al. 2003). The physicochemical properties of cold smoked salmon would describe the growth potential of *Listeria* spp.: The salt content (2%–8% water phase salt), phenol amount (2–15 ppm), pH (5.9–6.3), and water activity (0.95–0.98) of the product enable *Listeria* to multiply in the food at refrigeration temperatures (Hwang and Sheen 2009).

According to the EU regulation EC 2073/2005, predictive modeling is a preferred method to examine the growth potential of *L. monocytogenes* in RTE food. Several models exist in scientific literature to describe the growth rate and growth boundaries of *L. monocytogenes* and have been successfully validated for seafood products; however, they differ in complexity. This complexity arises through the parameters considered in those models such as the effect of storage temperature solely or in combination with pH, a_w, NaCl, and other environmental parameters relevant for certain food categories. The square-root model developed for cold smoked salmon only considers the temperature as an environmental parameter (Delignette-Muller et al. 2006). The remarkable growth behavior of *L. monocytogenes* at low (suboptimal) temperatures was emphasized in the review of McKeekin (2013), and the *dual lifestyle* (environmental to parasitic lifestyle) of the intracellular pathogen that needs to be considered for growth modeling approaches was discussed by McMeekin et al. (2013). Mejlholm et al. (2010) have evaluated the performance of six predictive growth rate models for *L. monocytogenes* based on accuracy and bias indices. According to their results the square-root model predicted a 70% faster growth rate in seafood (more than 86% were smoked salmon) than the observed one. The overestimation of the risk may lead, for instance, to excessive use of preservatives (Mejlholm et al. 2010). For the assessment of the growth of *L. monocytogenes* in smoked salmon the so-called Jameson effect (maximum population density) should be considered (Jameson 1962; McKeekin et al. 2013). Predictive microbiology in this sense is an essential tool for the quantitative microbial risk assessment (MQRA) in food and serves as a practical decision instrument for final outputs realized by the food industry.

All these properties require several preventive measures, funded knowledge, and applied HACCP regimes to control the occurrence, multiplication, and persistence of this bacterium in different manufacturing environments. This chapter aims at giving an overview of the presence and behavior of *L. monocytogenes* in fishery products at different processing levels and some current intervention strategies to control this bacterium during processing and in the final product.

2.2.2 Brief Overview of Methods Used for the Detection of *L. monocytogenes* Relevant for the Food Industry Sector

Due to the close relationship of *L. monocytogenes* to other *Listeria* species, the species-specific detection can be hampered by different methodological approaches. Further, the presence of other *Listeria* species is a useful indicator for the possible occurrence of the pathogenic *L. monocytogenes* species.

Conventional culture based methods (according to ISO; FSIS-USDA; FDA; NMKL) for the detection of *L. monocytogenes* in different food matrices remain widely accepted by the food industry and official control bodies. The introduction of combinations of new plating/enrichment media and different protocols has optimized the selectivity and identification of *Listeria* spp. (Zunabovic et al. 2011). Enrichment steps (nonselective and selective) are common to all culture-based reference methods obtaining a presumptive detection result after some days.

Further, the presence of *L. monocytogenes* can be masked during enrichment by closer related species, such as *L. innocua* or by the autochthonous microflora and specific antimicrobial components in the food matrix (Nero et al. 2009; Zitz et al. 2011).

Immunoassay-based methods basically rely on the natural binding affinity of antibodies (e.g., monoclonal, polyclonal, and recombinant) to antigens. These methods can be applied at different steps of enrichment or for the species confirmation of *L. monocytogenes*. However, enrichment-based detection with antibodies may be problematic because of environmental influences on virulence determinants of *L. monocytogenes* (Gasanov et al. 2005; Janzten et al. 2006). To encounter species-specific immune analysis, the detection based on structural components such as flagella, listeriolysin O (LLO) toxin, and protein p60 (encoded by *iap* gene) is preferable to whole-cell test formats (Churchill et al. 2006). To concentrate pathogens (e.g., *Listeria*) during enrichment, immunocapture techniques, including magnetic beads coated with specific antibodies, are used. Other immunoassay-based methods are enzyme-linked immunosorbent assays (ELISA) and latex agglutination assays. Several commercial kits are currently available in the market and are mainly based on conventional enzyme-linked immunosorbent assays (ELISA), Sandwich ELISA, Competitive ELISA, fluorescently labeled ELISA, and Latex Agglutination Assay (Churchill et al. 2006). ELISAs are very popular for the detection of foodborne pathogens due to their ease of use, generation of rapid results (approximately 30–50 h), and applicability to complex food matrices (e.g., fishery products) due to their high specificity coupled with a biochemical reaction

(Gasanov et al. 2005). An ELFA (enzyme-linked fluorescent assay) is more sensitive compared to ELISA due to the fluorescent labels that are conjugated to the antibodies. The detection is faster as the last step of the ELISA (the colorimetric reaction) can be left out; however, this advantage is connected to higher costs (Churchill et al. 2006). An example of a commercially available automated ELFA is the VIDAS® (Vitek Immuno Diagnostic Assay System, bioMerieux, France).

The time-consuming nature of conventional detection methods has led to much improved molecular methods tailored to the food industry needs. In this context PCR assays targeting various genes specific for *L. monocytogenes* (e.g., *hly*, *lmo0733*, *inl*A, *inl*B, *inl*C, 16S rRNA) can be named (Jadhav et al. 2012; Zunabovic et al. 2011). However, the limitations (e.g., estimation of the viability of the target organism, inhibitory substances, and complexity of matrices) of these techniques such as components interfering with the PCR reaction (e.g., cresols, fat, phenolics, and aldehydes in smoked salmon) (Simon et al. 1996), humic acid in soil sediments, heme in blood serum, and proteases in dairy products are often matrix related (Lantz et al. 1994). Frequently, preculture enrichment is indispensable for the enhancement of the sensitivity and detection of viable *L. monocytogenes* strains (Barocci et al. 2008). However, with regard to these optional procedures, there has been a lack of standardized and optimized procedures dealing with diverse food systems (Liu and Busse 2010). Martin et al. (2012) obtained significant differences (based on the Limit of Detection; LOD) between commercial DNA extraction kits in the detection of *L. monocytogenes* by real-time PCR in RTE meat products. The results were mainly influenced by the enrichment volume taken for extraction, the type of food matrix (fermented vs. cooked), and inoculum level.

Modifications of the nucleic acid based technology (e.g., Multiplex PCR, Real-time PCR) have emerged based on the simultaneous detection of pathogenic bacteria or the parallel detection of different bacterial species within the genus. These new and recently published approaches allow rapid, sensitive, and cost-effective analysis of *L. monocytogenes* in different samples (Table 2.2).

2.2.3 Prevalence and Transmission of *L. monocytogenes* in Fishery Products

The ubiquitous behavior of *L. monocytogenes* allows the organism to contaminate the fish at any point in the supply chain (e.g., aquatic areas, farms, processing units). A well-described food model is cold smoked salmon. Salmon (hot and cold smoked) comprises more than 30% of the European processed product market (Rotariou et al. 2014). It is a RTE food traditionally produced by salting, smoking (undergoing incomplete heat coagulation), trimming, or slicing the fish, followed by vacuum packaging (Jahncke and Herman 2001). The refrigerated shelf life (2°C–8°C recommended storage) is individually set at 3–8 weeks (Hwang and Sheen 2009; Rørvik 2000). The application of smoke and the salt content have been shown to have some inhibitory effect (reduction not elimination) on bacteria (Cornu et al. 2006; Rørvik 2000). This process has no significant effect on the inactivation of *L. monocytogenes*, but from a culinary aspect, this is inevitable in smoked fish. In this context various scientific literature and freeware programs deal with predictive growth response in fish matrices but always invest a lot of effort in fit-for-purpose growth models (Cornu et al. 2006).

It is generally considered that a single contamination source during fish processing is not likely to occur. Also it is well known that raw fishery products entering the company do not seem to be important contamination sources but the prevalence may vary (Jinneman et al. 2007; Rørvik, 2000; Rotariu et al. 2014; Thimothe et al. 2003; Zunabovic et al. 2012;). These variables can occur due to farm practices, the water, catching vessels, and hygiene conditions during catching and preprocessing. Hence, the sporadic occurrence in raw fish is generally considered low. Miettinen and Wirtanen (2005) also concluded that the position of *Listeria* on the fish carcass influences the actual prevalence. *Listeria* spp. was most often found in fish gills, followed by skin and viscera in rainbow trout samples. In another work *Listeria* spp. could be isolated from muscle (33%) and viscera samples (67%) with a high species variety including *L. monocytogenes*, *L. seeligeri*, *L. grayi*, and *L. welshimeri* (Jallewar et al. 2007).

Another study performed by Gudbjörnsdóttir et al. (2004) concluded that in the seafood sector almost all *Listeria* positive samples also included *L. monocytogenes* (91.1% of the positive samples) in contrast to other food types (meat and poultry) where *L. innocua* dominated. However, tropical fish were also seen to have an elevated presence of *L. innocua* serving as a hygiene indicator (Parihar et al. 2008).

TABLE 2.2

Novel PCR-Based Protocols for the Specific and Rapid Detection of *L. monocytogenes* in Different Food Matrices

Methodology	Sample Type	Target Species	Target Gene	Strain Evaluation	Reference
Multiplex-PCR	Processed meat products	*Listeria* spp., *L. innocua, L. ivanovii, L. grayi, L. monocytogenes, L. seeligeri, L. welshimeri*	*prs*; *oxidoreductase*; *lmo1030*; *lin0464*; *namA*; *lmo0333*; *scrA*	35 *Listeria* type strains; 26 non-*Listeria* strains	Ryu et al. (2013)
Multiplex real-time PCR (qPCR)	Boiled mussels, cephalopods, bivalves, meat, vegetables, RTE[a], food byproducts, environmental samples	*L. monocytogenes*; *Salmonella* spp.	*hlyA*; *invA*	2 *L. monocytogenes*; 9 non-*Listeria* strains	Garrido et al. (2013)
Real-time reverse-transcriptase PCR	Chilled pork	*L. monocytogenes*	*hlyA*	*L. monocytogenes*; CICC 21583; *L. innocua* 40 non-*Listeria* strains	Ye et al. (2012)
Quantitative real-time PCR (qPCR)	N/D[b]	*L. monocytogenes*	*hlyA*; TaqMan probe	18 *L. monocytogenes* strains; 10 *Listeria* spp.; 5 non-*Listeria* strains	Traunšek et al. (2011)
Combined enrichment real-time PCR	Salmon, pâte, green-veined cheese	*L. monocytogenes*	*prfA*	100 *L. monocytogenes*; 13 *L. innocua*, 3 *L. ivanovii*, 5 *L. seeligeri*, 6 *L. welshimeri*, 3 *L. grayi*; 29 non-*Listeria* isolates	Rossmanith et al. (2006)
Quantitative viable real-time PCR; inclusion of PMA	Environmental samples; swab samples	*L. monocytogenes*	Thermonuclease (*nuc*), *hlyA*	*L. monocytogenes* ATCC 19115	Martinon et al. (2012)

[a] RTE, Ready-to-eat.
[b] N/D, Not defined.

A critical step during cold smoked fish (e.g., salmon) processing is brining (liquid or dry-salting), where the occurrence of *L. monocytogenes* can increase significantly (Jinneman et al. 2007) because of the contamination of the recirculating brine solution. The brining solution mainly consists of water, salt, sugar, spices and flavorings, phosphates, protective bacteria, and chemical preservatives. Sodium nitrite is allowed in the United States (Jahncke and Herman 2001). However, according to a study of Rotariou et al. (2014) it was common practice in the smoked salmon industry to reuse the brine solution for several days. In contrast when applying crystal salt the resulting drips from the fish meat pose a risk of cross-contamination due to lack of splash protection. Also, the background flora influences the growth of inoculated *L. monocytogenes* strains notably (see also Table 2.1). Further, the environmental conditions can alter the virulence of *L. monocytogenes*. The increase of virulence in a low virulent *L. monocytogenes* strain grown in smoked salmon could be observed at elevated storage temperatures by Duodu et al. (2010).

Several prevalence studies were performed with RTE foods including fishery products (Table 2.3). *L. monocytogenes* was most often detected in smoked fish or in products with fish as ingredient. Unfortunately it is rarely mentioned which smoking process was applied (hot or cold) as often only the general term *smoked fish* is used (Lambertz et al. 2012). In an Estonian prevalence study *L. monocytogenes* was detected in about 8% of raw fish samples and in 5% of RTE fish samples, which differs from other published surveys (Kramarenko et al. 2013). Generally a higher incidence is expected in RTE fish products. According to the latest zoonosis report RTE fishery products were the most common samples exceeding the legal safety limit for *L. monocytogenes* (EFSA 2013).

Crustaceans (e.g., shrimp, lobster, prawns) in different convenience degrees have been demonstrated as sources of *Listeria* spp. in a plethora of publications with a highly variable prevalence. Despite cooking regimes (blanching, cooking) and freezing, *L. monocytogenes* could be detected in fresh, frozen, cooked, peeled shrimps, and lobsters at a prevalence rate between 0% and 50% (Wan Norhana et al. 2010). The quantitative levels have rarely been examined but are usually seen at low levels (Jinneman et al. 2007). However, it was observed that the growth rate of *L. monocytogenes* in cooked shrimps was better than in other fishery RTE products rising up to significant levels during chilled storage (Wan Norhana et al. 2010).

2.2.4 Occurrence of *Listeria* spp. in Manufacture Environments

The sporadic high prevalence in RTE fishery products shows that there is a need for tailored in-house control strategies to minimize the risk of human listeriosis. However, the step-forward toward the consumer should also be considered in risk assessments. Regarding the control of this bacterium, selection of appropriate reagents, proper cleaning, and disinfection procedures at frequent intervals should be seen as the primary option for food manufacture. Another crucial factor of process contamination is the job rotation during processing (Rørvik et al. 1997).

The probability for the increase of contamination is higher in the seafood sector than in meat processing because of the use of greater amounts of water during manufacture, forming moist environments favored by *Listeria* (Gudbjörnsdóttir et al. 2004). In the review of (Carpentier and Cerf 2011), conclusions have been drawn regarding the control of *Listeria* spp. in manufacturing environments. One of the main rules is to avoid water in sites where the product is exposed. Secondly, the floors (typically contaminated) should be cleaned and disinfected before the equipment in order to avoid dislodged bacteria from harborage sites on cleaned equipment surfaces. By following these rules, in addition to the improvement of hygiene and maintenance of equipment, it is possible to prevent contamination.

The rise of molecular subtyping methods, such as PFGE (Pulsed-field gel electrophoresis) or PCR-based fingerprinting (e.g., Rep-PCR, RAPD), provides in-depth information on the routes of contamination of *L. monocytogenes*. However, the methods differ in discrimination capacity and lack practical application in the food industry (Zunabovic et al. 2011). Therefore they are mainly performed by research laboratories for epidemiological purposes. Norton et al. (2001) investigated three smoked salmon production facilities by ribotyping and revealed unique contamination patterns for each processor with strains partly persisting up to 6 months. It could be shown that the environment served as primary source

TABLE 2.3

Overview of *L. monocytogenes* Prevalence Data in RTE[a] Fishery Products

Sample Type	Number of RTE Fish Samples	*L. monocytogenes* (%/N)	Presence of Other *Listeria* Species (%/N)	Sampling Point	Country	Reference
Gravad und (hot + cold) smoked fish; (80% *Salmo salar*)	558	14/221 (smoked) 14/200 (gravad) 0.5/55 (>100 CFU/g)	N/D[b]	Retail; FBO's[c]	National Survey: Sweden	Lambertz et al. (2012)
Smoked fish	3.222	17/1344 (cold smoked) 3.4/1.878 (hot smoked)	21/1.344 (cold smoked) 5.2/1.878 (hot smoked)	Retail	Great Britain	FSIS (2008)
Mayonnaise-deli-salads (fish as ingredient)	1.187	7/1187 (all variations) 10/471 (fish salad) 57/58 (smoked fish)	N/D[b]	FBO's[c]	Belgium	Uyttendaele et al. (2009)
Fishery products: Trout, mackerel, salmon, herring, eel	2.935	6/2.938 (smoked, cooked) 1.3/2.607 (<100 CFU[d]/g) 1.3/2.607 (>100 CFU[d]/g)	N/D[b]	Retail FBO's[c]	EU (Member States)	EFSA (2012)

[a] RTE, Ready-to-eat.
[b] N/D, Not described.
[c] FBO, Food business operator.
[d] CFU, Colony forming units.

of contamination as *L. monocytogenes* strains isolated from environmental samples and product samples were distinct from raw material isolates. It has been already described that *Listeria* can persist for more than 10 years in one processing unit gaining increased resistance to cleaning and disinfection with an ability to form biofilms (Carpentier and Cerf 2011). In an earlier study by Autio et al. (1999) the contamination sources were elucidated by PFGE during rainbow trout processing. Brining and post-brining areas (slicing, packaging) were most frequently contaminated by *L. monocytogenes*. Again, the recirculating brine solution (moisture environment) and its environment were dominated by few pulsotypes (specific indistinguishable PFGE strain cluster) and sporadic pulsotypes occurring in raw fish material but not during processing (Autio et al. 1999).

2.3 Strategies to Inhibit the Growth of *L. monocytogenes* and Other *Listeria* Species in RTE Fishery Products

L. monocytogenes is known to be widespread in the environment and therefore capable of surviving under adverse surroundings. Listeriosis (an illness caused by *L. monocytogenes*) is mainly caused by a food source contaminated somewhere in the food supply chain. One possible explanation of this widespread occurrence in food processing facilities is the formation of biofilms on various surfaces (see also Section 2.2.1) (Carpentier 2011). Recurrent *L. monocytogenes* strains have been identified in several investigations presuming a potential persistence in equipment, which may be linked to a resistance to disinfectants. In parallel to the adaption or tolerance of this microorganism to environmental stresses, the modern strategy of *mild food processing* in response to consumer demands for more *fresh* and *natural* food products allows the pathogen to survive and therefore to re-emerge as foodborne hazard. Further, the increased need for reformulated food (Havelaar et al. 2010), for example, reduction of salt concentration, implies parallel examination on the growth or adaptive behavior of *L. monocytogenes* in the reformulated food based on the combination of technological parameters and new food recipes is required (Stollewerk et al. 2012).

Thermal processing is the most commonly applied technique to control pathogenic bacteria in food. It is well known that several bacteria increase their resistance to heat after an exposure to elevated temperatures for a short time (Sergelidis and Abrahim 2009) or when other environmental stressors such as acidification occur in parallel or in a sequential way, leading to a so-called cross-protection (Skandamis et al. 2008). This adaptive response is well described for *L. monocytogenes* that are subjected to different technological hurdles. The highest heat resistance was observed after simultaneous exposure to acid and heat followed by osmotic shock. The parallel application results in more survivors than the sequential hurdle technology (Rajkovic et al. 2010).

Nonthermal processing including antimicrobial additives, in contrast, encompasses preservation techniques that are effective even at ambient temperatures. The application of nonthermal treatment and synergistic effects for the inactivation of foodborne pathogens has been extensively reviewed by Ross et al. (2003). Some of those technologies are discussed in relation to *Listeria* and fishery products in this chapter.

Ultra high pressure (UHP) or high pressure processing (HPP) of food products (mainly RTE) represents a nonthermal alternative of food preservation involving pressures of 5,000–11,000 bar (Brown 2011). Vegetative microorganisms are mainly inactivated, but endospores show higher resistances. Several studies have been conducted on the application of HPP in different food items with regard to changes in ingredients and properties such as protein, color, lipids, and contaminating bacteria (Gudbjornsdottir et al. 2010). Some research has been carried out in the area of surimi, salmon, slurries, and fish gels and the effect on endogenous enzymes but a lesser extent on microbial inactivation. The HPP inactivation of *L. monocytogenes* (Scott A) was more successful at a lower pressure (375 MPa; 5 min; 20°C) than for *Escherichia coli* (700 MPa) to achieve comparable destruction kinetics in a fish slurry (Ramaswamy et al. 2008). The destructive effect on *L. monocytogenes* has been measured depending on parameters, such as temperature, time, and pressure. Gudbjornsdottir et al. (2010) described the efficient inactivation of *L. innocua* in smoked salmon after reaching pressures between 700 and 900 MPa. However, the surviving or residual background microflora (e.g., lactic acid bacteria) also needs to be measured to estimate the real shelf life of a product. This was observed by Takahashi et al. (2012) after the synergistic treatment with biopreservatives (nisin and ε-polylysine) in minced tuna and salmon roe. *Listeria* cell numbers could be decreased below the detection level, but no significant reduction was found in total aerobic bacterial counts.

The application of antimicrobial packaging films using Chitosan (biopolymer) incorporated with nisin, sodium lactate (SL), sodium diacetate (SD), sodium benzoate (SB) or potassium sorbate (PS) show an inhibitory effect on the growth of *Listeria* in smoked salmon. These antimicrobials are approved by the United States and by the European Commission (DG SANCO 2013) at specific concentrations and food applications and are generally recognized as safe (GRAS). It could be shown that chitosan-coated films containing 4.5 mg/cm^2 (SL) alone, 4.5 mg/cm^2 SL in combination with 0.6 mg/cm^2 PS or 2.3 mg/cm^2 SL in combination with 500 IU/cm^2 had a high treatment effect against *L. monocytogenes* in salmon with long-term efficacy (Ye et al. 2008). The long-term inhibitory effect is of high importance when examining the shelf life at refrigerated temperatures where *Listeria* have enough time to regenerate and further multiply to significant levels. The synergistic antilisterial effect of SL and nisin in cold smoked rainbow trout was also confirmed for almost 30 days at 3°C without sensory changes (Nykänen et al. 2000).

Since the FDA (Food and Drug Administration) approval for "pulsed UV light in the production, processing and handling of food," numerous reports have been published about the decontamination of *Listeria* spp. in the area of food, food contact surfaces, and processing environments. For European Member States, the regulation EC 258/97 (Novel Food Regulation) adheres to the proof that the nutritional quality and chemical composition is not altered by new technologies. In principle, the high peak power (short pulse duration for higher energy deliverance) and the UV component of the broad spectrum of the flash (xenon lamps) are capable of inactivating bacteria and fungi (Rajkovic et al. 2010).

Cheigh et al. (2013) investigated the inactivation effects of intense pulsed light (IPL) on surface inoculated filleted shrimps, flatfish, and salmon at fluences of 0 and 17.2 J/cm^2/pulse. The decrease of inoculated (10^4 CFUs) viable *L. monocytogenes* cells was observed between 2.4- and 1.6-log reductions in the following order: shrimp, salmon, and flatfish fillets for 9800–3600 pulses (1960–720 s) at 0.70–1.75 mJ/cm^2/pulse. Because of the lamp, no significant influence of temperature increase on the product surface could be found. This side effect usually poses as a critical factor in IPL applications and has been reported by Ozer and Demirci (2006) for smoked salmon fillets where the surface temperature increased up to 100°C within 60 s of treatment time, resulting in visual changes of the product. The microbial inactivation applied to tuna carpaccio was lower per cm^2 for *L. monocytogenes* reaching levels up to 0.7 log CFU/cm^2 at the maximum energy level. At fluences of 2.1 J/cm^2 extended shelf life was not accomplished with the treated products but the microbial quality could be improved in a way to maintain the sensory quality over 1 week (Hierro et al. 2012).

Because of the perishable character of seafood products, preservative treatments applied to this food category should be carefully selected. Side effects resulting from physical treatments of seafood such as HPP and IPL necessitate the search for alternatives, for example, using biopreservation. This technique usually refers to the use of the natural or controlled microflora and/or its antimicrobial metabolites (Ghanbari et al. 2013) that contribute to food safety. *Carnobacterium*, *Enterococcus*, *Lactobacillus*, and *Lactococcus* are genera associated and isolated from fresh and seawater fish. A comprehensive overview of lactic acid bacteria (LAB) originating from various fish types was given by Ghanbari et al. (2013). Competitive LAB and bacteriocin-producing LAB as protective cultures for seafood applications have been shown to inhibit the growth of *Listeria* spp. through different model applications (Table 2.4). Especially for the application of antilisterially effective *Enterococcus* and *Bacillus* species, the non-pathogenicity (e.g., regarding horizontal transfer of plasmids) should be evidenced (Tomé et al. 2008). Tomé et al. (2008) investigated the antilisterial capacity of different bacterial strains (*Enterococcus faecium*, *Listeria curvatus*, *Listeria delbrueckii*, *Pediococcus acidilactici*) during cold smoked salmon production using *L. innocua* as a model strain. *E. faecium* inoculated in smoked salmon, in particular, showed the best biopreservation characteristics.

Next to various psychrotrophic, pathogenic, and spoilage bacteria surviving in fishery products, efforts have been made to inhibit the growth of *L. monocytogenes*. As this organism is a frequently occurring bacterial contaminant, the use of LAB isolated from different fish species proved to be successful in

TABLE 2.4

Studies Dealing with Growth Inhibition of *Listeria* spp. in Seafood

Matrix/Product Type	Bioprotective Strain(s)	Isolation Source	Study Focus	Reference
Simulated cold smoked fish system	*Carnobacterium* spp. and *Lactobacillus* spp. isolates; *C. divergens* V41; *C. piscicola* V1	Smoked Salmon	Growth inhibition at 4°C	Duffes et al. (1999)
Fish and shellfish products	*Enterococcus* isolates	Sea bass; sea bream	Characterization for inhibitory capacity	Chahad et al. (2012)
In vitro	*Lactobacillus casei* isolates; *Lactobacillus plantarum* isolates	Persian sturgeon; Beluga	Inhibitory capacity against 42 food-borne pathogens	Ghanbari et al. (2009)
Fish model system	*Carnobacterium maltaromaticum*[a]	*Surubim* fish fillets	Growth inhibition at 5°C 35 days; in combination with plant extract	dos Reis et al. (2011)

[a] Bacteriocin-producing and nonbacteriocin-producing strains.

lightly preserved fish products. However, there is still a huge need for research to further characterize strain-specific applications under various conditions including process and storage parameters.

2.4 Conclusions

The European Food Safety Authority (EFSA) has recently published technical specifications for the monitoring of temporal trends in zoonotic agents in animal and food populations at Community as well as Member State level. Regarding the microbiological safety of cold smoked salmon, *L. monocytogenes* has been selected as zoonotic agent for trend analysis and watching. Furthermore, it should be investigated to what extent cold smoked fish contributes to listeriosis cases at the European and worldwide level.

The specific virulence potential of *L. monocytogenes* in RTE fish needs to be assessed. The evaluation of natural antimicrobials applicable as ingredients in RTE fish and the type of application to eliminate the possibility of growth of the *dual lifestyle* pathogen should also be considered. In connection to this question, the propagation potential of low levels of *Listeria* species and *L. monocytogenes* in cold smoked salmon needs to be investigated using robust inoculation trials following Regulation (EC) No. 2073/2005, amended by Regulation No. 1441/2007. Several predictive models have been published with different complexities and therefore should be carefully applied in order not to over- or, even, underestimate the real risk of *L. monocytogenes*.

The increased consumption of RTE foods calls for food-specific measures to prevent contamination with this pathogen. In addition, there is a need for continued education of all related interest groups. Based on this multidisciplinary approach, further control measures can be established to ensure safer processing/handling of this food group. Because of changing processing technologies (e.g., nonthermal preservation), microorganisms can adapt to these conditions and therefore may be expected to reemerge in our products.

REFERENCES

Adrião, A., Vieira, M., Fernandes, I., Barbosa, M., Sol, M., Tenreiro, R.P., Chambel, L. et al. 2008. Marked intra-strain variation in response of *Listeria monocytogenes* dairy isolates to acid or salt stress and the effect of acid or salt adaptation on adherence to abiotic surfaces. *International Journal of Food Microbiology*. 123: 142–150.

Augustin, J.C., Bergis, H., Midelet-Bourdin, G., Cornu, M., Couvert, O., Denis, C., Huchet, V. et al. 2011. Design of challenge testing experiments to assess the variability of *Listeria monocytogenes* growth in foods. *Food Microbiology*. 28: 746–754.

Autio, T., Hielm, S., Miettinen, M., Sjöberg, A.M., Aarnisalo, K., Björkroth, J., Mattila-Sandholm, T., and Korkeala, H. 1999. Sources of *Listeria monocytogenes* contamination in a cold-smoked rainbow trout processing plant detected by pulsed-field gel electrophoresis typing. *Applied and Environmental Microbiology*. 65: 150–155.

Barocci, S., Calza, L., Blasi, G., Briscolini, S., De Curtis, M., Palombo, B., Cucco, L. et al. 2008. Evaluation of a rapid molecular method for detection of *Listeria monocytogenes* directly from enrichment broth media. *Food Control*. 19: 750–756.

Beaufort, A. 2011. The determination of ready-to-eat foods into *Listeria monocytogenes* growth and no growth categories by challenge tests. *Food Control*. 22: 1498–1502.

Beaufort, A., Cornu, M., Bergis, H., Lardeux, A.-L., and Lombard, B. 2008. Technical guidance document on shelf-life studies for *Listeria monocytogenes* in ready-to-eat foods. Community Reference Laboratory for *Listeria monocytogenes*. European Commission. http://ec.europa.eu/food/food/biosafety/salmonella/docs/shelflife_listeria_monocytogenes_en.pdf (accessed May 23, 2014).

Bertsch, D., Rau, J., Eugster, M.R., Haug, M.C., Lawson, P.A., Lacroix, C., and Meile, L. 2013. *Listeria fleischmannii* sp. nov., isolated from cheese. *International Journal of Systematic and Evolutionary Microbiology*. 63: 526–532.

Brown, M. 2011. Processing and food and beverage shelf life. In: *Food and Beverage Stability and Shelf Life*, Kilcast, D. and Subramaniam, P., eds., Woodhead Publishing, Philadelphia, PA, pp. 184–236.

Buchrieser, C., Rusniok, C., Kunst, F., Cossart, P., and Glaser, P. 2003. Comparison of the genome sequences of *Listeria monocytogenes* and *Listeria innocua*: Clues for evolution and pathogenicity. *FEMS Immunology and Medical Microbiology*. 35: 207–213.

Carpentier, B. and Cerf, O. 2011. Review—Persistence of *Listeria monocytogenes* in food industry equipment and premises. *International Journal of Food Microbiology*. 145: 1–8.

Carpentier, E. and Courvalin, P. 1999. Antibiotic resistance in *Listeria* spp. *Antimicrobial Agents and Chemotherapy*. 43: 2103–2108.

Chahad, O.B., Bour, M., Calo-Mata, P., Boudabous, A., and Barros-Velàzquez, J. 2012. Discovery of novel biopreservation agents with inhibitory effects on growth of food-borne pathogens and their application to seafood products. *Research in Microbiology*. 163: 44–54.

Cheigh, C.I., Hwang, H.J., and Chung, M.S. 2013. Intense pulsed light (IPL) and UV-C treatments for inactivating *Listeria monocytogenes* on solid medium and seafoods. *Food Research International*. 54: 745–752.

Churchill, R.L.T., Lee, H., and Hall, C. 2006. Detection of *Listeria monocytogenes* and the toxin listeriolysin O in food. *Journal of Microbiological Methods*. 64: 141–170.

Cornu, M., Beaufort, A., Rudelle, S., Laloux, L., Bergis, H., Miconnet, N., Serot, T., and Delignette-Muller, M.L. 2006. Effect of temperature, water-phase salt and phenolic contents on *Listeria monocytogenes* growth rates on cold-smoked salmon and evaluation of secondary models. *International Journal of Food Microbiology*. 106: 159–168.

Delignette-Muller, M.L., Cornu, M., Pouillot, R., and Denis, J.B. 2006. Use of Bayesian modelling in risk assessment: Application to growth of *Listeria monocytogenes* and food flora in cold-smoked salmon. *International Journal of Food Microbiology*. 106: 195–208.

DG SANCO. 2013. Food additives database. https://webgate.ec.europa.eu/sanco_foods/main pdf (accessed September 26, 2013).

dos Reis, F.B., de Souza, V.M., Thomaz, M.R.S., Fernandes, L.P., de Oliveira, W.P., and De Martinis, E.C.P. 2011. Use of *Carnobacterium maltaromaticum* cultures and hydroalcoholic extract of *Lippia sidoides* Cham. against *Listeria monocytogenes* in fish model systems. *International Journal of Food Microbiology*. 146: 228–234.

Duffes, F., Leroi, F., Boyaval, P., and Dousseta, X. 1999. Inhibition of *Listeria monocytogenes* by *Carnobacterium* spp. strains in a simulated cold smoked fish system stored at 48°C. *International Journal of Food Microbiology*. 47: 33–42.

Dufour, C. 2010. Application of EC Regulation No. 2073/2005 regarding *Listeria monocytogenes* in ready-to-eat foods in retail and catering sectors in Europe. *Food Control*. 22: 1491–1494.

Duodu, S., Holst-Jensen, A., Skjerdal, T., Cappelier, J.M., Pilet, M.F., and Loncarevic, S. 2010. Influence of storage temperature on gene expression and virulence potential of *Listeria monocytogenes* strains grown in a salmon matrix. *Food Microbiology*. 27: 795–801.

EFSA (European Food Safety Authority). 2012. The European Union summary report on trends and sources of zoonoses, zoonotic agents and food-borne outbreaks in 2010. *EFSA Journal*. 10: 2597.

EFSA (European Food Safety Authority). 2013. The European Union summary report on trends and sources of zoonoses, zoonotic agents and food-borne outbreaks in 2011. *EFSA Journal*. 11: 3129.

EFSA (Scientific Opinion of the Panel on Biological Hazards [BIOHAZ]). 2007. Request for updating the former SCVPH opinion on *Listeria monocytogenes* risk related to ready-to-eat foods and scientific advice on different levels of *Listeria monocytogenes* in ready-to-eat-foods and the related risk for human illness. *EFSA Journal*. 599: 1–42.

Fallah, A.A., Siavash Saei-Dehkordi, S., and Mahzounieh, M. 2013. Occurrence and antibiotic resistance profiles of *Listeria monocytogenes* isolated from seafood products and market and processing environments in Iran. *Food Control*. 34: 630–636.

Farber, J.M. and Peterkin, P.I. 1991. *Listeria monocytogenes*, a food-borne pathogen. *Microbiological Reviews*. 55: 476–511.

Food Standards Agency. 2009. General guidance for food business operators EC Regulation No. 2073/2005 on microbiological criteria for foodstuffs. Published on the website http://www.food.gov.uk/foodindustry/regulation/europeleg/eufoodhygieneleg/microbiolreg (accessed September 26, 2013).

FSIS. 2008. Food Survey Information Sheet 05/08. A microbiological survey of retail smoked fish with particular reference to the presence of *Listeria monocytogenes*. http://www.food.gov.uk/multimedia/pdfs/fsis0508.pdf (accessed July 2013).

Gandhi, M. and Chikindas, M.L. 2007. *Listeria*: A foodborne pathogen that knows how to survive. *International Journal of Food Microbiology*. 113: 1–15.

Garrido, A., Chapela, M.J., Román, B., Fajardo, P., Lago, J., Vieitesm, J.M., and Cabado, A.G. 2013. A new multiplex real-time PCR developed method for *Salmonella* spp. and *Listeria monocytogenes* detection in food and environmental samples. *Food Control*. 30: 76–85.

Gasanov, U., Hughes, D., and Hansbro, P.M. 2005. Methods for the isolation and identification of *Listeria* spp. and *Listeria monocytogenes*: A review. *FEMS Microbiology Reviews*. 29: 851–875.

Ghanbari, M., Jami, M., Domig, K.J., and Kneifel, W. 2013. Seafood biopreservation by lactic acid bacteria—A review. *LWT—Food Science and Technology*. 54: 315–324.

Ghanbari, M., Jami, M., Kneifel, W., and Domig, K.J. 2009. Antimicrobial activity and partial characterization of bacteriocins produced by lactobacilli isolated from Sturgeon fish. *Food Control*. 32: 379–385.

Graves, L.M., Helsel, L.O., Steigerwalt, A.G., Morey, R.E., Daneshvar, M.I., Roof, S.E., Orsi, R.H. et al. 2010. *Listeria marthii* sp. nov., isolated from the natural environment, Finger Lakes National Forest. *International Journal of Systematic and Evolutionary Microbiology*. 60: 1280–1288.

Gudbjornsdottir, B., Jonsson, A., Hafsteinsson, H., and Heinz, V. 2010. Effect of high-pressure processing on *Listeria* spp. and on the textural and microstructural properties of cold smoked salmon. *LWT—Food Science and Technology*. 43: 366–374.

Gudbjörnsdóttir, B., Suihko, M.L., Gustavsson, P., Thorkelsson, G., Salo, S., Sjöberg, A.M., Niclasen, O., and Bredholt, S. 2004. The incidence of *Listeria monocytogenes* in meat, poultry and seafood plants in the Nordic countries. *Food Microbiology*. 21: 217–225.

Havelaar, A.H., Brul, S., de Jong, A., de Jonge, R., Zwietering, M.H., and ter Kuile, B.H. 2010. Future challenges to microbial food safety. *International Journal of Food Microbiology*. 139: 79–94.

Hierro, E., Ganan, M., Barroso, E., and Fernández, M. 2012. Pulsed light treatment for the inactivation of selected pathogens and the shelf-life extension of beef and tuna carpaccio. *International Journal of Food Microbiology*. 158: 42–48.

Hof, H. 2003. History and epidemiology of listeriosis. *FEMS Immunology and Medical Microbiology*. 35: 199–202.

Hwang, C.A. and Sheen, S. 2009. Modelling the growth characteristics of *Listeria monocytogenes* and native microflora in smoked salmon. *Journal of Food Science*. 74: 125–130.

Jadhav, S., Bhave, M., and Palombo, E.A. 2012. Methods used for the detection and subtyping of *Listeria monocytogenes*. *Journal of Microbiological Methods*. 88: 327–341.

Jahncke, M.L. and Herman, D. 2001. Control of food safety hazards during cold-smoked fish processing, Chapter VI. *Journal of Food Science (Special Supplement)*. 66: 1104–1112.

Jallewar, P.K., Kalorey, D.R., Kurkure, N.V., Pande, V.V., and Barbuddhe, S.B. 2007. Genotypic characterization of *Listeria* spp. isolated from fresh water fish. *International Journal of Food Microbiology*. 114: 120–123.

Jameson, J.F. 1962. A discussion of the dynamics of *Salmonella* enrichment. *Journal of Hygiene*. 60: 193–207.

Janzten, M.M., Navas, J., Corujo, J., Moreno, R., López, V., and Martínez-Suárez, J.V. 2006. Specific detection of *Listeria monocytogenes* in foods using commercial methods: From chromogenic media to real-time PCR. *Spanish Journal of Agricultural Research*. 4: 235–247.

Jaroni, D., Ravishankar, S., and Juneja, V. 2010 Microbiology of ready to eat foods. In: *Ready to Eat Foods: Microbial Concerns and Control Measures*. A. Hwang and L. Huang, eds., CRC Press/Taylor & Francis Group, Boca Raton, FL, pp. 1–60.

Jinneman, K.C., Wekell, M.M., and Eklund, M.W. 2007. Incidence and behavior of *Listeria monocytogenes* in fish and seafood. In: *Listeria, Listeriosis and Food Safety*, Ryser, E.T. and Marth, E.H., eds., CRC Press, New York, pp. 617–653.

Kadam, S.R., den Besten, H.M.W., van der Veen, S., Zwietering, H.M., Moezelaar, R., and Abee, T. 2013. Diversity assessment of *Listeria monocytogenes* biofilm formation: Impact of growth condition, serotype and strain origin. *International Journal of Food Microbiology*. 165: 259–264.

Khelef, N., Lecuit, M., Buchrieser, C., Cabanes, D., Dussurget, O., and Cossart, P. 2006. *Prokaryotes—A Handbook on the Biology of Bacteria. Bacteria: Firmicutes, Cyanobacteria*, Vol. 4, Dworkin, M., ed., Springer, New York, pp. 404–476.

Kramarenko, T., Roasto, M., Meremäe, K., Kuningas, M., Põltsama, P., and Elias, T. 2013. *Listeria monocytogenes* prevalence and serotype diversity in various foods. *Food Control*. 30: 24–29.

Lambertz, S.T., Nilsson, C., Bradenmark, A., Sylvén, S., Johansson, A., Jansson, L.M., and Lindblad, M. 2012. Prevalence and level of *Listeria monocytogenes* in three categories of ready-to-eat foods in Sweden 2010. *International Journal of Food Microbiology*. 160: 24–31.

Lang-Halter, E., Neuhaus, K., and Scherer, S. 2013. *Listeria weihenstephanensis* sp. nov., isolated from the water plant *Lemna trisulca* taken from a freshwater pond. *International Journal of Systematic and Evolutionary Microbiology*. 63: 641–647.

Lantz, P.G., Hahn-Hägerdal, B., and Rådström, P. 1994. Sample preparation methods in PCR-based detection of food pathogens. *Trends in Food Science & Technology*. 5: 384–389.

Leclerq, A., Clermont, D., Bizet, C., Grimont, P.A.D., Le Fleche-Mateos, A., Roche S.M., Buchrieser, C. et al. 2010. *Listeria rocourtiae* sp. nov. *International Journal of Systematic and Evolutionary Microbiology*. 60: 2210–2214.

Liu, D. 2008. Epidemiology. In: *Handbook of Listeria monocytogenes*, Liu, D., ed., Taylor & Francis Group, Boca Raton, FL, pp. 27–59.

Liu, D. and Busse, H.J. 2010. *Listeria*, Chapter 15. In: *Molecular Detection of Food Borne Pathogens*, Liu, D., ed., CRC Press, Boca Raton, FL, pp. 207–220.

Martin, B., Garriga, M., and Aymerich, T. 2012. Pre-PCR treatments as a key factor on the probability of detection of *Listeria monocytogenes* and *Salmonella* in ready-to-eat meat products by real-time PCR. *Food Control*. 27: 163–169.

Martinon, A., Cronina, U.P., Quealy, J., Stapleton, A., and Wilkinson, M.G. 2012. Swab sample preparation and viable real-time PCR methodologies for the recovery of *Escherichia coli*, *Staphylococcus aureus* or *Listeria monocytogenes* from artificially contaminated food processing surfaces. *Food Control*. 24: 86–94.

McLauchlin, J. and Rees, C.E.D. 2009. Genus *Listeria* Pirie 1940a, 383AL. In: *Bergey's Manual of Systematic Bacteriology: The Firmicutes*, De Vos Paul, ed., Springer, New York, pp. 244–268.

McMeekin, T., Olley, J., Ratkowsky, D., Corkrey, R., and Ross, T. 2013. Predictive microbiology theory and application: Is it all about rates? *Food Control*. 29: 290–299.

Mejlholm, O., Gunvig, A., Borggaard, C., Blom-Hanssen, J., Mellefont, L., Ross, T., Leroi, F., Else, T., Visser, D., and Dalgaard, P. 2010. Predicting growth rates and growth boundary of *Listeria monocytogenes*—An international validation study with focus on processed and ready-to-eat meat and seafood. *International Journal of Food Microbiology*. 141: 137–150.

Midelet-Bourdin, G., Copin, S., Leleu, G., and Malle, P. 2010. Determination of *Listeria monocytogenes* growth potential on new fresh salmon preparations. *Food Control*. 21: 1415–1418.

Miettinen, H. and Wirtanen, G. 2005. Prevalence and location of *Listeria monocytogenes* in farmed rainbow trout. *International Journal of Food Microbiology*. 104: 135–143.

Nero, L.A., Mattos, M.R., Aguiar, M., Barros, F., Beloti, V., and Franco, B.D. 2009. Interference of raw milk autochthonous microbiota on the performance of conventional methodologies for *Listeria monocytogenes* and *Salmonella* spp. detection. *Microbiological Research*. 164: 529–535.

Norton, D.M., Mccamey, M.A., Gall, K.L., Scarlett, J.M., Boor, K.J., and Wiedmann, M. 2001. Molecular studies on the ecology of *Listeria monocytogenes* in the smoked fish processing industry. *Applied and Environmental Microbiology*. 67: 198–205.

Nykänen, N., Weckman, K., and Lapveteläinen, A. 2000. Synergistic inhibition of *Listeria monocytogenes* on cold-smoked rainbow trout by nisin and sodium lactate. *International Journal of Food Microbiology*. 61: 63–72.

Ozer, N.P. and Demirci, A. 2006. Inactivation of *Escherichia coli* O157:H7 and *Listeria monocytogenes* inoculated on raw salmon fillets by pulsed UV-light treatment. *International Journal of Food Science & Technology*. 41: 354–360.

Parihar, V.S., Barbuddhe, S.B., Danielsson-Tham, M.L., and Tham, W. 2008. Isolation and characterization of *Listeria* species from tropical seafoods. *Food Control*. 19: 566–569.

Rajkovic, A., Smigic, N., and Devlieghere, F. 2010. Contemporary strategies in combating microbial contamination in food chain. *International Journal of Food Microbiology* 141: 29–42.

Ramaswamy, H.S., Zaman, S.U., and Smith, J.P. 2008. High pressure destruction kinetics of *Escherichia coli* (O157:H7) and *Listeria monocytogenes* (Scott A) in a fish slurry. *Journal of Food Engineering*. 87: 99–106.

Rørvik, L.M. 2000. *Listeria monocytogenes* in the smoked salmon industry. *International Journal of Food Microbiology*. 62: 183–190.

Rørvik, L.M., Skjerve, E., Knudsen, B.R., Yndestad, M. 1997. Risk factors for contamination of smoked salmon with *Listeria monocytogenes*. *International Journal of Food Microbiology*. 37: 215–219.

Ross, A.I.V., Griffiths, M.W., Mittal, G.S., and Deeth, H.C. 2003. Combining nonthermal technologies to control foodborne microorganisms. *International Journal of Food Microbiology*. 89: 125–138.

Rossmanith, P., Krassnig, M., Wagner, M., and Hein, I. 2006. Detection of *Listeria monocytogenes* in food using a combined enrichment/real-time PCR method targeting the prfA gene. *Research in Microbiology*. 157: 763–771.

Rotariu, O., Thomas, D.J.I., Goodburn, K.E., Hutchison, M.L., and Strachan, N.J.C. 2014. Smoked salmon industry practices and their association with *Listeria monocytogenes*. *Food Control*. 35: 284–292.

Ruiz-Bolivar, Z., Neuque-Rico, M.C., Poutou-Piñales, R.A., Carrascal-Camacho, A.K. and Mattar, S. 2011. Antimicrobial susceptibility of *Listeria monocytogenes* food isolates from different cities in Colombia. *Foodborne Pathogens and Disease*. 8: 8.

Ryu, J., Park, S.H., Yeom, Y.S., Shrivastav, A., Lee, S.H., Kim, Y.R., and Kim, H.Y. 2013. Simultaneous detection of *Listeria* species isolated from meat processed foods using multiplex PCR. *Food Control*. 32: 659–664.

Sergelidis, D. and Abrahim, A. 2009. Adaptive response of *Listeria monocytogenes* to heat and its impact on food safety. *Food Control*. 20: 1–10.

Simon, M.C., Gray, D.I., and Cook, N. 1996. DNA extraction and PCR methods for the detection of *Listeria monocytogenes* in cold-smoked salmon. *Applied and Environmental Microbiology*. 62: 822–824.

Skandamis, P.N., Yoon, Y., Stopforth, J.D., Kendall, P.A., and Sofos, J.N. 2008. Heat and acid tolerance of *Listeria monocytogenes* after exposure to single and multiple sublethal stresses. *Food Microbiology*. 25: 294–303.

Srey, S., Jahid, I.K., and Ha, S.D. 2013. Review—Biofilm formation in food industries: A food safety concern. *Food Control*. 31: 572–585.

Stollewerk, K., Jofré, A., Comaposada, J., Arnau, J., and Garriga, M. 2012. The effect of NaCl-free processing and high pressure on the fate of *Listeria monocytogenes* and *Salmonella* on sliced smoked dry-cured ham. *Meat Science*. 90: 472–477.

Takahashi, H., Kashimura, M., Miya, S., Kuramoto, S., Koiso, H., Kuda, T., and Kimura, B. 2012. Effect of paired antimicrobial combinations on *Listeria monocytogenes* growth inhibition in ready-to-eat seafood products. *Food Control*. 26: 397–400.

Thimothe, J., Nightingale, K.K., Gall, K., Scott, V.N., and Wiedmann, M. 2003. Tracking of *Listeria monocytogenes* in smoked fish processing plants. *Journal of Food Protection*. 67: 328–341.

Tirpanalan, Ö., Zunabovic, M., Domig, K.J., and Kneifel, W. 2011. Mini review: Antimicrobial strategies in the production of fresh-cut lettuce products. In: *Science against Microbial Pathogens: Communicating Current Research and Technological Advances*, Méndez-Vilas, A., ed., Formatex Research Center, Badajoz, Spain, pp. 176–188.

Tomé, E., Gibbs, P.A., and Teixeira, P.C. 2008. Growth control of *Listeria innocua* 2030c on vacuum-packaged cold-smoked salmon by lactic acid bacteria. *International Journal of Food Microbiology*. 121: 285–294.

Traunšek, U., Toplak, N., Jeršek, B., Lapanje, A., Majstorović, T., and Kovač, M. 2011. Novel cost-efficient real-time PCR assays for detection and quantitation of *Listeria monocytogenes*. *Journal of Microbiological Methods*. 85: 40–46.

Troxler, R., von Graevenitz, A., Funke, G., Wiedemann, B., and Stock, I. 2000. Natural antibiotic susceptibility of *Listeria* species: *L. grayi*, *L. innocua*, *L. ivanovii*, *L. monocytogenes*, *L. seeligeri* and *L. welshimeri* strains. *Clinical Microbiology and Infection*. 6: 525–535.

Uyttendaele, M., Busschaert, P., Valero, A., Geeraerd, A.H., Vermeulen, A., Jacxsens, L., Goh, K.K., De Loy, A., Van Impe, J.F., and Devlieghere, F. 2009. Prevalence and challenge tests of *Listeria monocytogenes* in Belgian produced and retailed mayonnaise-based deli-salads, cooked meat products and smoked fish between 2005 and 2007. *International Journal of Food Microbiology*. 133: 94–104.

Vermeulen, A., Devlieghere, F., De Loy-Hendrickx, A., and Uyttendaele, M. 2011. Critical evaluation of the EU-technical guidance on shelf-life studies for *L. monocytogenes* on RTE-foods: A case study for smoked salmon. *International Journal of Food Microbiology*. 145: 176–185.

Wagner, M. and McLauchlin, J. 2008. Biology. In: *Handbook of Listeria monocytogenes*, Liu, D., ed., Taylor & Francis Group, Boca Raton, FL, pp. 3–25.

Walsh, D., Duffy, G., Sheridan, J.J., Blair, I.S., and McDowell, D.A. 2001. Antibiotic resistance among *Listeria*, including *Listeria monocytogenes*, in retail foods. *Journal of Applied Microbiology*. 90: 517–522.

Wan Norhana, M.N., Poole, S.E., Deeth, H.C., and Dykes, G.A. 2010. Prevalence, persistence and control of *Salmonella* and *Listeria* in shrimp and shrimp products: A review. *Food Control*. 21: 343–361.

Ye, K., Zhang, Q., Jiang, Y., Xu, X., Cao, J., and Zhou, G. 2012. Rapid detection of viable *Listeria monocytogenes* in chilled pork by real-time reverse-transcriptase PCR. *Food Control*. 25: 117–124.

Ye, M., Neetoo, H., and Chen, H. 2008. Effectiveness of chitosan-coated plastic films incorporating antimicrobials in inhibition of *Listeria monocytogenes* on cold-smoked salmon. *International Journal of Food Microbiology* 127: 235–240.

Yücel, N., Citak, S., and Önder, M. 2005. Prevalence and antibiotic resistance of *Listeria* species in meat products in Ankara, Turkey. *Food Microbiology*. 22: 241–245.

Zitz, U., Zunabovic, M., Domig, K.J., Wilrich, P.T., and Kneifel, W. 2011. Reduced detectability of *Listeria monocytogenes* in the presence of *Listeria innocua*. *Journal of Food Protection*. 74: 1282–1287.

Zunabovic, M., Domig, K.J., and Kneifel, W. 2011. Practical relevance of methodologies for detecting and tracing of *Listeria monocytogenes* in ready-to-eat foods and manufacture environments—A review. *LWT—Food Science and Technology*. 44: 351–362.

Zunabovic, M., Domig, K.J., Pichler, I., and Kneifel, W. 2012. Monitoring transmission routes of *Listeria* spp. in smoked salmon production with repetitive element sequence-based PCR techniques. *Journal of Food Protection*. 75: 504–511.

3

Fruit Juice Processing: Addressing Consumer Demands for Safety and Quality

Alonzo A. Gabriel

CONTENTS

3.1 Introduction

The market for fruit juices and fruit juice products has recently been benefiting from increasing consumer patronage due to the aggressive campaigns of the industry and the government and the desire of the consumers to lead healthful lifestyles. In a study conducted by the Food Marketing Institute (FMI) in the United States in 2008 and 2009, almost half of the consumers (41%–46%) stated that they buy products that cater to their health and wellness needs (Sloan, 2010). The beneficial effects of consuming fruits and fruit products are attributed to the presence of antioxidants, which have been pointed to alleviate a number of degenerative illnesses, including cardiovascular diseases, cancer, and aging (Kaur and Kapoor, 2001). Juice consumption may be considered a convenient means of meeting the recommended daily intake for fruits and vegetables as one-third to one-half cup of pure, undiluted juice is already equivalent to one serving of fruit or vegetable (Food and Nutrition Research Institute [FNRI], 1994; United States Department of Health and Human Services [USDHHS] and United States Department of Agriculture [USDA], 2005).

However, the fruit juice industry has also been challenged by a number of safety and quality issues related to microbiological contamination. As a consequence of improper practices throughout production, handling, storage, and processing, pathogenic and spoilage microorganisms may contaminate the product and increase the risks of disease outbreaks and product losses (Raybaudi-Masilia et al., 2009). Vojdani et al. (2008) summarized that from 1995 to 2005, 21 juice-associated outbreaks were reported to the US Centers for Disease Control and Prevention (CDC), resulting in 1366 illnesses. Most of these outbreaks involved the consumption of apple juice or cider and orange juice products. *Escherichia coli* O157:H7, -O111 (Shiga toxin-producing), and *Salmonella* spp. were reported as major causative agents of the outbreaks. To many food safety analysts and epidemiologists, such outbreaks were surprising as the inherent acidity of fruit juices had once been thought to be an effective controlling factor for these pathogens (Keller and Miller, 2006; Mazzotta, 2001).

Studies have shown that sublethal exposures to acidic environments are able to induce acid habituation (Goodson and Rowbury, 1989) or acid tolerance (Foster and Hall, 1990) in pathogens such as *E. coli* O157:H7, *Salmonella* spp., and *Listeria monocytogenes*, allowing them to exhibit both homologous and heterologous adaptations to subsequent acid and thermal stresses, respectively (Buchanan and Edelson, 1999; Koutsomanis and Sofos, 2004; Mazzotta, 2001; Ryu and Beuchat, 1998; Sharma et al., 2005). Food environments and processing conditions may, however, expose microorganisms to multiple stresses that could promote *stress hardening* (Skandamis et al., 2008). Aside from acidification, reduction of water activity (a_w) through addition of solutes or water evaporation and temperature control is also a commonly applied method for preventing the growth of microorganisms in food (Buchanan and Edelson, 1999; Tiganitas et al., 2009).

In response to the occurrences of juice-related outbreaks, the US Food and Drug Administration (USFDA) compelled manufacturers to comply with the juice Hazard Analysis and Critical Control Point (HACCP) by employing processing techniques capable of reducing populations of pertinent pathogens by 5 log cycles (Goodrich et al., 2005; USDA and USDHHS, 2001). Despite being a traditional method of pasteurization, thermal processing remains the primary means by which pathogens are eliminated from foods (Buchanan and Edelson, 1999) and a reliable and relatively affordable means of producing safe juice products (Mak et al., 2001). The negative effects of thermal processing on the sensory and nutritional qualities of juices have, however, prompted a number of studies to explore alternative processing methods including ultraviolet irradiation (Wright et al., 2000) and ultrasound exposure (Salleh-Mack and Roberts, 2007). Furthermore, the efficacies of a number of traditional and alternative natural antimicrobials as fruit juice additives are also being given attention (Raybaudi-Massilia et al., 2009). Hurdle technology–based juice processing using combinations of two or more physical or chemical treatments or combinations of physical and chemical treatments has also been studied (Akpomedaye and Ejechi, 1999; McNamee et al., 2010; Nguyen and Mittal, 2007; Noci et al., 2008). Hurdle technology involves the intelligent manipulation and combination of various intrinsic, extrinsic, and implicit food factors to come up with preservation techniques for food (Leistner and Gorris, 1995; Leistner and Gould, 2002).

Therefore, considering the market value of fruit juices and the challenges posed by microorganisms to the industry, this chapter was written to provide pertinent information related to processing such products for safety and quality optimization. Brief discussions on fruit juice consumption and composition are provided. A review on microbiological hazards, including infection outbreaks, government mitigation strategies, and traditional and novel processing technologies are also included in this chapter. Emphases are also given on conducting microbiological challenge studies as well as appropriate statistical tools for the establishment of precise process schedules.

3.2 Fruit Juices and Juice Consumption

The Codex Alimentarius of the Food and Agriculture Organization (FAO, 1992) defines juice as the unfermented but fermentable extract intended for direct consumption obtained by a mechanical process from sound, ripe fruits and preserved exclusively by physical means. The juice may be turbid or clear and may have been concentrated and later reconstituted with water suitable for the purpose of maintaining the essential composition and quality factors of the juice. Sugars or acidulants may be added but must be endorsed in the individual standard (Bates et al., 2001). In the market, juices are further classified according to the level of dilution and/or processing steps applied to the product. Bates et al. (2001) enumerated these classifications and explained the importance of declaring the product's commercial designation as prescribed by law. Products declared as *pure or 100% juice* are composed of nothing but extracted fruit juice that is not from concentrate. *Fresh squeezed* juices are those that have not been pasteurized and hence must constantly be refrigerated, while those marked *chilled, ready to serve* are usually held refrigerated and are made from concentrate and are usually pasteurized. Products declared as *not from concentrate* are single strength, and subjected to pasteurization after extraction, while those labeled *from concentrate* are reconstituted and pasteurized. Unpasteurized juice products that are single strength and immediately frozen after extraction are declared as *fresh frozen*. *Purees* contain pulp and are more viscous than juices, while *nectars* usually contain 20%–50% juice, sweetened and acidified, and may be clear or pulpy. Other juice products that are diluted with water include those declared as *juice drink*, *juice beverage*, and *juice cocktail*, which contain 10%–20% real juice. Products containing about 10% juice are designated *fruit+ade*, for example, lemonade.

Curtis (1997) discussed that in the United States, commercial juice production started with the advent of the pasteurization process for bottled grape juices in the late 1800s. Orange juice started to become a popular source of vitamin C in the 1940s when flash pasteurization protocols for frozen concentrates were established. With the ratification of the USFDA's Nutrition Labeling Education Act of 1990 (NLEA) and the aggressive promotions of the US Department of Agriculture (USDA) on the Food Guide Pyramid, consumer patronage of fruit juice and fruit juice–based beverages has continuously increased. Fruit juices, fruit-based beverages, and fruit-flavored drinks are continuously being given attention by functional food product developers in order to address the current consumer demands. Frederick (2009) recounted that in the past couple of years, there has been a significant shift in the way consumers approach beverages. Consumers have turned away from the traditional carbonated soft drinks and upgraded beverages as lifestyle items that offer various functionalities to meet their specific needs. Juice consumption is a convenient means of meeting the recommended daily intake for fruits and vegetables as one-third to one-half cup of pure, undiluted juice is already equivalent to one serving of fruit or vegetable (FNRI, 1994; USDHHS and USDA, 2005).

3.3 Composition and Physicochemical Properties

Southgate et al. (1995) explained that fruit juices rarely have constant composition and physicochemical characteristics due to the natural variations in the fruit composition and extraction procedures applied. Freshly extracted juices generally have similar moisture content as the source fruit. Minute amounts of nitrogenous substances including free amino acids and soluble cytoplasmic proteins are also included, with clarified juices having less than the cloudy ones. Fruit juices are nutritionally fat-free as very small amounts of lipids, most of which are essential flavor compounds, are found in the products. The extraction process also removes most plant cell wall polysaccharides, leaving only small amounts of soluble pectins in the expressed juice. Hence, the carbohydrates present in fruit juices are mainly mixtures of free sugars, the composition of which is also affected by other juice properties and processing and storage parameters. Curtis (1997) further explained that the sweetness of fruit juices is mainly due to the disaccharide sucrose and its monosaccharide components, glucose and fructose. The relative amounts of these sugars vary, depending on the fruit species. In citrus juice, the typical sucrose/fructose/glucose ratio was previously reported at 2:1:1, while that in apples ranges from 1:2:1 to 1:3:1. Furthermore, sorbitol has also been reported as one of the dominant sugars present in apple juice (Lea, 1995; Li Wan Po et al., 2002). The low pH and characteristic tartness of juices are due to the presence of organic acids, the most common of which are citric, malic, and tartaric acids (Curtis, 1997).

3.4 Microbial Safety of Fruit Juices

As the demands for fresh fruit and vegetable products increased, so did the number of occurrences of outbreaks of microbial illnesses due to the consumption of such products (Raybaudi-Massilia et al., 2009). Minimally processed fruits and vegetables, including unpasteurized fruit juices, may be inappropriately handled during production, harvesting, and storage and get contaminated with spoilage and pathogenic microorganisms. It has been previously estimated that 10%–30% of fruits and vegetables harvested in the United States are wasted due to mechanical injuries, physiological decays, and microbial spoilage caused by a wide variety of bacteria, yeasts, and molds (Harvey, 1978; Nguyen-The and Carlin, 1994).

Moreover, consumption of unpasteurized fruit juices has also become a significant public health concern as these products have been linked to a number of outbreaks of illnesses. In 1991, the occurrence of *E. coli* O157:H7 infections and hemolytic uremic syndrome was linked to traditionally pressed apple cider (Besser et al., 1993). Since then, a number of outbreaks linked to fruit juices caused by a number of pathogenic bacteria and parasites have been reported. In the CDC reports summarized by Vojdani et al. (2008), the five outbreaks caused by *Salmonella* spp. accounted for 52% of the reported illnesses and 63% of the reported hospitalizations. Apple juice products were reported vehicles in 10 of the 21 outbreaks and 8 outbreaks were reported to be due to orange juice consumption. Pineapple juice, fruit juice, and a mix of juices (apple, orange, grape, and pineapple mix) were implicated in the remaining three outbreaks. Furthermore, in 1999, 500 laboratory-confirmed cases of salmonellosis were reported to have been due to unpasteurized orange juice consumption in Australia (Luedtke and Powell, 2000).

The occurrences of such outbreaks due to fruit juice consumption have prompted the USFDA to sanction the juice-labeling regulation (USDA and USDHHS, 1998). This regulation aimed for processors of beverages containing juice or juice ingredients that have not been processed to achieve a 5 log reduction of a target pathogen population. The processors were required to use a warning label indicating the risks associated with drinking such beverages (Vojdani et al., 2008).

Vojdani et al. (2008) also recounted that despite the juice-labeling regulation, the number of reported juice outbreaks did not significantly decline. In fact, from 1999 through 2001, seven more outbreaks were reported by the CDC. In 1999, 398 cases of *Salmonella* Muenchen infection caused by orange juice consumption in 15 states, and in 2000, 88 more cases of *Salmonella* Enteritidis infection associated with orange juice were reported. Hence, in 2002, the USFDA implemented the federal juice HACCP, which compelled juice manufacturers to apply a processing or a combination of processing steps capable of reducing appropriate pathogens in fruit juice by 5 log cycles (USDA and USDHHS, 2001). The juice HACCP regulation became effective in January 2002, 2003, and 2004, for large, medium, and small businesses, respectively, and seemed to have an impact and reduced the occurrences of juice-associated outbreaks.

3.4.1 Microbial Stress Adaptation and Its Implication to Juice Safety

The implementation of juice HACCP from 2002 reduced but did not prevent the occurrence of juice-related outbreaks. The persistence of outbreaks indicated other challenges that the HACCP program was not able to address. These challenges included the noninclusion of retail establishments to the regulation, and noncompliance of processors (Vojdani et al., 2008). Furthermore, the application of generic pasteurization processes for juices that were established against a specific microorganism or a microorganism at a specific physiological state may contribute to the risk of having an outbreak. This is why the federal juice HACCP stipulated that a specific processing method should target the most resistant pathogen most likely to occur in the product (Mazzotta, 2001).

When fruit juice–related outbreaks were first reported, food microbiologists were surprised as it had long been thought that the inherent acidity of such products was effective enough to prevent pathogen survival (Keller and Miller, 2006; Mazzotta, 2001). Studies have, however, shown that pathogens like *E. coli* O157:H7 and *Salmonella* spp. are capable of developing adaptive mechanisms by undergoing genetic and physiologic changes that allow the cells to stay viable in acidic food environments (Foster and Hall, 1990; Goodson and Rowbury, 1989; Linton et al., 1996; Moat et al., 2002). A number of studies have also shown that adaptation of *E. coli* O157:H7, *Salmonella* spp., and *L. monocytogenes* to acidic conditions by exposure to gradually decreasing pH led to cross protection against thermal inactivation in fruit juices, milk, and chicken broth (Buchanan and Edelson, 1999; Ryu and Beuchat, 1998; Sharma et al., 2005). As acidification and heating are common means of controlling microorganisms in foods (Brown and Booth, 1991), the ability of the pathogen to develop increased resistance to these control factors raises safety concerns to products including fruit juices. The complex relationships between the stresses encountered by pathogenic microorganisms from both intrinsic food characteristics (pH, water

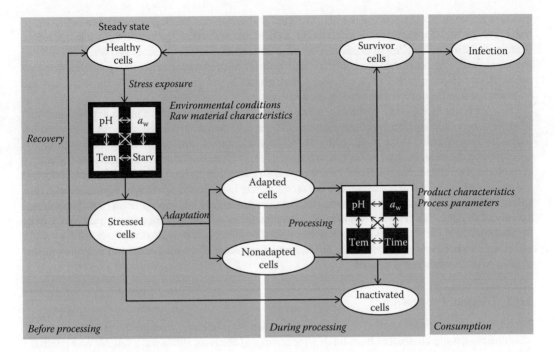

FIGURE 3.1 Influences of environmental stresses on subsequent resistance, adaptation, survival, and occurrence of foodborne infections. Healthy cells may be subjected to a number of stresses before the application of the processing step. Such stresses may include intrinsic food characteristics and extrinsic environmental conditions that may individually or interactively influence the induction of adaptive mechanisms toward subsequent food-processing steps. Similarly, the efficacy of the eventual food-processing step against pathogens may be dependent on individual or interactive influences of food and environmental parameters.

activity, nutrient availability, etc.) and extrinsic environmental conditions (temperature, etc.), and the induction of adaptive mechanisms to subsequent processing steps that lead to survival and eventual occurrences of diseases are summarized in Figure 3.1. Therefore, it becomes clear that in order to ensure the efficacy of a particular processing technology against pathogenic microorganisms in fruit juices, the possible influence of preprocessing conditions on microbial inactivation must be given consideration.

3.5 Fruit Juice Processing

The juice HACCP was implemented to guarantee safer fruit juice products by compelling manufacturers to apply processing steps that can effectively reduce pathogens in the products. Furthermore, the HACCP plan allowed manufacturers to examine their respective process flows and determine critical steps where possible contaminations are acquired and thus can be controlled. Complying with such a regulation is, however, not a simple feat as fruit juice processors have to deal with a number of challenges related to the product itself, the contaminating microorganisms, and other processing parameters. Nordby and Nagy (1980), Lea (1995), and Curtis (1997) explained that juice processing definitely requires more than fruit pressing and expressed juice collection. Apples appropriate for juice processing are sorted and washed prior to grinding and mechanical juice extraction by pressing. The initially extracted juice undergoes other treatments in order to increase production yield. Treatments with the enzymes cellulase and pectinase are conducted to degrade cell wall components and reduce product viscosity, respectively. In the production of clear juices, some enzymes and chemical-binding agents, together with ultrafiltration and centrifugation, are also used to treat the extracted juice and remove the haze or the suspended particles in cloudy juices. The haze is composed of polysaccharides, proteins, polyphenols, comminuted pulps, and polyvalent cations. The extracted and pretreated juices are then subjected to concentration through moisture removal. Such a process retards microbial and enzymatic spoilage through the reduction of water activity (a_w). Furthermore, concentration reduces the volume of the product for transportation, storage, and distribution. Apple juices can be concentrated to 70°Brix (Lea, 1995), while orange and lemon juices can be concentrated at 65°Brix and 40°Brix, respectively (Curtis, 1997). Aroma stripping can be done before or during concentration, where aromatic and flavor compounds are recovered and reintroduced into the concentrate or reconstituted juice.

The juice is finally subjected to the pasteurization process to inactivate contaminating pathogenic and spoilage microorganisms, and spoilage enzymes. The application of heat is possibly the most common means of pasteurization (Mak et al., 2001). However, a number of other methods are being studied as alternatives to heating, in order to curb the negative effects of heat on the nutritional and sensory quality of products. The USDA and USDHHS (2001) explained that while thermal pasteurization is an effective and proven technology that could achieve the recommended reduction in pathogen population and thus ensure the microbial safety of the juice, the USFDA does not believe it is appropriate to mandate thermal treatments to juice. The use of a nonthermal juice pasteurization, for as long as such a process attains the recommended pathogen reduction, may be applied to juices and provide consumers with a greater variety of safe products to choose form. The following are some of the traditional and emerging means of juice pasteurization being studied.

3.5.1 Thermal Processing

Despite being a traditional method of pasteurization, thermal processing remains the primary means by which pathogens are eliminated from foods (Buchanan and Edelson, 1999) and a reliable and relatively affordable means of producing safe juice products (Mak et al., 2001). The USFDA recognizes the efficacy of thermal pasteurization in reducing both pathogenic and spoilage organisms in fruit juices for product safety and shelf life stability (Donahue et al., 2004). Jay et al. (2005) discussed that thermal processing is applied to foods based on its ability to alter microbial cell structures and inactivate enzymes necessary for metabolism and growth. Horton et al. (2002) explained that microbial cell membranes and cell walls contain structural protein that can be easily denatured by heating. Many essential cellular processes are dependent on the integrity of the cell membranes and cell walls. Inactivation of metabolic enzymes would also mean the cessation of many biological reactions within the cells that could eventually lead to cell death.

In addition to heating time and temperature, the efficacy of thermal processing, however, has a limitation of being dependent on the inherent thermal resistance of target organisms and on the intrinsic food properties (Buchanan and Edelson, 1999; Jay et al., 2005). Such dependency can be characterized in order to quantify the relative contributions of each property to microbial inactivation and optimize a thermal process. The study conducted by Li Wan Po et al. (2002) in simulated apple juice demonstrated the dependence of the efficacy of heat inactivation of *E. coli* O157:H7 on some physicochemical parameters of the suspending medium. The study employed a rotatable central composite design of experiment to prepare simulated apple juice with various combinations of pH (2.82–4.06), titratable acidity (TA, 0.2%–0.74% w/v malic acid), and soluble solids (SS, 8.86°Brix–15.32°Brix). The heat inactivation studies were conducted at 53°C, 54°C, and 56°C. Results showed that the individual effects of pH and TA on the reduction of the test pathogen were significant at all heating temperatures. Increasing amounts of dissolved solids imparted protection to the cells, but such an effect was not statistically significant. Furthermore, Li Wan Po et al. (2002) also demonstrated that the physicochemical characteristics of the suspending medium could also interactively influence the inactivation of the test organism. The joint or interactive effects of pH and TA (pH × TA) and pH and SS (pH × SS) on the reduction of microbial population were found to be significant.

3.5.2 Ultraviolet Irradiation

A number of studies have determined the efficacy of ultraviolet (UV) irradiation as a means of eliminating pathogenic and spoilage microorganisms in fruit juices (Donahue et al., 2004; Franz et al., 2009; Wright et al., 2000). The Institute of Food Technologists (IFT) (IFT, 2000) discussed that

UV processing involves the use of radiation (100–400 nm) from the UV region of the electromagnetic spectrum. This UV range may be further divided and classified as UV-A (315–400 nm), UV-B (280–315 nm), UV-C (200–280 nm), and the vacuum UV range (100–200 nm). The UV-C range is germicidal and can effectively inactivate bacteria and viruses with maximum efficiency at wavelengths between 250 and 260 nm (Morgan, 1989). The UV-C range is germicidal due to its ability to break down the DNA of the microorganisms, impairing DNA transcription, replication, and compromising essential cellular functions (Unluturk et al., 2008). Furthermore, UV radiation causes significant damage in the cytoplasmic membrane and in the cellular enzyme activity, which could also contribute to microbial inactivation (Schenk et al., 2010). UV-C treatment was also reported to be effective in inactivating enzymes including polyphenol oxidase and peroxidase in apple juice (Falguera et al., 2011). The efficacy of UV processing has been previously reported to be affected by equipment design, treatment parameters, characteristics of the liquid being treated, and type of microorganism (Franz et al., 2009; Koutchma, 2009).

One limiting factor of UV-C treatment in fruit juices is the reduced penetration of UV-C due to the presence of dissolved organic solutes and compounds (Shama et al., 1996; Wright et al., 2000). The amount of soluble solids in liquid directly affects the penetration of UV-C (Guerrero-Beltrán and Barbosa-Cánovas, 2005). Hence, a clear juice like apple juice requires a lower UV dose to achieve effective reduction compared to orange juice and tropical juices (Keyser et al., 2008). This also explains the different reduction value obtained in the study of Oteiza et al. (2005) in the UV-C treatment of orange juice and multifruit juice samples. A radiation dose of 0.55 J/cm^2 was sufficient to achieve a 5 log reduction of *E. coli* ATCC 25922 and *E. coli* O157:H7 (EDL 933) in stirred 0.7 mm thick orange juice films. However, this dose only resulted in a 3 log reduction in multifruit juice under the same treatment conditions. Background microflora also contributes to the lowering of UV-C penetration in liquid (Oteiza et al., 2010; Shama et al., 1996). Wright et al. (2000) reported achieving the highest reduction (5.5 log CFU/mL) of *E. coli* O157:H7 in apple cider with very low initial yeasts and molds (1 log CFU/mL) treated with the highest radiation dose tested (61,005 µW s/cm^2). In order to overcome this limitation, the juice has to be exposed to UV as a thin film (Tran and Farid, 2004) and the flow rate and mixing of the liquid need to be increased, which can be accomplished by using the Dean vortex technology (Franz et al., 2009; Guerrero-Beltrán and Barbosa-Cánovas, 2005; Muller et al., 2011). Higher flow rates and mixing and reduced film thickness were reported to produce higher bactericidal effect (Koutchma et al., 2004; Oteiza et al., 2005). An increase in the reduction of *Lactobacillus plantarum* of approximately 2.5 log reduction was observed in orange juice treated with 7.7 kJ/L UV dose by increasing the flow rate (Muller et al., 2011).

The sensitivity of microorganisms to UV-C varies significantly, which can be explained by the differences in the thickness and chemical composition of the cell wall, which in turn determines the penetration rate of UV light into the cell (Montgometry, 1985), the cell size, which can alter the turbidity of the liquid food (Hansen, 1976), the genome size, repair ability (Tran and Farid, 2004), and the growth phase of the microorganism (Snowball and Hornsey, 1988). According to Gabriel (2012), a variation of sensitivity to UV-C light can also be observed between strains of the same species. In general, microorganisms follow an increasing order of sensitivity to UV-C light starting with gram-negative bacteria, gram-positive bacteria, viruses, fungi, spores, and cysts (Failly, 1994; Sastry et al., 2000). Table 3.1 summarizes the population reduction of pathogenic and spoilage microorganisms in fruit juices treated with UV-C light.

3.5.3 Ultrasound Processing

The potential of ultrasound processing as a juice pasteurization method has also been explored in some studies (Baumann et al., 2005; D'Amico et al., 2006; Patil et al., 2009; Salleh-Mack and Roberts, 2007). Ultrasound refers to pressure waves with a frequency of 20 kHz or more (Brondum et al., 1998; Butz and Tauscher, 2002). Patil et al. (2009) and Piyasena et al. (2003) discussed that, generally, ultrasound equipment uses frequencies from 20 kHz to 10 MHz and that those with frequencies between 20 and 100 kHz are called power ultrasound and are capable of inducing cavitation to inactivate microorganisms in foods. Cavitation is the process by which microbubbles are generated and collapsed within a liquid

TABLE 3.1

Reduction of Different Microorganisms in Fruit Juices Subjected to UV-C Treatment

Microorganisms	Fruit Juices	Reductions (log CFU)	Dosages (J/L)[a]	References
Escherichia coli K12	Apple	7.42	1,377	Keyser et al. (2008)
E. coli K-12 (ATCC 23716)[b]	Apple	5.80	3,333	Ukuku and Geveke (2010)
E. coli DH5α	Apple	6.00	7,714	Muller et al. (2011)
	Apple	ca. 4.00–5.00	8,100	Franz et al. (2009)
E. coli ATCC 25922	Pomegranate	6.15	34,400	Pala and Toklucu (2011)
	Orange	5.72	36,090	Pala and Toklucu (2013)
E. coli STCC 4201[c]	Orange	>6.00	13,550	Gayán et al. (2012)
E. coli cocktail (STCC 4201, STCC 471, ATCC 27325, ATCC 25922, and O157:H7 Chapman strain)[d]	Orange	>5.00	23,720	Gayán et al. (2012)
Lactobacillus plantarum	Apple	5.00	1,929	Muller et al. (2011)
Lactobacillus brevis LMG 11438	Apple	ca. 4.00	8,100	Franz et al. (2009)
Saccharomyces cerevisiae DSM 70478	Apple	ca. 4.00	9,643	Muller et al. (2011)
	Apple	ca. 3.00–4.00	16,200	Franz et al. (2009)
Alicyclobacillus acidoterrestris	Apple	ca. 4.00	9,643	Muller et al. (2011)

[a] Dosage = Total UV-C output per unit (W)/flow rate (L/s) (Keyser et al., 2008).
[b] Medium heated at 40.0°C.
[c] Medium heated at 60.0°C.
[d] Medium heated at 55.0°C.

medium. Bubble collapse within a medium can generate localized heating to as high as 5500°C and pressures of up to 100 MPa, and result in localized microbial inactivation. The pressure changes occurring from cavitations are the main mechanism for microbial cell disruption. Some critical factors influencing the efficacy of ultrasound processing include the type of microorganism, the frequency, the amplitude, the temperature, and the viscosity of the food (IFT, 2000; Sala et al., 1995).

Studies on the application of ultrasound as a single treatment to fruit juices are limited. The study of Salleh-Mack and Roberts (2007), which utilized water as the suspending medium, provides the closest data that can be used in assessing the efficacy of ultrasound in fruit juices as water is commonly added to juices. It was reported that 1.79 and 5.54 CFU/mL reductions in the population of *E. coli* ATCC 25922 were achieved by sonification at 24 kHz for 3 min and 10 min, respectively, at temperatures lower or equal to 30°C. In general, sonification alone is not very effective in the inactivation of microorganisms in food (Piyasena et al., 2003). Several reports indicate that sonification in conjunction with antimicrobials, heat, pressure, and UV-C resulted in significantly higher inactivation of microorganisms (Baumann et al., 2005; Ferrante et al., 2007, Gabriel, 2012).

3.5.4 Other Physical Means

Aside from the earlier enumerated methods of pasteurization, other physical methods for decontaminating juices of pathogenic and spoilage organisms also include high-pressure processing. Pressures of approximately 300–700 MPa for a few minutes can destroy pathogenic bacteria such as *Listeria*, *Salmonella*, *E. coli*, and *Vibrio* as well as spoilage microorganisms like yeasts and molds, while more resistant bacterial spores can survive pressures higher than 1000 MPa (Arroyo et al., 1999; Patterson, 2005; Teo et al., 2001). High pressure causes multiple damages in different parts of the cell, which could lead to cell death when the damages are beyond the ability of the cell to repair (Malone et al., 2002). In some cases, the damaged cells might recover if they are subjected to appropriate conditions (Bozoglu et al., 2004; Bull et al., 2004). Pressure-induced inactivation primarily accounts for the damage in the cell membrane and inactivation of key enzymes, including those involved in DNA replication and transcription (Hoover et al., 1989). At the protein level, pressure changes in the range of 100–300 MPa are

TABLE 3.2

Microbial Inactivation in Fruit Juices Using High-Pressure Treatment at Different Processing Conditions

Microorganisms	Fruit Juices	Pressure (MPa)	Temperature (°C)	Time (min)	Reduction (log CFU)	References
Listeria monocytogenes NCTC 11994	Apple	300	20	5	5.00	Jordan et al. (2001)
	Orange	300	20	5	3.00	Jordan et al. (2001)
	Apple, apricot, cherry, and orange	350	50	5	>8.00	Alpas and Bozoglu (2003)
Escherichia coli O157 C9490	Apple	500	20	5	5.00	Jordan et al. (2001)
	Orange	500	20	5	1.00	Jordan et al. (2001)
	Grapefruit	615	15	2	8.34	Teo et al. (2001)
	Apple	615	15	2	0.41	Teo et al. (2001)
Salmonella Typhimurium	Orange	400	25	10	7.04	Erkmen (2011)
Cryptosporidium parvum oocyst	Apple and orange juice	550		>1	>4.50	Slifko et al. (2000)

reversible but irreversible at pressures above 400 MPa (Tauscher, 1995). The level of microbial inactivation depends on the microbial parameters, which include the type of microorganism, its physiological state, and gram type; the physicochemical properties and temperature of the suspending juice; and the processing parameters, which include the time and magnitude of pressure treatment (Alpas et al., 2000; Benito et al., 1999; Patterson, 2005; Shigehisa et al., 1991).

Five decimal reductions of pathogenic microorganisms including *Listeria* spp., *Salmonella* spp., and *E. coli* in fruit juices have been reported to be achieved with the application of high pressure (Alpas and Bozoglu, 2003; Bull et al., 2004; Erkmen, 2011; Jordan et al., 2001; Teo et al., 2001). A summary of recent studies on the inactivation of pathogenic microorganisms in fruit juices using high pressure is presented in Table 3.2.

Pulsed electric field (PEF) processing of juices, which involves the application of pulses of high voltage (typically 20–80 kV/cm) to foods placed between two electrodes, has also been given some attention (Ayhan et al., 2002; IFT, 2000). PEF is reported to inactivate enzymes and microorganisms, both pathogenic and spoilage microorganisms (Sale and Hamilton, 1967; Zhang et al., 1994), without significant loss in flavor, color, and nutrients (Cserhalmi et al., 2006; Hodgins et al., 2002). However, PEF treatment does not have the same bactericidal effect on spores as it does on vegetative cells.

Ohmic heating, wherein the electric field energy is transformed into heat, occurs during PEF processing, causing an increase in temperature (Lindgren et al., 2002). However, temperature rise is relatively small, thereby reducing undesirable heat effects (Rodriguez et al., 2007). Also, as the increase in temperature is insignificant, microbial inactivation can primarily account for the field-induced impairment of the cell membrane and the contribution of ohmic heating is negligible (Jayaram et al., 1992; Pothakamury et al., 1995; Sale and Hamilton, 1967). PEF causes pore formation in the cell membrane in a process called electroporation, which causes cellular lysis followed by leakage of intracellular compounds. Similar to high-pressure processing, the damage caused by PEF is reversible or irreversible depending on the structural changes of the membrane (Tsong, 1990).

The efficiency of PEF processing depends on the electric field strength, treatment time, pulse width, pulse waveform, and temperature used in the process; conductivity and composition of the food material; and the microbial strain, growth phase, and initial inoculum size (Wouters and Smelt, 1997). Electric field strength, the number of pulses, and time and temperature of treatment directly affect the efficiency of microbial inactivation (Cserhalmi et al., 2006; Grahl and Märkl, 1996). However, increasing the pulse number is likely to result in significant heating. Mild electric field levels cause reversible pores in the membrane while drastic electric field intensity results in irreversible damage leading to cell death

(Ho et al., 1995; Tsong, 1990). On the other hand, larger microbial cells and cells in the exponential growth phase require lower field strength to be inactivated, compared to smaller cells and cells in the lag or stationary phase, respectively (Alvarez et al., 2000).

Several studies have already demonstrated the lethality of PEF on vegetative pathogenic microorganisms in fruit juices, which even reached more than 5 $\log_{10}$ reduction of microbial population. Evrendilek et al. (1999) reported 5.0 $\log_{10}$ and 5.4 $\log_{10}$ reductions of *E. coli* O157:H7 and *E. coli* 8739, respectively, in apple juice with the application of 29 kV/cm electric field for 172 µs at 42°C. Gupta et al. (2003) also achieved a 5 $\log_{10}$ reduction of *E. coli* K12 in apple juice with the application of 40 kV/cm and 100 pulses.

3.5.5 Antimicrobial Supplementation

Aside from the physical means previously described, studies on the efficacies of supplementing juices with traditional and alternative natural antimicrobial substances to achieve nonthermal pasteurization processes have also been given much attention. Raybaudi-Massilia et al. (2009) explained that antimicrobials may be classified as traditional or novel (also known as natural), depending on the origin of the substances. Traditional antimicrobials are (1) those that have been used for many years, (2) those that have been approved in many countries to be used in food processing, or (3) those that have been chemically synthesized. Traditional antimicrobials previously used as biostatics or biocides include organic acids like potassium sorbate (Walker and Phillips, 2008), malic acid (Raybaudi-Massilia et al., 2008), lactic acid (Uljas and Ingham, 1999), sodium benzoate, and fumaric and citric acids (Comes and Beelman, 2002).

Novel or alternative natural substances derived from animals, plants, and microorganisms have also been studied as antimicrobials in a number of fresh-cut produce and fruit juices. Cinnamon oil (Raybaudi-Massilia et al., 2006), clove oil (Nguyen and Mittal, 2007) lemon oil, oregano oil, carvacrol, cinnamaldehyde, eugenol, geraniol (Friedman et al., 2004), and citral (Ferrante et al., 2007) are some of the essential oils and active compounds of plant origin previously tested in fruit juices. Nisin, a peptide produced by a microorganism, has also been tested as an antimicrobial in juice (Yuste and Fung, 2004).

The specific targets of antimicrobials in the microbial cell structures include the cell wall, cell membrane, metabolic enzymes, transport proteins, and genetic systems (Raybaudi-Massilia et al., 2009). Davidson (2001) further explained that it is difficult to identify the specific action site of antimicrobials as many interacting reactions can occur simultaneously. For example, membrane-disrupting compounds can simultaneously cause cellular content leakage, interference with essential enzymatic or protein transport function, or adenosine triphosphate (ATP) loss.

3.5.6 Combination Processing: Hurdle Technology

Raybaudi-Massilia et al. (2009) further explained that most antimicrobial agents are only able to prevent microbial growth but are incapable of inducing cellular inactivation. Furthermore, the high treatment intensities required for physical processing to achieve microbial inactivation can result in adverse alterations in product quality (Ross et al., 2003) and may be too costly to have practical applications (Raso et al., 1998). Hence, better antimicrobial efficacies may be achieved when antimicrobial substances are combined with other compounds, or physical processing. In fact, a number of studies have looked into the efficacies of combining antimicrobial supplementation and physical treatments to inactivate microorganisms in fruit juice products. Akpomedaye and Ejechi (1999) combined mild heating and supplementation of plant extracts against fungi in fruit juices, while McNamee et al. (2010) combined PEF and a number of bacteriocins against spoilage and pathogen surrogate organisms in orange juice. Physical treatments can also be combined at milder settings such as in thermosonication (heating and ultrasound), manosonication (pressure and ultrasound), and manothermosonication (pressure, heating, and ultrasound) (Piyasena et al., 2003). These combinations of mild processes to inactivate microorganisms in foods are the bases of hurdle technology (Leistner and Gorris, 1995, Leistner and Gould, 2002).

Leistner and Gould (2002) explained that hurdle technology employs a deliberate manipulation and consideration of a number of preservative factors or techniques to control microorganisms in foods. Such factors include the intrinsic factors, which are inherent characteristics of foods (pH, a_w, composition, etc.), that contaminating microorganisms are inextricably exposed to. Processing and extrinsic factors are environmental conditions (processing temperature, storage temperature, relative humidity, etc.) that are deliberately applied to foods and effect preservative action during processing and storage. Finally, implicit factors are those that are related to the nature of the microorganisms that are present in the food, and their interaction with the food itself (injury, *stress hardening*, or adaptation). In hurdle technology, instead of applying one control method at an extreme setting resulting in a single mechanism of inactivation, multiple, mild preservative factors are applied, simultaneously assaulting different microbial homeostatic processes (Montville and Matthews, 2007).

However, the combination of two or more control factors may result in synergistic, additive, or antagonistic effects and hence must be thoroughly characterized before application to food. Conditions defining the boundaries of microbial growth or survival should be correctly defined in order for the industry to come up with products with the appropriate level of safety (Brocklehurst, 2004). Furthermore, the effects of intrinsic and extrinsic properties on microbial growth and inactivation can also be quantitatively defined in order to come up with precise or optimized processing technologies for specific foods (Li Wan Po et al., 2002; Silva et al., 1999). With appropriate statistical and mathematical tools, the influences of different products and processing factors on microbial responses to food may be quantitatively characterized.

3.6 Optimization Research and Response Surface Methodology

Product and process optimization studies have always been the focus of research and development efforts of any manufacturing industry (Verseput, 2000). Optimization research (OR) is undertaken in order to develop the best possible product or process in its class (Sidel and Stone, 1983). The response surface methodology (RSM) proposed by Box et al. (1978) and the multiple regression approach (MRA) described by Schutz et al. (1972) and Schutz (1983) are structured statistical approaches that can be used in OR. Between these two models, RSM seems to be the popularly used OR tool (Giovanni, 1983).

RSM is a collection of statistical and mathematical techniques useful for development, improvement, and optimization processes (Myers and Montgomery, 2002). It involves the use of quantitative data obtained from an appropriate experimental design to determine and simultaneously solve multivariate equations (Giovanni, 1983). Unlike classical OR tools that test one variable at a time, RSM can consider several variables at many different levels in a product or a process and at the same time account for the corresponding interactions among these variables and levels. Specifically, RSM enables the experimenters to generate and solve multivariate equations that (1) describe how the test variables affect the performance measure or quality characteristic or response; (2) determine the interrelationships among the test variables; and (3) describe the combined effect of all test variables on the response (Giovanni, 1983; Myers and Montgomery, 2002).

Giovanni (1983) explained that RSM is a four-step process that involves (1) identification of two or three variable (factors) critical to the product or the process being studied, (2) definition of the range of the levels of the previously identified variables, (3) identification of an experimental design to which the test combinations of experimental variables will be subjected, and (4) analyses and interpretation of the obtained data. Unlike the MRA model for OR where critical variables affecting a product or a process are not known beforehand, the critical variables should be established and systematically varied when an experimenter chooses to use RSM (Giovanni, 1983; Sidel and Stone, 1983).

An advantage of optimization over classical research methods is the application of experimental designs that limit the number of experiments needed to be performed while allowing possible interactions between the experimental variables (Adinarayana and Ellaiah, 2002) and providing sufficient information for acceptable results (Yu and Ngadi, 2004). A subset of the samples to be tested is selected from all possible samples that could be tested. While covering the range of variable levels identified by the experimenter, these designs emphasize those samples closest to the midpoints of the defined ranges, thereby decreasing the total number of samples that need to be tested. Such designs provide information on direct, pairwise interaction and curvilinear or quadratic variable effects on the response being measured (Verseput, 2000).

Sidel and Stone (1983) explained that the use of OR is not confined to the identification of the critical variables and the definition of their contribution to the response being studied. Data gathered from an OR could be processed and used to predict combinations of the identical critical variables that will yield a specific response. However, as in any scientific study, the results of a response surface analysis should not be extrapolated beyond the predefined limits (Giovanni, 1983). Any prediction made about a response that lies outside the defined region should be verified by experiments before putting reliance on it (Cochran and Cox, 1957).

3.6.1 Central Composite Designs of Experiment

In OR, experimental designs are employed to reduce the number of experiments needed to be performed without compromising the interactions between experimental variables and the statistical significance of the results (Adinaraya and Ellaiah, 2002; Yu and Ngadi, 2004). Cochran and Cox (1957) presented some of these experimental designs that could fit multivariate equations. Among the many designs of experiments used for RSM, central composite designs (CCD) are mostly considered for use because of certain advantages (Verseput, 2000).

Verseput (2000) explained that a CCD can be portioned into subsets and run sequentially to estimate the linear, two-factor interaction and quadratic variable effects. A CCD requires a minimum number of experiments to be run in order to provide much information on experiment variable effects and experimental error. Several varieties of CCD can be used in different experimental regions of interest and operability. The face-centered CCD (FCCCD), the central composite rotatable design (CCRD), and the inscribed CCD (ICCD) are the varieties presented by Verseput (2000). A noncentral composite design was described by Cochran and Cox (1957). The choice of a specific design depends on the availability of the resources and the operability of the defined variable setting combinations.

In a CCD, the region of interest is defined as the geometric area bound by the ranges of the variable combinations, while the region of operability is the geometric area that can be operationally achieved with acceptable safety that can produce a testable product or process. In a successful design of experiment, the region of operability will encompass the region of interest. This means that the experiments can be performed at all variable setting combinations in the region of interest defined by the variable ranges (Verseput, 2000). Figure 3.2 presents schematic representations of the different CCDs for two-variable experiments (x_1 and x_2). These figures show the coded variable settings ($\pm\alpha$, $\pm\Phi$, 0, ±1) and the variable setting combinations represented by the center, factorial, and axial or *star* points. Verseput (2000) discussed that the radius, designated by α, determines the geometry of the design region. An FCCCD of a two-variable study with an $\alpha = 1.00$ has a square design geometry (Figure 3.2a). A three-variable FCCCD with the same α value has cubical design geometry.

If the value of α is increased, the geometry of the design region becomes more circular or spherical. This is demonstrated by a CCRD (Figure 3.2b) where $\alpha = 1.4$ with the design region extended beyond

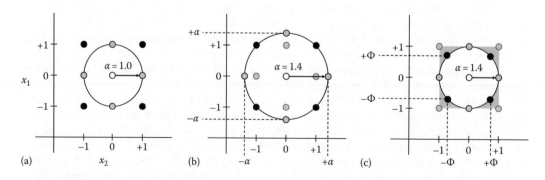

FIGURE 3.2 Varieties of the CCD of an experiment: two-variable (a) FCCCD, (b) RCCD, and (c) ICCD. Variable settings are represented by codes ($\pm\alpha$, $\pm\Phi$, 0, ±1) and variable combinations are represented by different points (O—center point, ◑—axial or *star* points, ●—factorial points).

the defined variable bounds. Thus, predicted responses at or near the axial points that would have been extrapolations in an FCCCD are now included in the CCRD region. Verseput (2000) defined rotatability as the equality of prediction error. In a CCRD, all points at the same radial distance (*r*) from the center point have the same magnitude of prediction error. In a CCRD involving more than two variables, though, the axial and factorial points lie on concentric spheres, thus giving them different prediction error magnitudes. A CCRD is usually employed when the experimenter does not know the location of the optimum point within the region of interest before the actual experiments are conducted. Among the varieties of CCD described by Cochran and Cox (1957) and Verseput (2000), the CCRD seems to be the more popular design for RSM practitioners.

The ICCD (Figure 3.2c) is another variety of CCD described by Verseput (2000) that is used when some points in the region of interest are not in the region of operability. In an ICCD, the actual design region is restricted to the defined variable ranges. The axial and factorial points are inscribed within the defined variable ranges while maintaining a proportional distance from the central point. As shown in Figure 3.2c, the axial points have been moved to the upper and lower bounds of the variable ranges while the factorial points have been set at ±Φ. In an ICCD, the experimenter is able to study the full ranges of the variables while excluding the points beyond the region of operability.

The studies reported by Gabriel et al. (2008) and Gabriel and Azanza (2009, 2010) demonstrate the application of the CCRD and RSM in the establishment of a predictive model for thermal inactivation in terms of decimal reduction times ($D_{72°C}$) of *Salmonella* Typhimurium (ATCC 13311) as a function of simulated citrus juice (SCJ) pH, TA, and SS. The $D_{72°C}$ value is equivalent to the unit of time (min) of heating that will result in the inactivation of *S.* Typhimurium in the SCJ by 1 log cycle (90%).

In these works, the established model was subjected to a multilevel validation process to assess its utility. Actual inactivation rates ($^aD_{72°C}$) were obtained in real citrus and other fruit juices and then compared to predicted values ($^pD_{72°C}$) (Table 3.3). The efficacy of the model in predicting the inactivation rates of other test organisms was also tested. Finally, the use of the model in establishing a pasteurization process was evaluated by applying a model-predicted process schedule on a citrus juice drink. The model exhibited acceptable predictive mathematical accuracy and bias and resulted in fail-safe predictions ($^pD_{72°C} > {}^aD_{72°C}$) when validated in select real citrus and noncitrus juices (Table 3.3). The model was also found to be applicable in predicting the inactivation of other reference strains in a citrus juice drink medium but was found to result in fail-dangerous ($^aD_{72°C} > {}^pD_{72°C}$) predictions in fruit juices that contained significant amounts of suspended pulp.

TABLE 3.3

Thermal Inactivation of *Salmonella* Typhimurium (ATCC 13311) in Real Juices

Real Juices	Physicochemical Properties			$D_{72°C}$ (s)^2	
	pH	TA (% Citric Acid)	SS (°Brix)	Predicted	Actual
Citrus juices (Gabriel and Azanza, 2010)					
Sweet orange	3.29	0.70	14.67	11.12[a]	7.25[b]
Blood orange	3.26	0.48	14.00	11.38[a]	7.64[b]
Grapefruit	3.12	1.23	10.00	9.40[a]	7.43[b]
Pomelo	3.19	0.63	14.33	9.80[a]	8.12[b]
Tangerine	3.77	0.41	12.13	20.23[a]	20.12[a]
Native orange drink	3.19	0.46	15.00	9.19[a]	6.29[b]
Other fruit juices (Gabriel, 2007)					
Guyabano (Soursop)	2.67	0.30	11.60	1.93[a]	1.63[b]
Pineapple	2.81	0.75	14.00	3.36[a]	2.81[b]
Guava	3.74	0.20	13.60	18.65[b]	20.66[a]
Tomato	3.66	0.38	6.38	18.58[b]	21.27[a]

[a,b] $D_{72°C}$ values on the same row followed by the same letter are not significantly different ($p > 0.05$).

Finally, the predictive model was used in the calculation of a thermal pasteurization process for a citrus juice drink with the recommended lethal rate of 99.999% (5 log reductions, 5D), based on the $^{P}D_{72°C}$ of *S.* Typhimurium. Microbial quality indicators including total aerobic plate and yeasts and molds significantly improved after heat pasteurization. Furthermore, the resulting pasteurized juice drink had color quality attributes that were comparable to that of the control, unpasteurized sample. The sensory quality parameters of both the control and heat-pasteurized samples were comparable (Gabriel and Azanza, 2009). These results validated the utility of the predictive model in establishing thermal processes for the tested fruit juice drink, and possibly of similar food systems.

3.7 Conclusion

The fruit juice industry might be benefiting from the increased health awareness of the consumers but has also been challenged by the occurrences of a number of outbreaks of microbial illnesses. To mitigate the public health risks, the USFDA implemented the federal juice HACCP program, which reduced but not completely eradicated juice-related outbreaks. Among the reasons of failure of the HACCP program to completely prevent outbreaks were factors related to manufacturers and consumers such as product mishandling and exemption from compliance to the regulation. Factors related to microorganisms such as stress adaptation and stress hardening may also contribute to the risk of outbreaks occurring. The HACCP program stipulates that in order to ensure safe products, juices should be processed such that the population of a target most resistant pathogenic species, or the most resistant physiological state of a pathogen likely to be present, will be reduced by 5 log cycles. The HACCP program also did not strictly recommend thermal processing as a means of decontaminating juices from pathogens, stating that other processing methods may be applied to products as long as the required 5 log reduction is met.

Consequently, the efficacies of a number of traditional and novel processing methods against microorganisms in juices, both physical and chemical means, either individually or combined have been reported. The applications of combinations of different, mild antimicrobial treatments as fruit juice–processing techniques are becoming popular and are the basis of *hurdle technology*, which aims to augment microbial inactivation and the retention of desirable product characteristics. The use of statistical and mathematical tools in characterizing the individual and combined effects of processing factors may be useful in the establishment of precise and accurate process schedules to optimize efficacy without compromising the sensory and nutritional quality of the product. Employment of statistical and mathematical tools, which involves the choice of an appropriate design of experiment (e.g., CCD) and regression analysis technique (e.g., RSM), allows a processor to develop and validate predictive models for the inactivation of a target pathogenic or spoilage organism. As such, unique process schedules for specific products may be established, avoiding the application of generic schedules that may result in under- or overprocessing and, respectively, compromise safety and quality of the product.

REFERENCES

Adinarayana, K. and Ellaiah, O. (2002). Response surface optimization of the critical medium components for the production of alkaline protease by newly isolated *Bacillus* sp. *Journal of Pharmacy and Pharmaceutical Sciences 5*, 272–278.

Akpomedaye, D. E. and Ejechi, B. O. (1999). The hurdle effect of mild heat and two tropical spice extracts on the growth of three fungi in fruit juices. *Food Research International 31*, 339–341.

Alpas, H. and Bozoglu, F. (2003). Efficiency of high pressure treatment for destruction of *Listeria monocytogenes* in fruit juices. *FEMS Immunology and Medical Microbiology 35*(3), 269–273.

Alpas, H., Kalchayanand, N., Bozoglu, F., and Ray, B. (2000). Interactions of high hydrostatic pressure, pressurization temperature and pH on death and injury of pressure-resistant and pressure sensitive strains of foodborne pathogens. *International Journal of Food Microbiology 60*, 33–42.

Alvarez, I., Raso, J., Palop, A., and Sala, F. J. (2000). Influence of different factors on the inactivation of *Salmonella* Senftenberg by pulsed electric fields. *International Journal of Food Microbiology 55*, 143–146.

Arroyo, G., Sanz, P. D., and Prestamo, G. (1999). Response to high pressure, low-temperature treatment in vegetables: Determination of survival rates of microbial populations using flow cytometry and detection of peroxidase activity using confocal microscopy. *Journal of Applied Microbiology 86*(3), 544–556.

Ayhan, Z., Zhang, Q. H., and Min, D. B. (2002). Effects of pulsed electric field processing and storage on the quality and stability of single strength orange juice. *Journal of Food Protection 65*, 1623–1627.

Bates, R. P., Morris, J. R., and Crandall, P. G. (2001). Principles and practices of small- and medium-scale fruit juice processing. *Food and Agriculture Organization Service Bulletin*, No. 146. Food and Agriculture Organization of the United Nations, Rome, Italy.

Baumann, A. R., Martin, S. E., and Feng, H. (2005). Power ultrasound treatment of *Listeria monocytogenes* in apple cider. *Journal of Food Protection 68*, 2333–2340.

Benito, A., Ventoura, G., Casadei, M., Robinson, T., and Mackey, B. (1999). Variation in resistance of natural isolates of *Escherichia coli* O157 to high hydrostatic pressure, mild heat, and other stresses. *Applied and Environmental Microbiology 65*(4), 1564–1569.

Besser, R. E., Lett, S. M., Webe, J. T., Doyle, M. P., Barrett, T. J., Wells, J. G., and Griffin, P. M. (1993). An outbreak of diarrhea and hemolytic uremic syndrome from *Escherichia coli* O157:H7 in fresh-pressed apple cider. *JAMA 269*, 2271–2220.

Box, G. E. P., Hunter, W. G., and Hunter, J. S. (1978). *Statistics of the Experimenters*. John Wiley & Sons, Inc., New York.

Bozoglu, F., Alpas, H., and Kalentuc, G. (2004). Injury recovery of foodborne pathogens in high hydrostatic pressure treated milk during storage. *FEMS Immunology and Medical Microbiology 40*(3), 243–247.

Brocklehurst, T. (2004). Challenge of food and the environment. In: *Modeling Microbial Response in Food*, R. C. McKellar and X. Lu (eds.), CRC Press, Boca Raton, FL, pp. 197–232.

Brondum, J., Egebo, M., Agerskov, C., and Busk, H. (1998). Online pork carcass grading with the autoform ultrasound system. *Journal of Animal Science 76*, 1859–1868.

Brown, M. H. and Booth, I. R. (1991). Acidulants and low pH. In: *Food Preservatives*, N. J. Russel and G. W. Could (eds.), Blackie, Glasgow, Scotland, pp. 22–43.

Buchanan, R. L. and Edelson, S. G. (1999). Effect of pH-dependent, stationary phase acid resistance on the thermal tolerance of *Escherichia coli* O157:H7. *Food Microbiology 16*, 447–458.

Bull, M. K., Zerdin, K., Howe, E., Goicoechea, D., Paramanandhan, P., Stockman, R., Sellahewa, J., Szabo, E. A., Johnson, R. L., and Stewart, C. M. (2004). The effect of high pressure on the microbial, physical and chemical properties of Valencia and Navel orange juice. *Innovative Food Science and Emerging Technologies 5*, 135–149.

Butz, P. and Tauscher, B. (2002). Emerging technologies: Chemical aspects. *Food Research International 35*, 279–284.

Cochran, W. G. and Cox, G. M. (1957). *Experimental Designs*, 2nd edn. John Wiley & Sons, Inc., New York.

Comes, J. E. and Beelman, R. B. (2002). Addition of fumaric acid and sodium benzoate as an alternative method to achieve a 5-log reduction of *Escherichia coli* O157:H7 population in apple cider. *Journal of Food Protection 65*, 476–483.

Cserhalmi, Z., Sass-Kiss, A., Tóth-Markus, M., and Lechner, N. (2006). Study of pulsed electric field treated citrus juices. *Innovative Food Science and Emerging Technologies 7*, 49–54.

Curtis, L. (1997). Juice up. Food Product Design. Available from: http://www.foodproductdesign.com/archive/1997/0797CS.html. Accessed June 28, 2010.

D'Amico, D. J., Silk, T. M., Wu, J., and Guo, M. (2006). Inactivation of microorganism in milk and apple cider treated with ultrasound. *Journal of Food Protection 69*, 556–563.

Davidson, P. M. (2001). Chemical preservatives and natural antimicrobial compounds. In: *Food Microbiology: Fundamentals and Frontiers*, 2nd edn., M. P. Doyle, L. R. Beuchat, and T. J. Montville (eds.), ASM Press, Washington, DC, pp. 593–627.

Donahue, D. W., Canitez, N., and Bushway, A. A. (2004). UV inactivation of *E. coli* O157:H7 in apple juice: Quality, sensory and shelf-life analysis. *Journal of Food Processing and Preservation 28*, 368–387.

Erkmen, O. (2011). Effects of high hydrostatic pressure on *Salmonella typhimurium* and aerobic bacteria in milk and fruit juices. *Romanian Biotechnological Letters 16*(5), 6540–6547.

Evrendilek, G. A., Zhang, Q. H., and Richter, E. R. (1999). Inactivation of *Escherichia coli* O157:H7 and *Escherichia coli* 8739 in apple juice by pulsed electric fields. *Journal of Food Protection 62*, 793–796.

Failly, J. (1994). Desinfection des eaux usees par rayonnement UV. *L'eau, l'industrie, les nuisances 32*, 58–60.

Falguera, V., Pagan, J., and Ibarz, A. (2011). Effect of UV irradiation on enzymatic activities and physico-chemical properties of apple juices from different varieties. *Food Science and Technology 44*, 115–119.

Ferrante, S., Guerrero, S., and Alzamora, M. S. (2007). Combined use of ultrasound and natural antimicrobials to inactivate *Listeria monocytogenes* in orange juice. *Journal of Food Protection 70*, 1850–1856.

Food and Agriculture Organization. (1992). Codex Alimentarius, Vol. 6, Fruit juices and related products. Food and Agriculture Organization of the United Nations, Rome, Italy.

Food and Nutrition Research Institute (FNRI, Philippines). (1994). *Food Exchange Lists for Meal Planning*, 3rd revision, FNRI Publication No. 57-ND8(3). Department of Science and Technology, Food and Nutrition Research Institute, Taguig, Philippines.

Foster, J. W. and Hall, H. K. (1990). Adaptive acidification tolerance response of *Salmonella* Typhimurium. *Journal of Bacteriology 172*, 771–778.

Franz, C. M. A. P., Specht, I., Cho, G., Graef, V., and Stahl, M. R. (2009). UV-C-inactivation of microorganisms in naturally cloudy apple juice using novel inactivation equipment based on Dean vortex technology. *Food Control 20*, 1103–1107.

Frederick, K. (2009). Booster shots for energy and health. *Food Technology 64*(9), 26–34.

Friedman, M., Henika, P. R., Levin, C. E., and Mandrell, R. E. (2004). Antibacterial activities of plant essential oils and their components against *Escherichia coli* O157:H7 and *Salmonella enterica* in apple juice. *Journal of Agricultural and Food Chemistry 52*, 6042–6048.

Gabriel, A. A. (2007). Predictive thermal inactivation model of *Salmonella* Typhimurium in citrus systems. An unpublished master's degree thesis presented to the Faculty of the College of Home Economics, University of the Philippines, Quezon City, Philippines.

Gabriel, A. A. (2012). Microbial inactivation in cloudy apple juice by multi-frequency *Dynashock* power ultrasound. *Ultrasonics Sonochemistry 19*, 346–351.

Gabriel, A. A. and Azanza, M. P. V. (2009). Quality of orange juice drink subjected to a predictive model-based pasteurization process. *Journal of Food Quality 32*, 452–468.

Gabriel, A. A. and Azanza, M. P. V. (2010). $D_{72°C}$ values of *Salmonella* Typhimurium in citrus juices: Predictive efficacy of a model. *Journal of Food Process Engineering 33*, 506–518.

Gabriel, A. A., Barrios, E. B., and Azanza, M. P. V. (2008). Modeling the thermal death *Salmonella* Typhimurium in citrus systems. *Journal of Food Process Engineering 31*, 640–657.

Gayán, E., Serrano, M. J., Monfort, S., Álvarez, I., and Condón, S. (2012). Combining ultraviolet light and mild temperatures for the inactivation of *Escherichia coli* in orange juice. *Journal of Food Engineering 113*(4), 598–605.

Giovanni, M. (1983). Response surface methodology and product optimization. *Food Technology 37*(11), 41–45.

Goodrich, R. M., Schneider, K. R., and Parish, M. E. (2005). The juice HACCP program: An overview. Food Safety and Toxicology Series, Institute Food and Agricultural Sciences, University of Florida, Gainesville, FL, FSHNO515, pp. 1–4.

Goodson, M. and Rowbury, R. J. (1989). Habituation to normal lethal acidity by prior growth of *Escherichia coli* at sublethal acid pH value. *Letters in Applied Microbiology 8*, 77–79.

Grahl, T. and Märkl, H. (1996). Killing of microorganisms by pulsed electric fields. *Applied Microbiology Biotechnology 45*, 148–157.

Guerrero-Beltrán, J. A. and Barbosa-Cánovas, G. V. (2005). Reduction of *Saccharomyces cerevisiae*, *Escherichia coli* and *Listeria innocua* in apple juice by ultraviolet light. *Journal of Food Process Engineering 28*, 437–452.

Gupta, B. S., Masterton, F., and Magee, T. R. A. (2003). Inactivation of *E. coli* K12 in apple juice by high voltage pulsed electric field. *European Food Research and Technology 217*, 434–437.

Hansen, R. (1976). Sterilization by use of intense ultraviolet radiation. *NordeuropAEisk Mejeri-Tidsskrift 42*, 60–67.

Harvey, J. M. (1978). Reduction of losses in fresh market fruits and vegetables. *Annual Review of Phytopathology 16*, 321.

Ho, S. Y., Mittal, G. S., Cross, J. D., and Griffiths, M. W. (1995). Inactivation of *Pseudomonas fluorescens* by high voltage electric pulses. *Journal of Food Science 60*, 1337–1340.

Hodgins, A. M., Mittal, G. S., and Griffiths, M. W. (2002). Pasteurization of fresh orange juice using low-energy pulsed electrical field. *Journal of Food Science 67*, 2294–2299.

Hoover, D. G., Metrick, K., Papineau, A. M., Farkas, D. F., and Knorr, D. (1989). Biological effects of high hydrostatic pressure on food microorganisms. *Food Technology 43*, 99–107.

Horton, H. R., Moran, L. A., Ochs, R. S., Rawn, J. D., and Scrimgeour, K. G. (2002). *Principles of Biochemistry*, 3rd edn. Prentice-Hall, Inc., Upper Saddle River, NJ.

Institute of Food Technologists (IFT). (2000). Kinetics of microbial inactivation for alternative food processing technologies. A report of the Institute of Food Technologists for the Food and Drug Administration of the US Department of Health and Human Services. Available at: http://www.fda.gov/Food/ScienceResearch/ResearchAreas/Safe PracticesforFoodProcesses/ucm100158.htm. Accessed July 22, 2010.

Jay, J. M., Loessner, M. J., and Golden, D. A. (2005). *Modern Food Microbiology*, 6th edn. Springer, New York.

Jayaram, S., Castle, G. S. P., and Pagaritis, A. (1992). Kinetics of sterilization of *Lactobacillus brevis* cells by the application of high voltage pulses. *Biotechnology and Bioengineering 40*, 1412–1420.

Jordan, S. L., Pascual, C., Bracey, E., and Mackey, B. M. (2001). Inactivation and injury of pressure-resistant strains of *Escherichia coli* O157:H7 and *Listeria monocytogenes* in fruit juice. *Journal of Applied Microbiology 91*, 463–469.

Kaur, C. and Kapoor, H. C. (2001). Antioxidants in fruits and vegetables—The millennium's health. *International Journal of Food Science and Technology 36*, 703–725.

Keller, S. E. and Miller, A. J. (2006). Microbiological safety of fresh citrus and apple juices. In: *Microbiology of Fruits and Vegetables*, G. M. Sapers, J. R. Gorny, and A. E. Yousef (eds.), CRC Press/Taylor & Francis Group, Boca Raton, FL, pp. 211–230.

Keyser, M., Müller, I. A., Cilliers, F. P., Nel, W., and Gouws, P. A. (2008). Ultraviolet radiation as non thermal treatment for inactivation of microorganisms in fruit juice. *Innovative Food Science and Emerging Technologies 9*, 348–354.

Koutchma, T. (2009). Advances in ultraviolet light technology for non-thermal processing of liquid foods. *Food and Bioprocess Technology 2*, 138–155.

Koutchma, T., Keller, S., Chirtel, S., and Parisi, B. (2004). Ultraviolet disinfection of juice products in laminar and turbulent flow reactors. *Innovative Food Science and Emerging Technologies 5*, 179–189.

Koutsomanis, K. P. and Sofos, J. N. (2004). Comparative acid stress response of *Listeria monocytogenes*, *Escherichia coli* O157:H7 and *Salmonella* typhimurium after habituation at different pH conditions. *Letters in Applied Microbiology 38*, 321–326.

Lea, A. G. H. (1995). Apple juice. In: *Production and Packaging of Non-Carbonated Fruit Juices and Fruit Beverages*, P. R. Ashurst (ed.), Blackie Academic & Professional, Great Britain, U.K., pp. 153–196.

Leistner, L. and Gorris, L. G. M. (1995). Food preservation by hurdle technology. *Trends in Food Science and Technology 6*, 41–46.

Leistner, L. and Gould, G. (2002). *Hurdle Technologies: Combinations of Treatments for Stability, Safety and Quality*. Kluwer Academic Publishers, New York, pp. 1–15.

Lindgren, M., Aronsson, K., Galt, S., and Ohlsson, T. (2002). Simulation of the temperature increase in pulse electric field (PEF) continuous flow treatment chambers. *Innovative Food Science and Emerging Technologies 3*(3), 233–245.

Linton, M., Smith, M. P., Chapkin, K. C., Baik, H. S., Bennett, G. N., and Foster, J. W. (1996). Mechanisms of acid resistance in enterohemorrhagic *Escherichia coli*. *Applied and Environmental Microbiology 62*, 3094–3100.

Li Wan Po, J. M., Piyasena, P., McKellar, R. C., Bartlett, F. M., Mittal, G. S., and Lu, X. (2002). Influence of simulated apple cider composition on the heat resistance of *Escherichia coli* O157:H7. *Lebensmittel-Wissenschaft und-Technologie 35*, 295–304.

Luedtke, A. and Powell, D. (2000). Fact sheet: A timeline of fresh juice outbreaks. The International Food Safety Network, Kansas State University, Manhattan, KS. Available at: http://foodsafety.k-state.edu/en/article-details.php?a=2&c=6&sc=37&id=427. Accessed July 22, 2010.

Mak, P. P., Ingham, B. H., and Ingham, S. C. (2001). Validation of apple cider pasteurization treatments against *Escherichia coli* O157:H7, *Salmonella*, and *Listeria monocytogenes*. *Journal of Food Protection 64*, 1679–1689.

Malone, A. S., Shelhammer, T. H., and Courtney, P. D. (2002). Effects of high pressure on the viability, morphology, lysis, and cell wall hydrolase activity of *Lactococcus lactis* subsp. *cremoris*. *Applied and Environmental Microbiology 68*(9):4357–4363.

Mazzotta, A. S. (2001). Thermal inactivation of stationary-phase and acid adapted *Escherichia coli* O157:H7, *Salmonella*, and *Listeria monocytogenes* in fruit juices. *Journal of Food Protection 64*, 315–320.

McNamee, C., Noci, F., Cronin, D. A., Lyng, J. G., Morgan, D. J., and Scannell, A. G. M. (2010). PEF based hurdle strategy to control *Pichia fermentans*, *Listeria innocua* and *Escherichia coli* K12 in orange juice. *International Journal of Food Microbiology 138*, 13–18.

Moat, A. G., Foster, J. W., and Spector, M. P. (2002). *Microbial Physiology*, 4th edn. Wiley-Liss, Inc., New York.

Montgometry, J. M. (1985). *Water Treatment: Principles and Design*. John Wiley & Sons, New York.

Montville, T. J. and Matthews, K. R. (2007). Growth, survival and death of microbes in food. In: *Food Microbiology: Fundamentals and Frontiers*, 3rd edn., M. P. Doyle and L. R. Beuchat (eds.), ASM Press, Washington, DC, pp. 3–22.

Morgan, R. (1989). Green light disinfections. *Dairy Industries International 54*, 33–35.

Muller, A., Stahl, M., Graef, V., Franz, C. M. A. P., and Huch, M. (2011). UV-C treatment of juices to inactivate microorganisms using Dean vortex technology. *Journal of Food Engineering 107*, 268–275.

Myers, R. H. and Montgomery, D. C. (2002). *Response Surface Methodology: Process and Product Optimization Using Designed Experiments*, 2nd edn. John Wiley & Sons, Inc., New York.

Nguyen, P. and Mittal, G. S. (2007). Inactivation of naturally occurring microorganisms in tomato juice using pulsed electric field (PEF) with and without antimicrobials. *Chemical Engineering and Processing 46*, 360–365.

Nguyen-The, C. and Carlin, F. (1994). The microbiology of processed fresh fruits and vegetables. *Critical Reviews in Food Science and Nutrition 34*, 371–401.

Noci, F., Riener, J., Walkling-Ribeiro, M., Cronin, D. A., Morgan, D. J., and Lyng, J. G. (2008). Ultraviolet irradiation and pulsed electric fields (PEF) in a hurdle strategy for the preservation of fresh apple juice. *Journal of Food Engineering 85*, 141–146.

Nordby, H. E. and Nagy, S. (1980). Processing oranges and tangerines. In: *Fruit and Vegetable Juice Processing Technology*, P. E. Nelson and D. K. Tressler (eds.), Avi Publishing Co., Westport, CT, pp. 35–96.

Oteiza, J. M., Giannuzzi, L., and Zaritzky, N. (2010). Ultraviolet treatment of orange juice to inactivate *E. coli* O157:H7 as affected by native microflora. *Food and Bioprocess Technology 3*(2), 603–614.

Oteiza, J. M., Peltzer, M., Gannuzzi, L., and Zaritzky, N. (2005). Antimicrobial efficacy of UV radiation on *Escherichia coli* O157:H7 in fruit juices of different absorptivities. *Journal of Food Protection 68*, 49–58.

Pala, C. U. and Toklucu, A. K. (2011). Effect of UV-C light on anthocyanin content and other quality parameters of pomegranate juice. *Journal of Food Composition Analysis 24*, 790–795.

Pala, C. U. and Toklucu, A. K. (2013). Microbial, physicochemical and sensory properties of UV-C processed orange juice and its microbial stability during refrigerated storage. *LWT-Food Science and Technology 50*, 426–431.

Patil, S., Bourke, P., Kelly, B., Frías, J., and Cullen, P. J. (2009). The effects of acid adaptation on *Escherichia coli* inactivation using power ultrasound. *Innovative Food Science and Emerging Technologies 10*, 486–490.

Patterson, M. F. (2005). Microbiology of pressure-treated foods. *Journal of Applied Microbiology 98*, 1400–1409.

Piyasena, P., Mohareb, E., and McKellar, R. C. (2003). Inactivation of microbes using ultrasound: A review. *International Journal of Food Microbiology 87*, 207–216.

Pothakamury, U. R., Monsalve-Gonzalez, A., Barbosa-Canovas, G. V., and Swanson, B. G. (1995). Inactivation of *Escherichia coli* and *Staphylococcus aureus* by pulsed electric field technology. *Food Research International 28*(2), 167–171.

Raso, J., Pagan, R., Codon, S., and Sala, F. J. (1998). Influence of temperature and pressure on the lethality of ultrasound. *Applied and Environmental Microbiology 64*, 465–471.

Raybaudi-Massilia, R. M., Mosqueda-Melgar, J., and Martín-Belloso, O. (2006). Antimicrobial activity of essential oils on *Salmonella* enteritidis, *Escherichia coli* and *Listeria innocua* in fruit juices. *Journal of Food Protection 69*, 1579–1580.

Raybaudi-Massilia, R. M., Mosqueda-Melgar, J., and Martín-Belloso, O. (2008). Antimicrobial activity of malic acid against *Listeria monocytogenes*, *Salmonella* enteritidis and *Escherichia coli* O157:H7 in apple, pear and melon juices. *Food Control 20*, 105–112.

Raybaudi-Massilia, R. M., Mosqueda-Melgar, J., Soliva-Fortuny, R., and Martín-Belloso, O. (2009). Control of pathogenic and spoilage microorganisms in fresh-cut fruits and fruit juices by traditional and alternative natural antimicrobials. *Comprehensive Reviews in Food Science and Food Safety 8*, 157–180.

Rodriguez, A. V., Moorillon, G. V., Zhang, Q. H., and Riva, E. R. (2007). Comparison of thermal processing and pulsed electric fields treatment in pasteurization of apple juice. *Food and Bioproducts Processing* 85(2), 93–97.

Ross, A. I. V., Griffiths, M. W., Mittal, G. S., and Deeth, H. C. (2003). Combining nonthermal technologies to control foodborne microorganisms. *International Journal of Food Microbiology 89*, 125–138.

Ryu, J. and Beuchat, L. R. (1998). Influence of acid tolerance responses on survival, growth, and thermal cross-protection of *Escherichia coli* O157:H7 in acidified media and fruit juices. *International Journal of Food Microbiology 45*, 185–193.

Sala, F., Burgos, J., Codon, S., Lopez, P., and Raso, J. (1995). *New Methods of Food Preservation*. Blackie Academic & Professional, London, U.K., pp. 176–204.

Sale, A. J. H. and Hamilton, W. A. (1967). Effects of high electric fields on microorganisms: I. Killing of bacteria and yeasts. *Biochimica et Biophysica Acta 148*, 781–788.

Salleh-Mack, S. Z. and Roberts, J. S. (2007). Ultrasound pasteurization: The effects of temperature, soluble solids, organic acids and pH on the inactivation of *Escherichia coli* ATCC 25922. *Ultrasonics Sonochemistry 14*, 323–329.

Sastry, S. K., Datta, K., and Worobo, R. W. (2000). Ultraviolet light. *Journal of Food Safety 65*, 90–92.

Schenk, M., Ilini, S. R., Guerrero, S., Blanco, G. A., and Alzamora, S. M. (2010). Inactivation of *Escherichia coli*, *Listeria innocua* and *Saccharomyces cerevisiae* by UV-C light: Study of cell injury by flow cytometry. *Food Science and Technology 44*, 191–198.

Schutz, H. G. (1983). Multiple regression approach to optimization. *Food Technology 37*(11), 46.

Schutz, H. G., Damrell, J. D., and Locke, B. H. (1972). Predicting hedonic ratings of raw carrot texture by sensory analysis. *Journal of Texture Studies 3*, 227–232.

Shama, G., Peppiatt, C., and Biguzzi, M. (1996). A novel thin film photoreactor. *Journal of Chemical Technology and Biotechnology 65*(1), 56–64.

Sharma, M., Adler, B. B., Harrison, M. D., and Beuchat, L. R. (2005). Thermal tolerance of acid-adapted and unadapted *Salmonella*, *Escherichia coli* O157:H7, and *Listeria monocytogenes* in cantaloupe juice and watermelon juice. *Letters in Applied Microbiology 41*, 448–453.

Shigehisa, T., Ohmori, T., Saito, A., Taji, S., and Hayashi, R. (1991) Effects of high hydrostatic pressure on characteristics of pork slurries and inactivation of microorganisms associated with meat and meat products. *International Journal of Food Microbiology 12*(2–3), 207–215.

Sidel, J. L. and Stone H. (1983). An introduction to optimization research. *Food Technology 37*(11), 36–38.

Silva, F. M., Gibbs, P., Vieira, M. C., and Silva, C. L. M. (1999). Thermal inactivation of *Alicyclobacillus acidoterrestris* spores under different temperature, soluble solids and pH conditions for the design of fruit processes. *International Journal of Food Microbiology 51*, 95–103.

Skandamis, P. N., Yoon, Y., Stopforth, J. D., Kendall, P. A., and Sofos, J. N. (2008). Heat and acid tolerance of *Listeria monocytogenes* after exposure to single and multiple sublethal stresses. *Food Microbiology 25*, 294–303.

Slifko, T. R., Raqhubeer, E., and Rose, J. B. (2000). Effect of high hydrostatic pressure on *Cryptosporidium parvum* infectivity. *Journal of Food Protection 63*(9), 1262–1267.

Sloan, A. E. (2010). Top ten functional food trends. *Food Technology 64*(4), 22–41.

Snowball, M. R. and Hornsey, I. S. (1988). Purification of water supplies using ultraviolet light. In: *Developments in Food Microbiology*, Vol. 3, R. K. Robinson (ed.), Elsevier Applied Science, London, U.K., pp. 171–192.

Southgate, D. A. T., Johnson, I. T., and Fenwick, G. R. (1995).Nutritional value and safety of processed fruit juices. In: *Production and Packaging of Non-Carbonated Fruit Juices and Fruit Beverages*, P. R. Ashurst (ed.), Blackie Academic & Professional, Great Britain, U.K., pp. 331–359.

Tauscher, B. (1995). Pasteurization of food by hydrostatic high pressure: Chemical aspects. *Zeitschrift für Lebensmitteluntersuchung und -Forschung A 200*, 3–13.

Teo, A. Y., Ravishankar, S., and Sizer, C. E. (2001). Effect of low-temperature, high-pressure treatment on the survival of *Escherichia coli* O157:7 and *Salmonella* in unpasteurized fruit juices. *Journal of Food Protection 64*(8), 1122–1127.

Tiganitas, A., Zeaki, N., Gounadaki, A. S., Drosinos, E. H., and Skandamis, P. N. (2009). Study of the effect of lethal and sublethal pH and a_w stresses on the inactivation or growth of *Listeria monocytogenes* and *Salmonella* Typhimurium. *International Journal of Food Microbiology 134*, 104–112.

Tran, M. T. T. and Farid, M. (2004). Ultraviolet treatment of orange juice. *Innovative Food Science & Emerging Technologies 5*, 495–502.

Tsong, T. Y. (1990). On electroporation of cell membranes and some related phenomena. *Bioelectrochemistry Bioenergetics 24*, 271–295.

Ukuku, D. O. and Geveke, D. J. (2010). A combined treatment of UV-light and radio frequency electric field for the inactivation of *Escherichia coli* K-12 in apple juice. *International Journal of Food Microbiology 138*, 50–55.

Uljas, H. E. and Ingham, S. C. (1999). Combination of intervention treatment resulting in a 5-$\log_{10}$-unit reductions in numbers of *Escherichia coli* O157:H7 and *Salmonella* Typhimurium DT104 organisms in apple cider. *Applied and Environmental Microbiology 65*, 1924–1929.

United States Department of Agriculture (USDA) and United States Department of Health and Human Services (USDHHS). (1998). Food labeling: Warning and notice: Labeling of juice products; Final rule, 21 CFR Part 101, 63 FR 37030-37056. US Food and Drug Administration, Washington, DC, July 8, 1998.

United States Department of Agriculture (USDA) and United States Department of Health and Human Services (USDHHS). (2001). Hazard analysis and critical control point (HACCP); Procedures for the safe and sanitary processing and importing of juice; Final rule, 21 CFR Part 120, 66 FR 6137-6202. US Food and Drug Administration, Washington, DC, January 19, 2001.

United States Department of Health and Human Services (USDHHS) and United States Department of Agriculture (USDA). (2005). *Dietary Guidelines for Americans*, 6th edn. Government Printing Office, Washington, DC.

Unluturk, S., Atilgan, M., Handan Baysal, A., and Tari, C. (2008). Use of UV-C radiation as a non-thermal process for liquid egg products (LEP). *Journal of Food Engineering 85*, 561–568.

Verseput, R. (2000). Digging into DOE: Selecting the right central composite design for response surface methodology applications. *Quality Digest*, QCI International. Available at: http://www.qualitydigest.com/june01/html/doe.html. Accessed July 22, 2010.

Vojdani, J. D., Beuchat, L. R., and Tauxe, R. V. (2008). Juice-associated outbreaks of human illness in the United States, 1995 through 2005. *Journal of Food Protection 71*, 356–364.

Walker, W. and Philips, C. A. (2008). The effect of preservatives on *Alicyclobacillus acidoterrestris* and *Propionibacterium cyclohexanicum* in fruit juice. *Food Control 19*, 974–981.

Wouters, P. C. and Smelt, J. P. P. M. (1997). Inactivation of microorganisms with pulsed electric fields: Potential for food preservation. *Food Biotechnology 11*, 193–229.

Wright, J. R., Sumner, S. S., Hackney, C. R., Pierson, M. D., and Zoecklein, B. W. (2000). Efficacy of ultraviolet light for reducing *Escherichia coli* O157:H7 in unpasteurized apple cider. *Journal of Food Protection 63*, 563–567.

Yu, L. J. and Ngadi, M. O. (2004). Textural and other quality properties of instant friend noodles as affected by some ingredients. *Cereal Chemistry 81*, 772–776.

Yuste, J. and Fung, D. Y. C. (2004). Inactivation of *Salmonella* Typhimurium and *Escherichia coli* O157:H7 in apple juice by a combination of nisin and cinnamon. *Journal of Food Protection 67*, 371–377.

Zhang, Q. H., Chang, F. J., and Barbosa-Cánovas, G. V. (1994). Inactivation of microorganisms in a semisolid model food using high voltage pulsed electric fields. *Lebensmittel-Wissenschaft und-Technologie 27*(6), 538–543.

4

Accumulation of Biogenic Amines in Foods: Hazard Identification and Control Options

Donatella Restuccia, Umile Gianfranco Spizzirri, Francesco Puoci, Ortensia Ilaria Parisi, Manuela Curcio, and Nevio Picci

CONTENTS

4.1 Introduction

Biogenic amines (BAs) are basic compounds of low molecular weight derived internally from plants, fruits, and vegetables. These organic bases can be divided into several groups according to their chemical structure—aromatic (tyramine and phenylethylamine), aliphatic (putrescine and cadaverine), or heterocyclic (histamine and tryptamine)—or in relation to the number of amino groups into monoamines (phenylethylamine and tyramine) and diamines (histamine, cadaverine, and putrescine). They can also be classified as volatile (phenylethylamine) and nonvolatile (histamine, cadaverine, putrescine, spermine, agmatine, tryptamine, and tyramine) BAs (EFSA, 2011). Figure 4.1 shows the chemical structures of the most important amines.

BAs execute a number of crucial functions in the physiology and development of eukaryotic cells. The most active BAs are histamine and tyramine. Polyamines such as putrescine, spermine, and spermidine also play essential roles in cell growth and differentiation via the regulation of gene expression and the

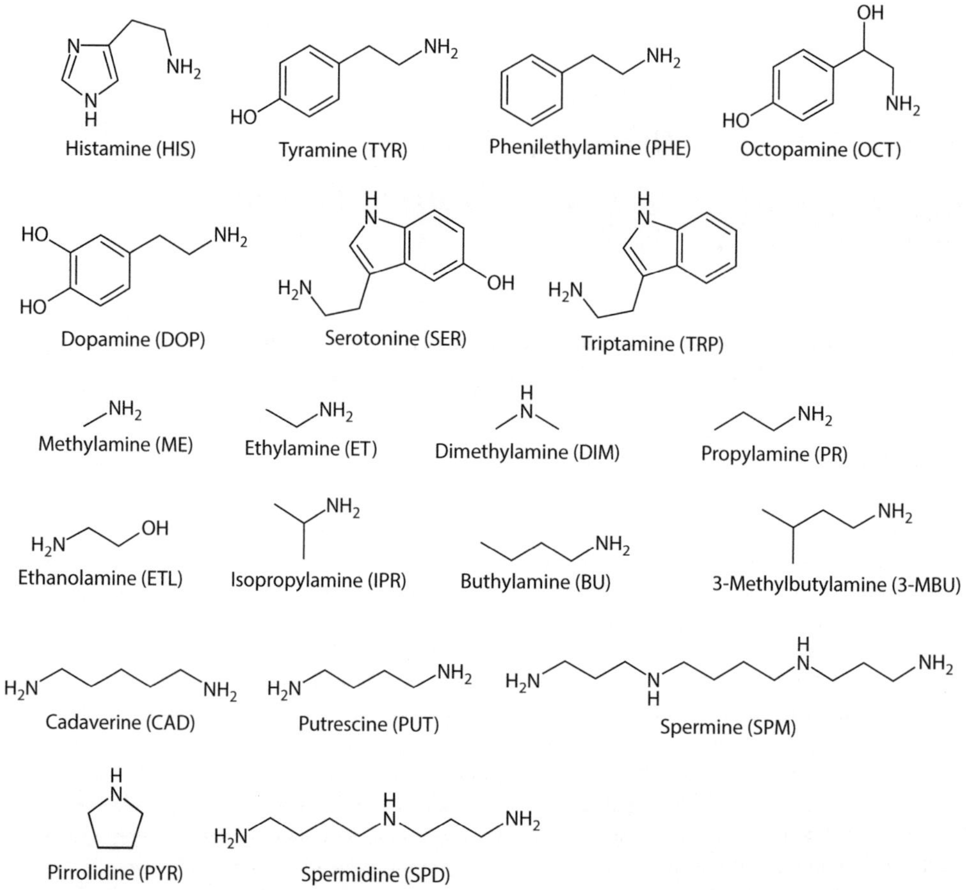

FIGURE 4.1 Chemical structures of biogenic amines.

modulation of signal transduction pathways. Low concentrations of BA are usually tolerated by the human body, and they are efficiently detoxified by mono- and diamine oxidase in the intestinal tract. However, these compounds can have adverse effects when present at high concentrations and pose a health risk for sensitive individuals.

While no upper limit of BA levels in foods is in place in legislation, some limits have been set only for histamine. In particular, the European Union's (EU) specific legislation, including microbial criteria for BA in foods, only covers histamine in fishery products; criteria for other BAs or other food products do not occur in any national legislation. Commission Regulation (EC) 2073/2005 on microbiological criteria for foodstuffs (as well as its amendments such as Regulation [EC] No. 1441/2007 and Regulation [EU] No. 365/2010) lays down food safety criteria for histamine in fishery products from fish species associated with a high amount of histidine between 100 mg/kg (m) and 200 mg/kg (M) and for fishery products that have undergone enzyme maturation treatment in brine, manufactured from fish species associated with a high amount of histidine between 200 mg/kg (m) and 400 mg/kg (M). Regulation (EC) No. 853/2004 lays down specific hygiene rules for foods of animal origin providing a possibility to set freshness criteria and limits for fishery products with regard to histamine and places the responsibility on food business operators to ensure that the limits with regard to histamine are not exceeded in the context of health standards for these products. In addition, for fish, the U.S. Food and Drug Administration (USFDA) proposed a guidance level of 50 ppm to be set, while the Australian and New Zealand Food Standards Code (Vol. 2, Standard 2.2.3, p. 22301, September 2002) states that "the level of histamine in fish or fish products must not exceed 200 mg/kg" (Bremer et al., 2003). The Codex Alimentarius provides

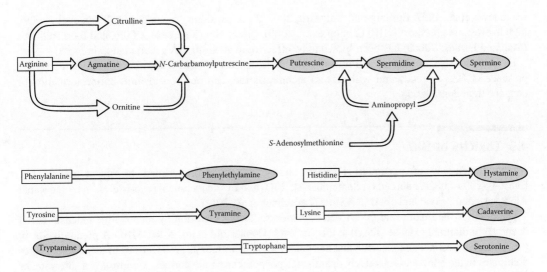

FIGURE 4.2 Generation of biogenic amines.

standards for some fish families' histamine levels as indicators for (1) decomposition (100 mg/kg) and (2) hygiene and handling (200 mg/kg).

Moreover, some authors have suggested other limits for BAs in foods, either for histamine alone (100 mg/kg) (Ten Brink et al., 1990) or as a sum of histamine, tyramine, putrescine, and cadaverine (900 mg/kg) (Shalaby, 1996; Valsamaki et al., 2000). Previous studies (Nout, 1994) considered that not all amines are equally toxic and that histamine, tyramine, and phenylethylamine were of major concern, proposing a limit for fermented foods of 50–100, 100–800, and 30 ppm for histamine, tyramine, and phenylethylamine, respectively, or a total of 100–200 ppm.

BAs originate in foods from the decarboxylation of the corresponding amino acid by the activity of exogenous enzymes released by various microorganisms (Figure 4.2). The presence of BAs in nonfermented foods generally indicates inadequate or prolonged storage; on the other hand, their presence in fermented foods could be unavoidable due to the diffusion of decarboxylases among lactic acid bacteria (LAB). It follows that the reasons for the determination of BAs in foods are twofold: first, their potential toxicity; second, the possibility of using them as food quality markers as their concentration can be related with the hygienic-sanitary quality of the process and with the freshness of the raw materials and the processed products.

As the consumption of food containing high concentrations of BAs may cause toxic reactions, considerable research has been undertaken in recent years to evaluate the presence of these compounds in various fermented, seasoned, or conserved foodstuffs (Russo et al., 2010) and efforts to reduce the occurrence of BAs deserve a priority and justify the challenge to the food industry to provide products with BA levels as low as possible. However, results indicate that the total amount of the different amines is strongly variable depending on the nature of the food, the microorganisms involved, and the environmental conditions. Considering that amine formation depends on multiple and complex variables, all of which interact, it follows that it is difficult to characterize the effects of each technological factor on aminogenesis during food fermentation, ripening, and/or storage. On the other hand, although the influence of these environmental factors is not always well characterized, they constitute the basis of the mitigation options to prevent or limit BA accumulation in foods. In particular, BA formation can be controlled by inhibiting microbial growth or inhibiting the decarboxylase activity of microbes (Wendakoon and Sakaguchi, 1995). It follows that the prevention of BA accumulation in food is generally achieved by temperature control, but also by using high-quality raw material, good manufacturing practice, nonamine-forming (amine-negative) or amine-oxidizing starter cultures for fermentation (Dapkevicius et al., 2000; Nieto-Arribas et al., 2009), and enzymes to oxidize amines (Dapkevicius et al., 2000). Moreover, microbial modeling to assess favorable conditions to delay BA formation

(Neumeyer et al., 1997; Emborg and Dalgaard, 2008a,b) in packaging techniques (Mohan et al., 2009), high hydrostatic pressure (HHP) (Bolton et al., 2009), irradiation (Kim et al., 2003), and food additives (Mah and Hwang, 2009a) has been shown to be effective in avoiding BA accumulation in foods.

In this context, this chapter will focus on the main aspects of hazard identification related with the presence of BAs in foods and will discuss techniques that can be used to limit amine formation or enhance their degradation.

4.2 Toxicity of BAs

Amine toxicity is of particular concern and it has been underlined by the Question No. EFSA-Q-2009-00829, adopted on September 21, 2011, and by the technical reports drafted by the Rapid Alert System for Food and Feed (RASFF) (Leuschner et al., 2013).

Oxidation is the main route of BA detoxification following ingestion by monoamine oxidase (MAO A and B) or diamine oxidase (DAO) at the gut level (Hungerford et al., 2010). MAO-A predominates in the stomach, intestine, and placenta and has polar aromatic amines as preferred substrates. MAO-B predominates in the brain and selectively deaminates nonpolar aromatic amines. Tyramine is a substrate for either form of MAO; MAO-A is responsible for the intestinal metabolism of tyramine, thereby preventing its systemic absorption (Azzaro et al., 2006). Tyramine, histamine, and phenylethylamine are also detoxified by N-methyltransferases or acetylation (Lehane et al., 2000; Broadley, 2010).

The severity of clinical symptoms depends on the amount and type of BA ingested and the correct functioning of the detoxification system. This last aspect could be influenced by human susceptibility, genetic disorder, and contemporary consumption of medication or BA-rich food or beverage. With regard to human susceptibility, it is clearly evidenced that certain physiological/pathological status can modify the sensitivity against BAs. For example, during pregnancy, there is a physiological increase of DAO production (up to 500-fold), which would explain the remissions of food intolerance (Mainz and Novak, 2007), while individuals affected by gastritis, irritable bowel syndrome, Crohn's disease, etc. are more sensible as the activity of oxidases is lower (Raithel et al., 2003). Moreover, several drugs are able to inhibit the enzymes involved in the detoxification process. Among them, MAO inhibitors that are used as antidepressant drugs could inhibit BA metabolism and increase tyramine sensitivity in a range from 2- to 5-fold for selegiline, 5- to 7-fold for moclobemide, and upto 13-fold for phenelzine (Bieck and Antonin, 1989; Zimmer et al., 1990). Other drugs, painkillers, antihypertensive drugs, mucolytics, antibiotics, antiarhythmic neuromuscular blocking, and agents reducing gut motility could act as MAO inhibitors (Mainz and Novak, 2007). Recently, isoflavones used as alternatives to hormone replacement therapy exhibited MAO inhibitory activity causing hypertensive crisis (Hutchins et al., 2005).

Considering that there are 1.3 billion people smoking worldwide, the evidence that tobacco smoke reduces MAO levels by up 40% together with other components present in cigarette proves that it could be one of the major risk factors for the onset of BA intoxication (WHO World Health Report, 2003; Broadley, 2010). Another important risk factor is the quantity of BA-rich food in a meal, which makes the prediction of the phenomenon of BA accumulation difficult. Moreover, foods rich in BAs like cheese are frequently consumed with alcoholic fermented beverages such as beer or wine, which increases the permeability of intestinal epithelial cells to Bas, enhancing the toxic effect (Silla-Santos et al., 1996). This phenomenon is referred to as the BA synergistic effect and may explain why the ingestion of aged cheese is more toxic than an equivalent of histamine administration in aqueous solution (Taylor and Summer, 1986).

BA intoxication is also called *scombroid poisoning* if it is caused by histamine or *cheese reaction* if it is related to tyramine, which is particularly abundant in aged cheese. The clinical signs of BA poisoning normally appear between 30 min to a few hours following BA ingestion and disappear within 24 h. Histamine poisoning is frequently misdiagnosed due to its typical symptoms that mimic those of allergy (Attaran and Probst, 2002). Histamine poisoning is characterized by headaches, vertigo, nausea and vomiting, enhanced secretion of gastric mucosa, gastrointestinal cramps, stomachache and diarrhea, hypotension, tachycardia, extrasystoles, itching, nose congestion, rhinorrhea, blepharitis, flushes, pruritus, and urticaria (Amon et al., 1999). Frequently, bass poisoning is a mild illness that is resolved by administration of antihistaminic combination drugs (Russel and Maretic, 1986). The *cheese reaction* is characterized by a release of catecholamines from the sympathetic nervous system and the adrenal medulla. This release may cause an increase in the mean arterial blood pressure ($\geq$180/120 mmHg) and heart rate because of peripheral vasoconstriction, resulting in hypertensive crisis, which is a more dangerous consequence. Tyramine ingestion of 125 mg/kg is necessary for the occurrence of this condition (Ladero et al., 2010). An ordinary meal contains 40 mg of tyramine; however, this value can increase drastically if the meal includes other fermented products.

It has been reported that the presence of 6 mg in one or two usual servings is thought to be sufficient to cause a mild adverse event while 10–25 mg will produce a severe adverse event in those using monoamine oxidase inhibitor (MAOI) drugs (McCabe-Sellers et al., 2006). For unmedicated adults, 200–800 mg of dietary tyramine is needed to induce a mild (30 mmHg) rise in blood pressure (Gardner et al., 1996). Polyamines such as pustrescine and cadaverine frequently present in food are considered not hazardous for human health (Bardocz, 1995). However, putrescine and cadaverine in particular may contribute to the dangerous synergistic effect that potentiates histamine or tyramine toxicity by inhibiting MAO, DAO, and hydroxymethyl transferase (Hui and Taylor, 1985; Bardócz, 1995; Straub et al., 1995). A positive correlation between cancer occurrence and polyamine consumption has been extensively reported (Gerner et al., 2004). For example, high putrescine levels in gastric carcinoma are caused by *Helicobacter pylori* (Shah and Swiatlo, 2008). Moreover, Theruvathu et al. (2005) demonstrated that polyamines stimulate the formation of mutagenic 1,N^2-propanodeoxyguanosine adducts from acetaldehyde, the first metabolite of ethanol, explaining the carcinogenicity of alcoholic beverage consumption, especially at the gastrointestinal tract level.

4.3 Hazard Identification

Generally speaking, food hazards fall into three categories:

1. Biological contamination: by organic materials present (e.g., animal, bird, insect remains) or from toxins produced from molds and bacteria.
2. Chemical and biochemical contamination: by chemicals deliberately introduced (e.g., pesticides), by accident (e.g., fuel), from cleaning chemicals, or actually produced through a process, and from biochemicals such as toxins produced by molds and fungi.
3. Physical contamination: by physical objects present in the raw materials (e.g., stones, glass, metallic parts), or picked up from plants (e.g., metallic components, glass), or accidentally dropped in by process operators/contractors (e.g., pens, tools).

Hazard identification provides the identification of the types of adverse health effects that can be caused by exposure to some agents as well as the characterization of the quality and weight of evidence supporting this identification. It is the process of determining whether exposure to a stressor can cause an increase in the incidence of specific adverse health effects and whether the adverse health effect is likely to occur in humans. The process examines the available scientific data for a given hazard and develops a weight of evidence to characterize the link between the negative effects and the considered agent. Hazard identification is based on experience and has a qualitative reasoning, providing a basis for identifying, evaluating, defining, and justifying the selection (and rejection) of control measures for reducing risk (Lee and Hathaway, 1998; Mortimore, 2000).

BAs are considered biochemical hazards as their occurrence in food is related with spoilage and fermentation although their presence has also been reported in raw materials of good hygienic quality such as putrescine in meat, fish, milk, and fruits. However, high concentrations of BA are generally produced in foods during fermentation. The main aim of lactic acid fermentation is the conversion of carbohydrates to lactic acid; therefore, the action of LAB is desirable for the production of fermented sausages, cheese, sour dough bread, and also for malolactic fermentation of wine. However, some of the bacteria involved in fermentation can produce BAs. Moreover, spoilage microorganisms may strongly contribute to BA formation in fermented foods even though food is not spoilt.

4.4 Conditions Influencing BA Formation

During food processing, amine formation is possible only when three conditions are realized: (1) the availability of free amino acids (precursors), (2) the existence of decarboxylase-positive microorganisms, and (3) the presence of suitable preconditions that allow bacterial growth, bacterial activity, decarboxylation, and synthesis (Anli and Bayram, 2008). In the set of metabolic pathways that break down amino acids into smaller units, several reactions are pointed, namely, decarboxylation, transamination, deamination, and desulfurization (Anli and Bayram, 2008).

As BA formation is a multifactorial phenomenon, all the factors bearing on the production of the substrate, the enzyme, and also their level of activity affect the type and amount of BA present in each case. Factors associated with the raw material (food composition, pH, handling conditions, etc.) such as the substrate source and reaction medium directly affect the availability of free amino acids (FAAs), whereas the presence of enzyme is closely connected to microbiological aspects (bacterial species and strain, bacterial growth, etc.). These factors are obviously interdependent and are further influenced by the technological processes associated with the types of food derivative and storage conditions. The combined action of these factors is what chiefly determines the final concentrations of BA, because it determines either directly or indirectly the presence and the activity of the substrate and enzyme.

4.4.1 Substrate Availability

FAAs play a fundamental role in the formation of amines in foods, in that they are the precursors of amines and, moreover, constitute a substrate for microbial growth (Ruiz-Capillas and Jiménez-Colmenero, 2005). Concentrations increase in association with proteolytic events taking place in the course of treatment and storage, due essentially to many of the microorganisms present. Microbial strains with high proteolytic enzyme activity also potentially increase the risk for BA formation in food systems, by increasing the availability of FAAs (Halàsz et al., 1994).

4.4.2 Microorganisms with Amino Acid Decarboxylase Activities

Many microorganisms possessing decarboxylase enzymes are able to produce amines and most of the amino acid decarboxylases require pyridoxal 5-1-phosphate (pyridoxal-P or PLP) as an essential coenzyme. Biodegradative decarboxylases do not show strict specificity toward the substrates and when the primary amino acid substrate is absent, structurally similar amino acids may be decarboxylated to form other BAs to provide pH homeostasis (Foster and Hall, 1991; Park et al., 1996). BA formation was suggested as a microbial strategy to survive in acidic environments or to supply alternative metabolic energy when bacterial cells are exposed to suboptimal substrate conditions (Cotter and Hill, 2003). In the pathway of amino acid catabolism, generally, the combined action of decarboxylases and a functional substrate/product transmembrane exchanger results in a proton motive force, which generates alkalinization of the cytoplasm (Wolken et al., 2006). The developing of a pH gradient across a membrane is associated with metabolic energy production and would be particularly important to microorganisms lacking a respiratory chain for generating high yields of adenosine triphosphate (ATP) (Vido et al., 2004). The decarboxylation pathways involve only two proteins: a decarboxylase and a transporter protein. The former converts the amino acid in the cytoplasm into the BA and carbon dioxide; the latter is responsible for the uptake of the amino acid from the medium and the excretion of the amine in an exchange process. Decarboxylase and transporter genes are generally organized in clusters located on the bacterial chromosome or on plasmids. A decarboxylase with a given amino acid specificity is adjacent to an exchanger with an equally strict amino acid/BA specificity, as was observed for histidine/histamine (Lucas et al., 2005), tyrosine/tyramine (Wolken et al., 2006), aspartate/alanine (Abe et al., 2002), ornithine/putrescine (Romano et al., 2012), and glutamate/g-aminobutyrate (Small and Waterman, 1998) decarboxylation systems.

The arginine deiminase pathway is the other type. Arginine is converted to ornithine. A very similar pathway has been described for agmatine, the decarboxylation product of arginine, which yields putrescine (Driessen et al., 1988; Griswold et al., 2004; Llácer et al., 2006). The pathway consists of one transport step and the sequential activities of three enzymes (Fernández and Zúñiga, 2006). As in the decarboxylation pathways, the transporter is responsible for the combined uptake and excretion of substrate and product, respectively, that is arginine/ornithine and agmatine/putrescine exchange. The first enzyme, arginine/agmatine deiminase, deiminates the substrate, yielding citrulline/carbamoylputrescine and ammonia. Next, the intermediate is phosphorylated from Pi by an ornithine/putrescine transcarbamylase, yielding carbamoylphosphate and the end product ornithine/putrescine. In both pathways, the final reaction is carried out by a carbamate kinase producing ATP, CO_2, and NH_3 from carbamoyl phosphate.

The production of BAs in vitro has been widely investigated during the last decades and both gram-positive (*Enterococcus faecalis, E. faecium, E. durans, E. hirae, Lactobacillus brevis, L. curvatus, L. buchneri, L. lactis, Leuconostoc mesenteroides, Streptococcus*) and gram-negative (*Enterobacter, Serratia liquefaciens, Escherichia coli, Samonella, Hafnia alvei, Citrobacter freundii, Morganella morganii, Klebsiella pneumoniae*, and *K. oxytoca*) bacteria have been found to possess the capacity to produce BAs (Okuzumi et al., 1982; Gingerich et al., 1999; Lehane and Olley, 2000; Marino et al., 2000; Kim et al., 2001; Ozogul and Ozogul, 2007; Lavizzari et al., 2010; Coton et al., 2012). Moreover, it has been reported that BAs can also be formed by yeast *Debaryomyces hansenii, Yarrowia lipolytica, Pichia jadinii, Geotrichum candidum, Candida incospicua*, and *Candida intermedia* (Wyder et al., 1999; Gardini et al., 2001, 2006; Roig-Sagués et al., 2002; Fernández et al., 2006). However, it is important to note that the ability to form BAs in vitro does not reflect the actual production in food, as environmental parameters (temperature, water activity, pH, precursor availability), microbiological factors (flora competition or presence of BA degraders), and optimal decarboxylase enzyme activity can all influence amine production.

Foods likely to contain high levels of BA include fish and fish products, dairy products, meat and meat products, fermented vegetables, soy products, alcoholic beverages such as wine and beer, fruits, nuts, and chocolate (Table 4.1). The most important BAs found in food are histamine, tyramine, putrescine, cadaverine, and phenylethylamine, which are products of the decarboxylation of histidine, tyrosine, ornithine,

TABLE 4.1
Most Common Biogenic Amines and the Produced Microorganisms in Foods

Food	Biogenic Amines	Produced Microorganism
Fish	Histamine	*Morganella morganii, Klebsiella pneumoniae, Hafnia alvei, Proteus mirabilis, Proteus vulgaris, Enterobacter clocae, Enterobacter aerogenes, Serratia fonticola, Serratia liquefacies, Citobacter freuidii, Clostridium* sp., *Pseudomonas fluorescens, Pseudomonas putida, Aeromonas* spp., *Plesiomonas shigelloides, Photobacterium* spp.
Cheese	Histamine	*Lactobacillus buchneri*
	Tyramine	*Enterococcus faecalis, Enterococcus faectum, Enterococcus durans, Enterococcus hirae, Lactobacillus brevis, Lactobacillus curvatus*
	Putrescine	Enterobacteriaceae, *Lactobacillus brevis*
Wine	Cadaverine	Enterobacteriaceae
	Histamine	*Oenococcus oeni, Lactobacillus hilgardii, Pediococcus parvulus*
	Tyramine	*Lactobacillus brevis, Lactobacillus hilgardii, Leuconostoc mesenteroides, Lactobacillus plantarum, Enterococcus faectum*
	Putrescine	*Lactobacillus brevis, Lactobacillus hilgardii, Leuconostoc mesenteroides, Lactobacillus plantarum, Oenococcus oeni, Lactobacillus buchneri, Lactobacillus zeae*
Meat	Histamine	Enterobacteriaceae, *Staphylococcus capitis*
	Tyramine	*Staphylococcus carnosus, Staphylococcus xylosus, Staphylococcus epidermidis, Staphylococcus saprophyticus, Lactobacillus brevis, Lactobacillus curvatus, Lactobacillus saket, Lactobacillus bavaricus, Carnobacterius divergens, Carnobacterius pisicola*
	Putrescine	Enterobacteriaceae, *Morganella morganii, S. liquefacies, Pseudomonas, Lactobacillus curvatus, Enterococcus*
	Cadaverine	Enterobacteriaceae

Source: Data from Ladero, V. et al., *Curr. Nutr. Food Sci.*, 6, 145, 2010.

lysine, and phenylalanine, respectively. Putrescine can also be formed through deimination of agmatine (Stratton et al., 1991; Lehane and Olley, 2000). Putrescine, a catabolic product of ornithine or arginine pathways, can be converted into spermidine, which can form spermine, as these three molecules are interconvertible (Pegg, 1986). Putrescine, spermidine, spermine, and cadaverine, also known as poly-amines, are mainly indicative of undesired microbial activity (Tabor and Tabor, 1985; Shalaby, 1996).

The capability to form BAs is generally considered a strain-specific characteristic rather than a spe-cies property. It is thus difficult to find precise correlations between BA contents and microorganism counts (Halász et al., 1994; Suzzi and Gardini, 2003; Standarová et al., 2008), suggesting that the genes encoding the BA-producing pathway may also be transferred by mobile elements (Marcobal et al. 2006a; Coton and Coton, 2009). However, in some cases, the genes encoding BA-producing enzymes (decarbox-ylation and transporter) have been sequenced and regulatory studies have been performed.

Amino acid decarboxylases are enzymes present in many microorganisms that may be either naturally present in food products or may be introduced by contamination before, during, or after food process-ing. In fermented foods, LAB can contribute to BA formation as they are present in raw materials, are part of the starter culture, or contaminate the product during its manufacture. To this regard, a number of authors have suggested the ability to produce BAs to be a negative trait when selecting a starter, sec-ondary, or adjunct culture to be used for making fermented products (Crow et al., 2001; Lonvaud-Funel, 2001). Between LAB, strains belonging to the genera *Lactobacillus, Enterococcus, Carnobacterium, Pediococcus, Lactococcus,* and *Leuconostoc* may be able to decarboxylate amino acids in several foods like cheese, fermented meat, vegetables, and beverages (Spano et al., 2010). Enterobacteriaceae belong-ing to the genera *Citrobacter, Klebsiella, Escherichia, Proteus, Salmonella,* and *Shigella* are associated with the production of considerable amounts of putrescine, cadaverine, and histamine in fish and meat products or, more generally, in spoiled food (Silla Santos, 1996; Bover'Cid et al., 2001a; Marino et al., 2003; Suzzi and Gardini, 2003). Putrescine synthesis was initially associated with gram-negative bac-teria, particularly members of the Enterobacteriacea (Marino et al., 2000; Chaves-López et al., 2006). However, recent reports cite the ability of certain LAB strains of *Lb. brevis* isolated from wine (Lucas et al., 2007) and *Lb. curvatus* and *E. faecalis* isolated from cheese to produce putrescine by agmatine deamination instead of the ornithine decarboxylation pathway; their contribution to putrescine produc-tion in food cannot, therefore, be ruled out. In fact, LAB are the main bacteria responsible for putrescine production in wine (Ancín-Azpilicueta et al., 2008).

Decarboxylase activity was observed in some species belonging to the genera *Micrococcus* and *Staphylococcus* in fermented sausages (de las Rivas et al., 2008). Although *Oenococcus oeni* was ini-tially considered the main bacterial species responsible for histamine accumulation in wine, recent stud-ies have shown that strains of *Lb. hilgardii* and *Pediococcus parvulus* may be effective as well (Lucas et al., 2005; Moreno-Arribas and Polo, 2008).

4.4.3 Technological Parameters

4.4.3.1 pH

The conditions in the reaction medium are a key factor in enzymatic activity, quite apart from their effect on the actual production of the enzyme through their impact on microbial growth. Amino acid decarboxylase activity is stronger in an acidic environment, the optimum pH being between 4.0 and 5.5 (Marcobal et al., 2006). There are two pH-related mechanisms acting simultaneously. One affects the growth by acidity, which inhibits the growth of microorganisms (Maijala et al., 1993). The other affects the production and activity of the enzyme because in low pH environments, bacteria are more stimulated to produce decarboxylase as a part of their defense mechanisms against acidity (Fernández et al., 2007). These opposing factors interfere with each other and the net result depends on their balance.

4.4.3.2 Oxygen

Oxygen supply also appears to have a significant effect on the biosynthesis of BA. *Enterobacter clo-acae* produces about half the quantity of putrescine in anaerobic compared with aerobic conditions, and *Klebsiella pneumoniae* synthesizes significantly less cadaverine but acquires the ability to produce putrescine under anaerobic conditions (Halàsz et al., 1994). Vacuum or modified atmosphere packaging (MAP), commonly used to prolong the shelf life of these kinds of products, entails altering O_2 levels and, hence, the growth and selection of some microorganism strains that can affect the production of BAs (Chong et al., 2011). However, it has been reported that the success of inhibition largely depends on the type of microflora, its environmental conditions such as temperature, and also the gas mix used in case of MAP; it may also be product-specific (Naila et al., 2010).

The redox potential of the medium also influences BA production. Conditions resulting in a reduced redox potential stimulate histamine production, and histidine decarboxylase activity seems to be inactivated or destroyed in the presence of oxygen.

4.4.3.3 Temperature

The quantitative production of BAs is usually reported to be temperature- and time-dependent (Zaman et al., 2009). Generally, the amine production rate increases with temperature. Conversely, BA accumulation is minimized at low temperatures through the inhibition of microbial growth and the reduction of enzyme activity. The optimum temperature for the formation of BAs by mesophilic bacteria has been reported to be between 20°C and 37°C, while the production of BA decreases below 5°C or above 40°C (Pachlovà et al., 2012).

4.4.3.4 NaCl and Sugars

The additives used in the preparation of fermented products are also important factors in the formation of BAs during this process, and their effect appears to depend on several factors (concentration, processing conditions, and others). In particular, sodium chloride plays an important role in microbial growth and therefore influences the activity of the amino acid decarboxylase. When sodium chloride content is 3.5%, the ability of *Lb. buchneri* to form histamine is partly inhibited, and when it is 5.0%, its formation is stopped (Maijala et al., 1994). This phenomenon can be attributed to reduced cell yield obtained in the presence of high sodium chloride concentration and to a progressive disturbance of the membrane located in microbial decarboxylase enzymes (Chander et al., 1989). However, the presence of sodium chloride activates tyrosine decarboxylase activity and inhibits histidine decarboxylase activity (Silla-Santos, 1996). Hence, it can be assumed that the effect of sodium chloride in inhibiting and stimulating BA production is strain-specific.

The presence of fermentable carbohydrate such as glucose can increase both growth and amino acid decarboxylase of bacteria. It has been demonstrated that the acidification produced by sugars in the fermentation process influences the formation of BAs (Arnol and Brown, 1978; Smith, 1980; Masson et al., 1996). The optimum glucose concentration for the formation of decarboxylase enzymes is 0.5%–2%, whereas concentrations above 3% inhibit it. This is related to the effect of reduced pH on microorganism growth and the activity of enzymatic systems (Masson and Montel, 1995; Hernàndez-Jover et al., 1997).

4.5 Control Options

The current control options to minimize BA occurrence in foods are mainly focused on the food-processing level, including raw material handling and the fermentation process, as they constitute the most important factors for amine accumulation in foods. During the storage of fermented food, the risk of BA increase is less important and is dependent on the previously mentioned stages.

Basically, the main approaches related with the accumulation delay of BAs in foods are particularly focused on (1) the assurance and maintenance of hygienic quality of raw materials and production processes in order to limit the contamination of food products with aminogenic microorganisms and (2) the implementation of specific conditions and/or production techniques aiming to inhibit (or eliminate) microorganisms with aminogenic potential or to prevent the growth and minimize the decarboxylase activity of microorganisms.

4.5.1 Hygienic Conditions

The presence of BAs in raw materials is associated with spoilage, suggesting poor hygienic practices, and may therefore indicate other food safety issues. Therefore, aminogenesis should be prevented in fresh commodities by improving food-handling standards through a preventative strategy from harvest/slaughtering to the consumer, implying that food quality and safety management relying on Hazard Analysis and Critical Control Point (HACCP) should be regarded as the primary approach (Hungerford, 2010). At the same time, good hygienic and manufacturing practices (GHP/GMP) along with proper cleaning and disinfection procedures should be carefully implemented from primary production.

Aminogenic organisms originating from animals (i.e., intestines, skin, fish gills, etc.) can be spread to other sites, surfaces, and equipment during handling of fresh raw materials (such as during degutting and filleting of fish, slaughtering, cutting, and mincing of meat, etc.) and consequently promote the accumulation of BAs during further processing and/or storage. Furthermore, any additional source of aminogenic microorganisms, decarboxylase enzymes, or preformed BAs should be minimized, for instance, assuring an optimal quality of water, brine for salting, as well as of spices and ingredients used during product manufacturing (Latorre-Moratalla et al., 2007; Innocente et al., 2009).

Moreover, the hygienic quality of raw materials and ingredients should also be assured in order to facilitate the dominance of starter bacteria from the early stages of fermentation (Maijala et al., 1995; Bover-Cid et al., 2001a; Novella-Rodríguez et al., 2004a,b). To this regard, a useful example is represented by the so-called low histamine technology (Bodmer et al., 1999) based on both the preventive approach (through GHP, GMP, and HACCP) and the implementation of specific technological measures for the manufacture of traditional alcoholic beverages (Bodmer et al., 1999).

4.5.2 Thermal Treatments and Freezing

Intervention strategies to improve the hygiene of raw materials should include, whenever possible, thermal treatments. In particular, pasteurization is the most common milk treatment used during cheesemaking to reduce the number of pathogenic and spoilage microorganisms and many authors found that its application is able to reduce the concentration of BAs in milk and dairy products. In fact, many decarboxylating bacteria such as Enterobacteriaceae and *Enterococcus* do not survive thermal treatment, this being the main explanation for the lower BA contents generally found in cheeses from pasteurized milk in comparison with those obtained from raw milk. Anyway, it should be kept in mind that deficient hygienic conditions can promote contamination with BA-producing microorganisms during the manufacture of dairy products after the thermal treatment has been accomplished. To this regard, the accumulation of elevated concentrations of BAs has been reported in cheeses derived from pasteurized milk (Linares et al., 2012; Loizzo et al., 2013).

As BA formation is temperature-dependent, it is decreased at low temperatures through the inhibition of microbial growth and the reduction of enzyme activity (Shalaby, 1996). In this sense, the time/temperature binomial is the most important risk factor for the formation of histamine and other BAs during the handling and storage of fresh commodities (e.g., meat and seafood products). For this reason, a strict adherence to the cold chain should be accomplished and low temperatures should be applied during storage to inhibit proteolytic and decarboxylase activity of bacteria (Rezae et al., 2007; Hernández-Orte et al., 2008). However, it should be kept in mind that proteolysis inhibition is not applicable for cheeses or fermented sausage manufacture because it is an essential process for coagulation and ripening.

Apart from mesophilic aminogenic organisms, which can be controlled by preventing time–temperature abuse, psychrotolerant bacteria are also relevant in relation to BA production in fish stored at chill temperature (Emborg and Dalgaard, 2006; Emborg et al., 2006). Freezing and temperatures near 0°C inhibit growth and activity of aminogenic organisms and therefore constitute the most effective way to prevent BA accumulation in fresh products, and thus also in raw materials (Ruiz-Capillas et al., 2010). In particular, it has been reported that, when thawed, frozen meat and fish are less susceptible to BA accumulation than unfrozen counterparts, probably because the microbiota are reduced to some extent as a result of the freezing process (Bover-Cid et al., 2006a, Flick and Granata, 2005).

However, it should be underlined that it is not always possible to control BA production through temperature alone, as some bacteria produce amines at temperatures below 5°C (Emborg and Dalgaard, 2006; Emborg et al., 2006).

4.5.3 High Hydrostatic Pressure

The application of pressure in the range of 100–1000 MPa is one of the most promising methods of food treatment and food preservation at room temperature and this nonthermal technology has the advantages of maintaining sensory and nutritional properties of foods. HHP is a preservation method that damages cell membranes of microorganisms, resulting in inactivation or sublethal injury (Rivas et al., 2008). It follows that when HHP is applied to raw material, a reduction in the number of bacteria may inhibit BA formation.

While reduction of BA concentrations has been shown in studies dealing with fish, meat, and milk hygiene (Novella-Rodríguez et al., 2002a; Latorre-Moratalla et al., 2007; Vidal-Carou et al., 2007; Ruiz-Capillas and Jimenez-Colmenero, 2010), other papers evidence an increased BA concentration after HHP application (Novella-Rodriguez et al., 2002b), underlining that the effects of high pressure processing on BA formation in foods need more investigation.

4.5.4 Starter Culture Selection

Fermentation represents a critical step in BA accumulation, and, generally speaking, the presence of these compounds in fermented foods cannot be totally avoided. In many cases, the accumulation of BAs in fermented foods has been attributed mainly to the activity of the nonstarter microflora. However, an indirect role of the starter LAB has been hypothesized as the peptidases released by the lysis of starter LAB could be essential in providing precursor amino acids (Valsamaki et al., 2000). It follows that, in the selection of starter cultures, the inability to form BAs but also their ability to grow well at the temperature intended for processing of the product and competitiveness in suppressing the growth of wild amine-producing microflora should be taken into consideration (Suzzi and Gardini, 2003). To this aim, principles of microbial ecology of food fermentation must be applied on the basis of product formulation (salt, sugar, preservatives, spices) and technological processing parameters (temperature, relative humidity, time, diameter/size, etc.) during the fermentation process. In particular, the inability to produce BAs should be considered an indispensable condition of strains intended to be used as starters and the European Food Safety Agency (EFSA) has recently introduced a system for a premarket safety assessment of selected taxonomic groups of microorganisms, leading to a *Qualified Presumption of Safety* (QPS) European equivalent of the Generally Recognized As Safe (GRAS) status (EFSA, 2007). However, it should be considered that starter cultures used in fermentation can also delay the formation of BAs (Bover-Cid et al., 2001b,c; Latorre-Moratalla et al., 2007; Mah and Hwang, 2009b) as they can be either amine-negative (not able to decarboxylate amino acid into BAs) or amine-oxidizing (oxidize BAs into aldehyde, hydrogen peroxide, and ammonia) bacteria as is proposed for fish sauce, sausage, or wine fermentation (Martuscelli et al., 2000; Fadda et al., 2001; Gardini et al., 2002; García-Ruiz et al., 2011; Zaman et al., 2011). The equilibrium between amines formed and degraded finally determines the BA level in food. Moreover, the complexity of the fermentation also enables the occurrence of a variety of additives, with synergistic or even antagonistic effects on ecology factors and a given ecological parameter can simultaneously show multiple targets or mechanisms of action. This is the case for the typical acidification accompanying nearly all, or most of, the fermentation processes. While a rapid and sharp decrease in pH is recognized as a key factor to reduce the growth of contaminating microbiota, it can also stimulate decarboxylation reactions in surviving microbiota as a response against unfavorable acidic environments (Vidal-Carou et al., 2007).

The protective performance of starter cultures to prevent BA accumulation will also be strongly conditioned by the adaptation of strains to the particular fermentation ecology, which can be better if strains are isolated from the same product or type of product. Standardized commercial preparations can be less effective than indigenous starters, although the influence is strain-dependent (Latorre-Moratalla et al., 2010). As an example, in fermented sausages, *Lb. sakei* and *Lb. curvatus* are well adapted to the meat fermentation environment, which make them good candidates for starter cultures as they are highly competitive to outgrow spontaneous fermenting flora and can efficiently inhibit gram-negative contaminating bacteria (Hugas and Monfort, 1997). Indeed, the literature confirms that starter cultures including decarboxylase-negative strains of *Lb. sakei* are the most protective as they reduce the overall amine accumulation by up to 95% in comparison with 30%–40% achieved with other commercial starters consisting of *Lb. plantarum* and *Pediococcus* spp. Mixed starter cultures, not only of LAB, but with other species involved in meat fermentation (e.g., Staphylococci) will contribute to control a wider variety of aminogenic microorganisms (Vidal-Carou et al., 2007).

4.5.5 Fermentation Parameters

Optimization of the technological conditions will favor proper implantation and development of the starter. Technological variables such as temperature and relative humidity of the ripening conditions (Maijala et al., 1995; Bover-Cid et al., 2001a) as well as the modification of the type or concentration of fermentable sugar or the addition of nontherapeutic antimicrobials (sulfite etc.) and additives (Bozkurt and Erkmen, 2002; Latorre-Moratalla et al., 2010) have been presented as appropriate measures to prevent the accumulation of BAs, not only during manufacture but also during storage of fermented products (Pinho et al., 2001; Suzzi and Gardini, 2003; Latorre-Moratalla et al., 2010). In particular, the temperature at which fermentation takes place influences the formation of BAs; indeed, it has been suggested that temperature could be a very useful parameter for preventing tyramine formation in dry sausage, chiefly by assuring conditions favorable to starter growth (Maijala et al., 1995). One explanation for the influence of processing temperature is that a higher fermentation temperature gives the starter culture the opportunity to outgrow nonstarter LAB. Moreover, it has been shown that additives and preservatives can reduce the formation of BAs by inhibiting bacterial growth and amine formation. Sodium sorbate, potassium sorbate, sodium hexametaphosphate at 2%, citric acid, succinic acid, D-sorbitol, and malic acid have been used to delay the formation of BAs in seafood and meat products. Naturally occurring specific inhibitory substances in spices and additives have also been shown to limit BA formation (curcumin, capsaicin, and piperine) as well as ginger, garlic, green onion, red pepper, clove, and cinnamon. However, although studies have shown the inhibitory effects of food additives and preservatives on amine accumulation, few authors have highlighted their potential negative effects. For example, the presence of preservatives has been reported to increase BA formation during sausage production (Komprda et al., 2004). Recently, it was found that curcumin inhibits DAO (Bhutani et al., 2009), which may inhibit BA reduction. When sodium sorbate and sodium hexametaphosphate were applied to sardines, a putrefactive odor was observed within 2 days at chill storage (Kang and Park, 1984). Other disadvantages of preservative use are a lack of available knowledge on their effectiveness against BAs in foods and the lack of consumer acceptance (Bjornsdottir, 2009). In summary, food additives and preservatives that work well in food require further investigation into the effectiveness in delaying BA production. At the same time, food additives that have shown a positive effect on delaying BA formation need to be tested in many food systems (such as cheese).

4.6 Degrading Methods

In the literature, attempts to destroy BAs once they are formed in the final product have also been described. In particular, irradiation may be an effective method acting by direct radiolysis of BAs (Mbarki et al., 2008) and by reducing the number of bacteria responsible for BA production (Kim et al., 2003) considering that the ionizing radiation inactivates the microorganisms by damaging the nucleic acid of cells (Farkas, 2006). However, although irradiation can be appropriate in eliminating BAs in foods once formed, it may pose some adverse effects on the aspects of food nutrition and organoleptic properties.

High temperature treatments have also been shown to be unsuitable to destroy formed BAs. In fact, these compounds are reported to be heat-stable (Tapingka et al., 2010) and cooking or prolonged exposure to heat will not eliminate the toxin (Shalaby, 1996; Duflos, 2009; Gonzaga et al., 2009).

For example, fish paste (Rihaakuru, Maldives local dish) is made through prolonged cooking (maximum 100°C), which eliminates all the potential bacteria responsible for histamine formation. However, Rihaakuru often contains high levels of histamine (>1000 ppm) as the histamine is believed to be formed in fish well before the cooking step and heat does not destroy histamine (Naila et al., 2010).

Other methods include the use of oxidizing microorganisms (BA-oxidizing bacteria) and enzymes such as DAO. BA-degrading bacteria could be introduced into a food-processing step to degrade the BAs in the food, or the bacteria could be used as a starter for fermented foods. These strategies are potential control measures where it is difficult to control BA levels through the traditional means of refrigeration, and to eliminate already formed BAs in food.

However, these methods are not according to the general principles of food hygiene that rely on prevention rather than elimination of problems after they appear and they have not been proven as effective and feasible.

4.7 Analytical Techniques for BA Determination in Foods

Detection of BAs is an important part of the investigation of incidents of potential food poisoning, for the verification of the food production processes (including HACCP) and as a measure of quality (freshness) of both raw materials and finished products (EFSA, 2011). The methods commonly used in the quantification of BAs can be divided into two groups: those based on the detection of BA themselves, and those based on the detection of the producer microorganisms.

Analytical methods for the determination of BAs in foods are reviewed by Önal (2007) and Sarkadi (2009). To obtain information on the levels of amines in cheese and to propose technological alternatives to reduce levels, a reliable methodology for the quantitative determination of bioactive amines must be available. All the methods used for amine determination involve two steps: amine extraction from the matrix and analytical determination. Reported extraction procedures consist of the use of acids (trichloracetic acid, hydrochloric, perchloric, thiodipropionic, or methansulfonic acids), solvents (petroleum ether, chloroform, or methanol), and filtration. The complexity of the varied food matrices is a critical consideration in obtaining adequate recoveries of all BAs. Depending on the complexity of the food matrix and the natural amount of free amino acids that can compete with the derivatizing agent, a further purification step might be needed prior to analytical determination. The analytical methods used for quantification of BAs are mainly based on chromatographic methods: thin layer chromatography, gas chromatography, capillary electrophoresis, and high-performance liquid chromatography (HPLC). HPLC is most often used for the analysis method of BAs. Due to low volatility and lack of chromophores of most Bas, the large majority of assays employ fluorimetric and UV detection with precolumn or postcolumn derivatization techniques. Liquid chromatographic methods with fluorescence or UV detection are reported in the literature for the determination of BAs, after derivatization with 6-aminoquinolyl-*N*-hydroxysuccinimidyl carbamate (Busto et al., 1996), benzoyl chloride (Huang et al., 1997), dabsyl chloride (Romero et al., 2000), dansyl chloride (Moret et al., 2005), and *o*-phthaldialdehyde (Alberto et al., 2002). It is well established that HPLC separation of the amines, postcolumn derivatization with *o*-phthalaldehyde, and fluorimetric detection is the method of choice (Novella-Rodrıguez et al., 2000).

More recently, a liquid chromatographic method with evaporative light-scattering detection (ELSD) was presented as a useful alternative to detect BAs in foods. The ELSD response is based on the amount of light scattered by analyte particles created by the evaporation of a solvent as it passes through a light beam. Therefore, the resulting signal corresponds to all compounds present in the sample that do not evaporate or decompose during evaporation of the solvent or mobile phase (Lucena et al., 2007). The use of ELSD permits the quantification of any solute less volatile than the solvent. However, the droplet size (and thus the response) is highly dependent on the flow of the nebulizing gas, the temperature of the evaporating tube, as well as the flow rate and the composition and physical characteristics of the mobile phase. The whole methodology, comprehensive of the homogenization–extraction process and the liquid chromatographic ELSD analysis, has been applied for the determination of BAs in different cheese samples, as their presence and relative amounts give useful information about freshness, level of ripening, and quality of storage (Restuccia et al., 2011; Spizzirri et al., 2013).

Traditional HPLC methods allow the determination of a wide variety of BAs, but the time required to carry out the analysis is relatively long, between 20 and 60 min/sample. Based on the HPLC principles, when sub-2 μm porous particles are used, speed and peak capacity can be extended to new limits. Ultra high-pressure liquid chromatography (UHPLC) is a new generation of chromatographic techniques, which in comparison with HPLC, operate at higher pressures, up to 15,000 psi, require less volume of sample, and capture detector signals at high data rates for fast-eluting peaks. Thus, UHPLC technology allows a drastic reduction on time analysis while increasing resolution and sensitivity (Nguyen et al., 2006). Latorre-Moratalla et al. (2009) proposed a rapid, precise, and versatile UHPLC method coupled with an online *o*-phthaldialdehyde postcolumn derivatization, which allows the determination of 12 BAs and polyamines in different food types. The method described by the authors is a reliable procedure to determine 12 BAs and polyamines usually present in food samples in less than 7 min of chromatographic elution, showing a satisfactory linearity, sensitivity, precision, and accuracy irrespective of the food matrix. The shortening of the run time was between 5- and 11-fold less in comparison with the conventional existing HPLC methods. This allows the analysis of a large number of samples, spending not only less time but also less solvent volumes, which is in agreement with environmental sustainability criteria.

In addition to the analytical methodologies previously described, several methods to detect the production of BAs through microorganisms have been developed (Linares et al., 2011). With respect to the detection of producer microorganisms, screening methods were initially based on the use of differential media containing a pH indicator to identify the BA-producer strains. However, these methods require the isolation and growth of the producer microorganism, and are therefore laborious and time-consuming.

The characterization of the genes encoding the decarboxylating enzymes led to the development of new methods based on polymerase chain reaction (PCR). A relationship between the presence of the gene encoding the decarboxylase and the capacity to synthesize BAs has been shown by several authors (Fernandez et al., 2004; Landete et al., 2005). A strain carrying aminoacyl-decarboxylase genes is a potential BA producer and its presence should be avoided in food. Therefore, these genes are good candidates for the design of specific primers for detecting BA-producer strains by PCR. Several sets of primers have been developed for the detection of tyramine-producing LAB (Fernandez et al., 2004; Coton and Coton, 2005). An important benefit of this technique is the detection of producer microorganisms before BAs themselves are detected in the sample; such detection can help to predict the likely accumulation of BAs in the final product. Three sets of primers have been developed for the detection of gram-positive histamine-producing bacteria (Coton and Coton, 2005; Landete et al., 2005), and two sets of primers have been proposed for the detection of gram-negative histamine-producing bacteria (de las Rivas et al., 2006). In addition, several sets of primers have been proposed for the detection of cadaverine- and putrescine-producing strains (de las Rivas et al., 2006). Although sensitive and specific under optimized conditions, conventional PCR has two drawbacks: the need to analyze the data by traditional endpoint analysis and the impossibility of template quantification. Real-time quantitative PCR (q-PCR) offers a potential alternative. This allows continuous monitoring of the PCR amplification process and, under appropriate conditions, the quantification of the target DNA present in the sample. A new set of primers has been designed for the detection of histamine-producing LAB by q-PCR (Lucas et al., 2008).

4.8 Conclusions

BAs are produced in food by amino acid decarboxylase-positive microorganisms and if ingested at high concentrations, can be detrimental for human health. They can accumulate in a wide variety of foods and beverages that have allowed microbial growth and activity during their manufacture, handling, and/or storage. The main prerequisites for the presence of BAs in foods include the availability of FAAs, the presence of microorganisms producing BA enzymes (mainly from raw materials and/or added starter cultures), and conditions allowing their growth (particularly temperature, pH), as well as conditions affecting the enzyme production and activity (particularly low pH). Fermentation provides the conditions indicated earlier allowing intensive microbial activity and therefore represents the key factor for

BA accumulation in foods. Unfortunately, a common rule cannot be defined for each type of product because each food type needs specific formulation and processing parameters, which have to be assessed on a case-by-case basis. Nowadays, the use of selected starter cultures is recognized as the most reliable approach to control the fermentation process, for both large-scale and traditional small productions of fermented foods and beverages. In this sense, microorganisms should be appropriately selected for each type of product and variety (i.e., substrate) taking into account their technological competence (competitiveness, influence on organoleptic characteristics, etc.) and safety requirements, including the inability to produce BAs.

Optimal intervention strategies to avoid the accumulation of BAs in foods should start from the accurate knowledge of the type of product and its productive process. As aminogenesis in fermented food and beverages is the result of complex phenomena affected by multiple factors, specific product/processor measures should be designed assessing and considering the particularities of the product, production process, and processing environment characteristics.

To this regard, as temperature is the major factor for controlling aminogenesis, traditionally, BA formation in food has been prevented, primarily by limiting microbial growth through chilling and freezing. However, secondary control measures to avoid amine synthesis or to reduce their levels once formed need to be considered as alternatives. Such approaches may include hydrostatic pressures, irradiation, controlled atmosphere packaging, or the use of food additives. Moreover, a variety of techniques can be combined together to control the microbial growth and enzyme activity during processing and storage for better shelf life extension and food safety.

REFERENCES

Abe, K., Onishi, F., Yagi, K. et al. 2002. Plasmid-encoded asp operon confers a proton motive metabolic cycle catalyzed by an aspartate-alanine exchange reaction. *Journal of Bacteriology* 184: 2906–2913.

Alberto, M.R., Arena, M.E., and Manca de Nadra, M.C.A. 2002. Comparative survey of two analytical methods for identification and quantification of biogenic amines. *Food Control* 13: 125–129.

Amon, U., Bangha, E., Küster, T., Menne, A., Vollrath, I.B., and Gibbs, B.F. 1999. Enteral histaminosis: Clinical implications. *Inflammation Research* 47: 291–295.

Ancín-Azpilicueta, C., González-Marco, A., and Jiménez-Moreno, N. 2008. Current knowledge about the presence of amines in wine. *Critical Reviews in Food Science and Nutrition* 48: 257–275.

Anl, R.E. and Bayram, M. 2008. Biogenic amines in wines. *Food Reviews International* 25: 86–102.

Arnol, S.H. and Brow, W.D. 1978. Histamine toxicity from fish products. *Advances in Food Research* 24: 113–154.

Attaran, R.R. and Probst, F. 2002. Histamine fish poisoning: A common but frequently misdiagnosed condition. *Emergency Medicine Journal* 19: 474–475.

Azzaro, A.J., Vandenberg, C.M., Blob, L.F. et al. 2006. Tyramine pressor sensitivity during treatment with the selegiline transdermal system 6 mg/24 h in healthy subjects. *Journal of Clinical Pharmacology* 46: 933–944.

Bardocz, S. 1995. Polyamines in food and their consequences for food quality and human health. *Trends in Food Science & Technology* 6: 341–346.

Bhutani, M.K., Bishnoi, M., and Kulkarni, S.K. 2009. Anti-depressant like effect of curcumin and its combination with piperine in unpredictable chronic stress-induced behavioral, biochemical and neurochemical changes. *Pharmacology Biochemistry and Behavior* 92: 39–43.

Bieck, P.R. and Antonin, K.H. 1989. Tyramine potentiation during treatment with MAO inhibitors: Brofaromine and moclobemide *vs* irreversible inhibitors. *Journal of Neural Transmission* 28(Suppl.): 21–31.

Bjornsdottir, K. 2009. Detection and control of histamine-producing bacteria in fish. DPhil thesis, Graduate Faculty of North Carolina State University, Raleigh, NC, p. 200. http://www.lib.ncsu.edu/theses/available/etd-03242009-101524/unrestricted/etd.pdf.

Bodmer, S., Imark, C., and Kneubühl, M. 1999. Biogenic amines in foods: Histamine and food processing. *Inflammation Research* 48: 296–300.

Bolton, G.E., Bjornsdottir, K., Nielsen, D., Luna, P.F., and Green, D.P. 2009. Effect of high hydrostatic pressure on histamine forming bacteria in yellowfin tuna and mahi-mahi skinless portions. *Institute of Food Technologists (IFT) Conference*, Anaheim, CA, 2009. Abstract nr. 006-05.

Bover-Cid, S., Hugas, M., Izquierdo-Pulido, M., and Vidal-Carou, M.C. 2000. Reduction of biogenic amine formation using a negative amino acid-decarboxylase starter culture for fermentation of fuet sausage. *Journal of Food Protection* 63: 237–243.

Bover-Cid, S., Hugas, M., Izquierd-Pulido, M., and Vidal-Carou, M.C. 2001a. Amino acid decarboxylase activity of bacteria isolated from fermented pork sausages. *International Journal of Food Microbiology* 66: 185–189.

Bover-Cid, S., Izquierdo-Pulido, M., and Vidal-Carou, M.C. 2001b. Effect of the interaction between a low tyramine-producing *Lactobacillus* and proteolytic staphylococci on biogenic amine production during ripening and storage of dry sausages. *International Journal of Food Microbiology* 65: 113–123.

Bover-Cid, S., Izquierdo-Pulido, M., and Vidal-Carou, M.C. 2001c. Effectiveness of a *Lactobacillus sakei* starter culture in the reduction of biogenic amine accumulation as a function of the raw material quality. *Journal of Food Protection* 64: 367–373.

Bozkurt, H. and Erkmen, O. 2002. Effects of starter cultures and additives on the quality of Turkish style sausage (sucuk). *Meat Science* 61: 149–156.

Bremer, P.J., Fletcher, G.C., and Osborne, C. 2003. Scombrotoxin in seafood. New Zealand Institute for Crop and Food Research Limited, A Crown Research Institute, Christchurch, New Zealand. Retrieved on March 16, 2011 at: www.crop.cri.nz/home/research/marine/pathogens/Scombrotoxin.pdf.

Broadley, K.H. 2010. The vascular effects of trace amines and amphetamines. *Pharmacology and Therapeutics* 125: 363–375.

Busto, O., Gulasch, J., and Borrull, F. 1996. Determination of biogenic amines in wine after precolumn derivatization with 6-aminoquinolyl-*N*-hydroxysuccinimidyl carbamate. *Journal of Chromatography A* 737: 205–213.

Chander, H., Batish, V.H., Babu, S., and Singh, R.S. 1989. Factors affecting amine production by a selected strain of *Lactobacillus bulgaricus*. *Journal of Food Science* 54: 940–942.

Chaves-López, C., De Angelis, M., Martuscelli, M., Serio, A., Paparella, A., and Suzzi, G. 2006. Characterization of the Enterobacteriaceae isolated from an artisanal Italian ewe's cheese (Pecorino Abruzzese). *Journal of Applied Microbiology* 101: 353–360.

Chong, C.Y., Abu Bakar, F., Russly, A.R., Jamilah, B., and Mahyudin, N.A. 2011. The effects of food processing on biogenic amines formation. *International Food Research Journal* 18: 867–876.

Coton, E. and Coton, M. 2005. Multiplex PCR for colony direct detection of Gram-positive histamine- and tyramine-producing bacteria. *Journal of Microbiology Methods* 63: 296–304.

Coton, E. and Coton, M. 2009. Evidence of horizontal transfer as origin of strain to strain variation of the tyramine production trait in *Lactobacillus brevis*. *Food Microbiology* 26: 52–57.

Coton, M., Delbès-Paus, C., Irlinger, F. et al. 2012. Biodiversity and assessment of potential risk factors of Gram-negative isolates associated with French cheeses. *Food Microbiology* 29: 88–98.

Cotter, P.D. and Hill, C. 2003. Surviving the acid test: Responses of Gram positive bacteria to low pH. *Microbiology and Molecular Biology Reviews* 67: 429–445.

Crow, V., Curry, B., and Hayes, M. 2001. The ecology of non-starter lactic acid bacteria (NSLAB) and their uses as adjuncts in New Zealand Cheddar. *International Dairy Journal* 11: 275–283.

Dapkevicius, M.L.N.E., Nout, M.J.R., Rombouts, F.M., Houben, J.H., and Wymenga, W. 2000. Biogenic amine formation and degradation by potential fish silage starter microorganisms. *International Journal of Food Microbiology* 57: 107–114.

de las Rivas, B., Marcobal, A., Carrascosa, A.V., and Munoz, R. 2006. PCR detection of foodborne bacteria producing the biogenic amines histamine, tyramine, putrescine, and cadaverine. *Journal of Food Protection* 69: 2509–2514.

de las Rivas, B., Ruiz-Capillas, C., Carrascosa, A.V., Curiel, J.A., Jimenez-Colmenero, F., and Muñoz, R. 2008. Biogenic amine production by Gram-positive bacteria isolated from Spanish dry-cured "chorizo" sausage treated with high pressure and kept in chilled storage. *Meat Science* 80: 272–277.

Driessen, A.J., Smid, E.J., and Konings, W.N. 1988. Transport of diamines by *Enterococcus faecalis* is mediated by an agmatine-putrescine antiporter. *Journal of Bacteriology* 170: 4522–4527.

Duflos, G. 2009. Histamine risk in fishery products. *Bulletin de l'Academie Veterinaire de France* 162: 241–246.

Emborg, J. and Dalgaard, P. 2006. Formation of histamine and biogenic amins in cold-smoked tuna—An investigation of psychrotolerant bacteria from samples implicated in cases of histamine fish poisoning. *Journal of Food Protection* 69: 897–906.

Emborg, J. and Dalgard, P. 2008a. Growth, inactivation and histamine formation of *Morganella psychrotolerans* and *Morganella morganii*—Development and evaluation of predictive models. *International Journal of Food Microbiology* 128: 234–243.

Emborg, J. and Dalgard, P. 2008b. Modelling the effect of temperature, carbon dioxide, water activity and pH on growth and histamine formation by *Morganella psychrotolerans*. *International Journal of Food Microbiology* 128: 226–233.

Emborg, J., Dalgaard, P., and Ahrens, P. 2006. *Morganella psychrotolerans* sp. nov., a histamine producing bacterium isolated from various seafoods. *International Journal of Systematic and Evolutionary Microbiology* 56: 2473–2479.

European Food Safety Authority (EFSA). 2007. Opinion of the Scientific Committee on a request from EFSA on the introduction of a Qualified Presumption of Safety (QPS) approach for assessment of selected microorganisms referred to EFSA. *The EFSA Journal* 587: 1–16.

European Food Safety Authority (EFSA). 2011. Scientific opinion on risk based control of biogenic amine formation in fermented foods. Panel on Biological Hazards. *The EFSA Journal* 9: 2393.

Fadda, S., Vignolo, G., and Oliver, G. 2001. Tyramine degradation and tyramine/histamine production by lactic acid bacteria and *Kocuria* strains. *Biotechnology Letters* 23: 2015–2019.

Farkas, J. 2006. Irradiation for better foods. *Trends in Food Science and Technology* 17: 148–152.

Fernández, M., del Río, B., Linares, D.M., Martín, M.C., and Álvarez, M.A. 2006. Real-time polymerase chain reaction for quantitative detection of histamine-producing bacteria: Use in cheese production. *Journal of Dairy Science* 89: 3763–3769.

Fernández, M., Linares, D.M., and Álvarez, M.A. 2004. Sequencing of the tyrosine decarboxylase cluster of *Lactococcus lactis* IPLA 655 and the development of a PCR method for detecting tyrosine decarboxylating lactic acid bacteria. *Journal of Food Protection* 67: 2521–2529.

Fernández, M., Linares, D.M., Rodríguez, A., and Alvarez, M.A. 2007. Factors affecting tyramine production in *Enterococcus durans* IPLA 655. *Applied Microbiology and Biotechnology* 73: 1400–1406.

Fernández, M. and Zúñiga, M. 2006. Amino acid catabolic pathways in lactic acid bacteria. *Critical Reviews in Microbiology* 32: 155–183.

Foster, J.W. and Hall, H.K. 1991. Inducible pH homeostasis and the acid tolerance response of *Salmonella* Typhimurium. *Journal of Bacteriology* 173: 5129–5135.

García-Ruiz, A., González-Rompinelli, E.M., Bartolomé, B., and Moreno- Arribas, M.V. 2011. Potential of wine-associated lactic acid bacteria to degrade biogenic amines. *International Journal of Food Microbiology* 148: 115–120.

Gardini, F., Martuscelli, M., Caruso, M.C. et al. 2001. Effects of pH, temperature and NaCl concentration on the growth kinetics, proteolytic activity and biogenic amine production of *Enterococcus faecalis*. *International Journal of Food Microbiology* 64: 105–117.

Gardini, F., Martuscelli, M., Crudele, M.A., Paparella, A., and Suzzi, G. 2002. Use of *Staphylococcus xylosus* as a starter culture in dried sausages: Effect on the biogenic amine content. *Meat Science* 61: 275–283.

Gardini, F., Tofalo, R., Belletti, N. et al. 2006. Characterization of yeasts involved in the ripening of Pecorino Crotonese cheese. *Food Microbiology* 23: 641–648.

Gardner, D.M., Shulman, K.I., Walker, S.E., and Tailor, S.A. 1996. The making of a user friendly MAOI diet. *Journal of Clinical Psychiatry* 57: 99–104.

Gerner, E.W. and Meyskens, F.L. 2004. Polyamines and cancer: Old molecules, new understanding. *Nature Reviews Cancer* 4: 781–792.

Gingerich, T.M., Lorca, T., Flick, G.J., Pierson, M.D., and McNair, H.M. 1999. Biogenic amine survey and organoleptic changes in fresh, stored, and temperature-abused bluefish (*Pomatomus saltatrix*). *Journal of Food Protection* 62: 1033–1037.

Gonzaga, V.E., Lescano, A.G., Huaman, A.A., Salmn-Mulanovich, G., and Blazes, D.L. 2009. Histamine levels in fish from markets in Lima, Peru. *Journal of Food Protection* 72: 1112–1115.

Griswold, A.R., Chen, Y.Y.M., and Burne, R.A. 2004. Analysis of an agmatine deiminase gene cluster in *Streptococcus mutans* UA159. *Journal of Bacteriology* 186: 1902–1904.

Halàsz, A., Baràth, A., Simon-Sarkadi, L., and Holzapfel, W. 1994. Biogenic amines and their production by microorganisms in food. *Trends in Food Science & Technology* 5: 42–49.

Hernàndez-Jover, T., Izquierdo-Pulido, M., Veciana-Nogues, M.T., Marinè-Font, A., and Vidal-Carou, M.C. 1997. Effect of starter cultures on biogenic amine formation during fermented sausage production. *Journal of Food Protection* 60: 825–830.

Hernández-Orte, P., Lapeña, A.C., Peña-Gallego, A. et al. 2008. Biogenic amine determination in wine fermented in oak barrels: Factors affecting formation. *Food Research International* 41: 697–706.

Hugas, M. and Monfort, J.M. 1997. Bacterial starter cultures for meat fermentation. *Food Chemistry* 59: 547–554.

Hui, J.Y. and Taylor, S.L. 1985. Inhibition of in vivo histamine metabolism in rats by foodborne and pharmacologic inhibitors of diamine oxidase, histamine *N*-ethyltransferase, and monoamine oxidase. *Toxicology and Applied Pharmacology* 81: 241–249.

Hungerford, J.M. 2010. Scombroid poisoning: A review. *Toxicon* 56: 231–243.

Hutchins, A.M., McIver, I.E., and Johnston, C.S. 2005. Hypertensive crisis associated with high dose soy isoflavone supplementation in a post-menopausal woman: A case report [ISRCTN98074661]. *BMC Women's Health* 5: 9–10.

Hwang, D.F., Chang, S.H., Shiua, C.Y., and Chai, T.J. 1997. High performance liquid chromatographic determination of biogenic amines in fish implicated in food poisoning. *Journal of Chromatography B* 693: 23–29.

Innocente, N., Marino, M., Marchesini, G., and Biasutti, M. 2009. Presence of biogenic amines in a traditional salted Italian cheese. *International Journal of Dairy Technology* 62: 154–160.

Kang, I.J. and Park, Y.H. 1984. Effect of food additives on the histamine formation during processing and storage of mackerel. *Bulletin of the Korean Fisheries Society* 17: 383–390.

Kim, J.H., Ahn, H.J., Jo, C., Park, H.J., Chung, Y.J., and Byun, M.W. 2004. Radiolysis of biogenic amines in model system by gamma irradiation. *Food Control* 15: 405–408.

Kim, J.H., Ahn, H.J., Kim, D.H., Jo, C., Yook, H.S., and Park, H.J. 2003. Irradiation effects on biogenic amines in Korean fermented soybean paste during fermentation. *Journal of Food Science* 68: 80–84.

Kim, S.H., Field, K.G., Morrissey, M.T., Price, R.J., Wei, C.I., and An, H. 2001. Source and identification of histamine-producing bacteria from fresh and temperature-abused albacore. *Journal of Food Protection* 64: 1035–1044.

Komprda, T., Smela, D., Pechova, P., Kalhotka, L., Stencl, J., and Klejdus, B. 2004. Effect of starter culture, spice mix and storage time and temperature on biogenic amine content of dry fermented sausages. *Meat Science* 67: 607–616.

Ladero, V., Calles, M., Fernández, M., and Alvarez, M.A. 2010. Toxicological effects of dietary biogenic amines. *Current Nutrition & Food Science* 6: 145–156.

Landete, J.M., Ferrer, S., and Pardo, I. 2005. Which lactic acid bacteria are responsible for histamine production in wine? *Journal of Applied Microbiology* 99: 580–586.

Latorre-Moratalla, M.L., Bosch-Fuste, J., Lavizzari, T., Bover-Cid, S., Veciana-Nogues, M.T., and Vidal-Caroua, M.C. 2009. Validation of an ultra-high pressure liquid chromatographic method for the determination of biologically active amines in food. *Journal of Chromatography A* 1216: 7715–7720.

Latorre-Moratalla, M.L., Bover-Cid, S., Aymerich, T., Marcos, B., Vidal-Carou, M.C., and Garriga, M. 2007. Aminogenesis control in fermented sausages manufactured with pressurized meat batter and starter culture. *Meat Science* 75: 460–469.

Latorre-Moratalla, M.L., Bover-Cid, S., Talon, R. et al. 2010b. Strategies to reduce biogenic amine accumulation in traditional sausage manufacturing. *Food Science and Technology* 43: 20–25.

Latorre-Moratalla, M.L., Bover-Cid, S., and Vidal-Carou, M.C. 2010a. Technological conditions influence aminogenesis during spontaneous sausage fermentation. *Meat Science* 85: 537–541.

Lavizzari, T., Breccia, M., Bover-Cid, S., Vidal-Carou, C., and Veciana-Nogues, M.T. 2010. Histamine, cadaverine, and putrescine production in vitro by Enterobacteriaceae and Pseudomonadaceae isolated from spinach. *Journal of Food Protection* 73: 385–389.

Lee, J.A. and Hathaway, S.C. 1998. The challenge of designing valid HACCP plans for raw food commodities. *Food Control* 9: 111–117.

Lehane, L. and Olley, J. 2000. Histamine fish poisoning revisited. *International Journal of Food Microbiology* 58: 1–37.

Leuschner, R.G.K., Aglika, H., Robinson, T., and Hugas, M. 2013. The Rapid Alert System for Food and Feed (RASFF) database in support of risk analysis of biogenic amines in food. *Journal of Food Composition and Analysis* 29: 37–42.

Linares, D.M., del Río, B., Ladero, V. et al. 2012. Factors influencing biogenic amines accumulation in dairy products. *Frontiers in Microbiology* 3: 1–10.

Linares, D.M., Martín, M.C., Ladero, V., Álvarez, M.A., and Fernández, M. 2011. Biogenic amines in dairy products. *Critical Reviews in Food Science and Nutrition* 51: 691–703.

Llácer, J.L., Polo, L.M., Tavarez, S., Alarcon, B., Hilario, R., and Rubio, V. 2006. The gene cluster for agmatine catabolism of *Enterococcus faecalis*. Studies of recombinant putrescine transcarbamylase and agmatine deiminase and a snapshot of agmatine deiminase catalysing its reaction. *Journal of Bacteriology* 189: 1254–1265.

Loizzo, M.R., Menichini, F., Picci, N., Puoci, F., Spizzirri, U.G., and Restuccia, D. 2013. Technological aspects and analytical determination of biogenic amines in cheese. *Trends in Food Science & Technology* 30: 38–55.

Lonvaud-Funel, A. 2001. Biogenic amines in wines: Role of lactic acid bacteria. *FEMS Microbiology Letters* 199: 9–13.

Lucas, P.M., Blancato, V.S., Claisse, O., Magni, C., Lolkema, J.S., and Lonvaud-Funel, A. 2007. Agmatine deiminase pathway genes in *Lactobacillus brevis* are linked to the tyrosine decarboxylation operon in a putative acid resistance locus. *Microbiology* 153: 2221–2230.

Lucas, P.M., Claisse, O., and Lonvaud-Funel, A. 2008. High frequency of histamine-producing bacteria in the enological environment and instability of the histidine decarboxylase production phenotype. *Applied Environmental Microbiology* 74: 811–817.

Lucas, P.M., Wolken, W.A.M., Claisse, O., Lolkema, J.S., and Lonvaud-Funel, A. 2005. Histamine producing pathway encoded on an unstable plasmid in *Lactobacillus hilgardii* 0006. *Applied and Environmental Microbiology* 71: 1417–1424.

Lucena, R., Cárdenas, S., and Valcárce, M. 2007. *Analytical and Bioanalytical Chemistry* 388: 1663–1672.

Mah, J.H. and Hwang, H.J. 2009a. Effects of food additives on biogenic amine formation in Myeolchijeot, a salted and fermented anchovy (*Engraulis japonicus*). *Food Chemistry* 114: 168–173.

Mah, J.H. and Hwang, H.J. 2009b. Inhibition of biogenic amine formation in a salted and fermented anchovy by *Staphylococcus xylosus* as a protective culture. *Food Control* 20: 796–801.

Maijala, R.L. 1994. Histamine and tyramine production by a *Lactobacillus* strain subjected to external pH decrease. *Journal of Food Protection* 57: 259–262.

Maijala, R.L., Eerola, S.H., Aho, M.A., and Hirn, J.A. 1993. The effects of GDL-induced pH decrease on the formation of biogenic amines in meat. *Journal of Food Protection* 56: 125–129.

Maijala, R.L., Eerola, S., Lievonen, S., Hill, P., and Hirvi, T. 1995. Formation of biogenic amines during ripening of dry sausages as affected by starter culture and thawing time of raw materials. *Journal of Food Science* 60: 1187–1190.

Maintz, L. and Novak, N. 2007. Histamine and histamine intolerance. *The American Journal of Clinical Nutrition* 85: 1185–1196.

Marcobal, A., de las Rivas, B., Moreno-Arribas, M.V., and Muñoz, R. 2006a. Evidence for horizontal gene transfer as origin of putrescine production in *Oenococcus oeni* RM83. *Applied and Environmental Microbiology* 72: 7954–7958.

Marcobal, A., Martin-Alvarez, P.J., Moreno-Arribas, M.V., and Muñoz, R. 2006b. A multifactorial design for studying factors influencing growth and tyramine production of the lactic acid bacteria *Lactobacillus brevis* CECT 4669 and *Enterococcus faecium* BIFI-58. *Research in Microbiology* 157: 417–424.

Marino, M., Maifreni, M., Moret, S., and Rondinini, G. 2000. The capacity of Enterobacteriaceae to produce biogenic amines in cheese. *Letters in Applied Microbiology* 31: 169–173.

Martuscelli, M., Crudele, M.A., Gardini, F., and Suzzi, G. 2000. Biogenic amine formation and oxidation by *Staphylococcus xylosus* strains from artisanal fermented sausages. *Letters in Applied Microbiology* 31: 228–232.

Masson, F. and Montel, M.C. 1995. Les amines biogenes dans les produits carnés. *Viandes et Produits Carnes* 16: 3–7.

Masson, F., Talon, R., and Montel, M.C. 1996. Histamine and tyramine production by bacteria from meat products. *International Journal of Food Microbiology* 32: 199–207.

Mbarki, R., Sadok, S., and Barkallah, I. 2008. Influence of gamma irradiation on microbiological, biochemical, and textural properties of bonito (*Sarda sarda*) during chilled storage. *Food Science and Technology International* 14: 367–373.

McCabe-Sellers, B.J., Staggs, C.G., and Bogle, M.L. 2006. Tyramine in foods and monoamine oxidase inhibitor drugs: A crossroad where medicine, nutrition, pharmacy, and food industry converge. *Journal of Food Composition and Analysis* 19: 58–65.

Mohan, C.O., Ravishankar, C.N., Gopal, T.K.S., Kumar, K.A., and Lalitha, K.V. 2009. Biogenic amines formation in seerfish (*Scomberomorus commerson*) steaks packed with O_2 scavenger during chilled storage. *Food Research International* 42: 411–416.

Moreno-Arribas, M.V. and Polom, M.C. 2008. Occurrence of lactic acid bacteria and biogenic amines in biologically aged wines. *Food Microbiology* 25: 875–881.

Moret, S., Smela, D., Populin, T., and Conte, L.S. 2005. A survey on free biogenic amine content of fresh and preserved vegetables. *Food Chemistry* 89: 355–361.

Mortimore, S. 2000. An example of some procedures used to assess HACCP systems within the food manufacturing industry. *Food Control* 11: 403–413.

Naila, A., Flint, S., Fletcher, G., Bremer, P., and Meerdink, G. 2010. Control of biogenic amines in food— Existing and emerging approaches. *Journal of Food Science* 75: 139–150.

Neumeyer, K., Ross, T., and McMeekin, T.A. 1997. Development of a predictive model to describe the effects of temperature and water activity on the growth of spoilage pseudomonads. *International Journal of Food Microbiology* 38: 45–54.

Nguyen, D.T.T., Guillarme, D., Rudaz, S., and Veuthey, J.L. 2006. Fast analysis in liquid chromatography using small particle size and high pressure. *Journal of Separation Science* 29: 1836–1848.

Nieto-Arribas, P., Poveda, J.M., Sesēna, S., Palop, L., and Cabezas, L. 2009. Technological characterization of *Lactobacillus* isolates from traditional Manchego cheese for potential use as adjunct starter cultures. *Food Control* 20: 1092–1098.

Nout, M.J.R. 1994. Fermented foods and food safety. *Food Research International* 27: 291–298.

Novella-Rodríguez, S., Veciana-Nogués, M.T., Roig-Sagues, A.X., Trujillo-Mesa, A.J., and Vidal-Carou, C. 2004a. Evaluation of biogenic amines and microbial counts throughout the ripening of goat cheeses from pasteurized and raw milk. *Journal of Dairy Research* 71: 245–252.

Novella-Rodríguez, S., Veciana-Nogués, M.T., Roig-Sagués, A.X., Trujillo-Mesa, A.J., and Vidal-Carou, M.C. 2004b. Comparison of biogenic amine profile in cheeses manufactured from fresh and stored (4°C, 48 hours) raw goats milk. *Journal of Food Protection* 67: 110–116.

Novella-Rodríguez, S., Veciana-Nogués, M.T., Saldo, J., and Vidal-Carou, M.C. 2002b. Effects of high hydrostatic pressure treatments on biogenic amine contents in goat cheeses during ripening. *Journal of Agricultural and Food Chemistry* 50: 7288–7292.

Novella-Rodríguez, S., Veciana-Nogués, M.T., Trujillo-Mesa, A., and Vidal-Carou, M.C. 2002a. Profile of biogenic amines in goat cheese made from pasteurized and pressurized milks. *Journal of Food Science* 67: 2940–2944.

Novella-Rodrıguez, S., Veciana-Nogues, M.T., and Vidal-Carou, M.C. 2000. Biogenic amines and polyamines in milks and cheeses by ion-pair high performance liquid chromatography. *Journal of Agricultural and Food Chemistry* 48: 5117–5123.

Okuzumi, M., Okuda, S., and Awano, M. 1982. Occurrence of psychrophilic and halophilic histamine-forming bacteria (N-group bacteria) on/in red meat fish. *Bulletin of the Japanese Society for the Science of Fish* 48: 799–804.

Önal, A., 2007. A review: Current analytical methods for the determination of biogenic amines in foods. *Food Chemistry* 103: 1475–1486.

Özogul, F. and Özogul, Y. 2007. The ability of biogenic amines and ammonia production by single bacterial cultures. *European Food Research and Technology* 225: 385–394.

Pachlovà, V., Bunka, F., Flasarovà, R., Vàlkovà, P., and Bunkovà, L. 2012. The effect of elevated temperature on ripening of Dutch type cheese. *Food Chemistry* 132: 1846–1854.

Park, Y., Bearson, B., Bang, S.H., Bang, I.S., and Foster, J.W. 1996. Internal pH crisis, lysine decarboxylase and the acid tolerance response of *Salmonella typhimurium*. *Molecular Microbiology* 20: 605–611.

Pegg, A.E. 1986. Recent advances in the biochemistry of polyamines in eukaryotes. *Biochemical Journal* 234: 249–262.

Pinho, O., Ferreira, I.M.P.L.V.O., Mendes, E., Oliveira, B.M., and Ferreira, M. 2001. Effect of temperature on evolution of free amino acid and biogenic amine contents during storage of Azeitão cheese. *Food Chemistry* 75: 287–291.

Raithel, M., Riedel, A., Küfner, M., Donhauser, N., and Hahn, E. 2003. Evaluation of gut mucosal diamine oxidase activity (DAO) in patients with food allergy and ulcerative colitis, idiopathic ulcerative colitis and Crohn's disease. *Gastroenterology* 124(Suppl. 1): A475.

Renata, G.K., Leuschner, A.H., Tobin, R., and Hugas, M. 2013. The Rapid Alert System for Food and Feed (RASFF) database in support of risk analysis of biogenic amines in food. *Journal of Food Composition and Analysis* 29: 37–42.

Restuccia, D., Spizzirri, U.G., Puoci, F. et al. 2011. A new method for the determination of biogenic amines in cheese by LC with evaporative light scattering detector. *Talanta* 85: 363–369.

Rezaei, M., Montazeri, N., Langrudi, H.E., Mokhayer, B., Parviz, M., and Nazarinia, A. 2007. The biogenic amines and bacterial changes of farmed rainbow trout (*Oncorhynchus mykiss*) stored in ice. *Food Chemistry* 103: 150–154.

Rivas, B., Gonzàlez, R., Landete, J.M., and Muñoz, R. 2008. Characterization of a second ornithine decarboxylase isolated from *Morganella morganii*. *Journal of Food Protection* 71: 657–661.

Roig-Sagués, A.X., Molina, A.P., and Hernandez-Herrero, M.M. 2002. Histamine and tyramine-forming microorganisms in Spanish traditional cheeses. *European Food Research and Technology* 215: 96–100.

Romano, A., Trip, H., Lonvaud-Funel, A., Lolkema, J.S., and Lucas, P.M. 2012. Evidence for two functionally distinct ornithine decarboxylation systems in lactic acid bacteria. *Applied and Environmental Microbiology* 78: 1953–1961.

Romero, R., Gazquez, D., Bagur, M.G., and Sanchez-Vinas, M. 2000. Optimization of chromatographic parameters for the determination of biogenic amines in wines by reversed-phase high-performance liquid chromatography. *Journal of Chromatography A* 871: 75–83.

Ruiz-Capillas, C. and Jiménez-Colmenero, F. 2004. Biogenic amines in meat and meat products. *Critical Reviews in Food Science and Nutrition* 44: 489–499.

Ruiz-Capillas, C. and Jiménez Colmenero, F. 2010. Biogenic amines in seafood products. In *Handbook of Seafood and Seafood Products Analysis*, L. Nollet and F. Todrà, eds., pp. 833–850. Taylor & Francis Group, LLC, Boca Raton, FL.

Russell, F.E. and Maretic, Z. 1986. Scombroid poisoning: Mini review with case histories. *Toxicon* 24: 967–973.

Russo, P., Spano, G., Arena, M. P., Capozzi, V., Grieco, F., and Beneduece, L. 2010. Are consumers aware of the risks related to biogenic amines in food? *Current Research, Technology and Education Topics in Applied Microbiology and Microbial Biotechnology* 2: 1087–1095.

Sarkadi, L.S. 2009. Biogenic amines. In *Process Induced Food Toxicants*, R.H. Stadler and D.R. Lineback, eds., John Wiley & Sons, Inc., Hoboken, NJ.

Shah, P. and Swiatlo, E. 2008. A multifaceted role for polyamines in bacterial pathogens. *Molecular Microbiology* 68: 4–16.

Shalaby, A.R. 1996. Significance of biogenic amines in food safety and human health. *Food Research International* 29: 675–690.

Silla Santos, M.H. 1996. Biogenic amines: Their importance in food. *International Journal of Food Microbiology* 29: 213–231.

Small, P.L.C. and Waterman, S.R. 1998. Acid stress, anaerobiosis and gadCB: Lessons from *Lactococcus lactis* and *Escherichia coli*. *Trends in Microbiology* 6: 214–216.

Smith, T.A. 1980. Amines in food. *Food Chemistry* 6: 169–200.

Spano, G., Russo, P., Lonvaud-Funel, A. et al. 2010. Risk assessment of biogenic amines in fermented food. *European Journal of Clinical Nutrition* 64: 95–100.

Spizzirri, U.G., Restuccia, D., Curcio, M., Parisi, O.I., Iemma, F., and Picci, N. 2013. Determination of biogenic amines in different cheese samples by LC with evaporative light scattering detector. *Journal of Food Composition and Analysis* 29: 43–51.

Standarová, E., Borkovcová, I., and Vorlová, L. 2008. The occurrence of biogenic amines in dairy products on the Czech market. *Acta Scientiarum Polonorum* 7: 35–42.

Stratton, J.E., Hutkins, R.W., and Taylor, S.L. 1991. Biogenic amines in cheese and other fermented foods: A review. *Journal of Food Protection* 54: 460–470.

Straub, B.W., Kicherer, M., Schilcher, S.M., and Hammes, W.P. 1995. The formation of biogenic amines by fermentation organisms. *Zeitschrift für Lebensmittel-Untersuchung und -Forschung A* 201: 79–82.

Suzzi, G. and Gardini, F. 2003. Biogenic amines in dry fermented sausages: A review. *International Journal of Food Microbiology* 88: 41–54.

Tabor, C.W. and Tabor, H. 1985. Polyamines in microorganisms. *Microbiological Reviews* 49: 81–99.

Tapingkae, W., Tanasupawat, S., Parkin, K.L., Benjakul, S., and Visessanguan, W. 2010. Degradation of histamine by extremely halophilic archaea isolated from high salt-fermented fishery products. *Enzyme and Microbial Technology* 46: 92–99.

Taylor, S.L. and Sumner, S.S. 1986. Determination of histamine, putrescine, and cadaverine. In *Seafood Quality Determination*, D.E. Kramer and J. Liston, eds., pp. 235–245. Elsevier, Amsterdam, the Netherlands.

Ten Brink, B., Damink, C., Joosten, H.M.L.J., and Huisin't Veld, J.H.J. 1990. Occurrence and formation of biologically active amines in foods. *International Journal of Food Microbiology* 11: 73–84.

Theruvathu, J.A., Jaruga, P., Nath, R.G., Dizdaroglu, R., and Brooks P.J. 2005. Polyamines stimulate the formation of mutagenic 1,N^2-propanodeoxyguanosine adducts from acetaldehyde. *Nucleic Acids Research* 33: 3513–3520.

Valsamaki, K., Michaelidou, A., and Polychroniadou, A. 2000. Biogenic amine production in Feta cheese. *Food Chemistry* 71: 259–266.

Vidal-Carou, M.C., Latorre-Moratallam, M.L., Veciana-Noguésm, M.T., and Bover-Cid, S. 2007. Biogenic amines: Risks and control. In *Handbook of Fermented Meat and Poultry*, F. Todrà, Y.H. Hui, I. Astiasarán, W.-K. Nip, J.G. Sebranek, E.T.F. Silveira, L.H. Stahnke, and R. Talon, eds., Chapter 43, pp. 455–468. Blackwell Publishing, Oxford, U.K.

Vido, K., Le Bars, D., Mistou, M.Y., Anglade, P., Gruss, A., and Gaudu, P. 2004. Proteome analyses of heme-dependent respiration in *Lactococcus lactis*: Involvement of the proteolytic system. *Journal of Bacteriology* 186: 1648–1657.

Wendakoon, C.N. and Sakaguchi, M. 1995. Inhibition of amino acid decarboxylase activity of *Enterobacter aerogenes* by active components in spices. *Journal of Food Protection* 58: 280–283.

Wolken, W.A., Lucas, P.M., Lonvaud-Funel, A., and Lolkema, J.S. 2006. The mechanism of the tyrosine transporter TyrP supports a proton motive decarboxylation pathway in *Lactobacillus brevis*. *Journal of Bacteriology* 188: 2198–2206.

Wyder, M.T., Bachmann, H.P., and Puhan, Z. 1999. Role of selected yeasts in cheese ripening: An evaluation in foil wrapped Raclette cheese. *Lebensmittel-Wissenschaft & Technologie* 32: 333–343.

Zaman, M.Z., Abdulamir, A.S., Abu Bakar, F., Selamat, J., and Bakar, J. 2009. A review: Microbiological, physiological and health impact of high level of biogenic amines in fish sauce. *American Journal of Applied Sciences* 6: 1199–1211.

Zaman, M.Z., Abu-Bakar, F., Jinap, S., and Bakar, J. 2011. Novel starter cultures to inhibit biogenic amines accumulation during fish sauce fermentation. *International Journal of Food Microbiology* 145: 84–91.

Zimmer, R., Puech, A.J., Philipp, F., and Korn, A. 1990. Interaction between orally administered tyramine and moclobemide. *Acta Psychiatrica Scandinavica* 360(Suppl.): 78–80.

5

Quantitative Microbial Risk Assessment Methods for Food Safety in RTE Fresh Vegetables

G.D. Posada-Izquierdo, G. Zurera, and Fernando Pérez-Rodríguez

CONTENTS

5.1 Introduction to Microbial Risk Assessment

Microbiological risk assessment (MRA) is a scientific-based process comprised of four steps: hazard identification, hazard characterization, exposure assessment, and risk characterization, which are defined by CAC (1999) as follows:

Hazard Identification: "The identification of biological agents capable of causing adverse health effects and which may be present in a particular food or group of foods."

Hazard Characterization: "The qualitative and/or quantitative evaluation of the nature of the adverse health associated with the hazard."

Exposure Assessment: "The qualitative and/or quantitative evaluation of the likely intake of a biological agent via food, as well as exposure from other sources if relevant."

Risk Characterization: "The process of determining the qualitative and/or quantitative estimation, including attendant uncertainties, of the probability of occurrence and severity of known or potential adverse health effects in a given population based on hazard Identification, Hazard Characterization and Exposure Assessment."

MRA is a relatively new tool in the field of food safety. In literature, it is possible to find risk assessments of *Salmonella* spp., *Listeria monocytogenes*, *Escherichia coli*, etc., in different food commodities (Pérez-Rodríguez et al. 2007a). MRA can be carried out in different manners according to the nature of the variables that are considered. MRA could be qualitative, quantitative, or semiquantitative. Often, the selection of one or another relies on the information available. The more information available, the more quantitative the assessment will be. In addition, a quantitative MRA could be point estimate or stochastic, the latter being preferable as it provides much more information than a point-estimate approach. In the first case, each variable is described by a single value, which is usually the mean of the variable or another representative statistic such as 5th and 95th percentile, mode, mean, or median (Lammerding and Fazil 2000).

In the exposure assessment phase, the exposure level to the food-borne pathogen at the time of consumption, associated with a specific food or food category, is estimated (FAO/WHO 1999). This step is mainly underpinned by the use of predictive microbiology models (see Section 5.3). Predictive models allow considering specific steps along the food chain where experimental observations are difficult or scarce (e.g., low levels after sterilization or levels at the moment of consumption) can be represented or considered with respect to their contribution to final exposure level and hence to final risk. Output variables in exposure assessment are prevalence and concentration of the microorganism, which are related to factors or input variables like temperature, pH, time, serving size, etc. All these input variables include an important source of variability and uncertainty, since in many cases, risk is derived from different possible scenarios for the different considered variables or factors (Pérez-Rodríguez et al. 2007b). For example, storage temperatures during food refrigeration usually vary in a wide range, including temperatures that support and do not support growth. This fact will greatly influence risk. Therefore, in order to take into consideration this inherent variability, probability distributions are used. By including probability distributions, a more complete and accurate estimate of risk can be obtained since all possible scenarios or combinations are contemplated. This information helps risk managers make better decisions concerning risk mitigation strategies, control measures, and food regulation. Probabilistic or stochastic MRA studies necessitate the application of different mathematical techniques to enable the use of probability distributions. The most used technique is Monte-Carlo analysis, which is a set of numerical methods that enable to operate with any type of distribution (Cullen and Frey 1999).

5.2 Hazard Identification in RTE Fresh Vegetables

In hazard identification, different types of scientific information and information source are used such as epidemiological data, microbiological surveys, consumption data, clinical studies, industrial data, international organization and governmental data, etc. in order to relate a specific microbiological hazard and a certain food or food category to a public health problem (CAC 1999). The result of this initial step will determine whether there is a need to undertake a whole MRA on the basis of the type of public health issue identified.

Ready-to-eat (RTE) fresh vegetables are foods widely demanded for their high nutritional value and easy use, which are frequently eaten raw. However, the can contain pathogens because the current industrial sanitizing treatments do not guarantee the total elimination of the pathogen when present (Abadias et al. 2008, Beuchat 2002, Parish et al. 2003). Although the prevalence of bacterial pathogens in vegetables is low as compared with other food products (Doyle and Erickson 2008, Francis et al. 1999, Garg et al. 1990, Gómez-López et al. 2008), RTE vegetables can represent a potential health risk due to the fact that no heat treatment is included in their production chain, and the only step intended to reduce the microbial load are washing and sanitizing treatments at the factory (Artés et al. 2009, Beuchat 2002).

Recent outbreaks linked to the consumption of RTE vegetables include cases of *Escherichia coli* O157:H7 in Denmark (2010), the Netherlands (2007), and Sweden (2005), and the recent and notorious *E. coli* O104 outbreak in Germany (2011) (Wu et al. 2011). Other fresh products such as romaine lettuce and seed sprouts have been related to food-borne outbreaks caused by *E. coli* O145 and O104 serotypes in the United States (CDC 2010) and Germany (Frank et al. 2011), respectively. The very young, elderly, and immunocompromised people are reported to be the most susceptible population groups (Ayers 2005, Schneider et al. 2009). For a quantitative MRA in RTE fresh vegetables, there is a need for a multitude of data on different aspects: the type of food product, the production chain, consumer habits, organism characteristics, ecology, human susceptibility, and strain virulence. While these data gaps are filled in to enable more reliable MRA studies, it is also necessary to continue contributing with predictive models describing microbial response of pathogenic bacteria in RTE fresh products along the food chain. These models can be especially useful at an operational level, helping food business operators and food safety authorities to design more effective control measures and risk management systems (e.g., HACCP). In the next sections, relevant aspects of the application of predictive microbiology models in risk assessment and risk management of RTE fresh vegetables will be discussed in more detail.

5.3 Predictive Microbiology

Predictive microbiology is a specialized branch of food microbiology, devoted to study and predict microbial behavior in foods through the application of mathematical functions, taking into account the effect of the most significant environmental factors (temperature, pH, water activity, etc.) (McMeekin et al. 1993). The process whereby these mathematical functions are derived is known as *modeling* (Ratkowsky et al. 1982, Roberts and Jarvis 1983). Besides, mathematical models are mathematical functions employing physical and chemical laws to describe, in mathematical terms, the behavior of a real system (Dym 2004).

In predictive microbiology, models are classified according to the type of response, model origin, model structure, and the type of mathematical function used for the model. For example, models can be classified into mechanistic and empirical ones, according to the type of knowledge used for its construction (Buchanan et al. 1997). The mechanistic models are those that are constructed by considering the biochemical and physiological mechanisms governing the microbial kinetics (McLauchlin et al. 2004). In turn, empirical models are mathematical equations exclusively derived from experimental data regardless of the underlying biochemical and physiological processes. In these cases, regression methods are applied to data to find the best-fit model representing experimental data. The log-logistic function and the modified equation of Gompertz are examples of empirical models (Buchanan et al. 1997, Haas et al. 1999). Because of the difficulty of developing mechanistic models due to the scarce knowledge about biological processes involved in the cell division, most models tend to be quasi-mechanistic, in which some of the parameters have certain biological meaning such as in the case of the model by Baranyi and Roberts (1994). In this model, the duration of the lag phase is based on the concentration of a limiting substrate when cells colonize a new habitat. The most widely used classification for kinetic models is the one based on the model structure, which leads to three levels of complexity or three types of models: primary, secondary, and tertiary models. Primary models are mathematical functions that describe the change in microbial concentration over time. In turn, secondary models are mathematical functions that relate intrinsic and extrinsic environmental factors such as pH, temperature, and water activity, with kinetic parameters, that is to say, growth rate, lag time, or maximum population density. Finally, tertiary models are computer applications integrating both primary and secondary models whereby end users can make predictions on microbial behavior (growth and inactivation) in foods under different environmental conditions. For more details on the topologies and applications of predictive microbiology models, readers are recommended to consult Pérez-Rodríguez and Valero (2013).

5.4 Quantifying and Modeling Microbial Response in the Ready-to-Eat Vegetable Industry: An Approach to Quantitative Risk Management

Food safety in the RTE vegetable industry is driven by two important factors. They are, on the one hand, the fact that the hygienization processes applied in industry are not fully effective in reducing microbial load in vegetables including pathogenic bacteria and viruses, and on the other hand, the fact that these products are preferably consumed raw without any additional culinary treatment (Francis et al. 1999). Hence, food safety in these products is achieved by applying an approach based on *hurdle theory*, in which different factors and preservation technologies are combined with the aim of reducing and/or inhibiting microbial growth along the food chain (Lee 2004). The key factors for the pathogen control in minimally processed vegetables are disinfection and storage temperatures since it is well known that pathogenic bacteria can survive from exposure to chlorine and subsequently grow if conditions are adequate for that (Ana et al. 2012, Delaquis et al. 2007, Franz et al. 2010, Legnani and Leoni 2004, Tromp et al. 2010). In relation to this, a quantitative microbial risk assessment (QMRA) study on *E. coli*

O157:H7 in leafy vegetables identified storage temperature and time together with contamination during the washing step as the most relevant risk factors (Danyluk and Schaffner 2011). The washing step with chlorinated water, where 20–200 ppm free chlorine is applied by the industry, may lead to reductions between 1 and 3 decimal logarithmic units (Aruscavage et al. 2006), which implies that bacteria could still be present after disinfection if initial levels are relatively high (Kolling and Matthews 2007, Zhao et al. 2000). Besides, Carrasco et al. (2007) reported that household temperature could support the growth of pathogens such as *L. monocytogenes, E. coli* O157:H7, and *Salmonella* spp. in vegetables, taking into account consumer behavior and habits in relation to RTE vegetables.

Quantifying the effect of food processes on bacterial growth and survival is crucial to establish effective preventive measures in RTE vegetables. The application of predictive models in a context of exposure assessment or risk assessment allows the identification of critical steps, risk factors, or process parameters. Tromp et al. (2010) pointed out that there is an increasing need to generate more quantitative data to undertake QMRA studies comprising the whole distribution chain. Moreover, Danyluk and Schaffner (2011), in relation to the risk by *E. coli* O157:H7 in leafy vegetables, stated that more information is required concerning lag time and growth of pathogenic bacteria and the impact of cross-contamination in the washing step.

5.5 Modeling the Disinfection of Pathogenic Bacteria in RTE Fresh Vegetables

Most disinfection models for pathogens have been developed for aqueous suspensions, with their main application in drinking water treatment and distribution systems. In those cases, predictive models may be developed following a *vitalistic* or *mechanistic* approach for deriving a mathematical function (Lambert and Johnston 2000). Disinfection compounds are reduced during the disinfection process due to interactions with other substances contained in water, like in the case of hypochlorite when contacting with organic matter (Winward et al. 2008) or dissipating off given their volatility. The Weibull model and its family of functions are able to consider nonstatic conditions for disinfection agents and determine inactivation over time by application of differential functions (Corradini and Peleg 2007):

$$\frac{dS(t)}{dt} = -b\big[C(t)\big] \cdot n \cdot \left[\frac{dS(t)}{-b\big[C(t)\big]}\right]^{n-1/n}$$

where

S is the logarithmic reduction of cells
$b[C(t)]$ is a coefficient dependent on the disinfection agent concentration varying over time (t)
n is another coefficient assumed to be constant for different values of $C(t)$

In the case of RTE vegetables, disinfection should be effective both for washing water and the vegetable being treated. The former case is crucial to minimize cross-contamination during washing (Gil et al. 2009), while the latter case has as objective to reduce the presence of pathogens in the food and improve quality and extend shelf life. From this, it seems clear that two types of models should be applied to have a complete representation for disinfection: one applied to the wash water and the other to the vegetable surface. For the first type of model, there exist abundant information and models developed for drinking water distribution systems that could be suitable for describing water disinfection in RTE vegetables. In the previous paragraph, a brief comment has been made on disinfection models in water distribution systems; however, this is an area in itself that would deserve a more thorough review including traditional and alternative technologies (WHO 2004). The disinfection method most widely studied corresponds to the use of hypochlorite in wash water, which is the most used method of treatment by the food industry in Spain and other countries (i.e., the United States). There are multiple studies assessing and quantifying chlorine efficiency in reducing microbial load including total microflora, hygiene indicators, and pathogenic microorganisms. These studies evidence a variable behavior, probably because there are several factors involved in its action mechanism that hampers their control, especially in experiments developed at pilot scale (Behrsing et al. 2000). This fact has led to this treatment be seldom described by means of deterministic mathematical models, that is, mathematical functions that relate factors involved (chlorine, temperature, etc.) and pathogen reduction over time (Pirovani et al. 2004). Hence, stochastic models are more generally applied, based on probability distributions accounting for observed variability and describing the frequency for the different efficacy levels or logarithmic reductions (Carrasco et al. 2010, Danyluk and Schaffner 2011, Franz et al. 2010, Tromp et al. 2010). It is more than evident that disinfection processes are key to mitigate risk, particularly if other process parameters are controlled and maintained at optimum values. However, it should also be noted that a total reduction of the microbial contamination present in vegetables is not feasible with the current technology (Beuchat et al. 2001, Lang et al. 2004). This is the reason why other events along the food chain of RTE vegetables such as cross-contamination or microbial growth are especially relevant for the occurrence of food-borne outbreaks. In this respect, the increasing development of large-scale food distribution chains all over the world can be an important factor that enables breaks in the cold chain and, consequently, increases the probability of growth of bacteria that survive hygienization processes.

5.6 Modeling the Growth of Pathogenic Bacteria in RTE Fresh Vegetables

Although there are many works dealing with the potential growth of pathogenic microorganisms in green leafy vegetables (Amanatidou et al. 1999, Francis and O'Beirne 2001, Ongeng et al. 2007, Sant'Ana et al. 2012a, Valero et al. 2006), few of them have developed predictive models (Crépet et al. 2009, Koseki and Isobe 2005, McKellar and Delaquis 2011, Sant'Ana et al. 2012b). Both models and observations hint that the microorganisms, *L. monocytogenes*, *Salmonella* spp., and *E. coli* O157:H7, are able to grow at high refrigeration temperatures, that is, 7°C–8°C, and even under modified atmosphere packaging conditions (McKellar and Delaquis 2011, Sant'Ana et al. 2012b). Table 5.1 showed a selection of significant works on *E. coli* O157:H7 growth and predictive models, considering different leafy vegetables and treatments. Studies have reported that *E. coli* O157:H7 is able to grow up to 2 logarithms in 3 days at 12°C and under modified atmosphere conditions (Luo et al. 2010). On the contrary, lower temperatures do not support growth of these pathogens or, in some cases, lead to their inactivation or destruction (Oliveira et al. 2010). The developed models have been secondary models intended to predict maximum growth rate, lag time, and maximum population density (MPD) as a function of storage temperature. Most of them are based on experiments carried out with microorganisms inoculated either in foods already treated or packaged or in raw vegetables, thus ignoring the effect of disinfection processes on the subsequent microbial growth (Koseki and Itoh 2001, McKellar and Delaquis 2011). The study performed by Koseki and Itoh (2001) evidenced an increased growth of *E. coli* O157:H7 in lettuce and cabbage when they were subjected to electrolyzed water treatment as compared to nontreated produces. The reviewed studies indicate that it is

TABLE 5.1

Growth Data and Models on *Escherichia coli* O157:H7 in Fresh-Cut Leafy Vegetables

Source	Food Matrix	Temperature (°C)	Commercial Treatment
Abdul-Raouf et al. (1993)	Lettuce	12	MAP[a]
Delaquis et al. (2002), McKellar and Delaquis (2011)	Lettuce	10	Heat treatment
Delaquis et al. (2007), McKellar and Delaquis (2011)	Lettuce	15	Nontreatment
Diaz and Hotchkiss (1996)	Lettuce	13	MAP
Francis and O'Beirne (2001)	Lettuce	8	Nontreatment
Li et al. (2001)	Lettuce	15	Chlorine
Luo et al. (2009)	Spinach	8 and 12	Nontreatment
Koseki and Isobe (2005)	Lettuce	10 and 15	Nontreatment
Posada-Izquiedo et al. (2013)	Lettuce	8, 13, 16	Chlorine and MAP
Smigic et al. (2009)	Broth	10, 12.5, 15	NEW[b] and MAP

[a] MAP, modified atmosphere packaging.
[b] NEW, neutralized electrolyzed water.

not yet clear whether chlorine reduces the growth capability of *E. coli* O157:H7 or growth is augmented through a reduction of competitive microflora (Delaquis et al. 2002). Nonetheless, the use and application of models not derived or validated under more realistic conditions resembling RTE vegetables could lead to significant biases in predictions and, in some cases, result in fail-dangerous predictions (i.e., underpredicting growth). A more extensive study on hygienization processes and their effect on subsequent microbial growth together with the development of suitable predictive models would help to better assess the risk posed by different pathogens along the food chain of RTE vegetables.

5.7 Importance of Quantifying Bacterial Transfer to RTE Fresh Vegetables

Cross-contamination phenomena when occurring at factory level can lead to the augmentation of the number of contaminated units at the end of the process. This is mainly because the water at the washing step can act as a transmission vehicle between contaminated and noncontaminated produces. In order to reduce risk by cross-contamination, industries apply disinfectant compounds (e.g., chlorine, electrolyzed water, etc.) at the washing step (López-Gálvez et al. 2009, Tomás-Callejas et al. 2012). Cross-contamination during washing results in the homogenization and/or redistribution of contamination over the processed lot (Wachtel and Charkowski 2002), specially when the disinfectant agent is reduced or is at an insufficient concentration due to either a major presence of organic matter or a deficient process control (Olaimat and Holley 2012). In this sense, it has been suggested that cross-contamination could be a determinant factor in the occurrence of food-borne outbreaks associated with RTE vegetables (Danyluk and Schaffner 2011, Pérez-Rodríguez et al. 2008, 2011, Roberts 1990). In spite of this, cross-contamination alone is not sufficient, and to end up with illness cases, temperature abuse or long periods of time with temperatures supporting growth are needed to reach risk levels at the moment of consumption (Rosset et al. 2004). Also in line with this statement, the risk assessment study carried out by Danyluk and Schaffner (2011) reported that when products were contaminated with *E. coli* O157:H7 at a prevalence of 0.1%, 99% cases originated at the washing step. Recent works performed at pilot scale demonstrate the capacity of *E. coli* O157:H7 to transfer from contaminated produces to final products through surfaces and wash water (Buchholz et al. 2012a,b). The study by Buchholz et al. (2012b) showed that initial levels of 2 log cfu/g *E. coli* O157:H7 resulted in a final contamination of 21.2 kg out of 78 kg lettuce initially not contaminated, where levels ranged between 2 and 3 log cfu/100 g. These results highlight the capacity of microorganisms to transfer from surfaces, equipment (shredder, conveyor belt, centrifuge, etc.), and wash water along the process, spreading contamination over a major number of final products.

Cutting and handling processes at the industrial level and subsequent steps during distribution, retailing, and consumption could be a cause of recontamination as evidenced by several works for both *E. coli* O157:H7 and *Salmonella* spp. (Gorman et al. 2002, Ravishankar et al. 2010, Wachtel and Charkowski 2002). In one of these works, it was demonstrated that *E. coli* O157:H7 could transfer from raw meat to lettuce leaves through cutting surfaces, reporting that 45% leaves ($n = 25$) in contact with the cutting table were contaminated with the pathogen (Wachtel et al. 2003). Despite the existence of quantitative data about bacterial transfer or cross-contamination of pathogens during the washing step, no predictive models have been developed yet (Pérez-Rodríguez et al. 2008). These models could be valuable tools in decision-making processes, quantitative risk management systems, and quantitative risk assessment; in addition, they could be applied to the establishment and selection of risk metrics, such as food safety objectives, performance objectives, performance criteria, or process parameters and other food safety criteria (see the following section).

5.8 Risk Management in RTE Fresh Vegetables Based on the Application of Food Safety Objectives

The food safety objective (FSO) is defined as "the maximum frequency and/or concentration of a hazard in a food at the time of consumption that provides or contributes to the appropriate level of protection (ALOP)" (ICMSF 2002). The FSO is proposed as a quantitative objective applied at the operational level on which risk management systems, hazard analysis and critical control points (HACCP) programs, or effective control measures are underpinned in order to assure a determined food safety level in the population. Likewise, other concepts have been defined related to FSO, which are encompassed in the so-called risk metrics:

- Performance Objective: "The maximum frequency and/or concentration of a hazard in a food at a specified step in the food chain before the time of consumption that provides or contributes to an FSO or ALOP, as applicable."
- Performance Criteria: "The effect in frequency and/or concentration of a hazard in a food that must be achieved by the application of one or more control measures to provide or contribute to a PO or an FSO."

To derive an FSO with a quantitative base, a risk assessment study is required, which enables to quantitatively relate ingested hazard doses and a public health objective expressed as the number of annual cases or other type of index such as the Disability Adjusted Life Years (DALY) (Havelaar et al. 2008, Zwietering 2005). There are some examples in literature on how probabilistic risk assessment models can be used to elucidate an FSO and other risk metrics, including attendant microbiological criteria, if needed (Delignette-Muller and Cornu 2008, Gkogka et al. 2013, Mejia et al. 2011). Consider the study on *Clostridium perfringens* in RTE meat or partly cooked meat, where a second-order model was developed to estimate the attendant risk (Crouch et al. 2009). Outcomes were used in this study to derive different risk metrics, control measures, and microbiological criteria based on a scenario analysis. In these cases, predictive models are crucial to precisely describe different processes and steps along the food chain and their impact on the final risk (Buchanan and Appel 2010).

The inequation proposed by ICMSF (2002) is a simplification and systematization of what is known as exposure assessment, a key step within the risk assessment methodology (CAC 1999). This adoption helps food business operators quantify the effect of different food processes on pathogenic microorganisms by means of three basic microbial processes and/or mathematical terms, that is, increments, reductions, and initial concentration. Sums for each term should result in values equal or lower than the established FSO. In that way, public health objectives related to an appropriate level of protection (ALOP) can be conveyed to operational levels in a more effective manner, guiding decision-making processes, HACCP systems, and in general quantitative risk management systems. In summary, this methodology based on the FSO concept and the inequation is proposed as an effective means to derive or select both performance objectives (POs) and performance criteria (PCs) (Membré et al. 2007, Zwietering 2005) and select and validate control measures (Schothorst et al. 2009) as well as food and process parameters (Gorris 2005). Once again, the quantification of food processes and development of suitable predictive models can be specially helpful for the RTE vegetable sector to carry out quantitative risk assessment studies and apply the FSO and the inequation proposed by ICMSF (2002) as valuable risk management tools. In the RTE vegetable industry, there are several food processes relevant to microbial risk where process parameters and control measures should be established to assure food safety. As evidenced in Figure 5.1, each process can be mathematically described by a predictive model that might be applied to assess the relative effect on the system capacity to reach specific FSOs or POs. Likewise, these predictive models can be used to identify critical steps or processes where controls should be applied to guarantee the compliance with FSOs, POs, or established microbiological criteria.

An example could be the application of inactivation models to determine the number of reductions and attendant process parameters required in the washing step to meet a specific food safety criterion (e.g., FSO or PO). In this example, initial concentration and increments produced during distribution, retailing, and consumption can also be estimated by predictive models or using quantitative data. If the example is applied to *E. coli* O157:H7 in RTE leafy vegetables, the terms in the aforementioned inequation could be defined as follows:

$$H_0 + \Sigma I - \Sigma R \leq \mathrm{FSO}$$

- $H_0 =$ initial concentration
- $\Sigma I = \Sigma I_{cc} + \Sigma I_{growth}$

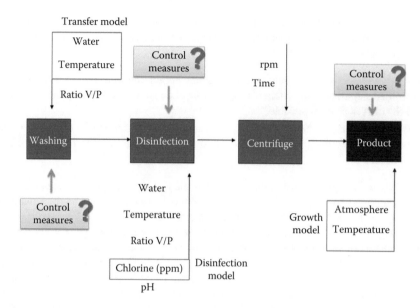

FIGURE 5.1 Scheme of operations and basic processes applied in the elaboration of RTE fresh vegetables and application of predictive models to derive process parameters and other control measures in different food steps.

- ΣI_{cc} stands for the sum of concentration increments (arithmetic units) of the pathogen because of cross-contamination during the washing step and ΣI_{growth} represents the sum of concentration increments (logarithmic units) of the pathogen due to growth during the whole food chain of RTE fresh vegetables
- $\Sigma R = \Sigma R_{chlorine}$, stands for the reductions (logarithmic units) of the pathogen produced by exposure to hypochlorite during washing

In order to define H_0, investigation and enumeration studies on the pathogen are needed. However, the prevalence for *E. coli* O157:H7 is usually so low that the number of samples required to obtain a reliable estimate of concentration of the pathogen is very high and unfeasible. Therefore, in these cases, a scenario analysis is the most appropriate approach. As regards increments (ΣI), due to both cross-contamination and growth, these can be obtained from either quantitative data from scientific literature or from the application of transfer and growth models, respectively. By applying both types of models, process parameters such as washing intensity, free chlorine level, storage temperature, or modified atmosphere can be related to specific increments and, consequently, to the capacity to meet a certain FSO or PO. Nevertheless, predictive models are limited in relation to type of pathogen (species, strain, serotype, etc.), food matrix, and commercial conditions, thereby affecting prediction accuracy. Reductions (ΣR) mainly occur at the washing step with chlorinated water. In this case, inactivation models could relate reductions to free chlorine level and treatment time. However, to the best of our knowledge, there are no secondary models specific to disinfection processes applied in the RTE fresh vegetable industry.

As a hypothetical example, the FSO for *E. coli* O157:H7 could be set to −5 log cfu/g. Although one could think that this value is low, other food safety objectives have also been set to low values such as the well-known criteria for *Clostridium botulinum* corresponding to 1 spore per 10^{12} containers. If a worst-case scenario is assumed for a product of fresh-cut lettuce, then the value for H_0 could be set to −2 log cfu/g as a sporadic high contamination level in raw lettuce entering the processing line and an abuse temperature of 8°C for 4 days during distribution, retailing, and household storage can be considered. To estimate ΣI_{growth}, a preexisting predictive model for *E. coli* O157:H7 growth in fresh-cut lettuce under MAP conditions that was subjected to chlorination (Posada-Izquierdo et al. 2013, in Table 2) was applied as follows:

$$\sqrt{\mu} = b\left(T - T_{min}\right) \qquad (5.1)$$

If variables for the square root model are defined as point-estimate values based on the worst-case scenario for growth described earlier, then

$$\mu = \left[0.10(8 - 5.13) \right]^2 = 0.26 \tag{5.2}$$

$$\Sigma I \sim 1 \text{ log cfu/g} = 0.26 \text{ log cfu/day} \cdot 4 \text{ days} \tag{5.3}$$

If these values are applied to the terms of the inequation proposed by ICMSF (2002), then

$$-5 \geq -2(H_0) + 1(\Sigma I) - \Sigma R \tag{5.4}$$

Therefore, the reductions required in the washing step would be

$$\Sigma R = 4 \text{ log cfu/g} \tag{5.5}$$

According to the calculations and variables, in order to meet the hypothetical FSO, a reduction of 4 log is required. Through predictive models, the relationship between reductions and process parameters are defined, which could help to determine what levels of free chlorine are more adequate to meet the established FSO (−5 log cfu/g). Also, combinations of different control measures might be applied based on the *hurdle theory*. These measures could be a better control of storage temperatures (< 8°C), resulting in $\Sigma I = 0.5$ according to the above secondary model, and a lower initial concentration of −3 log cfu/g, for example, consequence of a better control of raw material supplier. In such a case, ΣR needed to comply with the established FSO would be −2.5 log. The temperature value corresponding to this reduction would be used as the critical limit, a concept linked to HACCP systems.

Data used here have been given in a deterministic manner; however, variables could also be defined using probability distributions (i.e., stochastic approach), which would help to consider a greater number of scenarios (Rieu et al. 2007). In summary, Figure 5.2 represents predictive models and factors to be

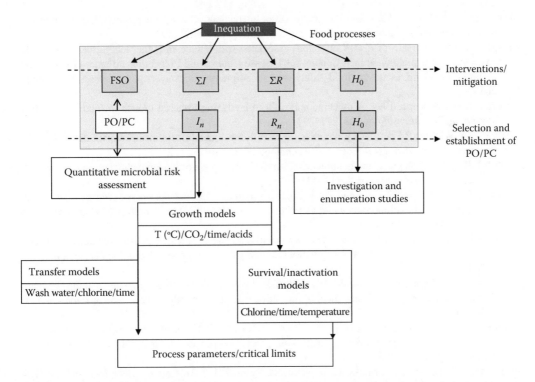

FIGURE 5.2 General scheme describing the application of the inequation proposed by ICMSF (2002) for quantitative risk management in RTE fresh vegetables based on the use of predictive microbiology models.

considered in quantitative studies based on the inequation proposed by ICMSF (2002) for the establishment and validation of PO, PC, and FSO along the different steps in the food chain of RTE fresh vegetables.

ACKNOWLEDGMENT

This work has been funded by MICINN through the projects AGL2010-20070 and AGL2008-03298, where G.D. Posada-Izquierdo is the holder of a predoctoral scholarship. Special thanks to the Research Group AGR-170 HIBRO of the *Plan Andaluz de Investigación, Desarrollo e Innovación* (PAIDI), International Campus of Excellence in the AgriFood Sector ceiA3.

REFERENCES

Abadias, M., J. Usall, M. Oliveira, I. Alegre, and I. Viñas. 2008. Efficacy of neutral electrolyzed water (NEW) for reducing microbial contamination on minimally processed vegetables. *International Journal of Food Microbiology* 123, 151–158.

Abdul-Raouf, U., L. Beuchat, and M. Ammar. 1993. Survival and growth of *Escherichia coli* O157:H7 on salad vegetables. *Applied and Environmental Microbiology* 59, 1999–2006.

Amanatidou, A., E. Smid, and L. Gorris. 1999. Effect of elevated oxygen and carbon dioxide on the surface growth of vegetable-associated micro-organisms. *Journal of Applied Microbiology* 86, 429–438.

Ana, A., M. Barbosa, M. Destro, M. Landgraf, and B. Franco. 2012. Growth potential of *Salmonella* spp. and *Listeria monocytogenes* in nine types of ready-to-eat vegetables stored at variable temperature conditions during shelf-life. *International Journal of Food Microbiology* 157, 52–58.

Artés, F., P. Gómez, E. Aguayo, V. Escalona, and F. Artés-Hernández. 2009. Sustainable sanitation techniques for keeping quality and safety of fresh-cut plant commodities. *Postharvest Biology and Technology* 51, 287–296.

Aruscavage, D., K. Lee, S. Miller, and J. Lejeune. 2006. Interactions affecting the proliferation and control of human pathogens on edible plants. *Journal of Food Science* 71, R89–R99.

Ayers, L.T. 2005. Outbreaks of *E. coli* Infections associated with lettuce and other leafy greens enteric diseases and Epidemiology Branch Centers for Disease Control and Prevention, Atlanta, GA. Available at: http://www.card.iastate.edu/food_safety/workshop4/presentations/ayers.pdf. Accessed November 3, 2013.

Baranyi, J. and T. Roberts. 1994. A dynamic approach to predicting bacterial-growth in food. *International Journal of Food Microbiology* 23, 277–294.

Behrsing, J., S. Winkler, P. Franz, and R. Premier. 2000. Efficacy of chlorine for inactivation of *Escherichia coli* on vegetables. *Postharvest Biology and Technology* 19, 187–192.

Beuchat, L.R. 2002. Ecological factors influencing survival and growth of human pathogens on raw fruits and vegetables. *Microbes Infection* 4, 413–423.

Beuchat, L.R., T. Ward, and C. Pettigrew. 2001. Comparison of chlorine and a prototype produce wash product for effectiveness in killing *Salmonella* and *Escherichia coli* O157:H7 on alfalfa seeds. *Journal of Food Protection* 64, 152–158.

Buchanan, R. and B. Appel. 2010. Combining analysis tools and mathematical modeling to enhance and harmonize food safety and food defense regulatory requirements. *International Journal of Food Microbiology* 139, 48–56.

Buchanan, R., R. Whiting, and W. Damert. 1997. When is simple good enough: A comparison of the Gompertz, Baranyi, and three-phase linear models for fitting bacterial growth curves. *Food Microbiology* 14, 313–326.

Buchholz, A., G. Davidson, B. Marks, E., Todd, and E. Ryser. 2012a. Quantitative transfer of *Escherichia coli* O157:H7 to equipment during small-scale production of fresh-cut leafy greens. *Journal of Food Protection* 75, 1184–1197.

Buchholz, A., G. Davidson, B. Marks, E. Todd, and E. Ryser. 2012b. Transfer of *Escherichia coli* O157:H7 from equipment surfaces to fresh-cut leafy greens during processing in a model pilot-plant production line with sanitizer-free water. *Journal of Food Protection* 75, 1920–1929.

CAC (Codex Alimentarius Commission). 1999. Principles and guidelines for the conduct of a Microbiological Risk Assessment. CAC/GL-30-1999. Secretariat of the Joint FAO/WHO Food Standards Programme, FAO, Rome, Italy.

Carrasco, E., F. Pérez-Rodríguez, A. Valero, R.M. García-Gimeno, G. Zurera. 2007. Survey of temperature and consumption patterns of fresh-cut leafy green salads: Risk factors for listeriosis. *Journal of Food Protection* 70, 2407–2412.

Carrasco, E., F. Pérez-Rodríguez, A. Valero, R. García-Gimeno, and G. Zurera. 2010. Risk assessment and management of *Listeria monocytogenes* in ready-to-eat lettuce salads. *Comprehensive Reviews in Food Science and Food Safety* 9, 498–512.

CDC (Centers for Disease Control and Prevention). 2010. Investigation update: Multistate outbreak of human *Salmonella* Montevideo infections. Available at: http://www.cdc.gov/salmonella/montevideo/index.html. Accessed November 20, 2013.

Corradini, M. and M. Peleg. 2007. A Weibullian model for microbial injury and mortality. *International Journal of Food Microbiology* 119, 319–328.

Crépet, A., V. Stahl, and F. Carlin. 2009. Development of a hierarchical Bayesian model to estimate the growth parameters of *Listeria monocytogenes* in minimally processed fresh leafy salads. *International Journal of Food Microbiology* 131, 112–119.

Crouch, E., D. Labarre, N. Golden, J. Kause, and K. Dearfield. 2009. Application of quantitative microbial risk assessments for estimation of risk management metrics: *Clostridium perfringens* in ready-to-eat and partially cooked meat and poultry products as an example. *Journal of Food Protection* 72, 2151–2161.

Cullen, A.C. and H.C. Frey. 1999. *Probabilistic Techniques in Exposure Assessment: A Handbook for Dealing with Variability and Uncertainty in Models and Inputs.* New York: Plenum.

Danyluk, M. and D. Schaffner. 2011. Quantitative assessment of the microbial risk of leafy greens from farm to consumption: Preliminary framework, data, and risk estimates. *Journal of Food Protection* 74, 700–708.

Delaquis, P., S. Bach, and L. Dinu. 2007. Behavior of *Escherichia coli* O157:H7 in leafy vegetables. *Journal of Food Protection* 70, 1966–1974.

Delaquis, S., S. Stewart, S. Cazaux, and P. Toivonen. 2002. Survival and growth of *Listeria monocytogenes* and *Escherichia coli* O157:H7 in ready-to-eat iceberg lettuce washed in warm chlorinated water. *Journal of Food Protection* 65, 459–464.

Delignette-Muller, M. and M. Cornu. 2008. Quantitative risk assessment for *Escherichia coli* O157:H7 in frozen ground beef patties consumed by young children in French households. *International Journal of Food Microbiology* 128, 158–164.

Diaz, C. and J. Hotchkiss. 1996. Comparative growth of *Escherichia coli* O157:H7, spoilage organisms and shelf-life of shredded iceberg lettuce stored under modified atmospheres. *Journal of the Science of Food and Agriculture* 70, 433–438.

Doyle, M. and M. Erickson. 2008. Summer meeting 2007—The problems with fresh produce: An overview. *Journal of Applied Microbiology* 105, 317–330.

Dym, C. 2004. *Principles of Mathematical Modeling.* London, U.K.: Academic Press.

FAO/WHO (Food Agriculture Organization/World Health Organization). 1999. Risk assessment of microbial hazard in foods. Report of the Joint FAO/WHO Expert Consultation, Geneva, Switzerland.

Francis, G. and D. O'Beirne. 2001. Food-borne pathogens: Effects of vegetable type, package atmosphere and storage temperature on growth and survival of *Escherichia coli* O157:H7 and *Listeria monocytogenes.* *Journal of Industrial Microbiology and Biotechnology* 27, 111–116.

Francis, G., C. Thomas, and D. O'Beirne. 1999. The microbiological safety of minimally processed vegetables. *International Journal of Food Science and Nutrition* 34, 1–22.

Frank, C., M.S. Faber, M. Askar, H. Bernard, A. Fruth, A. Gilsdorf, and D. Werber. 2011. Large and ongoing outbreak of haemolytic uraemic syndrome, Germany, May 2011. *Euro Surveillance: European Communicable Disease Bulletin* 16, 2–4.

Franz, E., S. Tromp, H. Rijgersberg, and H. Van der Fels-Klerx. 2010. Quantitative microbial risk assessment for *Escherichia coli* O157:H7, *Salmonella,* and *Listeria monocytogenes* in leafy green vegetables consumed at salad bars. *Journal of Food Protection* 73, 274–285.

Garg, N., J. Churey, and D. Splittstoesser. 1990. Effect of processing conditions on the microflora of fresh-cut vegetables. *Journal of Food Protection* 53, 701–703.

Gil, M., M. Selma, F. López-Gálvez, and A. Allende. 2009. Fresh-cut product sanitation and wash water disinfection: Problems and solutions. *International Journal of Food Microbiology* 134, 37–45.

Gkogka, E., M. Reij, L. Gorris, and M. Zwietering. 2013. The application of the Appropriate Level of Protection (ALOP) and Food Safety Objective (FSO) concepts in food safety management, using *Listeria monocytogenes* in deli meats as a case study. *Food Control* 29, 382–393.

Gómez-López, V., P. Ragaert, J. Debevere, and F. Devlieghere. 2008. Decontamination methods to prolong the shelf life of minimally processed vegetables, state-of-the-art. *Critical Reviews in Food Science and Nutrition* 48, 487–495.

Gorman, R., S. Bloomfield, and C. Adley. 2002. A study of cross-contamination of food-borne pathogens in the domestic kitchen in the Republic of Ireland. *International Journal of Food Microbiology* 76, 143–150.

Gorris, L. 2005. Food safety objective: An integral part of food chain management. *Assessment* 16, 801–809.

Haas, C., J. Rose, and C. Gerba. 1999. *Quantitative Microbial Risk Assessment*. New York: John Wiley & Sons, Inc.

Havelaar, A., E. Evers, and M. Nauta. 2008. Challenges of quantitative microbial risk assessment at EU level. *Trends in Food Science & Technology* 19, S26–S33.

ICMSF (International Commission on Microbiological Specification for Foods). 2002. *Microorganisms in Foods 7: Microbiological Testing in Food Safety Management*. New York: Kluwer Academic/Plenum Publishers.

Kolling, G.L. and K.R. Matthews. 2007. Influence of enteric bacteria conditioned media on recovery of *Escherichia coli* O157:H7 exposed to starvation and sodium hypochlorite. *Journal of Applied Microbiology* 103, 1435–1441.

Koseki, S. and S. Isobe. 2005. Prediction of pathogen growth on iceberg lettuce under real temperature history during distribution from farm to table. *International Journal of Food Microbiology* 104, 239–248.

Koseki, S. and K. Itoh. 2001. Prediction of microbial growth in fresh-cut vegetables treated with acidic electrolyzed water during storage under various temperature conditions. *Journal of Food Protection* 64, 1935–1942.

Lambert, R. and M. Johnston. 2000. Disinfection kinetics: A new hypothesis and model for the tailing of log-survivor/time curves. *Journal of Applied Microbiology* 88, 907–913.

Lammerding, A.M. and A. Fazil. 2000. Hazard identification and exposure assessment for microbial food safety risk assessment. *International Journal of Food Microbiology* 58, 147–157.

Lang, M., L. Harris, and L. Beuchat. 2004. Survival and recovery of *Escherichia coli* O157:H7, *Salmonella*, and *Listeria monocytogenes* on lettuce and parsley as affected by method of inoculation, time between inoculation and analysis, and treatment with chlorinated water. *Journal of Food Protection* 67, 1092–1103.

Lee, S. 2004. Microbial safety of pickled fruits and vegetables and hurdle technology. *Science* 4, 21–32.

Legnani, P. and E. Leoni. 2004. Effect of processing and storage conditions on the microbiological quality of minimally processed vegetables. *International Journal of Food Science and Technology* 39, 1061–1068.

Li, Y., R. Brackett, J. Chen, and L. Beuchat. 2001. Survival and growth of *Escherichia coli* O157:H7 inoculated onto cut lettuce before or after heating in chlorinated water, followed by storage at 5 or 15 degrees C. *Journal of Food Protection* 64, 305–309.

López-Gálvez, F., A. Allende, M. Selma, and M. Gil. 2009. Prevention of *Escherichia coli* cross-contamination by different commercial sanitizers during washing of fresh-cut lettuce. *International Journal of Food Microbiology* 133, 167–171.

Luo, Y., Q. He, and J. McEvoy. 2010. Effect of storage temperature and duration on the behavior of *Escherichia coli* O157:H7 on packaged fresh-cut salad containing romaine and iceberg lettuce. *Journal of Food Science* 75, 390–397.

Luo, Y., Q. He, J. McEvoy, and W. Conway. 2009. Fate of *Escherichia coli* O157:H7 in the presence of indigenous microorganisms on commercially packaged baby spinach as impacted by storage temperature and time. *Journal of Food Protection* 72, 2038–2045.

McKellar, R. and P. Delaquis. 2011. Development of a dynamic growth–death model for *Escherichia coli* O157:H7 in minimally processed leafy green vegetables. *International Journal of Food Microbiology* 151, 7–14.

McLauchlin, J., R. Mitchell, W. Smerdon, and K. Jewell. 2004. *Listeria monocytogenes* and listeriosis: A review of hazard characterisation for use in microbiological risk assessment of foods. *International Journal of Food Microbiology* 92, 15–33.

McMeekin, T., J. Olley, and T. Ross. 1993. *Predictive Microbiology: Theory and Application*. Somerset, U.K.: John Wiley & Sons, Ltd.

Mejia, Z., R. Beumer, and M. Zwietering. 2011. Risk evaluation and management to reaching a suggested FSO in a steam meal. *Food Microbiology* 138, 631–638.

Membré, J., J. Bassett, and L. Gorris. 2007. Applying the food safety objective and related standards to thermal inactivation of *Salmonella* in poultry meat. *Journal of Food Protection* 70, 2036–2044.

Olaimat, A. and R. Holley. 2012. Factors influencing the microbial safety of fresh produce: A review. *Food Microbiology* 32, 1–19.

Oliveira, M., J. Usall, C. Solsona, I. Alegre, I. Viñas, and M. Abadias. 2010. Effects of packaging type and storage temperature on the growth of foodborne pathogens on shredded "Romaine" lettuce. *Food Microbiology* 27, 375–380.

Ongeng, D., J. Ryckeboer, A. Vermeulen, and F. Devlieghere. 2007. The effect of micro-architectural structure of cabbage substratum and or background bacterial flora on the growth of *Listeria monocytogenes*. *International Journal of Food Microbiology* 119, 291–299.

Parish, M.E., L.R. Beuchat, T.V. Suslow, L.J. Harris, E.H. Garrett and J.N. Farber. 2003. Methods to reduce/eliminate pathogens from fresh and fresh-cut produce. *Food Science and Food Safety* 2, 161–173.

Pérez-Rodríguez, F., D. Campos, E. Ryser, A. Buchholz, G. Posada-Izquierdo, B. Marks, G. Zurera, and E. Todd. 2011. A mathematical risk model for *Escherichia coli* O157:H7 cross-contamination of lettuce during processing. *Food Microbiology* 28, 694–701.

Pérez-Rodríguez, F., R.M. Garcia-Gimeno, and G. Zurera. 2007a. Conceptual and methodological foundations for developing microbial risk assessment models. In *Food Microbiology Research Trends*, M.V. Palino, ed., pp. 93–129. New York: Nova Publisher.

Pérez-Rodríguez, F. and A. Valero. 2013. *Predictive Microbiology in Foods*, R.W. Hartel, ed. New York: Springer.

Pérez-Rodríguez, F., A. Valero, E. Carrasco, R. Garcia-Gimeno, and G. Zurera. 2008. Understanding and modeling bacterial transfer to foods: A review. *Trends in Food Science and Technology* 19, 131–144.

Pérez-Rodríguez, F., E.D. van Asselt., R.M. Garcia-Gimeno, G. Zurera, and M.H. Zwietering. 2007b. Extracting additional risk managers information from a risk assessment of *Listeria monocytogenes* in deli meats. *Journal of Food Protection* 70, 1137–1152.

Pirovani, M., A. Piagentini, D. Guemes, and S. Arkwright. 2004. Reduction of chlorine concentration and microbial load during washing-disinfection of shredded lettuce. *International Journal of Food Science and Technology* 39, 341–347.

Posada-Izquierdo, G., F. Pérez-Rodríguez, F. López-Gálvez, A. Allende, M. Selma, M. Gil, and G. Zurera. 2013. Modelling growth of *Escherichia coli* O157:H7 in fresh-cut lettuce submitted to commercial process conditions: Chlorine washing and modified atmosphere packaging. *Food Microbiology* 33, 131–138.

Ratkowsky, D., J. Olley, T. McMeekin, and A. Ball. 1982. Relationship between temperature and growth rate of bacterial cultures. *Journal Bacteriology* 149, 1–5.

Ravishankar, S., L. Zhu, and D. Jaroni. 2010. Assessing the cross contamination and transfer rates of *Salmonella enterica* from chicken to lettuce under different food-handling scenarios. *Food Microbiology* 27, 791–794.

Rieu, E., K. Duhem, E. Vindel, and M. Sanaa. 2007. Food safety objectives should integrate the variability of the concentration of pathogen. *Risk Analysis* 27, 373–386.

Roberts, D. 1990 Foodborne illness, sources of infection: Food. *Lancet* 336, 859–861.

Roberts, T. and B. Jarvis, 1983. Predictive modelling of food safety with particular reference to *Clostridium botulinum* in model cured meat systems. In *Food Microbiology: Advances and Prospects*, T.A. Roberts and F.A. Skinner, eds., pp. 85–95. New York: Academic Press.

Rosset, P., M. Cornu, V. Noël, E. Morelli, and G. Poumeyrol. 2004. Time-temperature profiles of chilled ready-to-eat foods in school catering and probabilistic analysis of *Listeria monocytogenes* growth. *International Journal of Food Microbiology* 96, 49–59.

Sant'Ana, A., M. Barbosa, M. Destro, M. Landgraf, and B. Franco. 2012a. Growth potential of *Salmonella* spp. and *Listeria monocytogenes* in nine types of ready-to-eat vegetables stored at variable temperature conditions during shelf-life. *International Journal of Food Microbiology* 157, 52–58.

Sant'Ana, A., B. Franco, and D. Schaffner. 2012b. Modeling the growth rate and lag time of different strains of *Salmonella enterica* and *Listeria monocytogenes* in ready-to-eat lettuce. *Food Microbiology* 30, 267–273.

Schneider, K.R., R.G. Schneider, M.A. Hubbard, and A. Chang. 2009. Preventing food-borne illness: *Escherichia coli* O157:H7. *Food Science and Human Nutrition* 31, 1–5.

Schothorst, M.V., M. Zwietering, T. Ross, R. Buchanan, and M. Cole. 2009. Relating microbiological criteria to food safety objectives and performance objectives. *Food Control* 20, 967–979.

Smigic, N., A. Rajkovic, E. Antal, H. Medic, B. Lipnicka, M. Uyttendaele, and F. Devlieghere. 2009. Treatment of *Escherichia coli* O157:H7 with lactic acid, neutralized electrolyzed oxidizing water and chlorine dioxide followed by growth under sub-optimal conditions of temperature, pH and modified atmosphere. *Food Microbiology* 26, 629–637.

Tomás-Callejas, A., F. López-Gálvez, A. Sbodio, F. Artés, F. Artés-Hernández, and T. Suslow. 2012. Chlorine dioxide and chlorine effectiveness to prevent *Escherichia coli* O157:H7 and *Salmonella* cross-contamination on fresh-cut Red Chard. *Food Control* 23, 325–332.

Tromp, S., H. Rijgersberg, and E. Franz. 2010. Quantitative microbial risk assessment for *Escherichia coli* O157:H7, *Salmonella enterica*, and *Listeria monocytogenes* in leafy green vegetables consumed at salad bars, based on modeling supply chain logistics. *Journal of Food Protection* 73, 1830–1840.

Valero, A., E. Carrasco, F. Pérez-Rodríguez, R. García-Gimeno, and G. Zurera. 2006. Growth/no growth model of *Listeria monocytogenes* as a function of temperature, pH, citric acid and ascorbic acid. *European Food Research and Technology* 224, 91–100.

Wachtel, M. and A. Charkowski. 2002. Cross-contamination of lettuce with *Escherichia coli* O157:H7. *Journal of Food Protection* 65, 465–470.

Wachtel, M., J. McEvoy, Y. Luo, A. Williams-Campbell, and M. Solomon. 2003. Cross-contamination of lettuce (*Lactuca sativa* L.) with *Escherichia coli* O157:H7 via contaminated ground beef. *Journal of Food Protection* 66, 1176–1183.

WHO (World Health Organization). 2004. Performance models. In *Water Treatment and Pathogen Control: Process Efficiency in Achieving Safe Drinking Water*, M.W. LeChevallier and K.-K. Au, eds., pp. 67–74. London, U.K.: IWA Publishing.

Winward, G., L. Avery, T. Stephenson, and B. Jefferson. 2008. Chlorine disinfection of grey water for reuse: Effect of organics and particles. *Water Research* 42, 483–491.

Wu, C., P. Hsueh, and W. Ko. 2011. A new health threat in Europe: Shiga toxin–producing *Escherichia coli* O104:H4 infections (review article). *Journal of Microbiology, Immunology and Infection* 44, 390–393.

Zhao, L., T. Montville, and D. Schaffner. 2000. Inoculum size of *Clostridium botulinum* 56A spore influences time-to-detection and percent growth-positive samples. *Journal Food Science* 65, 1369–1375.

Zwietering, M. 2005. Practical considerations on food safety objectives. *Food Control* 16, 817–823.

6

Mechanisms and Risks Associated with Bacterial Transfer between Abiotic and Biotic Surfaces

Stavros G. Manios, Anastasia E. Kapetanakou, Evangelia Zilelidou, Sofia Poimenidou, and Panagiotis N. Skandamis

CONTENTS

6.1 Routes and Vehicles of Bacterial Immigration in the Food Industry and the Domestic Environment

The ubiquitous nature of microbes renders their presence inevitable in the food-processing chain as well as in domestic kitchens. Their introduction into a sterile or well-disinfected environment and their establishment on an abiotic or biotic surface is termed contamination. The major contamination source in such environments derives from the incoming raw materials, which may be carriers of a large and diverse association of spoilage bacteria, such as pseudomonads, enterobacteria, lactic acid bacteria, enterococci, yeasts, and molds. From a food safety aspect, these raw materials may also serve as vehicles of pathogenic microorganisms, including *Listeria monocytogenes*, *Salmonella* spp., *Escherichia coli*, *Vibrio parahaemolyticus*, *Shigella*, and *Campylobacter*. These microorganisms may be part of the indigenous microflora of the food (of plant or animal origin), retrieved from the direct contact with the growth environment or introduced to the food during harvest and subsequent handling in the field and/or during slaughter and further-processing and storage. Similarly, the working staff in a food-processing environment or the cook in a domestic kitchen may also assist in the introduction and spreading of undesirable microbes, especially of pathogenic nature. In particular, food handlers may be potential disease vectors without experiencing any obvious symptoms. For instance, *Staphylococcus aureus* is commonly encountered on human skin and skin wounds, nose, or throat, and it can be easily transferred to foods or food-related surfaces. Other sources of undesirable contamination in the food industry or a kitchen environment include the water used for washing of the equipment, utensils, or the pets and pests that have access to a food preparation site, which may excrete a variety of food-poisoning microorganisms. For these reasons, it is of high importance that the hygienic level of the incoming raw materials as well as that of the food handlers should be strictly maintained and regularly validated, while all the food-related environments should also be protected from animals and pests.

Following entry to the processing environment, the microbial load of a potential contamination source may be easily spread to abiotic or biotic surfaces. These surfaces will further come into contact with the final product, facilitating the phenomenon of *cross-contamination*. This term is used to describe the transfer of microorganisms or viruses from a contaminated food, raw material, kitchen utensil, or person to other foods, whether it occurs directly or indirectly. Direct cross-contamination involves the direct transmission of an agent from raw foods to cooked or ready-to-eat (RTE) foods. For instance, the juices of a contaminated meat product may drip onto fresh vegetables or fruits, if they are stored improperly in the refrigerator. Indirect cross-contamination describes the transfer of an agent from food handlers, utensils, or equipment, which comes into direct contact with contaminated foods or raw materials, to foods that are ready for consumption. For example, the use of a knife or cutting board to cut fresh meat and further use of the same equipment to prepare a fresh-cut salad without prior sanitation may facilitate the transmission of the potential hazard from the meat to the salad. Figure 6.1 illustrates the potential routes of microbial contamination of a food-processing environment, as well as the identified vehicles that may further transfer this contamination to the final product.

Cross-contamination ranks at the top shelves of the greatest causes of food-borne illnesses in the world (Greig and Ravel 2009). It is usually caused due to improper handling and storage of raw

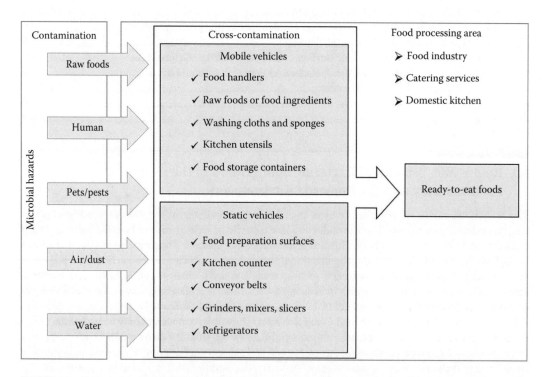

FIGURE 6.1 Routes of external microbiological contamination of food-processing environments and potential vehicles of further cross-contamination of RTE foods.

foods, and, therefore, it may likely occur in food-processing plants, catering services, or domestic kitchens. Because, in most cases, the microbial transmission is caused unintentionally, the precise description of the agent vehicle is rather difficult. In 2009, the Health Protection Agency of England and Wales reported that 29% of the outbreaks (28 out of 98) that occurred in the United Kingdom were attributed to cross-contamination cases (HPA 2009). Recently, the North Carolina Department of Health and Human services published a report of 100 salmonellosis cases associated with poor personal hygiene of the employees and/or inadequate dish machine temperatures in a hotel restaurant (NCDHHS 2013). Despite the fact that 87% of the sites where outbreaks occur are associated with foods prepared in households (van Asselt et al. 2008), the major problem of cross-contamination in these environments is that reporting of such sporadic cases is extremely rare. Therefore, identifying the sources of human infection and tracing the route of contamination of such nature are often complicated.

The key step that contributes to the prevention of food-poisoning via cross-contamination is to avoid or limit the possibilities of transfer of undesirable agents from contaminated surfaces to foods. In simple terms, this involves the identification of the potential vehicles and further keeping them away from foods that are not intended for further lethal processing. The most recognized vehicles of unintentional microbial transmission in a food environment are primarily divided into mobile and static. The first category comprises humans (employees or housewives), raw foods or food ingredients, washing cloths and sponges, kitchen utensils (e.g., knives, cutting boards, spoons, forks, spatulas), and food storage containers. The second category includes any surface that may come into contact with a contaminant, such as industrial surfaces, kitchen counters, static utensils (i.e., mixers, slicers, and grinders), and refrigerator shelves. As most of these may easily come into direct contact with foods that are ready for consumption, greater focus on the practice of improved and targeted hygiene measures in all food-processing sites should be followed in order to reduce the risk of food-borne illnesses. In the next paragraphs, emphasis will be given on the detailed description of the potential vehicles of cross-contamination in an industrial or domestic environment, in parallel with the measures that should be applied in order to limit any undesirable food-poisoning phenomena.

6.1.1 Food Handlers

Except for the external hazards that human hands may introduce in a food-processing environment, they are also regarded as one of the major vectors of microbial spread in this environment. In food industries and catering services, the need of rapid and automated work by the employees may lead to unintentional immigration of a microorganism from a contaminated surface or food to the handlers' hands, and further distribution to other surfaces or final products. For instance, in the food industry, the failure of a regular change of used gloves to new ones may lead to undesirable distribution of the microbial load of a contaminated batch to other batches or raw meat or, even worse, to the working surfaces. Indeed, Rodríguez et al. (2010) reported that handling of artificially contaminated ground meat may lead to the transfer and detection of the inoculated microorganism on RTE foods, even after wiping the hands with a paper towel. Similarly, in quick service restaurants, it is a common but hazardous practice to use the same pair of gloves to prepare meals that may contain uncooked meat products (i.e., bacon or turkey slices), eggs, and fresh vegetables, while in worst-case scenarios, the same food handler may also operate the cashier. Given that all food-related industries and restaurants should follow the established good manufacturing and good hygiene practices (GMPs and GHPs), cross-contamination is more controllable in such environments. In contrast, the insufficient knowledge of the consumers to prevent food-borne illnesses in the domestic environment increases the risk of food contamination via hands in houses (Redmond and Griffith 2003). Dharod et al. (2009) showed that the likelihood of food handlers' hands to be found contaminated with *S. aureus* was proportional to their awareness of food safety issues. The potential of hands to act as contamination vehicles was highlighted by Aldabe et al. (2011), who found that an infected person with *E. coli* O104:H4, was responsible for the hemolytic uremic syndrome that was expressed in two individuals in a household in France. Therefore, sufficient and regular hand washing should take place, in order to limit the risk of food-borne illnesses, especially in households, where the use of gloves by the food handlers is very rare.

6.1.2 Cleaning Sponges, Cloths, and Towels

Cleaning sponges and cloths require attention in households and restaurants, as they may serve as a reservoir of pathogenic bacteria. Beumer et al. (1996) revealed that 37% of analyzed kitchens cloths were found positive to *Listeria* spp., while in some cases, the populations of the pathogen exceeded 10^4 CFU/sample. A survey in the United States showed that 23 microbial species were isolated and identified from 140 different kitchen sponges and 13 species from 56 cleaning towels that were used in different households (Enriquez et al. 1997). Despite the fact that these means are intended for cleaning and sanitation procedures, several recent studies have revealed that they may also constitute important diffusers of pathogenic bacteria to surfaces and utensils, leading to cross-contamination of foods. A study of Scott and Bloomfield (1993) showed that the contamination level of food-related surfaces increased after washing with kitchen cloths, suggesting the translocation of the microorganisms from the cloths to the surfaces. Similarly, Kusumaningrum et al. (2003) tracked the potential of pathogens to be transferred from sponges to cutting surfaces and subsequently to fresh vegetables.

The problem arises from the fact that sponges and towels continually remain wet, with such an environment providing ideal surviving and/or growing conditions for microorganisms. In parallel, the repeated contact with food residues increases the levels of available nutrients, apart from the addition of new microorganisms (Erdogrul and Erbilir 2000; Rayner et al. 2004; Tate 2006). Kitchen sponges, especially, remain almost constantly wet and rich of residues, due to the large free surface that they have. Although humidity plays a significant role in the survival of microorganisms, extensive survival of *Campylobacter* on dry cloths has also been reported (Mattick et al. 2003). Even the coexistence of microorganisms with detergents has been found insufficient for their elimination, as these substances may not have any significant antimicrobial activity. In contrast, such sublethal stresses may activate the adaptation mechanisms of pathogens, and further render them more resistant to the same or other stresses, or affect their kinetic parameters on other foods. Manios et al. (2012) revealed that stresses encountered on the contamination route from kitchen sponges to food containers (i.e., commercial detergent, starvation on food container surface) reduced the time lag of *Salmonella* spp. in a simulated food environment.

In order to eliminate the risk of cross-contamination via kitchen sponges or cloths, several disinfection methods have been tested, including bleaching, dipping in vinegar or citric acid, boiling, and placing in the dishwasher or in microwave (Sharma et al. 2009; Tate 2006). Especially, microwave and dishwasher have been proved to be very effective in eliminating any microbial load in sponges, if treated properly (Sharma et al. 2009). In contrast, the efficacy of other natural or chemical disinfectants is affected by many factors such as duration of application, concentration, and temperature of the disinfectant (Rossi et al. 2012), which may not be easily standardized in a domestic environment. In any case, it is suggested to carefully dry the cloth or sponge after washing, in order to optimize the results of decontamination.

6.1.3 Kitchen Utensils and Storage Containers

The mobile equipment used in food industries, catering services, and households may also serve as a potential vehicle of bacterial contamination. Knives, spoons, forks, and cutting boards have been proposed to be the most common utensils that are usually found contaminated in such environments. On an industrial level, using the same knife or cutting board for different batches without proper sanitation may facilitate the distribution of undesirable microbes to all the processed batches. Therefore, it is suggested that these utensils should be regularly changed with sterile ones during working shifts. Similarly, in restaurants, using the same utensil/board for different ingredients may also affect the hygienic condition of the final product. Ravishankar et al. (2010) estimated the transfer rate of *Salmonella enterica* from poultry to lettuce through cutting boards and knives and found that improper sanitation caused 45.62% transfer of the pathogen to the vegetable. Following merely washing with tap water, Soares et al. (2012) observed that 1.8 log CFU/g of *Salmonella* Enteritidis, originating from inoculated chicken (5.1 log CFU/cm^2), was transferred to tomatoes via wooden cutting board. Such applications are more common in households, due to the lack of knowledge of the consumers about cross-contamination issues (Redmond and Griffith 2003). In a survey conducted by Phang and Bruhn (2011), less than half of the volunteers washed their cutting boards and knives properly before using them for other foods, while Klontz et al. (1995) revealed that 25% of the consumers in the United States did not find it necessary. In order to avoid this, a universal *multicolor rule* supports the use of boards and knives of the same color for each food category (i.e., green for produce, red for raw meat, yellow for poultry, tan for fish, blue for cooked foods, and white for dairy products). Especially for cutting boards, it is a fact that bacteria are able to penetrate their pores, where they remain protected from sanitation procedures. Thus, it is essential to follow effective antimicrobial applications on all kitchen utensils, giving special attention to cutting boards, which may require regular replacement with new ones when necessary.

In addition to cutting boards and knives, dishes and food containers may also constitute sources of contamination of cooked meals or food leftovers, respectively, if they are not treated properly (i.e., washed and dried). Mattick et al. (2003) studied the potential transfer from washed dishes to foods and found positive samples, regardless of the contact time, especially when dishes were towel-dried (compared with air-dried). The effectiveness of washing may vary with the application followed (de Jong et al. 2008) and the type of material used (Soares et al. 2012). Due to the variation of dishwashing techniques, this type of disinfection is not constantly effective. In contrast, a dishwasher may safely contribute to the removal of undesirable bacteria from dishes and food containers if the washing temperature and the amount of detergent are used to the suggested levels (Taché and Carpentier 2014).

6.1.4 Other Vehicles of Cross-Contamination

In the food-processing environments, there are also various sites or utensils that could *hide* and transmit microbes to foods. Faucet and spigots, sinks, kitchen counters, mixers, slicers, refrigerators, and other utensils (e.g., can opener, rubber spatulas, and scissors) are generally recognized as microbial reservoirs. The implementation of such spots in cross-contamination scenarios is attributed to the difficulty they pose for proper sanitation or their less significant role in food preparation, compared with other utensils such as cutting boards or knives. Chen et al. (2001) studied the bacterial transfer rates between contaminated hands and sterile spigots and vice versa, as well as the transmission of the contamination to lettuce. Results of their study highlighted the variability of transfer rates in the domestic environment as well as the fact that faucet spigots may be a significant source of cross-contamination. Therefore, emphasis should be given on faucets' effective disinfection with soap and water in order to prevent cross-contamination. The use of automated faucets (motion sensors) or those operated by foot pedals may likely eliminate this hazard. Similarly, the average populations of aerobic plate counts and enterobacteria found in domestic sinks reached 5.9 and 4.2 log CFU/sample in 150 different sinks (Chen et al. 2011). Kitchen counters are the most accessible spots of unintentional (accidental) contamination, while other equipment such as slicers and blenders require disassembling for proper disinfection. Even shopping bags have been reported to be potential vehicles of cross-contamination of foods (Carrasco et al. 2012). Azevedo et al. (2014) found that of 15 different household refrigerator handles, 33% and 13.3% were positive to Enterobacteriaceae and *E. coli*, respectively, while all samples (100%) were positive to coagulase positive *Staphylococcus*. As indicated in all potential cross-contamination vehicles, the proper application of disinfectants and good hygienic practices in any food-processing environment are the most effective tools to eliminate the risk of microbial immigration, establishment, and further formation of biofilm communities on food-related surfaces, thus ensuring the safety of the final product. The effective application of hygiene measures to control dispersal and establishment of microbial contamination in household or industrial environments requires systematic knowledge of the factors controlling bacterial attachment and subsequent biofilm formation.

6.2 Factors Affecting Attachment of Pathogens on Food Contact Surfaces—Biofilm Formation

During the processing of contaminated raw materials and food preparation, microorganisms entrapped in food residues are transmitted to the equipment plant surfaces. In the context of GMPs and GHPs, processing plant surfaces are cleaned and disinfected after use in order to remove food soil and eliminate the transferred microorganisms. However, the disinfection procedure may fail. Potential reasons of this failure could be inadequate disinfectant concentrations, the temperature of disinfectant application, the exposure time of disinfectant, the remaining soil on the surfaces, or the surface material (Herrera et al. 2007; Jullien et al. 2008; Langsrud et al. 2003; Van Houdt and Michiels 2010). Moreover, bacterial growth rate, surrounding nutrient status, attachment ability of the microorganisms, and cross-resistance or adaptation of the strains due to their ecological background may result in resistance and establishment of persistent strains. It is this deposit of bacterial cells on solid surfaces that leads to biofilm formation activity.

Biofilms are referred to in the literature as biologically active matrices of cells and extracellular substances in association with a solid surface (Bakke et al. 1984; Kumar and Anand 1998; Zobell 1943). Placed in the solid–liquid interface, bacterial cells, irreversibly attached to the food surface, are organized in multistructural communities, embedded into a glycocalyx (Kumar and Anand 1998). Capillary water channels are part of this porous structure of biofilms, and distribute water and nutrients to the included microorganisms (Poulsen 1999). During the last decades, research articles have revealed the multidisciplinary nature of biofilms, as they occur in industrial and clinical environments (Parsek and Fuqua 2004), gaining importance for safety and hygiene in the food industries, catering services, and medical supplies. In food equipment, biofilms may serve as a reservoir for food product contamination, due to the transfer phenomenon of microorganisms. Numerous outbreaks are related to food-borne pathogens, such as *L. monocytogenes*, *E. coli*, *S. enterica*, *Pseudomonas* spp., *Bacillus* spp., *Yersinia enterocolitica*, and *Campylobacter jejuni* (Chmielewski and Frank 2003; EFSA 2010; Shi and Zhu 2009).

There are three critical steps in the biofilm formation activity: (1) attachment, (2) microcolony and EPS production, and (3) maturation (Chmielewski and Frank 2003; Davey and O'Toole 2000). Bacterial cells may actively or passively adhere to the surfaces depending on the bacterial cells' surface properties, and the attachment can be reversible or irreversible. After reversible attachment, cells can still perform the Brownian motion and are easily removed or scan the surface to detect a better nutritional level for living (Chmielewski and Frank 2003; Poulsen 1999). Once cells are attached irreversibly, their removal becomes difficult and they are about to be established as a biofilm in the living form (Chmielewski and Frank 2003). The attachment and biofilm-forming capabilities of bacteria depend on the interaction of multiple factors, including the attachment surface, the presence of other bacteria, the temperature, the availability of nutrients and pH, the production of extracellular polysaccharides, and the cell-to-cell communication (Chmielewski and Frank 2003; Herrera et al. 2007; Rivas et al. 2007; Rode et al. 2007; Van Houdt and Michiels 2010).

6.2.1 Type of Surface

Any surface can support biofilm development where microorganisms are present (Kumar and Anand 1998). However, the properties of the attachment surface, along with the bacterial cell surface, affect the time required for attachment, and the strength of the forming biofilms (Kumar and Anand 1998; Poulsen 1999; Van Houdt and Michiels 2010).

Materials often used in the food environment are plastics, rubber, glass, cement, and stainless steel, and properties such as roughness, cleanability, disinfectability, and wettability determined by hydrophobicity may influence the attachment strength of the cells (Van Houdt and Michiels 2010). Bacterial adhesion is facilitated by high free energy and wet surfaces. Stainless steel and glass generally promote bacterial attachment due to their high free energy and hydrophilic nature, contrary to Teflon, nylon, buna-N rubber, and fluorinated polymers, which are considered hydrophobic (Chmielewski and Frank 2003; Shi and Zhu 2009). Twenty-five isolates of *Salmonella* spp. tested for adhesion abilities to four surface materials showed higher bacterial attachment to surfaces more positive in interfacial free energies (Chia et al. 2009).

Chavant et al. (2002) observed that the nature of the surface plants (hydrophilic stainless steel and hydrophobic polytetrafluoroethylene [PTFE]) and the temperature (8°C, 20°C, and 37°C) significantly affected the adhesion of *L. monocytogenes* LO28 and the colonization of the surfaces. Despite the consistent electronegative charge of bacteria, a strongly hydrophilic character was exhibited at 8°C, at any growth phase studied, resulting in low colonization of the hydrophobic material of PTFE. The stability of the biofilms varied. At 37°C, after the complete colonization of PTFE, a significant detachment of the pathogen was observed, while on stainless steel at 20°C, cells detached, but completely recolonized the surface 24 h later. In conclusion, this study suggested that PTFE surfaces could minimize the development of *L. monocytogenes* in cold rooms.

Studying the impact of material type among 8 commonly used surfaces in the kitchen environment, on the adhesion ability of 10 isolates of *L. monocytogenes*, Silva et al. (2008) observed that the best attachment was achieved at the threshold between hydrophobicity and hydrophilicity. Additionally, polypropylene and glass, despite the lower extent of adhesion, supported the viability of attached cells to 100%. This surface-dependent survival may play a significant role in cross-contamination scenarios, during the transfer of microorganisms between surfaces and food products.

Additionally, the rough appearance of stainless-steel surfaces due to cracks and crevices is capable of trapping bacteria, thus contributing to biofilm spread when entrapped bacteria escape by shear forces of the bulk fluid. Aluminum surfaces have larger crevices and sponge-like appearance, while nylon and Teflon are smooth (Kumar and Anand 1998). According to these data, it is of great importance and constitutes a significant challenge to develop and choose materials with improved food safety profiles to maintain the hygienic status of food environments (Pérez-Rodrígues et al. 2011; Van Houdt and Michiels 2010).

6.2.2 Microbial Strain

At first, bacterial strains adsorb reversibly at the conditioning abiotic surface. While bacterial surfaces are usually negatively charged, their adhesion is inhibited due to electrostatic repulsive forces. Therefore, the contribution of bacterial surface appendages is critical, as they provide cells with hydrophobicity. Repulsive forces are then overcome, and irreversible bonding is initiated. This process is usually completed within a few hours of contact (Chmielewski 2003; Kumar and Anand 1998; Poulsen 1999; Simoes 2010; Van Houdt 2010). Some of the bacterial surface appendages, known for their contribution to biofilm formation, are flagella, fimbriae, curli, and pili, and these are discussed in this section.

Flagella are helical structures extending out from the bacterial cytoplasm through the cell wall. They are flagellin protein threads that are capable of cell-to-surface interactions during biofilm formation. Flagella are responsible for bacterial motility, thus contributing to the growth and spread of biofilm development along surfaces. Initial attachment and biofilm maturation of *Pseudomonas fluorescens*, *L. monocytogenes*, *E. coli*, and *Y. enterocolitica* have been shown to be affected by the presence of these structures (Simões et al. 2010; Van Houdt and Michiels 2010), while the effect on *Salmonella* biofilm formation was contradictory (Giaouris et al. 2012).

Fimbriae are surface structures of gram-negative bacteria. Protein filamentous appendages are straight in shape; they do not contribute to motility and are implicated in biofilm formation by *E. coli*, including Shiga toxin–producing strains, *Klebsiella pneumoniae*, *Aeromonas caviae*, *Pseudomonas* spp., and *Vibrio* spp. (Simões et al. 2010; Van Houdt and Michiels 2010).

Curli fimbriae (or thin aggregative fimbriae) are amyloid cell-surface proteins and, together with cellulose, constitute the two main matrix components in *Salmonella* biofilms (Giaouris et al. 2012). The transcriptional activator CsgD regulates the coexpression of csgA-encoding curli and bcsA-encoding cellulose and leads to the formation of a hydrophobic network with tightly packed cells (Goulter-Thorsen et al. 2011; Jain and Chen 2007). As a wide range of environmental stress conditions influence CsgD transcription, it is evident that *Salmonella* and *E. coli* (EHEC and EPEC) pose the risk of transmission of pathogenicity due to the potential tolerance of well-organized biofilms to sanitizing agents (Jain and Chen 2007; Saldaña et al. 2009).

A conjugative plasmid encodes structures similar to fimbriae, which are referred to as pili. These appendages are involved in horizontal gene transfer, known as conjugation, and stimulate biofilm development and, in turn, dispersion of the plasmid from high-density biofilms through the conjugation. As transferred plasmid often contains resistance genes, the transmission of infections is of major importance for food safety when bacteria-producing pili are deposited on food contact surfaces (Van Houdt and Michiels 2010).

EPS is another factor, produced by bacteria, that contributes to biofilm formation by binding the cells and other particular materials together and to the surface (Van Houdt and Michiels 2010). Anchored to the surface and produced by the microorganism, these polyanionic extracellular substances contain polysaccharides, proteins, phospholipids, teichoic, and nucleic acids, and other polymeric substances, hydrated to 85%–95% water (Chmielewski and Frank 2003). Its role is critical and dependent on the microorganism to be attached, as it can serve as a conditioning film, with adhesive or anti adhesive properties, and determines the structure of the developed biofilm.

Ryu and Beuchat (2005) studied the contribution of EPS and curli production to the biofilm formation by *E. coli* O157:H7 on stainless steel. EPS was found to inhibit the initial attachment of the pathogen cells on stainless-steel coupons, while curli production had no effect on the attachment ability of the cells. However, curli facilitated the biofilm development, and the resistance of the pathogenic cells to chlorine increased when the cells produced biofilms, indicating the protected role of EPS and curli against stress. Additionally, EPS amounts were related to the persistence of the established surface bacteria and to the survivability of the cells during stressful conditions (Bremer et al. 2001; Habimana et al. 2009; Nakamura et al. 2013; Norwood and Gilmour 2001; Ryu and Beuchat 2005).

Chia et al. (2009) investigated the adherence properties of 25 strains of *Salmonella* to four types of surface material. Results revealed a strain-dependent attachment behavior, with adherence of higher numbers of *Salmonella* Sofia to all materials used. Because this was a multifactorial study, researchers concluded that the overall results could be attributed to the variability of bacterial surface appendages between different strains.

6.2.3 Temperature

Temperature constitutes a key factor for the attachment of bacterial cells to abiotic surfaces, as it plays a significant role in the microorganisms' adaptation, determines their ecological background, influences physiology in terms of inducing the expression of critical transcriptional regulators, and causes cell surface alterations. However, temperature is usually studied in combination with other environmental factors, such as the conditioning film present on the attachment surfaces, the pH of the surrounding media, osmosis conditions, and ecological parameters of the strains involved (Da Silva Meira et al. 2012; Dourou et al. 2011; Nascentes et al. 2012; Norwood and Gilmour 2001; Speranza et al. 2011).

Mai and Conner (2007) evaluated the attachment of *L. monocytogenes* to stainless steel under different nutrient (brain heart infusion [BHI] and minimal medium) and temperature (4°C, 20°C, 30°C, 37°C, and 42°C) conditions and found that the number of attached cells increased with increasing temperature, with the exception of 42°C, while at 20°C–42°C, the attached population was significantly higher under the influence of BHI nutrient composition compared to the minimal medium.

Nutrients and temperature have been shown to interact in the process of biofilm formation. In Kadam et al.'s study (2013), 143 strains of *L. monocytogenes* were tested for biofilm formation ability under 4 different temperatures (12°C, 20°C, 30°C, and 37°C) in rich, moderate, and poor nutrient media, and it was shown that biofilm formation decreased with decreasing temperature, while the serotype's effect was strongly correlated to the nutrient broth, but the origin of the strains (animal, meat, dairy industry, or human) did not affect biofilm formation.

Contrarily, it was demonstrated by Nilsson et al. (2011) that the origin of the strain affected biofilm formation in correlation with temperature, whereas clinical and environmental strains enhanced biofilm production at higher and lower temperatures, respectively. Serotype 1/2a produced more biofilm, among 95 strains screened, and biofilm production increased with increasing temperature at the most acidic or alkali growth conditions.

Thirty strains of *Salmonella* spp. tested for biofilm formation resulted in higher yields at 30°C for 24 h and at 22°C for 48 h (Stepanović et al. 2003). In the same study, microaerophilic and CO_2-rich conditions constituted the best environment for biofilm formation, which was in accordance with Giaouris and Nychas' (2006) study, which suggested the air–liquid interface as the optimum condition for biofilms. These observations lead to the inference of the multifactorial process that governs biofilm formation in food-processing plants.

6.2.4 Surface Conditioning

Most solid surfaces receive a net negative charge when immersed in water and thus accumulate nutrients by drawing organic material and ions into it, resulting in a conditioning surface (Poulsen 1999; Zottola and Sasahara 1994). This conditioning alters the physicochemical properties of the surface, affecting the subsequent attachment activity (Kumar and Anand 1998). Moreover, it is reported that bacterial attachment occurs most readily on surfaces that are coated by a conditioning film (Jullien et al. 2008; Simões et al. 2010), and this could be attributed to the fact that in the conditioning region, the accumulation of molecules will lead to a higher concentration of nutrients compared to the fluid phase (Kumar and Anand 1998).

Eighteen *L. monocytogenes* strains, with similar growth patterns at different temperatures and salinities, were compared for their adherence capabilities with or without the addition of 2%–5% NaCl. The results showed significant increase in attachment of all strains when NaCl was added, with the temperature of 37°C exhibiting the greatest enhancement, compared to 5°C or 15°C. Moreover, the addition of NaCl resulted in the formation of aggregate cells, tightly bound to the plastic surface, which could be a factor of persistence and increased risk in food-processing facilities (Jensen et al. 2007).

The importance of surface conditioning additionally to temperature was also observed by Dourou et al. (2011), who found that ground beef (solid substrate) exhibited the highest initial attachment, followed by fat-lean homogenate (liquid) and tryptic soy broth, at both temperatures studied (4°C and 15°C), with higher further attachment observed at 15°C. No difference was observed between stainless steel and high-density polyethylene. However, the negative effect of surface conditioning has also been reported, when limited attachment of *L. monocytogenes* was observed on yogurt-soiled surfaces, likely attributed to the inhibiting effect of low yogurt pH (Poimenidou et al. 2009).

6.2.5 Mixed-Species Biofilm Formation

In food-processing environments, microorganisms likely exist as multispecies cultures, where spatial and metabolic interactions between species lead to the development of multispecies biofilms (Giaouris et al. 2012). Depending on the secondary species, the biofilm cells of a pathogen may increase, decrease, or remain unaffected compared to single-species biofilms (Bremer et al. 2001; Habimana et al. 2009; van der Veen and Abee 2011). The composition of the surrounding nutrient medium has been shown to influence the contribution of each species to the mixed biofilms. More importantly, resistance to disinfectants such as peracetic acid and benzalkonium chloride was higher than that exhibited by respective monospecies biofilms (van der Veen and Abee 2011).

Mixed-species biofilms that occur on plant surfaces may also create a particular conditioning, which is encountered by the pathogens when they come into contact with the abiotic surfaces. Pathogen development may even be inhibited when natural *protective* biofilms occur. *Lactococcus lactis* is described as being efficient in controlling *L. monocytogenes* development, due to competition or bacteriocin production. Its existing biofilm may reduce the adhesion of the pathogen and EPS can inhibit the attachment, but this depends on the porosity of the biofilm structure (Habimana et al. 2009). A mixed-species biofilm may also play a significant role in the survival of attached bacteria, as indicated in the study of Bremer et al. (2001), in which *L. monocytogenes* cells were recoverable for longer periods when cultivated with *Flavobacterium* spp., compared to the pure culture.

The establishment of bacteria on surfaces and their ability to resist disinfection are considered among the fundamental mechanisms responsible for their persistence and reoccurrence in the food-processing environment. Persistence may result in *stress-hardened* cells, which are selected for certain food-related stresses and are difficult to be eradicated from the processing environment.

6.2.6 Persistence, Survival, and Stress Hardening of Attached Populations

Strains not removed by disinfection procedures may remain on the food contact facilities and inhabit a food plant for up to 12 years (Tompkin 2002). These strains are referred to as persistent and pose an increased risk of cross-contamination and transmission of infections to food products and consumers.

While Carpentier and Cerf (2011) report that persistence is not an endogenous ability of bacteria to survive, but only a result of improper cleaning, other researchers report that not only are harborage sites in food-processing facilities responsible for the persistence of bacterial cells, but other mechanisms also may lead to their establishment in food environments (Fox et al. 2011; Nakamura et al. 2013). Persistent cells are considered as populations of phenotypic variants, which exist in all bacterial populations studied to date and exhibit multidrug tolerance (Shah et al. 2006). Fox et al. (2011) investigated the physiological differences between persistent and presumed nonpersistent strains of *L. monocytogenes* and reported nonshared properties between them, with increased resistance of the persistent strains to quaternary ammonium compounds. A study of 77 isolates of *L. monocytogenes* by Nakamura et al. (2013) indicated additional attributes that differed between persistent and transient strains, where persistent strains produced higher levels of EPS and biofilms, and persisted more than benzalkonium chloride. Norwood and Gilmour (2001) studied the effect of different temperatures (4°C, 18°C, and 30°C) on the attachment of two *L. monocytogenes* strains, in mono- and mixed-cultures, containing *Staphylococcus xylosus* and *Pseudomonas fragi*. They showed that the optimum adherence condition was at 18°C in monoculture biofilms, and that significantly higher adherence was exhibited by the persistent *L. monocytogenes* strain.

When food contact surfaces are improperly cleaned, food soil remains in surfaces crevices and affects the present microorganism's survival and proliferation. Minced tuna, ground pork, and cabbage were used to compare the effect of food soiling on survival under desiccation of three food-borne pathogens, *L. monocytogenes*, *S.* Typhimurium, and *S. aureus* at 25°C through 30 days. *L. monocytogenes* exhibited the greatest survivability, irrespective of food components, while survival was dependent on food soil presence and food composition. Possibly, plant temperature has an effective role in survival, but this needs to be investigated (Helke et al. 1993; Takahashi et al. 2011).

The effect of relative humidity on biofilm formation and survival was shown to be of great importance for five food-borne pathogens (*L. monocytogenes*, *E. coli* O157:H7, *Salmonella* Typhimurium, *S. aureus*, and *Cronobacter sakazakii*). The RH may have further implications by means of transfer of biofilm cells to food products during food processing (Bae et al. 2012). The high level of water present in a biofilm contributes to the capillary forces holding the biofilm together, while dried biofilms are considered to be governed by weak adhesive forces in cell–cell and cell–surface communication (Rodríguez et al. 2007a). Thus, moist foods enhance the transfer and cross-contamination phenomena due to dried biofilms present, while high RH establishes wet biofilms on food contact surfaces.

Food-borne pathogens commonly undergo stressful conditions during food processing, such as exposure to heat, hydrogen peroxide, acid, and salt. Bacterial cells that survive adapt and obtain increased resistance, which is referred to as stress hardening. The arising stress protection is dependent on the stress encountered and the lethal factor applied, as adaptation to one type of stress can protect bacteria from the same or different types of lethal stresses (Lou and Yousef 1997; Skandamis et al. 2008; Tiganitas et al. 2009).

Several reports have demonstrated the impact of stress conditions in food-processing facilities on the hardening of microorganisms. This may result in an alteration of bacterial physiology and changes in surface hydrophobicity, surface charge, and subsequent attachment ability. Cell elongation and filamentation are two major physiology alterations that are reported, which may play a significant role in food safety issues (Jones et al. 2013; Visvalingam and Holley 2013). Moreover, bacterial cells may adapt to the stress they undergo and develop cross-protection mechanisms. Cold adaptation can lead to the elongation of mesophilic enteric organisms, and *E. coli* O157:H7, adapted and elongated at 6°C, was shown to attach better to both stainless-steel and glass surfaces, than cells grown at 37°C. Growth temperature of 15°C, however, facilitated attachment, compared to 6°C and 37°C, to both surfaces, while the glass surface resulted in better attachment compared to stainless steel (Visvalingam and Holley 2013).

One of the global goals of food markets is to extend the shelf life of food products with respect to the safety of consumers. Therefore, various antimicrobials are added to foods to prevent food-borne pathogens and spoilage bacteria growth during shelf life. Nevertheless, the stress-hardening potential has led to the concern that adaptation to antimicrobials may enhance the resistance of bacteria against other lethal conditions. Indeed, Yuan et al. (2012) showed that *S.* Typhimurium cells, adapted to organic acid salts such as sodium lactate and sodium acetate, exhibited increased resistance in simulated gastric fluid and heat treatment.

Attachment of bacterial cells may cause not only adaptation to stressful factors and cross-protection phenomena, but also susceptibility of the cells when exposed to new environments. The influence on the survival of *S.* Typhimurium under low temperatures and different a_w values showed that although attachment to beef surfaces prevented cell injury and death from hyperosmosis and low temperatures, planktonic cells were better adapted to osmotic stress and thus their viability increased compared to their attached counterparts (Kinsella et al. 2007).

6.3 Stochastic Description of Bacterial Transfer

Bacterial transfer could resemble a puzzle. The factors governing transfer are like pieces that depend upon each other and have to fit together for a clear picture of the whole puzzle (Figure 6.2). It is considered as an inherently random phenomenon. The uncertainty arises not only from the multitude of factors involved, some of which are still unknown, but also from the fact that the known factors cannot be easily controlled.

This uncertainty of bacterial transfer can be somehow captured by using a stochastic approach when studying this phenomenon. The first and simple step to quantify bacterial transfer is the calculation of

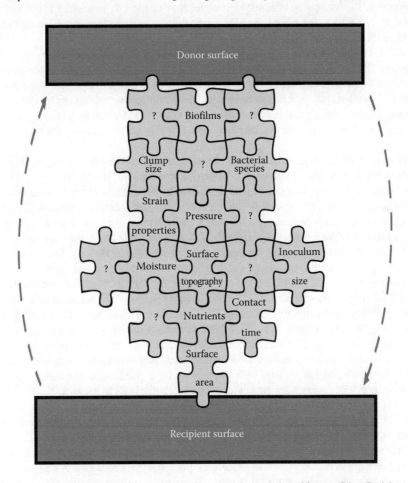

FIGURE 6.2 Schematic representation of bacterial transfer; a puzzle consisting of factors (Pérez-Rodríguez et al. 2008) that fit together to define the transportation of bacteria from one surface to another. Question mark pieces are the unknown factors involved in bacterial transfer.

bacterial transfer rate (Tr), alternatively called efficiency of transfer (EOT) (Pérez-Rodríguez et al. 2008; Rodríguez and McLandsborough 2007). Considering that the process involves two surfaces that come into contact, Tr is defined as the number of cells transferred to a surface (recipient surface) divided by the number of cells existing in the source of contamination (donor surface):

$$Tr(\%) = 100 \times \frac{\text{Number of bacteria on recipient surface}}{\text{Number of bacteria on donor surface}}$$

There are studies that also try to take into consideration the inherent variability of the transfer phenomenon in the representation of their results. These studies use transfer rates or their log transformation and probability distributions to describe their data. Other studies focus on the phenomenon and aim to provide a better understanding of the factors involved. Both types of studies will be presented in the following sections.

6.3.1 Transfer between Abiotic Surfaces and Foods of Plant Origin

Foods of plant origin such as fruits and vegetables have always been recognized as a valuable deposit of nutrients, a treasury of nature. This *pure* character of fresh produce, along with its RTE nature, has placed it in the center of a recent consumer trend that demands more natural, less chemically preserved products. However, the increasing consumption of fresh or minimally processed fruits and vegetables has raised safety issues as at the same time food-borne outbreaks linked to their consumption have also increased (Olaimat and Holley 2012).

The reasons that make plant-origin foods an important vehicle for food pathogens are well understood and documented (Berger et al. 2010). The extent to which different factors such as preharvest or postharvest practices contribute to their contamination has always been an important issue and extensive research has been undertaken in order to determine those contamination routes (Li-Cohen and Bruhn 2002). However, cross-contamination scenarios during cutting and shredding have also started to be considered when studies are conducted, as these scenarios can be determining for the safety of fresh produce.

Different studies have taken place to evaluate the transfer between abiotic surfaces and foods of plant origin. The studies that involve simulation of industrial processes for the production of fresh-cut vegetables focus more on the cross-contamination risk posed by different industrial equipment at different steps of the process. Often, this type of studies includes attempts toward mathematical representation and modeling of data. Buchholz et al. (2012a) used iceberg lettuce, romaine lettuce, and baby spinach to generate quantitative data regarding the transfer of *E. coli* O157:H7 from these vegetables to different pieces of industrial equipment used for the commercial processing of leafy greens. The data provided information regarding the pathogens' transferability between different processing surfaces, or for different contamination levels of the produce. Later on, the same researchers produced data on the transfer of *E. coli* O157:H7 from product-contaminated industrial equipment to fresh uninoculated lettuce batches. Buchholz et al. (2012b) indicated that up to 0.36% of even an initially low (2.0 log CFU/g) contaminated product can be transferred to noncontaminated products during processing. Similarly, during a former study by Pérez-Rodríguez et al. (2011), transfer data from the simulation of commercial processing of lettuce were obtained and transfer coefficients for each step of the processing line were estimated. According to the results, higher transfer took place from fresh produce to processing water and from equipment to lettuce. The researchers fitted the estimated coefficients in probability distributions in an attempt to better describe such stochastic phenomena as transfer events. There are also studies that focus on cross-contamination and transfer events that can occur in an industrial environment due to disinfection solutions used for the decontamination of fresh produce or even the water used for the washing and cleaning of fresh-cut vegetables (Luo et al. 2011). Poor water management in industries often leads to the introduction of pathogens in fresh produce processing lines and contamination of end products through a sequence of transfer events (Holvoet et al. 2012).

As already mentioned, kitchen and household equipment could be an ideal environment for pathogen transmission. Chai et al. (2008) used naturally contaminated vegetables with *C. jejuni* and

S. Typhimurium and simulated different handling scenarios that take place in the kitchen during food preparation. Transfer rates for all events were calculated and the data stressed out water as an important source of cross-contamination in the kitchen. Chen et al. (2001) used *Enterobacter aerogenes*, a non-pathogenic microorganism with similar attachment properties to *Salmonella* spp., and showed its transfer potential between different foods (including lettuce) and surfaces (e.g., cutting boards, hands). Great transfer variability was detected for events between foods and hands. Kusumaningrum et al. (2003) estimated the transfer rates of *S.* Enteritidis, *S. aureus*, and *C. jejuni* from kitchen sponges to stainless-steel surfaces and subsequently to cucumber slices, and highlighted the importance of extrinsic factors such as moisture or contact pressure on pathogen transfer. Indeed, moisture content and product composition have been investigated as factors contributing to bacterial transfer. Silagyi et al. (2009) showed that *E. coli* O157:H7 transferred more efficiently from stainless steel to products with higher moisture content (cantaloupe and other fresh produce). Interestingly, with respect to this, Moore et al. (2003) found no significant differences in the degree of bacterial transfer from contaminated stainless-steel surfaces to wet or dry lettuce. These authors presented their data both as arithmetic and geometric means of percent transfer in order to facilitate future readers to incorporate these data in risk assessment exercises. The work of Montville and Schaffner (2003) has been previously discussed. The impact of a surface's initial contamination on bacterial transfer was tested between gloves and lettuce and, as has been stated, percent transfer rates decreased with the increase of inoculum.

The investigation of surface topography on bacterial transfer was performed during a study conducted by Moore et al. (2007). Cucumber slices and four types of working surfaces commonly used in the kitchen for food preparation were chosen to assess the cross-contamination risk posed by different surface materials. Transfer of *S.* Typhimurium occurred at higher rates from stainless steel compared to other materials. The data approach chosen by the authors in this study is noteworthy. They claim that an estimation of bacterial transfer rates instead of bacterial numbers transferred would lead to a misinterpretation of the results, thus emphasizing the importance of a careful approach to experimental data. Soares et al. (2012) also stressed the importance of the food-handling surface. During their study of *S.* Enteritidis NAL+ transfer from chicken to cutting boards and subsequently to tomatoes, they also evaluated the efficiency of cleaning procedures, proposing stainless steel as a surface that is easier to clean and safer to use. In the same context, the study of Ravishankar et al. (2010) showed the significance of proper cleaning practices of food-handling surfaces and highlighted the importance of following the Food and Drug Administration (FDA) recommendations for washing practices in order to avoid the high risk of cross-contamination.

6.3.2 Transfer between Abiotic Surfaces and Foods of Animal Origin

Foods of animal origin (meat, fish, and dairy products) have been implicated several times in food-borne outbreaks. Food pathogens in such products will always be a major concern and an issue that needs control and rational interventions (Sofos and Geornaras 2010). The contribution of cross-contamination events to the dissemination of pathogens needs serious consideration in industrial environments that process foods of animal origin. On the other hand, incidents of food-poisoning at homes or restaurants due to food-handling and preparation practices have made the phrase *avoid cross-contamination* the most common instruction and an everyday *motto*. The topic of cross-contamination is indeed so significant for products of animal origin that it is questionable whether the risk posed by undercooked products is higher than the risk that cross-contamination events can have (Luber 2009).

As mentioned earlier, biofilms are a form of bacterial lifestyle that is today a major concern for food (especially meat) industries, a serious consideration for food safety authorities, and consequently an interesting topic for scientists. The biofilm formation and the *relationship* that will be developed *over time*, despite the *difficulties* between a number of bacteria and a food-handling surface, apparently play a significant role in bacterial transfer events. Pure *L. monocytogenes* biofilms have been shown to have different transfer potential to solid model foods compared to mixed-species biofilms (Midelet et al. 2006). This is of high significance for food preparation environments, where pathogens and spoilage microorganisms coexist and their behavior is highly dependable on their in-between interactions. It has been demonstrated that strong biofilm-forming bacteria also exhibit a higher transferability than weak biofilm formers. The increased survival attained by strong biofilm formers has been the argument for these controversial findings (Keskinen et al. 2008a). The same authors had similar findings when they tested the transfer of *L. monocytogenes* during the slicing of roasted turkey or salami and the impact of biofilm-forming ability or previous cell injury on transfer. The transfer of the pathogen was higher in absolute numbers for stronger biofilm formers. The transfer rate, though lower for strong biofilm-forming bacteria, led to a prolonged transfer in general, as opposed to weak biofilm formers (Keskinen et al. 2008b). Hansen and Vogel (2011) studied the transfer of *L. monocytogenes* from stainless steel to salmon products. The survival of pathogens during different desiccation treatments was investigated and the Weibull model was suitably fitted to the data. The authors also stressed on the difference of biofilm-forming compared to non-biofilm-forming *L. monocytogenes* cells regarding transfer efficacy. The impact of biofilm presence or absence was found to be stronger on transfer efficiency than the effect of any other stress that the cells might have been subjected to. The level of biofilm dryness has also been shown to be associated with the bacterial transfer of *L. monocytogenes* from stainless steel to bologna and hard salami (Rodríguez et al. 2007a). Capillary forces between the moister bologna or salami and the dry biofilm were thought to be responsible for the observations made; the dryness of the biofilm enhances the transfer efficiency of cells. These capillary forces, along with the decreased cell–cell and cell–surface adhesion upon drying, facilitate the propagation of cells to foods.

Food composition mainly in respect to moisture and fat content has been investigated as a contributing factor to bacterial transfer in studies involving foods of animal origin. As mentioned earlier, Kusumaningrum et al. (2003) simulated pathogen transfer from kitchen sponges to surfaces and subsequently to cucumber slices but also to roasted chicken with transfer rates from stainless-steel surfaces to chicken varying from 25% to 100%. The lower a_w of such a product as roasted chicken, compared to the high moisture contained in cucumber, as well as the difference in fat levels were thought to affect the transferability of the pathogens. It has been shown that the rate of bacterial transfer decreases in foods of high fat proportion (Vorst et al. 2006b). In addition, the same authors reported a logarithmic transfer of *L. monocytogenes* from inoculated slicer blade to turkey and bologna, while a linear transfer trend was observed for salami. Working in the same context, Luber et al. (2006) evaluated the risk of *Campylobacter* transfer from naturally contaminated chicken breasts or legs during handling operations in the kitchen. Interestingly, for the different parts of the same product, different transfer rates were reported. The authors described their data using transfer rates, unlike the proposed log-transformed transfer rates as a means for data representation. Regarding the impact of the formulation of a product of animal origin on bacterial transfer, it must also be noted that the food residues remaining on the surface of the cutting or slicing equipment can develop a layer that is associated with prolonged transfer of a pathogen (Lin et al. 2006; Vorst et al. 2006a). It has been discussed that the food residues that cover an abiotic surface might even change the surface's initial properties (hydrophobicity), thus impacting bacterial transfer (Midelet and Carpentier 2002).

When considering the risk of cross-contamination and bacterial transfer between abiotic surfaces and products of animal origin, the initial bacterial concentration has also been studied as a potential contributing factor. Rodríguez et al. (2007b) evaluated the impact of inoculum size on the transfer efficiency of *L. monocytogenes* from bologna to different surfaces. The authors found that the level of initial contamination does not play an important role in the transfer efficiency of the pathogen. The absolute numbers of transferred bacteria were higher when the inoculum size was higher, but the transfer rates did not significantly change. This is contradictory to other authors' results that report an inversely proportional correlation between inoculum size and transfer rate. Montville and Schaffner (2003) evaluated the impact of initial contamination on the transfer of *E. aerogenes* between a variety of foods and surfaces (chicken, hands, gloves, cutting boards, etc.). After 350 observations, the researchers found that there is a negative correlation between inoculum size and bacterial transfer; the higher the size, the slower the transfer rate. Fravalo et al. (2009) confirmed these observations by investigating the effect of contamination levels on *Campylobacter* transfer rates from chicken legs to cutting boards. For approximately 3.0 log CFU/g of *Campylobacter* on chicken skin, less than 10% transfer was observed, while for 1.0 log CFU/g, the transfer rate was more than 20%. Possible explanations for this correlation are the strongest attachment of bacteria in higher populations, as well as the clump size (Montville and Schaffner 2003; Pérez-Rodríguez et al. 2008).

A number of cross-contamination studies also take into account another important factor in terms of transferability—surface properties. The food-handling surface material, the roughness, and even the cracks and dents on the surface are parameters involved in this type of studies (Flores et al. 2006). Rodríguez and McLandsborough (2007) observed higher transfer efficiency of *L. monocytogenes* from stainless-steel than from polyethylene surfaces to bologna and American cheese. The roughness did not play an important role in transfer efficiency, but as the authors claim, it can affect *cleanability* and consequently bacterial transfer. Similarly, stainless steel has been found to be a *better* recipient surface compared to other materials (polyvinylchloride, polyurethane, polyethylene) when bacterial transfer from foods (beef, bologna) to surfaces was tested (Midelet and Carpentier 2002; Rodríguez et al. 2007b). Goh et al. (2014) estimated the transfer rates of *L. monocytogenes* from wooden or polyethylene cutting surfaces to cooled or hot chicken meat. Higher transfer rates from polyethylene (~90% for cool chicken after 30 min, while 50% for hot chicken) than from wooden surfaces (~70% for cool chicken after 30 min, while 20% for hot chicken) were reported. After 1 h, zero transfer was observed from wooden cutting boards. This, however, is somehow an artifact, as after 1 h, bacteria could not be detected on wood. It has been argued that the porosity of surfaces is also implicated in bacterial transfer. Surfaces in which bacteria can be easily absorbed or entrapped within the pores make them unavailable for subsequent transfer. In the same context, Tang et al. (2011) estimated the transfer rates between cooked chicken, cutting boards, and raw chicken and emphasized the importance of the surface material on which foods are handled during preparation in the kitchen. Furthermore, it is worth noting that the results of Goh et al. (2014) were also consistent with those of Tang et al. (2011) concerning the observed higher transfer rates to products of lower temperature—an observation that highlights temperature as another significant factor for bacterial transfer.

There are also studies with regard to other factors affecting bacterial transfer between foods of animal origin and inanimate surfaces. Such factors can be the contact time between the surfaces involved in the transfer event or, for example, the pressure applied on the surfaces (Dawson et al. 2007; Vorst et al. 2006b). The mechanical forces applied between the surfaces lead to the interruption of cell–cell or cell–surface bonds, or even the release of food exudates that will drift away internalized bacteria. All these possible reasons and explanations are always debatable as far as bacterial transfer is concerned. As Pérez-Rodríguez et al. (2008) state, the contribution of intrinsic factors on bacterial transfer (e.g., bacterial species or biofilm formation) could sometimes become insignificant due to the stronger impact of environmental factors. This, however, is the reality for every factor affecting bacterial transfer. There is always the possibility for a supposedly significant factor to finally not affect the outcome of a transfer event. The combination of factors in bacterial transfer has a much stronger impact compared to each factor alone. Hence, the picture of bacterial transfer is still vague and the missing pieces of information still remain the challenge for completing the bacterial transfer puzzle.

6.3.3 Modeling Bacterial Transfer

The recent deeper insight into the bacterial transfer phenomenon, along with the pronounced necessity of incorporating them in risk assessment studies, has implemented predictive microbiology with a new type of model: the bacterial transfer model. Modeling bacterial transfer can be considered as a challenge for predictive microbiologists. The significance of choosing the right model to describe a microbiological process is well known and the choice of the appropriate parameters as model inputs is decisive for the outcome of the prediction. This fact makes the construction of bacterial transfer models a complex task. As Pérez-Rodríguez et al. (2008) state, not only do unknown factors govern bacterial transfer, but also the known factors (contact time, bacterial strain properties, pressure, etc.) that are involved in cross-contamination events cannot be easily controlled. The challenge in modeling bacterial transfer in food-processing environments, such as the common house kitchen or a food industry, is the dynamic conditions that dominate those environments. At the very same time, a recipient surface can become the donor surface in terms of contamination. Furthermore, this event can take place in every possible direction. The contamination status changes constantly and each new generated pathway becomes the new starting point for more transfer scenarios.

As aforementioned, the first step to modeling transfer is the calculation of the transfer rates as the percentage of cells transferred from the source to the destination of contamination. It has also been mentioned that the multitude of factors that govern bacterial transfer lead to a high variability regarding this phenomenon and subsequently regarding the data obtained from relevant studies. Therefore, this variability, which is reflected on estimated transfer rates, can be depicted through probability distributions. For many authors, probability distributions can be constructed based on the log-transformed transfer rates instead of transfer rates. In fact, normal distribution is considered most suitable to best fit log-transformed transfer rates. Hoelzer et al. (2012) used available scientific data to produce probability distributions and models in order to describe the transfer of *L. monocytogenes* between foods and surfaces or during slicing, as well as the impact of sanitation methods on subsequent cross-contamination events. Normal distributions were also chosen to fit the log-transformed transfer data. In a similar study carried out by Chen et al. (2001) to determine transfer rates of *E. aerogenes* between food and surfaces used for food preparation, normal distribution was chosen as the most suitable and statistically convenient to represent the data. Similarly, Kusumaningrum et al. (2004) used the Anderson–Darling criteria and decided that the normal distribution gives the best fit to the log-transformed transfer rates. Alternatively, logistic distributions were ranked first when normal ones did not give the best results.

Research advances include attempts to construct models that mathematically describe cross-contamination scenarios. Usually, the models are empirical due to the lack of concrete knowledge on bacterial transfer phenomena. Interest has mainly been focused on pathogen transfer that takes place between cutting, shredding, or slicing equipment and foods. Particular interest in these processes is apparently attributed to the employed equipment (e.g., knife or slicer blades) or the type of foods involved (e.g., sliced RTE meat products, fresh-cut salads, and salmon) and consequently the risk implied. The models developed to describe these experimental data represent transfer as a function of cut/slice number or inoculation level. With respect to this, Pérez-Rodríguez et al. (2007) and Vorst et al. (2006b) agreed that when different transfer scenarios were investigated, the logarithmic decrease that was observed in the numbers of the microorganisms under study could be efficiently described by the log-linear model. The goodness of the log-linear model to fit the experimental data was only comparable to that of the Weibull model as reported by Pérez-Rodríguez et al. (2007).

The empirical models used so far have been proven sufficiently accurate and promising for risk assessment studies (Aarnisalo et al. 2007; Sheen 2008; Sheen and Hwang 2010). A more mechanistic approach was introduced by Møller et al. (2012) to treat their experimental data. They developed a model simulating the transfer of *S.* Typhimurium during the grinding of pork. The proposed model was an improved version of the model presented by Nauta et al. (2005), taking into account the tailing phenomenon that is observed after sequential slices during a grinding process. The authors claim that this type of model, though similar to empirical models such as that of Sheen and Hwang (2010), can more efficiently describe all the events that take place in a grinder. It also introduces the idea of the two environments existing in the grinder responsible for two different transfer rates occurring during the process. A mechanistic approach was also chosen by Aziza et al. (2006) to explain cross-contamination during cheese-smearing industrial operations. In this study, a binomial distribution was applied and the constructed model reflected the potential risk associated with this type of cheese manufacturing.

Although interesting, the approach of Aziza et al. (2006) was based on certain assumptions, for example, even bacterial distribution throughout the food matrix. Several authors (Hoelzer et al. 2012; Kusumaningrum et al. 2004; van Asselt et al. 2008) have made some of the following assumptions for the setup of the experiment and the subsequent buildup of the model: no bacteria are lost during processes, no cleaning of the equipment takes place in between use, no new contamination is introduced in the processing equipment for a certain number of operations, and the time after which the transfer event will take place does not have any impact on the event. Such assumptions seem to be necessary for the development of bacterial transfer models as the complexity and inherent variability of the phenomenon demands a simplification and sometimes a more linear representation of transfer pathways (Pérez-Rodríguez and Valero 2013; van Asselt et al. 2008).

Despite being uncertain and multidirectional as a phenomenon, bacterial transfer between foods and surfaces needs description and interpretation. This is in line with the apparent demand for more accurate risk assessment studies and trace-back investigations. Thus, the importance of considering bacterial transfer models in quantitative microbiological risk assessment (QMRA) becomes evident. Further investigation will lead to an abundance of available transfer data and the elimination of assumptions that are now necessarily being made. Additional information such as the educational level of handlers, their social behavior, or the frequency and the degree at which certain products can be initially contaminated with certain pathogens (Christensen et al. 2005; Ivanek et al. 2004; Mylius et al. 2007; Pérez-Rodríguez et al. 2008) would be also useful for QMRA.

6.4 Concluding Remarks

The transfer of bacteria from utensils and processing devices to foods and vice versa is critical for food safety. This chapter has sought to offer a detailed insight into the factors controlling this process. Frequent cross-contamination events along with accumulation of food soil on processing equipment and affinity of bacteria to surfaces are the first steps to the establishment of bacteria in the food-processing environment, either in household or industrial setup. In particular, it may result in the formation of biofilms and subsequent selection of stress- and sanitizer-resistant strains, which in turn may became persistent and difficult to eradicate. Induced stress-adaptation strategies assist bacteria in developing cross-tolerance to various inimical processing conditions and persist in the food-processing environ-ment. Therefore, compliance with hygiene rules and detailed cleaning and sanitation is the only measure that may limit the persistence and the establishment of bacteria that are strongly or even irreversibly associated with surfaces. The level of systematic knowledge obtained so far through studies targeted in simulation of bacterial transfer between various food contact surfaces and foods enabled the collection of massive data on the determination of the likelihood and easiness of bacterial transfer. In parallel, advancements in predictive modeling and stochastic assessment of bacterial behavior on foods may assist in the quantitative description of the bacterial transfer and cross-contamination events. Combining the available experimental data with stochastic modeling tools may enable the accurate simulation of bacte-rial transfer between abiotic and biotic surfaces and offers great assistance in the exposure assessment of modern QMRA approaches. Another potential approach at the level of tertiary modeling (i.e., software application) could be the development of integrated kinetic and probabilistic modeling software (Pin et al. 2011), which will combine growth/survival/inactivation and transfer models to give a prediction of bacterial dynamics throughout a whole food chain from raw materials to the final products and handling by the consumer.

REFERENCES

Aarnisalo, K., S. Sheen, L. Raaska, and M. Tamplin. (2007) Modelling transfer of *Listeria monocytogenes* dur-ing slicing of 'gravad' salmon. *International Journal of Food Microbiology* 118(1): 69–78.

Aldabe, B., Y. Delmas, G. Gault et al. (2011) Household transmission of haemolytic uraemic syndrome associ-ated with *Escherichia coli* O104:H4, south-western France, June 2011. *Eurosurveillance* 16: 3–5.

Azevedo, I., H. Albano, J. Silva, and Teixeira, P. (2014) Food safety in the domestic environment. *Food Control* 37: 272–276.

Aziza, F., E. Mettler, J.-J. Daudin, and M. Sanaa. (2006) Stochastic, compartmental, and dynamic modeling of cross-contamination during mechanical smearing of cheeses. *Risk Analysis* 26(3): 731–745.

Bae, Y., S. Baek, and S. Lee. (2012) Resistance of pathogenic bacteria on the surface of stainless steel depend-ing on attachment form and efficacy of chemical sanitizers. *International Journal of Food Microbiology* 153(3): 465–473.

Bakke, R., M. G. Trulear, J.A. Robinson, and W. G. Characklis. (1984) Activity of *Pseudomonas aeruginosa* in biofilms: Steady state. *Biotechnology and Bioengineering* 26(12): 1418–1424.

Berger, C. N., S. V. Sodha, R. K. Shaw et al. (2010) Fresh fruit and vegetables as vehicles for the transmission of human pathogens. *Environmental Microbiology* 12(9): 2385–2397.

Beumer, R. R., M. C. Te Giffel, E. Spoorenberg, and F. M. Rombouts. (1996) *Listeria* species in domestic environments. *Epidemiology and Infection* 117: 437–442.

Bremer, P. J., I. A. N. Monk, and C. M. Osborne. (2001) Survival of *Listeria monocytogenes* attached to stainless steel surfaces in the presence or absence of *Flavobacterium* spp. *Journal of Food Protection* 64(9): 1369–1376.

Buchholz, A. L., G. R. Davidson, B. P. Marks, E. C. D. Todd, and E. T. Ryser. (2012a) Quantitative transfer of *Escherichia coli* O157:H7 to equipment during small-scale production of fresh-cut leafy greens. *Journal of Food Protection* 75(7): 1184–1197.

Buchholz, A. L., G. R. Davidson, B. P. Marks, E. C. D. Todd, and E. T. Ryser. (2012b) Transfer of *Escherichia coli* O157:H7 from equipment surfaces to fresh-cut leafy greens during processing in a model pilot-plant production line with sanitizer-free water. *Journal of Food Protection* 75(11): 1920–1929.

Carpentier, B. and O. Cerf. (2011) Review-persistence of *Listeria monocytogenes* in food industry equipment and premises. *International Journal of Food Microbiology* 145(1): 1–8.

Carrasco, E., A. Morales-Rueda, and R. M. García-Gimeno. (2012) Cross-contamination and recontamination by *Salmonella* in foods: A review. *Food Research International* 45: 545–556.

Chai, L.-C., H.-Y. Lee, F. M. Ghazali et al. (2008) Simulation of cross-contamination and decontamination of *Campylobacter jejuni* during handling of contaminated raw vegetables in a domestic kitchen. *Journal of Food Protection* 71(12): 2448–2452.

Chavant, P., B. Martinie, T. Meylheuc, and M. Hebraud. (2002) *Listeria monocytogenes* LO28: Surface physicochemical properties and ability to form biofilms at different temperatures and growth phases. *Applied and Environmental Microbiology* 68(2): 728–737.

Chen, F. C., S. L. Godwin, and A. Kilonzo-Nthenge. (2011) Relationship between cleaning practices and microbiological contamination in domestic kitchens. *Food Protection Trends* 31: 672–679.

Chen, Y., K. M. Jackson, F. P. Chea, and D. W. Schaffner. (2001) Quantification and variability analysis of bacterial cross-contamination rates in common food service tasks. *Journal of Food Protection* 64(1): 72–80.

Chia, T. W. R., R. M. Goulter, T. McMeekin, G. A. Dykes, and N. Fegan. (2009) Attachment of different *Salmonella* serovars to materials commonly used in a poultry processing plant. *Food Microbiology* 26(8): 853–859.

Chmielewski, R. N. and J. F. Frank. (2003) Biofilm formation and control in food processing facilities. *Comprehensive Reviews in Food Science and Food Safety* 2(1): 22–32.

Christensen, B. B., H. Rosenquist, H. M. Sommer et al. (2005) A model of hygiene practices and consumption patterns in the consumer phase. *Risk Analysis* 25(1): 49–60.

Da Silva Meira, Q. G., I. de Medeiros Barbosa, A. J. Alves Aguiar Athayde, J. P. de Siqueira-Júnior, and E. L. de Souza. (2012) Influence of temperature and surface kind on biofilm formation by *Staphylococcus aureus* from food-contact surfaces and sensitivity to sanitizers. *Food Control* 25(2): 469–475.

Davey, M. E. and G. A. O'Toole. (2000) Microbial biofilms: From ecology to molecular genetics. *Microbiology and Molecular Biology Reviews* 64(4): 847–867.

Dawson, P., I. Han, M. Cox, C. Black, and L. Simmons. (2007) Residence time and food contact time effects on transfer of *Salmonella* Typhimurium from tile, wood and carpet: Testing the five-second rule. *Journal of Applied Microbiology* 102(4): 945–953.

de Jong, A. E. I., M. J. Nauta, and R. de. Jonge. (2008) Cross-contamination in the kitchen: Effect of hygiene measures. *Journal of Applied Microbiology* 105: 615–624.

Dharod, J. M., S. Paciello, A. Bermúdez-Millán et al. (2009) Bacterial contamination of hands increases risk of cross-contamination among low-income Puerto Rican meal preparers. *Journal of Nutrition Education and Behavior* 41: 389–397.

Dourou, D., C. S. Beauchamp, Y. Yoon et al. (2011) Attachment and biofilm formation by *Escherichia coli* O157:H7 at different temperatures, on various food-contact surfaces encountered in beef processing. *International Journal of Food Microbiology* 149(3): 262–268.

European Food Safety Authority (EFSA). (2010) The Community Summary Report on trends and sources of zoonoses, zoonotic agents and food-borne outbreaks in the European Union in 2008. *EFSA Journal* 8(1):1496 (410 pp.). doi:10.2903/j.efsa.2010.1496; Available online: www.efsa.europa.eu/efsajournal

Enriquez, C. E., R. Enriquez-Gordillo, D. I. Kennedy, and C. P. Gerba. (1997) Bacteriological survey of used cellulose sponges and cotton dishcloths from domestic kitchens. *Dairy, Food and Environmental Sanitation* 17: 2–24.

Erdogrul, O. and F. Erbilir. (2000) Microorganisms in kitchen sponges. *Internet Journal of Food Safety* 6: 17–22.

Flores, R. A., M. L. Tamplin, B. S. Marmer, J. G. Phillips, and P. H. Cooke. (2006) Transfer coefficient models for *Escherichia coli* O157:H7 on contacts between beef tissue and high-density polyethylene surfaces. *Journal of Food Protection* 69(6): 1248–1255.

Fox, E. M., N. Leonard, and K. Jordan. (2011) Physiological and transcriptional characterization of persistent and nonpersistent *Listeria monocytogenes* isolates. *Applied and Environmental Microbiology* 77(18): 6559–6569.

Fravalo, P., M.-J. Laisney, M.-O. Gillard, G. Salvat, and M. Chemaly. (2009) *Campylobacter* transfer from naturally contaminated chicken thighs to cutting boards is inversely related to initial load. *Journal of Food Protection* 72(9): 1836–1840.

Giaouris, E., N. Chorianopoulos, P. Skandamis, and G.-J. Nychas. (2012) Attachment and biofilm formation by *Salmonella* in food processing environments. In *Salmonella—A Dangerous Foodborne Pathogen*, eds. B. S. M. Mahmoud, pp. 157–180. Rijeka, Croatia: Intech.

Giaouris, E. D. and G.-J. E. Nychas. (2006) The adherence of *Salmonella* Enteritidis PT4 to stainless steel: The importance of the air-liquid interface and nutrient availability. *Food Microbiology* 23(8): 747–752.

Goh, S. G., A.-H. Leili, C. H. Kuan et al. (2014) Transmission of *Listeria monocytogenes* from raw chicken meat to cooked chicken meat through cutting boards. *Food Control* 37: 51–55.

Goulter-Thorsen, R. M., E. Taran, I. R. Gentle, K. S. Gobius, and G. Dykes. (2011) CsgA production by *Escherichia coli* O157:H7 alters attachment to abiotic surfaces in some growth environments. *Applied and Environmental Microbiology* 77(20): 7339–7344.

Greig, J. D. and A. Ravel. (2009) Analysis of foodborne outbreak data reported internationally for source attribution. *International Journal of Food Microbiology* 130: 77–87.

Habimana, O., M. Meyrand, T. Meylheuc, S. Kulakauskas, and R. Briandet. (2009) Genetic features of resident biofilms determine attachment of *Listeria monocytogenes*. *Applied and Environmental Microbiology* 75(24): 7814–7821.

Hansen, L. T. and B. F. Vogel. (2011) Desiccation of adhering and biofilm *Listeria monocytogenes* on stainless steel: Survival and transfer to salmon products. *International Journal of Food Microbiology* 146(1): 88–93.

Health Protection Agency (HPA). (2009) Electronic foodborne and non-foodborne gastrointestinal outbreak surveillance system. Report 2: Foodborne outbreaks in 2009. http://www.hpa.org.uk/webc/HPAwebFile/HPAweb_C/1287143129506.

Helke, D. M., E. B. Somers, and A. C. L. Wong. (1993) Attachment of *Listeria monocytogenes* and *Salmonella* Typhimurium to stainless steel and buna-N in the presence of milk and individual milk components. *Journal of Food Protection* 56(6): 479–484.

Herrera, J. J. R., M. L. Cabo, A. González, I. Pazos, and L. Pastoriza. (2007) Adhesion and detachment kinetics of several strains of *Staphylococcus aureus* subsp. *aureus* under three different experimental conditions. *Food Microbiology* 24(6): 585–591.

Hoelzer, K., R. Pouillot, D. Gallagher et al. (2012) Estimation of *Listeria monocytogenes* transfer coefficients and efficacy of bacterial removal through cleaning and sanitation. *International Journal of Food Microbiology* 157(2): 267–277.

Holvoet, K., L. Jacxsens, I. Sampers, and M. Uyttendaele. (2012) Insight into the prevalence and distribution of microbial contamination to evaluate water management in the fresh produce processing industry. *Journal of Food Protection* 75(4): 671–681.

Ivanek, R., Y. T. Gröhn, M. Wiedmann, and M. T. Wells. (2004) Mathematical model of *Listeria monocytogenes* cross-contamination in a fish processing plant. *Journal of Food Protection* 67(12): 2688–2697.

Jain, S. and J. Chen. (2007) Attachment and biofilm formation by various serotypes of *Salmonella* as influenced by cellulose production and thin aggregative fimbriae biosynthesis. *Journal of Food Protection* 70(11): 2473–2479.

Jensen, A., M. H. Larsen, H. Ingmer, and B. F. Vogel. (2007) Sodium chloride enhances adherence and aggregation and strain variation influences invasiveness of *Listeria monocytogenes* strains. *Journal of Food Protection* 70(3): 592–599.

Jones, T. H., K. M. Vail, and L. M. Mcmullen. (2013) Filament formation by foodborne bacteria under sublethal stress. *International Journal of Food Microbiology* 165(2): 97–110.

Jullien, C., T. Benezech, C. Le Gentil et al. (2008) Physico-chemical and hygienic property modifications of stainless steel surfaces induced by conditioning with food and detergent. *Biofouling* 24(3): 163–172.

Kadam, S. R., H. M. W. den Besten, S. van der Veen et al. (2013) Diversity assessment of *Listeria monocytogenes* biofilm formation: Impact of growth condition, serotype and strain origin. *International Journal of Food Microbiology* 165(3): 259–264.

Keskinen, L. A., E. C. D. Todd, and E. T. Ryser. (2008a) Transfer of surface-dried *Listeria monocytogenes* from stainless steel knife blades to roast turkey breast. *Journal of Food Protection* 71(1): 176–181.

Keskinen, L. A., E. C. D. Todd, and E. T. Ryser. (2008b) Impact of bacterial stress and biofilm-forming ability on transfer of surface-dried *Listeria monocytogenes* during slicing of delicatessen meats. *International Journal of Food Microbiology* 127(3): 298–304.

Kinsella, K. J., T. A. Rowe, I. S. Blair, D. A. Mcdowell, and J. J. Sheridan. (2007) The influence of attachment to beef surfaces on the survival of cells of *Salmonella enterica* serovar Typhimurium DT104, at different a(w) values and at low storage temperatures. *Food Microbiology* 24: 786–793.

Klontz, K. C., B. Timbo, S. Fein, and A. Levy. (1995) Prevalence of selected food consumption and preparation behaviours associated with increased risks of foodborne disease. *Journal of Food Protection* 58: 927–930.

Kumar, C. G. and S. K. Anand. (1998) Significance of microbial biofilms in food industry: A review. *International Journal of Food Microbiology* 42(1–2): 9–27.

Kusumaningrum, H. D., G. Riboldi, W. C. Hazeleger, and R. R. Beumer. (2003) Survival of foodborne pathogens on stainless steel surfaces and cross-contamination to foods. *International Journal of Food Microbiology* 85(3): 227–236.

Kusumaningrum, H. D., E. D. van Asselt, R. R. Beumer, and M. H. Zwietering. (2004) A quantitative analysis of cross-contamination of *Salmonella* and *Campylobacter* spp. via domestic kitchen surfaces. *Journal of Food Protection* 67(9): 1892–1903.

Langsrud, S., M. Sidhu, E. Heir, and A. L. Holck. (2003) Bacterial disinfectant resistance—A challenge for the food industry. *International Biodeterioration & Biodegradation* 51(4): 283–290.

Li-Cohen, A. E. and C. M. Bruhn. (2002) Safety of consumer handling of fresh produce from the time of purchase to the plate: A comprehensive consumer survey. *Journal of Food Protection* 65(8): 1287–1296.

Lin, C.-M., K. Takeuchi, L. Zhang et al. (2006) Cross-contamination between processing equipment and deli meats by *Listeria monocytogenes*. *Journal of Food Protection* 69(1): 71–79.

Lou, Y. and A. E. Yousef. (1997) Adaptation to sublethal environmental stresses protects *Listeria monocytogenes* against lethal preservation factors. *Applied and Environmental Microbiology* 63(4): 1252–1255.

Luber, P. (2009) Cross-contamination versus undercooking of poultry meat or eggs—Which risks need to be managed first? *International Journal of Food Microbiology* 134(1–2): 21–28.

Luber, P., S. Brynestad, D. Topsch, K. Scherer, and E. Bartelt. (2006) Quantification of *Campylobacter* species cross-contamination during handling of contaminated fresh chicken parts in kitchens. *Applied and Environmental Microbiology* 72(1): 66–70.

Luo, Y., X. Nou, Y. Yang et al. (2011) Determination of free chlorine concentrations needed to prevent *Escherichia coli* O157:H7 cross-contamination during fresh-cut produce wash. *Journal of Food Protection* 74(3): 352–358.

Mai, T. L. and D. E. Conner. (2007) Effect of temperature and growth media on the attachment of *Listeria monocytogenes* to stainless steel. *International Journal of Food Microbiology* 120(3): 282–286.

Manios, S. G., A. Fytros, and P. N. Skandamis. Effect of habituation on plastic, metal or glass surfaces in the presence of various food residues on survival and growth of *Salmonella*. Paper presented at the *99th International Association of Food Protection Meeting*, Providence, RI, July 22–25, 2012.

Mattick, K., K. Durham, M. Hendrix et al. (2003) The microbiological quality of washing-up water and the environment in domestic and commercial kitchens. *Journal of Applied Microbiology* 94: 842–848.

Midelet, G. and B. Carpentier. (2002) Transfer of microorganisms, including *Listeria monocytogenes*, from various materials to beef. *Applied and Environmental Microbiology* 68(8): 4015–4024.

Midelet, G., A. Kobilinsky, and B. Carpentier. (2006) Construction and analysis of fractional multifactorial designs to study attachment strength and transfer of *Listeria monocytogenes* from pure or mixed biofilms after contact with a solid model food. *Applied and Environmental Microbiology* 72(4): 2313–2321.

Møller, C. O. A., M. J. Nauta, B. B. Christensen, P. Dalgaard, and T. B. Hansen. (2012) Modelling transfer of *Salmonella* Typhimurium DT104 during simulation of grinding of pork. *Journal of Applied Microbiology* 112(1): 90–98.

Montville, R. and D. W. Schaffner. (2003) Inoculum size influences bacterial cross contamination between surfaces. *Applied and Environmental Microbiology* 69: 7188–7193.

Moore, C. M., B. W. Sheldon, and L.-A. Jaykus. (2003) Transfer of *Salmonella* and *Campylobacter* from stainless steel to romaine lettuce. *Journal of Food Protection* 66(12): 2231–2236.

Moore, G., I. S. Blair, and D. A. McDowell. (2007) Recovery and transfer of *Salmonella* Typhimurium from four different domestic food contact surfaces. *Journal of Food Protection* 70(10): 2273–2280.

Mylius, S. D., M. J. Nauta, and A. H. Havelaar. (2007) Cross-contamination during food preparation: A mechanistic model applied to chicken-borne *Campylobacter*. *Risk Analysis* 27(4): 803–813.

Nakamura, H., K.-I. Takakura, Y. Sone, Y. Itano, and Y. Nishikawa. (2013) Biofilm formation and resistance to benzalkonium chloride in *Listeria monocytogenes* isolated from a fish processing plant. *Journal of Food Protection* 76(7): 1179–1186.

Nascentes, N., E. Chiarini, M. Teresa, M. De Aguiar, and L. Augusto. (2012) PFGE characterisation and adhesion ability of *Listeria monocytogenes* isolates obtained from bovine carcasses and beef processing facilities. *Meat Science* 92(4): 635–643.

Nauta, M., I. van der Fels-Klerx, and A. Havelaar. (2005) A poultry-processing model for quantitative microbiological risk assessment. *Risk Analysis* 25(1): 85–98.

Nilsson, R. E., T. Ross, and J. P. Bowman. (2011) Variability in biofilm production by *Listeria monocytogenes* correlated to strain origin and growth conditions. *International Journal of Food Microbiology* 150(1): 14–24.

North Carolina Department of Health and Human Services (NCDHHS). (2013) Outbreak of *Salmonella* Typhimurium gastroenteritis associated with a hotel-based restaurant. Cumberland County, North Carolina, May 2013. http://foodpoisoningbulletin.com/wp-content/uploads/Final_Salmonella_Report_July_2013.pdf (accessed May 11, 2013).

Norwood, D. E. and A. Gilmour. (2001) The differential adherence capabilities of two *Listeria monocytogenes* strains in monoculture and multispecies biofilms as a function of temperature. *Letters in Applied Microbiology* 33(4): 320–324.

Olaimat, A. N. and R. A. Holley. (2012) Factors influencing the microbial safety of fresh produce: A review. *Food Microbiology* 32(1): 1–19.

Parsek, M. R. and C. Fuqua. (2004) Biofilms 2003: Emerging themes and challenges in studies of surface-associated microbial life. *Journal of Bacteriology* 186(14): 4427–4440.

Pérez-Rodrígues, D., P. Teixeira, R. Oliveira, and J. Azeredo. (2011) *Salmonella enterica* Enteritidis biofilm formation and viability on regular and triclosan-impregnated bench cover materials. *Journal of Food Protection* 74(1): 32–37.

Pérez-Rodríguez, F., D. Campos, E. T. Ryser et al. (2011) A mathematical risk model for *Escherichia coli* O157:H7 cross-contamination of lettuce during processing. *Food Microbiology* 28(4): 694–701.

Pérez-Rodríguez, F. and A. Valero. *Predictive Microbiology in Foods*. New York: Springer, 2013.

Pérez-Rodríguez, F., A. Valero, E. Carrasco, R. M. García, and G. Zurera. 2008. (2008) Understanding and modelling bacterial transfer to foods: A review. *Trends in Food Science & Technology* 19(3): 131–144.

Pérez-Rodríguez, F., A. Valero, E. C. D. Todd et al. (2007) Modeling transfer of *Escherichia coli* O157:H7 and *Staphylococcus aureus* during slicing of a cooked meat product. *Meat Science* 76(4): 692–699.

Phang, H. S. and C. M. Bruhn. (2011) Burger preparation: What consumers say and do in the home. *Journal of Food Protection* 74(10): 1708–1716.

Pin, C., G. Avendaño-Perez, E. Cosciani-Cunico et al. (2011) Modelling *Salmonella* concentration throughout the pork supply chain by considering growth and survival in fluctuating conditions of temperature, pH and a_w. *International Journal of Food Microbiology* 145(Suppl. 1): 96–102.

Poimenidou, S., C. A. Belessi, E. D. Giaouris et al. (2009) *Listeria monocytogenes* attachment to and detachment from stainless steel surfaces in a simulated dairy processing environment. *Applied and Environmental Microbiology* 75(22): 7182–7188.

Poulsen, L. V. (1999) Review: Article microbial biofilm in food processing. *Lebensmittel-Wissenschaft Und-Technologie* 32: 321–326.

Ravishankar, S., L. Zhu, and D. Jaroni. (2010) Assessing the cross contamination and transfer rates of *Salmonella enterica* from chicken to lettuce under different food-handling scenarios. *Food Microbiology* 27(6): 791–794.

Rayner, J., R. Veeh, and J. Flood. (2004) Prevalence of microbial biofilms on selected fresh produce and household surfaces. *International Journal of Food Microbiology* 95: 29–39.

Redmond, E. C. and C. J. Griffith. (2003) Consumer food handling in the home: A review of food safety studies. *Journal of Food Protection* 66: 130–161.

Rivas, L., N. Fegan, and G. A. Dykes. (2007) Attachment of Shiga toxigenic *Escherichia coli* to stainless steel. *International Journal of Food Microbiology* 115(1): 89–94.

Rode, T. M., S. Langsrud, A. Holck, and T. Møretrø. (2007) Different patterns of biofilm formation in *Staphylococcus aureus* under food-related stress conditions. *International Journal of Food Microbiology* 116(3): 372–383.

Rodríguez, A., W. R. Autio, and L. A. McLandsborough. (2007a) Effect of biofilm dryness on the transfer of *Listeria monocytogenes* biofilms grown on stainless steel to bologna and hard salami. *Journal of Food Protection* 70(11): 2480–2484.

Rodríguez, A., W. R. Autio, and L. A. McLandsborough. (2007b) Effects of inoculation level, material hydration, and stainless steel surface roughness on the transfer of *Listeria monocytogenes* from inoculated bologna to stainless steel and high-density polyethylene. *Journal of Food Protection* 70(6): 1423–1428.

Rodríguez, A., S. Crevecoeur, R. Dure, L. Delhalle, and G. Daube. Modeling microbial cross-contamination in quick service restaurants by means of experimental simulations with *Bacillus* spores. Paper presented at the *22nd International Symposium of Food Micro*, Copenhagen, Denmark, August 30–September 3, 2010.

Rodríguez, A. and L. A. McLandsborough. (2007) Evaluation of the transfer of *Listeria monocytogenes* from stainless steel and high-density polyethylene to bologna and American cheese. *Journal of Food Protection* 70(3): 600–606.

Rossi, E. M., D. Scapin, W. F. Grando, and E. C. Tondo. (2012) Microbiological contamination and disinfection procedures of kitchen sponges used in food services. *Food and Nutrition Sciences* 3: 975–980.

Ryu, J. and L. R. Beuchat. (2005) Biofilm formation by *Escherichia coli* O157:H7 on stainless steel: Effect of exopolysaccharide and Curli production on its resistance to chlorine. *Applied and Environmental Microbiology* 71(1): 247–254.

Saldaña, Z., J. Xicohtencatl-Cortes, F. Avelino et al. (2009) Synergistic role of Curli and cellulose in cell adherence and biofilm formation of attaching and effacing *Escherichia coli* and identification of Fis as a negative regulator of Curli. *Environmental Microbiology* 11(4): 992–1006.

Scott, E. and S. Bloomfield. (1993) An in-use study of the relationship between bacterial contamination of food preparation surfaces and cleaning cloths. *Letters in Applied Microbiology* 16: 173–177.

Shah, D., Z. Zhang, A. Khodursky et al. (2006) Persisters: A distinct physiological state of *E. coli*. *BMC Microbiology* 6: 53.

Sharma, M., J. Eastridge, and C. Mudd. (2009) Effective household disinfection methods of kitchen sponges. *Food Control* 20: 310–313.

Sheen, S. (2008) Modeling surface transfer of *Listeria monocytogenes* on salami during slicing. *Journal of Food Science* 73(6): 304–311.

Sheen, S. and C.-A. Hwang. (2010) Mathematical modeling the cross-contamination of *Escherichia coli* O157:H7 on the surface of ready-to-eat meat product while slicing. *Food Microbiology* 27(1): 37–43.

Shi, X. and X. Zhu. (2009) Biofilm formation and food safety in food industries. *Trends in Food Science & Technology* 20(9): 407–413.

Silagyi, K., S.-H. Kim, Y. M. Lo, and C. Wei. (2009) Production of biofilm and quorum sensing by *Escherichia coli* O157:H7 and its transfer from contact surfaces to meat, poultry, ready-to-eat deli, and produce products. *Food Microbiology* 26(5): 514–519.

Silva, S., P. Teixeira, R. Oliveira, and J. Azeredo. (2008) Adhesion to and viability of *Listeria monocytogenes* on food contact surfaces. *Journal of Food Protection* 71(7): 1379–1385.

Simões, M., L. C. Simões, and M. J. Vieira. (2010) A review of current and emergent biofilm control strategies. *LWT—Food Science and Technology* 43(4): 573–583.

Skandamis, P. N., Y. Yoon, J. D. Stopforth, P. Kendall, and J. N. Sofos. (2008) Heat and acid tolerance of *Listeria monocytogenes* after exposure to single and multiple sublethal stresses. *Food Microbiology* 25(2): 294–303.

Soares, V. M., J. G. Pereira, C. Viana, T. B. Izidoro, L. D. S. Bersot, and J. P. D. A. N. Pinto. (2012) Transfer of *Salmonella* Enteritidis to four types of surfaces after cleaning procedures and cross-contamination to tomatoes. *Food Microbiology* 30(2): 453–456.

Sofos, J. N. and I. Geornaras. (2010) Overview of current meat hygiene and safety risks and summary of recent studies on biofilms, and control of *Escherichia coli* O157:H7 in nonintact, and *Listeria monocytogenes* in ready-to-eat, meat products. *Meat Science* 86(1): 2–14.

Speranza, B., M. R. Corbo, and M. Sinigaglia. (2011) Effects of nutritional and environmental conditions on *Salmonella* sp. biofilm formation. *Journal of Food Science* 76(1): M12–M16.

Stepanović, S., I. Ćirković, V. Mijač, and M. Švabić-Vlahović. (2003) Influence of the incubation temperature, atmosphere and dynamic conditions on biofilm formation by *Salmonella* spp. *Food Microbiology* 20(3): 339–343.

Taché, J. and B. Carpentier. (2014) Hygiene in the home kitchen: Changes in behaviour and impact of key microbiological hazard control measures. *Food Control* 35: 392–400.

Takahashi, H., S. Kuramoto, S. Miya, and B. Kimura. (2011) Desiccation survival of *Listeria monocytogenes* and other potential foodborne pathogens on stainless steel surfaces is affected by different food soils. *Food Control* 22(3–4): 633–637.

Tang, J. Y. H., M. Nishibuchi, Y. Nakaguchi, F. M. Ghazali, A. Saleha, and R. Son. (2011) Transfer of *Campylobacter jejuni* from raw to cooked chicken via wood and plastic cutting boards. *Letters in Applied Microbiology* 52(6): 581–588.

Tate, N. J. (2006) Bacteria in household sponges: A study testing which physical methods are most effective in decontaminating kitchen sponges. *Saint Martin's University Biology Journal* 1: 65–74.

Tiganitas, A., N. Zeaki, A. S. Gounadaki, E. H. Drosinos, and P. N. Skandamis. (2009) Study of the effect of lethal and sublethal pH and a(w) stresses on the inactivation or growth of *Listeria monocytogenes* and *Salmonella* Typhimurium. *International Journal of Food Microbiology* 134(1–2): 104–112.

Tompkin, R. B. (2002) Control of *Listeria monocytogenes* in the food-processing environment. *Journal of Food Protection* 65(4): 709–725.

van Asselt, E. D., A. E. I. de Jong, R. de Jonge, and M. J. Nauta. (2008) Cross-contamination in the kitchen: Estimation of transfer rates for cutting boards, hands and knives. *Journal of Applied Microbiology* 105(5): 1392–1401.

van der Veen, S. and T. Abee. (2011) Mixed species biofilms of *Listeria monocytogenes* and *Lactobacillus plantarum* show enhanced resistance to benzalkonium chloride and peracetic acid. *International Journal of Food Microbiology* 144(3): 421–431.

Van Houdt, R. and C. W. Michiels. (2010) Biofilm formation and the food industry—A focus on the bacterial outer surface. *Journal of Applied Microbiology* 109: 1117–1131.

Visvalingam, J. and R. Holley. (2013) Adherence of cold-adapted *Escherichia coli* O157:H7 to stainless steel and glass surfaces. *Food Control* 30(2): 575–579.

Vorst, K. L., E. C. D. Todd, and E. T. Ryser. (2006a) Transfer of *Listeria monocytogenes* during slicing of turkey breast, bologna, and salami with simulated kitchen knives. *Journal of Food Protection* 69(12): 2939–2946.

Vorst, K. L., E. C. D. Todd, and E. T. Rysert. (2006b) Transfer of *Listeria monocytogenes* during mechanical slicing of turkey breast, bologna, and salami. *Journal of Food Protection* 69(3): 619–626.

Yuan, W., R. Ágoston, D. Lee, S.-C. Lee, and H.-G. Yuk. (2012) Influence of lactate and acetate salt adaptation on *Salmonella* Typhimurium acid and heat resistance. *Food Microbiology* 30(2): 448–452.

Zobell, C. E. (1943) The effect of solid surfaces upon bacterial activity. *Journal of Bacteriology* 46(1): 39–56.

Zottola, E. and K. C. Sasahara. (1994) Microbial biofilms in the food processing industry—Should they be a concern? *International Journal of Food Microbiology* 23(2): 125–148.

Section II

Detection of Food-Borne Pathogens

7

Molecular Techniques for Detection of Food-Borne Bacteria and for Assessment of Bacterial Quality

Robert E. Levin

CONTENTS

The polymerase chain reaction (PCR) constitutes a powerful method for rapidly detecting low numbers of pathogenic and toxigenic bacteria in complex food matrices. Recent developments have resulted in the ability to detect as low as 1 colony forming unit (CFU) of *Salmonella enterica* in a 25 g sample of ground beef within 4.5 h without enrichment. These recent innovations have the potential to eliminate most costly recalls and related illnesses by allowing raw food processors to detect the presence of pathogens in products well before shipment. This chapter deals in depth with these recent developments.

7.1 Conventional PCR

7.1.1 Introduction

The PCR requires the identification of a gene unique to the targeted organism and the DNA nucleotide sequence of that specific gene for the generation of primers. GenBank is an excellent source for gene sequences of known pathogens. A number of PCR primers have already been published for unique genes harbored by most recognized pathogens. Knowledge of the target gene sequence and the use of the Blast software program is all that one needs to identify suitable primer pairs.

7.1.2 Theory of Conventional PCR

The PCR uses repeated temperature cycling involving template denaturation, primer annealing, and the activity of DNA polymerase for extension of the annealed primers from the 3′-ends of both DNA strands (Figure 7.1a). This results in amplification of the specific target DNA sequence. The availability of the thermostable *Taq* DNA polymerase, from the extreme thermophile *Thermus aquaticus*, greatly facilitates repeated thermal cycling at ~95°C for template denaturation.

PCRs are usually performed in 0.5 or 0.2 mL thin-walled polyethylene PCR tubes containing a 20–50 mL total reaction volume. Variables that require optimization include components 5–9 in Table 7.1. Innis and Gelfand (1990) have discussed the optimization of PCRs in detail. Most thermal cycling of PCRs encompasses 35–40 cycles.

A typical thermal cycling protocol is presented in Table 7.2. After an initial denaturation step at 95°C, steps 2–4 are then sequentially performed for 35–40 cycles followed by a final extension (step 5) at 72°C to ensure that the final round of strand synthesis at high substrate concentration is completed. The sixth step involving reduction of the temperature to 4°C is used for convenient holding until agarose gel electrophoresis (AGE) is performed. The time required to traverse from one temperature to another is referred to as the ramp time and usually contributes significantly to the total thermal cycling time.

7.1.3 Agarose Gel Electrophoresis

The author routinely uses 1% agarose (electrophoresis grade, high melting) prepared with 0.05 M sodium borate buffer at pH 8.0 (Brody and Kern, 2004). In place of ethidium bromide that is carcinogenic, the author incorporates the noncarcinogenic fluorescent DNA dye Midori green into the agarose gel prior to casting. After 30 min at 75 V, the gels are visualized with an ultraviolet (UV) transilluminator (302 nm) for detection of green fluorescent bands. Stained amplicon bands are photographed with a digital camera. The relative fluorescence of the DNA bands is obtained with the public domain NIH image analysis program ImageJ. The relative fluorescent intensity of each band is plotted against the log of CFU/g of food.

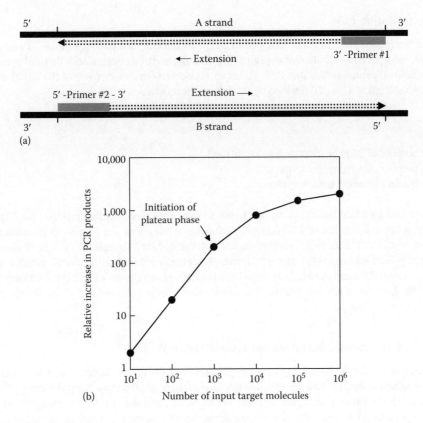

FIGURE 7.1 PCR amplification of DNA. (a) Amplification of a known target sequence with a set of two primers. (b) Dependence of PCR product accumulation on number of input targets from 35 amplification cycles.

TABLE 7.1

Typical PCR Components

1. Template DNA	1–10 μL
2. Tris-HCl (pH 8.3 at 20°C)	20 mM
3. KCl	25 mM
4. Triton X-100	0.1%
5. dNTP's (dATP, dCTP, dGTP, dTTP)	50 mM each
6. *Taq* polymerase	1.0 unit
7. MgCl$_2$	1.5 mM
8. Primers	10 pmol each

TABLE 7.2

Typical PCR Thermal Cycling Protocol

Step #	Process	Temperature (°C)	Time (min)
1	Initial denaturation	94	3
2	Denaturation	94	1
3	Annealing	60	1
4	Extension	72	1
5	Final extension	72	4–7
6	Terminate reactions and hold	4	—

7.1.4 Quantitative PCR

The author has found that the methodology described later is ideally suited for quantitative assessment of target DNA incorporating an internal standard (IS) with the use of a conventional thermal cycler.

The operational assumption is that PCR yields an exponential rate of increase of the initial number of target DNA molecules. This is described by the following equation:

$$P = T(2^n)$$

where

P is the number of PCR product molecules formed
T is the number of input target sequences
n is the number of amplification cycles

This exponential equation however does not apply to the entire amplification process. During the first few cycles, when the number of initial target molecules is very low, the rate of amplification can be anticipated to be low (Diaco, 1995). During the last few cycles, when the ratio of PCR product molecules to unreacted primer and *Taq* polymerase molecules has greatly increased, the overall amplification rate can be expected to decrease significantly from an exponential rate (Figure 7.1a). It is only along the linear portion of the plot that a reliable quantitative relationship between the number of input molecules and final products can be derived.

7.1.5 Use of an Internal Standard for Quantitative PCR

The application of an IS is based on the co-amplification of the target sequence and an internal DNA standard, which is amplified with equal efficiency as the target sequence by using the same primer pair. The IS contains the same primer binding sites as the target, and the two DNAs compete for reaction reagents to produce PCR products of different sizes, which can be separated in an agarose gel. The log ratio of intensities of amplified target DNA to IS is determined by the following equation given by Zachar et al. (1993):

$$-\mathrm{Log}\left(\frac{N_{n1}}{N_{n2}}\right) = \log\left(\frac{N_{01}}{N_{02}}\right) + n\log\left(\frac{\mathrm{eff}_1}{\mathrm{eff}_2}\right)$$

where

N_{n1} is the no. of resulting target molecules after amplification
N_{n2} is the no. of resulting IS molecules after amplification
N_{01} is the no. of initial target molecules
N_{02} is the no. of initial IS molecules
n is a given cycle
eff_1 is the efficiency of amplification of target molecules
eff_2 is the efficiency of amplification of IS molecules

If the efficiencies of amplification ($\mathrm{eff}_1/\mathrm{eff}_2$) are equal, the ratio of amplified products (N_{n1}/N_{n2}) is dependent on the log ratio of starting reactants (N_{01}/N_{02}). Even if the efficiencies of the two reactions are not equal, the values for N_{n1}/N_{n2} still hold assuming that $\mathrm{eff}_1/\mathrm{eff}_2$ is constant and amplification is in the exponential phase (Zachar et al., 1993). With this technique, varying amounts of target DNA are co-amplified with a constant amount of internal amplification standard (internal amplification control [IAC]). See Rupf et al. (1999) and Guan and Levin (2002) for details regarding synthesis of an IS. The resulting log ratio of fluorescence intensities of PCR products are plotted against the log of target CFU for construction of a standard curve (Figure 7.2) that is used for determining the number of CFU/g of food product. The advantage of this approach is that once the standard curve is generated, a single constant concentration of the internal amplification standard is added to each PCR vial. The operating assumption is that partial inhibition of amplification will impact both amplicons to the same extent. If extensive inhibition

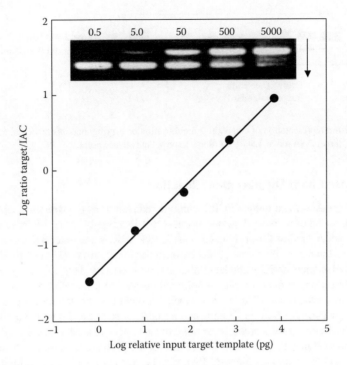

FIGURE 7.2 Standard curve for IAC. Amplification of 10-fold variations in amount (pg) of target DNA of 429 bp (upper bands) and constant amount (0.25 ng) of IAC (lower bands) of 222 pb. Numerical values indicate initial pg of target DNA. (From Zachar, V. et al., *Nucleic Acids Res.*, 8, 2917, 1993. With permission.)

of the PCR occurs so as to prevent the appearance of both DNA bands, then the assay is invalid. If only the IS is amplified, then clearly, a valid negative reaction has occurred.

7.1.6 Construction of a Calibration Curve for Quantitative PCR

Varying amounts of target DNA are co-amplified with a constant amount of the IS. The amplified co-PCR products are separated by electrophoresis in a 1.5% agarose gel (Figure 7.4). The gel images are then analyzed with NIH ImageJ software. To correct differences in the intensity of fluorescence of ethidium bromide-stained PCR fragments of different sizes, the intensity of the IS is multiplied by the ratio of the number of nucleotides in the target sequence to that of the IS. The log of the ratio of fluorescence intensity of the amplified target sequence to that of the IS is then plotted against the log of CFU to establish a standard curve, using a constant amount of IS and varying the number of CFU. It is important to note that the calibration curve is operational over no more than four log cycles of input target sequences (CFU).

Hübner et al. (1999) introduced the concept of *equivalence point* or equivalence in quantitative competitive PCR assays involving competitive amplification of an IS and the target DNA sequence. The IS consists of the same nucleotide sequence as the target sequence except that it contains a 22 bp insertion sequence to allow agarose gel distinction of the target and IS amplicons. This allows use of the same forward and reverse primer for both targets in the same PCR vial. The equivalence point is defined as that amount of each amplicon amplified to a visually equivalent extent resulting from equal starting concentrations of the IS and target DNA as illustrated in Figure 7.3. Here, a single concentration of the targeted DNA is added to a sequence of PCR tubes where the concentration of IS is varied (Figure 7.3). This approach suffers from the disadvantage that multiple PCR tubes must be run for each sample. It is important to note that competitive PCR involving an IS can only be used for determination of relative amounts of target and standard if the regression coefficient r^2 is better than 0.99 and the slope of the regression line is very close to unity reflecting equivalent amplification of equal amounts of initial target and IS (Raeymaekers, 1993; Hayward-Lester et al., 1995).

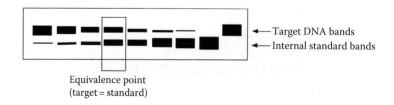

FIGURE 7.3 Idealized representation of competitive amplification of varying amounts of IS and a constant amount of target DNA. Lane 8, target DNA alone. Lane 9, IS alone. Lane 4, equivalence point.

7.1.7 Polymerase Chain Displacement Reaction

Harris et al. (2013) described a unique PCR technique referred to as a *polymerase chain displacement reaction (PCDR)*, which uses multiple nested primers that increased the sensitivity of normal quantitative PCR by 10-fold. A total of four primers are used. In PCDR, when extension occurs from the outer primer, it displaces the extension strand produced from the inner primer (IP) by utilizing a polymerase that has strand displacement activity and lacks 5′–3′ nuclease activity. This allows a greater than twofold increase of amplification product for each amplification cycle and therefore increases sensitivity and speed compared to conventional PCR. A dual-labeled probe of 22 nucleotides was used, which has a black hole quencher two (BH2) at the 3′-ends and a fluorophore at the 5′-end (Cy5). When there is no target present, the two ends of the probe come close to each other in a random coil, resulting in quenching of the fluorophore. Once there is template present for the probe to bind to, the distance between the quencher and fluorophore increases, allowing fluorescence. See Harris et al. (2013) for detailed diagrams comparing the amplification steps of conventional PCR and PCDR.

7.2 Real-Time PCR

7.2.1 Introduction

Real-time PCR (Rti-PCR) refers to the detection of PCR-amplified target DNA (amplicons) usually after each PCR cycle. The signal is readily followed on a computer screen where each point is automatically plotted and the extent of amplification is followed as a continual direct graphical plot (Figure 7.4). Computer software handles all of the preprogrammed calculations and plotting of data.

7.2.2 Advantages of Rti-PCR

Rti thermocyclers incorporating capillary sample air heating and cooling systems and thermoelectrically controlled blocks have greatly reduced ramp times. The use of shortened target DNA sequences (60–70 bp) in Rti-PCR results in more efficient amplification than standard PCR where amplicons are required to be at least 200 bp in length to allow detection by electrophoretic separation and also results in reduced extension times. Short PCR product yields are significantly improved by lower denaturation temperatures (Yap and McGee, 1991) so that a denaturation temperature of 90°C instead of 95°C is preferred. This also reduces ramp time. In addition, conventional PCR requires visualization of amplified products after AGE that usually involves additional 30–60 min. Rti-PCR eliminates this step through the use of a fluorometer built into the Rti-PCR thermal cycler that measures the intensity of fluorescence after each amplification cycle. Conventional PCR and the use of AGE is now frequently replaced with Rti-PCR systems by programmed generation of a thermal denaturation curve of the amplified product after the PCR that allows automatic calculation of the thermal denaturation temperature (T_m) of the amplified product value of the amplicon when SYBR green is the fluorescent reporter molecule. This is most useful in confirming the identity of a Rti-PCR amplicon.

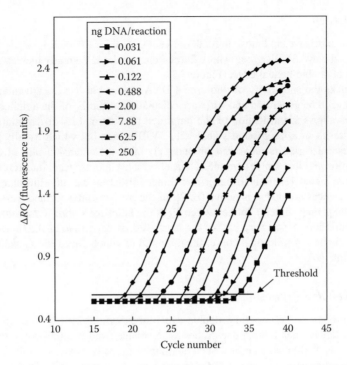

FIGURE 7.4 Appearance of Rti-PCR screen display of thermal cycling plots of target DNA amplification. Decreasing the quantity of target DNA results in an increased number of cycles required to detect fluorescence above the threshold level (shift of the plot to the right).

The quantitative range with conventional PCR is no more than four log cycles, whereas with Rti-PCR, an operational range of at least six to eight log cycles is usually achieved. Conventional thermocyclers can presently be acquired for $2000–$3000. In contrast, Rti-PCR systems presently range in price from $20,000 to $40,000. Rti-PCR can be performed with 16-well or up to 384-well microplates or with individual PCR tubes. Because of the significantly increased cost of reagents, particularly the fluorescent probes and dyes compared to reagents used with conventional PCR, the reaction volume is usually reduced from 50–100 mL to 10–20 mL.

In addition to detection of amplicons, Rti-PCR units can quantify amplified target DNA and differentiate alleles (determine point mutation or sequence variation). Allelic variation is assessed on the basis of T_m variations derived from analysis of melting curves of duplexes formed by fluorescent probes and amplicons. In addition, two or more different PCRs (multiplex PCR) amplifying different target sequences can be followed and quantified in the same PCR tube or well. Cockerill and Uhl (2002) extensively discussed the advantages of Rti-PCR.

Rti-PCR has great potential for the meat industry where massive recalls continue due to the presence of *Escherichia coli* O157:H7 and *Listeria monocytogenes* in various meat products. More recently, costly recalls in vegetables due to *E. coli* O157:H7 and outbreaks of salmonellosis from tomatoes have occurred. The present state of Rti-PCR technology should allow processors to detect product contamination from the production line in near *real time* to reduce such massive recalls by allowing detection prior to shipment of product.

7.2.3 Mechanisms of Rti-PCR

Rti-PCR depends on the emission of a UV-induced fluorescent signal that is proportional to the quantity of DNA that has been synthesized. Several fluorescent systems have been developed for this purpose and are discussed in the following.

7.2.3.1 SYBR Green

The simplest, least expensive, and most direct fluorescent system for Rti-PCR involves the incorporation into the PCR vial, the dye SYBR green whose fluorescence under UV greatly increases when bound to the minor groove of double-helical DNA (Figure 7.5).

SYBR green lacks the specificity of fluorescent DNA probes but has the advantage of allowing a DNA melting curve to be generated and software calculation of the T_m of the amplicon after the PCR (Figure 7.6). This allows identification of the amplified product and its differentiation from primer dimers that also result in a fluorescent signal with SYBR green but which usually have a lower T_m value. The fluorescent signal is measured immediately after the extension step of each cycle since thermal denaturation yielding single-stranded DNA eliminates fluorescence. Interference of the amplicon's signal by the signal resulting from primer-dimer formation can be eliminated by raising the temperature to a critical point that is above the T_m of the primer dimer formed (resulting in thermal denaturation of the primer dimers) but below the T_m of the amplicons prior to measuring the intensity of fluorescence emission. A software plot of the negative first derivative of the thermal denaturation plot yields a bell-shaped symmetrical curve, the midpoint of which yields the T_m value for the amplified product (Figure 7.6).

7.2.3.2 Dual-Labeled Probes

The TaqMan™ probes are proprietary double-dye probes synthesized by Perkin Elmer Applied Biosystems. A variety of such double-dye probes are available from a number of commercial sources and are referred to as dual-labeled probes. Such dual-labeled probes make use of the 5′–3′-exonuclease activity of *Taq* polymerase to produce a fluorescent signal. A custom-synthesized dual-labeled probe is incorporated into the PCR containing a sequence of nucleotides homologous to a specific nucleotide sequence of one strand of the amplicon internal to both primers. The probe harbors a fluorophore (reporter dye) such as 6-carboxyfluorescein (FAM) as the reporter dye at the 5′-end and 6-carboxytetramethyl-rhodamine (TAMRA) as the quenching dye at the 3′-ends or alternatively a *black hole quencher* at the 3′-ends (Figure 7.7), which are close enough to prevent emitted fluorescence of the reporter. A phosphate molecule is usually attached to the terminal 3′-thymine residue to prevent extension of the bound probe during amplification. The fluorescent emission spectrum of FAM is 500–650 nm. The fluorescent intensity of the quenching dye TAMRA changes very little over the course of PCR amplification.

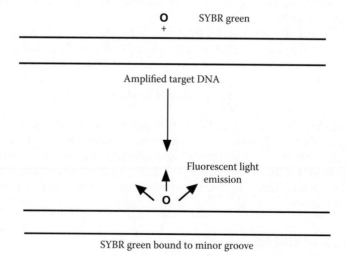

FIGURE 7.5 Mechanism of SYBR green fluorescence. SYBR green dye binds to the minor groove of the DNA double helix. The unbound dye emits little fluorescence that is enhanced eightfold when bound to double-stranded DNA.

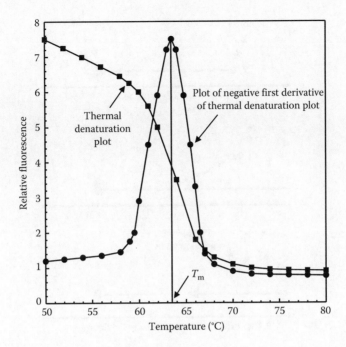

FIGURE 7.6 Thermal denaturation curve of double-stranded amplified target DNA with SYBR green I as reporter dye. As the double helix is denatured to the single-strand state, increasing amounts of SYBR green dissociate from the double-stranded DNA resulting in a linear decrease in fluorescence. A bell-shaped curve results from plotting the negative first derivative of the thermal denaturation plot with the apex coincident with the T_m value.

The intensity of TAMRA dye emission therefore serves as an IS with which to normalize the reporter (FAM) emission variation. Following each thermal denaturation step, the temperature is lowered to allow annealing of the probe to single-stranded amplicons. Increasing amounts of the single-stranded amplicons will bind increasing amounts of the probe. During primer extension, *Taq* polymerase cleaves the probe from the 5′ to the 3′ direction releasing the reporter dye which then emits fluorescence as a result of its increased distance from the quencher. Fluorescence is then measured following each extension stage of every cycle.

Black hole quenchers are molecular species that exhibit no inherent fluorescence themselves that are used in conjunction with fluorogenic reporter dyes with dual-labeled probes. Primer dimers are not detected by any of these dual-labeled probes that constitute an additional advantage in their use.

7.2.3.3 Fluorescent Resonance Energy Transfer

Fluorescence resonance energy transfer (FRET) involves the incorporation of two different oligonucleotide probes into the PCR. One probe (light donor) harbors a fluorescein label at its 3′-ends, and the other probe (light acceptor) is labeled with LightCycler Red 640 at its 5′-end. The sequences of the two probes are selected so that they can hybridize to the same strand of an amplicon in a head-to-tail orientation internal to both primers resulting in the two fluorescent dyes coming into close proximity to each other. Under UV, the fluorescein emits a green light that then excites the Red 640 dye because of their close proximity which in turn emits a red light that is proportional to the amount of amplicon present. The red emission is measured at 640 nm. This energy transfer is referred to as FRET and occurs when no more than one to five nucleotides separate the two dyes. Fluorescence is measured after each annealing step of every cycle. After annealing, the temperature is raised and the hybridization probes are displaced by the *Taq* polymerase. Primer dimers are not detected (see Levin, 2004, for a FRET diagram).

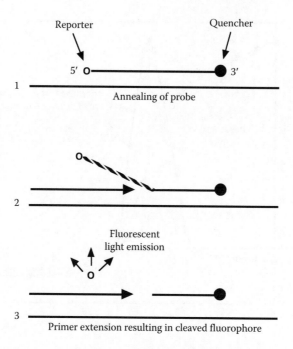

FIGURE 7.7   Mechanism of dual-labeled fluorescence probes. The 5′ nuclease activity of *Taq* DNA polymerase is utilized to cleave a dual-labeled probe during DNA amplification. (1) The probe is annealed to the target DNA sequence. The probe contains a reporter dye (O) at the 5′-end of the probe and a quencher dye (●) at the 3′-ends of the probe. (2) During PCR amplification, a complementary strand of DNA is synthesized, and the 5′ exonuclease activity of *Taq* polymerase excises the reporter dye. (3) Fluorescence of the reporter dye (O) occurs when it is separated from the quencher dye, and the intensity of fluorescence is measured.

An interesting variation of FRET involves the use of a single probe labeled at the 5′-end with a reporter dye LCRed64 or Cy5 and the addition of SYBR green I to the PCR mix (Loh and Yap, 2002). The labeled probe hybridizes to the homologous sequence of the amplified target strand, and then SYBR green I binds to the resulting double-stranded DNA so as to excite the reporter dye.

7.2.3.4 Molecular Beacons™

Molecular Beacons is the proprietary trade name of custom-synthesized nucleotide probes available from Stratagene with GC-rich complementary terminal nucleotides that form a hairpin configuration with a hybrid stem. A variety of such probes are available from a number of commercial sources. The probes harbor a reporter dye at one end and a quencher dye at the other end. The reporter dye or fluorophore is close to the quencher dye and is nonfluorescent in the hairpin configuration. The energy taken up by the fluorophore is transferred to the quencher and released as heat rather than being emitted as light. When the probe hybridizes to its homologous nucleotide sequence of the amplicons, a rigid double helix is formed that separates the quencher from the reporter dye resulting in emission of UV-induced fluorescence. Fluorescence is then measured after the annealing step of each cycle (see Levin, 2004, for a molecular beacon diagram).

7.2.4 Theory of Quantitative Real-Time PCR

The Rti-PCR can be divided into four characteristic sequential phases: phase one consists of a low level of exponential amplification that is masked by the background or threshold fluorescence; phase two consists of exponential amplification detectable above the threshold fluorescence; phase three consists of linear amplification efficiency and results in a steep increase in fluorescence; and phase four or the plateau phase consists of a decrease in the exponential rate of DNA amplification. During the plateau of DNA amplification, there is no direct quantitative relationship between the initial number of input target sequences and the amount of amplified DNA or level of fluorescence. The Ct value is used for quantitation of in-put target DNA and is defined as the number value that is defined as the number of cycles resulting in the amplification plot transecting the baseline fluorescence plot. The baseline plot is usually defined as 10× the standard deviation of the level of background fluorescence (Figure 7.4).

Quantitative real-time PCR (QRti-PCR) requires the design and use of proper controls for quantitation of the initial target sequences. Heid et al. (1996) was the first to develop such QRti-PCR methodology and made use of the Taqman dual-labeled probe reaction with FAM as the reporter dye and TAMRA as the quencher. The software calculates a value termed ΔRn or ΔRQ from the following: $\Delta Rn = (Rn^+) - (Rn^-)$, where Rn^+ = emission intensity of reporter/emission intensity of quencher at any given time in a reaction tube and Rn^- = emission intensity of reporter/emission intensity of quencher measured prior to PCR amplification in the same reaction tube. The ΔRn mean values are plotted on the Y-axis, and the number of cycles is plotted on the X-axis. During the early cycles, the ΔRn remains at baseline. When a sufficient amount of hybridization probe has been cleaved by the 5-nuclease activity of *Taq* polymerase, the intensity of reporter fluorescence emission increases. A threshold level of emission above the baseline is selected, and the point at which the amplification plot crosses the threshold is defined as Ct and is reported as the number of cycles at which the log phase of product accumulation is initiated (Figure 7.4). By setting up a series of wells containing a 4–5 log span of genomic DNA concentration (each concentration in triplicate wells), a series of amplification plots is generated by the software in real time (Figure 7.4). The amplification plots shift to the right as the quantity of input target DNA is reduced. Note that the flattened slopes and early plateaus do not influence the calculated Ct values. The Ct values decrease linearly with the log of increasing target quantity. A plot of the resulting Ct values on the Y-axis versus the log of the *ng* of input genomic DNA (or CFU) yields a straight line (Figure 7.8) that is then used as a standard curve for quantitation of samples with unknown levels of genomic DNA. This approach and the original nomenclature of Heid et al. (1996) have been universally adapted for QRti-PCR. However, this methodology does not address the issue that PCR inhibitors may be present in a DNA sample derived from a complex food product.

The detection of PCR inhibition and normalization of Ct values can be accomplished with the use of either of two methods: (1) splitting the sample into two parts, where one portion is subjected to QRti-PCR and the second portion is used to amplify an external control standard, and (2) quantitative comparative PCR using a normalization or *housekeeping* gene contained within the sample for RTi-PCR for correction of the observed Ct values. If equal amounts of nucleic acid are analyzed for each sample and if the amplification efficiency is identical for each sample, then the internal control (normalization gene or competitor) should give equal signals for all samples (Heid et al., 1996).

7.3 Isothermal Nucleic Acid Amplification

7.3.1 Introduction

Conventional and Rti-PCR methods require a thermal denaturation step at the end of each amplification cycle. The critical factor in isothermal amplification is the requirement for achieving separation of double-stranded DNA isothermally. Isothermal DNA amplification methods fall into one of two mechanistic categories, depending on whether they utilize enzyme-mediated denaturation or self-denaturation based on specially designed primers and probes. With a typical enzyme-based method, an RNA-degrading enzyme is used to initiate the displacement event by degrading the RNA region of a DNA–RNA heteroduplex. In the case of nucleic acid sequence-based amplification (NASBA), RNA regions are completely digested by RNase H so as to generate single-stranded DNA that can be annealed with another primer (Compton, 1991). In contrast, only partial RNA regions are cleaved to

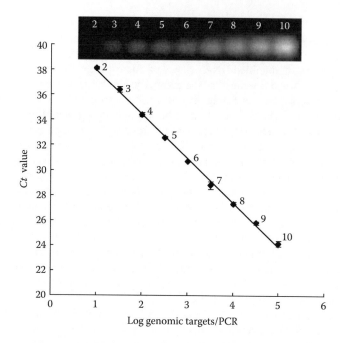

FIGURE 7.8 Relationship between the log number of genomic targets and relative Ct values from QRti-PCR amplification. Plotted values are the means and standard deviations derived from three independent assays. The straight line calculated by linear regression yielded a regression coefficient (R^2) of 0.998. Inset: image of Rti-PCR-amplified products of mixed bacterial broth culture from cod tissue. Lanes and corresponding plotted points designated 2–9: DNA from varying numbers of CFU per Rti-PCR (#2–9: 1.0×10^1, 3.2×10^1, 1.0×10^2, 3.2×10^2, 1.0×10^3, 3.2×10^3, 1.0×10^4, 3.2×10^4, and 1.0×10^5, respectively). Universal primers were used. (From Lee, J.-L. and Levin, R.E., *J. Fish. Sci.*, 1, 58, 2007.)

induce the displacement of one DNA strand in single-primer isothermal amplification (SPIA) (Kurn et al., 2005) and isothermal chimeric primer-initiated amplification of nucleic acids (ICAN) methods (Mukai et al., 2007). In certain enzyme-based methods, an RNase is used to initiate the displacement event by nicking or degrading the RNA region of *chimeric* DNA–RNA annealed primers. In addition, helicase enzymes can be used to unwind and separate double-stranded DNA so as to generate single-stranded templates isothermally (Vincent et al., 2004). In addition to using enzymes to promote denaturation during amplification, an autocycling denaturation process that does not rely on additional enzyme activities can occur with the use of specially designed primers and probes. Rolling circle amplification (RCA) utilizes a 5′–3′ exonuclease-deficient DNA polymerase to extend a primer on a circular template. This produces tandemly linked copies of the complementary template sequence (Fire and Xu, 1995; Lizardi et al., 1998). Loop-mediated amplification (LAMP) (Notomi et al., 2000) also employs a 5′–3′ exonuclease-deficient DNA polymerase and a set of four or six primers that recognize template sequences.

7.3.2 Detection of LAMP-Amplified DNA Products

A number of mechanisms are available for visual detection and quantification of LAMP-amplified products. These include the following: (1) visual observation of turbidity, (2) the addition of calcein plus Mn^{++} to LAMP reaction mixtures after amplification, (3) the addition of the metal ion indicator hydroxy naphthol blue (HNB) to reaction mixtures after amplification, (4) the incorporation of SYBR green into reaction mixtures, (5) the addition of ethidium bromide into LAMP reaction mixtures after amplification, (6) the addition of a FRET probe to reaction mixtures, (7) the addition of propidium iodide to reaction mixtures after amplification, and (8) the use of AGE and casting the gel with the addition of the fluorescent DNA dye ethidium bromide or the noncarcinogenic fluorescent DNA dye Midori green.

7.3.3 LAMP Primer Design

Several software programs are presently available online for the development of LAMP primers and include

1. LAMP designer (Premier Biosoft, International, Palo Alto, CA) http://www.premierbiosoft. com/tech_notes/Loop-Mediated-Isothermal-AmpliŽcation.html (Premier Biosoft, 2013).
2. Eiken Chemical Co., Ltd., Tokyo, Japan. PrimerExplorer V4 software http://primerexplorer. jp/e/v4_manual/index.html (Eiken Chemical, 2013).

Tomita et al. (2008) have emphasized that the following should be kept in mind in the design of LAMP primers: (1) both ends of the IPs should not be AT-rich, (2) the T_m value for each domain should be ~55°C–65°C, (3) the distance from the 5′-end of F2 to the 5′-end of F1 and the 5′-end of B2 to the 5′-end of B1 sites should be 40–60 bp, (4) the length of the amplified DNA region (from F2 to B2 sites inclusively) should not be >200 bp, and (5) HPLC-purified primers are recommended.

7.3.4 Calcein Plus Mn⁺⁺ for Visualization of LAMP-Amplified DNA Products

In the LAMP reaction solution, a huge amount of the amplification by-product pyrophosphate ion is produced. Tomita et al. (2008) described the use of Mn^{++} in conjunction with calcein as a fluorescence indicator of DNA amplification in LAMP reactions. Calcein strongly fluoresces under UV when forming chelation complexes with divalent metallic ions such as Ca^{++} and Mg^{++}. When Mn^{++} ions complex with calcein, fluorescence is quenched. Before the amplification reaction, calcein is combined with Mn^{++} ions, and UV fluorescence is quenched and the reaction solution is orange. Newly generated pyrophosphate binds preferably with Mn^{++}, and calcein then combines with Mg^{++} ions, resulting in enhanced fluorescence.

7.3.5 LAMP

LAMP utilizes a DNA polymerase isothermally at 60°C–65°C, a set of four primers that recognize a total of six distinct sequences on the target DNA. An IP containing sequences of the sense and antisense strands of target DNA initiates LAMP. The resulting strand displacement DNA synthesis primed by an outer primer releases a single strand of DNA. This serves as a template for DNA synthesis primed by the second inner and outer primers that hybridize to the other end of the target, which produces a stem-loop DNA structure. In subsequent LAMP cycling, one IP hybridizes to the loop on the product and initiates displacement DNA synthesis, yielding the original stem-loop DNA and a new stem-loop DNA with a stem twice as long. The cycling reaction continues with accumulation of ~10^9 copies of target DNA in less than 60 min. The final products are stem-loop DNA (see Premier Biosoft, 2013 and Notomi et al., 2000, for detailed diagrams of LAMP reactions.)

The presence of betaine or L-proline has been found to stimulate the rate of amplification and selectivity (Notomi et al., 2000). The identity of the DNA polymerase is a critical factor for efficient amplification. Bst polymerase was found best. Bst polymerase (derived from *Bacillus stearothermophilus*) has a helicase-like activity, making it able to unwind DNA strands and facilitating displacement. In addition, it lacks 5′–3′ DNase processing activity and will therefore not hydrolyze previously synthesized DNA strands it is displacing, which is an absolutely essential property of the enzyme. Its optimum functional temperature is between 60°C and 65°C, and it is denatured at temperatures above 70°C. These features make it useful in LAMP. The method is capable of yielding an unusually large amount of DNA, more than 500 mg/mL. At the completion of amplification, a white precipitate of magnesium pyrophosphate forms (Mori et al., 2001) that can be used to confirm amplification.

The yield of synthesized DNA can be quantified by measuring the intensity of fluorescence using ethidium bromide in agarose gels or in a Rti-PCR unit or alternatively by measuring the turbidity. Specificity of amplification can be readily distinguished from nonspecific amplification by different banding patterns on agarose gels. The final amplification products are mixtures of many different sizes of loop DNA, with several inverted repeats of the target sequence and cauliflower-like structures with multiple loops.

The LAMP reaction results in large amounts of pyrophosphate ion by-product. These ions react with Mg^{++} ions to form the insoluble product magnesium pyrophosphate. Since the soluble Mg^{++} ion concentration decreases as the LAMP reaction progresses, the LAMP reaction can be quantified by measuring the Mg^{++} ion concentration still remaining in the reaction solution. On the basis of this phenomenon, Goto et al. (2009) developed a simple colorimetric assay for detecting positive LAMP reactions based on the metal ion indicator HNB. The addition of 120 mM HNB to the LAMP reaction results in a color change from violet to sky blue as amplification proceeds. At 650 nm, the absorbance increases as the Mg^{++} concentration decreases. The assay is ideally suited for a microplate reader and an isothermal temperature of ~63°C. The reaction however could not be quantified on the basis of absorbance at 650 nm, although positive reactions were readily detected both visually and by a microplate reader. An excellent and complete colored animation of LAMP is available online (Eiken Chemical, 2013).

Nagamine et al. (2002) developed a method that accelerates the four-primer LAMP reaction system by using two additional primers, designated loop primers. Loop primers hybridize to the stem-loops, except for the loops that are hybridized by the IPs, and primer strand displacement DNA synthesis. The use of two such additional loop primers was found to reduce the reaction time of the four other LAMP primers by more than half. Most subsequently, developed LAMP assays thereafter have incorporated two loop primers so that the total number of LAMP primers now used is usually six.

7.3.6 LAMP Assay Detection of *Salmonella enterica* Serotypes

Hara-Kudo et al. (2005) reported on the development of a LAMP assay for *S. enterica* in whole egg enrichments that targeted the highly conserved *invA* gene of *S. enterica* serotypes. A total of 227 *Salmonella* strains representing 40 serotypes were detected while 62 strains of 23 bacterial species other than *Salmonella* were negative. The assay used six primers, two IPs, two outer primers, and two loop primers (Table 7.3). The LAMP reaction was performed at 65°C for 60 min. Bst polymerase was used for amplification. Amplified DNA was detected by fluorescence using the intercalating dye YO-PRO-1 iodide and by visual observation of turbidity. The level of detection was the DNA from ~2 CFU/LAMP assay (400 CFU/mL of enrichment). The level of detection with conventional PCR was ~16 CFU (3200 CFU/mL of enrichment).

Wang et al. (2008) reported on the development of a LAMP assay specific for *S. enterica* serotype targeting the *invA* gene. A temperature of 65°C for 60 min was used. The assay was 10-fold more sensitive than conventional PCR detecting 100 fg of DNA/reaction (DNA from ~20 CFU)/reaction. A set of two IPs and two outer primers (Table 7.3) were used to recognize six distinct sequences on the target DNA. The IPs are described as forward IP and backward inner primer (BIP). The forward IP consisted of the complementary sequence of F1 (22 nt), a T–T–T–T linker, and F2 (20 nt). The BIP consisted of BIC (21 nt), a T–T–T–T linker, and the complementary sequence of B2C (20 nt). The outer primers were F3 and B3, which are located outside of the F2 and B2 regions. The authors presented a detailed diagram of the primer locations and their design.

7.3.7 Isothermal and Chimeric-Primer Amplification of Nucleic Acids

Mukai et al. (2007) developed an efficient method of isothermally amplifying DNA termed ICAN. This method allows the amplification of targeted DNA under isothermal conditions at ~55°C for 60 min. The method requires a pair of 5′-DNA–RNA-3′ chimeric primers that possess ~3 RNA (ribose nucleotide residues) at the 3′-ends, thermostable RNase H that can cleave the RNA portion of the

TABLE 7.3

LAMP Primers

Genus, Species, or Strain		Primers (5′–3′)	Gene	References
Listeria monocytogenes	FIP	CGT-GTT-TCT-TTT-CGA-TTG-GCG-TCT- TTT-TTT-CAT-CCA-TG-CAC-CAC-C	*hlyA*	Tang et al. (2011)
	BIP	CCA-CGG-AGA-TGC-AGT-GAC-AAA-TGT- TTT-GGA-TTT-CTT-CTT-TTT-CTC-CAC-AAC		
	F3	TTG-GGC-AC-AAA-CTG-AAG-C		
	B3	GCT-TTT-ACG-AGA-GCA-CCT-GG		
	LF	TAG-GAC-TTG-CAG-GCG-GAG-ATG		
	LB	GCC-AAG-AAA-AGG-TTA-CAA-AGA-TGG		
L. monocytogenes	FOP (F3)	ATC-ACT-CTG-GAG-GCT-ACG	*hlyA*	Wang et al. (2011)
	BOP (B3)	TTC-CCA-CCA-TTC-CCA-AG		
	FIP	TGA-ACA-ATT-TCG-TTA-CCT-TCA-GGA-TTG- CTC-AAT-TCA-ACA-TCT-CTT		
	BIP	TAG-CTC-ATT-TCA-CAT-CGT-CCA-TCA-GTG- CAT-TCT-TTG-GCG-TAA		
	BLP	TTG-CCA-GGT-AAC-GCA-AGA-AAT		
Escherichia coli (STEC)	stx1-F3	TGA-TTT-TTC-ACA-TGT-TAC-CTT-TC	*stx1* (*vt1*)	Wang et al. (2012)
	stx1-B3	TAA-CAT-CGC-TCT-TGC-GAC		
	stx1-FIP	CCT-GCA-ACA-CGC-TGT-AAC-GTC-AGG-TAC- AAC-AGC-GGT-TA		
	stx1-BIP	AGT-CGT-ACG-GGG-ATG-CAG-ATA-GTG-AGG- TTC-CAC-TAT-GC		
	stx1-LF	GTA-TAG-CTA-CTG-TCA-CCA-GAC-AAT-G		
	stx1-LB	AAA-TCG-CCA-TC-GTT-GAC-TAC-TTC-T		
	stx2-F3	CGC-TTC-AGG-CAG-ATA-CAG-AG	*stx2* (*vt2*)	
	Ssstx2-B3	CCC-CCT-GAT-GAT-GGC-AAT-T		
	stx2-FIP	TTC-GCC-CCC-AGT-TCA-GAG— TGA-GTC-AGG- CAC-TGT-CTG-AAA-CT		
	stx2-BIP	TGC-TTC-CGG-AGT-ATC-GGG-GAG-CAG-TCC- CCA-GTA-TCG-CTG-A		

(Continued)

TABLE 7.3 (Continued)

LAMP Primers

Genus, Species, or Strain	Primers (5'–3')	Gene	References
stx2-LF	GCG-TCA-TCG-TAT-ACA-CAG-GAG-C		
stx2-LB	GAT-GGT-GTC-AGA-GTG-GGG-AGA-A		
eae-F3	TGA-CTA-AAA-TGT-CCC-CGG	eae	
eae-B3	CGT-TCC-ATA-ATG-TTG-TAA-CCA-G		
eae-FIP	GAA-GCT-GGC-TAC-CGA-GAC-TCC-CAA-AAG-CAA-CAT-GAC-CGA		
eae-BIP	GCG-ATC-TCT-GAA-CGG-GGA-TTC-CTG-CAA-CTG-TGA-CGA-AG		
eae-LF	GCC-GCA-TAA-TTT-AAT-GCC-TTG-TCA		
eae-LB	ACG-CGA-AAG-ATA-CCG-CTC-T		
E. coli (STEC) VT1-FIP	GCT-CTT-GCC-ACA-GAC-TGC-ACA-TTC-GTT-GAC-TAC-TTC-TTA-TCT-GG	vt1 (stx1)	Hara-Kudo et al. (2007)
VT1-BIP	ctg-TGA-CAG-CTG-AAG-CT-TAC-GCG-AAA-TCC-CCT-CTG-AAT-TTG-CC		
VT1-F3	GCT-ATA-CCA-CGT-TAC-AGC-GTG		
VT1-B3	ACT-ACT-CAA-CCT-TCC-CCA-GTT-C		
VT1-Loop F	AGG-TTC-GGC-TAT-GCG-ACA-TTA-AAT		
VT2-FIP	GCT-CTT-GAT-GCA-TCT-CTG-GTA-CAC-TCA-CTG-GTT-TCA-TCA-TAT-CTG-G		
VT2-BIP	CTG-TCA-CAG-CAG-AAG-CCT-TAC-GGA-GGA-AAT-TCT-CCC-TGT-ATC-TGC-C		
VT2-F3	CAG-TTA-TAC-CAC-TCT-GCA-ACG-TG		
VT2-B3	CTG-ATT-CGC-CGC-CAG-TC		
VT2-Loop F2	TGT-ATT-ACC-ACT-GAA-CTC-CAT-TAA-CG		
VT2-Loop F2	GC-ATT-TCC-ACT-AAA-CTC-CAT-TAA-CG		
E. coli (STEC) STX1-F3	GAA-CTG-GGG-AAG-GTT-GAG	stx1 (vt1)	Sasitharan et al. (2013)
STX1-B3	CAC-GGA-CTC-TTC-CAT-CTG		
STX1-FIP	TCC-CAG-AAT-TGC-ATT-AAT-GCT-TCC-GTC-CTG-CCT-GAT-TAT-CAT-GG		

E. coli (Commensurate)	STX1-BIP	AGC-GTG-GCA-TTA-ATA-CTG-AAT-TGT-ACA-TAG-AAG-GAA-ACT-CAT-CAG-AT	malB	Hill et al. (2008)
	STX1-LF	TCT-TCC-TAC-ATG-AAC-AGA-GTC-TTG-T		
	STX1-LF	CAT-CAT-CAT-GCA-TCG-GGA-GTT		
	FOP F3	GCC-ATC-TCC-TGA-TGA-CGC		
	BOP B3	ATT-TAC-CGC-AGC-CAG-ACG		
	BIP (B1+B2c)	CTG-GGG-CGA=GGT-CGT-GGT-ATT-CCG-ACA-AAC=ACC-ACG-AAT-T		
	FIP(1c+F2)	CAT-TTT-GCA-GCT-GTA-CGC-TCG-CAG-GCC-ATC-ATG-AAT-GTT-GCT		
	Loop F	CTT-TGT-AAC-AAC-CTG-TCA-TCG-ACAS		
	LOOP B	ATC-AAT-CTC-GAT-ATC-CAT-GAA-GGT-G		
Salmonella enterica	FIP	GAC-GAC-TGG-TAC-TGA-TCG-ATA-GTT-TTT-CAA-CGT-TTC-CTG-CGG	invA	Hara-Kudo et al. (2005)
	BIP	CCG-GTG-AAA-TTA-TCG-CCA-CAC-AAA-ACC-CAC-CGC-CAG-G		
	F3	GGC-GAT-ATT-GGT-GTT-TAT-GGG-G		
	B3	AAC-GAT-AAA-CTG-GAC-CAC-GG		
	Loop F	GAC-GAA-AGA-GCG-TGG-TAA-TTA-AC		
	Loop B	GGG-CAA-TTC-GTT-ATT-GGC-GAT-AG		
Salmonella enterica[a]	OF	CCT-GAT-CGC-ACT-GAA-TAT-C	invA	Kim et al. (2011)
	OR	CGA-AAG-AGC-GTG-GTA-ATT-AAC		
	IF	CGA-TGA-CTG-ACT-ATA-CAA-GUA-CGC-TGG-CGA-TAT-TGG-TGT-TTA-TG		
	IR	CTA-GTA-CAT-GAA-GCT-AAA-GAC-CGC-AGG-AAA-CGT-TGA-A		
Salmonella enterica	FRET Probe	FAM-CGT-TCT-ACA-TTG-ACA-GAA-TCC-TCA-G-DABCYL	invA	Wang et al. (2008)
	FIP F2	CCC-AGA-TCC-CCG-CAT-TGT-TGA-TTT-TTC-CGC-CCC-ATA-TTA-TCG-CTA-T		
	BIP B2	GAC-CAT-CAC-CAA-TGG-TCA-GCA-TTT-TAT-TGG-CGG-TAT-TTC-GGT-GGG		
	FOP	GTT-CAA-CAG-CTG-CGT-CAT-GA		
	BOP	CGT-TAT-TGC-CGG-CAT-CAT-TA		

(Continued)

TABLE 7.3 (*Continued*)

LAMP Primers

Genus, Species, or Strain		Primers (5′–3′)	Gene	References
Campylobacter jejuni	CJ-FIP (F1c-F2)	ACA-GCA-CCG-CCA-CCT-ATA-GTA-GAA-GCT-TTT-TTA-AAC-TAG-GGC	cj0414	Yamazaki et al. (2008)
	CJ-BIP (9B1-B2c)	AGG-CAG-CAG-AAC-TTA-CGC-ATT-GAG-TTT-GAA-AAA-ACA-TTC-TAC-CTC-T		
	CJ-F3 (F3)	GCA-AGA-CAA-TAT-TAT-TGA-TCG-C		
	CJ-B3 (B3c)	CTT-TCA-CAG-GCT-GCA-CTT		
	CJ-LF (LFc)	CTA-GCT-GCT-ACT-ACA-GAA-CCA-C		
	CJ-LB (LB)	CAT-CAA-GCT-TCA-CAA-GGA-AA		
Campylobacter coli	CC-FIP (F1c-F2)	AAG-AGA-TAA-ACA-CCA-TGA-TCC-CAG-TCA-TGA-ATG-AGC-TTA-CTT-TAG-C		
	CC-BIP (B1-B2c)	GCG-GCA-AAG-ACT-TAT-GAT-AAA-GCT-ACC-GCC-ATT-CCT-AAA-ACA-AG		
	CC-F3 (F3)	TGG-GAG-CGT-TTT-TGA-TCT		
	CC-B3 (B3c)	AAT-CAA-ACT-CAC-CGC-CAT		
	CC-LF (LFc)	CCA-CTA-CAG-CAA-AGG-TGA-TG		
	CC-LB (LB)	CCA-CGA-TAG-CCT-TTA-TGG-A		

^a Underlined nucleotides are ribonucleotides (RNA).

elongated strand, and a DNA polymerase such as BcaBEST that lacks 5′–3′-exonucleaes activity with strong strand displacing activity. The ICAN reaction mechanism is based on the ability of RNase H to introduce a nick at the 5′-RNA and DNA-3′ junction of an elongated strand synthesized from the chimeric primer. The DNA polymerase then amplifies the region encompassed by both primers. Since BcaBEST DNA polymerase lacks 5′–3′ exonuclease activity, it is able to initiate DNA elongation with concomitant strand displacement from a nicked site at every cleavage of the 3′-ends of the RNA portion of the elongated strand. The complementary strand is then synthesized starting from another chimeric primer on the displaced single-stranded DNA. This displaced strand can then serve as a template. Three distinct bands were observed by AGE. One band (large) possessed both of the primer sequences, another band (midsized) possessed only one primer (forward or reverse), and the third band (small) possessed no primer sequence. The authors concluded that two amplification mechanisms were involved *multipriming* and *template switching* (see Mukai et al., 2007, for diagrams of these two mechanisms). The ICAN technique yielded ~7× greater amplification than the PCR when comparable DNA primers were used.

7.3.8 Isothermal Chain Amplification

Jung et al. (2010) developed a new and highly sensitive and specific isothermal chain amplification (ICA) system as an improvement of a combination of LAMP (utilizing outer and inner primers) and ICAN (the DNA–RNA chimeric primers and the RNase H concept) for isothermal amplification of targeted DNA. This isothermal target and signaling probe amplification (iTPA) system employed four primers: an outer F primer and outer R primer and an inner F primer and inner R primer in addition to a probe labeled at the 5′-end with FAM and with DABCYL at the 3′-ends. Two main displacement events are mediated by the DNA degrading activity of RNase H within an RNA–DNA heteroduplex and the strand displacement activity of DNA polymerase during DNA extension. The two different displacement events are designed to occur simultaneously and lead to two different amplification routes, thereby producing notably high levels of isothermal amplification yielding three distinguishable amplicons. The method is ideally suited for quantification of the initial number of targets and is able to detect a single gene copy. A detailed diagram of the mechanism for this iTPA mechanism is presented by Jung et al. (2010).

Kim et al. (2011) extended this iTPA system for detection of low numbers of *Salmonella* sp. by targeting the *invA* gene conserved in all serotypes of *S. enterica*. The assay employs two forward primers and two reverse primers plus a DNA–RNA–DNA chimeric FRET probe that is hybridized with the target DNA, and the RNA region of the duplex is specifically cleaved by RNase H. The FRET probe recognizes five distinct regions of the *invA* target sequence and is labeled at the 5′-end with FAM and with DABCYL at the 3′-ends. The iTPA procedure not only amplifies the target DNA but also the FRET probe, resulting in enhanced sensitivity. Samples of chicken meat and several other foods were homogenized with 225 mL of Luria-Bertani (LB) broth after seeding and were then incubated for enrichment. Enrichment samples (100 mL) were then processed for DNA extraction. Amplification reactions were incubated at 58°C for 60 min. The limit of detection was ~2 CFU/sample following nonselective enrichment. The entire assay was completed within 24 h.

7.3.9 LAMP Assay Detection of *Listeria monocytogenes*

Tang et al. (2011) reported on the development of a LAMP assay for detection of *L. monocytogenes* in poultry that targeted the *hlyA* gene that encodes the virulence factor listeriolysin O present only in *L. monocytogenes*. A set of six primers (Table 7.3) targeting eight regions of the gene was used. Incubation of the LAMP assay was at 65°C for 40 min. Amplified DNA was detected by the observation of turbidity, by the addition of calcein plus Mn++ ions to LAMP tubes after amplification, and by AGE. With calcein and Mn++ addition, the color of the LAMP solution turned green in DNA-amplified tubes and remained orange in tubes without amplification. The LAMP assay was applied to 60 chicken samples for detection of naturally contaminating *L. monocytogenes*. Samples (25 g) were homogenized with

225 mL of Half Frazer's broth followed by incubation at 30°C for 24 h. Enrichment samples (100 mL) were then inoculated into 10 mL of Frazer's broth and incubated for 24 h followed by DNA extraction, DN purification with a kit, and LAMP assay. Among the 60 samples, 7 were detected as positive for *L. monocytogenes* via culture and LAMP techniques while 5 were detected as positive via the PCR. The LAMP assay was 100-fold more sensitive than the PCR assay.

Wang et al. (2011) developed a LAMP assay for detection of *L. monocytogenes* seeded into poultry tissue. A total of five primers (Table 7.3) two outer, two inner, and one loop primer targeted the *hlyA* gene. Detection of amplified DNA was by visual observation of turbidity and AGE. The LAMP assay involved incubation at 63°C for 20 min. Chicken frankfurters (25 g) were added to 225 mL of 0.9% NaCl and homogenized with various numbers of *L. monocytogenes* followed by differential centrifugation of 10 mL. After cell lysis, a spin column was used for DNA purification. The limit of detection of the LAMP assay was ~20 CFU/g of chicken frankfurter. A comparative 35-cycle PCR assay yielded a limit of detection of ~2000 CFU/g.

7.3.10 LAMP Assay Detection of Shiga Toxin-Producing Strains of *Escherichia coli*

A LAMP assay for detection of verotoxin toxin (VT, STX)-producing strains of *E. coli* in ground beef and radish sprouts was developed by Hara-Kudo et al. (2007). The assay utilized primers for VT1 and six primers for VT2 (Table 7.3). After seeding, samples (25 g) were incubated in 225 mL of modified EC broth at 41°C for 18 h. The sensitivity of detection with pure cultures was the DNA from 0.7 to 2.2 CFU/LAMP assay. The LAMP assays were conducted at 65°C for 60 min. When 10 samples (25 g) each of ground beef and radish sprouts were inoculated with ~7 CFU/25 g and ~28 CFU/25 g, respectively, the PCR assay was unable to detect VT-producing *E. coli* from 5/10 of the seeded samples. In contrast, The LAMP assay succeeded in detecting 9/10. The LAMP assay was at least 10-fold more sensitive than a comparable PCR assay. In addition, VT-producing *E. coli* were detected in 10/10 seeded beef samples with both the PCR and LAMP assays and in 9/10 samples by selective agars. It is of notable significance that with four naturally contaminated beef samples, all were positive by the LAMP assay for VT-producing *E. coli* while selective agar cultivation following of enrichments was negative for all four samples. Detection of positive LAMP assays was by visual assessment of turbidity and by AGE.

Wang et al. (2012) developed a series of LAMP assays for detection of shiga toxin-producing strains of *Escherichia coli* (STEC) strains seeded into ground beef containing 23% fat. The LAMP assays targeted three genes, *stx1*, *stx2*, and *eae* (intimin enterocyte attachment gene). A different set of six primers was used for each of these three genes (Table 7.3). When seeded with 1–2 CFU/25 g sample, the LAMP assays (65°C for 60 min) achieved positive detection after 6–8 h of enrichment at 42°C. The LAMP assays were somewhat superior to the corresponding PCR assays.

Sasitharan et al. (2013) reported on the development of a LAMP assay for detection of *E. coli* O157:H7 strains. A set of four primers was used that targeted the *stx1* gene (Table 7.3). Incubation of LAMP assays was at 62°C for 60 min. The extent of amplification was assessed by visual observation using SYBR green or calcein. AGE indicated a similar level of detection sensitivity with LAMP and conventional PCR. However, the quantitative yield of amplified DNA was ~100-fold greater with LAMP, allowing visual confirmation of amplification.

7.3.11 LAMP Assay Detection of Commensal Strains of *Escherichia coli* as an Index of Sanitation

Presumptive tests for *E. coli* as an index of sanitation of water supplies, foods, and environmental samples require a 24 h incubation period. Hill et al. (2008) developed a LAMP assay for *E. coli* strains utilizing a total of three sets of primers (outer, loop, and inner) targeting the *malB* gene (Table 7.3). This gene is highly conserved across diverse lineages of *E. coli* and is not shared by other Gram-negative bacteria except *Shigella* spp. Incubation of LAMP reactions was at 66°C for 60 min. The assay was capable of detecting 10 gene copies per LAMP reaction tube. Amplified DNA was detected in agarose gels and by direct visual observation visually with propidium iodide or SYBR green added after amplification with and without UV enhancement.

7.3.12 LAMP Assays for Detection of *Campylobacter jejuni* and *Campylobacter coli*

Yamazaki et al. (2008) report on the development of LAMP assays for rapid detection of *C. jejuni* and *C. coli*. LAMP reactions were performed at 60°C for 60 min in a real-time turbidimeter. For detection of *C. jejuni*, six primers were developed for the putative oxidoreductase gene *cj0414* (Table 7.3). For detection of *C. coli*, six primers were developed for the putative heavy metal transporter gene *gufA* (Table 7.3). Positive amplification was assessed with the turbidimeter and by visual observation of turbidity.

Yamazaki et al. (2009) utilized the same LAMP systems for comparison with conventional culture methods to detect *C. jejuni* and *C. coli* in naturally contaminated chicken meat samples. A total of 144 samples were enriched for LAMP. The limit of sensitivity was the DNA from 7.9 and 3.8 CFU per LAMP reaction tube for *C. jejuni* and *C. coli*, respectively, which were equivalent to 213 and 103 CFU/mL of enrichment broth. Among the 144 samples, 68 were culture positive and 67 were LAMP positive for campylobacters. In addition, the LAMP assays identified 57 *C. jejuni*-positive samples, 5 *C. coli*-positive samples, and 5 samples positive for both species. The one sample that was culture positive and LAMP negative yielded a *C. jejuni* culture.

7.4 Removal of PCR Inhibitors from Foods

7.4.1 Introduction

Most foods can be expected to contain a number of potent PCR inhibitors. A number of substances in ground beef are potent inhibitors of the PCR (Abu Al-Soud and Radstrom, 2000; Bhaduri and Cottrell, 2001) and include fats, proteins, enzymes, antibiotics, organic and inorganic chemicals, and polysaccharides (Rossen et al., 1992; Wilson, 1997). Heme-containing agents such as hemoglobin and myoglobin have been found to be potent inhibitors of the PCR (Rossen et al., 1992; Akane et al., 1994; Bélec et al., 1998; Abu Al-Soud and Radstrom, 2000). In addition, ground beef has been found to contain lipids (Rossen et al., 1992) that are PCR inhibitors. In plant tissues, various compounds including polysaccharides, phenolic compounds, and chlorophyll can inhibit PCR amplification (Wilson, 1997; Lee and Levin, 2011). Chlorophyll in particular, derived from leafy vegetables, such as lettuce and spinach, is present in significant quantities and is a notably powerful inhibitor of the PCR. Shellfish, particularly oysters, contain a notably high level of DNase activity (Wang and Levin, 2010). Mechanisms involving PCR inhibition from food components include (1) binding of inhibitor to target DNA, (2) binding of inhibitor to *Taq* polymerase, (3) protease digestion of *Taq* polymerase, and (4) DNase digestion of target DNA. For these reasons, the use of the PCR for detection of food-borne pathogenic bacteria is usually preceded by enrichment cultivation that is time-consuming unless appropriate methodology is invoked for removal of large amounts of such PCR inhibitors.

A number of commercial resins and miniature spin columns and DNA purification kits are available for purifying DNA from small samples. These however are not suitable for purifying microbial DNA derived directly (without enrichment), for example, from a 25 g sample of ground beef, where the total amount of PCR inhibitors such as myoglobin and hemoglobin is quite large.

7.4.2 Use of Activated Carbon for Removal of PCR Inhibitors from Foods

A gram of activated carbon (1–2 mm particle size) reportedly contains ~800–1500 m^2 (0.2–0.37 acres) of internal surface derived from its extensive submicroscopic pores (Bansal and Goyal, 2005). This surface area is contained predominantly within micropores that have effective diameters smaller than 2 nm allowing the uptake and binding of 45% of its weight of various solutes. However, the

TABLE 7.4

Recovery of *S. enterica* ser. Enteritidis after Treatment with 4.6 g Activated Carbon Coated with 0.05 g Milk Proteins at pH 5.0 and 7.0

Initial (CFU/mL)[a]	Mean Percent Recovery (Uncoated, PB pH 7.0)[b]	Mean Percent Recovery (Coated, 0.05 M Acetate Buffer pH 5.0)[b]	Mean Percent Recovery (Coated, 0.05 M PBS pH 7.0)[b]
111	2.0 ± 0.3	97.2 ± 0.2	98.2 ± 1.3
116	0.0 ± 0.5	98.3 ± 0.4	98.3 ± 1.2
133	1.2 ± 1.0	73.1 ± 1.8	96.6 ± 1.6
115	0.5 ± 0.5	63.3 ± 5.7	88.7 ± 2.5
100	1.0 ± 1.2	65.8 ± 4.1	98.5 ± 3.5
126	1.8 ± 0.8	98.7 ± 1.2	94.0 ± 1.7
Overall mean	1.1 ± 0.8	82.7 ± 17.1	95.7 ± 2.0

Source: Opet, N. and Levin, R., *J. Microbiol. Methods*, 94, 69, 2013a.

[a] CFU/mL in a total of 30.0 mL of 0.05 M PBS (pH 7.0).

[b] Means of duplicate plate counts.

problem with the use of activated carbon for removal of PCR inhibitors from foods is twofold: (1) it will bind DNA following lysis of cells and (2) it will bind all bacterial cells quite tenaciously, resulting in their loss. To overcome this problem, Abolmaaty et al. (2007) coated activated carbon particles with cells of *Pseudomonas fluorescens*. Later, Luan and Levin (2012) used the clay bentonite (~1–2 μm particle size) to coat the carbon particles. In addition, since bentonite particles are negatively charged, they will repel bacterial cells that are also negatively charged. More recently, Opet and Levin (2013a) reported on the efficacy of coating activated carbon particles with milk proteins (Table 7.4). The pH of a cell suspension of *S. enterica* in contact with coated carbon particles was found to be a critical factor, with a pH of 7.0 resulting in a notably higher recovery of cells compared to a pH of 5.0. This pH effect is most probably due to neutralization of positive charges on the milk proteins by hydroxyl groups.

7.4.3 Use of Beta-Cyclodextrin for Removal of Fat from Ground Beef

A major obstacle to the direct application of the PCR to meat products such as ground beef is the presence of fat at a level as high as 27% or more. Bacterial cells tend to concentrate in the fat phase of such food products (Caplan et al., 2013), so that by removal of the fat, a majority of targeted bacterial cells are also removed. The use of beta-cyclodextrin has proven to be an ideal method for selective removal of fat (triglycerides) without a loss in bacterial cells. Beta-cyclodextrin is a cyclic polysaccharide compound in which seven glucose monomers are linked to form a conical toroid structure with a hollow interior (Figure 7.9a). The exterior of the beta-cyclodextrin molecule is hydrophilic. In contrast, the interior (0.62–0.78 nm diam.) that has a depth of 0.78 nm (Szejtli, 1998) is hydrophobic.

These structural features mean that the hydrophobic ends of molecules such as the methyl group (0.40 nm, diam.; Chang, 2001) of triglycerides (fats) can be captured inside the internal aspect of the beta-cyclodextrin toroid (Figure 7.9b). This results in a highly favorable sequence of physical–chemical events: (1) fat globules are disrupted, (2) the methyl groups of individual triglyceride molecules are captured by the internal aspect of the beta-cyclodextrin toroids, and (3) bacterial cells are forced into the aqueous phase so as to result in noninjurious separation of all bacterial cells from the fat content of the product. A typical protocol for ground beef of 15% fat or less that results in the PCR detection of 3 CFU of *Salmonella* per gram within 4.5 h without enrichment is given by Opet and Levin (2013b).

The major disadvantage in the use of beta-cyclodextrin is the cost. An alternative to beta-cyclodextrin is starch. However, conventional starch products such as corn starch or potato starch are not suitable because of their viscosity and rapid gelling properties. However, soluble starch at a level of 10% is quite suitable. Amylose in aqueous solution occurs as an alpha helix having an outside diameter of 0.97 nm

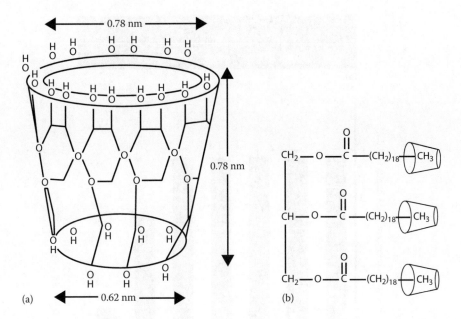

FIGURE 7.9 (a) Toroid structure of beta-cyclodextrin. (b) Envisioned interaction of triglyceride (fat molecule) with beta-cyclodextrin.

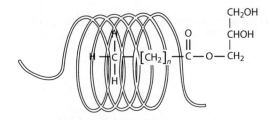

FIGURE 7.10 Alpha helix of amylose and starch–lipid inclusion. Each turn consists of six glucose molecules. The internal channel is hydrophobic and allows the entry of a terminal methyl group and chain of an individual fatty acid of a triglyceride to enter.

and an inner hydrophobic core of 0.74 nm (Lopez et al., 2012) that is more than large enough to accommodate the entry of the terminal methyl group of an individual fatty acid derived from a triglyceride (Figure 7.10).

7.5 PCR Amplification of DNA Sequences Only from Live Cells

7.5.1 Introduction

Conventional PCR, Rti-PCR, and LAMP will amplify target sequences from DNA derived from both live and dead cells. Nogva et al. (2003) were the first to introduce the use of a DNA-intercalating molecule into dead cells to prevent DNA amplification.

7.5.2 Ethidium Bromide Monoazide and Propidium Monoazide

Ethidium bromide monoazide (EMA) and propidium monoazide (PM) are flat planar molecules having the ability to enter dead cells with damaged membranes and intercalating into the DNA double helix.

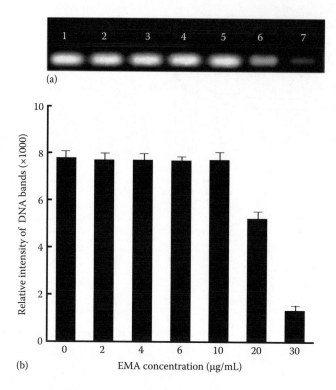

(a)

(b)

FIGURE 7.11 Optimization of the maximum amount of EMA not inhibiting the PCR from viable cells of *Escherichia coli* O157:H7. Cell suspensions (3.2×10^7 CFU) were treated with different concentrations of EMA. (a) A typical agarose gel image of PCR-amplified products. Lanes 1–7, varying concentrations of EMA (0, 2, 4, 6, 10, 20, and 30 µg/mL of EMA, respectively). (b) Bar graphs of the mean relative fluorescence intensities of DNA bands derived from triplicate agarose gel bands.

PM is required at a much higher level of concentration than EMA to inhibit amplification (Lee and Levin, 2009). The photo-inducible azide group allows these agents to covalently cross-link the two opposing DNA strands so as to prevent thermal denaturation of the DNA thereby preventing amplification. The operational basis of this selective methodology is based on the extent of damage of the cytoplasmic membrane accompanying cell death. With all such studies, it is important to first determine the maximum amount of EMA or PM that will not enter viable cells and inhibit PCR amplification. Figure 7.11 represents such a typical study and indicates that 10 mg/mL of EMA failed to enter viable cells of *E. coli* O157:H7. In contrast, 20 mg/mL of EMA resulted in a significant (35%) level of PCR inhibition and 30 mg/mL resulted in 85% inhibition. Therefore, the concentration of EMA in all such studies with *E. coli* 157:H7 should not exceed 10 mg/mL.

Elevated temperatures (80°C–100°C) will destroy vegetative bacterial cells and result in extensive damage to the cytoplasmic membrane. Subjecting bacterial cells to sodium hypochlorite (NaClO) similarly results in extensive damage to the cytoplasmic membrane and cell death. The higher the concentration of NaClO, the greater the damage to the cytoplasmic membrane. The level of EMA or PM required to enter dead cells depends on the extent of damage to the cytoplasmic membrane. Figure 7.12a illustrates that destruction of *E. coli* O157:H7 by 1000 ppm NaClO required 0.4 mg/mL of EMA to completely inhibit amplification. In contrast, destruction by 1 ppm NaClO required 0.6 mg/mL of EMA to completely inhibit amplification, reflecting much less membrane damage by 1 ppm NaClO.

Destruction of bacterial cells by UV irradiation, gamma radiation, or x-rays results in little or no membrane damage. EMA and PM will therefore not effectively distinguish live cells from cells destroyed by these mechanisms of irradiation.

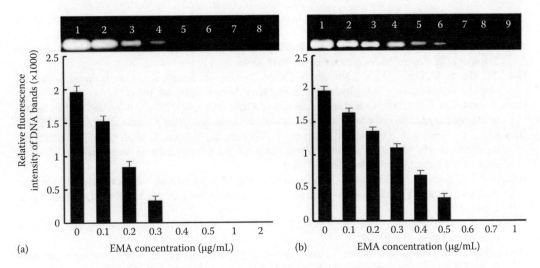

FIGURE 7.12 Determination of minimum amount of EMA to inhibit amplification of DNA from cells of *Escherichia coli* 0157:H7 killed with 1000 and 1 ppm NaClO. (a) 9×10^3 CFU killed by 1000 ppm NaClO and treated with different concentrations of EMA. Top: typical agarose gel image of PCR-amplified products. Lanes 1–8, varying concentrations of EMA (0, 0.1, 0.2, 0.3, 0.4, 0.5, 1, and 2 μg/mL); bottom: bar graphs of fluorescence intensity of corresponding DNA bands derived from PCR with respect to corresponding concentrations of EMA. (b) 9×10^3 CFU killed by 1 ppm NaClO treated with different concentrations of EMA. Top, typical agarose gel image of PCR-amplified products. Lanes 1–9, varying concentrations of EMA (0, 0.1, 0.2, 0.3, 0.4, 0.5, 0.6, 0.7, 0.8, and 1 μg/mL); bottom, bar graphs of fluorescence intensity of corresponding DNA bands derived from PCR with respect to corresponding concentrations of EMA.

REFERENCES

Abolmaaty, A., Gu, W., Witkowsky, R., and Levin, R.E. (2007) The use of activated charcoal for the removal of PCR inhibitors from oyster samples. *Journal of Microbiological Methods* 68: 349–352.

Abu Al-Soud, W. and Radstrom, P. (2000) Purification & characterization of PCR-inhibitory components in blood cells. *Journal of Clinical Microbiology* 39: 485–493.

Akane, A., Matsubara, K., Nakamura, H., Takahashi, S., and Kimura, K. (1994) Identification of the heme compound copurified with deoxyribonucleic acid (DNA) from bloodstains, a major inhibitor of polymerase chain reaction (PCR) amplification. *Journal of Forensic Sciences* 39: 362–372.

Bansal, R. and Goyal, M (2005). *Activated Carbon Adsorption*. Taylor & Francis Group, New York, 497pp.

Bélec, L., Authier, J., Eliezer-Vanerot, M.C., Piédouillet, C., Mohamed, A.S., and Gherardi, R.K. (1998) Myoglobin as a polymerase chain reaction (PCR) inhibitor: A limitation for PCR from skeletal muscle tissue avoided by the use of *Thermus thermophilus* polymerase. *Muscle & Nerve* 21: 1064–1067.

Bhaduri, S. and Cottrell, B. (2001) Sample preparation methods for PCR detection of *Escherichia coli* O157:H7, *Salmonella* Typhimurium and *Listeria monocytogenes* on beef chuck shoulder using a single enrichment medium. *Molecular and Cell Probes* 15: 267–274.

Brody, J.R. and Kern, S.E. (2004) Sodium boric acid: A Tris-free, cooler conductive medium for DNA electrophoresis. *Biotechniques* 36: 214–216.

Caplan, Z., Melilli, C., and Barbano, D. (2013) Gravity separation of fat, somatic cells, and bacteria in raw and pasteurized milks. *Journal of Dairy Science* 96: 2011–2019.

Chang, R. (2001) *Physical Chemistry for the Chemical and Biological Sciences*, 3rd edn. University Science Books, Sausalito, CA, p. 681.

Cockerill III, F. and Uhl, J. (2002) Applications and challenges of real-time PCR for the clinical microbiology laboratory. In Reischl, U., Wittwer, C., and Cockerill, F., eds., *Rapid Cycle Real-Time PCR—Methods and Application*. Springer Verlag, New York, pp. 3–27.

Compton, J. (1991) Nucleic acid based sequence-amplification. *Nature* 350: 91–92.

Diaco, R. (1995) Practical considerations for the design of quantitative PCR assays. In Innis, M.A., Gelfand, D.H., and Sinsky, J.J., eds., *PCR Protocols: A Guide to Methods and Applications*. Academic Press, New York, pp. 84–108.

Eiken Chemical. 2013. Eiken genome site. http://loopamp.eiken.co.jp/e/lamp/anim.html. Accessed August 29, 2013.

Fire, A. and Xu, S. (1995) Rolling replication of short DNA circles. *Proceedings of the National Academy of Sciences of the United States of America* 92: 4641–4645.

Goto, M., Honda, E., Ogura, A., and Nonoto, A. (2009) Colorimetric detection of loop-mediated isothermal amplification reaction by using hydroxy naphthol blue. *Biotechniques* 46: 167–172.

Guan, J. and Levin, R.E. (2002) Quantitative detection of *Escherichia coli* O157:H7 in ground beef by immunomagnetic separation and competitive polymerase chain reaction. *Food Biotechnology* 16: 155–166.

Hara-Kudo, Y., Nemoto, J., Ohtsuka, K., Segawa, Y., Takatori, K., Kojima, T. et al. (2007) Sensitive and rapid detection of vero toxin-producing *Escherichia coli* using loop-mediated isothermal amplification. *Journal of Medical Microbiology* 56: 398–406.

Hara-Kudo, Y., Yoshino, M., Kojima, T., and Ikedo, M. (2005) Loop-mediated isothermal amplification for the rapid detection of *Salmonella*. *Federation of European Microbiological Societies Letters* 253: 155–161.

Harris, C., Sanchez-Vargas, I., Olsen, K., Alphey, L., and Fu, G. (2013) Polymerase chain displacement reaction. *Biotechniques* 54: 93–97.

Hayward-Lester, A., Ofner, P., Sabatini, S., and Doris, P. (1995) Accurate and absolute quantitative measurement of expression by single-tube RT-PCR and HPLC. *Genome Research* 5: 494–499.

Heid, C., Stevens, J., Livak, K., and Williams, P. (1996) Real time quantitative PCR. *Genome Research* 6: 986–994.

Hill, J., Beriwal, S., Chandra, I., Paul, V., Kapil, A., Singh, T. et al. (2008) Loop-mediated isothermal amplification assay for rapid detection of common strains of *Escherichia coli*. *Journal of Clinical Microbiology* 46: 2800–2904.

Hübner, P., Studer, E., and Lüthy, J. (1999) Quantitative competitive PCR for the detection of genetically modified organisms in food. *Food Control* 10: 353–358.

Innis, M. and Gelfand, D. (1990) Optimization of PCRs. In Innis, M.A., Gelfand, D.H., Sninsky, J.J., and White, T.J., eds., *PCR Protocols: A Guide to Methods and Applications*. Academic Press, New York, pp. 3–20.

Jung, C., Chung, J., Kim, U., Kim, M., and Park, H. (2010) Isothermal target and signaling probe amplification method, based on a combination of an isothermal chain amplification technique and a fluorescence resonance energy transfer cycling probe technology. *Analytical Chemistry* 82: 5937–5943.

Kim, J., Jahng, M., Lee, G., Lee, K., Chae, H., Lee, J. et al. (2011) Rapid and simple detection of the *invA* gene in *Salmonella* spp. by isothermal target and probe amplification. *Letters in Applied Microbiology* 52: 399–405.

Kurn, N., Chen, P., Heath, J., Kopf-Sill, A., Sephen, K., and Wang, S. (2005) Novel isothermal, linear nucleic acid amplification systems for highly multiplexed applications. *Clinical Chemistry* 51: 1973–1981.

Lee, J.-L. and Levin, R.E. (2007) Rapid quantification of total bacteria on cod fillets by using real-time PCR. *Journal of Fisheries Sciences* 1: 58–67.

Lee, J.-L. and Levin, R.E. (2009) A comparative study of the ability of EMA and PMA to distinguish viable from heat killed mixed bacterial flora from fish fillets. *Journal of Microbiological Methods* 76: 93–96.

Lee, J.-L. and Levin, R.E. (2011) Detection of 5 CFU/g of *E. coli* O157:H7 on lettuce using activated charcoal and real-time PCR without enrichment. *Food Microbiology* 28: 562–567.

Levin, R.E. (2004) The application of real-time PCR to food and agricultural systems: A review. *Food Biotechnology* 18(1): 97–133.

Lizardi, P., Huang, X., Zhu, Z., Bray-Ward, P., Thomas, D., and Ward, D. (1998) Mutation detection and single-molecule counting using isothermal rolling-circle amplification. *Nature Genetics* 19: 225–232.

Loh, J. and Yap, E. (2002) Rapid and specific detection of *Plesiomonas shigelloides* directly from stool by light-Cycler PCR. In Reischl, U., Wittwer, C., and Cockerill, F., eds., *Rapid Cycle Real-Time PCR: Methods and Applications, Microbiology and Food Analysis*. Springer Verlag, Berlin, Germany, pp. 161–169.

Lopez, C., de Vries, A., and Marrink, S. (2012) Amylose folding under the influence of lipids. *Carbohydrate Research* 364: 1–7.

Luan, C. and Levin, R. (2012) Use of activated carbon coated with bentonite for increasing the sensitivity of PCR detection of *Escherichia coli* O157:H7 in Canadian oyster (*Crassostrea gigas*) tissue. *Journal of Microbiological Methods* 72: 67–72.

Mori, Y., Nagamine, K., Tomita, N., and Notomi, T. (2001) Detection of loop-mediated isothermal amplification reaction by turbidity derived from magnesium pyrophosphate formation. *Biochemical Biophysical Research Communications* 289: 150–154.

Mukai, H., Uemori, T., Takeda, O., Kobayashi, E., Yamamoto, J., Nisiwaki, K. et al. (2007) An efficient isothermal DNA amplification system using three elements of 5′-DNA-RNA-3′ chimeric primers, RNaseH and strand-displacing DNA polymerase. *Journal of Biochemistry* 141: 273–281.

Nagamine, K., Hase, T., and Notomi, T. (2002) Accelerated reaction by loop-mediated isothermal amplification using loop primers. *Molecular and Cellular Probes* 16: 223–229.

Nogva, H.K., Drømtorp, S.M., Nissen, H., and Rudi, K. (2003) Ethidium monoazide for DNA-based differentiation of viable and dead bacteria by 5′-nuclease PCR. *Biotechniques* 34: 804–813.

Notomi, T., Okayama, H., Masubuchi, H., Yonekawa, T., Watanabe, K., Amino, N. et al. (2000) Loop-mediated isothermal amplification of DNA. *Nucleic Acids Research* 28(12): e63, 7pp.

Opet, N. and Levin, R. (2013a) Efficacy of coating activated carbon with milk proteins to prevent binding of bacterial cells from foods for PCR detection. *Journal of Microbiological Methods* 94: 69–72.

Opet, N. and Levin, R. (2013b) Use of β-cyclodextrin and activated carbon for quantification of *Salmonella enterica* ser. Enteritidis from ground beef by conventional PCR without enrichment. *Food Microbiology* 38: 75–79.

Premier Biosoft. 2013. LAMP designer site (Premier Biosoft, International, Palo Alto, CA). http://www.premierbiosoft.com/tech_notes/Loop-Mediated-Isothermal-Amplification.html. Accessed September 17, 2013.

Raeymaekers, L. (1993) Quantitative PCR: Theoretical considerations with practical implications. *Analytical Biochemistry* 214: 582–585.

Rossen, L., Nørskov, P., Holmstrøm, K., and Rasmussen, O. (1992) Inhibition of PCR by components of food samples, microbial diagnostic assays and DNA-extraction solutions. *International Journal of Food Microbiology* 17: 37–45.

Rupf, S., Merte, K., and Eschrich, K. (1999) Quantification of bacteria in oral samples by competitive polymerase chain reaction. *Journal of Dental Research* 78: 850–856.

Sasitharan, D., Delvam, M., and Paul, W. (2013) Loop mediated isothermal amplification for diagnosis of *Escherichia coli* O157:H7 and viewpoints on its progression into realistic point of care. *International Journal of Environmental and Biological Science* 1: 244–247.

Szejtli, J. (1998) Introduction and general overview of cyclodextrin chemistry. *Chemical Review* 98: 1743–1753.

Tang, M., Zhou, S., Zhang, X.Y., Pu, J., Ge, Q.L., Tang, X. et al. (2011) Rapid and sensitive detection of *Listeria monocytogenes* by loop-mediated isothermal amplification. *Current Microbiology* 63: 511–516.

Tomita, N., Mori, Y., Kanda, H., and Notomi, T. (2008) Loop-mediated isothermal amplification (LAMP) of gene sequences and simple visual detection of products. *Nature protocols* 3: 877–882.

Vincent, M., Xu, Y., and Kong, H. (2004) Helicase dependent isothermal DNA amplification. *European Microbiological Organization Reports* 5: 795–800.

Wang, F., Jiang, L., and Ge, B. (2012) Loop-mediated isothermal amplification assays for detecting shiga toxin-producing *Escherichia coli* in ground beef and human stools. *Journal of Clinical Microbiology* 50: 91–97.

Wang, L., Shi, L., Alam, M., Geng, Y., and Li, L. (2008) Specific and rapid detection of food borne *Salmonella* by loop-mediated isothermal amplification method. *Food Research International* 451: 69–74.

Wang, S. and Levin, R.E. (2010) Interference of real-time PCR quantification of *Vibrio vulnificus* by a novel DNase from the eastern oyster (*Crassostrea virginica*). *Food Biotechnology* 24: 121–134.

Wang, X., Geng, F., Zhang, X., Wang, Y., Ma, X., Su, X. et al. (2011) A loop-mediated isothermal amplification assay for rapid detection of *Listeria monocytogenes*. *Journal of Food Safety* 31: 546–552.

Wilson, I. (1997) Inhibition and facilitation of nucleic acid amplification. *Applied and Environmental Microbiology* 63: 3741–3751.

Yamazaki, W., Taguchi, M., Kawai, T., Kawasu, K., Sakata, J., Inoue, K. et al. (2009) Comparison of loop-mediated isothermal amplification assay and conventional culture methods for detection of *Campylobacter jejuni* and *Campylobacter coli* in naturally contaminated chicken samples. *Applied and Environmental Microbiology* 75: 1597–1603.

Yamazaki, W., Yamaguchi, M., Isibashi, M., Kitazato, M., Nukina, M., Misawa, N., and Inoue, K. (2008) Development and evaluation of a loop-mediated isothermal amplification assay for rapid and simple detection of *Campylobacter jejuni* and *Campylobacter coli*. *Journal of Medical Microbiology* 57: 444–451.

Yap, E. and McGee, J. (1991) Short PCR product yields improved by lower denaturation temperatures. *Nucleic Acids Research* 19:1713.

Zachar, V., Thomas, A., and Goustin, A. (1993) Absolute quantification of target DNA: A simple competitive PCR for efficient analysis of multiple samples. *Nucleic Acids Research* 8: 2917–2018.

8

Recent Developments in Molecular-Based Food-Borne Pathogen Detection

Azlin Mustapha and Prashant Singh

CONTENTS

8.1 Introduction

Modern day molecular methods are being increasingly used for food applications. Such techniques often demonstrate high sensitivity and selectivity for target organisms, and some are even equipped to detect single molecules of target nucleic acids. With increasing numbers of infectious pathogen genomes being sequenced, these available genetic data can be exploited for the development of species-specific diagnostic assays to detect important pathogenic bacteria in foods. Molecular methods are also quite useful for detection of and discrimination among closely related pathogens that cannot be differentiated using culture methods.

Polymerase chain reaction (PCR)–based methods for detecting pathogens in foods have become more readily available and more widely used in the food industry. The most commonly used dual-labeled probe in real-time PCR assays is the TaqMan® probe, which relies on the 5′ exonuclease activity of Taq DNA polymerase to cleave a fluorescent dye from a quencher molecule and generate a signal during real-time PCR amplification. This process requires a TaqMan® probe with a relatively high melting temperature, which, in turn, reduces the allelic discrimination capabilities of the probe and the PCR amplification efficiency (Pierce and Wangh 2011). Another major disadvantage of symmetrical real-time PCR assay is that the hybridizing probe has to compete with the single-stranded complementary DNA fragment generated at the denaturation step of the PCR cycle. Towards the end of the cycle, the PCR amplicon accumulates and annealing of the single-stranded amplicons generated at the denaturation step dominates the probe hybridization. Thus, only a fraction of amplicons are detected by the probe (Pierce and Wangh 2011). Another limitation of conventional real-time PCR reaction is that it undergoes a self-limiting saturation after the exponential phase (30–40 cycles), and increasing the number of cycles does not lead to an increase in signal strength. Also, the signal dramatically decreases for samples with a low DNA copy number, leading to an increase in standard deviations among replicates and reducing the robustness of the assay.

While commercial development of PCR-based diagnostics has progressed significantly in recent years, currently approved PCR standard test kits to date are still categorized as high or moderate complexity under the Clinical Laboratory Improvement Amendments (Holland and Kiechle 2005). Efficient nucleic acid preparation from complex sample types, such as whole blood, stool, and foods, requires highly skilled personnel to manually perform multiple processing steps in a dedicated laboratory space, batched testing with next-day reporting of results, and costly reagents and instrumentation. These features are some of the major hurdles that still preclude PCR-based assays from being classified as a *simple test* based on official recommendations (Holland and Kiechle 2005).

This review aims to highlight some of the more recent developments in molecular techniques that can be applied for pathogen detection in foods. Table 8.1 provides a summary of the advantages and drawbacks of the methods discussed. Although many of them are still not commercially available, these methods have a high potential to be applied to foods for improved and robust pathogen detection, especially when coupled with robotic liquid handling devices and plate loaders.

TABLE 8.1

Summary of Methods

	Advantage	Application	Multiplex Capability	Disadvantage
Locked nucleic acid	Higher affinity for DNA and RNA, higher base mismatch discrimination property, higher resistance toward nuclease attack	Dual-labeled probe for qPCR, fluorescent probe for FISH assays, capture probe, oligonucleotide for microarray		High synthesis cost
Peptide nucleic acids	Higher affinity for DNA and RNA, more sensitive to nucleic acid mismatches, less sensitive to variations in salt concentration, resistant to nucleases and peptidases, higher capabilities to penetrate biological materials	PNA-FISH, PNA probes for qPCR; blocker probe, PNA clamps		High synthesis cost
Linear-after-the-exponential PCR	Remains in linear phase longer, higher multiplexing capability, better allelic differentiation capabilities, increased probe signal strength, single-cell detection capability	Multiplex assay for pathogen and SNP detection	Yes	Complex experimental design
High-resolution melting analysis	Simple, high sensitivity, low cost, nondestructive method	SNP genotyping, pathogen detection, presequence screening, methylation analysis	No	Lower resolution for A–T or T–A mutations
Helicase-dependent isothermal DNA amplification	Isothermal DNA amplification, helps to eliminate the use of expensive PCR instrument	Can be used for the development of point-of-care diagnostic assays	Yes	
Microfluidics PCR	Complete assay on a single chip, requires a lesser amount of sample and analytes, which helps to cut down the running cost, very high ramp rate	Rapid pathogen detection assays	Yes, but mostly in uniplex format	Still in development phase for the detection of food pathogens
Digital PCR	Single molecule detection capability, detection of rare alleles, absolute quantification of copy number variants	Detection of pathogens at very low concentration	Chip format— yes Droplet PCR—no	Very high instrument cost

8.2 Locked Nucleic Acid

8.2.1 Chemical Structure

Locked nucleic acid (LNA) is a nucleic acid analog that contains one or more LNA nucleotide monomers with a bicyclic furanose unit locked in an RNA mimic sugar confirmation. In recent years, a number of structural and conformational isomers of LNA have been synthesized with β-D-ribose and α-L-LNA configurations (Figure 8.1a) being the most commonly used in the area of research and diagnostics (Vester and Wengel 2004). The structural difference between a DNA or an RNA molecule and an LNA molecule is the introduction of an additional 2′-C,4′-C-oxymethylene link (Figure 8.1a) (Singh et al. 1998). Oligonucleotides containing LNA bases are synthesized according to standard DNA synthesis procedures. These oligonucleotides are also compatible with modified and unmodified nucleotide bases, such as DNA, RNA, and 2′*O*-Me-RNA, and conform to Watson–Crick base pairing (Petersen et al. 2000).

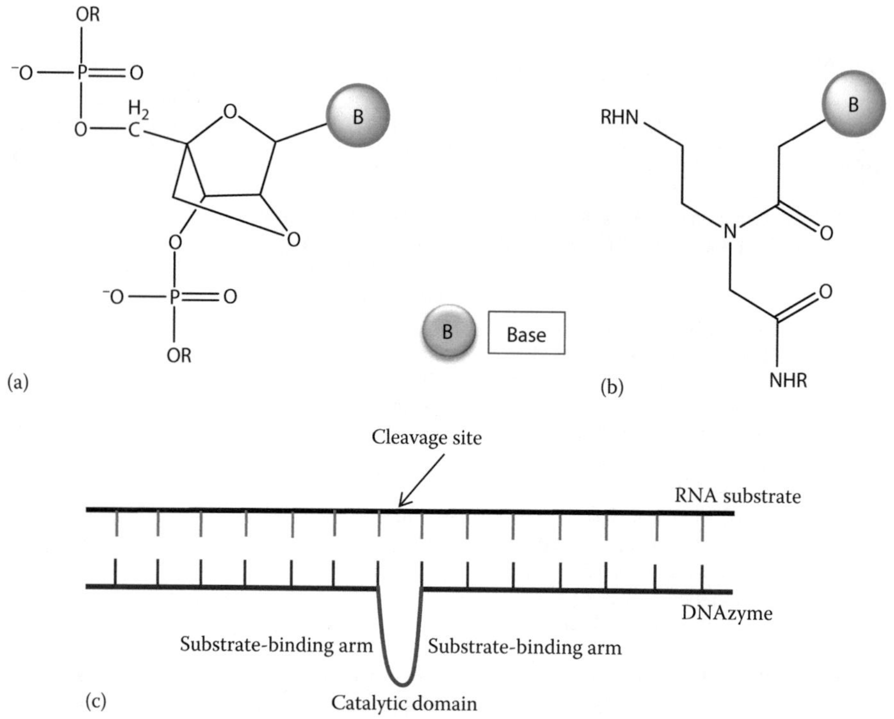

FIGURE 8.1 Structure of LNA (a) and PNA (b), and schematic representation of DNAzyme–substrate complexes (c).

8.2.2 Properties and Advantages

The robust structure of LNA molecules confers several superior properties to an LNA oligonucleotide that make it a powerful tool in the field of molecular biology research. Compared to natural oligonucleotides, an LNA molecule possesses a very high affinity toward complementary DNA, RNA, and LNA molecules. The reason for the high affinity is its favorable entropy and a more efficient stacking of the nucleotide bases within the molecule (Vester and Wengel 2004). LNA molecules possess higher base mismatch discrimination properties, whereby a single base mismatch in an LNA oligonucleotide would have a greater destabilizing effect during hybridization than natural oligonucleotides. During synthesis, LNA molecules have been modified to such an extent that they are not recognized by natural enzyme systems as substrates, making them resistant to the action of nucleases present in a biological system. Thus, LNA molecules are more stable, less toxic, and resistant to nuclease attack in biological systems.

8.2.3 Applications of LNA

8.2.3.1 Single Base Polymorphism Detection

A single base polymorphism (SNP) in a gene sequence can confer many phenotypic properties in living organisms, including antibiotic resistance in bacteria. There is an increasing demand for the development of rapid molecular assays for the detection of SNPs. LNA-based probes have better SNP discrimination capabilities, making them much more suited for SNP detection protocols. Letertre et al. (2003) compared minor groove binding (MGB) and LNA probes for the detection of staphylococcal enterotoxin gene using a TaqMan assay. The results obtained from both the assays were equivalent, suggesting that an LNA probe is a promising alternative to an MGB probe and can be used for the development of rapid pathogen detection assays for food applications.

8.2.3.2 Capture Probes

A eukaryotic mRNA molecule has a poly-A tail at the 3′ end, allowing for a poly-T sequence to be used as a capture probe in the process of isolating mRNA. A 20-mer capture probe with alternating LNA and DNA monomers has been shown to increase the yield of mRNA by 30- to 50-fold. A capture probe consisting of LNA molecules also works efficiently under low-salt buffer conditions, which is an added advantage (Jacobsen et al. 2004).

An LNA oligonucleotide and capture probe have been used for gene expression profiling by microarray (Castoldi et al. 2007). The incorporation of LNA molecules in a capture probe helped to achieve a higher hybridization affinity, leading to a more reproducible signal and improved base differentiation capabilities of SNPs. Lezar and Barros (2010) reported the development of an oligonucleotide microarray that could identify a total of 32 fungi species and their potential to produce mycotoxins in a single assay.

8.2.3.3 LNA-Fluorescence In Situ Hybridization

LNA oligonucleotide technology has also been successfully linked to fluorescence in situ hybridization (FISH). LNA probes for FISH have the property of a high binding affinity, which results in a shorter hybridization time. An LNA/DNA mixmer with every second nucleotide being an LNA has been shown to generate better results than DNA oligonucleotides (Silahtaroglu et al. 2003). Kubota et al. (2006) studied the improvement of FISH efficiency by using an LNA/DNA probe. The application of a short probe comprising LNA bases displayed a remarkable improvement in in-situ hybridization without compromising probe specificity.

8.2.3.4 LNAzyme

LNAzymes are catalytically active oligonucleotides containing LNA molecules that hybridize to target molecules in a sequence-specific manner (Dolinšek et al. 2013). Breaker and Joyce (1994) first reported catalytically active DNA oligonucleotides (DNAzymes) that were capable of cleaving RNA molecules. A DNAzyme is made up of a 15-nucleotide catalytic domain, which is flanked on both sides by two substrate-binding arms (Figure 8.1c). The sequence of the catalytic domain cannot be changed without losing the catalytic activity, but the sequence of their substrate-binding arms can be changed and be substituted with LNA molecules. The incorporation of LNA molecules in an LNAzyme sequence has been reported to significantly increase the efficiency of cutting RNA molecules, especially those that form secondary structures. The improved catalytic activity of the LNAzyme is due to its high binding affinity, which helps it to better hybridize with complex RNA structures. LNAzymes also show better base differentiation capabilities (Dolinšek et al. 2013) and can be used in a single or multiplex format for specific degradation of targeted RNA molecules. The method can be combined with in vitro transcription and reverse transcription and can be coupled with many molecular methods.

8.3 Peptide Nucleic Acid

8.3.1 Chemical Structure

Peptide nucleic acids (PNAs) are DNA analogs that mimic DNA structure. Their unique chemical properties have been explored for the development of many diagnostic assays. In a PNA molecule, the negatively charged sugar–phosphate backbone is replaced by a neutral polyamide backbone composed of *N*-(2-aminoethyl) glycine units (Nielsen et al. 1991) (Figure 8.1b). Even after compositional differences in the backbone of PNA and DNA molecules, the distance between the nucleotide bases in PNA remains the same as that of natural DNA molecules. The similarity in nucleotide distance between DNA and PNA allows these molecules to base pair with each other in accordance with Watson–Crick rules (Egholm et al. 1993).

8.3.2 Properties and Advantages

The lack of electrostatic repulsions in PNA molecules allows for PNA/DNA and PNA/RNA base pairing with higher thermal stability as compared to the base pairing in DNA/DNA and DNA/RNA duplexes. Given the higher affinity of PNA for DNA and RNA, a much more sensitive, selective, and structurally robust assay can be designed for the detection of targets when PNA molecules are used as probes. Since PNA-based probes have a higher affinity for DNA and RNA molecules, this also allows for the use of shorter PNA probes with higher affinity. Short probes are preferred for applications such as the detection of SNPs. Another important property is that the molecules that make up PNA-based probes are more sensitive to mismatches (Stender et al. 2002). A single base pair mismatch in PNA/DNA duplexes has a greater destabilizing effect when compared with what would occur in DNA/DNA complexes of the same length.

Yet another outstanding property of a PNA-based probe is that unlike DNA–DNA hybridization, a PNA–DNA hybridization is less sensitive to variations in salt and ionic concentrations. PNA–DNA binding can even be achieved at lower salt concentrations, which are detrimental for DNA-based probes because secondary structures of DNA and RNA molecules destabilize at these lower salt concentrations. On the other hand, the destabilization of a DNA or an RNA secondary structure at lower salt concentrations makes them more accessible and leads to improved binding of a PNA probe to a targeted site (Yilmaz et al. 2006).

A further advantage of a PNA probe is that it is naturally resistant to nucleases and peptidases. Since the PNA polyamide backbone is an unnatural structure, PNA molecules are not recognized by naturally occurring enzyme systems, which, in turn, allow them to be resistant to enzymatic degradation and be more stable and suitable for biological applications (Demidov et al. 1994). Owing to their superior performance in various experimental conditions, analytical methods based on PNA probes are more robust and offer higher sensitivity, especially in complex matrices.

8.3.3 Applications of PNA

8.3.3.1 Food Safety

One of the most routinely used methods that uses PNA probes for pathogen identification is PNA-FISH. In PNA-FISH, a PNA probe labeled with a fluorescent dye and targeted to bind to a specific DNA or RNA sequence can be used for the detection of pathogens in food samples, in biological tissues and fluids, in culture media, in filters, on slides, or in solution (Yu and Dunn 1996; Valdivia et al. 1998). The advantage of PNA-FISH in comparison with DNA-FISH is that PNA probes have a higher ability to penetrate recalcitrant biological structures, such as the membranes of gram-positive cells (Brehm-Stecher et al. 2007). Owing to their higher binding affinity toward a target DNA/RNA molecule, robust chemical structure, resistance toward biological enzymes, and independence from solution ionic strength, PNA probes are able to achieve a higher sensitivity than a DNA probe.

Brehm-Stecher et al. (2005) designed a PNA-FISH assay using fluorescein-labeled *Listeria* genus-specific PNA probe targeting the 16S ribosomal subunit to detect *Listeria* in food samples. A PNA-FISH method was described by Lehtola et al. (2005) for the detection of *Campylobacter coli*, *C jejuni*, and *Campylobacter lari*, which are important food- and water-borne pathogens. A PNA-FISH method was also used for the detection of *Escherichia coli* in a biofilm obtained from pipe samples in a drinking water network (Juhna et al. 2007). In this study, the PNA-FISH assay performed better than a culture-based method, which failed to detect viable but nonculturable (VNBC) *E. coli* from the biofilm samples. Azevedo et al. (2003) developed a PNA-FISH assay targeting the helix 6 region of *Helicobacter pylori* 16S rRNA for the detection of this pathogen in water biofilm samples. Lehtola et al. (2006) described a high-affinity PNA-FISH assay for detecting *Mycobacterium avium*, *M. avium* subsp. *avium*, *M. avium* subsp. *paratuberculosis*, and *M. avium* subsp. *silvaticum* in potable water biofilms. A PNA-FISH procedure for the detection of *Acinetobacter* spp. and *Pseudomonas aeruginosa* showed a 100% sensitivity and specificity for *Acinetobacter* spp. and 100% sensitivity and 95% specificity for *P. aeruginosa* (Peleg et al. 2009).

8.3.3.2 PCR Probes

Light-up probes (LightUp Technologies, Göteborg, Sweden) and LightSpeed probes (Boston Probes, Bedford, MA) are fluorescently labeled PNA probes. Light-up probes are labeled with the highly fluorescent dye, thiazole orange. In the absence of a complementary target, these probes do not fluoresce but, upon hybridization, become fluorescent. Light-up probes have been used for species-specific detection of *Salmonella* (Isacsson et al. 2000). On the other hand, LightSpeed probes are dual-labeled PNA probes that are labeled with fluorescent and quencher dyes (Stender et al. 2002). These probes are also known as linear PNA beacons and have been found to be useful for hybridizations in alcoholic solutions (Fiandaca et al. 2005).

Blocker probe is a PNA-based probe used for PCR applications. This is an unlabeled PNA probe that is designed to bind and block the binding site on nontarget DNA, thus reducing nonspecific signals from closely related species. Owing to a higher discriminatory capability of PNA probes, these blocker probes also help to improve the detection of point mutations. Stender et al. (2002) reported a chemiluminescent in situ hybridization method for the identification and enumeration of *E. coli* from municipal water. A blocker probe was added into this assay at a higher concentration to compete and bind to a *P. aeruginosa* target sequence, thus reducing nonspecific signals.

PNA clamps allow for a stronger PNA–DNA binding than DNA–DNA binding, and DNA polymerase does not recognize PNA–DNA complexes. These two properties of PNA are used in designing PNA clamps to block amplification from a set of nontarget DNA from closely related bacteria or for the detection of mutations that exist at lower frequencies. Binding of PNA clamps inhibits the competitive amplification of one set of DNA/allele, thus making the conditions favorable for another set of DNA/allele (Ørum et al. 1993). This strategy can be used for the detection of pathogens that exist in low numbers.

8.4 Linear-after-the-Exponential Polymerase Chain Reaction

8.4.1 Description

Linear-after-the-exponential polymerase chain reaction (LATE-PCR) is an advanced form of asymmetrical PCR in which both the primers are added at different concentrations, and a low temperature molecular beacon probe is used for the detection of targets. Unlike other asymmetrical PCR assays, LATE-PCR uses an improved primer and probe design criteria, which ensures a high level of sensitivity and specificity (Pierce and Wangh 2011). During a LATE-PCR amplification, the PCR reaction uses both primers equally, amplifying both strands of the amplicons in equal proportions during the initial exponential phase. After the limiting primer is exhausted, the PCR reaction turns asymmetric using only the excess primer and producing only single-stranded amplicons. These single-stranded amplicons, produced by the extension of the excess primers in the linear phase of the PCR reaction, are freely available to the probe for hybridization, thus eliminating competition between the probe and the complementary DNA strand (Figure 8.2a). This elimination of competition allows for the use of low-temperature probes for the detection of amplicons generated in the PCR reaction (Pierce and Wangh 2011). Low melting temperature probes are easier to design, have better allelic differentiation capabilities, and lower background fluorescence, which increases the signal strength by 80%–250%. Because the amplicon detection occurs after completion of the PCR reaction, the addition of extra probes has no inhibitory effect on the reaction itself (Sanchez et al. 2004). The use of a low-temperature probe also allows for higher multiplexing capabilities over a wide range of temperatures and detection channels of the real-time PCR (Rice et al. 2013). Another advantage of this approach is, unlike conventional PCR, which undergoes self-limiting saturation, LATE-PCR remains in the linear phase, which, in turn, helps to achieve higher sensitivity (Sanchez et al. 2004; Pierce and Wangh 2011).

8.4.2 Applications of LATE-PCR

Recently, a number of LATE-PCR assays have been published for the detection of clinical pathogens. Rice et al. (2013) reported a highly multiplexed single-tube assay for the detection of bacterial and fungal

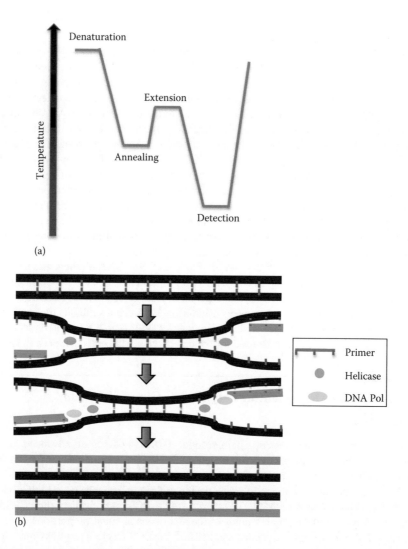

(a)

(b)

FIGURE 8.2 Schematic representation of LATE-PCR amplification (a). During HDA (b), helicase unbinds dsDNA allowing for annealing of primers. Both primers are then extended by two DNA polymerases (Pol), amplifying the targeted DNA sequence.

pathogens associated with sepsis, including *Klebsiella* spp., *Acinetobacter baumannii*, *Staphylococcus aureus*, *Enterobacter* spp., *Enterococcus* spp., *P. aeruginosa*, coagulase-negative staphylococci, and *Candida* spp. The limit of detection of this assay varied between 100 and 1000 copies of genomic DNA. The assay developed in this study was further validated using clinical samples at the University of California (Davis, CA). The assay demonstrated a sensitivity of ≥97.8% and 91.4% in monoplex and multiplex format, respectively (Gentile et al. 2013).

Food sources, especially meat, are becoming a source of *Clostridium difficile* infections. A single-tube multiplex LATE-PCR assay targeting all the toxigenic genes (*tcd*A, *tcd*B, *tcd*C, and *cdt*B) was developed for the detection of *C. difficile* (Pierce et al. 2012). The single tube LATE-PCR assay developed in this study generated more information than any other commercial assay. Foot-and-mouth disease (FMD) is a highly infectious disease that affects cloven-hoofed animals, namely beef cattle, goat, and sheep. Pierce et al. (2010) reported a LATE-PCR assay for the detection of the FMD virus, the RNA virus agent of this disease.

8.5 High-Resolution Melting Analysis

8.5.1 Description

Melt curve analysis of amplicons generated in a qPCR reaction using high-resolution dyes is known as high-resolution melting curve analysis (HRMA). The initial real-time application of this technique utilized the SYBR® Green intercalating dye for monitoring the accumulation of amplicons and relied on the melt curve for checking the specificity of the amplified PCR product (Morrison et al. 1998). DNA denaturation or melting has been used for some time to study the nucleotide base composition and DNA structure of molecules. The advent of HRMA qPCR technology, its ease of use, simplicity, high sensitivity, low cost, and nondestructive nature make this a tool of choice in the field of SNP genotyping and pathogen detection. In widening the scope of this technology, its application in the areas of clinical samples and food safety has also increased steadily.

8.5.2 Instrumentation

The instrument required for HRMA comes in two formats: (1) an HRMA instrument alone or (2) a qPCR system combined with HRMA. A combination of a qPCR instrument with HRMA allows for both quantification and genotyping of the target in one assay. The most sensitive system currently available in the market is the HR-1 (Idaho Technology Inc., Salt Lake City, UT) (Vossen et al. 2009). The HR-1 instrument is a capillary-based system and can analyze only one sample at a time. Apart from the hardware, the software for HRMA is also a vital part of the system.

8.5.3 HRMA Dyes

The dyes used for HRMA are a class of intercalating dyes that are also known as saturating dyes. The SYBR Green I–based master mixes still remain the most widely used for performing qPCR, but these have several drawbacks that make them unsuitable for HRMA. SYBR Green I has been shown to affect the melting temperature of amplicons, exhibit preferential binding to certain amplicons, and bind only to the minor groove of DNA (Gudnason et al. 2007). Saturating dyes are another class of intercalating dyes that do not show any preferential binding to amplicons and bind to the target DNA to completely saturate the amplicon without inhibiting the PCR reaction in the process, thus generating a melt curve of higher resolution (Wittwer et al. 2003). LCGreen® was the first saturating dye made available in the market (Wittwer et al. 2003). At present, many saturating dyes are available in the market that can be used for HRMA, including LCGreen® (Idaho Technology Inc., Salt Lake City, UT), Syto9® (Invitrogen, Carlsbad, CA), EvaGreen® (Biotium, Hayward, CA), and LightCycler® 480 ResoLight Dye (Roche, Indianapolis, IN). Each one of these dyes differs from one another in terms of molecular size, excitation spectrum, and emission spectrum. They also tend to show a variation in melt curve peaks obtained for the same amplicons, thus requiring PCR optimization for each.

8.5.4 HRMA Process

HRMA is a simple process that is a slight modification from a conventional melting curve assay and does not require any additional technical skills to perform. No additional modifications to an existing real-time PCR assay are required for HRMA. Because the HRMA process is a very sensitive method for studying DNA sequence polymorphism, it can help to cut down on sequencing costs. HRMA can be performed by directly adding a saturating dye in a qPCR master mix or the saturating dye can be added after the completion of a PCR. HRMA can be performed using a real-time PCR instrument or a specific HRMA instrument, such as HR-1 (Idaho Technology, Salt Lake City, UT).

8.5.5 Advantages and Drawbacks

Unlike other techniques, such as single-strand conformation analysis (SSCA), denaturing high-pressure liquid chromatography (dHPLC), denaturing gradient gel electrophoresis (DGGE), and capillary electrophoresis, that are used for genotyping, HRMA is a nondestructive method. Because of this, the amplicons generated at the completion of the assay can be subsequently analyzed by other methods, such as sequencing and gel electrophoresis.

One of the limitations of HRMA is its lower resolution for the differentiation of some variants, such as A–T or T–A changes (Vossen et al. 2009). The difference in the melt curve of these variants is subtle, but due to recent developments in this field, resolution limits of some instruments have improved (Gundry et al. 2008).

8.5.6 Applications of HRMA

8.5.6.1 Mutation Detection

HRMA was initially developed for the detection of sequence variations in DNA, but the process was later refined and also used for genotyping applications (Wittwer et al. 2003). Because HRMA is more efficient when used for smaller sized amplicons, it is useful for SNP genotyping, whereby amplicons of a smaller size (70–150 bp) are typically preferred (Vossen et al. 2009). During the process of HRMA, its software automatically scores the genotypes.

8.5.6.2 Presequence Screening

HRMA is a tool of choice for reducing the number of samples for sequencing. The process is suitable for screening longer genes in which multiple regions/exons need to be screened. The presequence screening may lead to a significant cost saving (Provaznikova et al. 2008) by eliminating variants that do not need to be sequenced in a sample.

8.5.6.3 Food Safety

Ajitkumar et al. (2012) reported an HRMA assay for the rapid identification of mastitis pathogens. A conserved primer was used to amplify the hyper-variable region (V5–V6) of the 16S rRNA gene for all the pathogens. The assay enabled identification of nine common mastitis pathogens, including *E. coli*, *Streptococcus agalactiae*, *Streptococcus dysgalactiae*, *Streptococcus uberis*, *S. aureus*, *Klebsiella pneumoniae*, *Mycoplasma bovi*, *Arcanobacterium pyogenes*, and *Corynebacterium bovis*.

Yang et al. (2009) described a universal PCR coupled with HRMA for the identification of 100 clinically relevant bacterial pathogens, which included Class A and Class B biothreat agents. Cheng et al. (2006) described a broad range real-time PCR targeting the 16S rRNA gene coupled with HRMA for the identification of 25 clinically important bacteria. Most of the bacterial species targeted in this study were common food-borne pathogens.

Zeinzinger et al. (2012) reported a triplex HRMA assay for rapid identification and simultaneous subtyping of frequently isolated *Salmonella enterica* serovars. The assay was based on a screening of SNPs present in the *fljB*, *gyrB*, and *ycfQ* genes and the generation of specific melt curve profiles based on these SNPs in the targeted regions. The study evaluated 417 *Salmonella* isolates comprising 46 different serotypes. It allowed for the differentiation of 37 serotypes using a close-tube triplex HRMA.

Ong et al. (2010) developed an HRMA assay for the detection of mutations in the *rpoB*, *inhA*, *ahpC*, and *katG* genes and/or promoter regions that confer resistance toward rifampin and isoniazid in *Mycobacterium tuberculosis*. Members of the genus *Enterococcus* are opportunistic pathogens and cases of vancomycin-resistant enterococci (VRE) are increasing. Gurtler et al. (2012) reported an HRMA assay for simultaneous identification of *Enterococcus* species and vancomycin resistance genotyping.

8.5.6.4 Microbial Population Dynamics in Foods

Apart from food safety, HRMA assays are also being used for studying bacterial community profiles in food samples. Porcellato et al. (2012a) compared the applicability of HRMA to other methods to study the population dynamics of fermented dairy foods. The HRMA results obtained in that study were in agreement with PCR-DGGE results. In another study (Porcellato et al. 2012b), an HRMA was used to study the nonstarter lactic acid bacterial population diversity of Norvegia cheese by targeting the V1 and V3 variable regions of the 16S rRNA gene and a $(GTG)_5$ primer-based repetitive DNA. HRMA clustering analysis showed a high discriminatory power, which was comparable to gel-based methods.

8.6 Helicase-Dependent Isothermal DNA Amplification

8.6.1 Description

Helicase-dependent isothermal amplification or helicase-dependent amplification (HDA) is a method that employs DNA helicase and DNA polymerase enzymes for the amplification of DNA at an isothermal temperature. An HDA assay uses a DNA helicase enzyme to separate two complementary strands of a double-stranded DNA (dsDNA). This action of the DNA helicase enzyme generates single-stranded DNA (ssDNA) molecules that act as templates for primer binding (Figure 8.2b). The use of a helicase enzyme at this step also helps to eliminate the high-temperature denaturation step required in conventional PCR amplification. The ssDNA generated after a DNA helicase action is stabilized by coating them with ssDNA-binding proteins (SSBPs). Sequence-specific primers anneal to the ssDNA generated in the last step and are extended by DNA polymerase, amplifying the target DNA and generating a dsDNA. This dsDNA acts as a substrate for DNA helicase in the next round of DNA amplification, resulting in an exponential amplification of the DNA (Vincent et al. 2004).

Initial HDA experiments were performed using *E. coli* UvrD helicase (Runyon and Lohman 1989) because the molecule possesses the ability to unwind blunt-ended DNA fragments. In addition to *E. coli* UvrD helicase, the reaction mixture in HDA assays also consisted of the T4 gene 32 protein or the RB 49 gene 32 protein acting as SSBPs (Casas-Finet and Karpel 1993; Desplats et al. 2002), Klenow fragment of DNA polymerase I, and the accessory protein, MutL (An et al. 2005). The DNA amplification was performed by incubating the reaction mixture at 37°C for 1 h, and the process was called mesophilic HDA (mHDA) (Vincent et al. 2004).

Subsequent HDA experiments utilized a thermostable form of UvrD helicase (Tte-UvrD) isolated from *Thermoanaerobacter tengcongensis* (An et al. 2005). This thermostable Tte-UvrD helicase was characterized and found to be stable and active in a temperature range of 45°C–65°C. It also possessed the ability to unbind blunt-ended DNA and dsDNA with a 3′- or 5′-ssDNA tail, without the need of MutL and SSBP. This thermostable Tte-UvrD helicase enzyme was used for the development of a thermophilic version of HDA, which is called tHDA (An et al. 2005). Performing an HDA at a higher temperature (tHDA) increases the efficiency and sensitivity of the reaction. A detection limit of 10 bacterial genomic DNA was achieved using tHDA, which was much better when compared to the detection limit of mHDA (1000 bacterial genomic DNA copies). The tHDA also does not require MutL and SSBP for amplification, helping to keep the process simple and cutting the cost of reagents (An et al. 2005).

However, the amplification of long sequences of DNA using HDA was not initially possible because of the low processivity of UvrD helicase. A new HDA protocol with high processivity and speed was reported using T7 bacteriophage replication machinery (Jeong et al. 2009). The process employs a hexameric T7 DNA helicase, which has a higher processivity and fidelity. Helicase-dependent amplification using T7 DNA helicase even allowed the amplification of whole genomes using single primer pairs (Jeong et al. 2009).

8.6.2 Applications of HDA

Chow et al. (2008) reported the development of a disposable isothermal HAD assay coupled with a single-use vertical-flow device for the detection of toxigenic *C. difficile*. An asymmetrical amplification was performed, and the amplified product and the internal amplification control (IAC) were detected using digoxigenin (DIG)- and fluorescein isothiocyanate (FITC)-labeled probes. A sensitivity of 20 copies of *C. difficile* genomic DNA per reaction or 10^4 colony forming units (CFU/g) of fecal sample was achieved using this assay. This HDA assay demonstrated 100% sensitivity and specificity, can be performed in less than 1.5 h, serves as a good alternative to PCR and enzyme immunoassay (EIA) methods, and can be used as a point-of-care diagnostic tool.

Frech et al. (2012) developed an isothermal HDA diagnostics assay for the detection of *S. aureus*. The assay generated biotinylated HDA amplicons, which were detected by hybridization on silicon chips using a specific immobilized capture probe. A sensitivity of 2 CFU per HDA reaction was achieved using this assay. The assay worked efficiently for swab samples with greater than 100 *S. aureus* CFU per swab.

8.7 Microfluidics PCR

8.7.1 Description

Microfluidics is an area of science that deals with behavior, precise control, and manipulation of fluid in an environment at a very small scale (small volume and size and low energy consumption) (Whitesides 2006). The application of microfluidic technology in the area of bioanalytical methods is quite promising and allows for miniaturization and integration of analytical methods. This miniaturization and its integration with other related analytical assays allow for increased reaction efficiency and the performance of a complete assay on a single chip. In microfluidic qPCR, multiple uniplex qPCR reactions are performed in nanoliter volumes and at high density on a microfluidic chip (Ishii et al. 2013). The microfluidic technology was initially applied for the development of conventional PCR, but was subsequently integrated with real-time detection technologies, capillary electrophoresis, and other techniques.

8.7.2 Advantages

Microfluidic assays offer several advantages, such as the requirement of a lesser amount of sample, which is very important when working with low initial DNA concentration samples, consumption of a lesser amount of analytes, which helps to cut down on the running cost, and the use of rapid thermal conductivity that allows for rapid heating and cooling. These factors allowed for the development of rapid detection methods by reducing the turn-around time of the assay, resulting in portability and integration of PCR with pre- and post-PCR processing modules on a single chip, which, in turn, facilitate automation of microfluidic assays (Whitesides 2006). All of these unique sets of capabilities of microfluidic technologies can be used for the development of single-use, self-contained, disposable miniaturized PCR chips that can be applied for on-site testing of multiple pathogens on a single chip, helping to reduce cross-contamination and risks of exposure to biohazard agents (Park et al. 2011).

8.7.3 Microfluidics PCR Design

8.7.3.1 Small Volume

High ramp rate and rapid thermal cycling are the most important features of microfluidic PCR. The high ramp rate and rapid heat transfer in microfluidic PCR are achieved because of the very small reaction volume of the PCR, and they are also facilitated by a high surface area to reaction volume ratio. The smaller reaction volume in microfluidic PCR also offers the advantage of a uniform temperature distribution across the reaction volume, thus increasing the efficiency of the reaction and leading to a higher yield (Park et al. 2011).

8.7.3.2 Device

The two most important criteria that are considered while selecting the material for the fabrication of microfluidic devices are the cost and performance of the material. The most commonly used polymeric materials for the fabrication of the devices are polydimethylsiloxane (PDMS) (Thorsen et al. 2002), polymethylmethacrylate (PMMA) (Hataoka et al. 2004), and polycarbonate (PC) (Hashimoto et al. 2004). Owing to its nontoxic, inert nature, higher thermal stability, and compatibility with PCR, PDMS has emerged as a material of choice for the fabrication of microfluidic PCR chips (Whitesides 2006). Two designs are currently being used for the fabrication of microfluidic PCR (Park et al. 2011). First is the stationary phase reaction chamber system, which works in the same manner as a conventional PCR system where the reaction volume is kept stationary and the PCR is performed by altering the temperature in the reaction chamber. The second design is called a continuous flow system. In this design, the PCR reaction mixture flows across various fixed temperature zones on the chip for PCR cycling. In comparison with the stationary phase design, the continuous flow approach provides faster thermal cycling (Park et al. 2011).

8.7.3.3 Post-PCR Analysis

The PCR product obtained at the completion of a microfluidic PCR can be analyzed either by conventional off-line methods, which require transferring the PCR product for analysis, or the chip itself can be integrated with on-chip detection techniques, such as capillary electrophoresis (Lagally et al. 2004), DNA hybridization microarray, fluorescence real-time detection, and electrochemical methods (Park et al. 2011).

8.7.3.4 Applications of Microfluidics PCR

Ishii et al. (2013) reported the development of a microfluidic quantitative PCR assay for the detection of multiple food- and water-borne pathogens. This microfluidic assay used 20 TaqMan probes for species-specific detection of 10 enteric pathogens, their virulence genes, and a fecal indicator bacterium. One of the most important criteria for performing microfluidic qPCR is the detection chemistry and qPCR conditions for all targets in the assay, which should be kept the same for every primer pair used. The microfluidic qPCR assay in this study was performed using a BioMark HD reader (Fluidigm, South San Francisco, CA). The assay allowed for the quantification of as low as two copies of target molecules per reaction. The PCR reaction efficiency of all targets in this assay ranged between 90% and 110%. The detection limit for spiked fecal samples ranged between 10^2 and 10^4 cells.

Ramalingam et al. (2010) fabricated a disposable microfluidic chip, which was preloaded with different primer pairs for the simultaneous real-time detection multiple waterborne pathogens (*Aeromonas hydrophila*, *K. pneumoniae*, *S. aureus*, and *P. aeruginosa*). PCR amplification was conducted using multiple parallel microreactors in a uniplex format. During fabrication of this microfluidic chip, different microreactors of the chip were deposited with different lyophilized primer pairs for the individual detection of pathogens in each microreactor. A loading channel was used to distribute the PCR mix with template DNA into each individual microreactor. The flow of liquid PCR mix dissolved the lyophilized primers and also purged out the air through a venting port. Amplification on the chip was measured using EvaGreen dye in a prototype real-time PCR, which was integrated to the chip. Using this assay, the lowest limit of detection (LLOD) of 1610 copies and 51 CFU/mL was achieved for *S. aureus* and *E. coli*, respectively.

8.8 Digital PCR

8.8.1 Description

Digital PCR (dPCR) is another advancement in the area of microfluidic PCR technology. In dPCR, the sample to be analyzed is diluted, divided, and distributed into hundreds of separate reaction chambers. The reduction in reaction volume leads to an increase in the rate of chemical reaction, an exponential increase in the ramp rate due to the increase in surface area to volume ratio, and a reduction in the amount of sample required (Markey et al. 2010). This sample distribution strategy provides each reaction chamber with either one copy of the target molecule or none. After PCR amplification, the number of positive and negative reaction chambers can be counted, and the number of target molecules in the sample can be accurately calculated. Initially, the hypothesis was tested using a 384-well plate with a reaction volume of 5 μL per partition or per well (Baker 2012). However, recent advancements in the area of nanofabrication and microfluidic technology allowed for the creation of chips that can generate millions of nanoliter or even picoliter size reaction chambers on a single chip. The sensitivity and resolution power of dPCR assays are determined by the number of reaction chambers present on the chip—the higher the number, the greater the sensitivity and resolution of the chip.

8.8.2 Advantages

Digital PCR uses the same set of primers and probes that is used for qPCR assays, but dPCR assays offer much higher sensitivity and precision. Using a conventional qPCR assay, it is hard to differentiate between samples that have less than a twofold gene expression or copy number difference. Using standard qPCR reactions also makes it challenging to detect rare alleles that exist at a frequency of less than 1%. On the other hand, a dPCR assay can measure much smaller gene expression differences and identify alleles that exist at a frequency of one in thousands. The other advantage of dPCR is that it does not require calibration and internal controls that are necessary for a qPCR assay (Baker 2012).

8.8.3 Designs

8.8.3.1 Chip Format

Fluidigm (San Francisco, CA) was the first company to commercialize dPCR in a chip format. The Fluidigm dPCR instrument consists of a system to mix the sample with the reagent, distribute the sample mixture on a chip, perform thermocycling, and read the results of each reaction chamber separately. The company manufactures two models of dPCR instruments, the simpler EP1 version can perform only endpoint detection, while the high-end model, BioMark HD system, can also perform qPCR.

8.8.3.2 Droplet Format

The droplet PCR is another format for performing dPCR. In digital droplet PCR, the sample is mixed with reagents, and the instrument generates tiny discrete aqueous droplets of nanometer to micrometer ranges, dispersed in a continuous oil phase. These tiny droplets are stabilized using oil and other stabilizing agents. The droplets are transferred to a thermocycler; PCR amplification is performed whereby each droplet acts as a separate reaction chamber and results are read separately for each droplet. The droplet PCR cannot perform qPCR assay but it generates high-quality digital PCR data owing to the generation of a higher number of droplets. This high-throughput droplet PCR offers a shorter amplification time and allows for the detection of even a single copy of RNA or virion (Park et al. 2011; Baker 2012).

8.8.4 Applications of dPCR

Gonzales et al. (2011) studied the development of a high-throughput OpenArray® qPCR assay (Life Technologies, Carlsbad, CA) for identification and virulence characterization of *E. coli* O157 and non-O157 serotypes using a panel of 28 potential target genes. A detection limit of 10^3–10^4 CFU/mL was achieved using this assay. The results obtained were comparable to results obtained from conventional real-time PCR. However, OpenArray allowed simultaneous quantification of a much higher number of targets in a single assay.

Zhu et al. (2012) reported a microfluidic emulsion PCR assay for the detection of a single cell of *E. coli* O157 in a high background microflora of 10^6 *E. coli* K12 cells. Because the emulsion PCR performs amplification of each DNA molecule in separate droplets, it allows for the quantification of pathogens at very low concentrations. The assay was performed with varying cell densities of *E. coli* O157 (0.5–0.00005 cells per droplet), while keeping the concentration of *E. coli* K12 cell concentration constant (50 cells per droplet). Specific fluorescently labeled primers were incorporated for each target. This assay was successfully able to detect *E. coli* O157 at all concentrations. The standard curve constructed using obtained data showed a correlation coefficient (R^2) of 0.99.

Sciancalepore et al. (2013) reported the development of a fast, culture-independent, microdroplet-based multiplex PCR method targeting two genes (tyrosine decarboxylase and agmatine deiminase) for the detection of biogenic amine producing bacteria. This microfluidic assay allowed for the integration of DNA isolation and PCR amplification on a single platform. PCR results of serially diluted cell suspensions (300, 150, 30, and 15 bacterial cells) showed no reduction in the amplification yield even at the lowest cell concentration. The total time needed for performing this assay was 32 min, which was far less than a standard qPCR assay.

8.9 Conclusions and Perspectives

Over the past decade, significant advances have been made in the area of nucleic acid–based detection technologies. These advances have revolutionized the technique of pathogen detection in the clinical, environmental, and food sectors. Despite the great contribution of PCR and PCR-related techniques in the field of food safety and pathogen detection, these methods are quite complex and expensive and require skilled labor to perform. With the onset of the new technologies described in this chapter, multiple considerations are required in order to fully exploit them as diagnostic tools for food industry application. These considerations include robustness, accuracy and sensitivity, rapidity, the requirement for much smaller samples and less dependence on highly skilled labor to operate, as well as be cost-effective in order to warrant their use. A powerful detection tool should have high-throughput capability, require a short analysis time, offer a low detection limit, and be portable.

Some of the methods highlighted in this chapter have great potential to be used as powerful tools for detecting food-borne pathogens, not only for routine pathogen testing by the food industry but also for epidemiological investigations in the event of an outbreak. At the high rate of research development and understanding of these powerful molecular techniques, it would be expected that many of these technologies will be developed and readily applied for pathogen detection in foods in the next 10–15 years.

REFERENCES

Ajitkumar, P., H. W. Barkema, and J. De Buck. 2012. Rapid identification of bovine mastitis pathogens by high-resolution melt analysis of 16S rDNA sequences. *Veterinary Microbiology* 155(2): 332–340.

An, L., W. Tang, T. A. Ranalli, H.-J. Kim, J. Wytiaz, and H. Kong. 2005. Characterization of a thermostable UvrD helicase and its participation in helicase-dependent amplification. *Journal of Biological Chemistry* 280(32): 28952–28958.

Azevedo, N. F., M. J. Vieira, and C. W. Keevil. 2003. Establishment of a continuous model system to study *Helicobacter pylori* survival in potable water biofilms. *Water Science and Technology* 47(5): 155–160.

Baker, M. 2012. Digital PCR hits its stride. *Nature Methods* 9(6): 541.

Breaker, R. R. and G. F. Joyce. 1994. A DNA enzyme that cleaves RNA. *Chemistry and Biology* 1(4): 223–229.

Brehm-Stecher, B. F., J. J. Hyldig-Nielsen, and E. A. Johnson. 2005. Design and evaluation of 16S rRNA-targeted peptide nucleic acid probes for whole-cell detection of members of the genus *Listeria*. *Applied and Environmental Microbiology* 71(9): 5451–5457.

Brehm-Stecher, B. F. and E. A. Johnson. 2007. Rapid methods for detection of *Listeria*. In *Listeria, Listeriosis, and Food Safety*, eds. E. Marth and E. Ryser, 3rd edn., pp. 257–281. Marcel Dekker, New York.

Casas-Finet, J. R. and R. L. Karpel. 1993. Bacteriophage T4 gene 32 protein: Modulation of protein-nucleic acid and protein-protein association by structural domains. *Biochemistry* 32(37): 9735–9744.

Castoldi, M., V. Benes, M. W. Hentze, and M. U. Muckenthaler. 2007. miChip: A microarray platform for expression profiling of microRNAs based on locked nucleic acid (LNA) oligonucleotide capture probes. *Methods* 43(2): 146–152.

Cheng, J.-C., C.-L. Huang, C.-C. Lin et al. 2006. Rapid detection and identification of clinically important bacteria by high-resolution melting analysis after broad-range ribosomal RNA real-time PCR. *Clinical Chemistry* 52(11): 1997–2004.

Chow, W. H. A., C. McCloskey, Y. Tong et al. 2008. Application of isothermal helicase-dependent amplification with a disposable detection device in a simple sensitive stool test for toxigenic *Clostridium difficile*. *Journal of Molecular Diagnostics* 10(5): 452–458.

Demidov, V. V., V. N. Potaman, M. D. Frank-Kamenetskil et al. 1994. Stability of peptide nucleic acids in human serum and cellular extracts. *Biochemical Pharmacology* 48(6): 1310–1313.

Desplats, C., C. Dez, F. Tétart, H. Eleaume, and H. M. Krisch. 2002. Snapshot of the genome of the pseudo-T-even bacteriophage RB49. *Journal of Bacteriology* 184(10): 2789–2804.

Dolinšek, J., C. Dorninger, I. Lagkouvardos, M. Wagner, and H. Daims. 2013. Depletion of unwanted nucleic acid templates by selective cleavage: LNAzymes open a new window for detecting rare microbial community members. *Applied and Environmental Microbiology* 79(5): 1534–1544.

Dunn, D. A., L. Valdivia, and H. Yu. Polymeric peptide probes and uses thereof. WIPO Patent 1996036734, issued November 22, 1996.

Egholm, M., O. Buchardt, L. Christensen et al. 1993. PNA hybridizes to complementary oligonucleotides obeying the Watson-Crick hydrogen-bonding rules. *Nature* 365(6446): 566–568.

Fiandaca, M., K Oliveira, and H. Stender. 2005. Hybridization of PNA probes in alcohol solutions. WIPO patent application PCT/US2005/019570, filed June 3, 2005.

Frech, G. C., D. Munns, R. D. Jenison, and B. J. Hicke. 2012. Direct detection of nasal *Staphylococcus aureus* carriage via helicase-dependent isothermal amplification and chip hybridization. *BMC Research Notes* 5(1): 430.

Gentile, N. L., A. M. Dillier, G. V. Williams et al. 2013. Verification of monoplex and multiplex linear-after-the-exponential PCR gene-specific sepsis assays using clinical isolates. *Journal of Applied Microbiology* 114(2): 586–594.

Gonzales, T. K., M. Kulow, D.-J. Park et al. 2011. A high-throughput open-array qPCR gene panel to identify, virulotype, and subtype O157 and non-O157 enterohemorrhagic *Escherichia coli*. *Molecular and Cellular Probes* 25(5): 222–230.

Gudnason, H., M. Dufva, D. D. Bang, and A. Wolff. 2007. Comparison of multiple DNA dyes for real-time PCR: Effects of dye concentration and sequence composition on DNA amplification and melting temperature. *Nucleic Acids Research* 35(19): e127.

Gundry, C. N., S. F. Dobrowolski, Y. Ranae Martin et al. 2008. Base-pair neutral homozygotes can be discriminated by calibrated high-resolution melting of small amplicons. *Nucleic Acids Research* 36(10): 3401–3408.

Gurtler, V., D. Grando, B. C. Mayall, J. Wang, and S. Ghaly-Derias. 2012. A novel method for simultaneous *Enterococcus* species identification/typing and *van* genotyping by high resolution melt analysis. *Journal of Microbiological Methods* 90(3): 167–181.

Hashimoto, M., P.-C. Chen, M. W. Mitchell, D. E. Nikitopoulos, S. A. Soper, and M. C. Murphy. 2004. Rapid PCR in a continuous flow device. *Lab on a Chip* 4(6): 638–645.

Hataoka, Y., L. Zhang, Y. Mori, N. Tomita, T. Notomi, and Y. Baba. 2004. Analysis of specific gene by integration of isothermal amplification and electrophoresis on poly (methyl methacrylate) microchips. *Analytical Chemistry* 76(13): 3689–3693.

Holland, C. A., and F. L. Kiechle. 2005. Point-of-care molecular diagnostic systems-past, present and future. *Current Opinion in Microbiology* 8(5): 504–509.

Isacsson, J., H. Cao, L. Ohlsson et al. 2000. Rapid and specific detection of PCR products using light-up probes. *Molecular and Cellular Probes* 14(5): 321–328.

Ishii, S., T. Segawa, and S. Okabe. 2013. Simultaneous quantification of multiple food-and waterborne pathogens by use of microfluidic quantitative PCR. *Applied and Environmental Microbiology* 79(9): 2891–2898.

Jacobsen, N., P. S. Nielsen, D. C. Jeffares et al. 2004. Direct isolation of poly (A)+ RNA from 4 M guanidine thiocyanate-lysed cell extracts using locked nucleic acid-oligo (T) capture. *Nucleic Acids Research* 32(7): e64–e64.

Jeong, Y.-J., K. Park, and D.-E. Kim. 2009. Isothermal DNA amplification in vitro: The helicase-dependent amplification system. *Cellular and Molecular Life Sciences* 66(20): 3325–3336.

Juhna, T., D. Birzniece, S. Larsson et al. 2007. Detection of *Escherichia coli* in biofilms from pipe samples and coupons in drinking water distribution networks. *Applied and Environmental Microbiology* 73(22): 7456–7464.

Kubota, K., A. Ohashi, H. Imachi, and H. Harada. 2006. Improved in situ hybridization efficiency with locked-nucleic-acid-incorporated DNA probes. *Applied and Environmental Microbiology* 72(8): 5311–5317.

Lagally, E. T., J. R. Scherer, R. G. Blazej et al. 2004. Integrated portable genetic analysis microsystem for pathogen/infectious disease detection. *Analytical Chemistry* 76(11): 3162–3170.

Lehtola, M. J., C. J. Loades, and C. William Keevil. 2005. Advantages of peptide nucleic acid oligonucle-otides for sensitive site directed 16S rRNA fluorescence in situ hybridization (FISH) detection of *Campylobacter jejuni, Campylobacter coli* and *Campylobacter lari. Journal of Microbiological Methods* 62(2): 211–219.

Lehtola, M. J., E. Torvinen, I. T. Miettinen, and C. William Keevil. 2006. Fluorescence in situ hybridization using peptide nucleic acid probes for rapid detection of *Mycobacterium avium* subsp. *avium* and *Mycobacterium avium* subsp. *paratuberculosis* in potable-water biofilms. *Applied and Environmental Microbiology* 72(1): 848–853.

Letertre, C., S. Perelle, F. Dilasser, K. Arar, and P. Fach. 2003. Evaluation of the performance of LNA and MGB probes in 5′-nuclease PCR assays. *Molecular and Cellular Probes* 17(6): 307–311.

Lezar, S. and E. Barros. 2010. Oligonucleotide microarray for the identification of potential mycotoxigenic fungi. *BMC Microbiology* 10(1): 87.

Markey, A. L., S. Mohr, and P. J. R. Day. 2010. High-throughput droplet PCR. *Methods* 50(4): 277–281.

Morrison, T. B., J. J. Weis, and C. T. Wittwer. 1998. Quantification of low-copy transcripts by continuous SYBR Green I monitoring during amplification. *Biotechniques* 24(6): 954–958.

Nielsen, P. E., M. Egholm, R. H. Berg, and O. Buchardt. 1991. Sequence-selective recognition of DNA by strand displacement with a thymine-substituted polyamide. *Science* 254(5037): 1497–1500.

Ong, D. C. T, W.-C. Yam, G. K. H. Siu, and A. S. G Lee. 2010. Rapid detection of rifampicin-and isoniazid-resistant *Mycobacterium tuberculosis* by high-resolution melting analysis. *Journal of Clinical Microbiology* 48(4): 1047–1054.

Ørum, H., P. E. Nielsen, M. Egholm, R. H. Berg, O. Buchardt, and C. Stanley. 1993. Single base pair mutation analysis by PNA directed PCR clamping. *Nucleic Acids Research* 21(23): 5332–5336.

Park, S., Y. Zhang, S. Lin, T.-H. Wang, and S. Yang. 2011. Advances in microfluidic PCR for point-of-care infectious disease diagnostics. *Biotechnology Advances* 29(6): 830–839.

Peleg, A. Y., Y. Tilahun, M. J. Fiandaca et al. 2009. Utility of peptide nucleic acid fluorescence in situ hybrid-ization for rapid detection of *Acinetobacter* spp. and *Pseudomonas aeruginosa. Journal of Clinical Microbiology* 47(3): 830–832.

Petersen, M., C. B. Nielsen, K. E. Nielsen et al. 2000. The conformations of locked nucleic acids (LNA). *Journal of Molecular Recognition* 13(1): 44–53.

Pierce, K. E., H. Khan, R. Mistry, S. D. Goldenberg, G. L. French, and L. J. Wangh. 2012. Rapid detection of sequence variation in *Clostridium difficile* genes using LATE-PCR with multiple mismatch-tolerant hybridization probes. *Journal of Microbiological Methods* 91(2): 269–275.

Pierce, K. E., R. Mistry, S. M. Reid et al. 2010. Design and optimization of a novel reverse transcription linear-after-the-exponential PCR for the detection of foot-and-mouth disease virus. *Journal of Applied Microbiology* 109(1): 180–189.

Pierce, K. E. and L. J. Wangh. 2011. LATE-PCR and allied technologies: Real-time detection strategies for rapid, reliable diagnosis from single cells. In *PCR Mutation Detection Protocols*, eds. B. D.M. Theophilus and R. Rapley, pp. 47–66. Humana Press, New York.

Porcellato, D., H. Grønnevik, K. Rudi, J. Narvhus, and S. B. Skeie. 2012a. Rapid lactic acid bacteria identifica-tion in dairy products by high-resolution melt analysis of DGGE bands. *Letters in Applied Microbiology* 54(4): 344–351.

Porcellato, D., H. M. Østlie, K. H. Liland, K. Rudi, T. Isaksson, and S. B. Skeie. 2012b. Strain-level character-ization of nonstarter lactic acid bacteria in Norvegia cheese by high-resolution melt analysis. *Journal of Dairy Science* 95(9): 4804–4812.

Provaznikova, D., T. Kumstyrova, R. Kotlin et al. 2008. High-resolution melting analysis for detection of MYH9 mutations. *Platelets* 19(6): 471–475.

Ramalingam, N., Z. Rui, H.-B. Liu et al. 2010. Real-time PCR-based microfluidic array chip for simultaneous detection of multiple waterborne pathogens. *Sensors and Actuators B: Chemical* 145(1): 543–552.

Rice, L. M., A. H. Reis, B. Ronish et al. 2013. Design of a single-tube, endpoint, linear-after-the-exponential-PCR assay for 17 pathogens associated with sepsis. *Journal of Applied Microbiology* 114(2): 457–469.

Runyon, G. T. and T. M. Lohman. 1989. *Escherichia coli* helicase II (*uvr*D) protein can completely unwind fully duplex linear and nicked circular DNA. *Journal of Biological Chemistry* 264(29): 17502–17512.

Sanchez, J. A., K. E. Pierce, J. E. Rice, and L. J. Wangh. 2004. Linear-after-the-exponential (LATE)–PCR: An advanced method of asymmetric PCR and its uses in quantitative real-time analysis. *Proceedings of the National Academy of Sciences of the United States of America* 101(7): 1933–1938.

Sciancalepore, A. G., E. Mele, V. Arcadio et al. 2013. Microdroplet-based multiplex PCR on chip to detect foodborne bacteria producing biogenic amines. *Food Microbiology* 35(1): 10–14.

Silahtaroglu, A. N., N. Tommerup, and H. Vissing. 2003. FISHing with locked nucleic acids (LNA): Evaluation of different LNA/DNA mixmers. *Molecular and Cellular Probes* 17(4): 165–169.

Singh, S. K., A. A. Koshkin, J. Wengel, and P. Nielsen. 1998. LNA (locked nucleic acids): Synthesis and high-affinity nucleic acid recognition. *Chemical Communications* 4: 455–456.

Stender, H., M. Fiandaca, J. J. Hyldig-Nielsen, and J. Coull. 2002. PNA for rapid microbiology. *Journal of Microbiological Methods* 48(1): 1–17.

Thorsen, T., S. J. Maerkl, and S. R. Quake. 2002. Microfluidic large-scale integration. *Science* 298(5593): 580–584.

Valdivia, L., Y. U. Hong, and D. A. Dunn. 1998. Polymeric peptide probes and uses thereof. European Patent EP 0871770, issued October 21, 1998.

Vester, B. and J. Wengel. 2004. LNA (locked nucleic acid): High-affinity targeting of complementary RNA and DNA. *Biochemistry* 43(42): 13233–13241.

Vincent, M., Yan X., and H. Kong. 2004. Helicase-dependent isothermal DNA amplification. *EMBO Reports* 5(8): 795–800.

Vossen, R. H. A. M., E. Aten, A. Roos, and J. T. den Dunnen. 2009. High-resolution melting analysis (HRMA): More than just sequence variant screening. *Human Mutation* 30(6): 860–866.

Whitesides, G. M. 2006. The origins and the future of microfluidics. *Nature* 442(7101): 368–373.

Wittwer, C. T., G. H. Reed, C. N. Gundry, J. G. Vandersteen, and R. J. Pryor. 2003. High-resolution genotyping by amplicon melting analysis using LCGreen. *Clinical Chemistry* 49(6): 853–860.

Yang, S., P. Ramachandran, R. Rothman et al. 2009. Rapid identification of biothreat and other clinically relevant bacterial species by use of universal PCR coupled with high-resolution melting analysis. *Journal of Clinical Microbiology* 47(7): 2252–2255.

Yilmaz, L. Ş., H. E. Ökten, and D. R. Noguera. 2006. Making all parts of the 16S rRNA of *Escherichia coli* accessible in situ to single DNA oligonucleotides. *Applied and Environmental Microbiology* 72(1): 733–744.

Zeinzinger, J., A. T. Pietzka, A. Stöger et al. 2012. One-step triplex high-resolution melting analysis for rapid identification and simultaneous subtyping of frequently isolated *Salmonella* serovars. *Applied and Environmental Microbiology* 78(9): 3352–3360.

Zhu, Z., W. Zhang, X. Leng et al. 2012. Highly sensitive and quantitative detection of rare pathogens through agarose droplet microfluidic emulsion PCR at the single-cell level. *Lab on a Chip* 12(20): 3907–3913.

9

Nanobiotechnology for the Detection of Food-Borne Pathogens

Niamh Gilmartin and Barry Byrne

CONTENTS

9.1 Introduction: Background and Driving Forces

A study in 2011 estimated that in the United States between the years 2000 and 2008, 9.4 million episodes of food-borne illness were caused by 31 major pathogens resulting in 55,961 hospitalizations and 1,351 deaths (Scallan et al. 2011). Similarly, in the European Union (EU), a total of 5,648 food-borne outbreaks were reported in 2011 resulting in 69,553 human cases, 7,125 hospitalizations, and 93 deaths (EFSA [European Food Safety Authority] 2013). This report also showed that in 2011 campylobacteriosis caused by the pathogen *Campylobacter* was the highest reported food-borne-related disease in humans (220,209 confirmed cases), followed by salmonellosis (95,548 confirmed cases), verotoxigenic *Escherichia coli* (VTEC) infection (9,485), and yersiniosis (7,017 confirmed cases). *Listeria* was seldom detected above the legal safety limit from ready-to-eat foods, resulting in only 1476 reported cases of listeriosis. Such statistics highlight the widespread prevalence of food-borne illness despite strict regulations governing food both at legislative and industrial levels. To protect public health, the monitoring for the presence of food-borne pathogens is critical, and EU commission regulation (EC) No. 2073/2005 sets maximum levels for certain contaminants in food, as summarized in Table 9.1.

The contamination of food by pathogenic organisms can have serious economic consequences and presents a considerable risk to human health even at low concentrations ($\approx$10–100 cells). To ensure the safety of the food we eat, it is imperative that these food-borne pathogens are detected at all stages of the food chain and that detection methods can meet the challenge of detection at such low concentrations. Due to the low numbers of pathogens that may be present in a sample, detection is difficult without highly sensitive detection systems. Culture methods remain the *gold standard* for detection, and they are the primary methods cited in the legislation for the determination of food-borne pathogens. While culture-based methods are relatively cost-effective, have good sensitivity, and ultimately provide the end user with useful qualitative and quantitative information, they are severely restricted by analysis time. For example, colony count estimation can range from 24 h for *E. coli* to 7 days for *Listeria monocytogenes*,

TABLE 9.1

Permitted Limits of Major Food-Borne Pathogens in EU

Pathogen	Acceptable Limits in Food
Salmonella	Absent within a set weight of food during shelf life
E. coli (VTEC/STEC)	50 CFU/g minced meat during processing (value set for *E. coli*)
	1000 CFU/g ready-to-eat fruit and vegetables (value set for *E. coli*)
L. monocytogenes	100 CFU/g ready-to-eat foods during their shelf life
Campylobacter	No defined limit

Source: European commission regulation (EC) No. 2073/2005 on microbial criteria for foodstuffs, 2005.

which is not conducive to rapid analysis. Furthermore, such lengthy analysis times present significant difficulties for quality control of semiperishable foods. These culture-based methods are also hampered by the need for a pre-enrichment step especially in cases where the pathogen of interest is present only in trace amounts, as is often the case in many food-borne illnesses where pathogens enter viable but nonculturable states due to environmental stress (Dreux et al. 2007). The consequence of this phenomenon is the underestimation of pathogen numbers and the ability of the cells to regain the potential to subsequently cause infection, thus posing a health risk.

Several sensitive immunological or molecular-based assays have been developed for food-borne pathogen detection (Adler et al. 2008; Byrne et al. 2009; Velusamy et al. 2010). A comparison of conventional culture methodology to enzyme-linked immunosorbent assay (ELISA), polymerase chain reaction (PCR), and lateral flow immunoprecipitation technologies for the detection of *Salmonella*, *Campylobacter*, *Listeria*, and *E. coli* O157:H7 in processed meat showed a percent agreement ranging from 80% to 100% depending on the pathogen detected and the method used (Bohaychuk et al. 2005). These assays have significant advantages over culture-based methods as they are rapid and can detect viable but nonculturable pathogens; however, the low pathogen numbers typically present in food samples and matrix effects from the food are limiting the detection level in these assays and more sensitive assays are required.

Even though state-of-the-art techniques for detecting microorganisms enable the quantification of very low concentrations of bacteria, to date it has been difficult to obtain successful results in real samples in a simple, reliable, and rapid manner. Advances in nanomaterial research have made it possible to achieve enhanced sensitivity, an improved response time, and increased potential for device portability through the use of nanomaterials. The combination of advances in nanotechnology and biosensors has resulted in increased focus on the use of nanobiotechnology for the development of sensitive assays for food-borne pathogen detection in food matrices (Gilmartin and O'Kennedy 2012). Nanostructured materials are typically between 1 and 100 nm in size and display unique properties not traditionally observed in their bulk counterparts, and the properties of the most common nanoparticles are shown in Table 9.2. Enhanced electrical/optical properties and increased surface area render them suitable for applications such as *label-free* detection and in the development of biosensors with enhanced sensitivities and improved response times (Yang et al. 2008; Zhao et al. 2009).

In this chapter, key developments in the area of nanobiotechnology for the detection of bacterial pathogens, such as *E. coli*, *L. monocytogenes*, and *Salmonella* species and *Campylobacter*, will be discussed and the potential and challenges of applying such technologies for high-throughput screening in complex food sample matrices will be evaluated.

9.2 Optical Detection

9.2.1 Colorimetric Detection

One of the first uses of nanoparticles in biosensing was that of colloidal gold in lateral flow immunoassays, which are one of the most important products of the diagnostics industry. In the presence of a pathogen, pathogen-specific antibodies labeled with gold nanoparticles (AuNP) result in a visible

TABLE 9.2

Description of Nanomaterials Commonly Used in Nanobiotechnology

Nanomaterial	Typical Diameter	Description
Quantum dots	2–10 nm	QDs are colloidal semiconducting fluorescent nanoparticles consisting of a semiconductor material core (normally cadmium mixed with selenium or tellurium), which has been coated with an additional semiconductor shell (usually zinc sulfide).
Magnetic nanoparticles	1–100 nm	These are made of compounds of magnetic elements such as iron, nickel, and cobalt and can be manipulated using magnetic fields.
Carbon nanotubes		CNTs are allotropes of carbon consisting of graphene sheets rolled up into cylinders. Multiwalled nanotubes (MWNTs) are essentially a number of concentric single-walled nanotubes (SWNTs).
SWNTs	0.4–3 nm	
MWNTs	2–100 nm	
AuNPs	5–110 nm	These are made of gold and may take the form of spheres, cubes, hexagons, rods, or nanoribbons.

Source: Adapted from Gilmartin, N. and O'Kennedy, R., *Enzyme Microb. Technol.*, 50(2), 87, 2012.

increase in color in the detection zone of the lateral flow devices (LFDs). Typical run times are less than 10 min, and the speed and simplicity of these methods have resulted in the development of commercial lateral flow immunoassay (LFIA)-based tests (e.g., Singlepath®, Duopath®, and RapidChek®) in the preliminary screening for pathogen contamination such as *Salmonella, Campylobacter, Listeria,* and *E. coli* O157:H7 in foods. The reported limit of detection (LOD) of bacteria using such strips ranges from 10^4 to 10^7 colony forming unit (CFU)/mL without an enrichment step (Hossain et al. 2012).

With the focus of research in nanobiotechnology shifting to highly sensitive detection and the exploitation of the properties of nanoparticles for this purpose, fewer colorimetric assays using nanoparticles are reported. However, the simplicity and low cost of colorimetric assays have led to the development of *on-slide* detection of the food-borne pathogen, *Campylobacter jejuni*, achieved using dual enlargement of AuNPs, first by Au atoms and, subsequently, by silver atoms. This simple and cost-effective detection format resulted in a strong color intensity that can easily be recognized by the unaided eye or measured by an inexpensive flatbed scanner (Cao et al. 2011). A PCR-based colorimetric assay, using single-stranded DNA (ssDNA) probes and nonfunctionalized gold nanoparticles, was also developed for the detection of *Salmonella* species (Prasad and Vidyarthi 2011). The simple, low-cost, and sensitive method relies upon the fact that in the absence of pathogen DNA, the ssDNA probe prevents the aggregation of AuNPs, resulting in a red color, whereas in the presence of pathogen DNA, dsDNA, using the ssDNA probe, is formed; therefore, the aggregation of the AuNPs occurs and the color changes to blue. The sensitivity and specificity of the colorimetric assay when compared to the conventional culture method were 89.15% and 99.04%, respectively, when food samples were tested (Figure 9.1).

9.2.2 Chemiluminescence and Fluorescence Detection

Fundamental drawbacks of traditional fluorophore-based assays include broad absorption and emission profiles, which reduce the possibility of multiplexing; however, quantum dots (QDs), with their very narrow emission bands, may overcome this. Using QDs, Zhao and colleagues detected three food-borne pathogenic bacteria, *Salmonella* Typhimurium, *Shigella flexneri*, and *E. coli* O157:H7, to a detection limit of 10^3 CFU/mL in milk and apple juice (Zhao et al. 2009).

It is the highly fluorescent properties of QDs that have helped achieve the ultimate goal in the detection of pathogens, namely, the detection of a single cell. Hahn et al. used streptavidin-conjugated QDs to detect a single *E. coli* O157:H7 bacterial pathogen, showing sensitivity improvements two orders of magnitude greater than conventional fluorescent dyes (Hahn et al. 2005). Here, *E. coli* cells were labeled with biotinylated anti-*E. coli* O157:H7 antibodies, which could complex to the streptavidin-conjugated QDs, thus facilitating their detection. While the detection of a single cell is in itself a significant achievement, single-cell detection in complex food matrix is even more remarkable. Zhao et al. detected a single

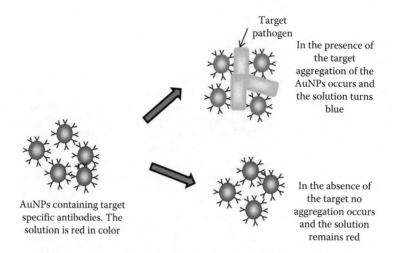

Target pathogen

In the presence of the target aggregation of the AuNPs occurs and the solution turns blue

AuNPs containing target specific antibodies. The solution is red in color

In the absence of the target no aggregation occurs and the solution remains red

FIGURE 9.1   Colorimetric detection using aggregation of gold nanoparticles.

E. coli O157 cell in a spiked ground beef sample using fluorescent nanoparticles conjugated to anti-*E. coli* O157 antibodies (Zhao et al. 2004). The capture of a single *Salmonella* bacterium was also achieved using nanoparticles; in this case, a gold/silica heterostructured nanorod scaffold functionalized with anti-*Salmonella* antibodies and a fluorescent dye (Fu et al. 2008).

A highly sensitive chemiluminescence-based assay utilizing the enhanced signal amplification properties was demonstrated by Wang et al. (2011). In this example, a sandwich complex was formed using *Salmonella* sp.–specific antibodies and antibody-coated AuNPs. Silver was then deposited around the AuNPs, resulting in the signal amplification of the chemiluminescence-based assay and a detection limit of 5 CFU/mL.

9.2.3 Light Scattering Detection

Surface-enhanced Raman scattering (SERS) greatly amplifies optical signals and therefore is one of the most sensitive diagnostic approaches available. A sensitive method combining immunomagnetic separation (IMS) and SERS was developed to enumerate *E. coli*. Magnetic spherical nanoparticles labeled with anti–*E. coli* antibodies were used to separate and concentrate the *E. coli* cells, which later interacted with the Raman labels (labeled gold nanorods) giving rise to a LOD of 8 CFU/mL (Guven et al. 2011). The multiplex detection using SERS of *S.* Typhimurium and *Staphlococcus aureus* pathogens was achieved using silicia-coated magnetic probes functionalized with pathogen-specific antibodies for capture followed by detection using pathogen-specific SERS probes (Wang et al. 2011). Detection limits of 10^3 CFU/mL were demonstrated by this study in complex food matrices (milk, spinach solution, and peanut butter), clearly demonstrating how such nanobiotechnology-based approaches can be successfully applied for sensitive pathogen detection in multiple food samples. Finally, the ability to distinguish between six species within the *Listeria* genus (including the human pathogen *L. monocytogenes*) based on a bacterial sample's SERS spectrum using a portable low-cost Raman spectral acquisition unit (Green et al. 2009) offers great potential for cost-effective, rapid, and sensitive assay for pathogen detection in food.

9.3 Electrical Detection

Electrical detection methods are appealing because of their low cost, low power consumption, ease of miniaturization, and potential multiplexing capability. Electrochemical sensors can be based on potentiometry, amperometry, voltammetry, coulometry, AC conductivity, or capacitance measurements. One of the advantages of using electrochemical detection is the portability and therefore accessibility of these systems to a variety of environments.

In the same way as nanoparticles amplified signals in optical-based detection of food pathogens, significant signal amplification can also be achieved through the use of nanoparticles in electrical detection systems. Lin et al. demonstrated a 13-fold increase in the signal achieved by combining the effects of ferrocenedicarboxylic acid and AuNPs in a disposable amperometric immunosensing strip for the rapid detection of *E. coli* O157:H7 (Lin et al. 2008). The detection limit of this assay was approximately 6 CFU/strip in phosphate buffered saline (PBS) buffer and 50 CFU/strip in milk. As a consequence of the coupling of anodic stripping voltammetry with copper-enhanced AuNP-labeled pathogen-specific antibodies, a highly sensitive electrochemical amplification immunoassay for *Salmonella* Typhi was created, resulting in a detection limit as low as 98.9 CFU/mL (Dungchai et al. 2008).

Over the past two decades, the benefits of label-free detection systems, such as a reduction in the resources required for assay development, simplified assay design, and reduced cost and assay times, have been recognized as offering many advantages over labeled systems. This is also true in the field of food-borne pathogen detection, and the harnessing of the electrical properties of nanomaterials such as carbon nanotubes has resulted in the development of numerous label-free detection systems. Zelada-Guillén et al. demonstrated the use of single-walled carbon nanotubes (SWCNTs) as excellent ion-to-electron transducers and covalently immobilized aptamers as biorecognition elements for the detection of *E. coli* CECT 675 cells, which are used as a model organism acting as a nonpathogenic surrogate for the pathogenic *E. coli* O157:H7 (Zelada-Guillén et al. 2010). The assay was carried out on real samples in a couple of minutes in a direct, simple, and selective way at concentration levels as low as 6 CFU/mL in complex matrices such as milk or 26 CFU/mL in apple juice. The use of electrical detection for multiplex samples is a challenge. The simultaneous, multiplexed determination of pathogenic bacteria in milk samples using a disposable electrochemical immunosensor was achieved by using pathogen-specific antibodies conjugated with nanocrystals with different releasable metal ions (Viswanathan et al. 2012). The multiwall carbon nanotube–polyallylamine-modified screen printed electrode (MWCNT-PAH/SPE)-based immunosensor showed an LOD of 400 cells/mL for *Salmonella*, 400 cells/mL for *Campylobacter*, and 800 cells/mL for *E. coli*.

9.4 Preconcentration Using Magnetic Nanoparticles

Sample preparation and the reduction of matrix effects have benefited from improvements in magnetic separation technology. Magnetic particles are coupled to a biorecognition element such as antibodies, aptamers, or nucleic acids, and the application of a magnetic field separates the target analyte from the matrix. Immunocapture of pathogenic bacteria is the most common method used, with capture efficiencies between 85% and 97% as observed in a buffer (Dudak et al. 2009) and 94% capture efficiency of *E. coli* O157:H7 cells as observed in beef extract (Varshney et al. 2005; Huang et al. 2011). Guven et al. (2011) observed a decrease in immunocapture efficiency when greater than 10^5 cells were present, and although this presents a difficulty in terms of accurate enumeration of the pathogen, this level far exceeds the permissible levels in food and thus does not represent a real limitation to the use of magnetic nanoparticles. Preconcentration of pathogenic bacteria via nanomagnetic separation coupled with sensitive detection methods such as chemiluminescence, *real-time* PCR, and QDs resulted in lower detection limits and quicker assays.

While magnetic nanoparticles have advantages in terms of rapid assay development, the challenges regarding robust surface chemistry to attach the antibody molecules to the surface still need to be overcome if the use of nanomagnetic particles is to become more widespread.

9.5 Future Trends

The upward trend of outbreaks of food-borne illness caused by pathogens such as *Campylobacter* and toxin-producing *E. coli* indicates the necessity to monitor food-borne contaminants throughout the food chain from production, processing, and distribution to the point of sale. Advances in assay portability and ease of use are required to meet this goal, and several on-site assays have been described. The in situ detection of *E. coli* O157:H7 and *S.* Typhimurium in real samples to a level of 3 and 15 CFU/mL was achieved using an immuno-AuNP network-based ELISA biosensor integrated with a sample concentration step based on immunomagnetic separation (Cho and Irudayaraj 2013). Further advances in pathogen detection were achieved by Wang et al. using a bionanotechnology-driven approach to first detect the presence of a pathogen and then its subsequent destruction (Wang et al. 2010). Antibody-conjugated oval-shaped AuNPs were used for the label-free detection of *S.* Typhimurium. Following detection, the bacteria–AuNP conjugate was exposed to near-infrared radiation, resulting in a decrease in cell viability due to photothermal lysis. This represents a significant advance in the use of nanobiotechnology for food safety, and if the health concerns of nanomaterials can be addressed, then this is a promising approach for the future.

Many current approaches for pathogen detection using nanoparticles rely on the use of antibodies to capture the pathogen; however, the use of aptamers in this regard is gaining in popularity. Aptamers are synthesized nucleic acid or peptide molecules that bind to a specific target and have numerous advantages over antibodies including improved stability, low cost, and ease of production. Aptamers were recently isolated for a number of pathogens including *E. coli* O157:H7 (Lee et al. 2012), *C. jejuni* (Dwivedi et al. 2010), and *S.* Typhimurium (Dwivedi et al. 2013), using a whole-cell systemic evolution of ligands by exponential enrichment (SELEX) method. The use of aptamers as an alternative to antibodies for food-borne pathogen detection is an emerging field; therefore, there are a limited number of reports of aptamer capture linked to nanomaterials. The potential of an aptamer-based capture and detection using nanomaterials was demonstrated by Ming et al. They showed the detection of *Salmonella paratyphi A* by harnessing using SWNTs and DNAzyme-labeled aptamer detection probes with an observed detection limit of 10^3 CFU/mL (Yang et al. 2013). Similarly, Zelada-Guillén et al. coupled SWNTs and aptamers as biorecognition elements for the electrical detection of *E. coli* CECT 675 cells (Zelada-Guillén et al. 2010). By harnessing the advantages over antibodies of aptamers as biorecognition elements and the increased sensitivity possible with nanomaterials, this field offers huge potential for the future of highly sensitive food-borne pathogen detection.

In the search for ultrasensitive detection methods, biobarcode assays (BCA) show great promise. These assays allow signal amplification by using target-specific antibody or aptamer-coated nanoparticle to concentrate the analyte and a second target-specific coated nanoparticle decorated with short stretches of DNA called DNA barcodes (Figure 9.2). The presence of the multiple copies of the DNA barcode per nanoparticle allows the detection of any analyte, even if only tiny amounts are present (Hearty et al. 2010). These assays offer highly sensitive detection capabilities and have significant potential for multiplexing.

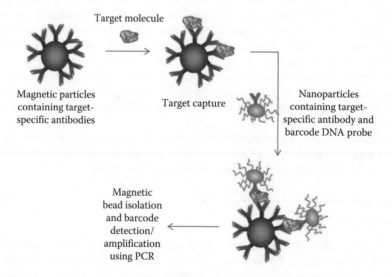

FIGURE 9.2   Principle of a biobarcode-based assay.

Initial studies in this area targeted prepared DNA or toxins of food-borne pathogens. In their DNA biobarcode assay, Zhang et al. avoided the use of PCR techniques by using a fluorescently labeled barcode DNA in a 1:100 target-specific DNA probe-to-barcode ratio for the detection of *Salmonella enterica* serovar. A second target-specific DNA probe was attached to magnetic nanoparticles and both probes were mixed with the DNA from *Salmonella* Enteritidis, thus forming a sandwich assay configuration. A magnetic field was applied to separate the sandwich from the unreacted materials, and the biobarcode DNA was released from the AuNPs and measured by fluorescence. The large number of barcode DNA per binding event means substantial amplification of the signal, and in this case, the detection limit of this BCA for the detection of *S.* Enteritidis was 1 ng/mL (Zhang et al. 2009). The detection limit was further improved to 0.5 ng/mL by using electrochemical detection rather than fluorescence, with lead sulfide (PbS) or cadmium sulfide (CdS) as nanotracers instead of fluorescein. This biobarcoded electrochemical biosensor was also used for the simultaneous detection of *Bacillus anthracis* and *S.* Enteritidis, with an observed detection limit of 50 pg/mL for *B. anthracis* (Zhang et al. 2010).

Similar biobarcode assay for toxins produced from food pathogens was reported for the detection of semicarbazide with an LOD of 8 pg/mL (Tang et al. 2011) and aflatoxin B-1 with an LOD 0.2 ng/mL (Tang et al. 2010). The use of a BCA assay in real food samples, namely, plain milk, green bean sprouts, and raw eggs showed an LOD of 13–26 CFU/mL for heat-killed and whole cell *S.* Typhimurium (Pratiwi et al. 2013). Future efforts in this area will need to focus on optimization of the DNA extraction method, but the ultrasensitivity of this assay shows great promise for the future of food-borne pathogen detection.

Lab-on-a-chip technology has recently been suggested as a perfect medium for portable and real-time medical diagnostics. Surely, there is no reason why one cannot use lab-on-a-chip for detecting food-borne pathogens. Lab-on-a-chip is a device that integrates several laboratory functions onto one small platform, typically only millimeters or centimeters in size. A lab-on-a-chip normally involves the handling of very small fluid volumes; this introduces the area of microfluidics that deals with the behavior, precise control, and manipulation of fluids that are constrained to a small submillimeter scale. Further, lab-on-a-chip technology utilizes a network of channels and wells that are etched onto glass, silicon, or polymer chips to build mini-laboratories. Pressure or electrokinetic forces move small volumes in a finely controlled manner through the channels. Lab-on-a-chip enables sample handling, mixing, dilution, electrophoresis, staining, and detection on a single integrated system. The main advantages of lab-on-a-chip are ease of use, speed of analysis, low sample and reagent consumption, and high reproducibility due to standardization, and automation. In summary, the lab-on-a-chip-based biosensor is a perfect medium to make portable and real-time biosensing of food-borne pathogens possible.

A recent lab-on-a-chip system for the label-free analysis of the highly toxic Staphylococcal enterotoxin B (SEB), using a biological semiconductor transducer, demonstrated an LOD of 5 ng/mL (Yang et al. 2010a). SEB samples were loaded into the device using a peristaltic pump, and the change in resistance resulting from antibody–antigen interactions was continuously monitored and recorded. This lab-on-a-chip system permits rapid detection and semiautomated operation, thus reducing the user handling of the sample. Yang et al. (2010b) also developed a lab-on-a-chip device by combining anti-SEB antibody–coated CNTs, enhanced chemiluminescence, and a cooled charge-coupled device (CCD) detector. SEB in buffer, soy milk, apple juice, and meat baby food was assayed with a detection limit of 0.01 ng/mL using the CCD detector, which was more sensitive than the conventional ELISA. The potential of lab-on-a-chip technology for the development and parallel processing of multiple analytes simultaneously was demonstrated by García-Aljaro et al. (2010). They reported CNT-based immunosensors for *E. coli* O157:H7 (bacterium) and T7 bacteriophage (virus). The transduction element consisted of SWCNTs, which were functionalized with specific antibodies, aligned in parallel bridging two gold electrodes to function as a chemiresistive biosensor. The detection limits demonstrated were 10^3–10^5 CFU/mL for *E. coli* O157:H7 and 10^3 pfu/mL for T7 bacteriophage. Although lab-on-a-chip technology offers huge advantages to contaminant detection in food, the matrix effects and complications presented by this matrix still need to be overcome. The incorporation of a rapid sample preparation protocol is crucial to reduce detection time, increase sensitivity, and thus avail of the power of lab-on-a-chip technology to position immunoassay-based assays at the forefront of screening procedures for food safety.

9.6 Conclusion

The small size of the nanomaterials has resulted in large surface-to-volume ratios and is responsible for superior electronic and optical properties, when compared with the bulk material. Their use in nano-biotechnology applications has facilitated the interaction with an increased number of target molecules, leading to the development of biosensors with enhanced sensitivities and improved response times, which is essential for the detection of food-borne pathogens. Nanomaterials have also proven invaluable in the signal amplification of detection assays resulting in the detection of a single cell in some foods; however, the matrix effects of some foods continue to hamper the sensitivity of some assays. Nanobiotechnology, including rapid tests; immunosensor-based applications, and nucleic acid–based assays, have revolutionized the field of food pathogen detection by reducing analysis times and facilitating the positive identification of the pathogen of interest, for example, determining the species of a bacterial pathogen, in a sensitive and specific manner. The ability to translate these technologies into assay configurations, whereby multiple targets can be analyzed simultaneously, is a major advantage, especially given that multiple pathogens may be potentially responsible for a single case of food-borne poisoning.

Despite the clear advantages of the promising approaches demonstrated to date, much work still needs to be done before they become a viable alternative to conventional methods. The main challenges to overcome are the sample matrix effects of the food on the sensitivity of the assays, the time it takes to do the assays, and the amount of sample preparation required. Certainly in terms of matrix effects and sample preparation the pathogen specific magnetic nanoparticles, used to concentrate the pathogen, partially addresses this concern but their widespread adoption in the food industry has not materialized. The rapid nature of the nanobiotechnology-based approaches to food pathogen detection offers huge potential for both cost and time saving for the food industry; however, in order for increased uptake of these methods by the food industry, the validation against the conventional methods is paramount.

The continued growth of the food sector relies heavily on the ability to produce food that is safe to consume. The monitoring of contamination in industrial food-producing plants is not only a legislative requirement but is paramount to building consumer confidence, and as a result, it is essential to provide companies with analytical techniques that allow them to monitor food-borne contamination in a fast, safe, and efficient manner. With advances in nanobiotechnology, microfluidics, and miniaturized devices, portable, rapid, sensitive, easy-to-use diagnostic tools for in situ food contaminant detection are fast becoming a reality. Due to the low level of bacterial pathogens ($\approx$10–100 cells) in food that can have an adverse effect on human health, it is critical that monitoring of food-borne pathogens occurs at all stages of the food chain to ensure the safety of the food we eat. As a consequence, *farm-to-fork* monitoring of food for contaminants will reduce the number of food-borne incidents, thus increasing consumer confidence in the quality and safety of food and therefore strengthening the competiveness of the industry.

ACKNOWLEDGMENTS

The financial support of Science Foundation Ireland (SFI, grant no. 10/CE3/B1821) and EU seventh Framework Grant No. FP7-SME-2011-3-286713 is gratefully acknowledged.

REFERENCES

Adler, M, R Wacker, and CM Niemeyer. 2008. Sensitivity by combination: Immuno-PCR and related technologies. *Analyst* 133(6): 702–718.

Bohaychuk, VM, GE Gensler, RK King, JT Wu, and LM McMullen. 2005. Evaluation of detection methods for screening meat and poultry products for the presence of foodborne pathogens. *J. Food Prot.* 68(12): 2637–2647.

Byrne, B, E Stack, N Gilmartin, and R O'Kennedy. 2009. Antibody-based sensors: Principles, problems and potential for detection of pathogens and associated toxins. *Sensors* 9(6): 4407–4445.

Cao, C, LC Gontard, LL Thuy Tram, A Wolff, and DD Bang. 2011. Dual enlargement of gold nanoparticles: From mechanism to scanometric detection of pathogenic bacteria. *Small* 7(12): 1701–1708.

Cho, I-H and J Irudayaraj. 2013. In-situ immuno-gold nanoparticle network ELISA biosensors for pathogen detection. *Int. J. Food Microbiol.* 164(1): 70–75.

Dreux, N, C Albagnac, M Federighi, F Carlin, CE Morris, and C Nguyen-the. 2007. Viable but non-culturable *Listeria monocytogenes* on parsley leaves and absence of recovery to a culturable state. *J. Appl. Microbiol.* 103(4): 1272–1281.

Dudak, FC, IH Boyaci, A Jurkevica, M Hossain, Z Aquilar, HB Halsall, CJ Seliskar, and WR Heineman. 2009. Determination of viable *Escherichia coli* using antibody-coated paramagnetic beads with fluorescence detection. *Anal. Bioanal. Chem.* 393(3): 949–956.

Dungchai, W, W Siangproh, W Chaicumpa, P Tongtawe, and O Chailapakul. 2008. *Salmonella typhi* determination using voltammetric amplification of nanoparticles: A highly sensitive strategy for metalloimmunoassay based on a copper-enhanced gold label. *Talanta* 77(2): 727–732.

Dwivedi, HP, RD Smiley, and L-A Jaykus. 2010. Selection and characterization of DNA aptamers with binding selectivity to *Campylobacter jejuni* using whole-cell SELEX. *Appl. Microbiol. Biotechnol.* 87(6): 2323–2334.

Dwivedi, HP, RD Smiley, and L-A Jayjus. 2013. Selection of DNA aptamers for capture and detection of *Salmonella* Typhimurium using a whole-cell SELEX approach in conjunction with cell sorting. *Appl. Microbiol. Biotechnol.* 97(8): 3677–3686.

EFSA (European Food Safety Authority), ECDC (European Centre for Disease Prevention and Control). 2013. The European Union summary report on trends and sources of zoonoses, trends and sources of zoonoses, zoonotic agents and food-borne outbreaks in 2011. *EFSA J.* 11(4): 1–250.

Fu, J, B Park, G Siragusa, L Jones, R Tripp, Y Zhao, and Y-J Cho. 2008. An Au/Si hetero-nanorod-based biosensor for *Salmonella* detection. *Nanotechnology* 19(15): 155502.

García-Aljaro, C, LN Cella, DJ Shirale, M Park, FJ Muñoz, MV Yates, and A Mulchandani. 2010. Carbon nanotubes-based chemiresistive biosensors for detection of microorganisms. *Biosens. Bioelectron.* 26(4): 1437–1441.

Gilmartin, N and R O'Kennedy. 2012. Nanobiotechnologies for the detection and reduction of pathogens. *Enzyme Microb. Technol.* 50(2): 87–95.

Green, GC, ADC Chan, and BS Luo. 2009. Identification of species using a low-cost surface-enhanced Raman scattering system with wavelet-based signal processing. *IEEE Trans. Instrum. Meas.* 58(10): 3713–3722.

Guven, B, N Basaran-Akgul, E Temur, U Tamer, and IH Boyaci. 2011. SERS-based sandwich immunoassay using antibody coated magnetic nanoparticles for *Escherichia coli* enumeration. *Analyst* 136(4): 740–748.

Hahn, MA, JS Tabb, and TD Krauss. 2005. Detection of single bacterial pathogens with semiconductor quantum dots. *Anal. Chem.* 77(15): 4861–4869.

Hearty, S, P Leonard, and R O'Kennedy. 2010. Barcodes check out prostate cancer a 7-nm light pen makes its mark. *Nat. Nanotechnol.* 5: 9–11.

Hossain, SMZ, C Ozimok, C Sicard, SD Aguirre, MM Ali, Y Li, and JD Brennan. 2012. Multiplexed paper test strip for quantitative bacterial detection. *Anal. Bioanal. Chem.* 403(6): 1567–1576.

Huang, H, C Ruan, J Lin, M Li, LM Cooney, WF Oliver, Y Li, and A Wang. 2011. Magnetic nanoparticle based magnetophoresis for efficient separation of *E. coli* O157:H7. *Trans. ASAE* 54(3): 1015–1024.

Lee, YJ, SR Han, J-S Maeng, Y-J Cho, and S-W Lee. 2012. In vitro selection of *Escherichia coli* O157:H7-specific RNA aptamer. *Biochem. Biophys. Res. Commun.* 417(1): 414–420.

Lin, Y-H, S-H Chen, Y-C Chuang, Y-C Lu, TY Shen, CA Chang, and C-S Lin. 2008. Disposable amperometric immunosensing strips fabricated by Au nanoparticles-modified screen-printed carbon electrodes for the detection of foodborne pathogen *Escherichia coli* O157:H7. *Biosens. Bioelectron.* 23(12): 1832–1837.

Prasad, D and AS Vidyarthi. 2011. Gold nanoparticles-based colorimetric assay for rapid detection of *Salmonella* species in food samples. *World J. Microbiol. Biotechnol.* 27(9): 2227–2230.

Pratiwi, FW, P Rijiravanich, M Somasundrum, and W Surareungchai. 2013. Electrochemical immunoassay for *Salmonella* Typhimurium based on magnetically collected Ag-enhanced DNA biobarcode labels. *Analyst* 138(17): 5011–5018.

Scallan, E, RM Hoekstra, FJ Angulo, RV Tauxe, M-A Widdowson, SL Roy, JL Jones, and PM Griffin. 2011. Foodborne illness acquired in the United States—Major pathogens. *Emerg. Infect. Dis.* 17(1): 7–15.

Tang, D, Yongliang Y, R Niessner, M Miró, and D Knopp. 2010. Magnetic bead-based fluorescence immunoassay for aflatoxin B1 in food using biofunctionalized rhodamine B-doped silica nanoparticles. *Analyst* 135(10): 2661–2667.

Tang, Y, L Yan, J-j Xiang, W-z Wang, and H-y Yang. 2011. An immunoassay based on bio-barcode method for quantitative detection of semicarbazide. *Eur. Food Res. Technol.* 232(6): 963–969.

Varshney, M, LJ Yang, XL Su, and YB Li. 2005. Magnetic nanoparticle-antibody conjugates for the separation of *Escherichia coli* O157:H7 in ground beef. *J. Food Prot.* 9: 1804–1811.

Velusamy, V, K Arshak, O Korostynska, K Oliwa, and C Adley. 2010. An overview of foodborne pathogen detection: In the perspective of biosensors. *Biotechnol. Adv.* 28(2): 232–254.

Viswanathan, S, C Rani, and JA Ho. 2012. Electrochemical immunosensor for multiplexed detection of food-borne pathogens using nanocrystal bioconjugates and MWCNT screen-printed electrode. *Talanta* 94: 315–319.

Wang, S, AK Singh, D Senapati, A Neely, H Yu, and PC Ray. 2010. Rapid colorimetric identification and targeted photothermal lysis of *Salmonella* bacteria by using bioconjugated oval-shaped gold nanoparticles. *Chemistry* 16(19): 5600–5606.

Wang, Y, S Ravindranath, and J Irudayaraj. 2011. Separation and detection of multiple pathogens in a food matrix by magnetic SERS nanoprobes. *Anal. Bioanal. Chem.* 399(3): 1271–1278.

Wang, Z, N Duan, J Li, J Ye, S Ma, and G Le. 2011. Ultrasensitive chemiluminescent immunoassay of *Salmonella* with silver enhancement of nanogold labels. *Luminescence* 26(2): 136–141.

Yang, H, H Li, and X Jiang. 2008. Detection of foodborne pathogens using bioconjugated nanomaterials. *Microfluid. Nanofluid.* 5: 571–583.

Yang, M, Z Peng, Y Ning, Y Chen, Q Zhou, and L Deng. 2013. Highly specific and cost-efficient detection of *Salmonella* Paratyphi A combining aptamers with single-walled carbon nanotubes. *Sensors* 13(5): 6865–6881.

Yang, M, S Sun, HA Bruck, Y Kostov, and A Rasooly. 2010a. Electrical percolation-based biosensor for real-time direct detection of staphylococcal enterotoxin B (SEB). *Biosens. Bioelectron.* 25(12): 2573–2578.

Yang, M, S Sun, Y Kostov, and A Rasooly. 2010b. Lab-on-a-chip for carbon nanotubes based immunoassay detection of staphylococcal enterotoxin B (SEB). *Lab on a Chip* 10(8): 1011–1017.

Zelada-Guillén, GA, SV Bhosale, J Riu, and FX Rius. 2010. Real-time potentiometric detection of bacteria in complex samples. *Anal. Chem.* 82(22): 9254–9260.

Zhang, D, DJ Carr, and EC Alocilja. 2009. Fluorescent bio-barcode DNA assay for the detection of *Salmonella enterica* serovar Enteritidis. *Biosens. Bioelectron.* 24(5): 1377–1381.

Zhang, D, MC Huarng, and EC Alocilja. 2010. A multiplex nanoparticle-based bio-barcoded DNA sensor for the simultaneous detection of multiple pathogens. *Biosens. Bioelectron.* 26(4): 1736–1742.

Zhao, XJ, LR Hilliard, SJ Mechery, YP Wang, RP Bagwe, SG Jin, and WH Tan. 2004. A rapid bioassay for single bacterial cell quantitation using bioconjugated nanoparticles. *Proc. Natl. Acad. Sci. U.S.A.* 101(42): 15027–15032.

Zhao, Y, M Ye, Q Chao, N Jia, Y Ge, and H Shen. 2009. Simultaneous detection of multifood-borne pathogenic bacteria based on functionalized quantum dots coupled with immunomagnetic separation in food samples. *J. Agric. Food Chem.* 57(2): 517–524.

10

Molecular Nanotechnology: Rapid Detection of Microbial Pathogens in Food

Swati Vyas and Vandana B. Patravale

CONTENTS

10.1 Introduction

Pathogenic microorganisms are a critical concern in food industries and water management facilities due to rapid microbial load buildup and toxic effects on public health. Important contamination sources in the dairy products industry are the surroundings and the stagnancy occurring in milk-processing equipment encouraging microbial growth (Brooks and Flint 2008). Online process control is necessary to ensure that under the best possible hygienic circumstances of milking, pasteurization, and cooling equipment, the somatic cell count is less than 1500 per mL of milk according to the guidelines in the National Dairy Code slated by the Canadian Dairy Information Centre (CDIC) (Code, National Dairy 2005). Infections are also caused from inadvertent consumption of untreated drinks, including apple juice and unpasteurized raw milk. Swift monitoring and correct analysis of food products are vital to ensure the safety of the end consumers. Concomitantly, the recognition of pathogens in food, water, and air has become a considerable issue for researchers. Moreover, conventional methods necessitate the presence of complex instrumentation facilities and skilled personnel and are not amenable to be used on-site. From infected food and contaminated water to outbreaks and pandemics, pathogenic agents cause detrimental diseases in unsuspecting consumers.

Rapid analytical time is indisputably key to a successful detection technology and is especially critical for directing instantaneous actions toward treating unfortunate victims of pathogenic attacks and curbing spread. Furthermore, simultaneous recognition and identification of diverse species of microbes is essential. Innovations leading to on-site deployable detection tests have become an imperative requisite in constant preemptive monitoring for preventing and containing outbreaks. Compact biosensing devices are already being developed with attractive traits such as sensitivity, specificity, robustness, and high-speed readouts, to mediate timely pathogenic detection (Jianrong et al. 2004). Several scientists have worked to build such devices and with due success, keeping in view such requisite properties by applying core nanotechnology-based techniques centric to the use of nanomaterials to accomplish the recognition of infectious pathogens, especially multicomponent opaque media like milk and other dairy products (Arora et al. 2011). A vast range of nanomaterials from silver nanoparticles and their surface plasmonic waves to liposomes and even viruses such as bacteriophages for the detection of microbial toxins, surface antigens, and nucleic acids can be accomplished with remarkable sensitivity and limits of detection (Arora et al. 2013). Biorecognition can be achieved by either immunosensing, enzymatic, or nucleic acid detection. Immunological sensors rely on the antigen–antibody reactions specific for target cells. The bioconjugates thus formed can be analyzed and quantified using a multitude of sensing technologies such as fluorescence, electrical conductance, cantilevers, surface plasmon resonance, and immunomagnetic separations (Pei et al. 2013). Biosensor probes can also detect nucleic acid sequences through hybridization augmented using labels, redox chemistry, and intercalating agents as strategies and pinpoint the identity of the relevant pathogenic species present in samples (Hahm 2013). Successful research findings have encouraged the implementation of nanoscale diagnostics for the prevention and containment of infectious diseases in varied sectors worldwide, averting outbreaks and conserving public health and wellness.

Even though conventional techniques can be authentically used for the detection of pathogens, they involve prolonged processes and extensive protocols engaging either the culturing methods or the identification of specific metabolites and cellular multiplication cycles. In this chapter, we describe contemporary biosensing technologies founded upon micro- and nanofabrication. These technologies permit a tremendous expansion of detection limits, consequently shortening the time needed for yielding confirmatory results with little or no sample preparation.

10.2 Discrepancies in Contemporary Methods

Traditional biomolecular diagnostic methodologies are nonetheless employed, in some cases even as gold standard, in research laboratories all over the globe to detect disease-causing agents with a reasonable level of sensitivity and reproducibility. Undeniably, the majority of these techniques are ineffective on sites such as airports, the food industry and food distribution centers, or in underdeveloped states where resources are inadequate, as they frequently require complicated, costly instrumentation facilities and skilled personnel. Moreover, the elevated cost and diminutive shelf life of some materials such as enzymes restrain the utilities of standard microbial identification techniques in developing countries. Even with their accuracy and sensitivity, current technologies like enzyme-linked immunosorbent assay (ELISA) and polymerase chain reaction (PCR) need elaborate sample preparations and have prolonged result delivery times that hinder rapid response and disease therapy. Conventionally, the occurrence of most microbes including bacteria, protozoa, and fungi has been identified using optical microscopes, typically post growth in pure culture (Petrenko and Sorokulova 2004). Typically, an aliquot from the infected sample is visually examined in the field of the microscope for the presence of the microbial pathogen. Additionally, for bacteria or fungi, the following verification concerns the appearance of growth patterns in differential media and through other biochemical testing procedures (Jeppesen 1995). However, there are various discrepancies in conventional techniques; for instance, samples contain a large number of pathogens and typically require a day-long incubation to produce results. Another major limitation of culturing is that some pathogens fail to grow and their detection becomes difficult (Rompre et al. 2002). In case of virus particles that display sizes in the nanoscale range, high-capacity electron microscopes become indispensable for visualization, and growing viruses is all the more complex before investigation (Colliex and Mory 1994).

10.3 Leading Nanotechnology-Based Methods: Transition from Recognition of Whole-Cell Microbes to Smaller Biomolecules

Distinctive luminescent, electrical, and magnetic properties of nanomaterials that can aid rapid detection of microbial pathogens can be used to build quicker, sensitive, and inexpensive detection bioassays. Although accuracy, sensitivity, and speed are critical parameters, many research efforts have been directed toward fabricating nanosystems that are economical, vigorous, reliable, and reproducible. Such systems can be easily made the point of care for applications in rustic areas of developing countries. Recent progress in nanofabrication has presented distinctive advantages for developing biosensors for pathogenic identification in many ways. Biosensor probes produced with equivalent or smaller measurements of a microbial cell can offer high sensitivity and low limits of detection. For example, well-designed probes such as nanotubes, nanoparticles, nanowires, and nanomechanical devices are used for identifying pathogens (Palchetti and Mascini 2008). Additionally, this technology can incorporate numerous processes sequentially for uni-step biosensing or in parallel for high-throughput monitoring (Gehring and Tu 2011). An important outcome of nanofabrication technology is the creation of nanofluidic devices, termed lab-on-a-chip (Henderson 2007). These systems present a suitable platform for biomolecular sensing in both immunological recognition and nucleic acid detection. Nanodevices normally consist of channels and biosensing compartments with dimensions of a few to hundreds of nanometers (Vo-Dinh 2002). Therefore, they entail diminutive quantities of samples and reagents. The small dimensions offer high surface to volume ratio, facilitating localization of target molecules in the *sensor* area. Additionally, quick mass transfer in the channels significantly shortens analysis time. Because a biosensor is typically made of nanomaterials, the nanosurface can be readily functionalized to specifically confine target microbial cells under constant flow circumstances (Kurkina and Balasubramanian 2012). This chapter will describe contemporary efforts of detecting pathogens using nanofabricated devices. Moreover, by using novel approaches, nanotechnology holds the potential to develop bioassays that can be conducted in dense media such as blood and milk, with no sample preparation, yielding rapid and reliable results in uncomplicated and end user–friendly configurations as described in the following sections.

10.4 Means of Detection at Nanoscale

10.4.1 Surface Markers

The microbial display of components on their cellular surfaces as a means for recognition has been widely researched, where the surface pharmacology varies depending upon microbial species. Most of these surface components are proteins, carbohydrates, lipoproteins, lipopolysaccharides, and peptides that are sensed through capture by specific antibodies. Immunoassays that may employ fluorescent, enzyme, and radiolabels or label-free methods utilize such specific antibodies, often monoclonal, that can trap antigenic microbes by binding to precise cellular structures on the microbial surface.

Salmonella enterica is a notorious food-borne pathogen that displays surface antigenic proteins such as H:gm flagellar protein and SefA fimbrial protein, recognized by the innate and adaptive immune systems (Nhan et al. 2011). These proteins are restricted to *Sa. enterica* and the closely associated group D *Salmonella*, and are pivotal contributors in *Salmonella* infections. Typically, human *Salmonella* infections are caused due to bacterial exposure from consumption of food products such as infected poultry and livestock, common sources harboring *Salmonella* spp. (Stevens et al. 2009). Likewise, microbial diseases such as enteritis, arthritis, plague, and lymphadenitis are instigated in humans due to the prevalence of bacteria from the genus *Yersinia* (Leo and Skurnik 2011). This particular group of microbes causes infection by binding to host cells and releasing proteins intrinsically that deter and disrupt cellular functions such as phagocytosis in macrophages. *Yersinia enterocolitica* harbors such a protein, Yop51, a protein tyrosine phosphatase, which is also a surface epitope and is recognizable as a distinguishing element (Starnbach and Bevan 1994).

Viral food-borne infectious agents such as noroviruses frequently spread through contaminated foods spiked by infected food handlers causing gastroenteritis in humans. Genetic materials of noroviruses are housed in capsid protein amino acid sequences. The monomeric protein of virus-like particles emitted by baculovirus upon capsid protein expression is composed of a shell domain and a protruding, domain or the P domain (Lindesmith et al. 2012). Within the subdivisions of the P domain, namely P1 and P2, the P2 subdomain is involved in interactions with neutralizing antibodies and can be rapidly detected by them. Opportunist human pathogens like fungi and fungal-like organisms account for significant large-scale food production damages by infecting roots of major food crops such as rice, wheat, maize, potatoes, and soybeans (*Pythium* and *Fusarium* spp.) and decaying post harvest stock caused by the *Geotrichum, Mucor,* and *Rhizopus* fungal species (Thornton and Wills 2013). Yeast-like fungi such as *Geotrichum candidum* commonly found in tomatoes harbor an immunogenic high molecular weight (~200 kDa) extracellular polysaccharide surface antigen in the arthroconidia and hyphae regions (Thornton and Wills 2013). Recombinant cell wall galactomannoprotein from *Aspergillus fumigatus* is yet another surface marker helping the confirmation of pulmonary aspergillosis by immunoassays.

Vibrio cholerae variants display surface phenotypes that differ in their susceptibility to antibody-dependent recognition. Surface epitopes comprising capsular polysaccharides and lipopolysaccharides in these species bind to antibodies and initiate antibody-mediated lytic cycle, and this property renders them feasible as indicators of bacterial infection (Attridge and Holmgren 2009). Brucellosis is another food-borne pathological condition that causes febrile illness in humans and the causative microbes are present in infected dairy products. The outer membrane proteins of *Brucella* spp. such as r31CSP (31 kDa cell surface protein of *Brucella abortus*) are recognizable elements and monoclonal antibodies generated against them in murine models have shown zero cross reactivity with *Yersinia, Salmonella,* and *V. cholerae,* and have been implicated in the detection of infected samples (Kumar et al. 2007). The cell wall structural component (lipopolysaccharide) of *Brucella* spp. has a polysaccharide scaffold that is highly precise and specific to the microbe, and anti-lipopolysaccharide antibodies have been identified in milk, and serum samples of infected animals, paving the way for use of this isolated component in differential serological diagnosis of brucellosis.

10.4.2 Nucleic Acid Microarrays and Real-Time PCR

Traditional microbial cultures even today are the gold standard for the detection of food-borne pathogens, of which 31 species are majorly estimated to cause approximately 10 million illnesses, hospitalizations, and deaths annually (Scallan et al. 2011). These techniques are accurate, but undoubtedly time-consuming, and do not provide quantitative data. High-throughput, microarray platforms are high-speed detection tools that can screen thousands of samples without compromising on the efficiency, accuracy, and sensitivity. Rapid detection of food-borne illnesses is a cause for concern, and microarray technology offers fast, accurate, reliable, and sensitive equipment for quick identification and screening of thousands of samples in a very short span of time (Seidel and Niessner 2008).

Microarrays employ antibodies or nucleic acid probes for the specific detection of microbial variants with accurate cataloging for microbial phenotypes and genotypes (Gehring et al. 2013). Apart from detecting whole-cell microbes and nucleic acids, this technology can also be similarly applied to identifying toxins unleashed in food samples from pathogens. Although this technology often vastly employs labels such as fluorescent probes, label-free detection strategies have also been developed for rapid quantitative detection (Figure 10.1).

Escherichia coli is designated as a major food adulterant present in meat products by the Food Safety and Inspection Service of the United States Department of Agriculture (Federal, Register 2012).

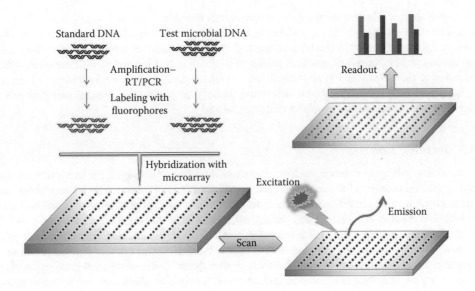

FIGURE 10.1 Nucleic acid microarray technology.

Antibody microarrays have been recently developed by Hegde et al. for the detection of various target groups including six important non-O157 serogroups, O26, O45, O103, O111, O121, and O145 serogroups of *E. coli* (Hegde et al. 2013). This technology requires minimum antibody concentrations and facilitates rapid identification of *E. coli* bacteria in food samples, thus increasing food safety. Microarrays can provide significant assistance in tracking sources of contamination in case of outbreaks and additionally distinguish possible epidemic-associated strains. *Salmonella* spp., an important pathogenic group for disease estimation, can be detected with fast high-throughput microarray assays that are reliable, sensitive, and accurate. Recently, Guo et al. designed microarrays that detect O-antigen-specific genes of *Salmonella* that can identify all serotypes (Guo et al. 2013). Detection of enteropathogenic bacteria such as *Yersinia* and emerging food contaminant (Hanifian and Khani 2012), Hepatitis E virus (Wacheck et al. 2012), which frequently appear in slaughter animals, is consequently critical for the safety of food in farming and processed meat industries. Lateral flow immunoassays (LFIA) based on antibodies can capture immobilized recombinant antigens of Hepatitis E virus and outer protein D of *Yersinia* spp. (Wutz et al. 2013). LFIAs have been implemented as the basis of identification using microarrays and can rapidly aid detection of potentially present food pathogens functioning as a promising analytical tool.

Nucleic acids are indispensable components that form a precise blueprint as a decipherable facet for mapping a particular species and are of tremendous importance in recognizing variants, genotypes, and phenotypes of an organism, thus enabling detection. Greatly accurate and sensitive amplification and electrophoresis techniques such as PCR are available for identification and these can pinpoint the occurrence of microbial genetic material in suspected food samples (Quigley et al. 2011). Real-time PCR technique is a more advanced method that displays high specificity, sensitivity, and reasonable rapidity for the simultaneous detection of microbial nucleic acids belonging to different species (Maurin 2012). Anthrax spores, a recent cause of global concern, harbor virulent plasmids and genetic markers in the chromosomal region. Rapid detection of these markers by active, real-time recognition of short DNA stretches can be achieved by pyrosequencing, a powerful identification tool that can readily uncover anthrax spores in spiked liquids (milk, water, juices) and processed meat with accuracy (Amoako et al. 2013). The methodology has immense applications to counter potential food-borne bioterrorism activities and defense sectors. Applications of real-time PCR additionally include identification of other microbes like *Salmonella* spp. (Margot et al. 2013) and *E. coli* (Ozpinar et al. 2013), commonly found in infected food products with minimal false-positive results and a lower limit of detection. Interestingly, the Crespo-Sempere group has developed a fast detection method using real-time PCR enabling the quick identification of *Alternaria* spp., a fungus commonly found in fruits, grains, and vegetables (tomatoes), and that can distinguish between viable and nonviable populations (Crespo-Sempere et al. 2013). The results from amplified DNA can be quantified, giving an estimation of the amount of damage caused by mycotoxins and phytotoxic metabolites produced by the fungus. Vine and vinegar are often infected by microbes of the *Acetobacter* spp. (*Acetobacter cerevisiae* and *Acetobacter malorum*) (Valera et al. 2013). Real-time PCR is a fast method to detect and quantify the presence of contaminating microbes in regular wine and vinegar samples with an efficiency of >80%.

10.4.3 Secreted Toxins

Disease-causing pathogens that commonly contaminate food secrete damaging protein toxins that assist the pathogens in sustaining a full-blown infection in the host species. Protein toxins influence host pathogen interactions by posttranslational mechanisms that modify host cell endogenous functions that are either inhibited or activated to serve pathogenic growth and spread. Although in recent times, a thorough mechanistic understanding of toxin-assisted pathogenic infections has undoubtedly helped in designing vaccines and therapies to prevent disease occurrences in humans, toxins are also a powerful means for the identification of infectious agents in food. Bioterrorism threats and epidemic outbreaks of pathogenic diseases through spiked food have stressed upon the need for rapid on-site biosensing detection tools for quick identification (Anderson and Bokor 2012). Biosensors are devices capable of converting signals generated upon the detection of pathogenic byproducts in samples such as protein toxins, or chemicals, into amplified signal indicators that can be easily visually read and quickly interpreted. Biosensing devices comprise molecular probes as fundamental detector elements, which capture analytes such as microbial toxins from viable bacterial or fungal cells (Banerjee and Bhunia 2010). Nucleic acids and peptides are attractive molecular probes for biosensors because they are compliant in developing tertiary structures and their interaction with toxins can be mapped quantitatively by surface plasmon resonance, magnetoelastic immunosensors, amperometric immunosensors, surface acoustic waves, and quartz crystal microbalances.

The virulence potential and antibiotic resistance properties of various serotypes of Shiga toxin, a by-product of *E. coli* bacteria, are an effective means to sense the bacteria from meat samples, one of the main sources that harbor bacteria (Momtaz and Jamshidi 2013). Oceanic seafood emerging in European waters is commonly infected with shellfish-poisoning toxins such as tetrodotoxin, a powerful neurotoxin that can cause paralysis in humans (Campbell et al. 2013). It is difficult to accurately identify the presence of this particular toxin by sensitive chromatographic methods such as high-performance liquid chromatography as established by the Association of Analytical Communities; however, newer techniques such as surface plasmon resonance optical biosensors can specifically identify the toxins and serve as an effective screening tool. Enterotoxins from *Staphylococcus aureus* are deadly pathogenic toxins and frequently cause food poisoning globally (Brizzio et al. 2013). Physiological distresses resulting from these enterotoxins are significant and include diarrhea and vomiting. Rapid screening immunoassays based on multiplex PCR are effective in reporting the identification of typical enterotoxin genes (sea, seb, sec, sed, and see), especially in disease occurrences.

10.5 Nanomaterials: Sensitivity and Specificity in Detection

10.5.1 Metallic Nanoparticles: Au, Ag, Si

Distinct optical properties are exhibited by metallic nanoparticles that are a function of their composition and specific geometry. In the area of medical diagnostics and imaging, metallic nanoparticles composed of gold, silver, and recently silica have found much utility for rapid screening of biological and food samples for early diagnosis as well as to trace out possible sources of microbial contamination during epidemic outbreaks. These nanoparticles when conjugated with affinity ligands such as antigenic surface epitopes and antitoxin antibodies work efficiently as biological chemical sensors in rapid detection (Kaittanis et al. 2010). Metal nanoparticles offer the flexibility to be tailored via surface modification chemistry, and their merger with promising detection technologies has resulted in economical, faster, sensitive, and accurate diagnostic bioassays.

Gold nanoparticles have been used as labels for qualitative and quantitative detection of various target molecules in combination with fluorescence, amperometry, surface plasmon resonance, Raman scattering, and magnetic force methods (Cao et al. 2011). Cost-efficient techniques employ colorimetric analyses for rapid identification of microbial presence manifested as strips that require only a few microliters of sample for testing. Such a rapid on-site bioassay has been developed for the identification of *Salmonella* Typhimurium, and uses oval-shaped gold nanoparticles conjugated with anti-*Salmonella* antibody (Wang et al. 2010). The detection technique is label-free and has a low limit of detection with high selectivity over other pathogens, and exploits a change in color occurring due to the formation of nanoaggregates of the bacterium with gold nanoparticles.

LFIAs are rapid, convenient, cost-effective, and user-friendly detection methods that are based on porous nitrocellulose membrane strips and colored nanoparticles such as gold nanoparticles and have been vastly employed in point-of-care testing, clinical diagnosis, and testing for agricultural and environmental pollutants. Lately, LFIA using gold nanoparticles as a probe was developed for sensitive detection of the food-borne *E. coli* O127:H7 strain that showed detectable limits as low as 0.4 nM (Rastogi et al. 2012).

Silver nanoparticles were originally known for their antimicrobial properties and were used as new-generation antimicrobials to combat infections. Silver nanoparticles can be fabricated using different methodologies as spherical, cylindrical, and triangular nanoparticles that are amenable to functionalization with target ligands such as peptides, antibodies, oligonucleotides, and small molecules. Often coupled with surface plasmon resonance, scanning transmission electron microscopy, and amperometric detection methods, these nanoparticles are readily employable for rapid microbial identification. Likewise, surface-enhanced Raman spectroscopy (SERS) is another frequent detection tool combined with silver nanoparticles. Silver nanoparticles are versatile and have been employed as detection aids for microbes like *E. coli*, *Salmonella* spp., *Listeria monocytogenes*, and *St. aureus* (Sundaram et al. 2013). Biosorbed silver nanospheres bearing polyclonal antibodies against *E. coli* displayed accurate and specific bacterial detection in milk and apple juice samples determined by SERS within 1 h (Naja et al. 2010). The adaptability of this method to detect other pathogen species by incorporating the relevant antibody makes it effective and versatile. Silver biopolymer nanosubstrates use silver as a SERS label for amplified signal detection of various pathogen species as a reliable tool for rapid detection and classification of pathogens present in food. Silver nanoparticles are entrapped within polymer matrices and enhance mapping by Raman scattering.

Another methodology developed using 100 nm silica-coated particles as a magnetic probe for SERS-based detection can identify pathogens such as serogroups of *Salmonella* and *St. aureus* in varied food matrices such as spinach solution and peanut butter with specificity and high sensitivity (Wang et al. 2011). Likewise, magnetic silica–coated nanoparticles have been employed as substrates for the immobilization of antibodies coupled with fluorescent quantum dots using surface chemistry (Zhao et al. 2009). The nanosystem is capable of attaching to antigenic surface epitopes *Sa.* Typhimurium, *Shigella flexneri*, and *E. coli* O157:H7, enabling the detection of these microbes using fluorescence and magnetic separation selectively and accurately. Although this particular technology requires 2 h to produce the result, it is nevertheless sensitive and effective in microbial content monitoring and identification.

10.5.2 Quantum Dots

In the recent decade, the field of nanotechnology has seen the development and wide application of discrete semiconductor nanocrystals as versatile nanomaterials in various areas such as engineering, food, biopharmaceuticals, and radio and image contrast agents. These nanomaterials popularly called quantum dots are luminescent homogeneous particles generally composed of inorganic binary alloys such as cadmium selenide, indium arsenide, zinc sulfide, cadmium sulfide, and indium phosphide (Cheki et al. 2013). Quantum dots are fluorophores with several advantages such as narrow emission, broad excitation spectra, photostability, and high quantum yield. Quantum dots can distinctly display different colors depending upon the particle size and material composition, and, therefore, their emission spectra color can be tuned, enabling the usage of quantum dots for simultaneous detection of multiple analyte components. Moreover, differentially sized quantum dots can be coexcited with a single excitation wavelength. Quantum dots have been widely applied as fluorescent biomarkers and labels for the rapid detection of various biological entities such as proteins, peptides, microbes, and endogenous biomolecules. For instance, Zhu et al. have reported Cry1Ab protein detection in genetically modified maize MON810 by QD-FLISA (Zhu et al. 2011). An important food-borne pathogen responsible for a high mortality rate (approximately 30%), *L. monocytogenes*, which uses two surface proteins called Internalin A and Internalin B to infect and invade host cells, has been detected by a highly sensitive, rapid, and reproducible bioassay using quantum dot–based fluorescence immunoassays (Tully et al. 2006). These immunoassays employ three cloned peptide fragments (MR23, 35, and 45 kDa) of Internalin B as the basis for detection.

10.5.3 Nanotubes, Nanofibers, and Nanowires

Nanofibers are nanostructured materials with exceptional biochemical properties and are potential candidates to engineer biosensors that work on immunochromatography and immunoassay formats with high sensitivity, fast response, and less cost. Generally fabricated from biocompatible materials, nanofiber-based biosensors have been designed using nanosized fibers of nitrocellulose, polyvinylidene fluoride, polyethersulfone, and others. These materials possess outstanding binding capacity to separate and confine biomolecules and microbial cells from food and clinical samples. Nanofibers are fabricated using electrospinning, a popular, versatile, and cost-effective technique for synthesis. Nanofibers have a high surface area and mass transfer rate, thereby enhancing the biochemical binding and signal-to-noise ratio of the biosensor. For instance, nanofibers produced by electrospinning embedded in nitrocellulose membrane have been functionalized with antibodies and employed in biosensors based on capillary separation and conductometric immunoassays for the detection of pathogenic bacteria and viruses, specifically *E. coli* O157:H7 and bovine viral diarrhea virus (Luo et al. 2010). The biosensor displays a short detection time of 8 min, with a low limit of detection indicating high sensitivity of the biosensor system. In an immunoassay developed by Arumugam and coworkers, the detection of *E. coli* O157:H7 was enabled by oligonucleotide probes imprinted on vertically aligned carbon nanofibers constructed within a simple and robust multiplexed biosensor (Arumugam et al. 2009). Vertically aligned carbon nanofibers possess interesting characteristics such as better electrical and thermal conductivities, a broad electrochemical potential window, higher mechanical strength, biocompatibility, and flexible surface chemistry.

The combination of well-designed polymers with carbon nanotubes is of growing significance owing to its ease of construction and its capacity to include conducting materials within porous polymeric systems to develop electrochemical biosensing devices. For example, a disposable electrochemical biosensor for the real-time measurements of *E. coli* O157:H7, *Campylobacter*, and *Salmonella* was fabricated by immobilizing a blend of anti-*E. coli*, anti-*Campylobacter*, and anti-*Salmonella* antibodies on the surface of the multiwall carbon nanoelectrodes modified with polyallylamine (Viswanathan et al. 2012). The method is accurate and sensitive, and illustrates the practicability of multiplexed detection of bacterial presence in milk samples. Fungal species such as *Botrytis cinerea*, a plant pathogenic fungus, can be identified using immunosensors fabricated with multiwalled carbon nanotubes (Fernandez-Baldo et al. 2009). Similarly, *Staphylococcal* Enterotoxin B (Yang et al. 2010), *Salmonella* Paratyphi A (Yang et al. 2013), and other *Salmonella* serovars (Amaro et al. 2012) have been successfully detected with lab-on-a-chip immunoassay systems based on carbon nanotubes.

Nanomaterials such as silicon nanowires have been useful for label-free field effect transistor applications as biosensors for fast and sensitive recognition of diverse significant analytes such as bacteria and viruses. Rapid, real-time, sensitive, and specific detection of *Bacillus cereus* can be achieved with direct charge transfer disposable immunosensors for on-field identification (Pal et al. 2007). The method utilizes the phenomenon of a direct charge electron transport in a voltage-controlled switch layout for bacterial recognition. This technology employs polyclonal antibodies as the biological sensor for capture and polyaniline nanowires as the electrical transducer. Polyaniline nanowires work as electron transfer labels in the immunoassay, facilitating signal intensification.

10.5.4 Magnetic Nanoparticles

The identification of pathogens present in food matrices by immunomagnetic separation techniques has received much interest. These techniques are based on cell separation by use of antibody-conjugated beads comprising a magnetic core that can be isolated by the application of an external magnetic field. The magnetic beads functionalized with antibodies work specifically to detect a particular target microbe. Immunomagnetic separation–based methods have been used in combination with recognition tools such as cell culturing, ELISA, and PCR. Dynabeads™ of 2.8 mm dimensions (Invitrogen Dynal A.S., Oslo, Norway) conjugated to antibodies can specifically capture *L. monocytogenes* by isolating the target microbe with an efficiency of 7%–23%, and are commercially accessible (Yang et al. 2007). Quantum dots as adjunct fluorophores have been used successfully to identify *E. coli* O157:H7 pure cultures, *Salmonella* serovar Typhimurium from chicken carcass wash water, and *Sa.* Typhimurium, *Sh. flexneri*, and *E. coli* O157:H7 from apple juice and milk (Zhao et al. 2009). Based on their size, iron oxide particles generally demonstrate diverse properties when a magnetic field is applied. Superparamagnetic properties of the particles increase upon reducing their size to a nanometric range, thereby making nano-sized particles more efficient in immunomagnetic separation than micron-sized ones (Yavuz et al. 2006). Additionally, nanoparticles ≤ 100 nm have surfaces more amenable to attachment with ligands and demonstrate lower sedimentation rates indicating better stability with minimized magnetic dipole–dipole interactions (Gupta and Gupta 2005). Magnetic nanoparticles (~30 nm) in conjunction with microfluidic chip and microelectrode–based impedance immunosensor can rapidly detect *L. monocytogenes* in food samples (Kanayeva et al. 2012). The microfluidic chips and interdigitated microelectrodes enhance sensitivity of the label-free immunoassay. Moreover, a combination with label-free infrared spectroscopy imparts specificity by aid of fingerprinting through microbial strain-specific antibodies (*E. coli* O157:H7, *Sa.* Typhimurium) for on-site recognition and monitoring of pathogens in food matrices such as milk and spinach extracts within 30 min (Ravindranath et al. 2009).

10.5.5 Fluorescent Polymeric Nanoparticles

Polymeric nanoparticles have been used as versatile nanocarriers of fluorophores for detection. For instance, immunoassays based on polystyrene nanoparticles embedded with Eu(III) ions (fluorescent marker) have been developed by the Hewlett group for sensitive recognition of biological targets such as anthrax protective agent (PA) in samples spiked with PA in conjunction with ELISA (Tang et al. 2009). The assay sensitivity is very high, 0.01–100 ng/mL, about 100-fold more than conventional ELISA with 100% accuracy. Valanne et al. built an immunoassay using the same polystyrene nanoparticle system, with high-affinity monoclonal antibodies for directly detecting adenovirus serotype 2 and nasopharyngeal aspirates through sandwich-based immunoreactions with sensitivity 800-fold over traditional immunoassay methods (Valanne et al. 2005). Udomsangpetch et al. reported a rapid diagnostic on-site test for the detection of the malarial parasite, *Plasmodium falciparum*, by employing sensitive polystyrene nanoparticles conjugated with antigen (or antibody) for screening malarial antibodies (or antigens) (Polpanich et al. 2007). Another test for the identification of microbes involves visual and spectroscopic detection and employs agar nanoparticles embedded with phospholipids and the colored polymer, polydiacetylene (Silbert et al. 2006). Surface interaction of the nanoparticles with cell membrane components of the bacteria triggers a rapid color change from blue to red along with an intense fluorescent emission. This technology is universal for gram-negative and gram-positive bacteria including *Sa. enterica* serovar Typhimurium, *B. cereus*, and *E. coli*.

10.5.6 Viruses

Viruses that have found tremendous application in the microbial detection of food are mainly bacteriophages or phages, and these can infect numerous host cell types. Phages are bacterial viruses composed of genetic materials enclosed within a shielding protein coat, and remain in a metabolically inactive condition until they accidentally contact a potential host cell. If the host bacterial cell is viable and appropriate, the phages will bind to the cell, introducing their nucleic acid material and commence a parasitic capture, also termed the phage lytic cycle, consequently beginning massive phage replications by exploiting host cell energy and machinery. Recognizing the host preferences of phages paves their utilization in the detection and monitoring of target pathogens occurring in food and water. Phages can be biotechnologically and genetically engineered for applicability in identification; however, natural bacteriophages coupled with novel detector aids can be used for screening and identification. Bacterial cells that fall within the phages' range of potential hosts have recognition elements or *sensors* that are identified by phages as an opportunity for capture. The absence of recognition receptors prevents the phages from attacking and sets them off in search for infecting other hosts. A few phages can invade a vast assortment of bacteria belonging to different genus and species, and there are others that infect a limited variety of bacterial hosts. The phages' inherent survival and parasitic replication tendency forms an important identification tool for the rapid detection of bacterial pathogens in foods. Phage assays based on either bioengineered phages that are genetically modified to recognize specific species or on wild-type phages have been applied for screening food matrices, water, and clinical samples, and comprise a wide range of strategies for detection. There is extensive literature on the use of virus-based techniques for host species detection, and further information may be referred to Smartt et al. (2012).

10.5.7 Lipidic Nanosystems

Nanosystems comprised of lipids such as liposomes have been well researched for application in the rapid detection of microbial contamination. LFIAs using test strips have been employed for the quick identification of cholera toxin in matrices of food samples (Ahn and Durst 2008). These are based on ganglioside liposomes that are readily captured by immobilized antibodies present in the analyte. The test zone or the analyte zone displays a colored band that can be quantitatively measured with a computer-assisted scanner. The test retains sensitivity and reproducibility at varying pH and salt concentrations. Liposome nanovesicles coated with anti-*E. coli* O157:H7 antibodies and loaded with sulphorhodamine B dye have been combined with immunomagnetic beads in an immunoassay format to detect *E. coli* (DeCory et al. 2005). Likewise, the detection of *Salmonella* spp. using fluorescence immunoassays based on sulphorhodamine B–encapsulated immunoliposomes can yield rapid results with low detection limits in various food and water samples such as tap water, apple juice, milk, meat, and cheese (Shin and Kim 2008). Portable colorimetric immunoassays comprising methylene blue–loaded immunoliposomes tagged with anti-*Salmonella* common structural antigen antibodies have been similarly developed in a nitrocellulose strip format for *Salmonella* spp. identification (Ho et al. 2008). This assay eliminates false, positive results that may surface due to cross reactivity with *E. coli* and *Listeria* spp. Blue-green algae or cyanobacteria produce toxic microcystins (cyclic heptapeptides) that contaminate drinking water and seafood. Fluorescent immunoliposomes have been employed as tracer molecules in a sensitive rapid screening test based on immunochromatography for detecting and monitoring microcystins (Khreich et al. 2010) (Figure 10.2).

10.6 Promising Innovations in Nanobiotechnology

10.6.1 Nanoarrays

Nanomaterial characteristics employed for pathogen identification can be tuned by controlling the composition, size, shape, and surface chemistry. Specifically, engineering the nanoparticles using their structural parameters, composition, self-assembly, size, and binding properties can directly relate to apposite modifications in the electronic, emissive, absorptive, light scattering, and conductive properties. Moreover, contemporary research has seen a sudden increase in the development of surface patterning methods that assure the production of nanoarrays of pathogen biological targets. SERS coupled with silver nanoscale array substrates can detect pathogenic bacteria, wherein the substrate consists of a support layer of 500 nm silver film on a glass slide and a deposit of silver nanorod array for obtaining spectra toward rapidly detecting microbes such as whole-cell bacteria, *Sa. aureus*, *Staphylococcus epidermidis*, *Sa. typhimurium*, generic *E. coli*, *E. coli* O157:H7, *E. coli* DH 5a, and bacteria mixtures with minimum

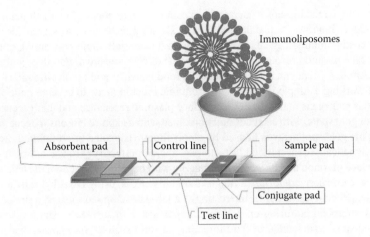

FIGURE 10.2 Fluorescently tagged immunological liposomes as nanodetectors in nitrocellulose strip–based testing.

sample preparation (Chu et al. 2008). An important advantage of this method is its ability to differentiate between Gram types, different species, their mixture, and strains and, furthermore, between viable and nonviable cells based on spectral variations. The rapid and precise identification of fundamental nucleic acids such as DNA has grabbed global focus for diagnosing an assortment of diseases and detecting detrimental microbial load in solid and liquid foods. Nanoarray platforms fabricated by a photolithography and composed of high–molecular weight polystyrene and anodic aluminum oxide membranes (200 nm pore diameter) can detect oligodeoxynucleotides assisted by surface plasmon resonance with enhanced sensor response capabilities (Celen et al. 2011).

10.6.2 Nano-Cantilevers

Contemporary trends in research pertaining to biosensing have been directed to developing label-free bioanalytical devices that are of small size, are portable, and can work with rapidity. This is particularly factual for nanosized cantilever biosensors that can endure alteration in mechanical characteristics upon specific adherence to biological molecular entities. A range of approaches are available to design and fabricate cantilevers catering to their applications such as the detection of proteins, DNA molecules, bacteria, and viruses depending upon the target biophysical properties. In the recent decade, the manufacture and developmental studies of biosensors based on cantilevers for biochemical research and clinical applications have been performed, and this field has greatly expanded the extent of nano-cantilever sensors. Biomolecular events can be easily monitored in situ and observed using cantilever-based devices. Nano-cantilevers are robust devices with high sensitivity and selectivity for detecting physical, chemical, and biological entities by capturing variations in cantilever bending or in resounding frequency; they are of two types: static and dynamic (Hwang et al. 2009). Although a majority of static cantilever sensors can sense in real time precise targets existing in an aqueous milieu, protein recognition by dynamic cantilevers is conducted mainly in air or in vacuum. Several biomolecular cantilever detection techniques can be classified depending upon the targets: proteins, DNA molecules, and biological particles. For instance, a sensitive, simple, and rapid cantilever-based sensor has been developed for identifying *V. cholerae* O1as the target pathogen in food and water that causes cholera (Sungkanak et al. 2010). Likewise, *L. monocytogenes*, a known food-borne pathogen with high mortality rate, can be successfully recognized from milk within 1 h (Sharma and Mutharasan 2013).

10.6.3 Nanosurface Plasmon Resonance

Surface plasmon resonance imaging can sense the incidence of biomolecules, cells, and pathogenic microbes attached to the sensor surface by analyzing the variations in the refractive indices. The variations in reflectivity arising on the biosensor surface from cellular and biomolecular attachments is determined pre and post analyte binding (Narsaiah et al. 2012). Existing techniques to screen for microbial contamination use costly immunoreagents or reliance upon substantially time-consuming microbial growth in cultures. Surface plasmon resonance imaging is a label-free, sensitive, portable, multiplex pathogen identification technology that can detect microbes in food matrices and predict the safety of food.

An excellent working example is a fusion microfluidic biochip made to execute multiplexed recognition of unicellular pathogens that utilizes both surface plasmon resonance and fluorescence imaging and holds an array of gold spots, with each dot functionalized with a capture biosensor agent targeting a definite microbe. This gold spot array is housed by a polydimethylsiloxane microfluidic flow compartment that conveys magnetically concentrated test samples and can effectively image antibody-bound *E. coli* O157:H7 by surface plasmon and fluorescence imaging with quantification (Zordan et al. 2009). Another biosensor detects *Fusarium culmorum*, a fungus found in cereals, using DNA hybridization and surface plasmonic waves (Pascale et al. 2013). Interestingly, a lab-on-a-chip biosensor developed by Bouguelia et al. combines microbial culturing on protein microarrays with surface plasmon resonance imaging for the rapid and specific detection of food-infecting bacteria such as *Sa. enterica* serovar Enteritidis, *Streptococcus pneumoniae*, and *E. coli* O157:H7 (Bouguelia et al. 2013). Human noroviruses, responsible for causing gastroenteritis outbreaks and conventionally identified using time-consuming techniques, have morphology similar to feline calcivirus, both of which can be detected in intact form using a surface plasmon resonance image and antibody capture technique with high sensitivity (Yakes et al. 2013). Consumer fresh whole-fat milk and powdered milk preparations are often contaminated with *Cronobacter* spp., and these can be detected in real time using surface plasmon biosensors grafted with poly(2-hydroxyethyl methacrylate) on the chip surface by identifying bacteria-positive dairy samples with low limits of detection (Rodriguez-Emmenegger et al. 2011).

10.6.4 Fiber Optics

Optical biosensors are promising tools for the detection of microbial pathogens, heavy metals, pesticides, and toxic materials in foods to verify safety for consumption. The core components of fiber optic–based biosensors consist of an optical light source, optical transmission fibers, immobilized biorecognition elements such as antibodies or microbes, optical probes (fluorescent indicators) for transduction, and an optical identification method to decide the sensitivity and limit of detection. Fiber optic biosensors are equipped with useful parts such as cuvette holders, collimating lenses, flow insertion arrangements with pump, capillary tubings, and optical filters, all of which aid the measurements of luminescence, phosphorescence, absorbance, fluorescence, and reflectance (Narsaiah et al. 2012). Fiber optic biosensors are of two types: intrinsic and extrinsic. In intrinsic sensors, contact with the bioanalyte takes place inside an aspect of the optical fiber; and contrastingly, in extrinsic sensors, the optical fiber couples light to and from the area where the optical light ray is affected by bioanalyte interactions. The light is directed into the sample via a single fiber, concomitantly reflected post interaction with the biochemical zone, and collected in the detector region via another fiber optic cable. An important benefit concerning optical biosensors based on optical fibers is that it allows sample analysis to be performed over elongated spaces enabling on-field examinations. An effective tool in food pathogen detection, fiber optic technology has been well researched and offers promising label-free methods for the rapid screening of various disease-causing agents in food (Figure 10.3 and Table 10.1).

10.7 Efficacy of Nanomaterials in Detection

Interesting technological advances have emerged in nanotechnology pathogen recognition in foods. The application of nanoparticles as biomarkers or labels in combination with innovative detection methods has resulted in enhancements in selectivity, sensitivity, and simultaneous multiplexing capacities. Metallic nanoparticles comprising gold, silver, or silica have unique optical and electronic properties, and can be modulated using parameters such as their size and composition. Upon functionalizing with affinity ligands, these nanomaterials have found significant implications as chemical and biological

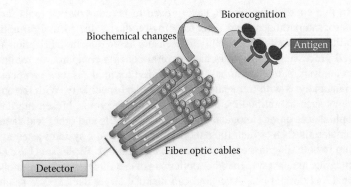

FIGURE 10.3 Fiber optic technology for identification of pathogens.

TABLE 10.1

Classification of Fiber Optic Technology Given by Monk and Walt (2004) and Implications in
Food Screening for Pathogen Detection

Biorecognition Element	Analytes	Food Matrices	References
Enzymes	Aflatoxins from *Aspergillus* spp.	Dry fruits (pistachios, raisins, and prunes)	Rojas-Duran et al (2007)
Nucleic acids	*Brettanomyces bruxellensis* (yeast)	Alcoholic beverages (wine, beer)	Cecchini et al. (2012)
Antigen–antibody	*Sa. typhimurium*	Meat (ground pork)	Ko and Grant (2006)
	L. monocytogenes and *Listeria ivanovii*	Soft cheese, sliced meats, and seafood	Mendonca et al. (2012)
Whole cells	*Bacillus badius*	Milk	Verma et al. (2010)
Biomimetics	Flavin (mimics flavoprotein enzymes)	Glucose-based matrices	Tognalli et al. (2009)

sensors. The characteristics of the nanomaterials used for pathogen detection and parameters controlling them can be tuned by varying the size, shape, constitution, and surface modification chemistries. Moreover, contemporary times have witnessed an expansion toward the fabrication of surface-modeling techniques using chemistry and hold great potential in building numerous, high-throughput, multiplex, sensitive, and rapid arrays of microbial targeting ligands.

Nanosystems with interesting capabilities have been developed wherein, for instance, nanodevices of sizes close to 50 or 100 nm along with any surface functionalizing coatings of capture biomolecules have been fabricated to identify targets that are still smaller such as lipidic cellular components, or larger like whole cells, or even having similar proportions such as viruses. It can be anticipated that nanodevices of a certain size can intermingle differently depending upon the variability they encounter in terms of size and can elicit corresponding response signals unique to a specific kind of interaction. Undoubtedly, these nanodevices have complete capabilities of being translated into rapidly sensing point-of-care aids enabling quick and sensitive on-field pathogenic identification of food and water samples.

10.8 Summary

The safety and quality of foods are a foremost concern to the food industry. The sensitive detection of contaminants such as disease-causing microbial agents and pesticides is an important requisite in guaranteeing food quality and safety. Traditional techniques of identifying harmful agents such as microbial culturing are inefficient in terms of speed, sensitivity, specificity, reproducibility, and sometimes even accuracy. Such methods currently employed to detect pathogens are additionally labor-intensive. In this context, biosensors have proved to be constructive tools for the monitoring of food- and waterborne pathogens and their toxic by-products, due to their simplicity, sensitivity, specificity, reliability, and on-site applicability. Biosensors based on nanofabrication are materializing in the area of food processing. Biosensors have several benefits compared to traditional techniques including rate per analysis, sample volumes, time required for analysis, and prospective implications for multiplex variant analysis with tremendous sensitivity and specificity. With the application of bio-recognition elements such as antibodies, nucleic acids, and enzymes, biosensing devices have been applied for the rapid detection of biological targets in viable cells and other real samples. Biosensors that employ nanomaterials such as metallic nanoparticles and lipidic systems as energy transmittance tools are promising in both diagnostic and theranostic applications. Biosensors based on surface plasmon resonance imaging, nanoarrays, and fiber optics can detect their targets accurately and selectively using photophysical and optical properties and can identify target nucleic acid sequences, polymorphic regions, and gene expression, proteins, pesticides, and others. Nanosized detection systems are undoubtedly emerging as revolutionary methods in the area of food and environmental safety and monitoring. In summary, current nanotechnology trends have presented techniques that are apposite for scale-up to on-site applicable devices ensuring rapidity, reliability, and sensitive identification of pathogenic food contaminants in samples.

REFERENCES

Ahn, S. and R. A. Durst. 2008. Detection of cholera toxin in seafood using a ganglioside-liposome immunoassay. *Anal Bioanal Chem* 391(2):473–478. doi: 10.1007/s00216-007-1551-1.

Amaro, M., S. Oaew, and W. Surareungchai. 2012. Scano-magneto immunoassay based on carbon nanotubes/gold nanoparticles nanocomposite for *Salmonella enterica* serovar *typhimurium* detection. *Biosens Bioelectron* 38(1):157–162. doi: 10.1016/j.bios.2012.05.018.

Amoako, K. K., T. W. Janzen, M. J. Shields, K. R. Hahn, M. C. Thomas, and N. Goji. 2013. Rapid detection and identification of *Bacillus anthracis* in food using pyrosequencing technology. *Int J Food Microbiol* 165(3):319–325. doi: 10.1016/j.ijfoodmicro.2013.05.028.

Anderson, P. D. and G. Bokor. 2012. Bioterrorism: Pathogens as weapons. *J Pharm Pract* 25(5):521–529. doi: 10.1177/0897190012456366.

Arora, P., A. Sindhu, N. Dilbaghi, and A. Chaudhury. 2011. Biosensors as innovative tools for the detection of food borne pathogens. *Biosens Bioelectron* 28(1):1–12. doi: 10.1016/j.bios.2011.06.002.

Arora, P., A. Sindhu, H. Kaur, N. Dilbaghi, and A. Chaudhury. 2013. An overview of transducers as platform for the rapid detection of foodborne pathogens. *Appl Microbiol Biotechnol* 97(5):1829–1840. doi: 10.1007/s00253-013-4692-5.

Arumugam, P. U., H. Chen, S. Siddiqui, J. A. Weinrich, A. Jejelowo, J. Li, and M. Meyyappan. 2009. Wafer-scale fabrication of patterned carbon nanofiber nanoelectrode arrays: A route for development of multiplexed, ultrasensitive disposable biosensors. *Biosens Bioelectron* 24(9):2818–2824. doi: 10.1016/j.bios.2009.02.009.

Attridge, S. R. and J. Holmgren. 2009. *Vibrio cholerae* O139 capsular polysaccharide confers complement resistance in the absence or presence of antibody yet presents a productive target for cell lysis: Implications for detection of bactericidal antibodies. *Microb Pathog* 47(6):314–320. doi: 10.1016/j.micpath.2009.09.013.

Banerjee, P. and A. K. Bhunia. 2010. Cell-based biosensor for rapid screening of pathogens and toxins. *Biosens Bioelectron* 26(1):99–106. doi: 10.1016/j.bios.2010.05.020.

Bouguelia, S., Y. Roupioz, S. Slimani, L. Mondani, M. G. Casabona, C. Durmort, T. Vernet, R. Calemczuk, and T. Livache. 2013. On-chip microbial culture for the specific detection of very low levels of bacteria. *Lab Chip* 13(20):4024–4032. doi: 10.1039/c3lc50473e.

Brizzio, A. A., F. A. Tedeschi, and F. E. Zalazar. 2013. Multiplex PCR strategy for the simultaneous identification of *Staphylococcus aureus* and detection of staphylococcal enterotoxins in isolates from food poisoning outbreaks. *Biomedica* 33(1):122–127. doi: 10.1590/s0120-41572013000100015.

Brooks, J. D. and S. H. Flint. 2008. Biofilms in the food industry: Problems and potential solutions. *Int J Food Sci Technol* 43(12):2163–2176. doi: 10.1111/j.1365-2621.2008.01839.x.

Campbell, K., P. Barnes, S. A. Haughey, C. Higgins, K. Kawatsu, V. Vasconcelos, and C. T. Elliott. 2013. Development and single laboratory validation of an optical biosensor assay for tetrodotoxin detection as a tool to combat emerging risks in European seafood. *Anal Bioanal Chem* 405(24):7753–7763. doi: 10.1007/s00216-013-7106-8.

Cao, X., Y. Ye, and S. Liu. 2011. Gold nanoparticle-based signal amplification for biosensing. *Anal Biochem* 417(1):1–16. doi: 10.1016/j.ab.2011.05.027.

Cecchini, F., M. Manzano, Y. Mandabi, E. Perelman, and R. S. Marks. 2012. Chemiluminescent DNA optical fibre sensor for *Brettanomyces bruxellensis* detection. *J Biotechnol* 157(1):25–30. doi: 10.1016/j.jbiotec.2011.10.004.

Celen, B., G. Demirel, and E. Piskin. 2011. Micro-array versus nano-array platforms: A comparative study for ODN detection based on SPR enhanced ellipsometry. *Nanotechnology* 22(16):165501. doi: 10.1088/0957-4484/22/16/165501.

Cheki, M., M. Moslehi, and M. Assadi. 2013. Marvelous applications of quantum dots. *Eur Rev Med Pharmacol Sci* 17(9):1141–1148.

Chu, H., Y. Huang, and Y. Zhao. 2008. Silver nanorod arrays as a surface-enhanced Raman scattering substrate for foodborne pathogenic bacteria detection. *Appl Spectrosc* 62(8):922–931. doi: 10.1366/000370208785284330.

Code, National Dairy. 2005. *Production and Processing Regulations.* ed. Canadian Dairy Information Center. Department of Agriculture, Agricultural Marketing Service, United States Government.

Colliex, C. and C. Mory. 1994. Scanning transmission electron microscopy of biological structures. *Biol Cell* 80(2–3):175–180.

Crespo-Sempere, A., N. Estiarte, S. Marin, V. Sanchis, and A. J. Ramos. 2013. Propidium monoazide combined with real-time quantitative PCR to quantify viable *Alternaria* spp. contamination in tomato products. *Int J Food Microbiol* 165(3):214–220. doi: 10.1016/j.ijfoodmicro.2013.05.017.

DeCory, T. R., R. A. Durst, S. J. Zimmerman, L. A. Garringer, G. Paluca, H. H. DeCory, and R. A. Montagna. 2005. Development of an immunomagnetic bead-immunoliposome fluorescence assay for rapid detection of *Escherichia coli* O157:H7 in aqueous samples and comparison of the assay with a standard microbiological method. *Appl Environ Microbiol* 71(4):1856–1864. doi: 10.1128/aem.71.4.1856-1864.2005.

Federal, Register. 2012. *Shiga Toxin-Producing Escherichia coli in Certain Raw Beef Products.* ed. USDA Food Safety and Inspection Service.

Fernandez-Baldo, M. A., G. A. Messina, M. I. Sanz, and J. Raba. 2009. Screen-printed immunosensor modified with carbon nanotubes in a continuous-flow system for the *Botrytis cinerea* determination in apple tissues. *Talanta* 79(3):681–686. doi: 10.1016/j.talanta.2009.04.059.

Gehring, A., C. Barnett, T. Chu, C. DebRoy, D. D'Souza, S. Eaker, P. Fratamico et al. 2013. A high-throughput antibody-based microarray typing platform. *Sensors (Basel)* 13(5):5737–5748. doi: 10.3390/s130505737.

Gehring, A. G. and S. I. Tu. 2011. High-throughput biosensors for multiplexed food-borne pathogen detection. *Annu Rev Anal Chem (Palo Alto Calif)* 4:151–172. doi: 10.1146/annurev-anchem-061010-114010.

Guo, D., B. Liu, F. Liu, B. Cao, M. Chen, X. Hao, L. Feng, and L. Wang. 2013. Development of a DNA microarray for molecular identification of all 46 *Salmonella* O serogroups. *Appl Environ Microbiol* 79(11):3392–3399. doi: 10.1128/aem.00225-13.

Gupta, A. K. and M. Gupta. 2005. Synthesis and surface engineering of iron oxide nanoparticles for biomedical applications. *Biomaterials* 26(18):3995–4021. doi: 10.1016/j.biomaterials.2004.10.012.

Hahm, J. I. 2013. Biomedical detection via macro- and nano-sensors fabricated with metallic and semiconducting oxides. *J Biomed Nanotechnol* 9(1):1–25.

Hanifian, S. and S. Khani. 2012. Fate of *Yersinia enterocolitica* during manufacture, ripening and storage of Lighvan cheese. *Int J Food Microbiol* 156(2):141–146. doi: 10.1016/j.ijfoodmicro.2012.03.015.

Hegde, N. V., C. Praul, A. Gehring, P. Fratamico, and C. Debroy. 2013. Rapid O serogroup identification of the six clinically relevant Shiga toxin-producing *Escherichia coli* by antibody microarray. *J Microbiol Methods* 93(3):273–276. doi: 10.1016/j.mimet.2013.03.024.

Henderson, E. 2007. BioForce Nanosciences, Inc. *Nanomedicine (Lond)* 2(3):391–396. doi: 10.2217/17435889.2.3.391.

Ho, J. A., S. C. Zeng, W. H. Tseng, Y. J. Lin, and C. H. Chen. 2008. Liposome-based immunostrip for the rapid detection of *Salmonella. Anal Bioanal Chem* 391(2):479–485. doi: 10.1007/s00216-008-1875-5.

Hwang, K. S., S. M. Lee, S. K. Kim, J. H. Lee, and T. S. Kim. 2009. Micro- and nanocantilever devices and systems for biomolecule detection. *Annu Rev Anal Chem (Palo Alto Calif)* 2:77–98. doi: 10.1146/annurev-anchem-060908-155232.

Jeppesen, C. 1995. Media for *Aeromonas* spp., *Plesiomonas shigelloides* and *Pseudomonas* spp. from food and environment. *Int J Food Microbiol* 26(1):25–41.

Jianrong, C., M. Yuqing, H. Nongyue, W. Xiaohua, and L. Sijiao. 2004. Nanotechnology and biosensors. *Biotechnol Adv* 22(7):505–518. doi: 10.1016/j.biotechadv.2004.03.004.

Kaittanis, C., S. Santra, and J. M. Perez. 2010. Emerging nanotechnology-based strategies for the identification of microbial pathogenesis. *Adv Drug Deliv Rev* 62(4–5):408–423. doi: 10.1016/j.addr.2009.11.013.

Kanayeva, D. A., R. Wang, D. Rhoads, G. F. Erf, M. F. Slavik, S. Tung, and Y. Li. 2012. Efficient separation and sensitive detection of *Listeria monocytogenes* using an impedance immunosensor based on magnetic nanoparticles, a microfluidic chip, and an interdigitated microelectrode. *J Food Prot* 75(11):1951–1959. doi: 10.4315/0362-028x.jfp-11-516.

Khreich, N., P. Lamourette, B. Lagoutte, C. Ronco, X. Franck, C. Creminon, and H. Volland. 2010. A fluorescent immunochromatographic test using immunoliposomes for detecting microcystins and nodularins. *Anal Bioanal Chem* 397(5):1733–1742. doi: 10.1007/s00216-009-3348-x.

Ko, S. and S. A. Grant. 2006. A novel FRET-based optical fiber biosensor for rapid detection of *Salmonella typhimurium*. *Biosens Bioelectron* 21(7):1283–1290. doi: 10.1016/j.bios.2005.05.017.

Kumar, S., U. Tuteja, and H. V. Batra. 2007. Generation and characterization of murine monoclonal antibodies to genus-specific 31-kilodalton recombinant cell surface protein of *Brucella abortus*. *Hybridoma (Larchmt)* 26(4):211–216. doi: 10.1089/hyb.2007.009.

Kurkina, T. and K. Balasubramanian. 2012. Towards in vitro molecular diagnostics using nanostructures. *Cell Mol Life Sci* 69(3):373–388. doi: 10.1007/s00018-011-0855-7.

Leo, J. C. and M. Skurnik. 2011. Adhesins of human pathogens from the genus *Yersinia*. *Adv Exp Med Biol* 715:1–15. doi: 10.1007/978-94-007-0940-9_1.

Lindesmith, L. C., K. Debbink, J. Swanstrom, J. Vinje, V. Costantini, R. S. Baric, and E. F. Donaldson. 2012. Monoclonal antibody-based antigenic mapping of norovirus GII.4–2002. *J Virol* 86(2):873–883. doi: 10.1128/jvi.06200-11.

Luo, Y., S. Nartker, H. Miller, D. Hochhalter, M. Wiederoder, S. Wiederoder, E. Setterington, L. T. Drzal, and E. C. Alocilja. 2010. Surface functionalization of electrospun nanofibers for detecting *E. coli* O157:H7 and BVDV cells in a direct-charge transfer biosensor. *Biosens Bioelectron* 26(4):1612–1617. doi: 10.1016/j.bios.2010.08.028.

Margot, H., R. Stephan, S. Guarino, B. Jagadeesan, D. Chilton, E. O'Mahony, and C. Iversen. 2013. Inclusivity, exclusivity and limit of detection of commercially available real-time PCR assays for the detection of *Salmonella*. *Int J Food Microbiol* 165(3):221–226. doi: 10.1016/j.ijfoodmicro.2013.05.012.

Maurin, M. 2012. Real-time PCR as a diagnostic tool for bacterial diseases. *Expert Rev Mol Diagn* 12(7):731–754. doi: 10.1586/erm.12.53.

Mendonca, M., N. L. Conrad, F. R. Conceicao, A. N. Moreira, W. P. da Silva, J. A. Aleixo, and A. K. Bhunia. 2012. Highly specific fiber optic immunosensor coupled with immunomagnetic separation for detection of low levels of *Listeria monocytogenes* and *L. ivanovii*. *BMC Microbiol* 12:275. doi: 10.1186/1471-2180-12-275.

Momtaz, H. and A. Jamshidi. 2013. Shiga toxin-producing *Escherichia coli* isolated from chicken meat in Iran: Serogroups, virulence factors, and antimicrobial resistance properties. *Poult Sci* 92(5):1305–1313. doi: 10.3382/ps.2012-02542.

Monk, D. J. and D. R. Walt. 2004. Optical fiber-based biosensors. *Anal Bioanal Chem* 379(7–8):931–945. doi: 10.1007/s00216-004-2650-x.

Naja, G., P. Bouvrette, J. Champagne, R. Brousseau, and J. H. Luong. 2010. Activation of nanoparticles by biosorption for *E. coli* detection in milk and apple juice. *Appl Biochem Biotechnol* 162(2):460–475. doi: 10.1007/s12010-009-8709-6.

Narsaiah, K., S. N. Jha, R. Bhardwaj, R. Sharma, and R. Kumar. 2012. Optical biosensors for food quality and safety assurance—A review. *J Food Sci Technol* 49(4):383–406. doi: 10.1007/s13197-011-0437-6.

Nhan, N. T., E. Gonzalez de Valdivia, M. Gustavsson, T. N. Hai, and G. Larsson. 2011. Surface display of *Salmonella* epitopes in *Escherichia coli* and *Staphylococcus carnosus*. *Microb Cell Fact* 10:22. doi: 10.1186/1475-2859-10-22.

Ozpinar, H., B. Turan, I. H. Tekiner, G. Tezmen, I. Gokce, and O. Akineden. 2013. Evaluation of pathogenic *Escherichia coli* occurrence in vegetable samples from district bazaars in Istanbul using real-time PCR. *Lett Appl Microbiol* 57(4):362–367. doi: 10.1111/lam.12122.

Pal, S., E. C. Alocilja, and F. P. Downes. 2007. Nanowire labeled direct-charge transfer biosensor for detecting *Bacillus* species. *Biosens Bioelectron* 22(9–10):2329–2336. doi: 10.1016/j.bios.2007.01.013.

Palchetti, I. and M. Mascini. 2008. Electroanalytical biosensors and their potential for food pathogen and toxin detection. *Anal Bioanal Chem* 391(2):455–471. doi: 10.1007/s00216-008-1876-4.

Pascale, M., F. Zezza, and G. Perrone. 2013. Surface plasmon resonance genosensor for the detection of *Fusarium culmorum*. *Methods Mol Biol* 968:155–165. doi: 10.1007/978-1-62703-257-5_12.

Pei, X., B. Zhang, J. Tang, B. Liu, W. Lai, and D. Tang. 2013. Sandwich-type immunosensors and immunoassays exploiting nanostructure labels: A review. *Anal Chim Acta* 758:1–18. doi: 10.1016/j.aca.2012.10.060.

Petrenko, V. A. and I. B. Sorokulova. 2004. Detection of biological threats. A challenge for directed molecular evolution. *J Microbiol Methods* 58(2):147–168. doi: 10.1016/j.mimet.2004.04.004.

Polpanich, D., P. Tangboriboonrat, A. Elaissari, and R. Udomsangpetch. 2007. Detection of malaria infection via latex agglutination assay. *Anal Chem* 79(12):4690–4695. doi: 10.1021/ac070502w.

Quigley, L., O. O'Sullivan, T. P. Beresford, R. P. Ross, G. F. Fitzgerald, and P. D. Cotter. 2011. Molecular approaches to analysing the microbial composition of raw milk and raw milk cheese. *Int J Food Microbiol* 150(2–3):81–94. doi: 10.1016/j.ijfoodmicro.2011.08.001.

Rastogi, S. K., C. M. Gibson, J. R. Branen, D. E. Aston, A. L. Branen, and P. J. Hrdlicka. 2012. DNA detection on lateral flow test strips: Enhanced signal sensitivity using LNA-conjugated gold nanoparticles. *Chem Commun (Camb)* 48(62):7714–7716. doi: 10.1039/c2cc33430e.

Ravindranath, S. P., L. J. Mauer, C. Deb-Roy, and J. Irudayaraj. 2009. Biofunctionalized magnetic nanoparticle integrated mid-infrared pathogen sensor for food matrixes. *Anal Chem* 81(8):2840–2846. doi: 10.1021/ac802158y.

Rodriguez-Emmenegger, C., O. A. Avramenko, E. Brynda, J. Skvor, and A. B. Alles. 2011. Poly(HEMA) brushes emerging as a new platform for direct detection of food pathogen in milk samples. *Biosens Bioelectron* 26(11):4545–4551. doi: 10.1016/j.bios.2011.05.021.

Rojas-Duran, T., I. Sanchez-Barragan, J. M. Costa-Fernandez, and A. Sanz-Medel. 2007. Direct and rapid discrimination of aflatoxigenic strains based on fibre-optic room temperature phosphorescence detection. *Analyst* 132(4):307–313. doi: 10.1039/b610789c.

Rompre, A., P. Servais, J. Baudart, M. R. de-Roubin, and P. Laurent. 2002. Detection and enumeration of coliforms in drinking water: Current methods and emerging approaches. *J Microbiol Methods* 49(1):31–54.

Scallan, E., R. M. Hoekstra, F. J. Angulo, R. V. Tauxe, M. A. Widdowson, S. L. Roy, J. L. Jones, and P. M. Griffin. 2011. Foodborne illness acquired in the United States—Major pathogens. *Emerg Infect Dis* 17(1):7–15.

Seidel, M. and R. Niessner. 2008. Automated analytical microarrays: A critical review. *Anal Bioanal Chem* 391(5):1521–1544. doi: 10.1007/s00216-008-2039-3.

Sharma, H. and R. Mutharasan. 2013. Rapid and sensitive immunodetection of *Listeria monocytogenes* in milk using a novel piezoelectric cantilever sensor. *Biosens Bioelectron* 45:158–162. doi: 10.1016/j.bios.2013.01.068.

Shin, J. and M. Kim. 2008. Development of liposome immunoassay for *Salmonella* spp. using immunomagnetic separation and immunoliposome. *J Microbiol Biotechnol* 18(10):1689–1694.

Silbert, L., I. Ben Shlush, E. Israel, A. Porgador, S. Kolusheva, and R. Jelinek. 2006. Rapid chromatic detection of bacteria by use of a new biomimetic polymer sensor. *Appl Environ Microbiol* 72(11):7339–7344. doi: 10.1128/aem.01324-06.

Smartt, A. E., T. Xu, P. Jegier, J. J. Carswell, S. A. Blount, G. S. Sayler, and S. Ripp. 2012. Pathogen detection using engineered bacteriophages. *Anal Bioanal Chem* 402(10):3127–3146. doi: 10.1007/s00216-011-5555-5.

Starnbach, M. N. and M. J. Bevan. 1994. Cells infected with *Yersinia* present an epitope to class I MHC-restricted CTL. *J Immunol* 153(4):1603–1612.

Stevens, M. P., T. J. Humphrey, and D. J. Maskell. 2009. Molecular insights into farm animal and zoonotic *Salmonella* infections. *Philos Trans R Soc Lond B Biol Sci* 364(1530):2709–2723. doi: 10.1098/rstb.2009.0094.

Sundaram, J., B. Park, Y. Kwon, and K. C. Lawrence. 2013. Surface enhanced Raman scattering (SERS) with biopolymer encapsulated silver nanosubstrates for rapid detection of foodborne pathogens. *Int J Food Microbiol* 167(1):67–73. doi: 10.1016/j.ijfoodmicro.2013.05.013.

Sungkanak, U., A. Sappat, A. Wisitsoraat, C. Promptmas, and A. Tuantranont. 2010. Ultrasensitive detection of *Vibrio cholerae* O1 using microcantilever-based biosensor with dynamic force microscopy. *Biosens Bioelectron* 26(2):784–789. doi: 10.1016/j.bios.2010.06.024.

Tang, S., M. Moayeri, Z. Chen, H. Harma, J. Zhao, H. Hu, R. H. Purcell, S. H. Leppla, and I. K. Hewlett. 2009. Detection of anthrax toxin by an ultrasensitive immunoassay using europium nanoparticles. *Clin Vaccine Immunol* 16(3):408–413. doi: 10.1128/cvi.00412-08.

Thornton, C. R. and O. E. Wills. 2013. Immunodetection of fungal and oomycete pathogens: Established and emerging threats to human health, animal welfare and global food security. *Crit Rev Microbiol.* doi: 10.3109/1040841x.2013.788995.

Tognalli, N. G., P. Scodeller, V. Flexer, R. Szamocki, A. Ricci, M. Tagliazucchi, E. J. Calvo, and A. Fainstein. 2009. Redox molecule based SERS sensors. *Phys Chem Chem Phys* 11(34):7412–7423. doi: 10.1039/b905600a.

Tully, E., S. Hearty, P. Leonard, and R. O'Kennedy. 2006. The development of rapid fluorescence-based immunoassays, using quantum dot-labelled antibodies for the detection of *Listeria monocytogenes* cell surface proteins. *Int J Biol Macromol* 39(1–3):127–134. doi: 10.1016/j.ijbiomac.2006.02.023.

Valanne, A., S. Huopalahti, T. Soukka, R. Vainionpaa, T. Lovgren, and H. Harma. 2005. A sensitive adenovirus immunoassay as a model for using nanoparticle label technology in virus diagnostics. *J Clin Virol* 33(3):217–223. doi: 10.1016/j.jcv.2004.11.007.

Valera, M. J., M. J. Torija, A. Mas, and E. Mateo. 2013. *Acetobacter malorum* and *Acetobacter cerevisiae* identification and quantification by Real-Time PCR with TaqMan-MGB probes. *Food Microbiol* 36(1): 30–39. doi: 10.1016/j.fm.2013.03.008.

Verma, N., S. Kumar, and H. Kaur. 2010. Fiber optic biosensor for the detection of Cd in milk. *J Biosens Bioelectron* 1:102.

Viswanathan, S., C. Rani, and J. A. Ho. 2012. Electrochemical immunosensor for multiplexed detection of food-borne pathogens using nanocrystal bioconjugates and MWCNT screen-printed electrode. *Talanta* 94:315–319. doi: 10.1016/j.talanta.2012.03.049.

Vo-Dinh, T. 2002. Nanobiosensors: Probing the sanctuary of individual living cells. *J Cell Biochem Suppl* 39:154–161. doi: 10.1002/jcb.10427.

Wacheck, S., C. Werres, U. Mohn, S. Dorn, E. Soutschek, M. Fredriksson-Ahomaa, and E. Martlbauer. 2012. Detection of IgM and IgG against hepatitis E virus in serum and meat juice samples from pigs at slaughter in Bavaria, Germany. *Foodborne Pathog Dis* 9(7):655–660. doi: 10.1089/fpd.2012.1141.

Wang, S., A. K. Singh, D. Senapati, A. Neely, H. Yu, and P. C. Ray. 2010. Rapid colorimetric identification and targeted photothermal lysis of *Salmonella* bacteria by using bioconjugated oval-shaped gold nanoparticles. *Chemistry* 16(19):5600–5606. doi: 10.1002/chem.201000176.

Wang, Y., S. Ravindranath, and J. Irudayaraj. 2011. Separation and detection of multiple pathogens in a food matrix by magnetic SERS nanoprobes. *Anal Bioanal Chem* 399(3):1271–1278. doi: 10.1007/s00216-010-4453-6.

Wutz, K., V. K. Meyer, S. Wacheck, P. Krol, M. Gareis, C. Nolting, F. Struck et al. 2013. New route for fast detection of antibodies against zoonotic pathogens in sera of slaughtered pigs by means of flow-through chemiluminescence immunochips. *Anal Chem* 85(10):5279–5285. doi: 10.1021/ac400781t.

Yakes, B. J., E. Papafragkou, S. M. Conrad, J. D. Neill, J. F. Ridpath, W. Burkhardt, 3rd, M. Kulka, and S. L. Degrasse. 2013. Surface plasmon resonance biosensor for detection of feline calicivirus, a surrogate for norovirus. *Int J Food Microbiol* 162(2):152–158. doi: 10.1016/j.ijfoodmicro.2013.01.011.

Yang, H., L. Qu, A. N. Wimbrow, X. Jiang, and Y. Sun. 2007. Rapid detection of *Listeria monocytogenes* by nanoparticle-based immunomagnetic separation and real-time PCR. *Int J Food Microbiol* 118(2): 132–138. doi: 10.1016/j.ijfoodmicro.2007.06.019.

Yang, M., Z. Peng, Y. Ning, Y. Chen, Q. Zhou, and L. Deng. 2013. Highly specific and cost-efficient detection of *Salmonella* Paratyphi A combining aptamers with single-walled carbon nanotubes. *Sensors (Basel)* 13(5):6865–6881. doi: 10.3390/s130506865.

Yang, M., S. Sun, Y. Kostov, and A. Rasooly. 2010. Lab-on-a-chip for carbon nanotubes based immunoassay detection of Staphylococcal Enterotoxin B (SEB). *Lab Chip* 10(8):1011–1017. doi: 10.1039/b923996k.

Yavuz, C. T., J. T. Mayo, W. W. Yu, A. Prakash, J. C. Falkner, S. Yean, L. Cong et al. 2006. Low-field magnetic separation of monodisperse Fe_3O_4 nanocrystals. *Science* 314(5801):964–967. doi: 10.1126/science.1131475.

Zhao, Y., M. Ye, Q. Chao, N. Jia, Y. Ge, and H. Shen. 2009. Simultaneous detection of multifood-borne pathogenic bacteria based on functionalized quantum dots coupled with immunomagnetic separation in food samples. *J Agric Food Chem* 57(2):517–524. doi: 10.1021/jf802817y.

Zhu, X., L. Chen, P. Shen, J. Jia, D. Zhang, and L. Yang. 2011. High sensitive detection of Cry1Ab protein using a quantum dot-based fluorescence-linked immunosorbent assay. *J Agric Food Chem* 59(6): 2184–2189. doi: 10.1021/jf104140t.

Zordan, M. D., M. M. Grafton, G. Acharya, L. M. Reece, C. L. Cooper, A. I. Aronson, K. Park, and J. F. Leary. 2009. Detection of pathogenic *E. coli* O157:H7 by a hybrid microfluidic SPR and molecular imaging cytometry device. *Cytometry A* 75(2):155–162. doi: 10.1002/cyto.a.20692.

11

Detection of Mycotoxin-Producing Molds and Mycotoxins in Foods

Alicia Rodríguez, María J. Andrade, Mar Rodríguez, and Juan J. Córdoba

CONTENTS

11.1 Molds and Their Effects in Foods

Molds are found in a wide range of foods as microbial contaminants due to their capacity to use a variety of substrates and to their relative tolerance to low pH, low water activity, and low temperature (Garcia et al., 2009).

It has been reported that molds growing on foods can have beneficial effects on the products by contributing to flavor development as in ripened foods (Martín et al., 2006) or in *Monascus*-fermented products by producing bioactives metabolites (Shi and Pan, 2011). Some molds isolated from foods or their purified enzymes have been reported to be beneficial because of their biological and industrial applications (Benito et al., 2004; Karimi and Zamani, 2013). However, molds may have a negative impact on foods due to spoilage, causing economic losses worldwide (Dagnas and Membré, 2012). In addition, some molds could synthesize mycotoxins. Although the growth of molds on foods is not necessarily associated with the formation of mycotoxins (Fernández-Cruz et al., 2010), in many cases, their presence leads to the accumulation of these metabolites in foods, which is considered a hazard for human health as they may cause autoimmune illnesses or neurological disorders and some mycotoxins even have teratogenic, carcinogenic, and mutagenic effects (García et al., 2009).

Mycotoxins are produced by several mold genera, mainly *Penicillium, Aspergillus, Fusarium,* and *Alternaria* (Niessen, 2007; Martín et al., 2013). The most frequently found mycotoxins in foods are aflatoxins, ochratoxin A, patulin, fumonisins, trichothecenes (T-2 toxin, HT-2 toxin, neosolaniol, monoacetoxy scirpenol, diacetoxyscirpenols, deoxynivalenol [DON], nivalenol, 3- and 15-acetoxynivalenol), zerealonone, and the *Alternaria* toxins (alternariol, alternariol monomethyl ether, altertoxin, tenuazonic acid, and altenuene) (Sudakin, 2003; Eriksen and Pettersson, 2004; Bhat et al., 2010; Fernández-Cruz et al., 2010; Bulder et al., 2012; Marín et al., 2013; Shephard et al., 2013). These mycotoxins have been detected in a wide variety of agricultural products such as nuts, fruits and derived juices, cereals, spices, cacao and its derivates, coffee, tea, wine, milk, and ripened foods, including cheese and meat products, as a consequence of surface growth of a mycobiota population (Mansfield et al., 2008; Marín et al., 2008, 2013; Hashem and Alamri, 2010). All the mentioned mycotoxins excepting patulin are the most commonly found mycotoxins in cereals and their derivates (Marín et al., 2013). Patulin is most likely to contaminate fruit juices, mainly apple juice (Marín et al., 2013). The occurrence of mycotoxin contamination in foods is influenced by several factors including climatic condition, harvest season, and storage or ripening conditions.

In order to protect human health from risks associated with mycotoxins, many countries have adopted regulations or guidelines regarding maximum limits of mycotoxins in food. This maximum permitted or recommended level of mycotoxins varies in regulations for different countries (Table 11.1). Concerning the European Union, the Commission Regulation (EC) No. 1881/2006 amended by several later Commission regulations set the maximum levels for aflatoxins, ochratoxin A, patulin, fumonisins (sum of B1 and B2), trichothecenes, and zerealonone in several foodstuffs (Table 11.1). To protect the health of infants and young children, this Regulation lays down maximum levels of all the mycotoxins in

TABLE 11.1

Summary of the Maximum Levels Permitted or Recommended for Aflatoxins, Ochratoxin A, Zearalenone, Fumonisins, and Patulin in Different Foodstuffs

Mycotoxin	Country	Maximum Limits (μg/kg)[a]	Foodstuffs[b]
Total aflatoxins	USA	20	Brazil nuts, peanuts, and peanut products, pistachio products
	UE	4–15	Nuts, peanuts, pistachios, dried fruit, cereals, spices
	Japan	10	All
Aflatoxin B1	UE	2–12	Nuts, peanuts, pistachios, dried fruit, cereals, spices
Aflatoxin M1	USA	0.5	Milk
	China	0.5	Milk and milk products
	UE	0.05	Milk and milk products
Ochratoxin A	UE	2–80	Unprocessed cereals and derived products, dried vine fruit, coffee, wine, grape juice, spices, liquorice
	Russia	5	Cereals and cereal products
	China	5	Cereals, milled products from cereals, legumes, pulses
Zearalenone	UE	50–400	Cereals and cereal products
Fumonisins	UE	800–4000	Maize and derived products
	USA	2000–4000	Corn and corn products
Patulin	UE	25–50	Fruit juices, fermented drinks derived from apples, solid apple products
	China	50	Fruit products containing apple or hawthorn
	USA	50	Apple juice and its derivates

Sources: European Commission, *Official J. Eur. Union,* L 364, 5, 2006; European Mycotoxins Awareness Network, Mycotoxins legislation worldwide, 2012, http://www.mycotoxins.org/, accessed September 9, 2013.

[a] When the maximum limits are different depending on the food products, the range of maximum levels is specified.

[b] Other than products for infants and young children.

food mentioned here for this population group. Other worldwide countries including the United States, Japan, China, and Russia have established specific maximum limits for these mycotoxins in different food products too (Table 11.1).

11.2 Methods to Detect Mycotoxin-Producing Molds

The detection and quantification of toxigenic molds in raw materials, preprocessed, and final foods are critical for the production of safe foods. The incorporation of rapid methods to monitorize the presence of mycotoxin-producing molds within Hazard Analysis Critical Control Point (HACCP) systems would allow taking appropriate corrective actions to avoid risks associated with the mycotoxin accumulation in foods. Thus, methods to detect mycotoxin-producing molds are of great utility in preventive food safety systems as they can increase the safety of processed foods and avoid economic losses as a consequence of eliminating foods contaminated with mycotoxins. However, methods to detect the accumulation of mycotoxins in foods are only useful in avoiding the consumption of those foods that are contaminated. In addition, the last methods may have the emerging problem of masked mycotoxins, which are compounds generated in mycotoxin-contaminated commodities by plant metabolism or by food processing that coexist together with the native toxins (Galaverna et al., 2009; Berthiller et al., 2013).

Different methods to detect mycotoxin-producing molds in foods have been reported. These methods can be classified into those based on mold characterization by conventional microbiological methods and those based on mold detection by molecular techniques. The first kind of methods needs a subsequent evaluation of mycotoxin production as only mold characterization does not give any information about the toxicity of the studied mold strains. In Sections 11.3 and 11.4, the conventional microbial method of detection of toxin production and molecular methods based on nucleic acid analysis will be analyzed and discussed.

11.3 Detection of Toxic Molds by Conventional Microbiological Methods and Analysis of Mycotoxin Production by Emerging Analytical Techniques

Detection of mycotoxin-producing molds by this method includes a multistep process comprised of sampling, culture, isolation, and characterization by microbiological methods and evaluation of mycotoxin production including toxin extraction, cleanup procedures, and finally qualitative or quantitative analysis of mycotoxins.

11.3.1 Isolation and Characterization of Mycotoxin-Producing Molds by Microbiological Analysis

Traditionally, culture media suitable for enumeration of mycotoxin-producing molds, such as Malt Extract agar, Potato Dextrose agar, Dichloran 18% glycerol (DG18) agar, or Rose Bengal Chloramphenicol agar have been used (Beuchat, 2003). These microbiological methods allow isolation and characterization of mold strains from foods. This kind of detection is time-consuming (minimum 5 days of incubation) and labor-intensive, needs morphological and physiological tests, and often requires mycological expertise.

11.3.2 Evaluation of Mycotoxin Production by Analytical Methods

After the isolation and characterization of mold strains by microbiological methods, it is necessary to complement this analysis by evaluating the production of mycotoxins by analytical methods. In this analysis, the sampling procedure is the largest source of variability associated with the mycotoxin test and the most crucial step in obtaining reliable results (Köppen et al., 2010). The difficulty of preparing a representative sample depends on several factors, for instance, the particle size or the number of particles per unit mass or even the complexity of the culture media or food matrices. Some strategies to reduce sampling errors consist of increasing the laboratory sample size, the degree of sample comminution, the subsample size, and the number of aliquots quantified (Whitaker et al., 2009). All these factors make it impossible to use only one technique for the determination of all mycotoxins. For this reason, a large number of procedures for the detection of mycotoxins and improvements in sampling were developed (Cigić and Prosen, 2009; Turner et al., 2009; Köppen et al., 2010; Shephard et al., 2013).

11.3.2.1 Sample Pretreatment Methods

This is a very important step for the determination of mycotoxins. The most used techniques include cleanup steps, which are vital for a successful protocol. The extraction method used to remove mycotoxins from the biological matrix depends on the chemical structure of the mycotoxin, and can affect the sensitivity of the results (Turner et al., 2009). The four most important methods used in cleaning up mycotoxin samples are solvent extraction (SE), supercritical fluid extraction (SFE), solid phase extraction (SPE), and the use of immunoaffinity columns (IA).

SE involves exploiting the different solubility levels of the mycotoxin in aqueous phase and in immiscible organic phase, to extract it into one solvent leaving the rest of the matrix in the other. The choice of the appropriate solvent depends on which culture medium or matrix food is being used, and which mycotoxins are being determined. Solvents such as acetonitrile, chloroform, methanol, ethyl acetate, hexane, and cyclohexane are commonly used (Cigić and Prosen, 2009; Turner et al., 2009).

SFE uses a supercritical fluid, such as CO_2, to extract the required compound from the matrix. This technique is not suitable for routine analysis due to its high costs and the need for specialized equipment (Holcomb et al., 1996).

SPE is generally performed by passing aqueous samples through a solid sorbent in a column. The mycotoxin is eluted from the solid medium with an appropriate organic solvent and a washing step is done with an organic solvent to remove the interferences (Hayat et al., 2012). A variation of the SPE is the solid phase microextraction (SPME), which employs fibers coated with different stationary phases on which analytes are sorbed from liquid or gaseous samples (Cigić and Prosen, 2009).

IA is a method of purification or concentration of mycotoxins with immunoaffine columns, which has to be accompanied by the following analysis. The sample extract passes through the column (with specific antibodies immobilized), the mycotoxin binds to the antibodies, and the other substances go through the column. The mycotoxin is washed off the column and can be analyzed with the help of different methods (Urusov et al., 2010).

11.3.2.2 Analysis of Mycotoxins by Chromatographic Methods

After pretreatment methods (cleanup), mycotoxins should be free of any metabolites that could interfere in their analysis. Traditionally, the separation of mycotoxins is achieved by using the following chromatographic methods whose advantages and disadvantages are also described. In addition, some characteristics of these methods are summarized in Table 11.2.

11.3.2.2.1 Thin-Layer Chromatography

Thin-layer chromatography (TLC) is a low-cost and rapid analytical technique that offers the ability to screen large numbers of samples yielding qualitative or semiquantitative estimations by visual inspection (Cigić and Prosen, 2009; Turner et al., 2009). It is a technique for screening that is useful for separation, assessment of purity, and identification of the mycotoxin, thanks to its simplicity and rapidity.

11.3.2.2.2 High-Performance Liquid Chromatography

There are several methods based on high-performance liquid chromatography (HPLC) whose differences lie in the coupled detector used. The most commonly found detection methods are ultraviolet (UV), diode array (DA), fluorescence (FL), single mass spectrometric (MS), and tandem mass spectrometric (MS/MS). Although a number of mycotoxins have natural fluorescence (e.g., ocratoxin A and aflatoxins) and can be detected directly in HPLC-FLD, some of them do not have a suitable chromophore, and their determination requires derivatization (Shephard et al., 2013).

TABLE 11.2

Type of Quantification, Required Equipment, and Assay Time of the Main Analytical Methods for Toxigenic-Producing Mold Detection

Methodology	Quantification	Equipment	Time of Analysis
Microbial characterization and production of mycotoxins in culture media or foods	No	—	>5 days
Analysis of mycotoxins by			
TLC	Semiquantitative	UV lamp	Previous cleanup + 30 min
HPLC-MS/MS	Yes	HPLC	Previous cleanup + 20–30 min
UHPLC-MS/MS	Yes	UHPLC	Previous cleanup + 4–15 min
LC/ESI-QTOF-MS/MS	Yes	QTOF-MS/MS and HPLC-ESI	Previous cleanup + 4–8 min
ELISA	Yes	ELISA reader	Previous cleanup + 1–2 h
Flow-through immunoassays	Semiquantitative	—	5 min
Lateral flow or strip test	Semiquantitative	—	5–10 min
Fluorescent polarization immunoassays	Semiquantitative	Fluorescence reader	5–10 min
Surface plasmon resonance	Yes	SPR system	25 min
Fiber-optic immunosensor	Yes	Spectrometer	Previous cleanup + 10–60 min
Aptamer-based methods	Yes	Dependent detection method	10 min to 2 h
Infrared spectroscopy	Yes	Infrared spectroscopy	5 min
Multispectral and hyperspectral imaging	No	Light source, wavelength dispersive device, and area detector	20–30 min

Sources: Elaborated with data from Beuchat, L.R., Media for detecting and enumerating yeasts and moulds, in *Handbook of Culture Media for Food Microbiology*, Elsevier Science B.V., Amsterdam, the Netherlands, 2003, pp. 369–385; Zheng, M.Z. et al., *Mycopathologia*, 161, 261, 2006; Krska, R. and Molinelli, A., *Anal. Bioanal. Chem.*, 393, 67, 2009; Hayat, A. et al. *Food Control*, 26, 401, 2012; Beltrán, E. et al., *Anal. Chim. Acta*, 783, 39, 2013; Liao, C.D. et al., *J. Agric. Food Chem.*, 61, 4771, 2013; Sirhan, A.Y. et al. *Food Control*, 31, 35, 2013.

Using HPLC-MS/MS, the simultaneous determination of mycotoxins belonging to different chemical families can be performed. Liao et al. (2013) developed a rapid and accurate HPLC-MS method (15 min) to detect 26 mycotoxins in different food matrices.

The introduction of ultra-HPLC (UHPLC) has provided additional advantages in the determination of mycotoxins in foodstuffs. This is the result of using columns packed with sub-2 μm particles, which allow obtaining narrower peaks than those obtained with traditional HPLC columns (Beltrán et al., 2013).

More recently, the liquid chromatography and electrospray ionization quadrupole time-of-flight mass spectrometry (LC/ESI-QTOF-MS/MS) provides a high sample throughput, with high resolution, which allows the acquisition of the full range mass spectral data instead of just a single ion. This method provides an efficient tool for the detection of mycotoxins in foods, such as aflatoxins (Sirhan et al., 2013).

11.3.2.2.3 Gas Chromatography

The main problem of gas chromatography (GC) is that only thermally stable and volatile products can be analyzed. However, most of the mycotoxins are not volatile and they have to be derivatized for being analyzed using GC (Cigić and Prosen, 2009). Normally, this system is coupled to MS, flame ionization detector (FID), or Fourier transform infrared spectroscopy (FTIR) detection techniques. However, the use of GC detection is not expected for routine analysis due to cheaper and faster alternatives such as the previously described HPLC methods (Turner et al., 2009).

11.3.2.2.4 Capillary Electrophoresis

Capillary electrophoresis (CE) allows significant separation of a large number of mycotoxins, but it has never had as much popularity as HPLC. Achieving low enough detection limits might present a serious problem in CE; therefore, the mycotoxins for which CE methods have been developed are mostly those for which fluorescence detection can be used (Cigić and Prosen, 2009).

11.3.2.3 Analysis of Mycotoxins by Immunological Methods

Immunoassays are based on the ability of a specific antibody to distinguish the three-dimensional structure of a specific mycotoxin from other molecules. Due to the fact that mycotoxins are low-molecular substances that do not have their own immunogenicity, it is necessary to bind them to a carrier protein before immunization. Commercial immunological techniques for mycotoxins are based on specific monoclonal and polyclonal antibodies produced against the toxin. This means that only commercial kits have been developed for certain mycotoxins (European Mycotoxins Awareness Network, 2011). These techniques are fast as the results are obtained within 2 h to a few minutes. The disadvantage of these kits lies in the fact that they are for single use, which can increase the cost of screening.

11.3.2.3.1 Enzyme-Linked Immunosorbent Assay

Commercially available enzyme-linked immunosorbent assay (ELISA) kits for the detection of mycotoxins are normally based on a direct competitive assay format that uses either a primary antibody specific for the target molecule or a conjugate of an enzyme and the required target. The complex formed will then interact with a chromogenic substrate to give a measurable result in 1–2 h (Turner et al., 2009).

11.3.2.3.2 Membrane-Based Immunoassays

The membrane methods of immunoanalysis for mycotoxins are based on a membrane that contains immobilized antibodies. There are two technologies: immunofiltration or flow-through assay and immunochromatography or lateral flow test. Although flow-through immunoassays have been developed for mycotoxins, they are not as commercially successful as test strips (Krska and Molinelli, 2009).

Immunochromatography, such as the flow-through immunoassays, consists of qualitative or semi quantitative tests. The lateral flow test assay is rapid (5 min), easy-to-use, and suitable for testing mycotoxins in the field. However, so far, major restrictions are matrix dependence, lack of appropriate specific antibodies, and, therefore, selectivity and sensitivity problems (Krska and Molinelli, 2009). Some of these problems are being improved with the development of new technologies such as the detector-free semiquantitative strip (DFQ-strip) (Zhang et al., 2013).

11.3.2.3.3 Fluorescent Polarization Immunoassays

Fluorescent polarization immunoassay (FPI) is based on the competition between the mycotoxin and a mycotoxin-fluorescein tracer for a mycotoxin-specific antibody. Polarization is a measure of the orientation of the fluorescence emission from both horizontal and vertical directions but not a direct measure of fluorophore concentration. The method is simple to use and can be field-portable; however, matrix effects may exist (Zheng et al., 2006).

11.3.2.3.4 Surface Plasmon Resonance–Based Immunoassay

The surface plasmon resonance–based immunoassay (SPR) is designed as an inhibition assay. A fixed amount of mycotoxin-specific antibody is mixed with a sample containing an unknown amount of mycotoxin, forming a complex. The sample is then passed over a sensor surface to which mycotoxins have been immobilized. The amount of noncomplexed antibodies is determined as they bind to the immobilized mycotoxin on the sensor surface. The SPR method has considerable advantages: a very small sample volume is necessary for mycotoxin analysis, the chip can be regenerated, and it is user-friendly. However, sensitivity may be an obstacle for some SPR systems and analysis cost is quite high (Urusov et al., 2010).

11.3.2.3.5 Fiber-Optic Immunosensor

For the detection of mycotoxins, one format of fiber-optic sensor uses the evanescent wave effect, in which an evanescent wave is generated at the interface between an optical fiber and an outside lower refractive index material. If the proper wavelength is selected, fluorescent molecules in this region can absorb energy and a portion of the fluorescence is coupled back into the fiber and can be detected. By immobilizing antibodies to the surface of an optical fiber, fluorescent materials from the bulk solution can be achieved. However, this method has some limitations in sensitivity and can be influenced by the solvents, because they change the medium refraction index (Zheng et al., 2006).

11.3.2.4 Aptamer-Based Assays for Detection of Mycotoxins

Aptamers are single-stranded oligonucleotides of DNA or RNA sequences produced by an in vitro selection process using systematic evolution of ligands by exponential enrichment (SELEX). Due to their unique three-dimensional structures, aptamers can bind to their target molecules with high affinity and specificity with different spatial structures and folding patterns (Ellington and Szostak, 1990). Aptamers possess significant advantages over antibodies as they are chemically synthesized, immunization of animals is not required, and aptamers are not susceptible to denaturation in the presence of solvents commonly used in the extraction of mycotoxins. Furthermore, the function of immobilized aptamers can be easily regenerated and aptamers can be reused (Hayat et al., 2012). Aptamer sequences of mycotoxins such as aflatoxins, ochratoxin A, fumonisin B1, zearalenona, and deoxynivlenol have been selected (Yang et al., 2013). The combination of aptamer recognition technique and novel nanomaterials was developed and used in various optical and electrochemical analytical methods for mycotoxin analysis (Hayat et al., 2012).

11.3.2.5 Analysis of Mycotoxins by Indirect Methods

These methods are based on the measurement and interpretation of a parameter or compound in the sample that change in relation to the mycotoxin concentration. Most of these methods are also noninvasive, meaning less sample manipulation and faster throughput (Cigić and Prosen, 2009).

Various methods using infrared spectroscopy have been developed, such as FTIR in diffuse reflection or attenuated total reflection, and near-infrared spectroscopy (NIR). NIR was applied to determine aflatoxins and ocratoxin A (Cigić and Prosen, 2009).

11.3.2.5.1 Proteomic Analysis for Investigating Food-Borne Mycotoxins

Mycotoxin production in foods is influenced by many factors such as availability of nutrients (carbon source as glucose, starch, fat, and lactate), temperature, and water activity. In conditions where mycotoxins are being produced, some proteins are selectively produced. Thus, an increase in the number of some proteins has been related with the production of several mycotoxins. This is the case of trichothecene C-3 acetyltransferase, a protein involved in trichothecene biosynthesis (Giacometti et al., 2013). A comparative proteomic analysis between OTA-producing and OTA-nonproducing strains has allowed associating some proteins to the production of ochratoxin A by *Aspergillus carbonarius* (Crespo-Sempere et al., 2011). The specific detection of these proteins by proteomic analysis allows the detection of mycotoxin production. In addition, proteomic analysis will raise future knowledge of mycotoxin production at the cellular level (Giacometti et al., 2013). Different methods are used for the quantitative proteomic characterization of fungi such as those based on one-dimensional and two-dimensional electrophoresis with poststaining, gel-free techniques based on in vitro or in vivo protein targeting or label-free approaches, for instance, spectral counting or approaches based on precursor signal intensity (Giacometti et al., 2013).

11.3.2.5.2 Metabolomics for Investigating Food-Borne Mycotoxins

Quite a different approach for the indirect detection of mycotoxin-producing molds is the detection of secondary metabolite production. Some correlation between the release of these compounds and the biosynthesis of mycotoxins seems to exist, and can be used as taxonomic markers for toxigenic and nontoxigenic mold species differentiation (Magan and Evans, 2000). The determination of these compounds may be performed with classical methods such as GC-MS, LC-MS/MS analysis, or LC high-resolution mass spectrometric (Nielsen et al., 2009), or with modern sensor arrays, the so-called electronic nose (EN) system for volatile metabolites. This measurement approach using a relatively sophisticated EN system requires 20–25 min/sample (Sahgal et al., 2007).

11.3.2.6 Machine Vision Techniques for Mycotoxin Detection in Foods

The machine vision systems are nondestructive spectral techniques that are available for the detection of mycotoxins in foods including thermal, x-ray, and color imaging (Teena et al., 2013). Multispectral and hyperspectral imaging techniques combine spectral and spatial imaging analyses to detect biochemical changes in food resulting from mold growth; however, both techniques are slightly different. The multispectral imaging technique has been used for detecting aflatoxins and fumonisins (Firrao et al., 2010; Kalkan et al., 2011). Fluorescence hyperspectral imaging has been used to detect aflatoxin (Hruska et al., 2013; Yao et al., 2013).

When the total time of analysis of mycotoxin-producing molds in foods by conventional microbial methods and the subsequent evaluation of mycotoxin production is taken into consideration, it can be concluded that this methodology is highly time-consuming (higher than 5 days), even when emerging techniques for mycotoxin analysis are used (Table 11.2). Therefore, this methodology is not appropriate for the routine analysis of mycotoxin-producing molds. However, the techniques described for mycotoxin analysis are very useful to detect exclusively the accumulation of mycotoxins in foods. Thus, some of these techniques could be used for the verification of hazard due to mycotoxin accumulation in HACCP systems.

11.4 Detection of Mycotoxin-Producing Molds by Molecular Methods

An alternative to mold characterization plus evaluation of mycotoxins for detecting mycotoxin-producing molds in foods could be the detection of genes involved in the biosynthesis of these metabolites by nucleic acid–based methods. Among these methods, PCR-based techniques have been the subject of considerable focus and, more particularly, real-time quantitative PCR (qPCR). In the last few years, new molecular techniques, such as loop-mediated isothermal amplification (LAMP), are being developed to detect toxigenic molds related to food. Regarding studies about the expression of toxin biosynthetic genes (genomic analysis), molecular methods including reverse transcription (RT)-PCR, RT-qPCR, microarrays, and RNA sequencing have been reported. The procedures followed in these methods and their utility for detecting mycotoxin-producing molds in foods will be discussed in the following sections.

11.4.1 Nucleic Acid Extraction of Mycotoxin-Producing Molds

Before setting up molecular methods to quantify mycotoxin-producing molds in foods, it is necessary to have adequate nucleic acid extraction methods available, as the sample quality is probably the most important aspect to ensure the reproducibility and sensitivity of the analysis and the correct detection and quantification of toxigenic molds (Hayat et al., 2012). The main factor affecting the DNA extraction from food samples is the presence of their components, which may reduce the sensitivity of the amplification reaction (Demeke and Jenkins, 2010). They act as inhibitors to PCR-based methods as they can interfere in cell lysis, contribute to nucleic acid degradation, or inhibit DNA polymerase (Wilson, 1997). Additionally, when a PCR-based protocol is developed to detect molds, the efficiency of the lysis of their cells and the nucleic acid purification has a critical importance due to the composition of the mold cell wall, which is a complex structure composed of chitin, glucans, lipids, and other polymers that are extremely resistant (Hayak et al., 2012; Leite et al., 2012). Thus, the effect of different steps in the DNA and RNA extraction procedures such as type of dilution or extraction buffer, method for breakage of the mold cell wall, type of commercial DNA extraction kit, DNA extraction solvents, and DNA precipitation alcohols should be carefully evaluated before selection.

In spite of the fact that it has been reported that there is no single DNA extraction method suitable for all mold species (Hayat et al., 2012), Luque et al. (2012b) developed an accurate method to obtain DNA from different ochratoxigenic, aflatoxigenic, and patulin-producing mold species directly from foods to be used in conventional PCR methods (Luque et al., 2011, 2012a, 2013b). Furthermore, Rodríguez et al. (2012d) optimized a mold DNA extraction method from different food matrices to be used in advanced molecular techniques such as qPCR. The latter is widely used to obtain a good

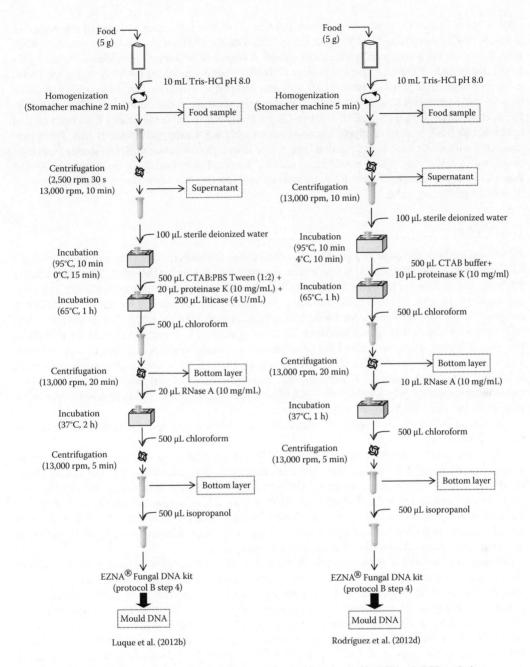

FIGURE 11.1 Detailed protocols for mold DNA extraction from foods to be used in PCR and qPCR analysis.

yield of DNA from foods for quantifying six kinds of toxigenic molds belonging to different species and genera and even nontoxigenic molds by qPCR (Rodríguez et al., 2011a,b, 2012a,b,c,d,e,f,g,h; Bernáldez et al., 2013, 2014). For the optimization of both DNA extraction methods, CTAB as extraction buffer and the EZNA® Fungal DNA Kit were used. The main difference between both methods consists of additional steps for sample cleaning in the protocol optimized by Luque et al. (2012b). Figure 11.1 shows a diagram of both these procedures. These protocols should be considered as a prior step to PCR or qPCR analyses to obtain a high quality of mold DNA to be used for the detection of mycotoxin-producing molds in foods.

RNA-based methods are more reliable than DNA-based PCR for detecting and quantifying toxigenic molds, as the former methods only detect live molds. The literature on protocols for extracting RNA from food-associated toxigenic molds is still scarce. A considerable number of studies have been published that include different extraction buffers (Trizol® or buffers provided for RNA extraction protocols), several physical grinding methods using mortar and pestle or bead beating, and purification using well-known commercial kits in this field or organic solvents such as chloroform or ethanol (Schmidt-Heydt et al., 2009; Leite et al., 2012; Lozano-Ojalvo et al., 2013). All these studies have been carried out to extract RNA from mold mycelium; however, no mold RNA extraction methods from foods contaminated with molds have been optimized yet. Nevertheless, the extraction of RNA directly from food commodities is not easy. Thus, considerable effort is required to optimize a unique protocol for the extraction and purification of mold RNA from foods to be used in the evaluation of gene expression in mycotoxin-producing molds.

11.4.2 PCR

PCR-based methods appear to be a good tool to detect toxigenic molds early in order to control or reduce mold mass and consequently prevent mycotoxins entering the food chain (Dao et al., 2005; Hayat et al., 2012). For the detection of toxigenic molds by PCR, unique DNA sequences have to be selected as primer binding sites. As genes involved in the mycotoxin biosynthesis are supposed to be exclusively present in toxigenic molds, they may be chosen for designing primers useful for a sensitive and specific detection PCR method. Many of these genes have been identified and their DNA sequences have been published. However, complete information about the genes involved in the biosynthetic pathway of certain mycotoxins is not available yet. Additionally, PCR techniques on the basis of unique conserved genes, such as internal transcribed spacer (ITS) and intergenic spacer (IGS) regions of rDNA units, or unique bands from random amplified polymorphic DNA (RAPD) analysis (Sartori et al., 2006) have been reported as suitable for detecting food-related toxigenic molds.

A review of PCR methods developed to detect the most relevant toxigenic molds in foods is summarized in Table 11.3.

In the detection of patulin-producing molds, PCR protocols have been developed using primers designed on the basis of the isoepoxidon dehydrogenase (*idh*) gene, involved in the biosynthesis of the mycotoxin. Most of the reported methods to detect patulin-producing molds were designed to only detect specific patulin-producing *Penicillium* (Paterson et al., 2000), *Aspergillus* (Varga et al., 2003), or *Byssochlamys* (Hosoya et al., 2012) species. However, Luque et al. (2011) developed a PCR method on the basis of the *idh* gene for detecting patulin-producing molds from different food origins regardless of the genus (*Penicillium*, *Aspergillus*, and *Emericella*) and species.

Although the ochratoxin A biosynthetic pathway has not been completely elucidated (Färber and Geisen, 2004), different PCR assays based on genes involved in this biosynthesis have been developed for the detection of specific ochratoxigenic mold species. Thus, Dao et al. (2005) designed two primer pairs based on the polyketide synthase gene: one specifically for detecting *Aspergillus ochraceus* and the other for detecting several producing species mainly belonging to the *Aspergillus* genus (*A. ochraceus*, *A. carbonarius*, *Aspergillus melleus*, *Aspergillus sulfureus*, and *Penicillium verrucosum*). For *Penicillia*, Bogs et al. (2006) reported a PCR method for detecting the two ochratoxigenic species *P. verrucosum* and *Penicillium nordicum*. It was based on the ochratoxin A polyketide synthase (*otapks*PN) and the ochratoxin A nonribosomal peptide synthetase (*otanps*PN) genes from *P. nordicum* located in a partial gene cluster of the ochratoxin A biosynthetic genes (Karolewiez and Geisen, 2005). Recently, a PCR method based on the *otanps*PN gene capable of detecting ochratoxin A–producing molds regardless of genus and species has been described (Luque et al., 2013b). Considering PCR protocols based on multicopy sequences, several authors have used the ITS sequence for detecting some of the main ochratoxigenic species belong to *Aspergillus* contaminating food (Patiño et al., 2005; Sardiñas et al., 2011).

Regarding the detection of aflatoxigenic molds, most of the developed PCR protocols have used more than one primer pair because aflatoxin precursor genes are involved in the biosynthesis of other mycotoxins such as sterigmatocystin by nonaflatoxigenic molds (Levin, 2012). Therefore, Geisen (1996)

TABLE 11.3

PCR Methods Described for the Detection of the Main Food-Related Toxigenic Molds

Targeted Mycotoxins	Producing Mold Genus	Gene Target	References
Ochratoxin A	*Penicillium*	*otanps*PN, *otapks*PN	Bogs et al. (2006)
	Aspergillus	ITS	Patiño et al. (2005)
		ITS	Sardiñas et al. (2011)
		RAPD-PCR	Sartori et al. (2006)
	Aspergillus, *Penicillium*, *Emericella*	*otanps*PN	Luque et al. (2013a)
	Aspergillus, *Penicillium*	PKS	Dao et al. (2005)
	Aflatoxins, sterigmatocystin	*aflR*, *omt-1*	Somashekar et al. (2004)
		nor-1, *ver-1*, *omt-1*, *aflR*	Chen et al. (2002)
	Aspergillus	*ver-1*, *omt-1*, *aflR*	Del Fiore et al. (2010)
		aflR, *aflS*, *aflD*, *aflO*, *aflQ*	Degola et al. (2007)
		aflR	Manonmani et al. (2005)
	Aspergillus, *Penicillium*, *Rhizopus*	*omt-1*	Luque et al. (2012a)
	Aspergillus	*nor-1*, *ver-1*, *omt-1*	Geisen (1996)
Patulin	*Byssochlamys*	*idh*	Hosoya et al. (2012)
	Aspergillus	*idh*	Varga et al. (2003)
	Penicillium	*idh*	Paterson et al. (2000)
	Aspergillus, *Penicillium*, *Emericella*	*idh*	Luque et al. (2011)
Fumonisins	*Fusarium*	fum5	Bluhm et al. (2002)
		fum1, fum13	Ramana et al. (2011)
		IGS	Jurado et al. (2006)
Trichothecenes	*Fusarium*	*tri6*	Bluhm et al. (2002)
		tri5, *tri6*	Ramana et al. (2011)
		tri3, *tri5*, *tri7*	Quarta et al. (2006)
		IGS	Jurado et al. (2005)
		IGS	Jurado et al. (2006)

reported a PCR method based on aflatoxin biosynthetic genes capable of detecting the sterigmatocystin-producing *Aspergillus versicolor* as well as the aflatoxigenic *Aspergillus flavus* and *Aspergillus parasiticus*. Most of the PCR primers designed for detecting food-related aflatoxigenic molds have been based on genes involved in the biosynthesis of aflatoxins such as the norsolorinic acid reductase (*nor-1*) encoding gene (Geisen, 1996), the versicolorin A dehydrogenase (*ver-1*) encoding gene (Geisen, 1996), the sterigmatocystin *O*-methyltransferase (*omt-1*) encoding gene (Geisen, 1996; Somashekar et al., 2004), and the aflatoxin regulatory (*aflR*) gene (Somashekar et al., 2004; Manonmani et al., 2005). These PCR methods targeted specific aflatoxin-producing species belonging to *Aspergillus*, mainly *A. flavus* and *A. parasiticus*. For detecting aflatoxigenic mold species belonging to the *Aspergillus* genus as well as to the *Penicillium* or *Rhizopus* genera, a PCR protocol using the *omt-1* gene as the target has been reported (Luque et al., 2012b).

PCR assays based on primers targeted to mycotoxin biosynthetic pathway genes have been described for fumonisin- (Bluhm et al., 2002) and trichothecene-producing (Bluhm et al., 2002; Nicholson et al., 2004) *Fusarium* species. For detecting trichothecene-producing *Fusarium*, primers have been designed mainly toward the trichodiene synthase encoding gene *tri5* and the regulatory gene *tri6*. Regarding the detection of fumonisin-producing *Fusarium* species, PCR protocols were based on *fum5* gene encoding a polyketide synthase. Other authors have designed specific primers to detect trichothecene- and fumonisin-producing species of *Fusarium* by using multicopy target sequences including the IGS region (Jurado et al., 2005, 2006).

Within the PCR-based methods an interesting strategy is the use of multiplex PCR methods capable of detecting different toxigenic molds in a single amplification reaction using more than one primer pair, thus requiring smaller amounts of reagents as well as less time and effort of analysis (Hayat et al., 2012). Several approaches relying on multiplex PCR have been developed to detect mycotoxin-producing molds. Most of them have been designed to detect mold species producing the same mycotoxin, mainly for aflatoxigenic molds as described earlier (Chen et al., 2002; Sartori et al., 2006; Del Fiore et al., 2010; Ramana et al., 2011). Several multiplex PCR assays based on genes involved in mycotoxin biosynthesis have been reported to simultaneously detect trichothecene- and fumonisin-producing *Fusarium* species (Bluhm et al., 2002; Ramana et al., 2011). In the same genus, Quarta et al. (2006) have developed a multiplex PCR method for the detection of different trichothecene chemotypes. A PCR assay for the simultaneous detection of food-related patulin, ochratoxin A, and aflatoxin-producing molds has been described by Luque et al. (2013a).

11.4.3 Real-Time qPCR

Among the PCR-based techniques, qPCR is considered a method of choice for the detection and quantification of pathogen microorganisms and, specifically, of mycotoxin-producing molds. One of its major advantages is that it is faster than conventional culture-based and conventional PCR methods (Postollec et al., 2011). It is also highly sensitive, specific, and enables the simultaneous detection of different mycotoxin-producing molds (Suanthie et al., 2009; Rodríguez et al., 2012c). Moreover, qPCR offers the possibility of quantifying mycotoxin-producing molds through the measurement of their target gene copy numbers or their colony-forming units (cfu) (Mulè et al., 2006; Rodríguez et al., 2012e,h), and it does not require postamplification manipulations, hence limiting the risk of contamination (Postollec et al., 2011).

The correct choice of target sequencing for designing primers is essential for the development of new qPCR protocols for the detection and quantification of mycotoxin-producing strains belonging to different mold species and genera (Niessen, 2007; Atoui et al., 2012). qPCR methods for the detection of mycotoxin-producing molds target, in general, genes involved in the mycotoxin biosynthetic pathways, although some authors have used as target constitutive genes. A brief description of the main qPCR methods to quantify mycotoxin-producing molds in food samples is given here and summarized in Table 11.4.

Regarding constitutive genes, the single-copy ones such as the *calmodulin* and β-*tubulin* genes have been successfully used to detect and quantify by qPCR ochratoxigenic species in foods (Mulè et al., 2006; Morello et al., 2007) or to discriminate between species of *Fusarium* in wheat seeds (Yin et al., 2009). However, the sensitivity of the assay can be considerably improved when multicopy sequences, such as ribosomal DNA regions, are used. In this way, several qPCR protocols based on spacers of rDNA, IGS, and ITS have been reported for the detection of mycotoxin-producing *Fusarium*, *Aspergillus*, and *Alternaria* species (Yli-Mattila et al., 2008; Suanthie et al., 2009; Sardiñas et al., 2011; Edwards et al., 2012; Pavón et al., 2012b; Crespo-Sempere et al., 2013). Finally, the elongation factor 1 α (EF1α) appears to be consistent in *Fusarium*, and was used by Nicolaisen et al. (2009) as target for developing a qPCR method to quantify 11 individual *Fusarium* species that produce different kinds of trichothecenes and fumonisins in cereals. However, from the point of view of food safety, the utility of these regions as target sequences for the development of qPCR procedures to detect and quantify mycotoxin-producing molds, regardless of genera or species, is limited as the same mycotoxin is produced by different mold species or even genera.

The use of toxin biosynthetic genes as target sequence allows their use in an adequate design of qPCR methods to quantify toxigenic molds. Taking into account the toxin biosynthetic genes identified and their DNA sequences, a considerable number of qPCR methods to detect and quantify mycotoxin-producing molds have been published. Most of the qPCR methods designed for *Aspergillus* and *Penicillium* genera detection pay particular attention to the enumeration of aflatoxin- and ochratoxin A-producing molds in foods. In the previous section corresponding to PCR, a detailed description of the main genes involved in both mycotoxins was provided. Several qPCR methods for detecting aflatoxin-producing molds have been previously reported. Thus, the detection of *Aspergillus* spp. that produce aflatoxins has been performed using qPCR assays based on the *nor-1* gene in maize, pepper, and paprika (Mayer et al., 2003) and in stored peanuts (Passone et al., 2010). Rodríguez et al. (2012e) have developed a qPCR assay

TABLE 11.4

qPCR Methods Described for the Detection of the Main Food-Related Toxigenic Molds

Targeted Mycotoxins	Producing Mold Genus	Gene Target	References
Ochratoxin A	*Penicillium*	*pks*	Geisen et al. (2004)
		pks	Schmidt-Heydt and Geisen (2007)
	Aspergillus	Calmodulin	Mulè et al. (2006)
		β-tubulin	Morello et al. (2007)
		ITS	González-Salgado et al. (2009)
		pks	Atoui et al. (2007)
		β-ketosynthase and acyltransferase domain	Selma et al. (2009)
		pks	Castellá and Cabañes (2011)
	Aspergillus, Penicillium, Emericella	*nps*	Rodríguez et al. (2011b)
Aflatoxin	*Aspergillus*	ITS	Sardiñas et al. (2011)
		nor-1	Mayer et al. (2003)
		nor-1	Passone et al. (2010)
	Aspergillus, Penicillium, Rhizopus	*omt-1*	Rodríguez et al. (2012e)
Patulin	*Penicillium, Aspergillus, Emericella*	*idh*	Rodríguez et al. (2011a)
Sterigmatocystin	*Aspergillus, Emericella*	*fluG*	Rodríguez et al. (2012b)
Verrucosidin	*Penicillium*	SVr1 probe	Rodríguez et al. (2012a)
Cyclopiazonic acid	*Penicillium, Aspergillus*	*dmaT*	Rodríguez et al. (2012h)
Trichothecenes	*Fusarium*	β-tubulin	Yin et al. (2009)
		ITS (HT-2 and T-2)	Edwards et al. (2012)
		IGS, ITS	Yli-Mattila et al. (2008)
		EF1α	Nicolaisen et al. (2009)
		Tri12	Nielsen et al. (2012)
		Tri5	Fredlund et al. (2010)
Fumonisins	*Fusarium*	EF1α	Nicolaisen et al. (2009)
		FUM1	Marín et al. (2010)
		FUM2, FUM21	Lazzaro et al. (2012)
Zearalenone	*Fusarium*	PKS13	Atoui et al. (2012)
Tenuazonic acid, alternariol, alternariol mono-methyl ether, and altenuene	*Alternaria*	ITS	Crespo-Sempere et al. (2013)
		ITS	Pavón et al. (2012b)

based on the *omt-1* gene for the quantification of aflatoxin producers belonging to different mold species in various food commodities (peanuts, spices, and ripened products). Regarding the quantification of ochratoxigenic molds in foods, several qPCR methods to detect *Aspergillus* and *Penicillium* spp. from different food products have been reported. Thus, several qPCR approaches for quantifying ochratoxin A–producing *A. carbonarius* and *Aspergillus niger* in grapes have been carried out (Atoui et al., 2007; Selma et al., 2009; Castellá and Cabañes, 2011). Regarding ochratoxin A–producing *Penicillia*, Geisen et al. (2004) and Schmidt-Heydt et al. (2008) have designed qPCR systems targeting the *pks* gene for the quantificacion of *P. nordicum* and *P. verrucosum*, respectively, in wheat. Recently, Rodríguez et al. (2011b) developed a new qPCR based on the *otanpsPN* gene capable of detecting and quantifying ochratoxin A–producing molds regardless of species or genera to be used in different foods.

Although patulin is a mycotoxin produced by a variety of molds (Selmanoglu, 2006; Niessen, 2007) in a wide range of food products, only one qPCR protocol using the *idh* gene as a target for primer design has been developed so far to quantify *Penicillium, Aspergillus,* and *Emericella* producers of this mycotoxin in fruits, cooked meat products, and ripened products (Rodríguez et al., 2011a).

Recently, original qPCR assays to quantify cyclopiazonic acid-, sterigmatocystin-, and verrucosidin-producing molds were developed (Rodríguez et al., 2012a,b,h). In the case of cyclopiazonic acid, it has been reported that the enzyme dimethylallyl tryptophan synthase (DMAT) is implicated in its production (Chang et al., 2009). Thus, the *dmaT* gene encoding this enzyme was used by Rodríguez et al. (2012h) for designing a protocol to quantify cyclopiazonic acid–producing molds in foods. In the case of verrucosidin, no information is available about verrucosidin-encoding genes in *Penicillium polonicum* and other verrucosidin-producing species to be used for designing specific primers and probes. An alternative used by Rodríguez et al. (2012a) in the development of a qPCR protocol to quantify producers of this mycotoxin was the use of a DNA probe related to verrucosidin-producing *P. polonicum* obtained after a differential molecular screening procedure reported by Aranda et al. (2002). Regarding the quantification of sterigmatocystin-producing molds, the *fluG* gene, which seems to be involved in the biosynthesis of sterigmatocystin (Seo et al., 2006; Shwab and Keller, 2008), was used as target sequence for the development of a qPCR method to quantify sterigmatocystin-producing molds in foods (Rodríguez et al., 2012b).

With regard to qPCR for quantifying mycotoxin-producing *Fusarium*, there are a wide number of published assays based on genes directly involved in the three types of mycotoxins (trichothecenes, fumonisins, and zearalenone). Focusing on the quantification of type A and type B trichothecene-producing molds, Fredlund et al. (2010) and Yli-Mattila et al. (2008) optimized qPCR to detect type A trichothecene–producing *Fusarium* in cereal grains, and Nielsen et al. (2012) reported qPCR assays to identify and quantify the 3ADON-, 15ADON-, and NIV-producing *Fusarium*. On the other hand, Marín et al. (2010) and Lazzaro et al. (2012) optimized qPCR methods based on FUM1, and FUM2 and FUM21, respectively, to quantify fumonisin-producing *Fusarium* in cereals. Regarding the quantification of *Fusarium* species that produce zearalenone, a qPCR based on the polyketide synthase gene PKS13 involved in this mycotoxin biosynthesis has been reported to quantify *Fusarium graminearum* and *Fusarium culmorum* in maize (Atoui et al., 2012).

A special note must be made in this section on multiplex qPCR developed to quantify mycotoxin-producing molds at the same time in a single reaction. Based on this method, Yi et al. (2009) developed multiplex qPCR assays for the simultaneous detection of different mycotoxin-producing *Fusarium* species in foods. Suanthie et al. (2009) optimized a multiplex qPCR to quantify toxigenic *Aspergillus*, *Penicillium*, and *Fusarium* and Rodríguez et al. (2012c) developed a multiplex qPCR to quantify aflatoxin-, ochratoxin A-, and patulin-producing molds in different kinds of foods.

Finally, using the extracted RNA as template, several systems for monitoring the mycotoxin production based on RT-qPCR have been developed. The resulting cDNA can be amplified by the specific qPCR genes. RT-qPCR is highly sensitive and allows the quantification of small changes in gene expression; thus, it could be an appropriate indicator for determining the risk from specific toxigenic species. Concretely, this technique has been widely used to study the expression of genes involved in different types of mycotoxins under specific environmental conditions. In this way, several RT-qPCR methods have been described to analyze the punctual expression of genes involved in the aflatoxin biosynthetic pathway (Degola et al., 2007; Abdel-Hadi et al., 2010; Lozano-Ojalvo et al., 2013), ochratoxin A biosynthetic pathway (Schmidt-Heydt and Geisen, 2007), fumonisin biosynthetic pathway (Fanelli et al., 2012, 2013; Lazzaro et al., 2012; Marín et al., 2013), trichothecenes biosynthetic pathway (Marín et al., 2010; Kulik et al., 2012), and a RT-qPCR has also been recently designed to detect and quantify viable *Alternaria* spp. (Pavón et al., 2012a).

11.4.4 Controls and Normalization of PCR and qPCR

To avoid false negative and false positive results produced by using molecular techniques, an adequate control of the efficiency of the reaction is a fundamental aspect in these assays (Hoorfar et al., 2004). Several controls are recommended to correctly interpret the results of molecular techniques such as processing positive control (PPC), processing negative control (PNC), amplification positive control (APC), nontemplate control (NTC), and internal amplification control (IAC) (Rodríguez-Lázaro et al., 2007). Focusing on IAC, the guidelines for PCR testing of food-borne pathogens have proposed its presence for PCR-based diagnostic tests (OECD, 2007; ISO:22174, 2005; ISO:22119, 2011). The use of IAC is an adequate strategy to assess the validity of the PCR results and avoid false negative results derived from PCR inhibitors (Hoorfar et al., 2004). It is a nontarget DNA sequence that is included in the same sample reaction together with the target sequence for simultaneous amplification (Rodríguez et al., 2012h). Therefore, when an IAC is incorporated in a PCR test, the corresponding amplification product should always be obtained, even though the target sequence is not present.

There are two main strategies for an IAC to be included in a PCR-based method: competitive IAC and noncompetitive IAC (Rodríguez-Lázaro et al., 2007). Both strategies are useful if they have been well optimized. Although IAC for PCR-based protocols for detecting and quantifying food-related mycotoxin-producing molds is not yet extensively used, Rodríguez et al. (2012h) designed a qPCR including a competitive IAC to quantify CPA-producing molds and Rodríguez et al. (2012a) developed a qPCR with a noncompetitive IAC to quantify verrucosidin-producing molds in foods.

During gene expression analysis, normalization must be performed to correct for differences in mRNA target in relation to the amount of total cDNA added to each reaction in the relative quantification assays (Livak and Schmittgen, 2001). rRNA or mRNA from housekeeping genes can serve for normalization. All the reference genes should be validated for the stability of their expression in the specific study conditions (Postollec et al., 2011). Generally, it is suggested to use several housekeeping genes and calculate a normalization factor from the geometric mean of their expression levels (Wong and Medrano, 2005).

Therefore, those protocols of PCR or qPCR selected to be applied for the detection of mycotoxin-producing molds should have an IAC that allows an accurate detection avoiding false negatives.

In most of the previous PCR or qPCR methods, the limit of detection in foods ranged from 1 to 3 log cfu/g, cfu/mL, or cfu/cm^2. This limit is lower that the level of log 3 cfu/g, cfu/mL, or cfu/cm^2 of toxigenic molds, which could be considered safe, as there is no production of mycotoxins in foods below that level (Rodríguez et al., 2012e). In addition, these methods are reasonably less time-consuming. They can be completed in 5–7 h analysis. Thus, qPCR protocols could be appropriate for the routine detection of mycotoxin-producing molds in raw materials and preprocessed foods, allowing the possibility of corrective actions to avoid the accumulation of mycotoxins in processed foods.

11.5 Future Perspectives: New Molecular Methods

In the last decade, new molecular techniques have emerged for proteomic, metabolomic, and genomic analysis. Some of them have been used to detect mycotoxin-producing molds as well as gene expression studies or to prevent mycotoxin production at the posttranscriptional level of gene expression. These procedures are described as the future new molecular methods to detect mycotoxin-producing molds.

LAMP of DNA has emerged as an alternative to the use of PCR-based methods in food safety testing. LAMP is a relatively novel method that amplifies nucleic acids under isothermal conditions (Notomi et al., 2000). This method employs a DNA polymerase and a set of four specially designed primers that recognize a total of six distinct sequences on the target DNA (Notomi et al., 2000). Its advantages over PCR-based techniques are shorter reaction time, no need for specific equipment, high sensitivity and specificity, as well as comparably low susceptibility to inhibitors present in foods, which enables the detection of the pathogens, and shorter time of analysis and preparation (Niessen et al., 2013).

This technique has been already used for detecting several mycotoxin-producing *Fusarium* and *Aspergillus* species in several food commodities. Thus, Niessen and Vogel (2010) and Denschlag et al. (2012) reported LAMP assays for the detection of *Fusarium cerealis, F. culmorum*, and *F. graminearum*, the major producers of the DON in model barley samples. Luo et al. (2012) designed three sets of LAMP primers to identify *A. flavus, Aspergillus nomius*, and *A. parasiticus* in peanuts, coffee beans, and Brazil nuts. Finally, Storari et al. (2013) developed LAMP assays based on the *Acpks* and *Anpks* genes for the identification of ochratoxin A–producing isolates of *A. carbonarius* and *A. niger*, respectively. Recently, a real-time LAMP assay was developed for quantifying gushing-inducing *Fusarium* spp. (Denschlag et al., 2013). This last approach that designed a set of LAMP primers based on mycotoxin biosynthetic genes could be used for detecting and quantifying toxigenic molds in the near future.

As described in the previous section, RNA can be used as a template to study the mycotoxin biosynthesis gene expression under specific environmental conditions as these genes are induced and not expressed constitutively (Peplow et al., 2003; Price et al., 2005). The use of RT-qPCR only allows the analysis of small changes in the expression of a specific gene; however, in order to predict whether myco-toxin biosynthesis may be possible under certain environmental conditions in food samples, monitoring the whole pathway of the genes would be appropriate. A certain expression pattern can be expected to indicate the onset of mycotoxin biosynthesis within the mold (Schmidt-Heydt and Geisen, 2007). In the last few years, some of these patterns have been developed; while Schmidt-Heydt and Geisen (2007) reported a microarray carrying oligonucleotides of the fumonisins, aflatoxin, ochratoxin A, trichothe-cene (types A and B), and patulin biosynthesis pathway, Kristensen et al. (2007) developed a microar-ray for the detection and identification of 14 species of *Fusarium*. A study of the relationship between aflatoxin gene cluster expression, environmental factors, growth, and toxin production by *A. flavus* was performed using a microarray including 30 genes of this mycotoxin biosynthetic pathway (Abdel-Hadi et al., 2012). It was demonstrated that a strain of *Fusarium kyushuense* was able to produce aflatoxins B1 and G1 by using a microarray covering oligonucleotides specific for the aflatoxin pathway genes of *A. flavus* (Schmidt-Heydt et al., 2009).

Finally, *RNA sequencing* is an emerging and promise technique used to get information about a sam-ple's RNA content in different matrices such as foods. This approach is not widely used yet. However, Rokas et al. (2012), Lin et al. (2013), and Xiao et al. (2013) used this technique for genomic studies in toxigenic *Aspergillus fumigatus*, *A. flavus*, and *Fusarium* spp. The use of this technique will allow examining the expression level of mRNA and obtaining information about the genetic material relat-ing to RNA in the mold when toxigenic species grow on the surface of different foods under the specific environmental conditions of their processing. This information will be of great utility in the efficient use of the reverse genetic tool RNA interference as a control method to minimize mycotoxin production in foods.

11.6 Conclusions

The presence of mycotoxin-producing molds during food processing and storage poses hazard for the consumers as this can lead to the accumulation of mycotoxins in food products. It is necessary to apply rapid methods to detect mycotoxin-producing molds in HACCP systems that allow taking corrective actions to avoid mycotoxin accumulation in prepared foods. Microbiological conventional methods and subsequent evaluation of mycotoxins are not suitable for the routine detection of mycotoxin-producing molds in foods, as this combination is highly time-consuming. However, the analysis of mycotoxins with emerging techniques on the basis of proteomics and metabolomics could be very useful to exclu-sively detect mycotoxins in foods in the verification process of HACCP systems. The detection of genes involved in mycotoxins biosynthesis by nucleic acid–based methods such as conventional PCR and qPCR described earlier is a good alternative for the rapid and sensitive detection of mycotoxin-producing molds in raw materials and preprocessed foods. Accurate mold DNA extraction methods from food such as those proposed in the present work (Figure 11.1) should be considered as previous steps for PCR and qPCR analysis with primers and methods for the corresponding investigated mycotoxin-producing molds. The selected PCR and qPCR protocols should incorporate an IAC to avoid false negatives in the analysis of toxigenic molds. Furthermore, new protocols on the basis of genomic analysis such as LAMP-loop, RT-PCR, RT-qPCR, microarrays, and RNA sequencing are being optimized to detect toxi-genic molds or to study the expression of genes involved in mycotoxin production. Most of these methods could be applied to the analysis of mycotoxin-producing molds in foods with the purpose of increasing food safety throughout food processing and storage.

ACKNOWLEDGMENTS

This work was funded by project AGL2007-64639, AGL2010-21623, and Carnisenusa CSD2007-00016, Consolider Ingenio 2010 of the Spanish Comision Interministerial de Ciencia y Tecnología, and GRU10162 of the Junta de Extremadura and FEDER.

REFERENCES

Abdel-Hadi, A., Carter, D., Magan, N. Temporal monitoring of the *nor-1* (aflD) gene of *Aspergillus flavus* in relation to aflatoxin B1 production during storage of peanuts under different water activity levels. *Journal of Applied Microbiology* 109 (2010): 1914–1922.

Abdel-Hadi, A., Schmidt-Heydt, M., Parra, R., Geisen, R., Magan, N. A systems approach to model the relationship between aflatoxin gene cluster expression, environmental factors, growth and toxin production by *Aspergillus flavus*. *Journal of the Royal Society Interface* 9 (2012): 757–767.

Aranda, E., Rodríguez, M., Benito, M.J., Asensio, M.A., Córdoba, J.J. Molecular cloning of verrucosidin-producing *Penicilllium polonicum* genes by differential screening to obtain a DNA probe. *International Journal of Food Microbiology* 76 (2002): 55–61.

Atoui, A., El Khoury, A., Kallassy, M., Lebrihi, A. Quantification of *Fusarium graminearum* and *Fusarium culmorum* by real-time PCR system and zearalenone assessment in maize. *International Journal of Food Microbiology* 154 (2012): 59–65.

Atoui, A., Mathieu, F., Lebrihi, A. Targeting a polyketide synthase gene for *Aspergillus carbonarius* quantification and ochratoxin A assessment in grapes using real-time PCR. *International Journal of Food Microbiology* 115 (2007): 313–318.

Beltrán, E., Ibáñez, M., Portolés, T. et al. Development of sensitive and rapid analytical methodology for food analysis of 18 mycotoxins included in a total diet study. *Analytica Chimica Acta* 783 (2013): 39–48.

Benito, M.J. Rodríguez, M., Martín, A., Aranda, E., Córdoba, J.J. Effect of the fungal protease EPg222 in the sensory characteristics of dry fermented sausage "salchichón" ripened with commercial starter cultures. *Meat Science* 67 (2004): 497–405.

Bernáldez, V., Córdoba, J.J., Rodríguez, M., Cordero, M., Polo, L., Rodríguez, A. Effect of *Penicillium nalgiovense* as protective culture in processing of dry-fermented sausage "salchichón." *Food Control* 32 (2013): 69–76.

Bernáldez, V., Rodríguez, A., Martín, A., Lozano, D., Córdoba, J.J. Development of a multiplex qPCR method for simultaneous quantification in dry-cured ham of an antifungal-peptide *P. chrysogenum* strain used as protective culture and aflatoxin-producing moulds. *Food Control* 36 (2014): 257–265.

Berthiller, F., Crews, C., Dall'Asta, C. et al. Maskedmycotoxins: A review. *Molecular Nutrition and Food Research* 57 (2013): 165–186.

Beuchat, L.R. Media for detecting and enumerating yeasts and moulds. In *Handbook of Culture Media for Food Microbiology*, pp. 369–385. Amsterdam, the Netherlands: Elsevier Science B.V., 2003.

Bhat, R., Rai, R.V., Karim, A.A. Mycotoxins in food and feed: Present status and future concerns. *Comprehensive Reviews in Food Science and Food Safety* 9 (2010): 57–81.

Bluhm, B.H., Flaherty, J.E., Cousin, M.A., Woloshuk, C.P. Multiplex polymerase chain reaction assay for the differential detection of trichothecene- and fumonisin-producing species of *Fusarium* in cornmeal. *Journal of Food Protection* 65 (2002): 1955–1961.

Bogs, C., Battilani, P., Geisen, R. Development of a molecular detection and differentiation system for ochratoxin A producing *Penicillium* species and its application to analyse the occurrence of *Penicillium nordicum* in cured meats. *International Journal of Food Microbiology* 107 (2006): 39–47.

Bulder, A.S., Arcella, D., Bolger, M. et al. Fumonisins (addendum). In *Safety Evaluation of Certain Food Additives and Contaminants*, pp. 325–327. WHO food additives series 65, prepared by the 74th meeting of the Joint FAO/WHO Expert Committee on Food Additives (JECFA). Geneva, Switzerland: WHO, 2012.

Castellá, G., Cabañes, F.J. Development of a real time PCR system for detection of ochratoxin A-producing strains of the *Aspergillus niger* aggregate. *Food Control* 22 (2011): 1367–1372.

Chang, P.K., Horn, B.W., Dorner, J.W. Clustered genes involved in cyclopiazonic acid production are next to the aflatoxin biosynthesis gene cluster in *Aspergillus flavus*. *Fungal Genetics and Biology* 46 (2009): 176–182.

Chen, R.S., Tsay, J.G., Huang, Y.F., Chiou, R.Y. Polymerase chain reaction-mediated characterization of moulds belonging to the *Aspergillus flavus* group and detection of *Aspergillus parasiticus* in peanut kernels by a multiplex polymerase chain reaction. *Journal of Food Protection* 65 (2002): 840–844.

Cigić, I.K., Prosen, H. An overview of conventional and emerging analytical methods for the determination of mycotoxins. *International Journal of Molecular Sciences* 10 (2009): 62–115.

Crespo-Sempere, A., Estiarte, N., Marín, S., Sanchis, V., Ramos, A.J. Propidium monoazide combined with real-time quantitative PCR to quantify viable *Alternaria* spp. contamination in tomato products. *International Journal of Food Microbiology* 165 (2013): 214–220.

Crespo-Sempere, A., Gil, J.V., Martínez-Culebras, P.V. Proteome analysis of the fungus *Aspergillus carbonarius* under ochratoxin A producing conditions. *International Journal of Food Microbiology* 147 (2011): 162–169.

Dagnas, S., Membré, J.M. Predicting and preventing mold spoilage of food products. *Journal of Food Protection* 76 (2012): 538–551.

Dao, H.P., Mathieu, F., Lebrihi, A. Two primer pairs to detect OTA producers by PCR method. *International Journal of Food Microbiology* 36 (2005) 215–220.

Degola, F., Berni, E., Dall'Asta, C. et al. A multiplex RT-PCR approach to detect aflatoxigenic strains of *Aspergillus flavus*. *Journal of Applied Microbiology* 103 (2007): 409–417.

Del Fiore, A., Reverberi, M., De Rossi, P., Tolani, V., Fabbri, A.A., Fanelli, C. Polymerase chain reaction-based assay for the detection of aflatoxigenic fungi on maize kernels. *Quality Assurance and Safety of Crops and Foods* 2 (2010): 22–27.

Demeke, T., Jenkins, R. Influence of DNA extraction methods, PCR inhibitors and quantification methods on real-time PCR assay of biotechnology-derived traits. *Analytical and Bioanalytical Chemistry* 396 (2010): 1977–1990.

Denschlag, C., Vogel, R.F., Niessen, L. *Hyd5* gene-based detection of the major gushing-inducing *Fusarium* spp. in a loop-mediated isothermal amplification (LAMP) assay. *International Journal of Food Microbiology* 156 (2012): 189–196.

Denschlag, C., Vogel, R.F., Niessen, L. *Hyd5* gene based analysis of cereals and malt for gushing-inducing *Fusarium* spp. by real-time LAMP using fluorescence and turbidity measurements. *International Journal of Food Microbiology* 162 (2013): 245–251.

Edwards, S.G., Imathiu, S.M., Ray, R.V., Back, M., Hare, M.C. Molecular studies to identify the *Fusarium* species responsible for HT-2 and T-2 mycotoxins in UK oats. *International Journal of Food Microbiology* 156 (2012): 168–175.

Ellington, A., Szostak, J.W. *In vitro* selection of RNA molecules that bind specific ligands. *Nature* 346 (1990): 818–822.

Eriksen, G.S., Pettersson, H. Toxicological evaluation of trichothecenes in animal feed. *Animal Feed Science and Technology* 114 (2004): 205–239.

European Commission. Commission Regulation (EC) No. 1881/2006 of 19 December 2006 setting maximum levels for certain contaminants in foodstuffs. *Official Journal of the European Union* L 364 (2006): 5–24.

European Mycotoxins Awareness Network. 2011. Commercial immunological kits for the analysis of mycotoxin. http://www.mycotoxins.org/ (accessed July 14, 2013).

European Mycotoxins Awareness Network. 2012. Mycotoxins legislation worldwide. http://www.mycotoxins.org/ (accessed September 9, 2013).

Fanelli, F., Iversen, A., Logrieco, A.F., Mulè, G. Relationship between fumonisin production and FUM gene expression in *Fusarium verticillioides* under different environmental conditions. *Food Additives and Contaminants—Part A Chemistry, Analysis, Control, Exposure and Risk Assessment* 30 (2013): 365–371.

Fanelli, F., Schmidt-Heydt, M., Haidukowski, M., Geisen, R., Logrieco, A., Mulè, G. Influence of light on growth, fumonisin biosynthesis and FUM1 gene expression by *Fusarium proliferatum*. *International Journal of Food Microbiology* 153 (2012): 148–153.

Färber, P., Geisen, R. Analysis of differentially-expressed ochratoxin A biosynthesis genes of *Penicillium nordicum*. *European Journal of Plant Pathology* 110 (2004): 661–669.

Fernández-Cruz, M.L., Mansilla, M.L., Tadeo, J.L. Mycotoxins in fruits and their processed products: Analysis, occurrence and health implications. *Journal of Advanced Research* 1 (2010): 113–122.

Firrao, G., Torelli, E., Gobbi, E., Raranciuc, S., Bianchi, G., Locci, R. Prediction of milled maize fumonisin contamination by multispectral image analysis. *Journal of Cereal Science* 52 (2010): 327–330.

Fredlund, E., Gidlund, A., Pettersson, H., Olsen, M., Borjesson, T. Real-time PCR detection of *Fusarium* species in Swedish oats and correlation to T-2 and HT-2 toxin content. *World Mycotoxin Journal* 3 (2010): 77–88.

Galaverna, G., Dall'Asta, C., Mangia, M., Dossena, A., Marchelli, R. Masked mycotoxins: An emerging issue for food safety. *Czech Journal of Food Sciences* 27 (2009): S89–S92.

Garcia, D., Ramos, A.J., Sanchis, V., Marín, S. Predicting mycotoxins in foods: A review. *Food Microbiology* 26 (2009): 757–769.

Geisen, R. Multiplex polymerase chain reaction for the detection of potential aflatoxin and sterigmatocystin producing fungi. *Systematic and Applied Microbiology* 19 (1996): 388–392.

Geisen, R., Mayer, Z., Karolewiez, A., Färber, P. Development of a real time PCR system for detection of *Penicillium nordicum* and for monitoring ochratoxin A production in foods by targeting the ochratoxin polyketide synthase gene. *Systematic and Applied Microbiology* 27 (2004): 501–507.

Giacometti, J., Tomljanovic, A.B., Josic, D. Application of proteomics and metabolomics for investigation of food toxins. *Food Research International* 54 (2013): 1042–1051.

González-Salgado, A., Patiño, B., Gil-Serna, J., et al. Specific detection of *Aspergillus carbonarius* by SYBR® Green and TaqMan® quantitative PCR assays based on the multicopy ITS2 region of the rRNA gene. *FEMS Microbiology Letters* 295 (2009): 57–66.

Hashem, M., Alamri, S. Contamination of common spices in Saudi Arabia markets with potential mycotoxin-producing fungi. *Saudi Journal of Biological Sciences* 17 (2010): 167–175.

Hayat, A., Paniel, N., Rhouati, A., Marty, J.-L., Barthelmebs, L. Recent advances in ochratoxin A-producing fungi detection based on PCR methods and ochratoxin A analysis in food matrices. *Food Control* 26 (2012): 401–415.

Holcomb, M., Thompson, H.C., Cooper, W.M. SFE extraction of Aflatoxins (B1, B2, G1, and G2) from corn and analysis by HPLC. *Journal of Supercritical Fluids* 9 (1996): 118–121.

Hoorfar, J., Malorny, B., Abdulmawjood, A., Cook, N., Wagner, M., Fach, P. Practical considerations in design of internal amplification controls for diagnostic PCR assays. *Journal of Clinical Microbiology* 42 (2004): 1863–1968.

Hosoya, K., Nakayama, M., Matsuzawa, T., Imanishi, Y., Hitomi, J., Yaguchi, T. Risk analysis and development of a rapid method for identifying four species of *Byssochlamys*. *Food Control* 26 (2012): 169–173.

Hruska, Z., Yao, H., Kincaid, R. et al. Fluorescence imaging spectroscopy (FIS) for comparing spectra from corn ears naturally and artificially infected with aflatoxin producing fungus. *Journal of Food Science* 78 (2013): 1313–1320.

ISO:22174. Anonymous. Microbiology of food and animal feeding stuffs. Polymerase chain reaction (PCR) for the detection of food-borne pathogens. General requirements and definitions. Geneva, Switzerland: International Organization for Standardization, 2005.

ISO:22119. Anonymous. Microbiology of food and animal feeding stuffs. Real-time Polymerase chain reaction (PCR) for the detection of food-borne pathogens. General requirements and definitions. Geneva, Switzerland: International Organization for Standardization, 2011.

Jurado, M., Vázquez, C., Marín, S., Sanchís, V., González-Jaén, M.T. PCR-based strategy to detect contamination with mycotoxigenic *Fusarium* species in maize. *Systematic and Applied Microbiology* 29 (2006): 681–689.

Jurado, M., Vázquez, C., Patiño, B., González-Jaén, M.T. PCR detection assays for the trichothecene-producing species *Fusarium graminearum, Fusarium culmorum, Fusarium poae, Fusarium equiseti* and *Fusarium sporotrichioides*. *Systematic and Applied Microbiology* 28 (2005): 562–568.

Kalkan, H., Beriat, P., Yardimci, Y., Pearson, T.C. Detection of contaminated hazelnuts and ground red chili peppers flakes by multispectral imaging. *Computer and Electronics in Agriculture* 77 (2011): 28–34.

Karimi, K., Zamami, A. *Mucor indicus*: Biology and industrial perspectives: A review. *Biotechnology Advances* 31 (2013): 466–481.

Karolewiez, A., Geisen, R. Cloning a part of ochratoxin A biosynthetic gene cluster of *Penicilllium nordicum* and characterization of the ochratoxin polyketide synthase gene. *Systematic and Applied Microbiology* 28 (2005): 588–597.

Köppen, R., Koch, M., Slegel, D., Merkel, S., Maul, R., Nehls, I. Determination of mycotoxins in foods: Current state of analytical methods and limitations. *Applied Microbiology and Biotechnology* 86 (2010): 1595–1512.

Kristensen, R., Gauthier, G., Berdal, K.G., Hamels, S., Remacle, J., Holst-Jensen, A. DNA microarray to detect and identify trichothecene- and moniliformin-producing *Fusarium* species. *Journal of Applied Microbiology* 102 (2007): 1060–1070.

Krska, R., Molinelli, A. Rapid test strips for analysis of mycotoxins in food and feed. *Analytical and Bioanalytical Chemistry* 393 (2009): 67–71.

Kulik, T., LŁojko, M., Jestoi, M., Perkowski, J. Sublethal concentrations of azoles induce tri transcript levels and trichothecene production in *Fusarium graminearum*. *FEMS Microbiology Letters* 335 (2012): 58–67.

Lazzaro, I., Susca, A., Mulé, G. et al. Effects of temperature and water activity on FUM2 and FUM21 gene expression and fumonisin B production in *Fusarium verticillioides*. *European Journal of Plant Pathology* 134 (2012): 685–695.

Leite, G.M., Magan, N., Medina, A. Comparison of different bead-beating RNA extraction strategies: An optimized method for filamentous fungi. *Journal of Microbiological Methods* 88 (2012): 413–418.

Levin, R.E. PCR detection of aflatoxin producing fungi and its limitations. *International Journal of Food Microbiology* 156 (2012): 1–6.

Liao, C.D., Wong, J.W., Zhang, K., Hayward, D.G., Lee, N.S., Trucksess, M.W. Multi-mycotoxin analysis of finished grain and nut products using High-Performance Liquid Chromatography–Triple-Quadrupole Mass Spectrometry. *Journal of Agricultural and Food Chemistry* 61 (2013): 4771–4782.

Lin, J.Q., Zhao, X.X., Zhi, Q.Q., Zhao, M., He, Z.M. Transcriptomic profiling of *Aspergillus flavus* in response to 5-azacytidine. *Fungal Genetics and Biology* 56 (2013): 78–86.

Livak, K.J., Schmittgen, T.D. Analysis of relative gene expression data using real-time quantitative PCR and the $2^{(-\Delta\Delta CT)}$ method. *Methods* 25 (2001): 402–408.

Lozano-Ojalvo, D., Rodríguez, A., Bernáldez, V., Córdoba, J.J., Rodríguez, M. Influence of temperature and substrate conditions on the omt-1 gene expression of *Aspergillus parasiticus* in relation to its aflatoxin production. *International Journal of Food Microbiology* 166 (2013): 263–269.

Luo, J., Vogel, R.F., Niessen, L. Development and application of a loop mediated isothermal amplification assay for rapid identification of aflatoxigenic molds and their detection in food samples. *International Journal of Food Microbiology* 159 (2012): 214–224.

Luque, M.I., Andrade, M.J., Rodríguez, A., Bermúdez, E., Córdoba, J.J. Development of a multiplex PCR method for the detection of patulin-, ochratoxin A- and aflatoxin-producing moulds in foods. *Food Analytical Methods* 6 (2013a): 1113–1121.

Luque, M.I., Andrade, M.J., Rodríguez, A., Rodríguez, M., Córdoba, J.J. Development of a protocol for efficient DNA extraction of patulin-producing moulds from food for sensitive detection by PCR. *Food Analytical Methods* 5 (2012b): 684–694.

Luque, M.I., Córdoba, J.J., Rodríguez, A., Núñez, F., Andrade, M.J. Development of a PCR protocol to detect ochratoxin A producing moulds in food products. *Food Control* 29 (2013b): 270–278.

Luque, M.I., Rodríguez, A., Andrade, M.J., Gordillo, R., Rodríguez, M., Córdoba, J.J. Development of a PCR protocol to detect patulin producing moulds in food products. *Food Control* 22 (2011): 1831–1838.

Luque, M.I., Rodríguez, A., Andrade, M.J., Martín, A., Córdoba, J.J. Development of a PCR protocol to detect aflatoxigenic molds in food products. *Journal of Food Protection* 75 (2012a): 85–94.

Magan, N., Evans, P. Volatiles as an indicator of fungal activity and differentiation between species, and the potential use of electronic nose technology for early detection of grain spoilage. *Journal of Stored Products Research* 36 (2000): 319–340.

Manonmani, H.K., Anand, S., Chandrashekar, A., Rati, E.R. Detection of aflatoxigenic fungi in selected food commodities by PCR. *Process Biochemistry* 40 (2005): 2859–2864.

Mansfield, M.A., Jones, A.D., Kuldau, G.A. Contamination of fresh and ensiled maize by multiple *Penicillium* mycotoxins. *Postharvest Pathology and Mycotoxins* 98 (2008): 330–336.

Martín, A., Córdoba, J.J., Aranda, E., Córdoba, M.G., Asensio, M.A. Contribution of a selected fungal population to the volatile compounds on dry-cured ham. *International Journal of Food Microbiology* 110 (2006): 8–18.

Marín, P., Magan, N., Vazquez, C., Gonzalez-Jaen, M.T. Differential effect of environmental conditions on the growth and regulation of the fumonisin biosynthetic gene FUM1 in the maize pathogens and fumonisin producers *Fusarium verticillioides* and *Fusarium proliferatum*. *FEMS Microbiology Ecology* 73 (2010): 303–311.

Marín, S., Hodzic, I., Ramos, A.J., Sanchís, V. Predicting the growth/no growth boundary and ochratoxin A production by *Aspergillus carbonarius* in pistachio nuts. *Food Microbiology* 25 (2008): 683–689.

Marín, S., Ramos, A.J., Cano-Sancho, G., Sanchis, V. Mycotoxins: Occurrence, toxicology, and exposure assessment. *Food and Chemical Toxicology* 60 (2013): 218–237.

Mayer, Z., Bagnara, A., Färber, P., Geisen, R. Quantification of the copy number of nor-1, a gene of the aflatoxin biosynthetic pathway by real-time PCR, and its correlation to the cfu of *Aspergillus flavus* in foods. *International Journal of Food Microbiology* 82 (2003): 143–151.

Morello, L.G., Sartori, D., de Oliveira Martínez, A.L. et al. Detection and quantification of *Aspergillus westerdijkiae* in coffee beans based on selective amplification of β-*tubulin* gene by using real-time PCR. *International Journal of Food Microbiology* 119 (2007): 270–276.

Mulè, G., Susca, A., Logrieco, A., Stea, G., Visconti, A. Development of a quantitative real-time PCR assay for the detection of *Aspergillus carbonarius* in grapes. *International Journal of Food Microbiology* 111 (2006): S28–S34.

Nicholson, P., Simpson, D.R., Wilson, A.H., Chandler, E., Thomsett, M. Detection and differentiation of trichothecene and enniatin-producing *Fusarium* species on small-grain cereals. *European Journal of Plant Pathology* 110 (2004): 503–514.

Nicolaisen, M., Supronienė, S., Nielsen, L.K., Lazzaro, I., Spliid, N.H., Justesen, A.F. Real-time PCR for quantification of eleven individual *Fusarium* species in cereals. *Journal of Microbiological Methods* 76 (2009): 234–240.

Nielsen, L.K., Jensen, J.D., Rodriguez, A., Jorgensen, L.N., Justesen, A.F. TRI12 based quantitative real-time PCR assays reveal the distribution of trichothecene genotypes of *F. graminearum* and *F. culmorum* isolates in Danish small grain cereals. *International Journal of Food Microbiology* 157 (2012): 384–392.

Nielsen, K.F., Mogensen, J.M., Johansen, M., Larsen, T.O., Frisvad, J.C. Review of secondary metabolites and mycotoxins from the *Aspergillus niger* group. *Analytical and Bioanalytical Chemistry* 395 (2009): 1225–1242.

Niessen, L. PCR-based diagnosis and quantification of mycotoxin producing fungi. *International Journal of Food Microbiology* 119 (2007): 38–46.

Niessen, L., Luo, J., Denschlag, C., Vogel, R.F. The application of loop-mediated isothermal amplification (LAMP) in food testing for bacterial pathogens and fungal contaminants. *Food Microbiology* 36 (2013): 191–206.

Niessen, L., Vogel, R.F. Detection of *Fusarium graminearum* DNA using a loop mediated isothermal amplification (LAMP) assay. *International Journal of Food Microbiology* 140 (2010): 183–191.

Notomi, T., Okayama, H., Masubuchi, H. et al. Loop-mediated isothermal amplification of DNA. *Nucleic Acids Research* 28 (2000): e63.

Organization Economic Cooperation and Development (OECD). OECD guidelines for quality assurance in molecular genetic testing, 2007. http://www.oecd.org/dataoecd/43/38839788.pdf (accessed May 20, 2014).

Passone, M.A., Rosso, L.C., Ciancio, A., Etcheverry, M. Detection and quantification of *Aspergillus* section Flavi spp. in stored peanuts by real-time PCR of *nor-1* gene, and effects of storage conditions on aflatoxin production. *International Journal of Food Microbiology* 138 (2010): 276–281.

Paterson, R.R.M., Archer, S., Kozzakiewicz, Z., Lea, A., Locke, T., O'Grady, E. A gene probe for the patulin metabolic pathway with potential use in novel disease control. *Biocontrol Science and Technology* 10 (2000): 509–512.

Patiño, B., González-Salgado, A., González-Jaén, M.T., Vázquez, C. PCR detection assays for the ochratoxin-producing *Aspergillus carbonarius* and *Aspergillus ochraceus* species. *International Journal of Food Microbiology* 104 (2005): 207–214.

Pavón, M.A., Gonzalez, I., Martin, R., Garcia, T. A real-time reverse-transcriptase PCR technique for detection and quantification of viable *Alternaria* spp. in foodstuffs. *Food Control* 28 (2012a): 286–294.

Pavón, M.T., González, I., Martín, R., García Lacarra, T. ITS-based detection and quantification of *Alternaria* spp. in raw and processed vegetables by real-time quantitative PCR. *Food Microbiology* 32 (2012b): 165–171.

Peplow, A.W., Tag, A.G., Garifullina, G.F., Beremand, M.N. Identification of new genes positively regulated by tri10 and a regulatory network for trichothecene mycotoxin production. *Applied and Environmental Microbiology* 69 (2003): 2731–2736.

Postollec, F., Falentin, H., Pavan, S., Combrisson, J., Sohier, D. Recent advances in quantitative PCR (qPCR) applications in food microbiology. *Food Microbiology* 28 (2011): 848–861.

Price, M.S., Conners, S.B., Tachdjian, S., Kelly, R.M., Payne, G.A. Aflatoxin conducive and non-conducive growth conditions reveal new gene associations with aflatoxin production. *Fungal Genetics and Biology* 42 (2005): 506–518.

Quarta, A., Mita, G., Haidukowski, M., Logrieco, A., Mulè, G., Visconti, A. Multiplex PCR assay for the identification of nivalenol, 3- and 15-acetyl-deoxynivalenol chemotypes in *Fusarium*. *FEMS Microbiology Letters* 259 (2006): 7–13.

Ramana, M.V., Balakrishna, K., Murali, H.C.S., Batra, H.V. Multiplex PCR-based strategy to detect contamination with mycotoxigenic *Fusarium* species in rice and fingermillet collected from southern India. *Journal of the Science of Food and Agriculture* 91 (2011): 1666–1673.

Rodríguez, A., Córdoba, J.J. Gordillo, R., Córdoba, M.G., Rodríguez, M. Development of two quantitative real-time PCR methods based on SYBR Green and TaqMan to quantify sterigmatocystin-producing molds in food. *Food Analytical Methods* 5 (2012b): 1514–1525.

Rodríguez, A., Córdoba, J.J., Werning, M.L., Andrade, M.J., Rodríguez, M. Duplex real-time PCR method with internal amplification control for quantification of verrucosidin producing molds in dry-ripened foods. *International Journal of Food Microbiology* 153 (2012a): 85–91.

Rodríguez, A., Luque, M.I., Andrade, M.J., Rodríguez, M., Asensio, M.A., Córdoba, J.J. Development of real-time PCR methods to quantify patulin-producing molds in food products. *Food Microbiology* 28 (2011a): 1190–1199.

Rodríguez, A., Rodríguez, M., Andrade, M.J., Córdoba, J.J. Development of a multiplex real-time PCR to quantify aflatoxin, ochratoxin A and patulin producing molds in foods. *International Journal of Food Microbiology* 155 (2012c): 10–18.

Rodríguez, A., Rodríguez, M., Luque, M.I., Justensen, A.F., Córdoba, J.J. Quantification of ochratoxin A-producing molds in food products by SYBR Green and TaqMan real-time PCR methods. *International Journal of Food Microbiology* 149 (2011b): 226–235.

Rodríguez, A., Rodríguez, M., Luque, M.I., Justesen, A.F., Córdoba, J.J. A comparative study of DNA extraction methods to be used in real-time PCR based quantification of ochratoxin A-producing molds in food products. *Food Control* 25 (2012d): 666–672.

Rodríguez, A., Rodríguez, M., Luque, M.I., Martín, A., Córdoba, J.J. Real-time PCR assays for detection and quantification of aflatoxin-producing molds in foods. *Food Microbiology* 31 (2012e): 89–99.

Rodríguez, A., Rodríguez, M., Martín, A., Delgado, J., Córdoba, J.J. Presence of ochratoxin A on the surface of dry-cured Iberian ham after initial fungal growth in the drying stage. *Meat Science* 92 (2012f): 728–734.

Rodríguez, A., Rodríguez, M., Martín, A., Núñez, F., Córdoba, J.J. Evaluation of hazard of aflatoxin B1, ochratoxin A and patulin production in dry-cured ham and early detection of producing moulds by qPCR. *Food Control* 27 (2012g): 118–126.

Rodríguez, A., Werning, M.L., Rodríguez, M., Bermúdez, E., Córdoba, J.J. Quantitative real-time PCR method with internal amplification control to quantify cyclopiazonic acid-producing molds in foods. *Food Microbiology* 32 (2012h): 397–405.

Rodríguez-Lázaro, D., Lombard, B., Smith, H. et al. Trends in analytical methodology in food safety and quality: Monitoring microorganisms and genetically modified organisms. *Trends in Food Science and Technology* 18 (2007): 306–319.

Rokas, A., Gibbons, J.G., Zhou, X., Beauvais, A., Latge, J.P. The diverse applications of RNA-seq for functional genomic studies in *Aspergillus fumigatus*. *Annals of the New York Academy of Sciences* 1273 (2012): 25–34.

Sahgal, N., Needham, R., Cabañes, F.J, Magan, N. Potential for detection and discrimination between mycotoxigenic and non-toxigenic spoilage moulds using volatile production patterns: A review. *Food Additives and Contaminants* 24 (2007): 1161–1168.

Sardiñas, N., Vázquez, C., Gil-Serna, J., Gónzalez-Jaén, M.T., Patiño, B. Specific detection and quantification of *Aspergillus flavus* and *Aspergillus parasiticus* in wheat flour by SYBR Green quantitative PCR. *International Journal of Food Microbiology* 145 (2011): 121–125.

Sartori, D., Furlaneto, M.C., Martins, M.K. et al. PCR method for the detection of potential ochratoxin-producing *Aspergillus* species in coffee beans. *Research in Microbiology* 157 (2006): 350–354.

Schmidt-Heydt, M., Geisen, R. A microarray for monitoring the production of mycotoxins in food. *International Journal of Food Microbiology* 117 (2007): 131–140.

Schmidt-Heydt, M., Hackel, S., Rufer, C.E., Geisen, R. A strain of *Fusarium kyushuense* is able to produce aflatoxin B1 and G1. *Mycotoxin Research* 25 (2009): 141–147.

Schmidt-Heydt, M., Richter, W., Michulec, M., Buttinger, G., Geisen, R. Comprehensive molecular system to study the presence, growth and ochratoxin A biosynthesis of *Penicillium verrucosum* in wheat. *Food Additives and Contaminants Part A* 25 (2008): 989–996.

Selma, M., Martínez-Culebras, P.V., Elizaquível, P., Aznar, R. Simultaneous detection of the main black aspergilli responsible for ochratoxin A (OTA) contamination in grapes by multiplex real-time polymerase chain reaction. *Food Additives and Contaminants Part A* 26 (2009): 180–188.

Selmanoglu, G. Evaluation of the reproductive toxicity of patulin in growing male rats. *Food and Chemical Toxicology* 44 (2006): 2019–2024.

Seo, J.A., Guan, Y., Yu, J.H. FluG-dependent asexual development in *Aspergillus nidulans* occurs via derepression. *Genetics* 172 (2006): 1535–1544.

Shephard, G.S., Berthiller, F., Burdaspal, P.A. et al. Developments in mycotoxin analysis: An update for 2011–2012. *World Mycotoxin Journal* 6 (2013): 3–30.

Shi, Y.C., Pan, T.M. Beneficial effects of *Monascus purpureus* NTU 568-fermented products: A review. *Applied Microbiology and Biotechnology* 90 (2011): 1207–1217.

Shwab, E.K., Keller, N.P. Regulation of secondary metabolite production in filamentous ascomycetes. *Mycological Research* 112 (2008): 225–230.

Sirhan, A.Y., Tan, G.H., Wong, R.C.S. Determination of aflatoxins in food using liquid chromatography coupled with electrospray ionization quadrupole time of flight mass spectrometry (LC-ESI-QTOF-MS/MS). *Food Control* 31 (2013): 35–44.

Somashekar, D., Rati, E.R., Anand, S., Chandrashekar. A. Isolation, enumeration and PCR characterization of aflatoxigenic fungi from food and feed samples in India. *Food Microbiology* 21 (2004): 809–813.

Storari, M., von Rohr, R., Pertot, I., Gessler, C., Broggini, G.A.L. Identification of ochratoxin A producing *Aspergillus carbonarius* and *A. niger* clade isolated from grapes using the loop-mediated isothermal amplification (LAMP) reaction. *Journal of Applied Microbiology* 114 (2013): 1193–1200.

Suanthie, Y., Cousin, M.A., Woloshuk, C.P. Multiplex real-time PCR for detection and quantification of mycotoxigenic *Aspergillus*, *Penicillium* and *Fusarium*. *Journal of Stored Products Research* 45 (2009): 139–145.

Sudakin, D.L. Trichothecenes in the environment: Relevance to human health. *Toxicology Letters* 143 (2003): 97–107.

Teena, M., Manickavasagan, A., Mothershaw, A., El Hadi, S., Jayas, S.D. Potential of machine vision techniques for detecting fecal and microbial contamination of food products: A review. *Food and Bioprocess Technology* 6 (2013): 1621–1634.

Turner, N.W., Subrahmanyam, S., Piletsky, S.A. Analytical methods for determination of mycotoxins: A review. *Analytica Chimica Acta* 632 (2009): 168–180.

Urusov, A.E., Zherdev, A.V. Dzantiev, B.B. Immunochemical methods of mycotoxin analysis (review). *Applied Biochemistry and Microbiology* 46 (2010): 253–266.

Varga, J., Rigó, K., Molnár, J. et al. Mycotoxin production and evolutionary relationships among species of *Aspergillus* section *Clavati*. *Antonie van Leeuwenhoek* 83 (2003): 191–200.

Whitaker, T.B., Trucksess, M.W., Weaver, C.M., Slate, A. Sampling and analytical variability associated with the determination of aflatoxins and ochratoxin A in bulk lots of powdered ginger marketed in 1-lb bags. *Analytical and Bioanalytical Chemistry* 395 (2009): 1291–1299.

Wilson, I.G. Inhibition and facilitation of nucleic acid amplification. *Applied and Environmental Microbiology* 63 (1997): 3741–3751.

Wong, M.L., Medrano, J.F. Real-time PCR for mRNA quantitation. *Biotechniques* 39 (2005): 75–85.

Xiao, J., Jin, X., Jia, X. et al. Transcriptome-based discovery of pathways and genes related to resistance against *Fusarium* head blight in wheat landrace Wangshuibai. *BMC Genomics* 14 (2013): 197.

Yang, X.H., Kong, W.J., Yang, M.H., Zhao, M., Ouyang, Z. Application of aptamer identification technology in rapid analysis of mycotoxins. *Chinese Journal of Analytical Chemistry* 41 (2013): 297–306.

Yao, H., Hruska, Z., Kincaid, R., Brown, R.L., Bhatnagar, D., Cleveland, T. Detecting maize inoculated with toxigenic and atoxigenic fungal strains with fluorescence hyperspectral imaging. *Biosystems Engineering* 115 (2013): 125–135.

Yin, Y., Liu, X., Ma, Z. Simultaneous detection of *Fusarium asiaticum* and *Fusarium graminearum* in wheat seeds using a real-time PCR method. *Letters in Applied Microbiology* 48 (2009): 680–686.

Yli-Mattila, T., Paavanen-Huhtala, S., Jestoi, M. et al. Real-time PCR detection and quantification of *Fusarium poae*, *F. graminearum, F. sporotrichioides* and *F. langsethiae* in cereal grains in Finland and Russia. *Archives of Phytopathology and Plant Protection* 41 (2008): 243–260.

Zhang, D., Li, P., Liua, W. et al. Development of a detector-free semiquantitative immunochromatographic assay with major aflatoxins as target analytes. *Sensors and Actuators B: Chemical* 185 (2013): 432–437.

Zheng, M.Z., Richard, J.L., Binder, J. A review of rapid methods for the analysis of mycotoxins. *Mycopathologia* 161 (2006): 261–273.

12

Electrochemical Biosensors for Food-Borne Pathogens

Subramanian Viswanathan

CONTENTS

12.1 Introduction

As agriculture and food technology advance and populations increase—the current global population is nearly 7 billion with 50% living in Asia—analytical and regulatory problems concerning food become very complex. Infections caused by food-borne pathogens are a frequent reminder of the complex food network that links us with animal, plant, and microbial populations around the world. The World Health Organization (WHO) estimates that some 2.2 million deaths occur annually due to food- and water-borne illnesses, and 1.9 million among them are children. Generally, the cooking process effectively kills pathogenic bacteria that are present in food; however, food styles have changed significantly in recent years, and more processed and ready-to-eat packaged foods are available, which increases the chance of exposure to pathogenic contamination. Although the safety of food has dramatically improved overall, progress is uneven and food-borne outbreaks from microbial contamination, chemicals, and toxins are common in many countries.

12.2 Emerging Food-Borne Pathogens

Food-borne diseases are well recognized but are considered to be at the emerging stage because they have only recently become more common. For example, outbreaks of salmonellosis have been reported for decades, but in the past 25 years, the incidence of the disease has increased in many continents. The most prominent food-borne pathogens include mycotoxins, exotoxins, and enterotoxins from *Escherichia coli* O157:H7 (*E. coli*), some strains of *Staphylococcus aureus*, *Shigella* spp., *Bacillus anthracis* (produces anthrax toxin), *Campylobacter jejuni*, *Clostridium perfringens*, *Clostridium botulinum* (produces a powerful paralytic toxin botulin), *Salmonella* spp., *Listeria monocytogenes*, *Vibrio cholera*, *Yersinia enterocolitica*, and *Coxiella burnetii* (Velusamy et al. 2010). Risk group includes infants, young children, pregnant women (and their fetuses), elderly and sick people, and others with weak immune systems, who get highly affected by pathogen outbreaks. A list of pathogens responsible for food-borne diseases is given in Table 12.1.

12.3 Food-Borne Pathogen Detection Methods

Food-borne diseases pose a considerable threat to human health and the economy of individuals, families, and nations (Bhunia 2008). Ever-increasing incidences of food-borne diseases have led to the employment of rapid and inexpensive method of analysis (Velusamy et al. 2010). In the food industry, the pathogen contamination is evaluated through periodic microbiological analysis such as culture, colony counting, immunological assays, fluorescence spectrophotometry, and polymerase chain reaction (PCR) (Ahmad et al. 2009). These methods do not allow easy, rapid monitoring because they involve complex analytical steps with expensive instrumentation, need well-trained operators, and, in some cases, increase the time of analysis. To overcome these limitations, an immediate need for the development of some alternative succeeded with an analytical approach.

12.4 Biosensors

Biosensors are a subgroup of chemical sensors that integrate biological sensing elements with physical transducers where the interactions between biological sensing elements and target molecules are directly converted into electronic signal (Figure 12.1). Biosensors represent a conceptually novel approach to real-time, on-site, simultaneous detection of multiple biohazardous agents (Bhadoria and Chaudhary 2011). Biosensor provides immediate interactive information about the sample being tested, enabling decision makers to take corrective measures and to quickly recognize impending threats. Biosensors represent a promising tool for food analysis because of their possibility to fulfill some demand that the classic methods of analysis do not attain (Arora et al. 2011). Several biosensors have been developed for the microbial analysis of food pathogens, including *E. coli*, *S. aureus*, *Salmonella*, and *L. monocytogenes*, as well as various microbial toxins. Biosensors have several potential advantages over other methods of analysis, including sensitivity in the range of ng/mL for microbial toxins and <100 colony-forming units (CFU)/mL for bacteria. Miniaturization enables biosensor integration into various stages of food production. Potential uses of biosensors for food microbiology include online process of microbial monitoring to provide real-time information on food production and analysis of microbial pathogens and their toxins in finished food. Rasooly and Herold (2006) reported the main biosensor approaches, technologies, instrumentation, and applications for food microbial analysis in their review (Rasooly and Herold 2006). There are various types of biosensors, including electrochemical, optical, piezoelectric, and acoustical sensors, that are used to serve the purpose of food safety. Among the different types of biosensors, the electrochemical

TABLE 12.1

List of Common Food-Borne Pathogens

Pathogens	Infective Dosage	Incubation Period	Symptoms	Possible Contaminants
Bacillus cereus	$>10^6$/g	30 min–15 h	Diarrhea, abdominal cramps, nausea, and vomiting (emetic type)	Meats, milk, vegetables, fish, rice, potatoes, pasta, and cheese
Campylobacter jejuni	400–500	1–7 days	Nausea, abdominal cramps, diarrhea, headache—varying in severity	Raw milk, eggs, poultry, raw beef, cake icing, water
Clostridium botulinum	$<10^{-9}$g	12–36 h	Nausea, vomiting, diarrhea, fatigue, headache, dry mouth, double vision, muscle paralysis, respiratory failure	Low-acid canned foods, meats, sausage, fish
Cl. Perfringens	$>10^8$	8–22 h	Abdominal cramps and diarrhea, some include dehydration	Meats and gravies
Cryptosporidium parvum	—	2–10 days	Watery diarrhea accompanied by mild stomach cramping, nausea, loss of appetite	Contaminated water or milk, person-to-person transmission
Escherichia coli	<10	2–4 days	Hemorrhagic colitis, possibly hemolytic uremic syndrome	Ground beef, raw milk
Giardia lamblia	—	1–2 weeks	Infection of the small intestine, diarrhea, loose or watery stool, stomach cramps, and lactose intolerance	*Giardia* is found in soil, food, water, or surfaces that have been contaminated with the feces from infected humans or animals
Hepatitis A	10–1000		Fever, malaise, nausea, abdominal discomfort	Water, fruits, vegetables, iced drinks, shellfish, salads
Listeria monocytogenes	<1000	2 days to 3 weeks	Meningitis, septicemia, miscarriage	Vegetables, milk, cheese, meat, seafood
Norwalk, Norwalk-like, or *Norovirus*	Unknown but presumed to be low	12–60 h	Nausea, vomiting, diarrhea, and abdominal cramps	Raw oysters/shellfish, water and ice, salads, frosting, person-to-person contact
Salmonellosis	15–20	12–24 h	Nausea, diarrhea, abdominal pain, fever, headache, chills, prostration	Meat, poultry, egg, milk products
Staphylococcus	—	1–6 h	Severe vomiting, diarrhea, abdominal cramping	Custard- or cream-filled baked goods, ham, tongue, poultry, dressing, gravy, eggs, potato salad, cream sauces, sandwich fillings
Shigella	<10	12–50 h	Abdominal pain, cramps, diarrhea, fever, vomiting, blood, and pus	Salads, raw vegetables, dairy products, poultry
Toxoplasma gondii	—	5–23 days	Mild illness (swollen lymph glands, fever, headache, and muscle aches)	Cat, rodent or bird feces, raw or undercooked food
Vibrio	>1 million	4 h to 4 days	Diarrhea, abdominal cramps, nausea, vomiting, headache, fever, and chills	Fish, shellfish
Yersiniosis	—	1–3 days	Enterocolitis, may mimic acute appendicitis	Raw milk, chocolate milk, water, pork, other raw meats

Source: Velusamy, V. et al., *Biotechnol. Adv.*, 28(2), 232, 2010.

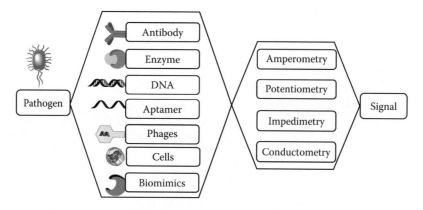

FIGURE 12.1   Schematic diagram of an electrochemical biosensor and its classification.

biosensors are the most common as a result of numerous advances leading to their well-understood biointeraction and detection process (Jiang et al. 2007; Lam et al. 2013).

12.5 Electrochemical Biosensors for Pathogens

In electrochemical biosensors, the variation in electron fluxes leads to the generation of an electrochemical signal, which is measured by the electrochemical detector (Caygill et al. 2010). Electrochemical biosensors are mainly based on the observation of current or potential changes due to interactions occurring at the transducer/biomolecule interface. These sensors are designed by coupling the biorecognition elements (e.g., antibodies, DNA, and receptors) to solid electrode surfaces (e.g., Pt-, Au-, Ag-, graphite-, or carbon-based conductors) or electrode arrays, which respond to applied electrical impulses such as potential or current. The versatility of electrochemical biosensors is reported with numerous applications in the area of food analysis (Viswanathan and Radecki 2008).

12.6 Classification Based on Biorecognition Elements

Biorecognition layers are often discussed as the most important functional part of electrochemical biosensor as they define the recognition specificity for pathogen detection (Bhadoria and Chaudhary 2011). They allow binding the specific pathogen of interest to the sensor for measurement with minimum interference from other components in complex mixtures. The popular bioprobes that have been employed on a biosensor surface for pathogen detection are antibodies, nucleic acids, bacteria phages, and most recently biomimics (Singh et al. 2013).

12.6.1 Antibodies

Antibodies are popular bioreceptors used in biosensors. The antigen is recognized as a foreign body. A specific antibody is generated to act against it by binding to it and operating to remove the antigen. Antibodies may be polyclonal, monoclonal, or recombinant, depending on their selective properties and the way they are synthesized (Pohanka 2009). By this specific recognition and interaction performed on the molecular level, antibodies and antigens can be exploited as a means for diagnostic testing. In this way, antibodies may serve as the basis for the biosensor detection system. Antibodies have been extensively explored as bioreceptors for pathogen detection and monitoring because of their ease of immobilization on a biosensor surface and their high level of specificity (kd ≈ 10^{-7}–10^{-11}) toward their target (Byrne et al. 2009). These antibodies have been successfully employed for the detection of pathogens, their spores, and toxins (Byrne et al. 2009; Viswanathan et al. 2006).

12.6.2 Enzymes

Biosensors making use of enzymes as bioreceptors is a well-established technique. Enzymes can be used to label antibodies much in the same fashion as in an enzyme-linked immunosorbent assay (ELISA). In the area of pathogen detection, using enzymes as biolabels not only provides biosensors with a high degree of specificity, but their catalytic activity can also amplify the pathogenic bacteria being detected and measured, allowing for sensitive analyses. The application of enzymes as labels has earned more popularity in immunoassay detection than other labels such as radioisotope and fluorescent tag. Enzymes offer the advantages of high sensitivity and the possibility of direct visualization. But there are a few limitations when using enzymes as labels, which include multiple assay steps and the possibility of interference from endogenous enzymes (Pividori et al. 2006). Antibodies/antibody fragments labeled to small enzymes can readily pass through cell membranes and by using electrodes it is then possible to observe cellular functions.

12.6.3 Nucleic Acids

Pathogenic bacteria have within them a unique signature in their deoxyribonucleic acid (DNA). All of the information contained in the DNA appears encoded in a series of amino acids and, as such, forms the identifying backbone of that structure. The recognition of these sequences is of fundamental importance to the control, reading, and detection of these molecular structures. The basic principle of a DNA biosensor is to detect the molecular recognition provided by the DNA probes and to transform it into the signal using a transducer. Current developments in DNA recognition have enhanced the power of DNA biosensors and biochips. In the case of pathogenic DNA detection, the identification of a target analyte's nucleic acid is achieved by matching the complementary base pairs that are often the genetic components of an organism. As each organism has unique DNA sequences, any self-replicating microorganism can be easily identified. Electrochemical biosensors based on nucleic acid as a biorecognition element are simple, rapid, and inexpensive, and hence, they are widely used in pathogen detection (Liao et al. 2007; Prasad and Vidyarthi 2009). DNA-based electrochemical assays were developed for bacterial pathogens such as *E. coli*, *Salmonella* spp., *C. jejuni*, and *S. aureus* (LaGier et al. 2007). Furthermore, direct detection of bacterial DNA was achieved using lysed cells rather than extracted nucleic acids, allowing streamlining of the process. The methods presented can be used to rapidly (3–5 h) screen environmental water samples for the presence of microbial contaminants and have the potential to be integrated into semiautomated detection platforms (LaGier et al. 2007). A multipathogen detection platform including *Salmonella*, *Listeria*, and *E. coli* has been developed by García et al. (2012). This method involves the immobilization of a thiolated capture probe that is able to hybridize with its complementary sequence (target). The hybridization event is detected using the ruthenium complex, $[Ru(NH_3)_5L]^{2+}$, where L is [3-(2-phenanthren-9-yl-vinyl)-pyridine] an electrochemical indicator. The combination of microelectromechanical systems (MEMS) technology to fabricate electrodes with a predetermined configuration and the use of a hybridization redox indicator, which interacts preferentially with dsDNA, results in the development of an approach that not only quantifies complementary target sequence but is also selective of *Salmonella* in the presence of other pathogens that can act as potential interferents (García et al. 2012).

A novel concept of metal-enhanced detection (MED) for the determination of DNA–DNA reactions could be applied for mismatch detection. The MED concept relies on the idea that metallic films deposited as a continuous layer or monolayer onto a solid electrode, or even electrostatically held, could greatly enhance the rate of electron transfer by reducing the distance between the donor and acceptor species and could lead to label-free assays during DNA hybridization reactions. The MED concept has been tested for the voltammetric detection of a gene sequence of *Microcystis* spp. The resulting biosensor involved the immobilization of a 17-mer DNA probe, which is complementary to a specific gene sequence of *Microcystis* spp. on a gold electrode via avidin–biotin chemistry. Electrochemical reduction and oxidation of DNA-captured Ag^+ ions provided the detection signals for the target gene sequence in the solution. A linear response of silver cathodic peak current with the concentration of the target oligonucleotide sequence was observed with a detection limit of 7×10^{-9} M (K'Owino et al. 2007).

Another example of an electrochemical genosensor based on 1-fluoro-2-nitro-4-azidobenzene modified octadecanethiol self-assembled monolayer has been fabricated for *E. coli* detection. The results of electrochemical response measurements investigated using methylene blue (MB) as a redox indicator reveal that this nucleic acid sensor has 60 s of response time, high sensitivity (0.5×10^{-18} M), and linearity of 0.5×10^{-18} to 1×10^{-6} M (Pandey et al. 2011a).

12.6.4 Aptamers

Numerous antibody and nucleic acid–based detection methods have been introduced to food-borne pathogen detection; however, the assay costs may limit the widespread use of such technologies. Thus, there is an unmet demand for new platforms of technology to improve the bacterial detection and identification in clinical practice. Aptamers are DNA or RNA molecules that can fold into a variety of structures (Chang et al. 2013; Ellington and Szostak 1990). They are small (i.e., 40–100 bases), synthetic oligonucleotides that can specifically recognize and bind to virtually any kind of target, including ions, whole cells, drugs, toxins, low–molecular weight ligands, peptides, and proteins. Aptamers can function as the biorecognition elements in biosensor applications. An aptamer attached to an electrode coated with single-walled carbon nanotubes interacts selectively with bacteria. The resulting electrochemical response is highly accurate and reproducible and starts at ultralow *E. coli* bacteria concentrations (single bacterium in 5 mL), thus providing a simple, selective method for pathogen detection (Zelada-Guillén et al. 2009).

12.6.5 Bacteriophages

Bioreceptors such as antibodies, enzymes, and nucleic acids are widely used as biomolecular recognition elements and they have both advantages and disadvantages when compared to one another. Recently, bacteriophages were utilized as biorecognition elements for the identification of various pathogenic bacteria (Singh et al. 2013). These powerful bacteriophages are viruses that bind to specific receptors on the bacterial surface in order to inject their genetic material inside the bacteria. These entities are typically of 20–200 nm in size. Phages recognize the bacterial receptors through their tail spike proteins. As the recognition is highly specific, it can be used for the typing of bacteria, thus opening the path for the development of specific pathogen detection technologies. Bacteriophage-based biorecognition has been combined with electrochemical detection methods to provide the specific detection of food-borne pathogens. Researchers have reported the application of phages as a biorecognition element for the detection of various pathogens such as *E. coli*, *S. aureus*, and *B. anthracis* spores by using different sensing platforms (Singh et al. 2013).

12.6.6 Cellular Bioreceptors

Most biosensor detection methods are based on nucleic acid hybridization or immunological techniques. Although these techniques are fast and specific on target pathogens or toxins, they often fail to provide any information on the biological and physiological activity of the target analyte and the identification of unknown hazardous agents. Mammalian cells, when used in cell-based biosensors (CBBs), can screen and detect analytes in a way that is physiologically pertinent and provide information about the biological function of the target (Velusamy et al. 2010). The cell-based microbial biosensor is an elaborate analytic device that uses microorganisms as recognition elements, mainly for application in environmental monitoring, food safety, military defense, and medicine (Xu and Ying 2011). CBBs, bioelectronic portable devices containing plant living cells, have been used for monitoring some physiological changes induced by pathogen-derived signal molecules called flagellin. Recently, screen-printed electrodes were adapted for the preparation of biosensors. The proton-sensitive thick films were printed using composite bulk modified with the addition of RuO_2 to measure the pH changes. Tobacco CBB can be used for the detection of flagellin, the virulence factor of bacterial pathogen. The detection of the electrochemical proton gradient across the plasma membrane serves as the analytical signal. The electrode response depends upon H^+ concentration in the extracellular solution. It can be conveniently observed on the surfaces of biosensors. The future development of these CBBs could draw advances in the selective monitoring of microbial pathogens and other physiologically active components. Moreover, this new method is much faster compared with traditional microbial testing (Oczkowski et al. 2007). The selection and immobilization of live cells are key steps that must first be addressed for microbial biosensors. Currently, genetically modified microorganisms play an increasingly significant role in improving the capacity of biosensors. Electrochemical transducers have been widely employed in microbial biosensors recently. Additionally, the microbial fuel cell (MFC), which has been mainly applied in biological oxygen demand (BOD) biosensors, is a promising technology. Xu and Ying (2011) reviewed recent developments of microbial biosensors with respect to their applications in food analysis, products in fermenting processes, and toxins in food.

12.6.7 Biomimetic Receptors

A receptor that is fabricated and designed to mimic a bioreceptor (antibody, enzyme, cell, or nucleic acids) is often termed a biomimetic receptor. Molecular imprinting is one of the techniques of producing artificial recognition sites by forming a polymer around a molecule, which can be used as a template. Though there are several methods, such as genetically engineered molecules and artificial membrane fabrication, the molecular imprinting technique has emerged as an attractive and highly accepted method for the development of artificial recognition agents. Molecular imprinted polymers (MIPs) can, in principle, be synthesized for any analyte molecule and are capable of binding target molecules with affinities and specificities on a par with biological recognition elements. The conducting polymers combine sensing properties with the ability to act as signal transducers for the biorecognition event (Travas-Sejdic et al. 2011). Chamberlain et al. (2012) reported a versatile carbohydrates-polypyrrole-coated microelectrode array-based platform to investigate carbohydrate-mediated protein and bacterial binding, with the objective of developing a generalizable method for screening inhibitors of host–microbe interactions. The platform's ability to analyze whole-cell binding was demonstrated using strains of *E. coli* and *Salmonella enterica* (Chamberlain et al. 2012).

Qi et al. (2013) synthesized bioimprinted films for selective bacterial detection. In their report, marine pathogen sulfate-reducing bacteria (SRB) were chosen as the template bacteria. Chitosan (CS) doped with reduced graphene sheets (RGSs) was electrodeposited on an indium tin oxide electrode, and the resulting RGSs–CS hybrid film served as a platform for bacterial attachment. A layer of nonconductive CS film was deposited to embed the pathogen, and acetone was used to wash away the bacterial templates. Electrochemical impedance spectroscopy measurements revealed that the charge transfer resistance (Rct) increased with increased SRB concentration between 1.0×10^4 and 1.0×10^8 CFU/mL. Hence, selectivity for bacterial detection can be improved if the bioimprinting technique is combined with other biorecognition elements (Qi et al. 2013).

12.7 Classification of Electrochemical Biosensors

The development of biosensors for monitoring food products is an interesting research topic. The main objective of biosensor researchers is to develop enhanced detection technologies with high levels of reliability, sensitivity, and selectivity with short assay times. These are critical factors for inspection of food products in industrial firms considering the short shelf life of products and low infection dose of pathogens in food samples. Efforts have been mainly focused on optimizing the biosensor transducer to improve detection sensitivity. The transducer plays an important role in the detection process of a biosensor. Biosensors can also be classified based upon the transduction methods they employ. Electrochemical-based detection methods are another possible means of transduction that has been used for the identification and quantification of food-borne pathogens. Electrochemical biosensors can be further classified into amperometric, potentiometric, impedimetric, and conductometric, based on the observed parameters such as current, potential, impedance, and conductance, respectively. Palchetti and Mascini (2008) reviewed food pathogen detection methods based on electrochemical biosensors, specifically amperometric, potentiometric, and impedimetric biosensors. The underlying principles and application of these biosensors are discussed with special emphasis on new biorecognition elements, nanomaterials, and lab-on-a-chip technology (Palchetti and Mascini 2008). Novel applications are also reviewed for the detection of food pathogens, as well as recent advances in electrochemical biosensors (Sadik et al. 2009).

12.7.1 Amperometric Method

Amperometric transduction is the most common electrochemical detection method that has been used for pathogen detection. In amperometric-based detection, the sensor potential is set at a value where the analyte produces current. Thus, the applied potential serves as the driving force for the electron transfer reaction, and the current produced is a direct measure of the rate of electron transfer. Clark and Lyons (1962) reported that oxygen electrodes possibly represent the basis for the simplest forms of amperometric biosensors, in which current was produced in proportion to the oxygen concentration. Many researchers have reported amperometric detection of food-borne pathogens such as *E. coli*, *Salmonella*, *L. monocytogenes*, and *C. jejuni* and have developed an amperometric detection method for determining the presence, the amount, and/or the concentration of an analyte in a microfluidic sensor, where the sensor is a part of an integrated sample acquisition system and/or analyte measurement device. In this method, the potential was applied and the related current measured at the integrated working electrode for no more than 10 s, nearly 2 s, and a relaxation time separating sequential amperometric measurements was longer than 1 s but shorter than 1 min.

In order to detect aflatoxin M1 in milk, an electrochemical immunosensor was fabricated by immobilizing the antibodies directly on the surface of the screen-printed electrodes (Micheli et al. 2005). The electrochemical technique chosen was chronoamperometry, performed at −100 mV. The results showed that using screen-printed electrodes, aflatoxin M1 was measured with a detection limit of 25 ppt and with a working range between 30 and 160 ppt (Micheli et al. 2005). A new method based on co-electropolymerizing staphylococcal protein A (SPA) and polypyrrole (PPy) to modify the sensing surface was developed and applied for the detection of *Salmonella* Typhimurium in a microamperometric immunosensor, which was an electrochemical system composed of two electrodes and fabricated based on MEMS technology (Sun et al. 2009). Electrochemical biosensor for the detection of *Aeromonas hydrophila*, an opportunistic pathogen that is recognized as an emerging food-borne hazard, was reported by Tichoniuk et al. (2010). The genosensor recognition layer was prepared using a mixed self-assembled monolayer (SAM) consisting of a thiolated single-stranded DNA probe (ssDNA). The voltammetric examination of the electroactive indicator, methylene blue (MB), was performed to investigate the influence of hybridization reaction time, concentration of target DNA fragments, and presence of noncomplementary DNA on the electrochemical response of the genosensor recognition interface. The biosensor enabled distinction between the DNA samples isolated from *A. hydrophila* (present at the concentration of 2.5 μg/cm^3) and other microbial DNA (Tichoniuk et al. 2010). Response surface methodology (RSM) is used for optimization of food-borne pathogen detection based on label-free electrochemical nucleic acid biosensors. *L. monocytogenes* amplicons obtained from food samples are used as models. The extent of hybridization is determined by using guanine oxidation signals obtained with differential pulse voltammetry. The ratio of electrochemical transductions after hybridization with complementary and noncomplementary targets is the response of the statistical analysis obtaining a detection limit of 267 pM (Urkut et al. 2011).

E. coli cells are isolated via immunomagnetic separation (IMS) and labeled with biofunctionalized electroactive polyaniline (immuno-PANI). Labeled cell complexes are deposited onto a disposable screen-printed carbon electrode (SPCE) sensor and pulled to the electrode surface by an external magnetic field, to amplify the electrochemical signal generated by PANI. Cyclic voltammetry is used to detect PANI and signal magnitude indicates the presence or absence of *E. coli*. As few as 7 CFU of *E. coli* (corresponding to an original concentration of 70 CFU/mL) were successfully detected on the SPCE sensor. The assay requires 70 min from sampling to detection, giving it a major advantage over standard culture methods in applications requiring high-throughput screening of samples and rapid results (Setterington and Alocilja 2011, 2012).

12.7.2 Potentiometric Method

The potentiometric biosensors monitor the accumulation of charge at zero current created by selective binding at the electrode surface (Bakker and Pretsch 2005). In potentiometric-based detection, the biorecognition process is converted into a potential signal. A high-resistance voltmeter is used to measure the electrical potential difference or electromotive force (EMF) between two electrodes at near-zero current. As potentiometry generates a logarithmic concentration response, the technique allows the detection of extremely small concentration changes (Zelada-Guillén et al. 2009).

12.7.3 Impedimetric Method

Electrochemical impedance spectroscopy (EIS) is playing an important role in biosensor development. In EIS measurements, a controlled AC electrical stimulus between 5 and 10 mV is applied over a range of frequencies, and this causes a current to flow through the biosensor, depending on the process. EIS biosensors have been exploited for bacterial detection by monitoring the changes in the solution–electrode interface due to the capture of microorganisms on the sensor surface. The capture of target analytes such as bacteria on the sensor usually increases the impedance due to the insulating properties. Bacteriophages have been used as a cross linkage between bacteria and electrode surface. The integration of impedance with biological recognition technology for the detection of pathogens has led to the development of impedance biosensors that are finding widespread use in recent years (Yang and Bashir 2008).

Lu et al. (2013) developed a label-free biosensor based on electrochemical impedance measurement for the determination of food-borne pathogenic bacteria. They also stated, in their previous work, how gold-tungsten wires (25 µm in diameter) were functionalized by coating them with polyethyleneimine-streptavidin-anti-*E. coli* antibodies to improve sensing specificity (Lu and Jun 2012). EIS has been used to detect and validate the resistance changes in a conventional three-electrode system in which $[Fe(CN)_6^{3-}]/[Fe(CN)_6^{4-}]$ served as the redox probe. The impedance data demonstrated a linear relationship between the increments of ΔRet and the logarithmic concentrations of *E. coli* suspension in the range of 10^3–10^8 CFU/mL. In addition, there were few changes of ΔRet when the sensor worked with *Salmonella*, which clearly evidenced the sensing specificity to *E. coli*. EIS was proven to be an ideal alternative to fluorescence microscopy for the enumeration of captured cells (Lu et al. 2013).

EIS is a widely used technique for probing bioaffinity interactions at the surfaces of electrically conducting polymers and can be employed to investigate the *label-free* detection of analytes via impedimetric transduction (Tully et al. 2008). Though EIS offers label-free detection compared to amperometry or potentiometry, its detection limits are inferior compared to traditional methods. Yang and Bashir (2008) used interdigitated microelectrodes as impedance sensors for the rapid detection of viable *Salmonella*. The impedance growth curves, impedance against bacterial growth time, were recorded at four frequencies (10 Hz, 100 Hz, 1 kHz, and 10 kHz) during the growth of *S.* Typhimurium. It was observed that impedance did not change until the cell number reached 10^5–10^6 CFU/mL. The greatest change in impedance was observed at 10 Hz. An equivalent electrical circuit, consisting of double-layered capacitors, a dielectric capacitor, and a medium resistor, was introduced and used for interpreting the change in impedance during bacterial growth. They reported that the detection times for 4.8×10^5 and 5.4×10^5 CFU/mL initial cell numbers were 9.3 and 2.2 h, respectively (Yang et al. 2004). The determination of *E. coli* using antibody–antigen binding method based on a covalently linked antibody on a conducting PANI film surface was reported. EIS was used to test the sensitivity and effectiveness of the sensor electrode by measuring the change in impedance values of electrodes before and after incubation with different concentrations of bacteria. As small a concentration as 10^2 CFU/mL of *E. coli* was successfully detected on the Au/PANI/Glu/antibody sensor with the upper detection limit of 10^7 CFU/mL. The mechanism of detection is very simple and rapid, giving it a major advantage over DNA-based techniques and other secondary antibody-based labeled methods (Chowdhury et al. 2012).

Nandakumar et al. (2008) developed a biosensor methodology based on Bayesian decision theory and electrochemical impedance spectroscopy for rapid and reliable detection of *S.* Typhimurium bacteria.

Furthermore, the technique has low computational complexity and is therefore well suited for use in hand-held pathogen detectors (Nandakumar et al. 2008). Biosensing electrodes have been fabricated by immobilization of *E. coli*-specific DNA probe onto dendritic cystine. The results of the electrochemical impedance spectroscopy studies reveal that this nucleic acid sensor exhibits a linear response to cDNA in the concentration range of 10^{-6}–10^{-14} M with a response time of 30 min (Pandey et al. 2011b). Pournaras et al. (2008) successfully developed a faradic impedimetric immunosensor based on electropolymerized polytyramine (Ptyr) films for the detection of *S.* Typhimurium in milk for the first time. Polyclonal anti-*Salmonella* was cross-linked, in the presence of glutaraldehyde vapors, on Ptyr-modified gold electrodes. In milk samples containing a low initial concentration of 10 CFU/mL *S.* Typhimurium, which actually defines the LOD of the immunosensors, signal changes of 33% and 88% were achieved after 3 and 10 h of incubation, respectively (Pournaras et al. 2008). Siddiqui et al. (2012) demonstrated that boron-doped ultrananocrystalline diamond microelectrode (UNCD) array (3×3 format, 200 μm diameter) improves signal reproducibility significantly and increases sensitivity by 4 orders of magnitude. This study marks the first demonstration of UNCD array-based biosensor that can reliably detect a model *E. coli* bacterium using EIS, positioning this technology for rapid adoption in point-of-use applications (Siddiqui et al. 2012).

12.7.4 Conductometric Method

Conductometric-based biosensors bond the relationship between conductance and a biorecognition technique. Most reactions involve a change in the ionic species concentration, which leads to a change in electrical conductivity or current flow. Normally, a conductometric biosensor consists of two metal electrodes separated by a certain distance and an AC voltage applied across the electrodes causes a current flow. During a biorecognition event, the ionic composition changes and the change in conductance between the metal electrodes is measured. Pal et al. (2008) developed a direct-charge transfer conductometric biosensor for the detection of *Bacillus cereus* in various food samples. The biosensor used the principle of a sandwich immunoassay, combined with an electron charge flow aided through conductive PANI, to generate an electronic signal. The biosensor was able to detect cell concentrations in the range of 35–88 CFU/mL in the food samples with a detection time of 6 min.

Recent technological improvements are still in progress for the development of advanced sensors to monitor specific biointeractions and different recognition events of biomolecules in solution and solid substrates. Some of the typical applications on screening of selected toxins and pathogens have been overviewed herein for environmental, agricultural, and food monitoring control with the emergence of important features of electrochemical sensor strategies (Erdem et al. 2012; Lubin and Plaxco 2010). Different modes of electrochemical-based food-borne pathogen detection are summarized in Table 12.2.

12.8 Nanometerials in Electrochemical Biosensors

According to the statement by the great scientist Arthur C. Clarke, "Any sufficiently advanced technology is indistinguishable from magic." This statement is particularly true in molecular biosensing based on nanotechnology where the detection limits are *magically* becoming smaller and smaller, even reaching zeptomolar concentrations in addition to opening up possibilities for ultrasensitive multiplexed detection. The whole area of biosensor development continues to be an extremely dynamic and growing area for scientific research (Pearson et al. 2000). Nanotechnology has recently become one of the most exciting forefront fields in biosensor fabrication (De Micheli et al. 2012). Nanotechnology is defined as the creation of functional materials, devices, and systems through the control of matter at the 1–100 nm scale. A wide variety of nanoscale materials of different sizes, shapes, and compositions are now available. The huge interest in nanomaterials is driven by their many desirable properties. The presence of nanomaterials in biosensors allows the use of many new signal transduction technologies in their manufacture. Because of their size, nanosensors, nanoprobes, and other nanosystems are revolutionizing the fields of chemical and biological analysis. The emergence of nanoscience and nanotechnology stimulates the scientists to fabricate devices for use in bioanalysis (via the integration of nanomaterials

TABLE 12.2

Different Electrochemical Method for Food-Borne Pathogen Detection

Pathogen	Bioreceptors	Technique	Limit of Detection (CFU/mL)	References
Escherichia coli	Antibody	Amperometric	78	Setterington and Alocilja (2011)
	Antibody-labeled	Voltammetry	400	Viswanathan et al. (2012)
	Antibody-labeled	Amperometric	81	Muhammad-Tahir and Alocilja (2004)
	Antibody	Impedimetric	—	Pournaras et al. (2008)
	Antibody	Impedimetric	10^1–10^7	Muñoz-Berbel et al. (2008)
	Antibody	Potentiometric	10	Ercole et al. (2003)
	Immunomagnetic	Voltammetry	6	Setterington and Alocilja (2012)
Salmonella typhimurium	Antibody	Voltammetry	400	Viswanathan et al. (2012)
	Antibody	Impedimetric	4.8 and 5.4×10^5	Yang et al. (2004)
	Antibody-labeled	Voltammetry	13	Freitas et al. (2014)
Bacillus cereus	Antibody	Conductometric	35–88	Pal et al. (2008)
	Immunomagnetic	Voltammetry	40	Setterington and Alocilja (2012)
Campylobacter jejuni	Antibody	Voltammetry	800	Viswanathan et al. (2012)
Vibrio parahaemolyticus	Antibody	Voltammetry	7.374×10^4	Zhao et al. (2007)

with biomolecules). Nanotechnology-based biorecognition devices are capable of rapid (in few minutes) and sensitive detection (even of single cell) of food-borne pathogens (Shinde et al. 2012). Sanvicens et al. (2009) provide a general overview of the progress, the limitations, and the future challenges of nanoparticle-based biosensors for the detection of pathogenic bacteria (Sanvicens et al. 2009). Polymeric nanoparticles, liposomes, vesicles, inorganic semiconducting, metallic and magnetic nanoparticles were conjugated with biomolecules such as antibodies, antibiotics, adhesion molecules, and complementary DNA sequences for the specific recognition of pathogens. Versatile chemistry, unique electrochemical properties, and strong ferromagnetic responses have made nanoparticle an efficient tool for various biodetector systems (Freitas et al. 2014). The development of nanoscale materials such as nanowires, nanofibers, nanoparticles, nanobelts or nanoribbons, and nanotubes has revolutionized clinical and molecular biology by their significant use as bioanalyzers and biodetectors, and nowadays it is widely used for the detection of contamination in food materials (Viswanathan et al. 2012). Viswanathan et al. (2012) described the fabrication of electrochemical immunosensors for multiple pathogens using nanocrystal-labeled anti-*E. coli*, anti-*Campylobacter*, and anti-*Salmonella* antibodies and a multiwall carbon nanotube screen-printed electrode (Figure 12.2).

Several label-free biosensors have been recently used for food-borne pathogen detection. Emerging nanomaterial-based sensors, particularly those developed by utilizing functionalized gold nanoparticles as a sensing component, potentially offer many desirable features needed for threat agent detection (Upadhyayula 2012). Maurer et al. (2012) developed a novel biosensor platform using an assembly of carbon nanotube, nanogold, and immobilized RNA capture elements designed for the rapid and selective detection of the bacterium *E. coli* (Maurer et al. 2012). Researchers are using two sets of nanoparticles, consisting of a group of gold or another material, along with a magnetic group. They expect that these two groups of materials will help them in developing an effective biosensor. Researchers conducted a study to demonstrate the use of target-specific DNA probes for salmonella. They are also focusing on the potential use of disposable carbon electrode–based detectors that generate an electrochemical signal (Hogan 2008). A novel material for electrochemical biosensing based on rigid conducting gold nanocomposites (nano-AuGEC) is reported (De Oliveira Marques et al. 2009). The spatial resolution of the immobilized thiolated DNA was easily controlled by merely varying the percentage of gold nanoparticles in the composition of the composite. As little as 9 fmol (60 pM) of synthetic DNA was detected in

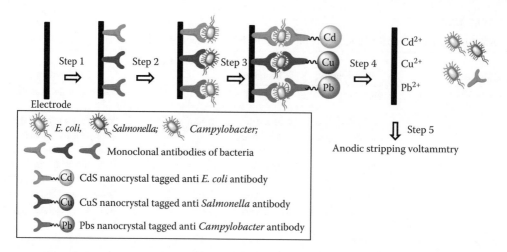

FIGURE 12.2 Outlines of multiplexed detection of pathogens using nanocrystal antibody conjugates and SPE. Step 1—Immobilization of antibodies on SPE, Step 2—Immunocapture, Step 3—NC–antibody conjugates immunobinding, Step 4—Dissolution of metal ions from NC, Step 5—Stripping voltammetric analysis of released metal ions. (From Viswanathan, S. et al., *Talanta*, 94, 315, 2012.)

hybridization experiments that used a thiolated probe (De Oliveira Marques et al. 2009). The synthesis and characterization of gold nanoparticles (AuNPs 20 nm sized) modified with k-casein derived peptides in order to monitor the peptide effect as bacterial adhesion inhibitor, and some aspects related to the stability of AuNP/peptide conjugates for a potential application in the design of an electrochemical biosensor for pathogen bacteria detection were also reported (Espinoza-Castañeda et al. 2013). Lam et al. (2012) demonstrated the performance of an effective, integrated platform for the rapid detection of bacteria that combines a universal bacterial lysis approach and a sensitive nanostructured electrochemical biosensor. The lysis is rapid, is effective at releasing intercellular RNA from bacterial samples, and can be performed with a simple, cost-effective device integrated with an analysis chip (Lam et al. 2012). A sensitive and selective genomagnetic assay for the electrochemical detection of food pathogens based on in situ DNA amplification with magnetic primers has been reported (Lermo et al. 2007).

Electrospinning is a versatile and cost-effective method for the fabrication of biocompatible nanofibrous materials. The novel nanostructure significantly increases the surface area and mass transfer rate, which improves the biochemical binding effect and sensor signal to noise ratio. Luo et al. (2010) developed an electrospinning method of nitrocellulose nanofibrous membrane and its antibody functionalization for the application of bacterial and viral pathogen detection. The capillary action of the nanofibrous membrane is further enhanced using oxygen plasma treatment. An electrospun biosensor is designed based on capillary separation and conductometric immunoassay. The silver electrode is fabricated using a spray deposition method, which is noninvasive for the electrospun nanofibers. The surface functionalization and sensor assembly process retain the unique fiber morphology. The antibody attachment and pathogen-binding effect is verified using the confocal laser scanning microscope (CLSM) and scanning electronic microscope (SEM). The electrospun biosensor exhibits linear response to both microbial samples, *E. coli* and bovine viral diarrhea virus (BVDV). The detection time of the biosensor is 8 min, and the detection limit is 61 CFU/mL and 10^3 CCID/mL for bacterial and viral samples, respectively. With the advantage of efficient antibody functionalization, excellent capillary capability, and relatively low cost, the electrospinning process and surface functionalization method can be implemented to produce nanofibrous capture membranes for different immunodetection applications (Luo et al. 2010). McGraw et al. (2011) developed a biosensor based on the use of nonwoven fiber membranes coated with a conductive polymer coating and functionalized with antibodies for biological capture. The study examines three methods: passive adsorption, gluteraldehyde cross-linking, and EDC/suffo-NHS cross-linking, for the immobilization of antibodies onto the three-dimensional conductive fiber surfaces in order to improve the specific capture of a target pathogen (McGraw et al. 2011).

Functionalized graphene oxide (GO) sheets were coupled with a signal amplification method based on the nanomaterial-promoted reduction of silver ions for the sensitive and selective detection of bacteria. A linear relationship between the stripping response and the logarithm of the bacterial concentration was obtained using an electrochemical technique for concentrations ranging from 10^2 to 10^8 CFU/mL (Wan et al. 2011). The objective of rapid and sensitive detection of food-borne infections was successfully achieved by using nanoparticles in very low doses.

12.9 Electronic Noses

Electronic noses have recently emerged as an interesting tool in the area of chemical and microbiological analysis of foods. Electronic noses are odor mappers that can discriminate several volatile compounds according to the electronic response (e.g., voltage, resistance, conductivity) arising from the array of sensors. Indeed, some volatile compounds can originate from biochemical processes of food as a consequence of technological treatments or product storage. Unpleasant odor may include substances originating from the metabolism of spoilage microorganisms, bacteria and fungi, which may naturally or accidentally contaminate the products prior to or during their production. The characteristic flavor of volatile compounds, the so-called fingerprint, may provide information about the safety and specific characteristics of food, acting sometimes as an indicator of process fault as well. The use of electronic noses can provide a rapid and accurate means of sensing the incidence of food-contaminant bacteria with little or no sample preparation.

Needham et al. applied an electronic nose for the early detection and differentiation of both bacteria (*Bacillus subtilis*) and fungi (*Penicillium verrucosum* and *Pichia anomala*) spoilage of bread analogues (Needham et al. 2005). Panigrahi et al. analyzed the headspace from fresh beef strip loins kept at 4°C for 10 days using polymer-based sensors (Panigrahi et al. 2006). Balasubramanian et al. (2008) detected the changes that occurred in the headspace from stored beef strip loins inoculated with *S.* Typhimurium using an electronic nose. The electronic nose could detect bacterial concentration as low as <10^2 CFU/mL, and it was also able to classify bacterial contamination independently of the *Alicyclobacillus* species (Gobbi et al. 2010).

12.10 Concluding Remarks and Future Directions

Biosensor technology has the potential to speed up the detection, increase specificity and sensitivity, enable high-throughput analysis, and be used for the monitoring of critical control points in food production. This chapter discusses the recent developments in food pathogen detection methods based on electrochemical biosensors. The underlying principles and application of these biosensors are discussed with special emphasis on new biorecognition elements and technology. The efficient merging of nanotechnology with naturally existing sophisticated biomolecular systems has been explored for the development of ultrasensitive electrochemical sensors. Nanomaterials have been used to achieve successful loading bioreceptors on the electrode surface, to promote electrochemical reaction, to impose barcode for biomaterials, and to amplify the signal of a biorecognition event. The resulting electrochemical nanobiosensors have been applied in areas of detection of pathogenic organisms, food safety, as well as clinical and environmental applications. At the same time, there are also encouraging promises in the development of electrochemical systems leading to further progress in the manufacture of portable devices.

REFERENCES

Ahmad, F., D. M. Tourlousse, R. D. Stedtfeld, G. Seyrig, A. B. Herzog, P. Bhaduri, and S. A. Hashsham. 2009. Detection and occurrence of indicator organisms and pathogens. *Water Environment Research* 81(10): 959–980.

Arora, P., A. Sindhu, N. Dilbaghi, and A. Chaudhury. 2011. Biosensors as innovative tools for the detection of food borne pathogens. *Biosensors and Bioelectronics* 28(1): 1–12.

Bakker, E. and E. Pretsch. 2005. Potentiometric sensors for trace-level analysis. *Trends in Analytical Chemistry* 24(3): 199–207.

Balasubramanian, S., S. Panigrahi, C. M. Logue, C. Doetkott, M. Marchello, and J. S. Sherwood. 2008. Independent component analysis-processed electronic nose data for predicting *Salmonella* Typhimurium populations in contaminated beef. *Food Control* 19(3): 236–246.

Bhadoria, R. and H. S. Chaudhary. 2011. Recent advances of biosensors in biomedical sciences. *International Journal of Drug Delivery* 3(4): 571–585.

Bhunia, A. K. 2008. Biosensors and bio-based methods for the separation and detection of foodborne pathogens. *Advances in Food and Nutrition Research* 54: 1–44.

Byrne, B., E. Stack, N. Gilmartin, and R. O'Kennedy. 2009. Antibody-based sensors: Principles, problems and potential for detection of pathogens and associated toxins. *Sensors (Switzerland)* 9(6): 4407–4445.

Caygill, R. L., G. E. Blair, and P. A. Millner. 2010. A review on viral biosensors to detect human pathogens. *Analytica Chimica Acta* 681(1–2): 8–15.

Chamberlain, J. W., K. Maurer, J. Cooper, W. J. Lyon, D. L. Danley, and D. M. Ratner. 2012. Microelectrode array biosensor for studying carbohydrate-mediated interactions. *Biosensors and Bioelectronics* 34(1): 253–260.

Chang, Y.-C., C.-Y. Yang, R.-L. Sun, Y.-F. Cheng, W.-C. Kao, and P.-C. Yang. 2013. Rapid single cell detection of *Staphylococcus aureus* by aptamer-conjugated gold nanoparticles. *Scientific Reports* 3: 1863.

Chowdhury, A. D., A. De, C. R. Chaudhuri, K. Bandyopadhyay, and P. Sen. 2012. Label free polyaniline based impedimetric biosensor for detection of *E. coli* O157:H7 bacteria. *Sensors and Actuators, B: Chemical* 171–172: 916–923.

Clark, L. C. and C. Lyons. 1962. Electrode systems for continuous monitoring in cardiovascular surgery. *Annals of the New York Academy of Sciences* 102(1): 29–45.

De Micheli, G., C. Boero, C. Baj-Rossi, I. Taurino, and S. Carrara. 2012. Integrated biosensors for personalized medicine. In *Proceedings of the 49th Design Automation Conference (DAC)*. IEEE, New York, pp. 6–11.

De Oliveira Marques, P. R. B., A. Lermo, S. Campoy, H. Yamanaka, J. Barbé, S. Alegret, and M. I. Pividori. 2009. Double-tagging polymerase chain reaction with a thiolated primer and electrochemical genosensing based on gold nanocomposite sensor for food safety. *Analytical Chemistry* 81(4): 1332–1339.

Ellington, A. D. and J. W. Szostak. 1990. In vitro selection of RNA molecules that bind specific ligands. *Nature* 346(6287): 818–822.

Ercole, C., M. Del Gallo, L. Mosiello, S. Baccella, and A. Lepidi. 2003. *Escherichia coli* detection in vegetable food by a potentiometric biosensor. *Sensors and Actuators B: Chemical* 91(1–3): 163–168.

Erdem, A., M. Muti, H. Karadeniz, G. Congur, and E. Canavar. 2012. Electrochemical biosensors for screening of toxins and pathogens. In D. P. Nikolelis (ed.), *Portable Chemical Sensors: Weapons against Bioterrorism*, NATO Science for Peace and Security Series A: Chemistry and Biology. Springer, Dordrecht, the Netherlands, pp. 323–334.

Espinoza-Castañeda, M., A. de la Escosura-Muñiz, G. González-Ortiz, S. M. Martín-Orúe, J. F. Pérez, and A. Merkoçi. 2013. Casein modified gold nanoparticles for future theranostic applications. *Biosensors and Bioelectronics* 40(1): 271–276.

Freitas, M., S. Viswanathan, H. P. A. Nouws, M. B. P. P. Oliveira, and C. Delerue-Matos. 2014. Iron oxide/gold core/shell nanomagnetic probes and CdS biolabels for amplified electrochemical immunosensing of *Salmonella* Typhimurium. *Biosensors and Bioelectronics* 51(0): 195–200.

García, T., M. Revenga-Parra, L. Añorga, S. arana, F. Pariente, and E. Lorenzo. 2012. Disposable DNA biosensor based on thin-film gold electrodes for selective *Salmonella* detection. *Sensors and Actuators, B: Chemical* 161(1): 1030–1037.

Gobbi, E., M. Falasconi, I. Concina, G. Mantero, F. Bianchi, M. Mattarozzi, M. Musci, and G. Sberveglieri. 2010. Electronic nose and *Alicyclobacillus* spp. spoilage of fruit juices: An emerging diagnostic tool. *Food Control* 21(10): 1374–1382.

Hogan, H. 2008. Biosensors that protect and serve. *Biophotonics International* 15(12): 30–31.

Jiang, X., J. Wang, Y. Ying, and Y. Li. 2007. Recent advances in biosensors for food safety detection. *Nongye Gongcheng Xuebao/Transactions of the Chinese Society of Agricultural Engineering* 23(5): 272–277.

K'Owino, I. O., S. K. Mwilu, and O. A. Sadik. 2007. Metal-enhanced biosensor for genetic mismatch detection. *Analytical Biochemistry* 369(1): 8–17.

LaGier, M. J., J. W. Fell, and K. D. Goodwin. 2007. Electrochemical detection of harmful algae and other microbial contaminants in coastal waters using hand-held biosensors. *Marine Pollution Bulletin* 54(6): 757–770.

Lam, B., J. Das, R. D. Holmes, L. Live, A. Sage, E. H. Sargent, and S. O. Kelley. 2013. Solution-based circuits enable rapid and multiplexed pathogen detection. *Nature Communications* 4: 2001.

Lam, B., Z. Fang, E. H. Sargent, and S. O. Kelley. 2012. Polymerase chain reaction-free, sample-to-answer bacterial detection in 30 minutes with integrated cell lysis. *Analytical Chemistry* 84(1): 21–25.

Lermo, A., S. Campoy, J. Barbé, S. Hernández, S. Alegret, and M. I. Pividori. 2007. In situ DNA amplification with magnetic primers for the electrochemical detection of food pathogens. *Biosensors and Bioelectronics* 22(9–10): 2010–2017.

Liao, J. C., M. Mastali, Y. Li, V. Gau, M. A. Suchard, J. Babbitt, J. Gornbein et al. 2007. Development of an advanced electrochemical DNA biosensor for bacterial pathogen detection. *Journal of Molecular Diagnostics* 9(2): 158–168.

Lu, L., G. Chee, K. Yamada, and S. Jun. 2013. Electrochemical impedance spectroscopic technique with a functionalized microwire sensor for rapid detection of foodborne pathogens. *Biosensors and Bioelectronics* 42(1): 492–495.

Lu, L. and S. Jun. 2012. Evaluation of a microwire sensor functionalized to detect *Escherichia coli* bacterial cells. *Biosensors and Bioelectronics* 36(1): 257–261.

Lubin, A. A. and K. W. Plaxco. 2010. Folding-based electrochemical biosensors: The case for responsive nucleic acid architectures. *Accounts of Chemical Research* 43(4): 496–505.

Luo, Y., S. Nartker, H. Miller, D. Hochhalter, M. Wiederoder, S. Wiederoder, E. Setterington, L. T. Drzal, and E. C. Alocilja. 2010. Surface functionalization of electrospun nanofibers for detecting *E. coli* O157:H7 and BVDV cells in a direct-charge transfer biosensor. *Biosensors and Bioelectronics* 26(4): 1612–1617.

Maurer, E. I., K. K. Comfort, S. M. Hussain, J. J. Schlager, and S. M. Mukhopadhyay. 2012. Novel platform development using an assembly of carbon nanotube, nanogold and immobilized RNA capture element towards rapid, selective sensing of bacteria. *Sensors (Switzerland)* 12(6): 8135–8144.

McGraw, S. K., E. C. Alocilja, K. J. Senecal, and A. G. Senecal. 2011. Antibody immobilization on conductive polymer coated nonwoven fibers for biosensors. In *NANOTECH*, Vol. 3. CRC Press, Boston, MA, pp. 56–59.

Micheli, L., R. Grecco, M. Badea, D. Moscone, and G. Palleschi. 2005. An electrochemical immunosensor for aflatoxin M1 determination in milk using screen-printed electrodes. *Biosensors and Bioelectronics* 21(4): 588–596.

Muhammad-Tahir, Z. and E. C. Alocilja. 2004. A disposable biosensor for pathogen detection in fresh produce samples. *Biosystems Engineering* 88(2): 145–151.

Muñoz-Berbel, X., N. Vigués, A. T. A. Jenkins, J. Mas, and F. J. Muñoz. 2008. Impedimetric approach for quantifying low bacteria concentrations based on the changes produced in the electrode–solution interface during the pre-attachment stage. *Biosensors and Bioelectronics* 23(10): 1540–1546.

Nandakumar, V., J. T. LaBelle, and T. L. Alford. 2008. Signal processing for rapid bacterial detection. In *42nd Asilomar Conference on Signals, Systems and Computers*, Pacific Grove, CA, October 26–29, 2008, pp. 1979–1981.

Needham, R., J. Williams, N. Beales, P. Voysey, and N. Magan. 2005. Early detection and differentiation of spoilage of bakery products. *Sensors and Actuators B: Chemical* 106(1): 20–23.

Oczkowski, T., E. Zwierkowska, and S. Bartkowiak. 2007. Application of cell-based biosensors for the detection of bacterial elicitor flagellin. *Bioelectrochemistry* 70(1): 192–197.

Pal, S., W. Ying, E. C. Alocilja, and F. P. Downes. 2008. Sensitivity and specificity performance of a direct-charge transfer biosensor for detecting *Bacillus cereus* in selected food matrices. *Biosystems Engineering* 99(4): 461–468.

Palchetti, I. and M. Mascini. 2008. Electroanalytical biosensors and their potential for food pathogen and toxin detection. *Analytical and Bioanalytical Chemistry* 391(2): 455–471.

Pandey, C. M., R. Singh, G. Sumana, M. K. Pandey, and B. D. Malhotra. 2011a. Electrochemical genosensor based on modified octadecanethiol self-assembled monolayer for *Escherichia coli* detection. *Sensors and Actuators, B: Chemical* 151(2): 333–340.

Pandey, C. M., G. Sumana, and B. D. Malhotra. 2011b. Microstructured cystine dendrites-based impedimetric sensor for nucleic acid detection. *Biomacromolecules* 12(8): 2925–2932.

Panigrahi, S., S. Balasubramanian, H. Gu, C. Logue, and M. Marchello. 2006. Neural-network-integrated electronic nose system for identification of spoiled beef. *LWT—Food Science and Technology* 39(2): 135–145.

Pearson, J. E., A. Gill, and P. Vadgama. 2000. Analytical aspects of biosensors. *Annals of Clinical Biochemistry* 37(Pt 2): 119–145.

Pividori, M. I., A. Lermo, S. Hemández, J. Barbé, S. Alegret, and S. Campoy. 2006. Rapid electrochemical DNA biosensing strategy for the detection of food pathogens based on enzyme-DNA-magnetic bead conjugate. *Afinidad* 63(521): 13–18.

Pohanka, M. 2009. Monoclonal and polyclonal antibodies production—Preparation of potent biorecognition element. *Journal of Applied Biomedicine* 7(3): 115–121.

Pournaras, A. V., T. Koraki, and M. I. Prodromidis. 2008. Development of an impedimetric immunosensor based on electropolymerized polytyramine films for the direct detection of *Salmonella* Typhimurium in pure cultures of type strains and inoculated real samples. *Analytica Chimica Acta* 624(2): 301–307.

Prasad, D. and A. S. Vidyarthi. 2009. DNA based methods used for characterization and detection of food borne bacterial pathogens with special consideration to recent rapid methods. *African Journal of Biotechnology* 8(9): 1768–1775.

Qi, P., Y. Wan, and D. Zhang. 2013. Impedimetric biosensor based on cell-mediated bioimprinted films for bacterial detection. *Biosensors and Bioelectronics* 39(1): 282–288.

Rasooly, A. and K. E. Herold. 2006. Biosensors for the analysis of food- and waterborne pathogens and their toxins. *Journal of AOAC International* 89(3): 873–883.

Sadik, O. A., A. O. Aluoch, and A. Zhou. 2009. Status of biomolecular recognition using electrochemical techniques. *Biosensors and Bioelectronics* 24(9): 2749–2765.

Sanvicens, N., C. Pastells, N. Pascual, and M. P. Marco. 2009. Nanoparticle-based biosensors for detection of pathogenic bacteria. *Trends in Analytical Chemistry* 28(11): 1243–1252.

Setterington, E. B. and E. C. Alocilja. 2011. Rapid electrochemical detection of polyaniline-labeled *Escherichia coli* O157:H7. *Biosensors and Bioelectronics* 26(5): 2208–2214.

Setterington, E. B. and E. C. Alocilja. 2012. Electrochemical biosensor for rapid and sensitive detection of magnetically extracted bacterial pathogens. *Biosensors* 2(1): 15–31.

Shinde, S. B., C. B. Fernandes, and V. B. Patravale. 2012. Recent trends in in-vitro nanodiagnostics for detection of pathogens. *Journal of Controlled Release* 159(2): 164–180.

Siddiqui, S., Z. Dai, C. J. Stavis, H. Zeng, N. Moldovan, R. J. Hamers, J. A. Carlisle, and P. U. Arumugam. 2012. A quantitative study of detection mechanism of a label-free impedance biosensor using ultrananocrystalline diamond microelectrode array. *Biosensors and Bioelectronics* 35(1): 284–290.

Singh, A., S. Poshtiban, and S. Evoy. 2013. Recent advances in bacteriophage based biosensors for food-borne pathogen detection. *Sensors* 13(2): 1763–1786.

Sun, J. Z., C. Bian, L. Qu, and S. H. Xia. 2009. Immobilizing antibody with electropolymerized staphylococcal protein A in micro amperometric immunosensor for detecting *Salmonella* Typhimurium. *Fenxi Huaxue/ Chinese Journal of Analytical Chemistry* 37(4): 484–488.

Tichoniuk, M., D. Gwiazdowska, M. Ligaj, and M. Filipiak. 2010. Electrochemical detection of foodborne pathogen *Aeromonas hydrophila* by DNA hybridization biosensor. *Biosensors and Bioelectronics* 26(4): 1618–1623.

Travas-Sejdic, J., H. Peng, H. H. Yu, and S. C. Luo. 2011. DNA detection using functionalized conducting polymers. *Methods in Molecular Biology (Clifton, N.J.)* 751: 437–452.

Tully, E., S. P. Higson, and R. O'Kennedy. 2008. The development of a 'labeless' immunosensor for the detection of *Listeria monocytogenes* cell surface protein, Internalin B. *Biosensors and Bioelectronics* 23(6): 906–912.

Upadhyayula, V. K. K. 2012. Functionalized gold nanoparticle supported sensory mechanisms applied in detection of chemical and biological threat agents: A review. *Analytica Chimica Acta* 715: 1–18.

Urkut, Z., P. Kara, Y. Goksungur, and M. Ozsoz. 2011. Response surface methodology for optimization of food borne pathogen detection in real samples based on label free electrochemical nucleic acid biosensors. *Electroanalysis* 23(11): 2668–2676.

Velusamy, V., K. Arshak, O. Korostynska, K. Oliwa, and C. Adley. 2010. An overview of foodborne pathogen detection: In the perspective of biosensors. *Biotechnology Advances* 28(2): 232–254.

Viswanathan, S. and J. Radecki. 2008. Nanomaterials in electrochemical biosensors for food analysis: A review. *Polish Journal of Food and Nutrition Sciences* 58(2): 157–164.

Viswanathan, S., C. Rani, and J.-a. A. Ho. 2012. Electrochemical immunosensor for multiplexed detection of food-borne pathogens using nanocrystal bioconjugates and MWCNT screen-printed electrode. *Talanta* 94: 315–319.

Viswanathan, S., L.-C. Wu, M.-R. Huang, and J.-A. Ho. 2006. Electrochemical immunosensor for cholera toxin using liposomes and poly(3,4-ethylenedioxythiophene)-coated carbon nanotubes. *Analytical Chemistry* 78(4): 1115–1121.

Wan, Y., Y. Wang, J. Wu, and D. Zhang. 2011. Graphene oxide sheet-mediated silver enhancement for application to electrochemical biosensors. *Analytical Chemistry* 83(3): 648–653.

Xu, X. and Y. Ying. 2011. Microbial biosensors for environmental monitoring and food analysis. *Food Reviews International* 27(3): 300–329.

Yang, L. and R. Bashir. 2008. Electrical/electrochemical impedance for rapid detection of foodborne pathogenic bacteria. *Biotechnology Advances* 26(2): 135–150.

Yang, L., Y. Li, C. L. Griffis, and M. G. Johnson. 2004. Interdigitated microelectrode (IME) impedance sensor for the detection of viable *Salmonella* Typhimurium. *Biosensors and Bioelectronics* 19(10): 1139–1147.

Zelada-Guillén, G. A., J. Riu, A. Düzgün, and F. X. Rius. 2009. Immediate detection of living bacteria at ultralow concentrations using a carbon nanotube based potentiometric aptasensor. *Angewandte Chemie International Edition* 48(40): 7334–7337.

Zhao, G., F. Xing, and S. Deng. 2007. A disposable amperometric enzyme immunosensor for rapid detection of *Vibrio parahaemolyticus* in food based on agarose/Nano-Au membrane and screen-printed electrode. *Electrochemistry Communications* 9(6): 1263–1268.

13

Novel Techniques for Preventing Bacterial Attachment to Foods and Food-Processing Surfaces

Yi Wang and Gary A. Dykes

CONTENTS

13.1 Introduction

Bacteria must attach to surfaces in order to move through the food chain and cause human disease or food spoilage. These surfaces may be abiotic, such as those associated with processing equipment, or biotic, such as those associated with food itself. Preventing bacterial attachment and removing already attached bacteria are important aspects of good hygiene and cleaning practice. The current practice to achieve this generally entails the use of synthetic surfactants. This approach is not always effective and suffers from the drawback of creating a potentially negative impact on the environment. For this reason, a number of alternative methods for controlling bacterial attachment are being investigated.

Bacterial attachment to surfaces is a complex process that is dependent on both the bacterial strain and the surface, and may be mediated by specific and/or nonspecific interactions. Specific interactions generally entail adhesin–receptor binding and often occur on biotic surfaces (i.e., foods). Nonspecific interactions are usually physicochemical in nature and tend to occur on abiotic surfaces (i.e., food-contacting surfaces). Novel approaches for controlling bacterial attachment generally aim to interfere with these interactions. For example, immunological techniques may be used to block specific interactions between bacterial cells and substratum surfaces, and antifouling paints may be used to modify the physicochemical properties of substratum surfaces and to reduce bacterial attachment. In this chapter, several novel techniques for preventing bacterial attachment are presented and discussed.

13.2 Blocking Specific Interactions

Specific interactions between bacterial cells and substratum surfaces are usually mediated by adhesin–receptor pairs. All naturally occurring bacteria are able to express adhesins on their surfaces. These may be membrane-bound outer membrane proteins, polysaccharides and cell surface appendages, or metabolites that are loosely bound to cell surfaces, such as biosurfactants and extracellular polymeric substances (EPS). Adhesins allow bacteria to specifically recognize and bind to a diverse spectrum of molecular motifs (i.e., receptors) on target surfaces. The binding of each adhesin is exclusive to a particular receptor in a lock-and-key fashion. For this reason, the expression of a given adhesin functions as an address indicator for the microbe and blocking this type of specific interactions may be an effective approach to control bacterial attachment. Techniques for preventing bacterial attachment by blocking the specific interactions are diverse, and several representative examples are described here and summarized in Table 13.1.

13.2.1 Competitive Blocking of Adhesin–Receptor Interactions Using Receptor Analogs

Exogenous receptor analogs that have the same binding preference as receptors can be used to competitively bind to and occupy the adhesin sites on the surface of bacterial cells and in turn reduce the likelihood of the bacteria attaching to a surface (Figure 13.1). Receptor analogs as antiattachment agents are used primarily against pathogenic bacteria that bind to glycoconjugates on animal cells using carbohydrate-specific attachment mechanisms (i.e., lectins). In these cases, the receptor analogs are usually saccharides with similar structures as the glycoprotein or glycolipid receptors. The use of mannose and its derivatives to competitively prevent *Escherichia coli* from attaching to the membrane of mammalian cells represents the first example of this approach (Ofek et al. 1977). Since this study was carried out, the sugar specificity of many bacteria has been determined and strategies for preventing the attachment of pathogens to mammalian tissues using receptor-like carbohydrates have been developed.

TABLE 13.1

Techniques of Preventing Bacterial Attachment by Blocking Specific Interactions

Techniques	Notes	References
Adhesin/receptor analogs	Not readily available, required in high concentrations, examinations for toxicity are needed	Goldhar et al. (1986), Kelly et al. (1999), Mysore et al. (1999)
Milk components	Readily available, cheap, nontoxic	Bitzan et al. (1998), Ofek et al. (2003), Ruiz-Palacios et al. (2003)
Immunization	Expensive, time-consuming, limited by sequence variation of adhesins	Moon and Bunn (1993)
Drugs for preventing biosynthesis of bacterial surface appendages	Not readily available, examinations for toxicity are needed	Pinkner et al. (2006), Cegelski et al. (2009)

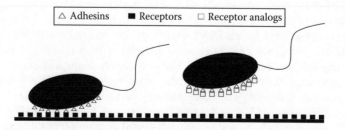

FIGURE 13.1   Schematic representation of the inhibition of bacterial attachment by receptor analogs.

Examples of clinical trials applying these techniques include the use of mannose and globotetraose to target type 1 fimbriae and P fimbriae of *E. coli* and prevent their attachment to the gastrointestinal tract and urinary tract of mammals (Svanborg-Edén et al. 1982; Goldhar et al. 1986), and the use of sialyl-3′-lactose to prevent the attachment of *Helicobacter pylori* to cultured human gastric cells and animal gastric tracts (Mysore et al. 1999). While these examples are in the medical sphere, these techniques have great potential for applications in the food industry and especially in meat and fresh produce processing. A limitation of the previous studies is that they do not provide information on the longevity of the attachment inhibitory ability of receptor analogs, which must be determined before putting them into use as antiattachment agents. Another drawback is that the concentrations of the carbohydrate receptor analogs required to give a significant inhibition of bacterial attachment are usually high (in the millimolar range) due to the low affinity of these saccharides for bacterial adhesins. A solution to this is to use multivalent adhesin inhibitors (i.e., attaching many copies of a saccharide to a suitable carrier). For example, a multiantennary α-mannosyl glycocluster was used as a vehicle for mannose to inhibit the attachment of type 1 fimbriated *E. coli* to erythrocytes (Lindhort et al. 1998).

An alternative approach to that described earlier is to use adhesin analogs that competitively bind to substratum surface receptors and act as inhibitors for bacterial attachment. An example of this is the use of synthetic peptides mimicking the sequence of a cell surface adhesin of *Streptococcus mutans* which mediates the attachment of the bacteria to a salivary protein on dental surfaces (Kelly et al. 1999). It is impractical, however, to use adhesin analogs to prevent bacterial attachment in the food industry because they are usually macromolecules that are not readily available, and again require relatively higher concentrations to provide a significant inhibitory effect. In addition, careful consideration must be given to their toxicity and immunogenicity.

13.2.2 Dairy Products: A Rich Source of Antiattachment Agents

Due to the safety concern raised by the potential application of adhesin/receptor analogs to prevent bacterial attachment in the food industry, prospecting for antiattachment agents in foodstuffs themselves has emerged as an approach to overcome this problem. A number of efficient antiattachment agents have been identified in milk products. Components of milk, such as proteins, saccharides, and fat globules, that can bind to bacterial adhesins and in turn block specific interactions have been identified. For example, sialylated glycoproteins in milk can bind to S fimbriated *E. coli* and *Mycoplasma pneumonia* (Ofek et al. 2003). Fucosylated saccharides can interact with surface adhesins of enteropathogenic *E. coli* (Cravioto et al. 1991) and *Campylobacter jejuni* (Ruiz-Palacios et al. 2003), prevent their attachment to different biotic surfaces, and in turn reduce the risk of infection. In addition, fat globules in milk can prevent the attachment of a range of bacteria by binding them because most of the fat globules carry glycoconjugates specific for bacterial adhesins. For example, L-lactoglobulin was found to bind *Listeria monocytogenes* cells (al-Makhlafi et al. 1994), bovine colostrum lipids can prevent the attachment of *H. pylori* and *Helicobacter mustelae* to immobilized glycolipids by binding the bacteria (Bitzan et al. 1998), and sialylated fat globules inhibit the attachment of S fimbriated *E. coli* to buccal epithelial cells (Schröten et al. 1992). In addition, milk glycoproteins can reconstitute the membrane of some biotic surfaces via their protein moieties. For example, lactoferrin, a common constituent of milk, has been shown to inhibit the attachment of *Actinobacillus actinomycetemcomitans*, *Prevotella intermedia*, and *Prevotella nigrescens* to fibroblast monolayers by reconstituting basement membranes (Alugupalli and Kalfas 1995, 1997). The benefits of using milk components as antiattachment agents are clear in that they are cheap, readily available, and toxicity is not an issue. While it may be possible to find suitable inhibitors for the attachment of particular pathogens in milk, the issue of bacterial strain specificity remains.

13.2.3 Antiattachment Vaccines

An appealing approach to block adhesin–receptor interaction is to use immunization with bacterial adhesins. Applying this technique to farm animals would invoke the production of adhesin-specific antibodies in the host and prevent bacterial attachment and subsequently infection. This in turn would prevent these pathogens from entering the food system. Blocking specific interactions using adhesin-based vaccines can be achieved either actively or passively. Active antiadhesin immunity prevents bacterial attachment by the stimulation of secretory IgA, but significant amounts of IgG will also be produced, which compromise the inhibitory effect of IgA on attachment. Passive antiadhesin immunity entails treatment of the host with an antiadhesin vaccine generated in another host. Target adhesins for vaccination are usually outer membrane proteins and fimbriae of bacteria. A good example of this approach is the use of the K88 fimbriae and related adhesins of enterotoxigenic *E. coli* to vaccinate farm animals and demonstrate that the antibodies that function to prevent bacterial attachment and infection are also produced in their milk (Moon and Bunn 1993). This technique can be used to improve safety and quality of dairy products and to reduce microbiological hazards of meat products by preventing bacterial attachment and infection in suckling calves.

A significant drawback of this technique is that the immune response is primarily directed to the major structure of the adhesin proteins, which often fails to result in effective antibody production due to substantial amino acid sequence variations. The applicability of the immunization technique is therefore largely limited. In addition, the safety of using immunological techniques has yet to be investigated.

13.2.4 Preventing the Biosynthesis of Adhesins

Another way of controlling specific interactions between bacteria and surfaces is to eliminate the biosynthesis of adhesins such as fimbriae. The biosynthesis of fimbriae is a complex process that involves the assembly of large heteropolymeric organelles via a chaperone usher–dependent pathway. Periplasmic chaperones escort structural components of an organelle to an usher located at the outer membrane where the actual synthesis of the organelle takes place. Disruption of the formation of fimbrial organelles at the assembly level in the chaperone usher pathway can effectively eliminate the biosynthesis of fimbriae. An example of this approach is the use of a family of synthetic chemicals, bicyclic 2-pyridones (also known as pilicides), which can chemically interfere in the chaperone usher pathway during the biosynthesis of type 1 and P fimbriae by *E. coli*. These chemicals reduce fimbriae-mediated attachment and biofilm formation in the strains tested by 90% (Pinkner et al. 2006). Recently, derivatives of bicyclic 2-pyridines were shown to inhibit the assembly of curli of *E. coli* strains and were therefore termed curlicides (Cegelski et al. 2009). Pillicides and curlicides target the key adhesins on bacterial surfaces responsible for their attachment to different surfaces and represent a new approach for the development of potential novel antiattachment agents. These compounds are, however, not readily available and need to be examined for toxicity before being applied in the food industry.

Taken together, strategies to inhibit bacterial attachment by blocking specific interactions seem to be promising and worth further development as they specifically target and inhibit the most direct virulence factor. The major drawback of these techniques also includes their specificity, which limits their application, as different bacteria carry different adhesins and one strain or species of bacteria may carry an array of different adhesins. Using combinations of different adhesin inhibitors and/or combinations of different techniques is therefore necessary to overcome this limitation. In addition, many of these techniques/adhesin inhibitors can be costly. For example, synthesizing receptor analogs or inhibitors for the biosynthesis of cell surface appendages and implementing the immunization technique are expensive and time-consuming. Furthermore, bacterial attachment is also mediated by factors other than adhesin–receptor interactions, such as physicochemical interactions. Therefore, preventing bacterial attachment by only blocking specific interactions is unlikely to be sufficient in most cases. More importantly, safety and longevity of some of the techniques need to be carefully determined before using them in the food industries.

13.3 Modification of Substratum Surfaces

While specific interaction-blocking techniques have a role to play in inhibiting bacterial attachment in the food system, the development strategies for preventing bacterial attachment in a more general fashion by altering nonspecific interactions is important. Modifying the physicochemical properties of substratum surfaces, and in turn interfering in the physicochemical interactions between bacterial cells and sub-stratum surfaces, is a promising approach to controlling bacteria in the food system. Physicochemical interactions between bacteria and surfaces include, but are not limited to, hydrophobic interactions (consisting of Lifshitz van der Waals interactions and hydrogen bonding), electrostatic interactions, steric interactions, and surface roughness/topology interactions. Altering these interactions so that they are unfavorable for bacterial attachment (i.e., turning attractive forces into repulsive forces) can prevent bacterial cells from approaching the substratum surface even before specific interactions can take place. The techniques described in this section are summarized in Table 13.2.

13.3.1 Antiattachment Coatings

Bacterial attachment, in most cases, happens in aqueous environments in which bacterial cells sense surface tension and will seek a less hydrophilic surface/interface to attach to in order to reduce the surface-free energy needed to maintain themselves. For this reason, hydrophobic interactions are generally a more important factor affecting bacterial attachment and many surface modification techniques for controlling attachment focus on reducing the hydrophobicity of substratum surfaces and in turn decreasing the preference of bacterial cells to attach to them. For example, poly(ethylene glycol) has been used as a coating material to prevent bacterial attachment to polyurethane surfaces (Park et al. 1998). The material can reduce the attachment of *Staphylococcus epidermidis* and *E. coli* by up to 2 log CFU/cm^2 by decreasing the hydrophobicity of polyurethane surfaces. Kregiel et al. (2013) reported that modifying polyvinyl chloride and silicone elastomer surfaces with reactive organo-silanes reduces the attachment of *Aeromonas hydrophila* by up to 2 log CFU/cm^2 by causing a change in surface tension energies. Roosjen et al. (2003) covalently coated glass surfaces with poly(ethylene oxide) brushes and demonstrated an ~98% reduction in the attachment of four foodborne pathogens (*S. epidermidis*, *Staphylococcus aureus*, *Streptococcus salivarius*, and *E. coli*). Due to the high mobility and extremely large exclusion volumes of poly(ethylene oxide), it is difficult for small particles (such as bacteria) to approach and establish themselves on the surface. A reduced Lifshitz van der Waals interaction energy between the bacterial cells and the substratum surface was noted after coating with poly(ethylene oxide) and this may also have contributed to the prevention of bacterial attachment.

The coating materials/techniques mentioned earlier are not universally useful as bacteria/surfaces are not mediated by nonspecific physicochemical interactions in all cases. For example, *Pseudomonas aeruginosa* can release surfactants that form hydrogen bonds with poly(ethylene oxide). This makes the ability of the poly(ethylene oxide) brushes to inhibit the attachment of *P. aeruginosa* weaker (~60% reduction) as compared to its effect on other microorganisms (Roosjen et al. 2003). The effects of surface coatings can therefore be compromised when there is a strong specific interaction between bacterial cells and the coating material.

While the surface-coating technique has been demonstrated to effectively prevent bacterial attachment in many cases, the stability of coating materials has raised concern. Koziarz and Yamazaki (1998) showed that six foodborne pathogens (*E. coli*, *Brevibacterium ammoniagens*, *P. aeruginosa*, *S. aureus*, *Salmonella* Typhimurium, and *Bacillus megaterium*) attach more efficiently and in higher numbers to hydrophobic polyester cloth than to hydrophilic cotton cloth, and polyvinyl alcohol coatings that reduce the hydrophobicity of polyester cloth can suppress the attachment of these bacteria by 1–2 log CFU/cm^2. However, polyvinyl alcohol coatings are not stable in the presence of detergents at a high temperature (e.g., autoclaving with 1% sodium dodecyl sulfate, which is a standard disinfection procedure) and therefore the value of this coating technique is greatly reduced. Nevertheless, solutions to these problems are available; for example, Koziarz and Yamazaki (1999) found that 1% acidified glutaraldehyde can stabilize polyvinyl alcohol coatings in the presence of detergents at a high temperature and does not affect its inhibitory effects on bacterial attachment.

Solutions of this type, however, often raise other concerns such as the toxicity of compounds (e.g., gluteraldehyde) that are used to overcome coating issues. In addition, testing the toxicity of coating materials can be time-consuming and expensive. For this reason, the use of food components as coating materials may represent a viable option. The use of a fish muscle protein, α-tropomyosin, as a surface coating material was found to significantly reduce attachment and biofilm formation by a wide range of

TABLE 13.2

Surface Modification Approaches to Prevent Bacterial Attachment

Techniques	Target Bacteria	Target Surfaces	Mechanisms	Notes	References
Poly(ethylene glycol) coating	*Staphylococcus epidermidis* and *Escherichia coli*	Polyurethane	Reducing hydrophobicity	—	Park et al. (1998)
Reactive organo-silane coating	*Aeromonas hydrophila*	Polyvinyl chloride and silicone elastomer	Changing surface tension energies	—	Kregiel et al. (2013)
Poly(ethylene oxide) brush coating	A range of foodborne pathogens	Glass	High mobility and large exclusion volumes, reducing Lifshitz van der Waals interaction energies	Less effective on attachment of *Pseudomonas aeruginosa*	Roosjen et al. (2003)
Polyvinyl alcohol coating	A range of foodborne pathogens	Polyester cloth	Reducing hydrophobicity	The coating material is not stable in the presence of detergents at a high temperature	Koziarz and Yamazaki (1998)
α-Tropomyosin	A range of foodborne pathogens	A range of abiotic surfaces	Reducing hydrophobicity, increasing negative charge	Fish muscle protein	Vejborg and Klemm (2008), Vejborg et al. (2008)
Plasma-assisted poly(ethylenediamine) coating	*Enterobacter sakazakii*	Stainless steel	Reducing hydrophobicity and surface tension, changing surface roughness/topology, a steric barrier provided by surface amino groups	Radio frequency plasma polymerization coating technique	Şen et al. (2009)
Argon plasma treatment	*S. epidermidis*	Polyethylene	Reducing hydrophobicity	—	Wang et al. (1995)
Nanosilver	*P. aeruginosa*	Poly(vinyl chloride)	Bactericidal and binding to proteins	—	Balazs et al. (2004)
Nanosilver	*E. coli* and *Staphylococcus aureus*	Facemask	Bactericidal and binding to proteins	—	Li et al. (2006)
Nano-TiO₂	*E. coli*	Reverse osmosis thin film-composite membrane	Production of active oxygen species that destroy bacterial outer membrane and inactivate surface proteins	Photocatalysis	Kim et al. (2003)
Electro-assisted method	Marine bacterial and *S. epidermidis*	Metal surfaces	Modification of the electrostatic properties	Remove attached bacteria cells and biofilms	van der Borden et al. (2004a,b), Wake et al. (2006)

foodborne pathogens on many different surfaces including glass, polystyrene, vinyl plastics, and stainless steel by up to 100-fold (Vejborg and Klemm 2008; Vejborg et al. 2008). α-Tropomyosin is a highly negatively charged and hydrophilic protein and, interestingly, as a biotic coating material, its ability to prevent bacterial attachment is purely physicochemical and not adhesin-specific (Vejborg and Klemm 2008). This nontoxic, cheap, and readily available coating material seems to have good potential for preventing bacterial attachment to hard surfaces in the food industry. Due to the inherent instability of physically absorbed protein coatings, α-tropomyosin is only appropriate for short-term use. A new technique that covalently binds α-tropomyosin to hard surfaces and extends the lifetime of the coating is currently under development (Klemm et al. 2010).

As the life time of surface coatings has raised concern, novel techniques such as plasma-assisted surface modification have been developed to overcome the challenge. Plasma is defined as a partially or wholly ionized gas with approximately equal numbers of positively and negatively charged particles. Plasmas coat materials on surfaces under high temperature conditions and result in a uniform spatial distribution and strong adhesion to the substrate. In addition, plasma coatings can be used to manipulate surface morphology, structure, chemical composition, and physicochemical properties of the materials for special needs. For example, Şen et al. (2009) used radio frequency plasma polymerization technique to polymerize monomeric ethylenediamine and coat the resultant polymer poly(ethylenediamine) on food-contacting surfaces and demonstrated a reduction in bacterial attachment. The attachment of *Enterobacter sakazakii* to stainless steel was used as a model system and a 99.74% reduction in attachment was observed upon coating the surface with poly(ethylenediamine). This reduction in attachment was reported to be associated with a reduction in substratum surface hydrophobicity and surface tension, a change in surface roughness and topology (from relatively rough to relatively flat), and exposure of amino groups on the surface that provided a steric barrier and repelled bacterial cells. Wang et al. (1995) demonstrated that argon plasma treatment reduced the attachment of *S. epidermidis* to polyethylene by reducing its surface hydrophobicity. In addition, exposure to plasmas has been shown to irreversibly damage bacterial cells, allowing *in situ* sterilization of substratum surfaces during the surface modification process. For instance, plasmas have been demonstrated to be effective against *E. coli*, *S. aureus*, *P. aeruginosa*, *Bacillus cereus*, *Bacillus subtilis*, and *Geobacillus stearothermophilus* (Roth et al. 2010; Sohbatzadeh et al. 2010; Sureshkumar et al. 2010; Zaaba et al. 2010). A significant drawback of this technique is that a high temperature (4,000–20,000 K) is required to initiate equilibrium plasmas, which is not appropriate for most of the substratum surfaces and the polymer coating materials.

In recent years, surface-grafted stimuli-responsive polymers, such as poly(*N*-isopropylacrylamide), have captured significant attention due to their ability to change physicochemical characteristics upon induction of environmentally triggered phase changes (Cordeiro et al. 2009). These materials are able to control biomolecular adsorption, bacterial cell attachment and release, and even cell functions, such as EPS production, by the attached cells. For example, attached cells can be released from the surface due to changes in the anchorage strength of cells caused by physicochemical changes in the surface upon induction of environmentally triggered phase changes (Ista et al. 2010).

Antiattachment coating techniques have a high potential to protect food-contact surfaces from bacterial attachment but are not extensively applied in the food industry. This could be because of the instability and toxicity of the coating materials. Most of the materials are polymers, which can be degraded by high temperatures, solvents, and high pressures and they may also release toxic monomers (e.g., polyethylene derivatives).

13.3.2 Application of Nanoparticles

Another approach to control biofouling is the application of nanoparticles. Nanoparticles are insoluble particles with a size smaller than 100 nm. Nanoparticles of antiattachment agents have a significantly larger surface-to-volume ratio and provide a more efficient means of preventing bacterial attachment and biofilm formation than many other methods. Nanoparticles are usually incorporated into polymeric substratum surfaces using three techniques: (1) nanoparticles are mixed with the polymer or are distributed in the matrix of the host polymer; (2) nanoparticles are produced during polymerization, and this technique is usually used for metal nanoparticles that are produced from metal salts by electronically active polymers during polymerization; and (3) dispersion of the nanoparticles in the monomer, in which case the polymerization is initiated in the presence of nanoparticles that are simultaneously trapped in the polymer network.

Nanoparticles are often made of metals that are cytotoxic to bacteria. Silver, for example, is a good candidate that is being extensively used in engineering systems. Hard surfaces with nanosilver incorporated into them can release silver particles or silver ions that can kill a broad range of bacteria and prevent them from attaching to the surface. Silver is also known to have a high affinity for proteins, which can block the specific interactions between bacterial cells and hard surfaces mediated by protein-based adhesin–receptor pairs. Balazs et al. (2004) found that silver completely inhibited the attachment of *P. aeruginosa* to poly(vinyl chloride) and efficiently prevented its colonization over 72 h. Li et al. (2006) reported that a facemask coated with silver had a 100% reduction in the attachment of viable *E. coli* and *S. aureus* after 48 h. In addition, silver is also effective in removing biofilms from surfaces. Biofilms create an environment that enhances antimicrobial resistance due to the production of EPS. The concentrations of antibiotics used to kill biofilm cells are usually 10–100 times higher than those used to kill planktonic cells. However, biofilms can absorb minerals and metals from the liquid phase that they are in contact with. In particular, EPS produced by gram-negative bacteria play an important role in metal biosorption. This makes silver effective in removing biofilms, especially when it is in nanoparticles that have a smaller size to penetrate biofilms. However, using nanosilver in the food industry is not appropriate in most cases due to its toxicity to humans.

Due to the toxicity of silver that makes it unsuitable for food manufacture, titanium dioxide (TiO_2) would be a better choice for surface nanoparticles and is widely used for food packaging. TiO_2 has photocatalytic effects that decompose organic chemicals and kill bacteria. Its photocatalysis generates various active oxygen species such as hydroxyl radicals and hydrogen peroxide by redox reactions. These active oxygen species can inactivate cell viability by destroying the outer membrane of bacterial cells and can inactivate functional proteins on cell surfaces, thereby preventing bacterial attachment. For example, an *E. coli* biofilm on a reverse osmosis thin film–composite membrane causes flux decline in industrial operations. Nano-TiO_2 was reported to effectively prevent the attachment of *E. coli* to the membrane (Kim et al. 2003). The advantages of using nano-TiO_2 are that it is relatively low in toxicity and cheap to produce.

Compared to antiattachment coatings, the application of nanoparticles has a dual effect. It not only prevents bacterial attachment but also kills bacterial cells. The common problem among these techniques, however, is that toxicity issues make many of them unsuitable for use in the food industry. In addition, the lifetime of the antiattachment effect induced by nanoparticles may not be long enough for industrial applications as the technique prevents bacterial attachment by releasing the particles. Released particles can also essentially be regarded as foreign bodies if they are released into foods, creating a regulatory issue.

13.3.3 Electro-Assisted Method

In addition to surface coatings and nanotechnology, the electro-assisted method is another approach to modify properties of electrically conductive surfaces to prevent bacterial attachment. Application of 50–100 mA/m^2 electric current can prevent the attachment of marine bacteria to titanium, and the effect can last for 2 years (Wake et al. 2006). The efficacy of this technique may be associated with the modification of the electrostatic properties (charge density/distribution) of metal surfaces. In addition to preventing bacterial attachment, the electro-assisted method was also reported to remove attached bacteria and biofilms from a metal surface. For example, 100 μA direct current detached more than 75% of attached cells or biofilms of *S. epidermidis* from stainless-steel surfaces (van der Borden et al. 2004a,b).

Surface modification is considered to be an effective approach for controlling bacterial attachment to foods and in food-related environments as its effects are general and not species/strain-specific, and no changes or addition to food formulation is required. The major challenges associated with many of these existing techniques are the toxicity and stability of surface-coating materials, which greatly limit their applicability.

13.4 Application of Phytochemicals

Due to the high cost of specific interaction-blocking techniques and the possible toxicity and instability of surface-coating materials, there is a strong interest in developing antiattachment agents that are cheap, safe, and readily available from natural sources. Phytochemicals that have been extensively studied for antimicrobial activities may represent prospective candidates as inhibitors of bacterial attachment. There is, however, very limited information regarding the antiattachment properties of plant materials. In this section, the most thoroughly studied plants/phytochemicals with respect to their antiattachment activities are presented and summarized in Table 13.3.

13.4.1 Modification of Bacterial Surface Structures/Properties by Phytochemicals

Many plant components prevent bacterial attachment to surfaces by altering the biological structures of cell surfaces, such as surface appendages, proteins, and cellular-bound EPS. For example, cranberry, a plant well characterized with respect to antiattachment properties, can prevent the attachment of a range of *E. coli* strains to a variety of surfaces by affecting cell surface macromolecules (Ofek et al. 2003). For example, it was found that cranberry juice can shorten the P fimbriae of an *E. coli* strain from ~148 to ~48 nm (Liu et al. 2006). Carvacrol, a terpenoid found in the essential oil of many plants, was found to inhibit the formation and development of a dual-species biofilm of *S. aureus* and *S.* Typhimurium on stainless steel by 1–3 log CFU/unit-biofilm (Knowles et al. 2005). The inhibitory effect was found to be associated with the protein level in the biofilms, which was reduced fivefold by carvacrol during biofilm development. Epigallocatechin gallate, the signature compound of green tea, was found to inhibit biofilm formation by *S. aureus* and *S. epidermidis* on plastic by approximately 50%–90% by reducing extracellular slime production (Blanco et al. 2005). A recent study showed that oolong tea extracts inhibited the attachment of *S. mutans* to hydroxyapatite by 50%–75% by coating the bacterial cells with oolong tea components (flavonoids, indolic compounds, and tannins) (Figure 13.2) and thereby reducing cell surface hydrophobicity (Wang et al. 2013). The coating activity is very likely to be associated with bacterial adhesins and/or outer-membrane proteins (Matsumoto et al. 1999) as these tea components may act as receptor analogs. It is reasonable to assume that this coating effect would happen to many foodborne pathogens as their attachment is often mediated by adhesins like *S. mutans*.

Some plant components, however, inhibit bacterial attachment not only by affecting extracellular structures but also by causing morphological damage. Extracts of tea-tree oil, for example, prevent

TABLE 13.3

Plant Extracts/Phytochemicals Used to Prevent Bacterial Attachment

Plants/Phytochemicals	Target Bacteria	Mechanisms	References
Cranberries	*Escherichia coli*	Affecting cell surface macromolecules	Ofek et al. (2003), Liu et al. (2006)
Carvacrol	*Staphylococcus aureus* and *Salmonella* Typhimurium	Reducing protein levels	Knowles et al. (2005)
Epigallocatechin gallate	*S. aureus* and *Staphylococcus epidermidis*	Reducing the production of cellular-bound slime	Blanco et al. (2005)
Oolong tea extracts	*Streptococcus mutans*	Cell surface coating	Matsumoto et al. (1999), Wang et al. (2013)
Tea-tree oil extracts	Methicillin-resistant *S. aureus*	Morphological destructions	Park et al. (2007)
Cinnamaldehyde	*E. coli*	Morphological destructions	Niu and Gilbert (2004)
Malaysian herbs	*Bacillus cereus*, *E. coli*, *Pseudomonas aeruginosa*, *Salmonella* Enteritidis, and *S. aureus*	Reducing cell surface hydrophobicity	Hui and Dykes (2012)
Garlic extracts, Florida plants	*P. aeruginosa*	Inhibition of quorum sensing	Rasmussen et al. (2005), Adonizio et al. (2008)

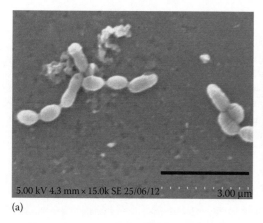
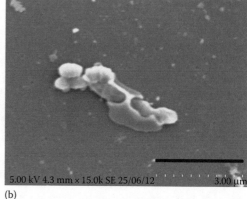

5.00 kV 4.3 mm × 15.0k SE 25/06/12 ┄┄┄┄┄┄┄ 3.00 μm

(a)

5.00 kV 4.3 mm × 15.0k SE 25/06/12 ┄┄┄┄┄┄┄ 3.00 μm

(b)

FIGURE 13.2   Scanning electron microscopic images of (a) untreated *Streptococcus mutans* cells and (b) oolong tea extract–treated *S. mutans* cells (scale bar: 3 μm).

biofilm formation by methicillin-resistant *S. aureus* on silicone by causing cellular damage (Park et al. 2007). Cinnamaldehyde, the compound that gives cinnamon its flavor and odor, reduces biofilms formed by *E. coli* on polystyrene by causing morphological damage (Niu and Gilbert 2004). Many researchers found that the inhibitory effect of some plants on attachment by some bacteria is due to a reduction in cell surface hydrophobicity. For example, Hui and Dykes (2012) screened the water extract of a range of Malaysian herbs for their effects on the attachment of five foodborne pathogens (*B. cereus*, *E. coli*, *P. aeruginosa*, *Salmonella* Enteritidis, and *S. aureus*) onto glass and stainless steel. Inhibitory effects were observed and suggested to be associated with the cell surface hydrophobicity reduction caused by the plant extracts. The reduction in hydrophobicity observed in this study may, however, be the result of the modification of cell surface biological factors such as flagella.

13.4.2 Interference with Quorum Sensing by Phytochemicals

In addition to altering bacterial cell surface structures, phytochemicals may also affect cell-to-cell interactions such as quorum sensing. Bacterial cells communicate and exchange information/cellular materials with quorum-sensing signal molecules. In some cases, this may be involved in biofilm formation by bacterial cells. Inactivating the path of quorum sensing may affect biofilm formation/development or the architecture of a formed biofilm. Quorum-sensing signal molecules are usually peptides, which makes many phytochemicals good candidates for blocking quorum-sensing systems as plant components, especially polyphenols, have a high affinity for proteins and peptides. For example, garlic extracts were found to reduce the biomass and change the architecture of *P. aeruginosa* biofilms in a flow system and to also reduce the tolerance of biofilm cells to antibiotic treatments such as tobramycin (Rasmussen et al. 2005). This was achieved by interrupting the quorum sensing of the bacteria mediated by *N*-acylhomoserine lactone molecules, which are produced by gram-negative bacteria and act as a transcriptional activator by interacting with DNA. The inhibitory effects of the extracts of three plants (*Conocarpus erectus*, *Callistemon viminalis*, and *Bucida buceras*) on biofilm formation by *P. aeruginosa* on plastic were also found to be associated with inhibition of quorum sensing (Adonizio et al. 2008). The plant extracts affected the quorum-sensing genes (*las* and *rhl*) and their respective signal molecules, and in turn inhibited the production of quorum-sensing regulated virulence factors (LasA protease, LasB elastase, and pyoverdin).

In summary, phytochemicals represent a good choice as antiattachment agents as they are natural, widespread, and safe for the food industry. In addition, they prevent bacterial attachment via different mechanisms, which gives them potential against a large variety of bacteria, makes them applicable in a range of circumstances, and therefore broadens their potential use. Limitations associated with using phytochemicals are also apparent. For example, many plant components are bactericidal and/or bacteriostatic and their use could be undesirable because it may result in resistant mutant strains. Precautions should therefore be taken to control phytochemical application at nonlethal levels. Another consideration is that bacterial attachment may be promoted by some phytochemicals, such as lectins, which are abundant in many plants. In addition, nondietary plant components may be toxic or may give undesirable flavor and careful examinations should be performed before using them in foodstuffs or on food-contact surfaces.

13.5　Microbial Therapy to Prevent Bacterial Attachment

Due to the high cost and safety problems of the aforementioned strategies to prevent bacterial attachment, much effort has been directed at another approach, namely, using nonharmful microbes to control the attachment of undesirable bacteria. Microbes that can effectively prevent the attachment of food pathogens are usually biosurfactant-producing probiotic bacteria and bacteriophage. In this section, several examples are described and discussed and these are summarized in Table 13.4.

13.5.1　Biosurfactant-Producing Bacteria

A number of nonpathogenic bacteria have been established to inhibit the attachment of human pathogens to a variety of surfaces. Most of the inhibiting bacteria were found to be biosurfactant producers. The activity of biosurfactants against bacterial attachment has long been demonstrated. Dairy *Streptococcus thermophilus* strains, for example, produce biosurfactants that prevent themselves from attaching to a Teflon surface (Busscher et al. 1994). As a product of microbial origin, biosurfactants have a number of advantages over synthetic surfactants. For example, they have low toxicity, strong surface activity, and emulsifying ability. In addition, they are biodegradable but stay effective under extreme environmental conditions (e.g., temperature, pH, and ionic strength). Using biosurfactants to prevent bacterial attachment therefore represents a potential approach for food industries to control bacteria, especially during food processing. Isolating biosurfactants from bacteria is time-consuming and introducing biosurfactant-producing bacteria into the food systems may be a more efficient approach. Among the biosurfactant-producing bacteria described, lactic acid bacteria have the best potential for application in food systems. For example, *Lactococcus lactis* can significantly prevent attachment and biofilm formation by pathogenic *B. cereus* in a silicone-based flow system (Ksontini et al. 2013). In addition, lactic acid bacteria can prevent the attachment of pathogens to biotic surfaces. For instance, *Lactobacillus* spp. can inhibit the attachment of a range of foodborne pathogens including *E. coli* O157:H7, *S.* Typhimurium, and *L. monocytogenes* to human intestinal cells (Johnson-Henry et al. 2007; Burkholder and Bhunia 2009; Lim and Im 2012; Abedi et al. 2013). However, most of the effective biosurfactants produced by lactic acid bacteria have yet to be identified and the mechanisms by which they prevent the attachment of pathogens are not well understood. It has been suggested that their inhibitory effect on attachment are due to surface activity. For example, biosurfactants produced by *L. lactis* can reduce the attachment of *S. aureus* and *S. epidermidis* to silicone rubber and the mechanism of this activity was found to be associated with a reduction in surface hydrophobicity and surface tension of silicone rubber (Rodrigues et al. 2004). In addition to producing biosurfactants, lactic acid bacteria may inhibit bacterial attachment by other mechanisms such as those acting as adhesin analogs that occupy the receptors on the substratum surface and/or by competitive growth (Wells and Mercenier 2008). Bacteria other than lactic acid

TABLE 13.4

Microbes That Can Prevent Bacterial Attachment

Microbes	Mechanisms	References
Lactic acid bacteria	Producing biosurfactants	Rodrigues et al. (2004), Ksontini et al. (2013)
Bacillus subtilis	Producing biosurfactants	Rivardo et al. (2009), Zeraik and Nitschke (2010)
Bacteriophages	Invading bacterial cells	Roy et al. (1993), Doolittle et al. (1995)

bacteria have also been studied for antiattachment activities. Biosurfactants produced by *B. subtilis*, for example, were found to inhibit the attachment of *S. aureus*, *L. monocytogenes*, and *Micrococcus luteus* to polystyrene surfaces by 63%–66% (Zeraik and Nitschke 2010) and prevent the biofilm formation by *E. coli* and *S. aureus* on polystyrene by more than 90% (Rivardo et al. 2009). The mechanism of the inhibitory effect of these bacteria on attachment and biofilm formation was also suggested to be due to the surface tension reducing activity of the biosurfactants.

13.5.2 Bacteriophage

Bacteriophages are viruses that infect bacteria and may act to prevent attachment and biofilm formation. For example, a Listeriaphage was found to prevent biofilm formation by *L. monocytogenes* on stainless steel and polypropylene (Roy et al. 1993). Bacteriophages infect bacteria by invading the bacterial cells. They attach to a specific receptor on the surface of bacteria, which usually includes lipopolysaccharides, teichoic acid, proteins, or even flagella. This specificity, however, limits the host range of bacteriophages to bacteria bearing such receptors. Some phages are more broad-spectrum and can infect multiple related species, while others are specific at a strain level, and this limits the use of phage to control biofilms. Formulating phage cocktails may provide a solution to this problem and improve their practical use.

Bacteriophages are more often used to detach existing biofilms. However, a common problem associated with using phages to remove bacterial biofilms is that they can only infect the cells on the surface of a biofilm but cannot access cells deep inside because of the EPS matrix associated with the biofilm. For example, phage E79 can infect the surface cells of a biofilm of *P. aeruginosa* but has no access to the inner cells (Cortés et al. 2011). Some phages, on the other hand, can infect both surface and inner cells by producing polysaccharide lyase enzymes that attack the EPS shield. For example, phage T4D was found to infect and lyse both surface and inner cells of an *E. coli* biofilm on polyvinylchloride surfaces in a modified Robbins device (Doolittle et al. 1995). The inability to produce polysaccharide lyase greatly limits the application of many effective phages against food-related pathogens. In addition, device-associated biofilms are often difficult to remove due to strong EPS production. For these reasons, genetic engineering of nonpolysaccharide lyase–producing bacteriophages to include this ability could be a potential direction for future research.

Using microbes to control bacterial attachment and biofilms is a new approach and worth further development as they are cheap and available. The challenge of this approach is that lactic acid bacteria may change the quality of some foods by fermentation (e.g., dairy products), and applying bacteriophages to biofilms may introduce bacterial virulence or toxin genes. Precautions need to be taken and further research needs to be performed to gain a better understanding of these techniques.

13.6 Conclusion

A substantial amount of research has been conducted and advances made in developing strategies to control bacterial attachment to biotic and abiotic surfaces. Based on our current understanding of these techniques, a universal approach that can prevent the attachment of all pathogenic bacteria to all food/food-related surfaces is still beyond reach. Strategies that address specific needs with respect to the control of the attachment of certain bacteria to certain surfaces have been developed. As these strategies are limited in scope, a better understanding of the mechanisms of bacterial attachment to surfaces is required in order to develop new and more advanced ones.

REFERENCES

Abedi, D., S. Feizizadeh, V. Akbari, and A. Jafarin-Dehkordi. 2013. In vitro anti-bacterial and anti-adherence effects of *Lactobacillus delbrueckii* subsp. *bulgaricus* on *Escherichia coli*. *Res Pharm Sci* 8:261–268.

Adonizio, A., K. F. Kong, and K. Mathee. 2008. Inhibition of quorum sensing-controlled virulence factor production in *Pseudomonas aeruginosa* by south Florida plant extracts. *Antimicrob Agents Chemother* 52:198–203.

Al-Makhlafi, H., J. McGuire, and M. Daeschel. 1994. Influence of preadsorbed milk proteins on adhesion of *Listeria monocytogenes* to hydrophobic and hydrophilic silica surfaces. *Appl Environ Microbiol* 60:3560–3565.

Alugupalli, K. R. and S. Kalfas. 1995. Inhibitory effect of lactoferrin on the adhesion of *Actinobacillus actinomycetemcomitans* and *Prevotella intermedia* to fibroblasts and epithelial cells. *APMIS* 103:154–160.

Alugupalli, K. R. and S. Kalfas. 1997. Characterization of the lactoferrin-dependent inhibition of the adhesion of *Actinobacillus actinomycetemcomitans*, *Prevotella intermedia* and *Prevotella nigrescens* to fibroblasts and to a reconstituted basement membrane. *APMIS* 105:680–688.

Balazs, D. J., K. Triandafillu, P. Wood et al. 2004. Inhibition of bacterial adhesion on PVC endotracheal tubes by RF-oxygen glow discharge, sodium hydroxide and silver nitrate treatments. *Biomaterials* 25:2139–2151.

Bitzan, M. M., B. D. Gold, D. J. Philpott et al. 1998. Inhibition of *Helicobacter pylori* and *Helicobacter mustelae* binding to lipid receptors by bovine colostrum. *J Infect Dis* 177:955–961.

Blanco, A. R., A. Sudano-Roccaro, G. C. Spoto, A. Nostro, and D. Rusciano. 2005. Epigallocatechin gallate inhibits biofilm formation by ocular staphylococcal isolates. *Antimicrob Agents Chemother* 49:4339–4343.

Burkholder, K. M. and A. K. Bhunia. 2009. *Salmonella enterica* serovar Typhimurium adhesion and cytotoxicity during epithelial cell stress is reduced by *Lactobacillus rhamnosus* GG. *Gut Pathog* 1:14–23.

Busscher, H. J., T. R. Neu, and H. C. van der Mei. 1994. Biosurfactant production by thermophilic dairy streptococci. *Appl Microbiol Biotechnol* 41:4–7.

Cegelski, L., J. S. Pinkner, N. D. Hammer et al. 2009. Small-molecule inhibitors target *Escherichia coli* amyloid biogenesis and biofilm formation. *Nat Chem Biol* 5:913–919.

Cordeiro, A. L., R. Zimmermann, S. Gramm et al. 2009. Temperature dependent physicochemical properties of poly(*N*-isopropylacrylamide-*co*-*N*-(1-phenylethyl) acrylamide) thin films. *Soft Matter* 5:1367–1377.

Cortés, M. E., J. C. Bonilla, and R. D. Sinisterra. 2011. Biofilm formation, control and novel strategies for eradication. In *Science against Microbial Pathogens: Communicating Current Research and Technological Advances*, ed. A. Méndez-Vilas, pp. 896–905. Extremadura, Spain: Formatex Research Center.

Cravioto, A., A. Tello, H. Villafan, J. Ruiz, S. del Vedovo, and J. R. Neeser. 1991. Inhibition of localized adhesion of enteropathogenic *Escherichia coli* to HEp-2 cells by immunoglobulin and oligosaccharide fractions of human colostrum and breast milk. *J Infect Dis* 163:1247–1255.

Doolittle, M. M., J. J. Cooney, and D. E. Caldwell. 1995. Lytic infection of *Escherichia coli* biofilms by bacteriophage-T4. *Can J Microbiol* 41:12–18.

Goldhar, J., A. Zilberberg, and I. Ofek. 1986. Infant mouse model of adherence and colonization of intestinal tissues by enterotoxigenic strains of *Escherichia coli* isolated from humans. *Infect Immun* 52:205–208.

Hui, Y. W. and G. A. Dykes. 2012. Modulation of cell surface hydrophobicity and attachment of bacteria to abiotic surfaces and shrimp by Malaysian herb extracts. *J Food Prot* 75:1507–1511.

Ista L. K., S. Mendez, and G. P. Lopez. 2010. Attachment and detachment of bacteria on surfaces with tunable and switchable wettability. *Biofouling* 26:111–118.

Johnson-Henry, K. C., K. E. Hagen, M. Gordonpour, T. A. Tompkins, and P. M. Sherman. 2007. Surface-layer protein extracts from *Lactobacillus helveticus* inhibit enterohaemorrhagic *Escherichia coli* O157:H7 adhesion to epithelial cells. *Cell Microbiol* 9:356–367.

Kelly, C. G., J. S. Younson, B. Y. Hikmat et al. 1999. A synthetic peptide adhesion epitope as a novel antimicrobial agent. *Nat Biotechnol* 17:42–47.

Kim, S. H., S. Y. Kwak, B. H. Sohn, and T. H. Park. 2004. Design of TiO_2 nanoparticle self-assembled aromatic polyamide thin-film-composite (TFC) membrane as an approach to solve biofouling problem. *J Memb Sci* 211:157–165.

Klemm, P., R. M. Vejborg, and V. Hancock. 2010. Prevention of bacterial adhesion. *Appl Microbiol Biotechnol* 88:451–459.

Knowles, J. R., S. Roller, D. B. Murray, and A. S. Naidu. 2005. Antimicrobial action of carvacrol at different stages of dual-species biofilm development by *Staphylococcus aureus* and *Salmonella enterica* serovar Typhimurium. *Appl Environ Microbiol* 71:797–803.

Koziarz, J., and H. Yamazaki. 1998. Immobilization of bacteria onto polyester cloth. *Biotechnol Techn* 12:407–410.

Kregiel, D., J. Berlowska, U. Mizerska, W. Fortuniak, J. Chojnowski, and W. Ambroziak. 2013. Chemical modification of polyvinyl chloride and silicone elastomer in inhibiting adhesion of *Aeromonas hydrophila*. *World J Microbiol Biotechnol* 29:1197–1206.

Ksontini, H., F. Kachouri, and M. Hamdi. 2013. Impact of *Lactococcus lactis* spp. *lactis* bio adhesion on pathogenic *Bacillus cereus* biofilm on silicone flowing system. *Indian J Microbiol* 53:269–275.

Li, Y., P. Leung, L. Yao, Q. W. Song, and E. Newton. 2006. Antimicrobial effect of surgical masks coated with nanoparticles. *J Hosp Infect* 62:58–63.

Lim, S. M. and D. S. Im. 2012. Inhibitory effects of antagonistic compounds produced from *Lactobacillus brevis* MLK27 on adhesion of *Listeria monocytogenes* KCTC3569 to HT-29 cells. *Food Sci Biotechnol* 21:775–784.

Lindhorst, T. K., C. Kieburg, and U. Krallmann-Wenzel. 1998. Inhibition of the type 1 fimbriae-mediated adhesion of *Escherichia coli* to erythrocytes by multiantennary alpha-mannosyl clusters: The effect of multivalency. *Glycoconjug J* 15:605–613.

Liu, Y., A. M. Black, L. Caron, and T. A. Camesano. 2006. Role of cranberry juice on molecular-scale surface characteristics and adhesion behavior of *Escherichia coli*. *Biotechnol Bioeng* 93:297–305.

Matsumoto, M., T. Minami, H. Sasaki, S. Sobue, S. Hamada, and T. Ooshima. 1999. Inhibitory effects of oolong tea extract on caries-inducing properties of *mutans streptococci*. *Caries Res* 33:441–445.

Moon, H. W. and T. O. Bunn. 1993. Vaccines for preventing enterotoxigenic *Escherichia coli* infections in farm animals. *Vaccine* 11:200–213.

Mysore, J. V., T. Wigginton, P. M. Simon, D. Zopf, L. M. Heman-Ackah, and A. Dubois. 1999. Treatment of *Helicobacter pylori* infection in rhesus monkeys using a novel anti-adhesion compound. *Gastroenterology* 117:1316–1325.

Niu, C. and E. S. Gilbert. 2004. Colorimetric method for identifying plant essential oil components that affect biofilm formation and structure. *Appl Environ Microbiol* 70:6951–6956.

Ofek, I., D. L. Hasty, and N. Sharon. 2003. Anti-adhesion therapy of bacterial diseases: Prospects and problems. *FEMS Immunol Med Microbiol* 38:181–191.

Ofek, I., D. Mirelman, and N. Sharon. 1977. Adherence of *Escherichia coli* to human mucosal cells mediated by mannose receptors. *Nature* 265:623–625.

Park, H., C. H. Jang, Y. B. Cho, and C. H. Choi. 2007. Antibacterial effect of tea-tree oil on methicillin-resistant *Staphylococcus aureus* biofilm formation of the tympanostomy tube: An in vitro study. *In Vivo* 21:1027–1030.

Park, K. D., Y. S. Kim, D. K. Han, Y. H. Kim, E. H. B. Lee, H. Suh, and K. S. Choi. 1998. Bacterial adhesion on PEG modified polyurethane surfaces. *Biomaterials* 19:851–859.

Pinkner, J. S., H. Remaut, F. Buelens et al. 2006. Rationally designed small compounds inhibit pilus biogenesis in uropathogenic bacteria. *Proc Natl Acad Sci USA* 103:17897–17902.

Rasmussen, T. B., T. Bjarnsholt, and M. E. Skindersoe et al. 2005. Screening for quorum-sensing inhibitors (QSI) by use of a novel genetic system, the QSI selector. *J Bacteriol* 187:1799–1814.

Rivardo, F., R. J. Turner, G. Allegrone, H. Ceri, and M. G. Matinotti. 2009. Anti-adhesion activity of two biosurfactants produced by *Bacillus* spp. prevents biofilm formation of human bacterial pathogens. *Appl Microbiol Biotechnol* 83:541–553.

Rodrigues, L., H. C. van der Mei, J. A. Teixeira, and R. Oliveira. 2004. Biosurfactant from *Lactococcus lactis* 53 inhibits microbial adhesion on silicone rubber. *Appl Microbiol Biotechnol* 66:306–311.

Roosjen, A., H. J. Kaper, H. C. van der Mei, W. Norde, and H. J. Busscher. 2003. Inhibition of adhesion of yeasts and bacteria by poly(ethylene oxide)-brushes on glass in a parallel plate flow chamber. *Microbiology* 149:3239–3246.

Roth, S., J. Feichtinger, and C. Hertel. 2010. Characterization of *Bacillus subtilis* spore inactivation in low-pressure, low-temperature gas plasma sterilization processes. *J Appl Microbiol* 108:521–531.

Roy, B, H. W. Ackermann, S. Pandian, G. Picard, and J. Goulet. 1993. Biological inactivation of adhering *Listeria monocytogenes* by Listeriaphages and a quaternary ammonium compound. *Appl Environ Microbiol* 59:2914–2917.

Ruiz-Palacios, G. M., L. E. Cervantes, P. Ramos, B. Chavez-Munguia, and D. S. Newburg. 2003. *Campylobacter jejuni* binds intestinal H(O) antigen (Fuc α1, 2Galβ1, 4GlcNAc), and fucosyloligosaccharides of human milk inhibit its binding and infection. *J Biol Chem* 278:14112–14120.

Schröten, H., A. Lethen, F. G. Hanisch et al. 1992. Inhibition of adhesion of S-fimbriated *Escherichia coli* to epithelial cells by meconium and feces of breast-fed and formula-fed newborns: Mucins are the major inhibitory component. *J Pediatr Gastroenterol Nutr* 15:150–158.

Şen, Y., U. Bağcı, H. A. Güleç, and M. Mutlu. 2009. Modification of food-contacting surfaces by plasma polymerization technique: Reducing the biofouling of microorganisms on stainless steel surface. *Food Bioprocess Technol* 5:166–175.

Sohbatzadeh, F., A. Hosseinzadeh Colagar, S. Mirzanejhad, and S. Mahmodi. 2010. *E. coli, P. aeruginosa*, and *B. cereus* bacteria sterilization using afterglow of non-thermal plasma at atmospheric pressure. *Appl Biochem Biotechnol* 160:1978–1984.

Sureshkumar A., R. Sankar, M. Mandal, and S. Neogi. 2010. Effective bacterial inactivation using low temperature radio frequency plasma. *Int J Pharm* 396:17–22.

Svanborg-Edén, C., R. Freter, L. Hagberg et al. 1982. Inhibition of experimental ascending urinary tract infection by an epithelial cell-surface receptor analogue. *Nature* 298:560–562.

Van der Borden, A. J., H. C. van der Mei, and H. J. Busscher. 2004a. Electric-current-induced detachment of *Staphylococcus epidermidis* strains from surgical stainless steel. *J Biomed Mater Res* 68B:160–164.

Van der Borden, A. J., H. van der Werf, H. C. van der Mei, and H. J. Busscher. 2004b. Electric current-induced detachment of *Staphylococcus epidermidis* biofilms from surgical stainless steel. *Appl Environ Microbiol* 70:6871–6874.

Vejborg, R. M., N. Bernbom, L. Gram, and P. Klemm. 2008. Antiadhesive properties of fish tropomyosins. *J Appl Microbiol* 105:141–150.

Vejborg, R. M. and P. Klemm. 2008. Blocking of bacterial biofilm formation by a fish protein coating. *Appl Environ Microbiol* 74:3551–3558.

Wake, H., H. Takahashi, T. Takimoto, H. Takayanagi, K. Ozawa, H. Kadoi, M. Okochi, and T. Matsunaga. 2006. Development of an electrochemical antifouling system for seawater cooling pipelines of power plants using titanium. *Biotechnol Bioeng* 95:468–473.

Wang, I., J. M. Anderson, M. R. Jacobs, and R. E. Marchant. 1995. Adhesion of *Staphylococcus epidermidis* to biomedical polymers: Contributions of surface thermodynamics and hemodynamic shear conditions. *J Biomed Mater Res* 29:485–493.

Wang, Y., S. M. Lee, and G. A. Dykes. 2013. Potential mechanisms for the effects of tea extracts on the attachment, biofilm formation and cell size of *Streptococcus mutans*. *Biofouling* 29:307–318.

Wells, J. M. and A. Mercenier. 2008. Mucosal delivery of therapeutic and prophylactic molecules using lactic acid bacteria. *Nat Rev Microbiol* 6:349–362.

Zaaba, S. K., T. Akitsu, H. Ohkawa et al. 2010. Plasma disinfection and the deterioration of surgical tools at atmospheric pressure plasma. *IEE J Trans Fundam Mater* 130:355–361.

Zeraik, A. E. and M. Nitschke. 2010. Biosurfactants as agents to reduce adhesion of pathogenic bacteria to polystyrene surfaces: Effect of temperature and hydrophobicity. *Curr Microbiol* 61:554–559.

Section III

Food Preservation and Intervention Techniques

14

Bacteriocins: The Natural Food Preservatives

Haider Khan, Steve H. Flint, and Pak-Lam Yu

CONTENTS

14.1 Introduction

The history of food preservation is as old as the mankind itself. A majority of foods are prone to spoilage by the contaminating microbes as they are a rich source of nutrients. Under favorable growth conditions (temperature, pH, and moisture availability), the spoilage and pathogenic microorganisms can easily grow in foods, thus rendering them unfit for use. Therefore, foods need to be preserved for future use and for transportation to distant places. The earliest methods of preservation included drying, salting, freezing (in cold regions), and fermentation of foods. With the advancement in knowledge, various novel food preservation methods emerged, which include canning, pasteurization, irradiation, and the use of chemical preservatives.

However, nowadays, there is an increasing consumer demand for minimally processed foods, which have not been subjected to harsh physical treatment, which often destroys the nutrients. Furthermore, many individuals are allergic to many chemical preservatives, and some chemical preservatives can even form carcinogenic by-products, for example, nitrosamines from nitrates (Kashani et al., 2012). Therefore, there is a growing interest among the researchers to search for safe and natural food preservatives, which have least undesirable effects.

Biopreservation of foods, which involves the use of safe microorganisms and/or their metabolites for extending the shelf life of foods, is one of the alternatives to harsh physical and chemical treatments (Galvez et al., 2007). The most important candidates in this regard are the lactic acid bacteria (LAB). Many species of LAB have been utilized as cultures for the preparation of fermented milk, meat, and vegetable products (such as yoghurt, cheese, sausages, and sauerkraut) for centuries and are, therefore, generally regarded as safe (GRAS). Among these LAB cultures, a large number of strains have now been known to produce ribosomally synthesized antimicrobial peptides and proteins known as bacteriocins. These bacteriocins are antagonistic to other closely related and sometimes unrelated bacterial strains and species. However, the producer strain is protected from its own bacteriocin by a dedicated immunity mechanism (Cleveland et al., 2001).

The first bacteriocin reported to be produced by a LAB strain was nisin, which is produced by strains of *Lactococcus lactis* subsp. *lactis* (*Streptococcus lactis*) (Mattick and Hirsch, 1947). It is a broad-spectrum antimicrobial found to be active against a large number of gram-positive bacteria that include *Listeria monocytogenes*, *Staphylococcus aureus*, *L. lactis* subsp. *lactis* and *cremoris*, *Lactobacillus del-brueckii* subsp. *bulgaricus*, *Clostridium*, and *Bacillus* spp. (Harris et al., 1992). Since the discovery of nisin, a large number of bacteriocins have been reported to be produced by LAB, and many of these have been found to be effective as preservatives against selected pathogenic and spoilage bacteria in a variety of food products. However, to date, nisin is the only bacteriocin that has been approved as a food preservative in more than 53 countries (Delves-Broughton, 2005; Mills et al., 2011).

14.2 Classification of Bacteriocins

Bacteriocins being proteinaceous are very heterogeneous in nature. Several attempts have been made to classify and group these heterogeneous molecules according to their primary amino acid structure and genetic characteristics. Based on the recent advances in bacteriocin research, a new classification scheme is being proposed that divides the bacteriocins into five classes (Figure 14.1). This classification scheme is a modified version of the scheme presented by Heng et al. (2007), which tries to encompass all the bacteriocins of gram-positive bacteria. The original scheme divided the bacteriocins into four classes,

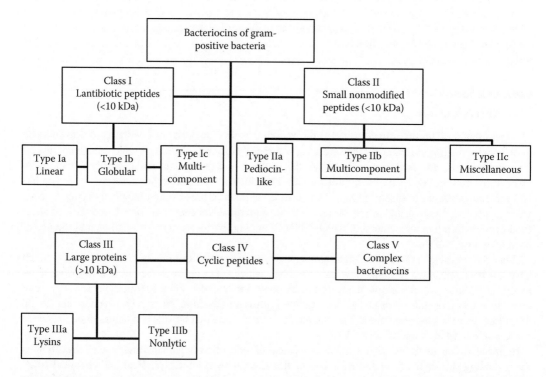

FIGURE 14.1 Classification scheme of bacteriocins produced by gram-positive bacteria. (Modified from Heng, N.C.K. et al., The diversity of bacteriocins in Gram-positive bacteria, in: *Bacteriocins: Ecology and Evolution*, M.A. Riley and M.A. Chavan, eds., Springer, Berlin, Germany, 2007, pp. 45–92.)

namely the posttranslationally modified lantibiotic peptides, nonmodified peptides, large proteins, and cyclic peptides. However, recent investigations have proved the existence of complex bacteriocins that require carbohydrate moiety in addition to the protein moiety for their activity. This necessitates the addition of a new class of complex bacteriocins to the Heng's scheme.

According to this classification scheme, the bacteriocins of gram-positive bacteria have been divided into five classes. The class I encompasses the small peptides (<10 kDa) termed as lantibiotics, which contain the unusual amino acids such as lanthionine and methyl lanthionine in their primary structure. These unusual amino acids are formed by posttranslational modification of normal amino acids. The lantibiotics are further subdivided into three subtypes: type Ia consisting of linear lantibiotics such as nisin, type Ib consists of compact globular peptides up to 19 amino acids in length such as mersacidin, and type Ic consisting of lantibiotics that require two or more modified peptides for their activity such as lacticin 3147.

The class II bacteriocins consist of small peptides (<10 kDa) that are not modified after translation. This is a very heterogeneous class and is further subdivided into three subtypes. Type IIa consists of *Listeria* active peptides, which all have a conserved motif YGNGV in their primary structure. The type IIb consists of bacteriocins that require the synergistic action of two or more nonmodified peptides for antimicrobial activity, whereas type IIc forms the repository for all the remaining nonmodified bacteriocins that do not belong either to subtype IIa or IIb.

In contrast to heat-stable peptides of classes I and II, the class III include heat-labile bacteriocins having molecular mass greater than 10 kDa and is further subdivided into lytic and nonlytic proteins. The class IV bacteriocins include the cyclic peptides. These peptides are synthesized as linear peptides and undergo posttranslational cyclization by head-to-tail ligation. Since head-to-tail ligation is a posttranslational modification, therefore, these peptides are assigned to a separate class.

Class V bacteriocins in the present scheme include all the bacteriocins that require a carbohydrate or lipid moiety in addition to the protein component for antimicrobial activity. The presence of such bacteriocins has long been disputed since no convincing data were available. But recent findings have revealed that sublancin produced by *Bacillus subtilis* (Oman et al., 2011) and glycocin F secreted by *Lactobacillus plantarum* KW30 (Stepper et al., 2011) are glycoproteins that require carbohydrate moiety for antimicrobial activity. The main characteristics, mode of action, and representative examples of different classes of bacteriocins are summarized in Table 14.1.

The bacteriocins belonging to classes I, II, and IV are mostly cationic peptides that are heat and pH stable, which are the desirable characteristics for use as food preservatives. Therefore, a large number of bacteriocins belonging to these three classes have been explored as food preservatives in a variety of food products. However, the most successful candidates in this regard are nisin, pediocin, and enterocin AS-48, which have been most extensively studied for their antimicrobial effects in food products.

14.3 Production, Purification, and Characterization of Bacteriocins

The journey for the discovery of a bacteriocin involves several steps. It starts from the screening for bacteriocin-producing strains from a large number of isolates. These strains are then grown in the most appropriate medium to produce the maximum amount of bacteriocin, which is then purified to homogeneity through a series of steps leading to identification and characterization.

LAB are fastidious organisms and require specific amino acids, vitamins, minerals, and a carbohydrate source for their growth (Bouguettoucha et al., 2011). These requirements are usually met by using complex media such as MRS, M17, Elliker lactic broth, and All Purpose Tween. These media contain large number of peptides that are in the molecular weight range of most bacteriocins (3000–6000 kDa). Therefore, the bacteriocin secreted in the medium in small quantities needs to be purified through a series of steps before being characterized at the molecular level (Carolissen-Mackay et al., 1997).

Since the bacteriocins are heterogeneous in nature, therefore, the purification strategy varies for each bacteriocin but usually involves an initial volume reduction step followed by final purification in a series of chromatographic steps that exploit the charge and hydrophobic properties of bacteriocins.

TABLE 14.1

Characteristic Features of Bacteriocin Classes and Their Examples

Class	Subtypes (If Any)	Main Characteristics	Mode of Action	Representative Bacteriocin of the Class			
				Name	Producing Species	Typical Target Bacteria	Food Applications
I (posttranslationally modified lantibiotics)	Ia (linear)	Elongated, screw-shaped, positively charged, amphiphilic, flexible molecules; mol. mass: 3–4 kDa; modified by two enzymes (LanB and C)	Receptor: lipid II; interfere with cell wall biosynthesis; form ion-permeable pores in the cytoplasmic membrane; membrane potential dissipation and cell death.	Nisin	*L. lactis* subsp. *lactis*	*L. lactis*, *Streptococcus thermophilus*, *Lactobacillus* spp., *L. monocytogenes*, *S. aureus*, *Bacillus cereus*, *Clostridium botulinum*	Processed cheese products, ricotta cheese, pasteurized milk and milk products, canned products, salad dressings, crumpets, sausages
	Ib (globular)	Globular and compact structure; neutral or negatively charged; mol. mass: 2–3 kDa; modified by a single enzyme (LanM)	Binds to lipid II interfering with cell wall biosynthesis.	Mersacidin	*Bacillus* sp.	*S. aureus*, *Staphylococcus simulans*	—
	Ic (multicomponent)	Two peptides act synergistically for antimicrobial activity; Each peptide is modified by separate LanM enzymes; A single LanT enzyme removes leader peptide and secrete the two peptides	One peptide component binds to lipid II forming a complex; the second peptide binds to this complex. The C-terminus of second peptide inserts into the lipid membrane forming ion leaking pores and cell death.	Lacticin 3147	*L. lactis* DPC3147	*L. lactis*, *L. monocytogenes*, *B. subtilis*, *S. aureus*, *Lactobacillus fermentum*	Yoghurt, cottage cheese, infant formula, sausages
II (posttranslationally unmodified peptides)	IIa (Pediocin-like)	Contain YGVNGVXC at their N-terminus; mol. mass: 3–7 kDa; peptide chain length 25–58 amino acids; cationic peptides	Receptor: Mannose phosphotransferase system (M-PTS); C-terminus inserts into the lipid membrane. Pores formed causing efflux of K+ ions, membrane potential dissipatior and cell death.	Pediocin PA-1/AcH	*Pediococcus acidilactici*	*Pediococcus pentosaceus*, *Lactobacillus helveticus*, *L. monocytogenes*	Meat products (sausages, meat sticks), cheddar cheese, munster cheese, liquid whole egg

Class	Characteristics	Mode of action	Bacteriocin	Producer	Sensitive organisms	Application
IIb (multicomponent)	Two individual peptides act synergistically for antimicrobial activity; cationic or amphiphilic peptides; mol. mass: 3–6 kDa; peptide chain length: 25–62 amino acids	Individual peptides interact using GxxxG motifs and penetrate cytoplasmic membrane. Form a transmembrane helix–helix structure resulting in pore formation, ion leakage, and cell death.	Enterocin 1071 (A and B)	*E. faecalis* BFE1071	*L. monocytogenes, Listeria innocua, Enterococcus faecalis, Enterococcus faecium*	Fish spread
IIc (miscellaneous)	Single-peptide non-pediocin-like bacteriocins; very heterogeneous group; mol. mass 3.5–7.5 kDa	Diverse modes of action. May be pore forming or inhibiting cell wall formation.	Enterocin EJ97	*E. faecalis* EJ97	*L. monocytogenes, S. aureus, Bacillus* spp., *E. faecalis, Geobacillus stearothermophilus*	Canned vegetable foods and drinks
III (large proteins >10 kDa) IIIa (lysins)	Heat labile; mol. mass 25–35 kDa; mature bacteriocin has 200–400 amino acids; have domain type structure, the N-terminus is the catalytic domain and C-terminus is the substrate recognition domain	Cleave within peptidoglycan moiety of cell wall of gram-positive bacteria causing cell lysis.	Enterolysin A	*E. faecalis*	*E. faecium, L. monocytogenes, L. lactis* subsp. *cremoris, Lactobacillus helveticus*	—
IIIb (nonlytic)	Heat labile; Mol. mass: 16–40 kDa; mature bacteriocin has 150–340 amino acids	Dysgalacticin uses mannose phosphotransferase system (M-PTS) as a receptor molecule interfering with glucose uptake and dissipating membrane potential; loss of ions cause cell death.	Dysgalacticin	*Streptococcus dysgalactiae ssp. equisimilis*	*Streptococcus pyogenes*	—
IV (cyclic bacteriocins)	N- and C-termini are ligated together forming a cyclic structure; molecular mass: 5.5–7.5 kDa; peptide chain 35–70 amino acid long	Permeabilize membranes of sensitive cells; no surface receptors required; the hydrophobic core inserts in the lipid membrane forming pores causing leakage of ions thus disrupting membrane potential and causing cell death.	Enterocin AS-48	*Enterococcus faecalis* subsp. *liquefaciens* S-48	*E. faecalis, E. faecium, Bacillus* spp., *S. aureus, L. monocytogenes, Escherichia coli*	Raw fruits and fruit juices, vegetable sauces, ready-to-eat salad, model sausages, canned fruits and vegetables, vegetable soups, and purees
V (complex bacteriocins)	Cysteine S-linked glycopeptides	Mode of action studies yet to be done.	Glycocin F	*L. plantarum* KW30	*L. plantarum, Lactobacillus brevis, L. delbrueckii* ssp. *lactis*	—

Ammonium sulfate precipitation, based on the salting-out principle, is most commonly used for initial volume reduction and concentration of bacteriocins. Other methods for initial volume reduction include the use of amberlite resins such as XAD-16, ultrafiltration, and lyophilization. The initial volume reduction is necessary since bacteriocins are produced in small quantities in large volumes of culture media. However, this initial step is not very selective, as proteins and peptides of the growth medium are also concentrated along with the bacteriocins. Therefore, further purification steps are necessary to separate bacteriocins from these contaminants and to purify them to homogeneity. Most of the bacteriocins are low-molecular-weight cationic molecules and contain hydrophobic amino acid residues; therefore, these properties are usually exploited to purify bacteriocins to homogeneity (Parada et al., 2007). Commonly used purification schemes include cation exchange chromatography and hydrophobic interaction chromatography followed by reverse-phase high-pressure liquid chromatography (RP-HPLC) as a final step. Gel filtration chromatography and other steps may also be included to further purify some of the bacteriocins, but RP-HPLC is usually the final step in the purification scheme, which purifies the bacteriocin to homogeneity and separates it from any remaining contaminants (Saavedra and Sesma, 2011). The purity is then confirmed by running the purified fraction in an SDS-PAGE gel, followed by in-gel activity testing. The presence of a single active band in the gel confirms that the bacteriocin has been purified to homogeneity, and the primary structure is then determined using techniques such as N-terminus sequencing and mass spectrometry.

The purification scheme described earlier is most commonly used and has been helpful in the purification and identification of many novel bacteriocins; however, there are several drawbacks. The scheme involves a number of steps, and some of them are quite time consuming and laborious. Furthermore, there are losses at each step, and, therefore, the greater the number of steps, the greater the loss in bacteriocin activity. Since the bacteriocins are produced in small quantities, the entire bacteriocin activity may be lost after final purification.

Alternative purification schemes have been suggested to either reduce the time or the number of purification steps thereby reducing the losses in activity. Novel methods have been developed that selectively separate bacteriocins from contaminants in minimal steps. One such novel procedure was developed by Yang et al. (1992), which exploits the cationic nature of bacteriocins. It is based on the principle that cationic bacteriocins are selectively adsorbed to the cells of the producer strain at neutral pH (maximum adsorption of about 90% at pH 6.0) and are desorbed at a pH of 2.0 (about 99% desorption). The procedure involves the production of bacteriocin in growth medium followed by heating the medium to about 70°C to kill the cells. The pH of the medium is then adjusted to 6.0, which results in the adsorption of the bacteriocin to heat-killed cells. The cells are then removed from the production broth and then resuspended in a small volume of saline buffer at pH 2 at a temperature of 4°C. This results in the desorption of cationic bacteriocins from the cells into the buffer. The bacteriocin-containing buffer is then dialyzed after the removal of cells, followed by lyophilization. The resulting bacteriocin is highly purified and may require only one more step such as RP-HPLC to purify bacteriocin to homogeneity. Yang et al. (1992) were able to recover more than 90% of the bacteriocin in the case of pediocin AcH, nisin, and leuconocin Lcm1. However, only 44% bacteriocin was recovered for sakacin A, indicating that this method, although very efficient, may not be suitable for all bacteriocins.

Another novel strategy was developed by Venema et al. (1997) for the purification of lactococcin B, which involves the addition of precooled ethanol at 4°C to the cell-free supernatant resulting in the precipitation of the bacteriocin. The precipitated bacteriocin can then be purified to homogeneity in one or two steps. This method has recently been used for the purification of sakacin C2 (Gao et al., 2010) and pediocin SA131 (Lee et al., 2010) in minimal steps.

A different approach for purifying the bacteriocin in minimal steps was used by Pingitore et al. (2009). They developed a simplified chemically defined medium for the production of salivaricin 1328 by *Lactobacillus salivarius* 1328. Since the defined medium was devoid of any contaminating peptides and proteins, a single ultrafiltration step was sufficient to purify and concentrate the bacteriocin from cell-free supernatant. This was evident by the presence of a single active band in a tricine SDS-PAGE gel.

Electrophoresis of purified bacteriocin using SDS-PAGE is usually the next step after the bacteriocin has been purified to homogeneity. The presence of a single active band in the gel as determined by the method developed by Bhunia et al. (1987) confirms the purity of the bacteriocin. The purified bacteriocin preparation can then be used for characterization at the amino acid level, or alternatively the active band can be eluted from the gel for further characterization.

The amino acid sequence of a bacteriocin can give a wealth of information about its function and mode of action and, also confirms the novelty or otherwise of the bacteriocin. It is, therefore, the ultimate task after purification. The classical method of choice usually is the N-terminal automated amino acid sequencing using Edman degradation method. The determination of the first few amino acids at the N-terminus is usually sufficient. Primers for PCR reaction can then be designed based on the amino acid sequence obtained, and the complete structural gene can then be amplified and sequenced. The complete amino acid sequence of the bacteriocin can then be deduced from the DNA sequence. Many novel bacteriocins have been characterized using this strategy, which include lactocyclicin Q (Sawa et al., 2009), plantaricin ASM1 (Hata et al., 2010), and lactococcin Q (Zendo et al., 2006). Recently, three novel bacteriocins have been identified and characterized by amino acid and DNA sequencing from *Lactobacillus sakei* D98 (Sawa et al., 2013).

To conclude, it can be said that with the development of new and improved methods for the purification of bacteriocins or innovative approaches for production and identification of bacteriocins, many novel bacteriocins may be identified and characterized at the molecular level thus adding to the database of existing bacteriocins.

14.4 Structure of Bacteriocins

14.4.1 Class I Bacteriocins (the Lantibiotics)

The lantibiotics group is one of the most extensively studied group of bacteriocins consisting of peptides (<5 kDa) having unusual amino acids such as lanthionine and methyl lanthionine in their primary structure, formed by extensive posttranslational modifications (Jack and Sahl, 1995). Indeed, the term lantibiotic is derived from L(anthionine)-containing antibiotics (McAuliffe et al., 2001).

The primary translation product of lantibiotics is a prepeptide consisting of a leader peptide at the N-terminus. The length of the leader peptide varies from 23 to 59 amino acids. No amino acid modification takes place in the leader peptide region; however, extensive modifications occur in the pro-peptide region while the leader peptide is still attached to the pro-peptide. Only three amino acids, serine, threonine, and cysteine, are involved in posttranslational modifications. Occasionally lysine, alanine, and isoleucine may also be posttranslationally modified (Sahl and Bierbaum, 1998).

The posttranslational modifications are brought about by specific enzymes. For lantibiotics such as nisin, two enzymes are involved that are encoded by genes designated as LanB and LanC. The enzyme encoded by the LanB gene is responsible for the dehydration of serine and threonine to dehydroalanine (Dha) and dehydrobutyrine (Dhb) respectively. The LanC protein then catalyzes cyclization of cysteine residues onto Dha or Dhb resulting in the formation of lanthionine and methyl lanthionine respectively (Chatterjee et al., 2005). The structures of selected lantibiotics of the three types are shown in Figure 14.2. The figure reveals that posttranslational modifications result in the formation of multiple thioether rings in the lantibiotic structure. These ringed structures and other posttranslational modifications have been considered to confer stability to the molecule against thermal and proteolytic degradation and also have a role in the antimicrobial activity of the lantibiotics (Sahl and Bierbaum, 1998; McAuliffe et al., 2001).

14.4.2 Class II Bacteriocins (the Unmodified Peptides)

In the early 1990s, bacteriocins such as leucocin A and pediocin PA-1 were described, which were found to be very active against the food pathogen *L. monocytogenes* (Hastings et al., 1991; Henderson et al., 1992). Later, many other peptide bacteriocins were discovered that had antilisterial properties. Studies of the primary structures of these bacteriocins revealed a consensus motif $YGNGV(X)C(X)_4C(X)V(X)_4A$ at their hydrophilic N-terminus with a more varying hydrophobic or amphiphilic C-terminus. These bacteriocins have, therefore, been grouped together, and to date, more than 20 pediocin-like bacteriocins

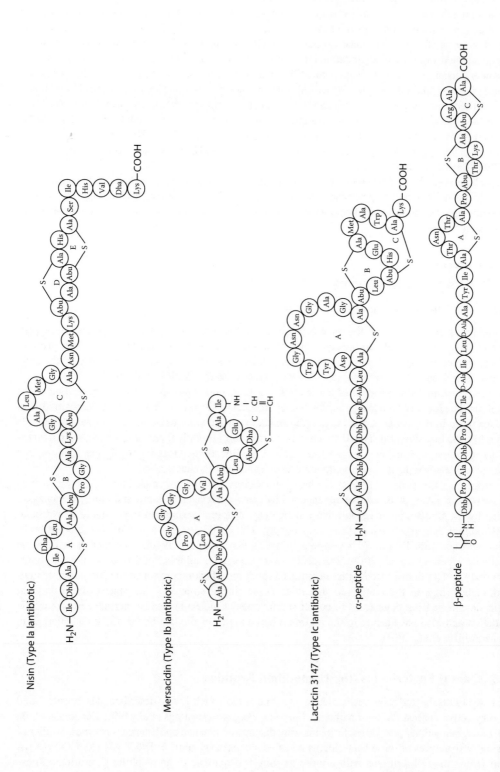

FIGURE 14.2 Structure of representative lantibiotics of subtypes of class I. Different rings of individual lantibiotic are indicated by capital letters. Abu–S–Ala, β-methyl lanthionine; Ala–S–Ala, lanthionine; Dha, dehydroalanine; Dhb, dehydrobutyrine. (Adapted from Nishie, M. et al., *Biocontrol Sci.*, 17, 1, 2012.)

have been described (Drider et al., 2006). In addition to the conserved motif, all the pediocin-like bacteriocins have a disulfide bond formed by the two cysteine residues at their N-terminus. Some bacteriocins (e.g., pediocin PA-1, sakacin G, and enterocin A) have an additional disulfide bond near their C-terminus (Figure 14.3). The presence of these disulfide bonds not only confers stability to the molecule but also enhances the antimicrobial activity of the bacteriocins (Fimland et al., 2000).

The multicomponent unmodified bacteriocins are 30–60 amino acid long cationic molecules having high isoelectric points, with amphiphilic or hydrophobic structures that are necessary for antimicrobial activity against the membranes of sensitive strains (Garneau et al., 2002). These bacteriocins require the complement of two or more peptides for antimicrobial activity. To date, only two-peptide multicomponent bacteriocins have been described.

In addition to the two subtypes of unmodified bacteriocins, type IIc includes all the one-peptide unmodified bacteriocins that do not have a conserved YGNGV motif at their N-terminus. They are very heterogeneous in nature and, therefore, have very diverse structures. The structures of representative bacteriocins of the three types of unmodified bacteriocins are presented in Figure 14.3.

14.4.3 Class III Bacteriocins (Large Proteins)

The majority of bacteriocins studied to date are heat-stable peptides, as discussed earlier. But in addition to these peptides, a few large proteinaceous (>10 kDa), heat-labile bacteriocins have also been described. A subgroup of these proteinaceous bacteriocins are the bacteriolysins, which are endopeptidases that lyse the cell walls of the sensitive strains in an enzymatic manner.

The primary structure of mature bacteriolysins consists of two domains, an N-terminus catalytic domain that belongs to the M37/M23 endopeptidase family, and a C-terminus substrate recognition domain. These two domains are connected by a threonine–proline-rich linker sequence (Cooper and Salmond, 1993; Simmonds et al., 1997; Nilsen et al., 2003; Maliničová et al., 2010).

14.4.4 Class IV Bacteriocins (the Circular Peptides)

The circular bacteriocins are distinct from linear bacteriocins as their N-terminus is ligated to C-terminus via an amide bond giving them a cyclic structure. They are ribosomally synthesized as linear peptides, which then undergo posttranslational cyclization after cleavage of the leader peptide. It has been postulated that the leader peptides play an important role in cyclization (Conlan et al., 2010). Due to the cyclic nature, they are resistant to many proteolytic enzymes and are also more stable at varying pH and temperature (Maqueda et al., 2008).

Enterocin AS-48 is the most extensively studied of the circular bacteriocins and was the first bacteriocin of this group to be purified to homogeneity and identified (Galvez et al., 1986, 1989). It, therefore, stands as the prototype of circular bacteriocins, and its schematic structure is shown in Figure 14.4.

The mature AS-48 molecule is a 70-amino acid globular protein consisting of five α-helices and a high proportion of hydrophobic amino acids (49%) present at the center of the globular structure. This hydrophobic core plays an important role in the antimicrobial activity, by inserting into the lipid bilayer of cell membrane (Sanchez-Barrena et al., 2003). Secondary structure prediction of other circular bacteriocins have shown a similar α-helical structure and a common saposin-like motif that helps in the interaction with the lipid membranes, thus making these circular peptides membrane active (Martin-Visscher et al., 2009).

The linear form of the AS-48 molecule has been shown to be 300-fold less active than the circular form and has also been found very sensitive to temperature changes. Similar results have also been observed for gassericin A, which indicate the importance of the circular form in conferring stability and greater potency against the sensitive strains (Montalban-Lopez et al., 2008).

14.5 Mode of Action of Bacteriocins

The bacteriocins of class I (lantibiotics), class II (unmodified peptides), and class IV (circular bacteriocins) are membrane-active peptides whose primary target is the cytoplasmic membrane of sensitive

Pediocin PA-1/AcH (Type IIa bacteriocin)

Lactococcin Q (Type IIb bacteriocin)

α-peptide

β-peptide

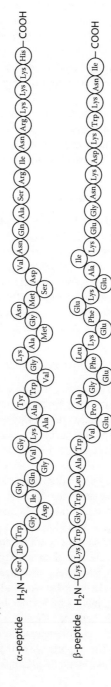

Lactococcin A (Type IIc bacteriocin)

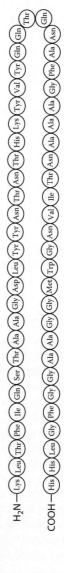

FIGURE 14.3 Structure of representative bacteriocins of subtypes of class II—the unmodified bacteriocins. (Adapted from Nishie, M. et al., *Biocontrol Sci.,* 17, 1, 2012.)

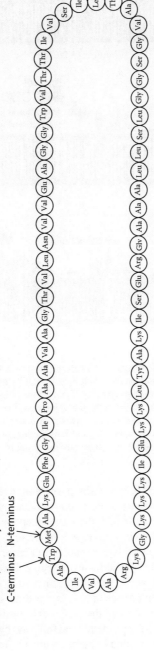

FIGURE 14.4 Schematic representation of the structure of the cyclic bacteriocin enterocin AS-48, a 70-amino acid peptide whose N-terminal amino acid (methionine) is ligated to C-terminal amino acid (tryptophan) forming an amide bond.

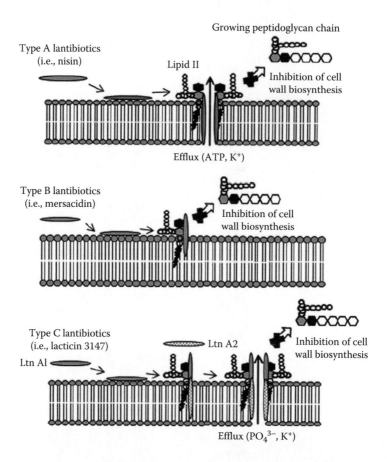

FIGURE 14.5 Proposed model for the mode of action of different subclasses of lantibiotics. *Type A* lantibiotics (i.e., nisin) initially interact with the membrane and then bind with lipid II, stabilizing the complex and making the poration complex in the target site. At the same time, they sequester lipid II that causes cell wall biosynthesis inhibition. *Type B* (i.e., mersacidin) lantibiotics interact with lipid II, resulting in the inhibition of cell wall biosynthesis. *Type C* two-component lantibiotics (i.e., lacticin 3147) interact with cell membrane by A1 peptide and followed by binding with lipid II. This triggers a conformational change of A1 peptide, whereupon a high-affinity binding site is generated for the A2 peptide, which is followed by pore formation. They also inhibit cell wall biosynthesis. (From Islam, M. et al., *Biochem. Soc. Trans.*, 40, 1528, 2012.)

bacteria. Since they are short peptides, they can easily pass through the cell walls of sensitive bacteria. However, they cannot pass through the outer protective membrane of gram-negative bacteria and are, therefore, usually not active against them although there are exceptions.

The lantibiotics have been further subdivided into three subgroups, and each subgroup has a different mode of action as evident from the graphical representation in Figure 14.5. The type Ia linear lantibiotics such as nisin, pep5, and epidermin have dual mode of action. The mode of action of nisin has been studied in detail. It is a cationic molecule and is, therefore, attracted by the negatively charged lipid membrane. On approaching the membrane, nisin uses lipid II (a precursor involved in cell wall biosynthesis) as a docking molecule prior to forming pores in the membrane. Initially, the N-terminal part binds to the carbohydrate moiety of lipid II in the target membrane followed by insertion of C-terminal part into the lipid phase of the membrane. The formation of pores results in the leakage of small molecules such as amino acids, potassium, and ATP (Figure 14.5). The increase in membrane permeability also causes complete dissipation of proton-motive force, that is, transmembrane potential ($\Delta\Psi$) and pH gradient (ΔpH) resulting in cell death (Bruno and Montville, 1993).

The binding of nisin to lipid II also interferes with the cell wall biosynthesis, which increases its efficacy against the sensitive cells. Furthermore, it also binds to negatively charged teichoic and lipoteichoic acids in the cell wall thus activating the autolytic cell wall enzymes resulting in the lysis of the cells (Chatterjee et al., 2005; Bierbaum and Sahl, 2009). This dual mode of action of nisin greatly enhances the bactericidal properties of nisin. In addition to the vegetative cells, nisin has also been found active against spores of *Bacillus* and *Clostridium* spp. Initial findings suggested the reaction of thiol groups in the spore wall with the Dha5 residue in nisin. However, recent experimental evidence shows that removal of the Dha5 residue from nisin does not affect its sporicidal activity. However, truncation of the C-terminus of the nisin molecule does interfere with the activity of nisin against spores, which indicates that nisin possibly acts on the membranes of germinating spores (Rink et al., 2007). The broad antimicrobial spectrum of nisin and its efficacy against the spores has made it very attractive as a food preservative in a wide range of food products.

The type Ib or globular lantibiotics such as mersacidin and actagardine are short peptides up to 19 amino acids in length, which do not form pores in the cytoplasmic membrane. They, however, form complex with lipid II thus preventing its incorporation into peptidoglycan units of cell wall, which results in slow lysis of the cell as shown in Figure 14.5 (Bauer et al., 2005).

The type Ic lantibiotics consist of two or more peptides that act synergistically to form pores in the membranes of sensitive cells. The mode-of-action studies of lacticin 3147, a two-peptide (A1 and A2) lantibiotic, have indicated that the A1 component, which is a short peptide like mersacidin, initially binds to lipid II. This facilitates interaction of the A2 component, which results in the formation of the complex of the two peptides with lipid II. The C-terminus of the A2 component then inserts into the membrane forming a pore resulting in the efflux of PO_4^{3-} and K^+ ions (Figure 14.5) resulting in cell death due to dissipation of membrane potential (Wiedemann et al., 2006).

The class II bacteriocins is a very diverse group of posttranslationally unmodified bacteriocins, which has been further subdivided into three subtypes. The type IIa, also known as pediocin-like bacteriocins, has been found active particularly against *L. monocytogenes*. In addition, they have also been found active against other gram-positive bacterial species belonging to the genera *Enterococcus*, *Lactococcus*, and *Clostridium*. Detailed studies of the mode of action of these listericidal peptides against the sensitive strains have indicated that the target of these peptides is the membrane of sensitive cells. They act on the membrane in two steps. The first step involves interaction of the peptide with a receptor in the membrane, and the second step is the formation of pores or channels (Figure 14.6).

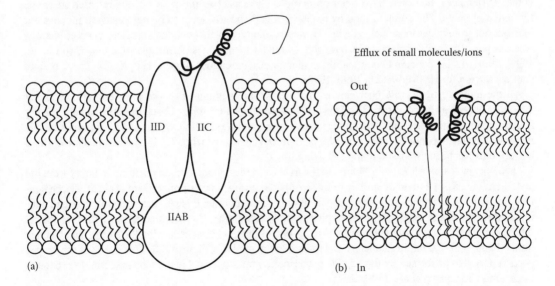

FIGURE 14.6 Schematic representation of mode of action of subclass IIa bacteriocins. (a) IIAB, IIC, and IID represent the subunits of the mannose phosphotransferase system (M-PTS). The bacteriocin recognizes the IIC and IID subunits and uses IIC subunit as a receptor. (b) The bacteriocin inserts into the membrane-forming pores resulting in the efflux of metabolites. (Adapted from Héchard, Y. and Sahl, H.G., *Biochimie*, 84, 545, 2002.)

Specific proteins of the mannose phosphotransferase system (M-PTS) act as receptors for the pediocin-like peptides. This system is responsible for import and translocation of sugars in the bacterial cells and is composed of three major components. One of the components of this system designated as enzyme II (EII) is in turn further composed of four subunits, IIA, IIB, IIC, and IID. The last two subunits (IIC and IID) are associated with the membrane where they form a complex through which the sugars enter the cell (Postma et al., 1993). Heterologous expression of the *mptC* gene (responsible for encoding IIC subunit) in the resistant *L. lactis* strains makes them sensitive to pediocin-like bacteriocins (Ramnath et al., 2004). Detailed studies have revealed that a 40-amino acid extracellular loop within the IIC subunit of sensitive strains acts as a receptor for the bacteriocins as evident from Figure 14.6 (Kjos et al., 2010).

After interaction with the receptor, the next step is the formation of pores in the membrane. This is achieved by insertion of the more hydrophobic or amphiphilic C-terminus region of the bacteriocin into the lipid membrane. After insertion, the more hydrophobic region faces the lipid bilayer, whereas the less hydrophobic region faces the lumen of the pore. The formation of the pores results in the efflux of small molecules particularly ions such as K^+ ions (Figure 14.6), which dissipates the membrane potential and deprives the cell of its energy resulting in death (Ennahar et al., 2000).

The type IIb bacteriocins require two or more peptides for their antimicrobial activity. To date, only two-peptide bacteriocins have been described. Studies have indicated that the individual peptides of a two-peptide bacteriocin are unstructured in aqueous solutions. Upon exposure to a hydrophobic membrane or membrane-mimicking environment, these peptides become organized and behave as a single antimicrobial entity (Hauge et al., 1999). The primary structure of all two-peptide bacteriocins contains GxxxG motifs, which play an important role in the interaction of the individual peptides with each other and penetration of the membranes (Nissen-Meyer et al., 2011). The individual peptides (α and β) of lactococcin G form a helical structure within the membrane of the sensitive strain with the two helices bonded by the GxxxG motifs. The positively charged C-terminus of the α-peptide helps in the penetration of the membrane of sensitive cell due to the membrane potential (negative inside). Within the membrane, the α and β peptides form a well-organized helix–helix structure with the C-termini of both peptides inside the cell (Figure 14.7). This results in the formation of pores with leakage of ions causing cell death (Moll et al., 1996; Oppegard et al., 2008).

Similar results have been obtained for plantaricin E/F and J/K (Fimland et al., 2008; Rogne et al., 2009). Differences, however, have been observed with respect to the type of ions released after the formation of pores. The pores formed by plantaricin E/F and J/K seem to be permeable to monovalent ions but not to divalent ions such as Mg^+. However, plantaricin E/F conducts cations more efficiently, whereas pores formed by plantaricin J/K seem to render membranes permeable to anions (Moll et al., 1999). Similarly, lactococcin G permeabilizes membranes for cations, such as Na^+, K^+, Li^+, Cs^+, Rb^+, and choline, but not for H^+ (Moll et al., 1996, 1998).

All the unmodified peptide bacteriocins that neither are pediocin-like nor have a multicomponent system have been grouped as type IIc. This is very heterogeneous group of bacteriocins that vary greatly in their primary structure and, therefore, have diverse modes of action. Some of the examples of this group include lactococcin A, lactococcin M, lacticin Q, and lactococcin 972. The mode of action of these bacteriocins has been discussed in the following lines.

Lactococcin A and lactococcin M bind to the M-PTS in the cell membrane resulting in pore formation and, therefore, act in a manner similar to those of pediocin-like bacteriocins (Diep et al., 2007). In contrast, lacticin Q does not bind to any receptor in the cell membrane. Instead, the positively charged peptide is attracted by the negatively charged phospholipid membrane. Upon interaction with the membrane, it forms an α helical structure and results in a flip-flop of the lipid bilayer causing the formation of a huge toroidal pore (4.6–6.6 nm in diameter) as depicted in the model in Figure 14.8. This is one of the largest pores reported to be formed by the peptide bacteriocins and is permeable to many macromolecules such as proteins (Yoneyama et al., 2009, 2011).

In opposition to the pore-forming activity of these bacteriocins, a different mode of action has been observed for lactococcin 972. This bacteriocin has been found to be active only against the dividing cells where it inhibits septum formation. No antimicrobial activity of this bacteriocin has been reported for nondividing cells of sensitive strains. Detailed analyses have shown that lactococcin 972 binds to

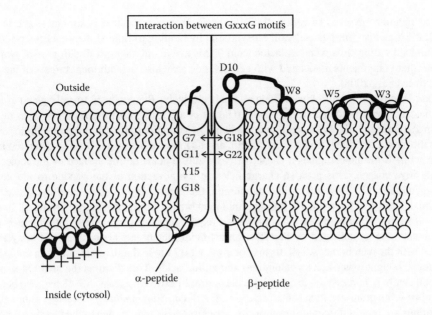

FIGURE 14.7 Schematic representation of mode of action of lactococcin G and its orientation in cytoplasmic membrane of sensitive cell. The cylinders indicate helical regions in the α and β peptides. The two peptides interact through G_7xxxG_{11}-motif in the α peptide and $G_{18}xxxG_{22}$-motif in the β peptide and form a transmembrane helix–helix structure. The highly positively charged flexible C-terminal end (shown by + signs) is forced through the cytosolic side of the membrane by the transmembrane potential (negative inside). The three tryptophan residues (W3, W5, and W8) at the N-terminal of the β peptide are near the outer membrane interface. (Adapted from Nissen-Meyer, J. et al., *Curr. Pharm. Biotechnol.*, 10, 19, 2009.)

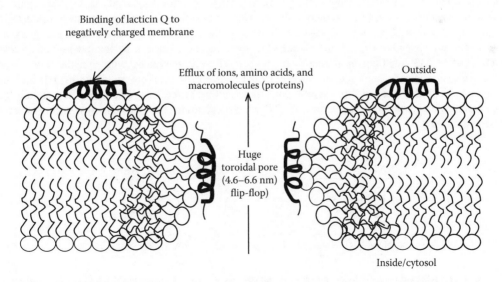

FIGURE 14.8 A model to depict mode of action of lacticin Q. The cationic lacticin Q molecules rapidly bind to the outer leaflet of the cell membrane and form huge toroidal pores (pore diameter, 4.6–6.6 nm) accompanied by lipid flip-flop. (Adapted from Yoneyama, F. et al., *Antimicrob. Agents Chemother.*, 55, 2446, 2011.)

lipid II, a precursor involved in cell wall biosynthesis. In this respect, it is similar to the lantibiotic mersacidin, which also inhibits cell wall biosynthesis by binding to lipid II. However, mersacidin also affects the nondividing cells whereas lactococcin 972 is active only against dividing cells, which indicates that other components associated with cell division may also be additional targets of lactococcin 972 (Martinez et al., 2008).

The class IV circular bacteriocins include all the bacteriocins that are head-to-tail ligated after translation. The examples include enterocin AS-48 (Galvez et al., 1989), uberolysin (Wirawan et al., 2007), gassericin A (Kawai et al., 1998), and leucocyclicin Q (Masuda et al., 2011). Among these, enterocin AS-48 is the most extensively studied bacteriocin, and its mode of action has been investigated in detail. The enterocin AS-48 molecule has a cluster of basic amino acids on the surface of the molecule that gives a positive charge. This positive charge helps in the attraction of the peptide to the negatively charged membrane. On approaching the membrane, the central hydrophobic core is exposed, which helps in the insertion of the molecule into the lipid membrane of the sensitive bacterial cell. In contrast to nisin and similar other peptides, AS-48 does not require any surface receptors and interacts directly with the membrane as indicated by its effect on artificial membranes, liposomes and bilayers. Upon interaction with the membrane, AS-48 forms pores (about 0.7 nm in diameter) resulting in the leakage of low-molecular-weight ions (such as potassium) and amino acids. This disrupts the membrane potential causing cell death. In contrast to most bacteriocins of gram-positive bacteria, AS-48 has also been found active against some gram-negative bacteria such as *E. coli* and *Salmonella typhimurium* although higher concentrations are required than for gram-positive bacteria. This may be due to the fact that AS-48 does not require any surface receptors for antimicrobial activity, and thus, higher concentrations can permeate the outer membrane as well as the inner membrane (Galvez et al., 1991; Sanchez-Barrena et al., 2003). Some other cyclic bacteriocins from LAB such as lactocyclicin Q and leucocyclicin Q also have weak antimicrobial activity against gram-negative bacteria, indicating that they may have a mode of action similar to enterocin AS-48.

In contrast to the peptide bacteriocins, the class III comprises heat-labile proteins having molecular mass >10 kDa. They have been further subdivided into lytic and nonlytic bacteriocins. The lytic bacteriocins act in an enzymatic manner, and their target is the peptidoglycan units of cell walls of sensitive gram-positive bacteria (Figure 14.9). The destruction of the cell wall results in the osmotic lysis of the cell (Malinicová et al., 2010).

The examples of nonlytic heat-labile bacteriocins include helveticin J (Joerger and Klaenhammer, 1986), streptococcin A-M57 (Heng et al., 2004), and dysgalacticin (Heng et al., 2006). The mode of action of dysgalacticin has been studied in detail. The dysgalacticin is a narrow-spectrum bacteriocin mainly active against *S. pyogenes*. Dysgalacticin binds to the phosphoenolpyruvate-dependent glucose- and mannose-PTS within the cell membrane of *S. pyogenes* as evidenced by the low-rate of glucose uptake (1.57 ± 1.13 nmol/mg protein/min) in the starved cells pretreated with dysgalacticin, while untreated cells showed a high rate of glucose uptake (32.8 ± 2.20 nmol/mg protein/min). The binding of dysgalacticin also dissipates membrane potential resulting in the loss of K$^+$ ions. To maintain the membrane potential, the cell uses its store of ATP, which depletes the cell of its energy resulting in death (Swe et al., 2009).

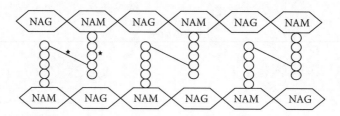

FIGURE 14.9 Schematic representation of peptidoglycan network structure in a gram-positive bacterial cell wall showing the specific target sites of lytic bacteriocins, that is, peptide chains of five amino acids (shown as ○) and cross-linking bridges (shown by horizontal lines). The target sites are marked with *. NAG, *N*-acetyl glucosamine; NAM, *N*-acetyl muramic acid.

14.6 Bacteriocins as Biopreservatives

Bacteriocins secreted by LAB are usually considered safe preservatives since they are active only against other bacteria and are harmless to eukaryotic cells. The exception is cytolysin secreted by *Enterococcus faecalis*, which has been found active against eukaryotic cells such as human, bovine, and horse erythrocytes, retinal cells, polymorphonuclear leukocytes, and human intestinal epithelial cells (Cox et al., 2005). Furthermore, bacteriocins being proteinaceous in nature are inactivated by the digestive proteases. Nisin when treated by the gastric enzyme trypsin is completely inactivated, and, therefore, any ingested nisin will not be harmful to beneficial microbes of the gut (Cleveland et al., 2001). Toxicological studies have shown that the lethal dose of nisin is 6950 mg/kg, which is far higher than that required for effective biopreservation (Balciunas et al., 2013). Similarly, pediocin AcH produced by *P. acidilactici* H was nonimmunogenic when tested in mice and rabbit and was also found sensitive to digestive proteolytic enzymes (Bhunia et al., 1990). These studies indicate the general safety of LAB bacteriocins to humans and their suitability for use as food preservatives.

Bacteriocins are added to the foods to extend shelf life, decrease economic losses due to food spoilage, substitute or reduce the level of added chemical preservatives, and reduce the intensity of harsh physical treatments thereby satisfying the consumer demand for minimally processed and lightly preserved foods. The biopreservative effect of bacteriocins in foods can be obtained in several ways. The simplest and easiest one is the addition of bacteriocin-producing bacterial strain to the product that grows and produces the bacteriocin in situ, thereby preserving the food against sensitive spoilage and pathogenic strains. Alternately, purified or semipurified preparations of bacteriocins can be added to food products to get the preservative effect.

14.6.1 Use of Live Bacterial Cultures for In Situ Bacteriocin Production

Many species and strains of LAB have been traditionally used as cultures for the preparation of fermented products of dairy, meat, and vegetable origin and are, therefore, of GRAS status. These cultures have now been hunted for bacteriocin producers, and bacteriocin-producing strains have been isolated from many traditional fermented dairy products such as *Leuconostoc mesenteroides* 406 isolated from Mongolian fermented mare's milk, airag (Wulijideligen et al., 2012), enterocin A–producing *E. faecium* MMRA isolated from Tunisian fermented milk drink, Rayeb (Rehaiem et al., 2012), enterocin KP–producing *E. faecalis* KP isolated from cheese (Isleroglu et al., 2012), and nisin A–producing *L. lactis* 40FEL3 isolated from Italian fermented foods (Dal Bello et al., 2012).

These bacteriocin-producing strains can be utilized as bacteriocin-producing cultures in fermented food products. If the bacteriocin-producing strain also has the proper technological properties such as high acid and flavor production, then it offers the extra advantage of being used as the sole culture for both fermentation and food preservation. However, if the bacteriocin-producing strain is not suitable as the main culture, then it can be utilized as an adjunct culture, along with the main starter provided that the bacteriocin producer does not interfere with the activity of the main starter culture. The bacteriocin-producing strains are even sometimes utilized for biopreservation of nonfermented foods provided they do not affect the quality of the food. A few examples of strains used for in situ bacteriocin production in fermented and nonfermented food products are shown in Table 14.2.

The table shows that *L. monocytogenes* is the most common organism that has been selected as the potential target of bacteriocins. This is due to the reason that this microbe has been reported to occur in diverse environments and successfully grow at a temperature range from 0°C to 50°C and also at a pH as low as 4.5 (Farber and Peterkin, 1991). This pathogen has been isolated from various food products; however, dairy and meat products are most commonly associated with listeriosis, and several outbreaks have been reported due to the consumption of contaminated dairy and meat products (Makino et al., 2005; de Castro et al., 2012; Lambertz et al., 2012; Hachler et al., 2013). Therefore, dairy and meat products have most commonly been selected as model foods to test the efficacy of bacteriocin-producing strains against *L. monocytogenes* (Table 14.2).

TABLE 14.2

Utilization of Bacteriocin-Producing Cultures for Food Preservation

Strain(s) Used for In Situ Bacteriocin Production	Bacteriocin Produced	Type of Food Preserved	Target Organisms	Reference(s)
P. acidilactici MCH14	Pediocin PA-1	Spanish dry-fermented sausages and frankfurters	*L. monocytogenes* *Clostridium perfringens*	Nieto-Lozano et al. (2010)
L. sakei CTC 494	Sakacin K	Dry fermented sausages	*L. monocytogenes*	Ravyts et al. (2008)
L. sakei	Sakacin P	Fermented sausages	*L. monocytogenes*	Urso et al. (2006)
L. plantarum EC52		Model sausage meat system	*L. monocytogenes* and *E. coli* 0157:H7	Diaz-Ruiz et al. (2012)
Lactococcus lactis subsp. *cremoris* 40FEL3	Nisin A	Cottage cheese	*L. monocytogenes*	Dal Bello et al. (2012)
L. lactis UL719	Nisin Z	Model cheese	*L. sakei*	Aly et al. (2012)
L. lactis DPC4275	Lacticin 3147	Smear-ripened cheese	*L. monocytogenes*	O'Sullivan et al. (2006)
L. lactis MG1614$_{Ent+}$	Enterocin A	Cottage cheese	*L. monocytogenes*	Liu et al. (2008)
E. faecalis A-48-32	Enterocin AS-48	Skimmed milk	*S. aureus*	Munoz et al. (2007)
P. acidilactici strain F	Pediocin	Skimmed milk; milk with 2% fat	*L. monocytogenes*	Somkuti and Steinberg (2010)
L. lactis IFPL 3593	Lacticin 3147	Semi-hard cheeses	*Clostridium tyrobutyricum*	Martinez-Cuesta et al. (2010)
Streptococcus macedonicus ACA-DC198	Macedocin	Kasseri cheese	*C. tyrobutyricum*	Anastasiou et al. (2009)
L. lactis subsp. *lactis* INIA 415	Nisin Z and lacticin 481	Ovine milk cheese	*Clostridium beijerinckii* INIA 63	Garde et al. (2011)
Lactobacillus curvatus CRL705	Lactocin 705 and lactocin AL705	Vacuum-packaged meat	*L. innocua*; *Brochothrix thermosphacta*	Castellano and Vignolo (2006)
E. faecium L50	Enterocins L50 (A and B); Enterocins P and Q	Alcoholic and nonalcoholic lager beers	*L. brevis*, *Pediococcus damnosus*	Basanta et al. (2008)
L. curvatus DF 126 and *L. plantarum* 423	Curvacin DF 126 and plantaricin 423	Ostrich meat salami	*L. monocytogenes*	Dicks et al. (2004)
L. lactis subsp. *lactis* CWBI B1410	Nisin	Traditional fermented fish (guedj)	Enteric bacteria	Diop et al. (2009)
L. plantarum 2.9	Plantaricin	Malted millet flour gruel	*B. cereus, E. coli, Salmonella enterica*	Valenzuela et al. (2008)
Lactobacillus amylovorus DCE 471	Amylorin L	Sourdough	*Leuconostocs, pediococci, and enterococci*	Leroy et al. (2007)
L. sakei 706	Sakacin A	Vacuum-packaged raw meat	*L. monocytogenes*	Jones et al. (2009)
Enterococcus mundtii	Mundticin	Smoked salmon	*L. monocytogenes*	Bigwood et al. (2012)

The utilization of bacteriocin-producing cultures is not only limited to laboratory experiments, but bacteriocin-producing bioprotective cultures are now commercially available as well. For example, an antilisterial culture is being marketed by Chr. Hansen, Denmark, as Bactoferm F-Lc for use in fermented sausages. This culture is a mixture of *P. acidilactici* and *L. curvatus*, which produce pediocin and saka-cin A respectively thus providing the preservative effect. Similarly, Danisco (Copenhagen, Denmark) is marketing HOLDBAC™ protective cultures for the prevention of spoilage and pathogenic contamination in fermented dairy, meat, poultry, and sea foods. These cultures include selected bacterial strains of *L. plantarum*, *Lactobacillus rhamnosus*, *L. sakei*, *Lactobacillus paracasei*, and *Propionibacterium freudenreichii* subsp. *shermanii*. These strains produce bacteriocins as well as other antimicrobials thus providing protection against *Listeria*, yeasts, and molds.

14.6.2 Bacteriocin Preparations as Food Preservatives

In addition to the application of bacteriocin-producing strains and cultures, purified and semipurified bacteriocin preparations have also been studied for their biopreservative efficacy in food products (Table 14.3). The bacteriocins may either be added as sole preservatives or be used in combination with other physical or chemical preservation methods as part of hurdle technology.

As evident from Table 14.3, essential oil, mild heat treatment, gamma irradiation, high hydrostatic pressure, and modified atmosphere storage have been applied as additional hurdles to extend the shelf life of food products. The use of multiple preservation treatments allows setting of each treatment at lower intensities than the level required if only a single preservation method is used. This results in more fresh-tasting, minimally processed, and nutritious products, and the use of safe and natural bacteriocins fits well in applying the concept of hurdle technology (Leistner, 2000).

A glance at Table 14.3 shows that nisin, pediocin PA-1/AcH, and enterocin AS-48 have been more extensively studied as compared to other bacteriocins, and their efficacy has been tested in a range of products of dairy, meat, and vegetable origin. These bacteriocins have properties that make them suitable candidates for use as food preservatives and have, therefore, attracted the attention of researchers.

Nisin is a 34-amino acid peptide with a molecular mass of 3.4 kDa. It is produced by many strains of *L. lactis* (*S. lactis*) and has a broad spectrum of activity against many species of gram-positive bacteria that include methicillin-resistant strains of *S. aureus*, *L. monocytogenes*, *E. faecalis*, *Lactobacillus* spp., and *Lactococcus* spp., and is also active against spores of *Bacillus* spp. and *Clostridium* spp. This broad spectrum has attracted the researchers to utilize nisin as a safe natural preservative in a variety of food products. It is the only bacteriocin to date that has been approved for use as food additive in over 53 countries, and its commercial preparation is available under the trade name of Nisaplin® (Liu and Hansen, 1990; Delves-Broughton, 2005).

The stability and activity of nisin in food products are mainly dependent upon the pH, and it is more active and heat stable at acidic pH (<3.5). The antimicrobial activity and heat stability of nisin markedly decrease at pH>6. Therefore, nisin is more suitable for use in foods at acidic pH. However, foods near to neutral pH can also be preserved with nisin provided a higher dose is used. For example, 1.25–2.5 mg/kg nisin level is considered sufficient for the preservation of high-acid tomato-based products, whereas 2.5–5.0 mg/kg nisin is required for preservation of low-acid canned vegetables (Delves-Broughton, 2005).

Pediocin is another wide-spectrum bacteriocin (active against many gram-positive bacteria) that has the potential to be used as food preservative particularly due to its strong antilisterial activity. It is a 44-amino acid peptide that has been reported to be produced by many strains of *P. acidilactici*. Since it has been isolated from different strains independently, therefore, it is known by different names such as pediocin PA-1, pediocin AcH, pediocin JD, and pediocin 5. However, they all have same structure and properties. Pediocin is a heat-stable peptide that is active over wide pH range (2–10); however, highest activity is achieved between pH 4 and 6.

A commercial pediocin-containing fermentate is also available as ALTA® 2341, which extends the shelf life of many products (Rodriguez et al., 2002) and is particularly found effective to control *L. monocytogenes* in meat and dairy products (Table 14.3).

While nisin and pediocin are effective only against gram-positive bacteria, the spectrum of activity of enterocin AS-48 not only includes most gram-positive bacteria but also extends to some gram-negative bacteria such as *E. coli*. Enterocin AS-48 is a cyclic peptide first reported to be produced by *E. faecalis* S-48 (Galvez et al., 1986). The spectrum of activity of enterocin AS-48 includes several pathogens such as *L. monocytogenes*, *S. aureus*, *Mycobacterium* spp., *B. cereus*, and *E. coli*. Furthermore, enterocin AS-48 is a heat-stable cyclic peptide that is active over a wide pH range (Galvez et al., 1989; Ananou et al., 2008). These properties have attracted researchers to test the effectiveness of this bacteriocin

TABLE 14.3

Utilization of Bacteriocin Preparations as Food Preservatives

Bacteriocin Utilized	Type of Food Preserved	Target Organisms	Additional Food Preservation Treatment (If Any)	Reference(s)
Nisin	Minas frescal cheese	*L. monocytogenes*		Malheiros et al. (2012)
	Whole and low-fat milk	*L. monocytogenes*	Extract of ginseng by-product	Kim et al. (2012)
	Brined white cheese	*L. innocua*	Heat treatment (63°C for 5 min)	Al-Holy et al. (2012)
	Soymilk	*L. monocytogenes*	Garlic shoot juice	Hsieh et al. (2011)
	Rainbow trout fillets	Natural microflora (e.g., LAB and psychrotrophic bacteria)	Essential oil (from *Zataria multiflora*)	Rahimabadi et al. (2013)
	Emulsion type sausages	Natural microflora (LAB, mesophilic and psychrotrophic bacteria), yeast, and mold	Modified atmosphere storage	Khajehali et al. (2012)
	Sausage casings	Spores of *Clostridium sporogenes* ATCC 3584		Wijnker et al. (2011)
	Raw meat	*L. monocytogenes*	Gamma radiation	Mohamed et al. (2011)
Nisin+pediocin	Cabbage, broccoli and mung bean sprouts	*L. monocytogenes*	Sodium acetate, citric acid, phytic acid, and potassium sorbate and EDTA	Bari et al. (2005)
Pediocin (ALTA 2341™)	Frankfurters	*L. monocytogenes*	Thermal pasteurization and vacuum packaging	Chen et al. (2004a)
	Frankfurters	*L. monocytogenes*	Vacuum packaging and postpackaging irradiation	Chen et al. (2004b)
	Bologna sausages	*L. monocytogenes*	Sodium diacetate, sodium lactate, and heat treatment	Grosulescu et al. (2011)
Pediocin K7	Shrikhand (traditional Indian dairy product)	*L. monocytogenes*		Jagannath et al. (2001)
Pediocin AcH	Chicken meat	*L. monocytogenes*		Goff et al. (1996)
Pediocin 5	Milk	*L. monocytogenes*		Huang et al. (1994)
Nisin+ enterocin AS-48	Rice pudding	*S. aureus*	High hydrostatic pressure	Pérez Pulido et al. (2012)
Enterocin AS-48	Commercial sauces	*S. aureus*	Mild heat and 2-nitro-1-propanol	Grande et al. (2012)
	Ready-to-eat salad	*L. monocytogenes*	Essential oils, fatty acids, Nisaplin (nisin)	Molinos et al. (2009)
	Canned corn, canned peas, coconut milk	*Geobacillus stearothermophilus*		Martinez-Viedma et al. (2009)
	Wheat flour dough	*Bacillus* spp.		Martinez-Viedma et al. (2011)
Hyicin 4244	Skimmed milk	*S. aureus*, *L. monocytogenes*		Duarte et al. (2013)
Sakacin C2	Milk (low fat and whole milk)	*E. coli* ATCC 25922	Lecithin, tween-80, ε-polylysine	Gao et al. (2013)

(Continued)

TABLE 14.3 (*Continued*)

Utilization of Bacteriocin Preparations as Food Preservatives

Bacteriocin Utilized	Type of Food Preserved	Target Organisms	Additional Food Preservation Treatment (If Any)	Reference(s)
Gassericins A and T	Custard cream	*Achromobacter denitrificans* AK1113, *B. cereus* AK1124, *L. lactis* subsp. *lactis* AK1155, *Pseudomonas fluorescens* AK1195	Glycine	Arakawa et al. (2009)
Bacteriocin HKT-9 (pediocin PA-1)	Vegetable salad	*Aeromonas hydrophila*, *S. aureus*		Kumar et al. (2012)
Bacteriocin PSY-2	Cod-fish fillets	*Staphylococcus* spp., *Pseudomonas* spp.		Sarika et al. (2012)
Enterocin EJ97	Canned corn, canned peas, coconut milk, coconut juice	*G. stearothermophilus*		Martinez- Viedma et al. (2010)
Bacteriocin RUC9	Iceberg lettuce	*L. monocytogenes*		Randazzo et al. (2009)
Amysin	Sliced bologna sausage	*L. monocytogenes*		Kaewklom et al. (2013)
Gassericin A	Custard cream	*B. cereus* AK1124 and *L. lactis* subsp. *lactis* AK1155	Glycine	Nakamura et al. (2013)
Bacteriocin KC24	UHT milk	*L. monocytogenes*		Han et al. (2013)
Plantaricin A	Paneer	*L. monocytogenes*		Singh et al. (2012)
Enterocin 4231	Dry fermented salami Púchov (Slovak product)	*Listeria innocua* Li1		Laukova and Turek (2011)
Piscicolin KH-1	Kamaboko (steamed surimi)	*E. faecium, Leuconostoc mesenteroides*		Hashimoto et al. (2011)

in food products of dairy, animal, and vegetable origin as evident from Table 14.3. It is also a suitable alternative to nisin especially in meat products where nisin may not be effective as a preservative due to binding to meat components (Rose et al., 1999) or in neutral or alkaline pH foods where nisin is not very effective (Cobos et al., 2001). Recently, a powder preparation of enterocin AS-48 has been obtained (Ananou et al., 2010), and further research may result in a commercial preparation of enterocin AS-48. Therefore, it is expected that a greater choice of safe and natural bacteriocins in utilizable form will be available in the near future.

14.7 Factors Affecting the Efficacy of Bacteriocins in Foods

The earlier account shows the potential of various bacteriocins to be used as food preservatives either produced in situ or ex situ. However, food systems represent a complex environment, and, therefore, various factors may influence the efficacy of bacteriocins. The live bacterial cultures added to a food product for in situ bacteriocin production may show reduced growth and bacteriocin production as compared to those in microbiological media due to unfavorable pH, a_w, temperature, or interference by food components such as spices and salts. The food-related factors such as food structure, buffering capacity, and additives may also negatively or positively affect bacteriocin production. The processing conditions such as freezing, thawing, and homogenization can also damage the cells of producer strain thus negatively affecting bacteriocin production. Sometimes the bacteriocin-producing strain may completely fail to produce the bacteriocin in the food due to loss of bacteriocin-producing trait, infection by bacteriophages, or antagonism of other bacteria toward the producer strain (Rodriguez et al., 2003).

The in situ bacteriocin production also depends on the interaction of the bacteriocin-producing strain with other microbes in the food environment, which include competition for nutrients, development of resistant strains, and production of other antimicrobial substances. The growth of bacteriocinogenic strain will be affected by these factors that, in turn, will also modify the food environment through consumption of nutrients and production of bacteriocin. Therefore, different factors dynamically interact with each other (Figure 14.10) ultimately dictating the food preservation potential of bacteriocinogenic strain in a complex food environment (Galvez et al., 2007).

The efficacy of purified or crude bacteriocin preparations added to foods is also affected by many factors. The composition and physicochemical properties of the food greatly influence the effectiveness of the bacteriocin. For example, nisin is more active in an acidic environment, and its solubility and efficacy are greatly reduced at pH above 6.0 (Liu and Hansen, 1990), and, therefore, higher concentrations of nisin are required for preservation of foods having pH above 6.0. Nisin also tends to bind to phospholipids and form a complex with glutathione (Rose et al., 2003) in meat products thus greatly reducing its efficacy.

Other food-related factors that may reduce the activity of bacteriocins include inactivation by enzymes, uneven distribution in the food matrix, interaction with food additives, and decreasing antimicrobial efficiency of the bacteriocin during the shelf life of food product.

In laboratory experiments, the antimicrobial effect of bacteriocins is tested against selected pathogenic and spoilage bacteria under controlled conditions. However, in real-life conditions, foods get contaminated with a range of microorganisms that include many species of gram-positive and gram-negative bacteria and may as well include fungal contamination. The efficacy of bacteriocins, therefore, may get reduced due to the presence of nonsensitive bacterial species such as gram-negative bacteria or due to the production of proteolytic enzymes. Furthermore, there are variations in the sensitivity of different strains of sensitive species, and appearance of bacteriocin-resistant strains can further aggravate the problem. Strains of *L. monocytogenes*, which have become resistant to nisin (Martinez et al., 2005), sakacin A (Dykes and Hastings, 1998), and plantaricin C19 (Rekhif et al., 1994), have been reported. More worryingly,

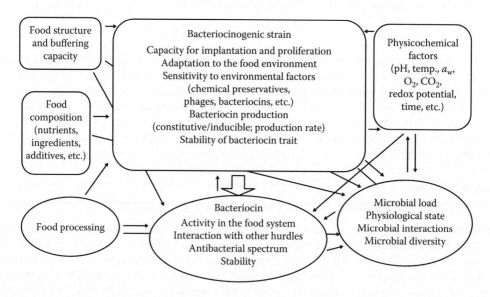

FIGURE 14.10 Influence of different factors on the efficacy of in situ bacteriocin production for biopreservation. (Reprinted from *International Journal of Food Microbiology*, 120, Galvez, A., Abriouel, H., Lopez, R.L., and Ben Omar, N., Bacteriocin-based strategies for food biopreservation, 51–70, Copyright (2007), with permission from Elsevier.)

the resistance of *L. monocytogenes* to any one pediocin-like bacteriocins can confer resistance against others bacteriocins of this class due to a general mechanism of resistance (Gravesen et al., 2002).

Other factors that can make the sensitive bacteria resistant to bacteriocins include formation of spores and biofilms (Kumar and Anand, 1998), thus limiting the efficacy of bacteriocins in foods.

14.8 Incorporation of Bacteriocins or Bacteriocin-Producing Cultures onto Packaging Films

Although the direct addition of bacteriocin-producing cultures or crude bacteriocin preparations to foods is a promising approach, it can also have certain disadvantages as discussed in the preceding paragraphs. Therefore, the concept of antimicrobial food packaging has evolved, which involves coating or incorporation of the antimicrobial onto edible or nonedible packaging films. The antimicrobial substance is slowly released from the film onto the surface of food thus providing long-term protection from any postprocessing contamination (Appendini and Joseph, 2002).

The packaging material used for the preparation of antimicrobial films may be nonedible such as polyethylene (PE), low-density polyethylene (LDPE), poly(vinyl chloride) (PVC), nylon, and even paper and cardboard. The antimicrobial substance either is simply spread onto the film or may involve use of binding agents for adherence to the packaging material. Edible antimicrobial films have also been synthesized, which have the advantage that they can be consumed along with the product. The matrix of such films may be polysaccharide based such as chitosan, alginate, κ-carrageenan, or protein based such as films made with wheat gluten, soy, zein, gelatin, whey and milk protein, and casein (Joerger, 2007).

A range of antimicrobials of natural and synthetic origin have been studied for their efficacy when incorporated onto packaging films; however, bacteriocins especially nisin is the most frequently incorporated antimicrobial onto packaging films (Joerger, 2007). Nisin-containing packaging films have been most frequently investigated against *L. monocytogenes* either as the sole antimicrobial (Ko et al., 2001; Cha et al., 2003; Mauriello et al., 2005) or in combination with other preservation treatments (Lungu and Johnson, 2005; Marcos et al., 2013).

In addition to nisin, antimicrobial films of other bacteriocins have also been successfully tested for the preservation of selected food products. The examples include incorporation of pediocin (ALTA 2351) into a cellulose-based film for the preservation of sliced ham (Santiago-Silva et al., 2009), lactocin 705 coating on a linear low density polyethylene (LLDPE) film (Massani et al., 2013), and LDPE films coated with enterocin 416K1, which efficiently reduced the *L. monocytogenes* counts on the surface of fresh soft cheeses and frankfurters (Iseppi et al., 2008).

Recently, a novel approach of incorporation of viable bacteriocin-producing strains onto packaging films has been investigated. While the antimicrobial activity of bacteriocin-containing films may decrease with the passage of time, the incorporation of bacteriocin-producing strains onto polymeric films can give a significant inhibition throughout the storage period. In an experiment, polyethylene terephthalate (PET) films were coated with either live culture of *Enterococcus casseliflavus* 416K1 or enterocin 416K1. The performance of these films was evaluated against *L. monocytogenes* in artificially contaminated seasoned cheese at 4°C. For the first 3 days, the trend of decrease in *L. monocytogenes* count was similar for both types of films. However, after 7 days, the decline induced by live bacteriocin-producing culture was significantly higher (decreased to 0.6 log, p < 0.05) than for enterocin-doped films (decreased to 2.6 log, p < 0.01), indicating the efficiency of live bacterial culture films (Iseppi et al., 2011). Similarly, when two LAB strains, which were producing bacteriocin-like substances, incorporated into alginate films along with nisin were found to significantly inhibit (p < 0.05) the growth of *L. monocytogenes* on the surface of smoked salmon for 28 days, the film containing nisin alone showed a bacteriostatic effect for only 14 days (Concha-Meyer et al., 2011). In another experiment, bacteriocin-producing *L. plantarum* strain was found to effectively control the growth of *L. innocua* when incorporated into bioactive cellulose-based films (Sanchez-Gonzalez et al., 2013). These recent experiments indicate the increasing interest of the researchers toward the development of films doped with viable bacteriocin-producing cultures due to their greater effectiveness as compared to films coated with bacteriocin preparations.

14.9 Heterologous Production and Bioengineering of Bacteriocins

Bacteriocin production by LAB is a desirable trait, and many LAB strains have been reported as natural bacteriocin producers. However, sometimes it is desirable to express the bacteriocin-producing trait in alternate hosts. The natural bacteriocin-producing strain may be a low producer, poorly adapted to a food environment, or may produce undesirable sensorial effects when grown in a food system. Sometimes resistance to antibiotics or possession of pathogenic characters will make the natural bacteriocin producer unsafe for use in foods. The transfer of bacteriocin-producing trait to a commercial starter strain may also be desirable to prevent the growth of pathogens or spoilage microorganisms during fermentation. Under all these circumstances, subcloning of bacteriocin-producing genes and their heterologous expression is an attractive strategy.

The successful bacteriocin production in alternate hosts depends on many factors that include the host strain selected, plasmid stability and copy number, transcription and translation signals, and the resistance of the host strain to the bacteriocin produced. If the host strain is sensitive to the produced bacteriocin, then the cognate immunity genes need to be transferred along with the structural gene for successful bacteriocin production (Rodriguez et al., 2003).

A diverse range of host species have been successfully modified to heterologously produce LAB bacteriocins, which even include distantly related species such as gram-negative bacterium *E. coli* and yeast cells. LAB are fastidious organisms that require complex media for growth. These media are rich in peptides, and, therefore, purification of bacteriocins from such media is a daunting task. However, *E. coli* can easily be grown in a minimal medium thus making it possible to develop faster and simpler procedures for the recovery of heterologously expressed bacteriocin from the culture supernatant. Many bacteriocins have, therefore, been expressed in *E. coli*, which include pediocin PA-1 (Liu et al., 2011b), enterolysin A (Nigutová et al., 2008), enterocin P (Gutiérrez et al., 2005), and nisin (Karakas-Sen and Narbad, 2012).

Many food-grade LAB strains have also been exploited as alternate hosts for heterologous bacteriocin production, making them useful as in situ bacteriocin producers for food preservation. Many enterocins produced by species of the genus *Enterococcus* have been known for their strong antilisterial activity. However, enterococci may transfer antibiotic resistance genes and sometimes also reported to be responsible for nosocomial infections such as bacteremia, endocarditis, or urinary tract infections (Franz et al., 2011). Therefore, heterologous expression of enterocin-producing genes in food-grade LAB is a good alternative to exploit the antilisterial property of these bacteriocins. Enterocin A is one of the most potent antilisterial bacteriocin (Drider et al., 2006), and the vector pENT02 encoding the genes for enterocin A production, immunity, and transport was introduced into a cheese starter culture strain *L. lactis* MG1614. The resulting transformant *L. lactis*$_{ENT+}$ when used as starter culture was able to reduce *L. monocytogenes* count below detectable limits within 2 days in cottage cheese without any compromise on the cheese quality when compared with a control (Liu et al., 2008). Similarly, the sec-dependent enterocin P structural gene was cloned in plasmid pLEB590 and was used to transform *L. lactis* MG1614. The purified bacteriocin concentration from the recombinant strain was 3.9-fold higher than from the natural enterocin P–producing strain and was found effective against *L. monocytogenes* thus showing the potential to be used for in situ bacteriocin production in dairy products (Liu et al., 2011a).

As discussed earlier, the development of resistance to bacteriocins has been observed for some bacteriocins. Therefore, recombinant strains producing two or more bacteriocins can be more effective when used for in situ bacteriocin production as the failure of one bacteriocin in case of the development of resistance can be compensated by the other. This can be more effective if the bacteriocins are unrelated and belong to different classes thus avoiding cross-resistance among sensitive bacterial species. Furthermore, the production of two bacteriocins can increase the spectrum of activity and effectiveness of recombinants under a wide range of environmental conditions such as pH of foods. With these objectives, a natural nisin producer *L. lactis* ESI 515 was transformed with the plasmid pF12160 carrying the genes for pediocin PA-1 production. The resulting recombinant strain was able to produce pediocin PA-1 without affecting its ability to produce nisin, showing its potential to be used for in situ bacteriocin production in cheese when used as starter culture (Reviriego et al., 2005).

A further advancement to the bacteriocin research is the bioengineering of structural genes to develop mutants having an enhanced activity, a wide antimicrobial spectrum, and an increased efficacy in a complex food environment. The lantibiotics and some class II bacteriocins have been particularly targeted for the development of bioengineered derivatives, and the bioengineering of lantibiotics has been recently covered in comprehensive reviews (Field et al., 2010; Cotter, 2012). Among the lantibiotics, nisin and lacticin 3147 have been the focus of bioengineering studies. Recently, a number of nisin derivatives have been obtained by a single mutation at serine 29. In contrast to natural nisin, these derivatives have enhanced activity against both gram-positive and gram-negative pathogens such as *E. coli* and *Salmonella enterica* serovar Typhimurium (Field et al., 2012). The bioengineering of lacticin 3147 resulted in mutants that were more resistant to heat and proteolytic enzymes as compared to native bacteriocin (Suda et al., 2010).

A few studies describe the bioengineering of class II bacteriocins. In one such study, an effort was made to increase the stability of pediocin PA-1. Bioengineered derivatives were obtained by site-directed mutagenesis where the methionine at position 31 of the mature peptide was replaced by a hydrophobic amino acid (alanine, isoleucine or leucine). The resulting derivatives were more stable and less prone to oxidation as compared to the native pediocin while the antimicrobial activity was not affected (Johnsen et al., 2000) and, therefore, may have greater efficacy when used in a complex food system.

Although utilization of bioengineered bacteriocins as food preservatives is a promising approach, however, it has legal implications. The use of genetically modified microorganisms (GMMs) is restricted by government regulations in many countries; however, self-cloning of LAB, which have GRAS status, attracts less stringent regulations. Self-cloning is

> the removal of nucleic acid sequences from a cell or an organism which may or may not be followed by reinsertion of all or part of that nucleic acid (or a synthetic equivalent) with or without prior enzymic or mechanical steps, into cells of the same species or into cells of phylogenetically closely related species which can exchange genetic material by natural physiological processes where the resulting micro-organism is unlikely to cause disease to humans, animals or plants.
>
> **Council Directive 98/81/EC (1998)**

Therefore, temporary introduction of plasmids, use of native LAB plasmids, or transfer of DNA between strains of same species or closely related species can be regarded as self-cloning, and the mutants, therefore, fall outside the jurisdiction of Contained Use legislation and are not regulated as GMMs. The examples of self-cloning include the development of nisin mutants by alteration of a single codon in the producing strain, and the resulting mutants were found to have enhanced activity against food pathogens such as *L. monocytogenes* (Field et al., 2008). Similarly, native plasmid (pMRC01) of *L. lactis* was modified by combining SOEing (splicing by overlap extension) PCR and gene knockout and/or replacement approaches, and mutants were obtained (Cotter et al., 2003). The success of these self-cloning approaches indicates that safe mutants can be generated secreting bioengineered bacteriocins that are more potent and effective than their natural counterparts. Such mutants can be used for in situ bacteriocin production providing more effective food preservation. Furthermore, purified or semipurified preparations of the bioengineered bacteriocins can also be made available in the future for use in foods. The engineered bacteriocins can, therefore, be regarded as the next generation of natural antimicrobials, which can be utilized for more efficient food preservation.

14.10 Conclusions and Future Trends

Bacteriocins produced by LAB are considered as safe and natural alternatives to harsh physical treatments and potentially harmful chemical preservatives. Therefore, many bacteriocins have been extensively studied for their antimicrobial efficacy in food products. However, the utilization of the majority of the bacteriocins is limited to laboratory experiments, and commercial preparations of only few bacteriocins are available, which include nisin, pediocin, and lacticin 3147. While nisin is the only bacteriocin that has been approved for use as food preservative, the preparations of pediocin and lacticin 3147 are available as fermentates of LAB having GRAS status and as such do not need prior approval for marketing.

The main reasons for the limited commercial exploitation of bacteriocins include higher processing costs and low yields at industrial scale and unavailability of cheap commercial media for optimum bacteriocin production. As an alternative, the utilization of bacteriocin-producing LAB strains is a cheap alternative. Since most LAB strains are GRAS and also have qualified presumption of safety status (Andreoletti et al., 2008), they can be easily utilized for in situ bacteriocin production, and many bacteriocin-producing cultures are already commercially available. This utilization of LAB strains is expected to grow with the coating or incorporation of bacteriocin-producing strains onto packaging films.

Another novel strategy recently employed to enhance the food preservation potential of bacteriocins is the bioengineering of their structural genes. Nisin and lacticin 3147 have been particularly targeted for bioengineering studies, and mutants have been developed that have greater potential to be used as more potent food preservatives than their natural counterparts.

To conclude, it can be said that what has been achieved is just the tip of the iceberg, and there is great potential in the application of bacteriocin-producing LAB strains or the bacteriocin preparations by the food industry. The development of various innovative approaches, such as those described in this chapter, can be commercialized and more widely adopted by the food industry for the preservation of minimally processed food products in a natural way.

REFERENCES

Al-Holy, M.A., A. Al-Nabulsi, T.M. Osaili, M.M. Ayyash, and R.R. Shaker. 2012. Inactivation of *Listeria innocua* in brined white cheese by a combination of nisin and heat. *Food Control* 23:48–53.

Aly, S., J. Floury, M. Piot, S. Lortal, and S. Jeanson. 2012. The efficacy of nisin can drastically vary when produced in situ in model cheeses. *Food Microbiology* 32:185–190.

Ananou, S., A. Munoz, A. Galvez, M. Martinez-Bueno, L. Maqueda, and E. Valdivia. 2008. Optimization of enterocin AS-48 production on a whey-based substrate. *International Dairy Journal* 18:923–927.

Ananou, S., A. Munoz, M. Martinez-Bueno et al. 2010. Evaluation of an enterocin AS-48 enriched bioactive powder obtained by spray drying. *Food Microbiology* 27:58–63.

Anastasiou, R., A. Aktypis, M. Georgalaki, M. Papadelli, L. De Vuyst, and E. Tsakalidou. 2009. Inhibition of *Clostridium tyrobutyricum* by *Streptococcus macedonicus* ACA-DC 198 under conditions mimicking Kasseri cheese production and ripening. *International Dairy Journal* 19:330–335.

Andreoletti, O., H. Bukda, and S. Buncic. 2008. Scientific opinion of the panel on biological hazards on a request from EFSA on the maintenance of the list of QPS microorganisms intentionally added to food or feed. *EFSA Journal* 923:1–48.

Appendini, P. and H.H. Joseph. 2002. Review of antimicrobial food packaging. *Innovative Food Science and Emerging Technologies* 3:113–126.

Arakawa, K., Y. Kawai, H. Iioka et al. 2009. Effects of gassericins A and T, bacteriocins produced by *Lactobacillus gasseri*, with glycine on custard cream preservation. *Journal of Dairy Science* 92:2365–2372.

Balciunas, E.M., F.A.C. Martinez, S.D. Todorov, B.D.G. de Melo Franco, A. Converti, and R.P. de Souza Oliveira. 2013. Novel biotechnological applications of bacteriocins: A review. *Food Control* 32:134–142.

Bari, M.L., D.O. Ukuku, T. Kawasaki, Y. Inatsu, K. Isshiki, and S. Kawamoto. 2005. Combined efficacy of nisin and pediocin with sodium lactate, citric acid, phytic acid, and potassium sorbate and EDTA in reducing the *Listeria monocytogenes* population of inoculated fresh-cut produce. *Journal of Food Protection* 68:1381–1387.

Basanta, A., J. Sanchez, B. Gomez-Sala, C. Herranz, P.E. Hernandez, and L.M. Cintas. 2008. Antimicrobial activity of *Enterococcus faecium* L50, a strain producing enterocins L50 (L50A and L50B), P and Q, against beer-spoilage lactic acid bacteria in broth, wort (hopped and unhopped), and alcoholic and non-alcoholic lager beers. *International Journal of Food Microbiology* 125:293–307.

Bauer, R., M.L. Chikindas, and L.M.T. Dicks. 2005. Purification, partial amino acid sequence and mode of action of pediocin PD-1, a bacteriocin produced by *Pediococcus damnosus* NCFB 1832. *International Journal of Food Microbiology* 101:17–27.

Bhunia, A.K., M.C. Johnson, and B. Ray. 1987. Direct detection of an antimicrobial peptide of *Pediococcus acidilactici* in sodium dodecyl sulfate-polyacrylamide gel electrophoresis. *Journal of Industrial Microbiology* 2:319–322.

Bhunia, A.K., M.C. Johnson, B. Ray, and E.L. Belden. 1990. Antigenic property of pediocin AcH produced by *Pediococcus acidilactici* H. *Journal of Applied Bacteriology* 69:211–215.

Bierbaum, G. and H.G. Sahl. 2009. Lantibiotics: Mode of action, biosynthesis and bioengineering. *Current Pharmaceutical Biotechnology* 10:2–18.

Bigwood, T., J.A. Hudson, J. Cooney et al. 2012. Inhibition of *Listeria monocytogenes* by *Enterococcus mundtii* isolated from soil. *Food Microbiology* 32:354–360.

Bouguettoucha, A., B. Balannec, and A. Amrane. Unstructured models for lactic acid fermentation—A review. 2011. *Food Technology and Biotechnology* 49:3–12.

Bruno, M.E.C. and T.J. Montville. 1993. Common mechanistic action of bacteriocins from lactic acid bacteria. *Applied and Environmental Microbiology* 59:3003–3010.

Carolissen-Mackay, V., G. Arendse, and J.W. Hastings. 1997. Purification of bacteriocins of lactic acid bacteria: Problems and pointers. *International Journal of Food Microbiology* 34:1–16.

Castellano, P. and G. Vignolo. 2006. Inhibition of *Listeria innocua* and *Brochothrix thermosphacta* in vacuum-packaged meat by addition of bacteriocinogenic *Lactobacillus curvatus* CRL705 and its bacteriocins. *Letters in Applied Microbiology* 43:194–199.

Cha, D.S., J.R. Chen, H.J. Park, and M.S. Chinnan. 2003. Inhibition of *Listeria monocytogenes* in tofu by use of a polyethylene film coated with a cellulosic solution containing nisin. *International Journal of Food Science and Technology* 38:499–503.

Chatterjee, C., M. Paul, L.L. Xie, and W.A. van der Donk. 2005. Biosynthesis and mode of action of lantibiotics. *Chemical Reviews* 105:633–683.

Chen, C.M., J.G. Sebranek, J.S. Dickson, and A.F. Mendonca. 2004a. Combining pediocin (ALTA 2341) with postpackaging thermal pasteurization for control of *Listeria monocytogenes* on frankfurters. *Journal of Food Protection* 67:1855–1865.

Chen, C.M., J.G. Sebranek, J.S. Dickson, and A.F. Mendonca. 2004b. Combining pediocin with postpackaging irradiation for control of *Listeria monocytogenes* on frankfurters. *Journal of Food Protection* 67:1866–1875.

Cleveland, J., T.J. Montville, I.F. Nes, and M.L. Chikindas. 2001. Bacteriocins: Safe, natural antimicrobials for food preservation. *International Journal of Food Microbiology* 71:1–20.

Cobos, E.S., V.V. Filimonov, A. Galvez et al. 2001. AS-48: A circular protein with an extremely stable globular structure. *FEBS Letters* 505:379–382.

Concha-Meyer, A., R. Schoebitz, C. Brito, and R. Fuentes. 2011. Lactic acid bacteria in an alginate film inhibit *Listeria monocytogenes* growth on smoked salmon. *Food Control* 22:485–489.

Conlan, F.B., A.D. Gillon, D.J. Craik, and M.A. Anderson. 2010. Circular proteins and mechanisms of cyclization. *Peptide Science* 94:573–583.

Cooper, V.J.C. and G.P.C. Salmond. 1993. Molecular analysis of the major cellulase (celV) of *Erwinia carotovora*: Evidence for an evolutionary "mix-and-match" of enzyme domains. *Molecular and General Genetics* 241:341–350.

Cotter, P.D. 2012. Bioengineering: A bacteriocin perspective. *Bioengineered* 3:313–319.

Cotter, P.D., C. Hill, and R.P. Ross. 2003. A food-grade approach for functional analysis and modification of native plasmids in *Lactococcus lactis*. *Applied and Environmental Microbiology* 69:702–706.

Council Directive. 1998. 98/81/EC of 26 October 1998 amending Directive 90/219/EEC on the contained use of genetically modified micro-organisms. *Official Journal of the European Communities* L 330:13–31.

Cox, C.R., P.S. Coburn, and M.S. Gilmore. 2005. Enterococcal cytolysin: A novel two component peptide system that serves as a bacterial defense against eukaryotic and prokaryotic cells. *Current Protein and Peptide Science* 6:77–84.

Dal Bello, B., L. Cocolin, G. Zeppa, D. Field, P.D. Cotter, and C. Hill. 2012. Technological characterization of bacteriocin producing *Lactococcus lactis* strains employed to control *Listeria monocytogenes* in cottage cheese. *International Journal of Food Microbiology* 153:58–65.

de Castro, V., J.M. Escudero, J.L. Rodriguez et al. 2012. Listeriosis outbreak caused by Latin-style fresh cheese, Bizkaia, Spain, August 2012. *Eurosurveillance* 17:8–10.

Delves-Broughton, J. 2005. Nisin as a food preservative. *Food Australia* 57:525–527.

Diaz-Ruiz, G., N. Ben Omar, H. Abriouel, M. Martinez-Canamero, and A. Galvez. 2012. Inhibition of *Listeria monocytogenes* and *Escherichia coli* by bacteriocin producing *Lactobacillus plantarum* EC52 in a meat sausage model system. *African Journal of Microbiology Research* 6:1103–1108.

Dicks, L.M.T., F.D. Mellett, and L.C. Hoffman. 2004. Use of bacteriocin-producing starter cultures of *Lactobacillus plantarum* and *Lactobacillus curvatus* in production of ostrich meat salami. *Meat Science* 66:703–708.

Diep, D.B., M. Skaugen, Z. Salehian, H. Holo, and I.F. Nes. 2007. Common mechanisms of target cell recognition and immunity for class II bacteriocins. *Proceedings of the National Academy of Sciences of the United States of America* 104:2384–2389.

Diop, M.B., R. Dubois-Dauphin, J. Destain, E. Tine, and P. Thonart. 2009. Use of a nisin producing starter culture of *Lactococcus lactis* subsp. *lactis* to improve traditional fish fermentation in Senegal. *Journal of Food Protection* 72:1930–1934.

Drider, D., G. Fimland, Y. Hechard, L.M. McMullen, and H. Prevost. 2006. The continuing story of class IIa bacteriocins. *Microbiology and Molecular Biology Reviews* 70:564–582.

Duarte, A.F.D., H. Ceotto, M.L.V. Coelho, M. Brito, and M.D.D. Bastos. 2013. Identification of new staphylococcins with potential application as food biopreservatives. *Food Control* 32:313–321.

Dykes, G.A. and J.W. Hastings. 1998. Fitness costs associated with class IIa bacteriocin resistance in *Listeria monocytogenes* B73. *Letters in Applied Microbiology* 26:5–8.

Ennahar, S., T. Sashihara, K. Sonomoto, and A. Ishizaki. 2000. Class IIa bacteriocins: Biosynthesis, structure and activity. *FEMS Microbiology Reviews* 24:85–106.

Farber, J.M. and P.I. Peterkin. 1991. *Listeria monocytogenes*, a food-borne pathogen. *Microbiological Reviews* 55:476–511.

Field, D., M. Begley, P.M. O'Connor et al. 2012. Bioengineered nisin A derivatives with enhanced activity against both Gram-positive and Gram-negative pathogens. *PLoS One* 7:e46884.

Field, D., P.M.O. Connor, P.D. Cotter, C. Hill, and R.P. Ross. 2008. The generation of nisin variants with enhanced activity against specific Gram-positive pathogens. *Molecular Microbiology* 69:218–230.

Field, D., C. Hill, P.D. Cotter, and R.P. Ross. 2010. The dawning of a 'Golden era' in lantibiotic bioengineering. *Molecular Microbiology* 78:1077–1087.

Fimland, G., L. Johnsen, L. Axelsson, M.B. Brurberg, I.F. Nes, V.G.H. Eijsink, and J. Nissen-Meyer. 2000. A C-terminal disulfide bridge in pediocin-like bacteriocins renders bacteriocin activity less temperature dependent and is a major determinant of the antimicrobial spectrum. *Journal of Bacteriology* 182:2643–2648.

Fimland, N., P. Rogne, G. Fimland, J. Nissen-Meyer, and P.E. Kristiansen. 2008. Three-dimensional structure of the two peptides that constitute the two-peptide bacteriocin plantaricin EF. *Biochimica et Biophysica Acta—Proteins and Proteomics* 1784:1711–1719.

Franz, C.M.A.P., M. Huch, H. Abriouel, W. Holzapfel, and A. Gálvez. 2011. Enterococci as probiotics and their implications in food safety. *International Journal of Food Microbiology* 151:125–140.

Galvez, A., H. Abriouel, R.L. Lopez, and N. Ben Omar. 2007. Bacteriocin-based strategies for food biopreservation. *International Journal of Food Microbiology* 120:51–70.

Galvez, A., G. Gimenezgallego, M. Maqueda, and E. Valdivia. 1989. Purification and amino acid composition of peptide antibiotic AS-48 produced by *Streptococcus* (*Enterococcus*) *faecalis* subsp. *liquefaciens* S-48. *Antimicrobial Agents and Chemotherapy* 33:437–441.

Galvez, A., M. Maqueda, M. Martinez-Bueno, and E. Valdivia. 1991. Permeation of bacterial-cells, permeation of cytoplasmic and artificial membrane-vesicles, and channel formation on lipid bilayers by peptide antibiotic AS-48. *Journal of Bacteriology* 173:886–892.

Galvez, A., M. Maqueda, E. Valdivia, A. Quesada, and E. Montoya. 1986. Characterization and partial purification of a broad-spectrum antibiotic AS-48 produced by *Streptococcus faecalis*. *Canadian Journal of Microbiology* 32:765–771.

Gao, Y.R., S.R. Jia, Q. Gao, and Z.L. Tan. 2010. A novel bacteriocin with a broad inhibitory spectrum produced by *Lactobacillus sake* C2, isolated from traditional Chinese fermented cabbage. *Food Control* 21:76–81.

Gao, Y.R., D.P. Li, and X.Y. Liu. 2013. Evaluation of the factors affecting the activity of sakacin C2 against *E. coli* in milk. *Food Control* 30:453–458.

Garde, S., M. Avila, R. Arias, P. Gaya, and M. Nunez. 2011. Outgrowth inhibition of *Clostridium beijerinckii* spores by a bacteriocin-producing lactic culture in ovine milk cheese. *International Journal of Food Microbiology* 150:59–65.

Garneau, S., N.I. Martin, and J.C. Vederas. 2002. Two-peptide bacteriocins produced by lactic acid bacteria. *Biochimie* 84:577–592.

Goff, J.H., A.K. Bhunia, and M.G. Johnson. 1996. Complete inhibition of low levels of *Listeria monocytogenes* on refrigerated chicken meat with pediocin AcH bound to heat-killed *Pediococcus acidilactici* cells. *Journal of Food Protection* 59:1187–1192.

Grande, B., M. Jose, H. Abriouel, R. Lucas, and A. Galvez. 2012. Increasing the microbial inactivation of *Staphylococcus aureus* in sauces by a combination of enterocin AS-48 and 2-nitropropanol, and mild heat treatments. *Food Control* 25:740–744.

Gravesen, A., M. Ramnath, K.B. Rechinger, N. Andersen, L. Jansch, Y. Hechard, J.W. Hastings, and S. Knochel. 2002. High-level resistance to class IIa bacteriocins is associated with one general mechanism in *Listeria monocytogenes*. *Microbiology-SGM* 148:2361–2369.

Grosulescu, C., V.K. Juneja, and S. Ravishankar. 2011. Effects and interactions of sodium lactate, sodium diacetate, and pediocin on the thermal inactivation of starved *Listeria monocytogenes* on bologna. *Food Microbiology* 28:440–446.

Gutiérrez, J., R. Criado, R. Citti, A. Martin, C. Herranz, I.F. Nes, L.M. Cintas, and P.E. Hernandez. 2005. Cloning, production and functional expression of enterocin P, a sec-dependent bacteriocin produced by *Enterococcus faecium* P13, in *Escherichia coli*. *International Journal of Food Microbiology* 103:239–250.

Hachler, H., G. Marti, P. Giannini et al. 2013. Outbreak of listerosis due to imported cooked ham, Switzerland 2011. *Eurosurveillance* 18:7–13.

Han, E.J., N.K. Lee, S.Y. Choi, and H.D. Paik. 2013. Short communication: Bacteriocin KC24 produced by *Lactococcus lactis* KC24 from kimchi and its antilisterial effect in UHT milk. *Journal of Dairy Science* 96:101–104.

Harris, L.J., H.P. Fleming, and T.R. Klaenhammer. 1992. Developments in nisin research. *Food Research International* 25:57–66.

Hashimoto, K., M.L. Bari, Y. Inatsu, S. Kawamoto, and J. Shima. 2011. Biopreservation of kamaboko (steamed surimi) using piscicolin KH1 produced by *Carnobacterium maltalomaticum* KH1. *Japanese Journal of Food Microbiology* 28:193–200.

Hastings, J.W., M. Sailer, K. Johnson, K.L. Roy, J.C. Vederas, and M.E. Stiles. 1991. Characterization of leucocin A-UAL 187 and cloning of the bacteriocin gene from *Leuconostoc gelidum*. *Journal of Bacteriology* 173:7491–7500.

Hata, T., R. Tanaka, and S. Ohmomo. 2010. Isolation and characterization of plantaricin ASM1: A new bacteriocin produced by *Lactobacillus plantarum* A-1. *International Journal of Food Microbiology* 137:94–99.

Hauge, H.H., D. Mantzilas, V.G.H. Eijsink, and J. Nissen-Meyer. 1999. Membrane-mimicking entities induce structuring of the two-peptide bacteriocins plantaricin E/F and plantaricin J/K. *Journal of Bacteriology* 181:740–747.

Héchard, Y. and H.G. Sahl. 2002. Mode of action of modified and unmodified bacteriocins from Gram-positive bacteria. *Biochimie* 84:545–557.

Henderson, J.T., A.L. Chopko, and P.D. van Wassenaar. 1992. Purification and primary structure of pediocin PA-1 produced by *Pediococcus acidilactici* PAC-1.0. *Archives of Biochemistry and Biophysics* 295:5–12.

Heng, N.C.K., G.A. Burtenshaw, R.W. Jack, and J.R. Tagg. 2004. Sequence analysis of pDN571, a plasmid encoding novel bacteriocin production in M-type 57 *Streptococcus pyogenes*. *Plasmid* 52:225–229.

Heng, N.C.K., N.L. Ragland, P.M. Swe, H.J. Baird, M.A. Inglis, J.R. Tagg, and R.W. Jack. 2006. Dysgalacticin: A novel, plasmid-encoded antimicrobial protein (bacteriocin) produced by *Streptococcus dysgalactiae* subsp. *equisimilis*. *Microbiology-SGM* 152:1991–2001.

Heng, N.C.K., P.A. Wescombe, J.P. Burton, R.W. Jack, and J.R. Tag. 2007. The diversity of bacteriocins in Gram-positive bacteria. In *Bacteriocins: Ecology and Evolution*, M.A. Riley, and M.A. Chavan, eds., pp. 45–92. Berlin, Germany: Springer.

Hsieh, Y.H., M. Yan, J.G. Liu, and J.C. Hwang. 2011. The synergistic effect of nisin and garlic shoot juice against *Listeria* spp. in soymilk. *Journal of the Taiwan Institute of Chemical Engineers* 42:576–579.

Huang, J., C. Lacroix, H. Daba, and R.E. Simard. 1994. Growth of *Listeria monocytogenes* in milk and its control by pediocin 5 produced by *Pediococcus acidilactici* UL5. *International Dairy Journal* 4:429–443.

Iseppi, R., S. de Niederhausern, I. Anacarso et al. 2011. Anti-listerial activity of coatings entrapping living bacteria. *Soft Matter* 7:8542–8548.

Iseppi, R., F. Pilati, M. Marini et al. 2008. Anti-listerial activity of a polymeric film coated with hybrid coatings doped with enterocin 416K1 for use as bioactive food packaging. *International Journal of Food Microbiology* 123:281–287.

Islam, M., J. Nagao, T. Zendo, and K. Sonomoto. 2012. Antimicrobial mechanism of lantibiotics. *Biochemical Society Transactions* 40:1528–1533.

Isleroglu, H., Z. Yildirim, M. Tokatli, N. Oncul, and M. Yildirim. 2012. Partial characterisation of enterocin KP produced by *Enterococcus faecalis* KP, a cheese isolate. *International Journal of Dairy Technology* 65:90–97.

Jack, R.W. and H.G. Sahl. 1995. Unique peptide modifications involved in the biosynthesis of lantibiotics. *Trends in Biotechnology* 13:269–278.

Jagannath, A., A. Ramesh, M.N. Ramesh, A. Chandrashekar, and M.C. Varadaraj. 2001. Predictive model for the behavior of *Listeria monocytogenes* Scott A in Shrikhand, prepared with a biopreservative, pediocin K7. *Food Microbiology* 18:335–343.

Joerger, M.C. and T.R. Klaenhammer. 1986. Characterization and purification of helveticin J and evidence for a chromosomally determined bacteriocin produced by *Lactobacillus helveticus* 481. *Journal of Bacteriology* 167:439–446.

Joerger, R.D. 2007. Antimicrobial films for food applications: A quantitative analysis of their effectiveness. *Packaging Technology and Science* 20:231–273.

Johnsen, L., G. Fimland, V. Eijsink, and J. Nissen-Meyer. 2000. Engineering increased stability in the antimicrobial peptide pediocin PA-1. *Applied and Environmental Microbiology* 66:4798–4802.

Jones, R.J., M. Zagorec, G. Brightwell, and J.R. Tagg. 2009. Inhibition by *Lactobacillus sakei* of other species in the flora of vacuum packaged raw meats during prolonged storage. *Food Microbiology* 26:876–881.

Kaewklom, S., S. Lumlert, W. Kraikul, and R. Aunpad. 2013. Control of *Listeria monocytogenes* on sliced bologna sausage using a novel bacteriocin, amysin, produced by *Bacillus amyloliquefaciens* isolated from Thai shrimp paste (Kapi). *Food Control* 32:552–557.

Karakas-Sen, A. and A. Narbad. 2012. Heterologous expression and purification of nisA, the precursor peptide of lantibiotic nisin from *Lactococcus lactis*. *Acta Biologica Hungarica* 63:301–310.

Kashani, H.H., H. Nikzad, S. Mobaseri, and E.S. Hoseini. 2012. Synergism effect of nisin peptide in reducing chemical preservatives in food industry. *Life Science Journal—Acta Zhengzhou University Overseas Edition* 9:496–501.

Kawai, Y., T. Saito, H. Kitazawa, and T. Itoh. 1998. Gassericin A: An uncommon cyclic bacteriocin produced by *Lactobacillus gasseri* LA39 linked at N- and C-terminal ends. *Bioscience Biotechnology and Biochemistry* 62:2438–2440.

Khajehali, E., S.S. Shekarforoush, A.H.K. Nazer, and S. Hoseinzadeh. 2012. Effects of nisin and Modified Atmosphere Packaging (MAP) on the quality of emulsion-type sausage. *Journal of Food Quality* 35:119–126.

Kim, W.J., K.Y. Min, K.T. Kim et al. 2012. Antimicrobial effect of the extracts of a ginseng by-product produced by subcritical water extraction, nisin, and their combination against *Listeria monocytogenes* in milk products. *Milchwissenschaft—Milk Science International* 67:370–373.

Kjos, M., Z. Salehian, I.F. Nes, and D.B. Diep. 2010. An extracellular loop of the mannose phosphotransferase system component IIC is responsible for specific targeting by class IIa bacteriocins. *Journal of Bacteriology* 192:5906–5913.

Ko, S., M.E. Janes, N.S. Hettiarachchy, and M.G. Johnson. 2001. Physical and chemical properties of edible films containing nisin and their action against *Listeria monocytogenes*. *Journal of Food Science* 66:1006–1011.

Kumar, C.G. and S.K. Anand. 1998. Significance of microbial biofilms in food industry: A review. *International Journal of Food Microbiology* 42:9–27.

Kumar, M., A.K. Jain, M. Ghosh, and A. Ganguli. 2012. Potential application of an anti-aeromonas bacteriocin of *Lactococcus lactis* ssp. *lactis* in the preservation of vegetable salad. *Journal of Food Safety* 32:369–378.

Lambertz, S.T., C. Nilsson, A. Bradenmark et al. 2012. Prevalence and level of *Listeria monocytogenes* in ready-to-eat foods in Sweden 2010. *International Journal of Food Microbiology* 160:24–31.

Laukova, A. and P. Turek. 2011. Effect of enterocin 4231 in Slovak fermented salami Puchov after its experimental inoculation with *Listeria innocua* Li1. *Acta Scientiarum Polonorum. Technologia Alimentaria* 10:423–431.

Lee, N.K., Park, Y.L., Park, Y.H., Kim, J.M., Nam, H.M., Jung, S.C., and Paik, H.D. 2010. Purification and characterization of pediocin SA131 produced by *Pediococcus pentosaceus* SA131 against bovine mastitis pathogens. *Milchwissenschaft—Milk Science International* 65:19–21.

Leistner, L. 2000. Basic aspects of food preservation by hurdle technology. *International Journal of Food Microbiology* 55:181–186.

Leroy, F., T. De Winter, M.R.F. Moreno, and L. De Vuyst. 2007. The bacteriocin producer *Lactobacillus amylovorus* DCE 471 is a competitive starter culture for type II sourdough fermentations. *Journal of the Science of Food and Agriculture* 87:1726–1736.

Liu, G.R., H.F. Wang, M.W. Griffiths, and P.L. Li. 2011a. Heterologous extracellular production of enterocin P in *Lactococcus lactis* by a food-grade expression system. *European Food Research and Technology* 233:123–129.

Liu, L., P. O'Conner, P.D. Cotter, C. Hill, and R.P. Ross. 2008. Controlling *Listeria monocytogenes* in cottage cheese through heterologous production of enterocin A by *Lactococcus lactis*. *Journal of Applied Microbiology* 104:1059–1066.

Liu, S.N., Y. Han, and Z.J. Zhou. 2011b. Fusion expression of pedA gene to obtain biologically active pediocin PA-1 in *Escherichia coli*. *Journal of Zhejiang University: Science B* 12:65–71.

Liu, W. and J.N. Hansen. 1990. Some chemical and physical-properties of nisin, a small-protein antibiotic produced by *Lactococcus lactis*. *Applied and Environmental Microbiology* 56:2551–2558.

Lungu, B. and M.G. Johnson. 2005. Potassium sorbate does not increase control of *Listeria monocytogenes* when added to zein coatings with nisin on the surface of full fat turkey frankfurter pieces in a model system at 4°C. *Journal of Food Science* 70:M95–M99.

Makino, S.I., K. Kawamoto, K. Takeshi et al. 2005. An outbreak of food-borne listeriosis due to cheese in Japan, during 2001. *International Journal of Food Microbiology* 104:189–196.

Malheiros, P.D., D.J. Daroit, and A. Brandelli. 2012. Inhibition of *Listeria monocytogenes* in minas frescal cheese by free and nanovesicle encapsulated nisin. *Brazilian Journal of Microbiology* 43:1414–1418.

Maliničová, L., M. Piknová, P. Pristaš, and P. Javorský. 2010. Peptidoglycan hydrolases as novel tool for anti-enterococcal therapy. In *Current Research, Technology and Education Topics in Applied Microbiology and Microbial Biotechnology* (Vol. 1), A. Méndez-Vilas, ed., pp. 463–472. Badajoz, Spain: Formatex Research Centre.

Maqueda, M., M. Sanchez-Hidalgo, M. Fernandez, M. Montalban-Lopez, E. Valdivia, and M. Martinez-Bueno. 2008. Genetic features of circular bacteriocins produced by Gram-positive bacteria. *FEMS Microbiology Reviews* 32:2–22.

Marcos, B., T. Aymerich, M. Garriga, and J. Arnau. 2013. Active packaging containing nisin and high pressure processing as post-processing listericidal treatments for convenience fermented sausages. *Food Control* 30:325–330.

Martinez, B., T. Bottiger, T. Schneider, A. Rodriguez, H.G. Sahl, and I. Wiedemann. 2008. Specific interaction of the unmodified bacteriocin lactococcin 972 with the cell wall precursor lipid II. *Applied and Environmental Microbiology* 74:4666–4670.

Martinez, B., D. Bravo, and A. Rodriguez. 2005. Consequences of the development of nisin-resistant *Listeria monocytogenes* in fermented dairy products. *Journal of Food Protection* 68:2383–2388.

Martinez-Cuesta, M.C., J. Bengoechea, I. Bustos, B. Rodriguez, T. Requena, and C. Pelaez. 2010. Control of late blowing in cheese by adding lacticin 3147-producing *Lactococcus lactis* IFPL 3593 to the starter. *International Dairy Journal* 20:18–24.

Martinez-Viedma, P.M., H. Abriouel, N. Ben Omar, R.L. Lopez, and A. Galvez. 2010. Effect of enterocin EJ97 against *Geobacillus stearothermophilus* vegetative cells and endospores in canned foods and beverages. *European Food Research and Technology* 230:513–519.

Martinez-Viedma, P.M., H. Abriouel, N. Ben Omar, R.L. Lopez, and A. Galvez. 2011. Inhibition of spoilage and toxigenic *Bacillus species* in dough from wheat flour by the cyclic peptide enterocin AS-48. *Food Control* 22:756–761.

Martinez-Viedma, P.M., H. Abriouel, N. Ben Omar, R.L. Lopez, E. Valdivia, and A. Galvez. 2009. Inactivation of *Geobacillus stearothermophilus* in canned food and coconut milk samples by addition of enterocin AS-48. *Food Microbiology* 26:289–293.

Martin-Visscher, L.A., X.D. Gong, M. Duszyk, and J.C. Vederas. 2009. The three dimensional structure of carnocyclin A reveals that many circular bacteriocins share a common structural motif. *Journal of Biological Chemistry* 284:28674–28681.

Massani, M.B., G.M. Vignolo, P. Eisenberg, and P.J. Morando. 2013. Adsorption of the bacteriocins produced by *Lactobacillus curvatus* CRL705 on a multilayer-LLDPE film for food-packaging applications. *LWT— Food Science and Technology* 53:128–138.

Masuda, Y., H. Ono, H. Kitagawa, H. Ito, F.Q. Mu, N. Sawa, T. Zendo, and K. Sonomoto. 2011. Identification and characterization of leucocyclicin Q, a novel cyclic bacteriocin produced by *Leuconostoc mesenteroides* TK41401. *Applied and Environmental Microbiology* 77:8164–8170.

Mattick, A.T.R. and A. Hirsch. 1947. Further observations on an inhibitory substance (nisin) from lactic streptococci. *Lancet* 2:5–8.

Mauriello, G., E. De Luca, A. La Storia, F. Villani, and D. Ercolini. 2005. Antimicrobial activity of a nisin-activated plastic film for food packaging. *Letters in Applied Microbiology* 41:464–469.

McAuliffe, O., R.P. Ross, and C. Hill. 2001. Lantibiotics: Structure, biosynthesis and mode of action. *FEMS Microbiology Reviews* 25:285–308.

Mills, S., C. Stanton, C. Hill, and R.P. Ross. 2011. New developments and applications of bacteriocins and peptides in foods. *Annual Review of Food Science and Technology* 2:299–329.

Mohamed, H.M., F.A. Elnawawi, and A.E. Yousef. 2011. Nisin treatment to enhance the efficacy of gamma radiation against *Listeria monocytogenes* on meat. *Journal of Food Protection* 74:193–199.

Molinos, A.C., H. Abriouel, R.L. Lopez, N. Ben Omar, E. Valdivia, and A. Galvez. 2009. Enhanced bactericidal activity of enterocin AS-48 in combination with essential oils, natural bioactive compounds and chemical preservatives against *Listeria monocytogenes* in ready-to-eat salad. *Food and Chemical Toxicology* 47:2216–2223.

Moll, G., H.H. Hauge, J. Nissen-Meyer, I.F. Nes, W.N. Konings, and A.J.M. Driessen. 1998. Mechanistic properties of the two-component bacteriocin lactococcin G. *Journal of Bacteriology* 180:96–99.

Moll, G., T.K. Ubbink, H.H. Hauge, J. Nissen-Meyer, I.F. Nes, W.N. Konings, and A.J.M. Driessen. 1996. Lactococcin G is a potassium ion-conducting, two-component bacteriocin. *Journal of Bacteriology* 178:600–605.

Moll, G.N., E. van den Akker, H.H. Hauge, J. Nissen-Meyer, I.F. Nes, W.N. Konings, and A.J.M. Driessen. 1999. Complementary and overlapping selectivity of the two-peptide bacteriocins plantaricin EF and JK. *Journal of Bacteriology* 181:4848–4852.

Montalban-Lopez, M., B. Spolaore, O. Pinato, M. Martinez-Bueno, E. Valdivia, M. Maqueda, and A. Fontana. 2008. Characterization of linear forms of the circular enterocin AS-48 obtained by limited proteolysis. *FEBS Letters* 582:3237–3242.

Munoz, A., S. Ananou, A. Galvez et al. 2007. Inhibition of *Staphylococcus aureus* in dairy products by enterocin AS-48 produced in situ and ex situ: Bactericidal synergism with heat. *International Dairy Journal* 17:760–769.

Nakamura, K., K. Kensuke, Y. Yasushi et al. 2013. Food preservative potential of gassericin A containing concentrate prepared from cheese whey culture supernatant of *Lactobacillus gasseri* LA39. *Animal Science Journal* 84:144–149.

Nieto-Lozano, J.C., J.I. Reguera-Useros, M.D. Pelaez-Martinez, G. Sacristan-Perez-Minayo, A.J. Gutierrez-Fernandez, and A.H. de la Torre. 2010. The effect of the pediocin PA-1 produced by *Pediococcus acidilactici* against *Listeria monocytogenes* and *Clostridium perfringens* in Spanish dry-fermented sausages and frankfurters. *Food Control* 21:679–685.

Nilsen, T., I.F. Nes, and H. Holo. 2003. Enterolysin A, a cell wall-degrading bacteriocin from *Enterococcus faecalis* LMG 2333. *Applied and Environmental Microbiology* 69:2975–2984.

Nigutová, K., L. Serenčová, M. Piknová, P. Javorský, and P. Pristaš. 2008. Heterologous expression of functionally active enterolysin A, class III bacteriocin from *Enterococcus faecalis* in *Escherichia coli*. *Protein Expression and Purification* 60:20–24.

Nishie, M., J.I. Nagao, and K. Sonomoto. 2012. Antibacterial peptides and bacteriocins: An overview of their diverse characteristics and applications. *Biocontrol Science* 17:1–16.

Nissen-Meyer, J., C. Oppegard, P. Rogne, H.S. Haugen, and P.E. Kristiansen. 2011. The two-peptide (Class-IIb) bacteriocins: Genetics, biosynthesis, structure, and mode of action. In *Prokaryotic Antimicrobial Peptides: From Genes to Applications*, D. Drider and S. Rebuffat, eds., pp. 197–212. New York: Springer.

Nissen-Meyer, J., P. Rogne, C. Oppegard, H.S. Haugen, and P.E. Kristiansen. 2009. Structure-function relationships of the non-lanthionine-containing peptide (class II) bacteriocins produced by gram-positive bacteria. *Current Pharmaceutical Biotechnology* 10:19–37.

Oman, T.J., J.M. Boettcher, H.A. Wang, X.N. Okalibe, and W.A. van der Donk. 2011. Sublancin is not a lanti-biotic but an S-linked glycopeptide. *Nature Chemical Biology* 7:78–80.

Oppegard, C., J. Schmidt, P.E. Kristiansen, and J. Nissen-Meyer. 2008. Mutational analysis of putative helix-helix interacting GxxxG-motifs and tryptophan residues in the two-peptide bacteriocin lactococcin G. *Biochemistry* 47:5242–5249.

O'Sullivan, L., E.B. O'Connor, R.P. Ross, and C. Hill. 2006. Evaluation of live-culture producing lacticin 3147 as a treatment for the control of *Listeria monocytogenes* on the surface of smear-ripened cheese. *Journal of Applied Microbiology* 100:135–143.

Parada, J.L., C.R. Caron, A.B.P. Medeiros, and C.R. Soccol. 2007. Bacteriocins from lactic acid bacteria: Purification, properties and use as biopreservatives. *Brazilian Archives of Biology and Technology* 50:521–542.

Pérez Pulido, R., J.T. del Arbol, M.J.G. Burgos, and A. Galvez. 2012. Bactericidal effects of high hydrostatic pressure treatment singly or in combination with natural antimicrobials on *Staphylococcus aureus* in rice pudding. *Food Control* 28:19–24.

Pingitore, E.V., E.M. Hebert, F. Sesma, and M.E. Nader-Macias. 2009. Influence of vitamins and osmolites on growth and bacteriocin production by *Lactobacillus salivarius* CRL 1328 in a chemically defined medium. *Canadian Journal of Microbiology* 55:304–310.

Postma, P.W., J.W. Lengeler, and G.R. Jacobson. 1993. Phosphoenolpyruvate—Carbohydrate phosphotransfer-ase systems of bacteria. *Microbiological Reviews* 57:543–594.

Rahimabadi, E.Z., M. Rigi, and M. Rahnama. 2013. Combined effects of *Zataria multiflora* boiss essential oil and nisin on the shelf-life of refrigerated rainbow trout (*Onchorynchus mykiss*) fillets. *Iranian Journal of Fisheries Sciences* 12:115–126.

Ramnath, M., S. Arous, A. Gravesen, J.W. Hastings, and Y. Hechard. 2004. Expression of mptC of *Listeria monocytogenes* induces sensitivity to class IIa bacteriocins in *Lactococcus lactis*. *Microbiology-SGM* 150:2663–2668.

Randazzo, C.L., I. Pitino, G.O. Scifo, and C. Caggia. 2009. Biopreservation of minimally processed iceberg lettuces using a bacteriocin produced by *Lactococcus lactis* wild strain. *Food Control* 20:756–763.

Ravyts, F., S. Barbuti, M.A. Frustoli et al. 2008. Competitiveness and antibacterial potential of bacteriocin-producing starter cultures in different types of fermented sausages. *Journal of Food Protection* 71:1817–1827.

Rehaiem, A., B. Martinez, M. Manai, and A. Rodriguez. 2012. Technological performance of the enterocin A producer *Enterococcus faecium* MMRA as a protective adjunct culture to enhance hygienic and sen-sory attributes of traditional fermented milk 'Rayeb'. *Food and Bioprocess Technology* 5:2140–2150.

Rekhif, N., A. Atrih, and G. Lefebvre. 1994. Selection and properties of spontaneous mutants of *Listeria monocytogenes* ATCC 15313 resistant to different bacteriocins produced by lactic acid bacteria strains. *Current Microbiology* 28:237–241.

Reviriego, C., A. Fernández, N. Horn, E. Rodríguez, M.L. Marín, L. Fernández, and J.M. Rodríguez. 2005. Production of pediocin PA-1, and coproduction of nisin A and pediocin PA-1, by wild *Lactococcus lactis* strains of dairy origin. *International Dairy Journal* 15:45–49.

Rink, R., J. Wierenga, A. Kuipers, L.D. Kluskens, A.J.M. Driessen, O.P. Kuipers, and G.N. Moll. 2007. Dissection and modulation of the four distinct activities of nisin by mutagenesis of rings A and B and by C-terminus truncation. *Applied and Environmental Microbiology* 73:5809–5816.

Rodriguez, J.M., M.I. Martinez, N. Horn, and H.M. Dodd. 2003. Heterologous production of bacteriocins by lactic acid bacteria. *International Journal of Food Microbiology* 80:101–116.

Rodriguez, J.M., M.I. Martinez, and J. Kok. 2002. Pediocin PA-1, a wide-spectrum bacteriocin from lactic acid bacteria. *Critical Reviews in Food Science and Nutrition* 42:91–121.

Rogne, P., C. Haugen, G. Fimland, J. Nissen-Meyer, and P.E. Kristiansen. 2009. Three-dimensional structure of the two-peptide bacteriocin plantaricin JK. *Peptides* 30:1613–1621.

Rose, N.L., P. Sporns, H.M. Dodd, M.J. Gasson, F.A. Mellon, and L.M. McMullen. 2003. Involvement of dehydroalanine and dehydrobutyrine in the addition of glutathione to nisin. *Journal of Agricultural and Food Chemistry* 51:3174–3178.

Rose, N.L., P. Sporns, M.E. Stiles, and L.M. McMullen. 1999. Inactivation of nisin by glutathione in fresh meat. *Journal of Food Science* 64:759–762.

Saavedra, L. and F. Sesma. 2011. Purification techniques of bacteriocins from lactic acid bacteria and other Gram-positive bacteria. In *Prokaryotic Antimicrobial Peptides—From Genes to Applications*, D. Drider and S. Rebuffat, eds., pp. 99–113. New York: Springer.

Sahl, H.G. and G. Bierbaum. 1998. Lantibiotics: Biosynthesis and biological activities of uniquely modified peptides from Gram-positive bacteria. *Annual Review of Microbiology* 52:41–79.

Sanchez-Barrena, M.J., M. Martinez-Ripoll, A. Galvez, E. Valdivia, M. Maqueda, V. Cruz, and A. Albert. 2003. Structure of bacteriocin AS-48: From soluble state to membrane bound state. *Journal of Molecular Biology* 334:541–549.

Sanchez-Gonzalez, L., J.I.Q. Saavedra, and A. Chiralt. 2013. Physical properties and antilisterial activity of bioactive edible films containing *Lactobacillus plantarum*. *Food Hydrocolloids* 33:92–98.

Santiago-Silva, P., N.F.F. Soares, J.E. Nobrega et al. 2009. Antimicrobial efficiency of film incorporated with pediocin (ALTA® 2351) on preservation of sliced ham. *Food Control* 20:85–89.

Sarika, A.R., A.P. Lipton, M.S. Aishwarya, and R.S. Dhivya. 2012. Isolation of a bacteriocin producing *Lactococcus lactis* and application of its bacteriocin to manage spoilage bacteria in high-value marine fish under different storage temperatures. *Applied Biochemistry and Biotechnology* 167:1280–1289.

Sawa, N., S. Koga, K. Okamura, N. Ishibashi, T. Zendo and K. Sonomoto. 2013. Identification and characterization of novel multiple bacteriocins produced by *Lactobacillus sakei* D98. *Journal of Applied Microbiology* 115:61–69.

Sawa, N., T. Zendo, J. Kiyofuji, K. Fujita, K. Himeno, J. Nakayama, and K. Sonomoto. 2009. Identification and characterization of lactocyclicin Q, a novel cyclic bacteriocin produced by *Lactococcus* sp. strain QU 12. *Applied and Environmental Microbiology* 75:1552–1558.

Simmonds, R.S., W.J. Simpson, and J.R. Tagg. 1997. Cloning and sequence analysis of zooA, a *Streptococcus zooepidemicus* gene encoding a bacteriocin-like inhibitory substance having a domain structure similar to that of lysostaphin. *Gene* 189:255–261.

Singh, A.K., M. Sandipan, M.D. Adhikari, R. Aiyagari, S. Mukherjee, and A. Ramesh. 2012. Fluorescence-based comparative evaluation of bactericidal potency and food application potential of anti-listerial bacteriocin produced by lactic acid bacteria isolated from indigenous samples. *Probiotics and Antimicrobial Proteins* 4:122–132.

Somkuti, G.A. and D.H. Steinberg. 2010. Pediocin production in milk by *Pediococcus acidilactici* in co-culture with *Streptococcus thermophilus* and *Lactobacillus delbrueckii* subsp. *bulgaricus*. *Journal of Industrial Microbiology and Biotechnology* 37:65–69.

Stepper, J., S. Shastri, T.S. Loo et al. 2011. Cysteine *S*-glycosylation, a new post-translational modification found in glycopeptide bacteriocins. *FEBS Letters* 585:645–650.

Suda, S., A. Westerbeek, P.M. O'Connor, R.P. Ross, C. Hill, and P.D. Cotter. 2010. Effect of bioengineering lacticin 3147 lanthionine bridges on specific activity and resistance to heat and proteases. *Chemistry and Biology* 17:1151–1160.

Swe, P.M., G.M. Cook, J.R. Tagg, and R.W. Jack. 2009. Mode of action of dysgalacticin: A large heat-labile bacteriocin. *Journal of Antimicrobial Chemotherapy* 63:679–686.

Urso, R., K. Rantsiou, C. Cantoni, G. Comi, and L. Cocolin. 2006. Technological characterization of a bacteriocin producing *Lactobacillus sakei* and its use in fermented sausages production. *International Journal of Food Microbiology* 110:232–239.

Valenzuela, A.S., G.D. Ruiz, N. Ben Omar et al. 2008. Inhibition of food poisoning and pathogenic bacteria by *Lactobacillus plantarum* strain 2.9 isolated from ben saalga, both in a culture medium and in food. *Food Control* 19:842–848.

Venema, K., M.L. Chikindas, J. Seegers, A.J. Haandrikman, K.J. Leenhouts, G. Venema, and J. Kok. 1997. Rapid and efficient purification method for small, hydrophobic, cationic bacteriocins: Purification of lactococcin B and pediocin PA-1. *Applied and Environmental Microbiology* 63:305–309.

Wiedemann, I., T. Böttiger, R.R. Bonelli et al. 2006. The mode of action of the lantibiotic lacticin 3147—A complex mechanism involving specific interaction of two peptides and the cell wall precursor lipid II. *Molecular Microbiology* 61:285–296.

Wijnker, J.J., E.A.W.S. Weerts, E.J. Breukink, J.H. Houben, and L.J.A. Lipman. 2011. Reduction of *Clostridium sporogenes* spore outgrowth in natural sausage casings using nisin. *Food Microbiology* 28:974–979.

Wirawan, R.E., K.M. Swanson, T. Kleffmann, R.W. Jack, and J.R. Tagg. 2007. Uberolysin: A novel cyclic bacteriocin produced by *Streptococcus uberis*. *Microbiology-SGM* 153:1619–1630.

Wulijideligen, T. Asahina, K. Hara, K. Arakawa, H. Nakano, and T. Miyamoto. 2012. Production of bacteriocin by *Leuconostoc mesenteroides* 406 isolated from Mongolian fermented mare's milk, Airag. *Animal Science Journal* 83:704–711.

Yang, R.G., M.C. Johnson, and B. Ray. 1992. Novel method to extract large amounts of bacteriocins from lactic acid bacteria. *Applied and Environmental Microbiology* 58:3355–3359.

Yoneyama, F., Y. Imura, K. Ohno, T. Zendo, J. Nakayama, K. Matsuzaki, and K. Sonomoto. 2009. Peptide lipid huge toroidal pore, a new antimicrobial mechanism mediated by a lactococcal bacteriocin, Lacticin Q. *Antimicrobial Agents and Chemotherapy* 53:3211–3217.

Yoneyama, F., K. Ohno, Y. Imura, M. Li, T. Zendo, J. Nakayama, K. Matsuzaki, and K. Sonomoto. 2011. Lacticin Q-mediated selective toxicity depending on physicochemical features of membrane components. *Antimicrobial Agents and Chemotherapy* 55:2446–2450.

Zendo, T., S. Koga, Y. Shigeri, J. Nakayama, and K. Sonomoto. 2006. Lactococcin Q, a novel two-peptide bacteriocin produced by *Lactococcus lactis* QU 4. *Applied and Environmental Microbiology* 72:3383–3389.

15

Use of Bacteriocins and Essential Oils for the Control of Listeria monocytogenes in Processed Foods

Corliss A. O'Bryan, Erin S. Whaley, Sara R. Milillo, Philip G. Crandall, Michael G. Johnson, and Steven C. Ricke

CONTENTS

15.1 *Listeria monocytogenes*

15.1.1 Discovery and Classification of *Listeria*

In 1924, E.G.D. Murray isolated Gram-positive rods from the blood of laboratory rabbits with peripheral monocytosis, and because he could not assign these microorganisms to any bacterial genus known at that time, he identified them as *Bacterium monocytogenes* (Murray et al. 1926). In 1927, Pirie isolated a Gram-positive bacterium from gerbils and suggested the genus *Listerella* for these organisms (Pirie 1940a). Both researchers independently deposited their cultures in the National Type Collection of the Lister Institute in London, where it was determined that the two organisms were the same. Murray and Pirie agreed to call the organism *Listerella monocytogenes* (Murray 1953) but the International Committee on Systematic Bacteriology rejected the genus name *Listerella* and so Pirie subsequently changed the genus name to *Listeria* (Pirie 1940b). The genus *Listeria* currently contains 10 species: *L. fleischmannii, L. grayi, L. innocua, L. ivanovii, L. marthii, L. monocytogenes, L. rocourtiae, L. seeligeri, L. weihenstephanensis*, and *L. welshimeri* (Collins et al. 1991). *L. monocytogenes* causes a severe foodborne disease in humans (den Bakker et al. 2010) and thus is of particular interest to food manufacturers.

Listeria monocytogenes is a Gram-positive, non-spore-forming, motile, facultatively anaerobic, rod-shaped bacterium. It is catalase positive and oxidase negative and expresses a beta hemolysin, which causes destruction of red blood cells. *L. monocytogenes* is actively motile by means of peritrichous flagella at room temperature (20°C–25°C), exhibiting a characteristic tumbling motility when viewed with light microscopy (Farber and Peterkin 1991). However, *L. monocytogenes* does not synthesize flagella at 37°C (Peel et al. 1988). Although *L. monocytogenes* is a mesophilic organism with an optimal growth temperature of 30°C–37°C, it has been reported to grow at temperatures as low as −0.4°C (Walker et al. 2008). *L. monocytogenes* readily multiplies in aerobic or microaerophilic environments with pH values between 5.6 and 9.6 (Low and Donachie 1997, Lungu et al. 2009). There are 13 serotypes of *L. monocytogenes* that can cause disease, but more than 90% of human isolates belong to only three serotypes: 1/2a, 1/2b, and 4b (Ward et al. 2004). *L. monocytogenes* serotype 4b strains are responsible for 33%–50% of sporadic human cases worldwide and for the majority of foodborne outbreaks in Europe and North America since the 1980s (Ward et al. 2004).

15.1.2 Foodborne Illness and *Listeria monocytogenes*

Food as a vehicle of infection with *L. monocytogenes* was unknown until an outbreak involving 41 cases (34 pregnancy-associated and 7 non-pregnancy-associated adults) occurred in the Maritime Provinces of Canada between March and September 1981 (Schlech et al. 1983). Of the 34 perinatal cases, there were 9 stillbirths, 23 live births of an ill infant with a subsequent 27% mortality rate, and 2 live births of a well infant; the adult mortality rate was 28.6%. Extensive case-control surveys were conducted and coleslaw consumption was found to be associated with illness, and coleslaw obtained from the refrigerator of one of the patients was shown to contain *L. monocytogenes* type 4b, the epidemic strain. Since 1981, several large outbreaks of listeriosis have been described and corresponding sources in food have been determined for each (Table 15.1). The particular source foods were associated epidemiologically

TABLE 15.1

Notable Outbreaks of Foodborne Listeriosis

Year(s)	Location	Food	# Cases	Mortality Rate, %	Miscarriages/ Stillbirths	Reference
1980–1981	Maritime Prov.	Coleslaw	124	34	9	Schlech et al. (1983)
1983	New England	Milk	49	29	7	Fleming et al. (1985)
1983–1984	Switzerland	Soft cheese	66	32		Bula et al. (1995)
1985	Western United States	Soft cheese	207	34	21	Linnan et al. (1988)
1992	France	Pork rillettes	120	32	10	Goulet et al. (1998)
1998–1999	United States	Hot dogs, deli meats	108	21	4	CDC (1999)
1999	Finland	Butter	25	24		Lyytikainen et al. (2000)
2000	United States	Deli turkey meat	30	23	3	Olsen et al. (2005)
2000–2001	United States	Soft cheese	12	N/A	5	CDC (2001)
2005	Switzerland	Tomme cheese	10	30	2	Bille et al. (2006)
2006	Czech Rep.	Cheese	75	16	4	Vit et al. (2007)
2007	Massachusetts	Milk	5	60	1	CDC (2007)
2006–2007	Germany	Scalded sausages	16	31		Winter et al. (2009)
2008	Canada	Ready-to-eat meats	57	40		PHAC (2008)
2008	Canada	Hard cheese	38	5	3	Gaulin et al. (2012)
2009–2010	Europe	Curd cheese	34	24		Fretz et al. (2010)
2011	United States (28 states)	Cantaloupe	147	22	1	CDC (2012a)
2012	United States (14 states)	Soft cheese	22	18	1	CDC (2012b)
2013	United States (5 states)	Soft cheese	6	17	1	CDC (2013)

in each instance, and the implicated strain of *Listeria* was isolated from the food that was determined to be the source. Although the majority of listeriosis cases are sporadic, the study of outbreaks allows identification of source in most cases, which is difficult to do in sporadic cases (Cartwright et al. 2013). Listeriosis can have an incubation period of as much as 70 days and therefore investigation of sporadic cases of listeriosis often does not lead to a direct product isolate–human isolate link (Farber and Peterkin 1991). If a food vehicle can be determined in sporadic cases, it is most likely to be a category of product (e.g., soft cheese) and not a specific brand (Barnes et al. 1989). However, active surveillance for *Listeria*, linked with laboratory studies of the refrigerator contents of patients with listeriosis, has proven that sporadic cases also have sources in food (Riedo et al. 1994, Vázquez-Boland et al. 2001). Strains of *L. monocytogenes* found in the refrigerators of patients were identical to strains isolated from the blood or spinal fluid of the patient. Hot dogs and delicatessen meats appear to be frequent sources of infection, as well as unpasteurized cheese products, particularly soft cheeses.

15.1.3 Association of *Listeria monocytogenes* with Processed Foods

The demands of modern life have led to an increase in ready-to-eat (RTE) food products, with convenience ranking as one of the key factors in the success of a food product (Weststrate et al. 2002, Crandall et al. 2011). In general, RTE foods include (1) products containing raw ingredients, (2) foods that are susceptible to post-lethality contamination and have an extended shelf life under refrigeration, and (3) food products that are intended to be consumed without further cooking. Epidemiological data suggest that RTE foods are more likely to be associated with outbreaks of listeriosis than other food products (Bell and Kyriakides 2005).

Listeria is capable of growing in a broad range of temperatures including refrigeration (e.g., 1°C–45°C), wide range of pH conditions (e.g., 4.3–9.5), relatively low water activity (>0.90), and high salt concentrations (up to 10%), enabling survival and growth in many different food processing environments (Farber and Peterkin 1991). Many studies have demonstrated the ability of *L. monocytogenes* to colonize, multiply, and persist in the food processing environment and on food processing equipment over extended periods (Kabuki et al. 2004, Lappi et al. 2004).

Overall, *Listeria* is a very adaptable pathogen that is capable of survival even after freezing, surface dehydration, and spray chilling; however, *Listeria* can easily be killed with proper cooking (Junttila et al. 1988). Data from facilities producing RTE foods intended to be consumed without further cooking has demonstrated that recontamination from the processing environment after the cook process is the primary means of contamination of processed RTE foods, and not a failure of processing (Tompkin 2002, Kornacki and Gurtler 2007). For foods that are processed and considered RTE, such as processed meats and pasteurized dairy products, a *zero tolerance* for the presence of *L. monocytogenes* has been in place in the United States for more than 15 years (Shank et al. 1996). Under refrigerated storage, *L. monocytogenes* can survive and grow to a dangerous level, increasing the likelihood of disease if consumed (Doyle 1988, Glass and Doyle 1989).

15.2 Food Antimicrobials

Food antimicrobials are defined as compounds used to inhibit microorganisms by extending the lag phase of growth (Davidson and Harrison 2002). Bacterial inhibition or complete cell death may also be achieved by the use of food antimicrobials (Nastro and Finegold 1972). Most of the antimicrobials and sanitizing agents currently used in food manufacturing have been in use for 50–100 plus years (Davidson and Harrison 2002). Benzoic acid and its salts, for example, were one of the first groups of antimicrobials approved for use in foods (Davidson and Harrison 2002). Antimicrobials are often classified as traditional or naturally occurring (Davidson et al. 2013). In the United States, many traditional antimicrobials are approved for use in food products including acetic acid in baked goods and condiments, benzoic acid in beverages and margarine, lactic acid in meats and fermented foods, nitrate and nitrite in cured meats, sorbic acid in wines, and sulfites in fruit and potato products (Davidson et al. 2013). Some antimicrobials are designated by the U.S. Food and Drug Administration (FDA) to be generally recognized as safe (GRAS), which allows it to be used in a specific application without additional regulatory approval (Montville and Chikindas 2013).

Food antimicrobials have been used to inhibit spoilage organisms and prolong the shelf life and quality of foods for many years, some of which can also be used for the additional purpose of controlling the growth of specific pathogens (Davidson and Harrison 2002, Ricke 2003, Ricke et al. 2005, Li et al. 2011, Sofos et al. 2013). Examples of this include the use of nitrite to inhibit the growth of *Clostridium botulinum* in cured meats (Johnston et al. 1969), specific organic acid sprays used to reduce pathogens on beef carcass surfaces (Castillo et al. 2001), nisin and lysozyme used to inhibit *C. botulinum* in pasteurized processed cheese (Hughey and Johnson 1987, Delves-Broughton et al. 1996), and lactate and diacetate used to inhibit *L. monocytogenes* in processed meats (Perumalla et al. 2008).

Consumers have voiced an increased interest in food products that are minimally processed, have reduced amounts of synthetic additives, and are perceived as more natural (Gould 1996, Van Loo et al. 2010, 2011, 2012). Supermarkets all over the world are striving to meet consumer demand and offer a wider selection of such food products (Lockie et al. 2002). These consumer demands have led processors to develop alternative food processing and preservation methods. There are natural antimicrobials readily available from plants, animals, and microorganisms where they constitute part of host defense mechanisms against (other) microorganisms (Brul and Coote 1999, Joerger 2003, Berghman et al. 2005, Sirsat et al. 2009, Ricke and Wideman 2013). The use of natural antimicrobials as food preservatives is considered helpful for avoiding the excessive physical processing of food to ensure microbial safety, which often alters organoleptic properties of food, such as occurs with canning. Natural antimicrobials, such as essential oils and herbs, are traditionally known for their antimicrobial properties and used in different indigenous practices. As many of these compounds are safe to consume at certain levels of use, their application in food as natural preservatives could be a preferred option for many food manufacturers. This chapter will concentrate on the use of two such classes of natural antimicrobials: bacteriocins and essential oils, alone and in combination.

15.3 Bacteriocins

Many bacteria create metabolic products that act as growth inhibitors against spoilage and pathogenic bacteria (Daeschel 1989). Many Gram-positive bacteria produce cationic, amphiphilic, membrane-permeabilizing small peptides that have antimicrobial activity against a wide range of microorganisms, collectively referred to as bacteriocins (Table 15.2) (Jenssen et al. 2006). For example, *Lactobacillus* spp.

TABLE 15.2

Antilisterial Effects of Various Bacteriocins Used in Food Systems

Food	Bacteriocin or Producer	Results	Reference
Fermented semidry sausage	*Pediococcus acidilactici*	2 Log reduction	Berry et al. (1990)
Fresh beef	Pediocin	0.5–2.0 Log reduction	Nielsen et al. (1990)
Frankfurters	*Pediococcus acidilactici*	Growth inhibition	Berry et al. (1991)
Cottage cheese	Nisin	Growth inhibition	Ferreira and Lund (1996)
Ricotta cheese	Nisin	Growth inhibition	Davies et al. (1997)
Nonfat milk	Nisin	4–6 Log reduction	Jung et al. (1992)
Half and half	Nisin	<1 Log reduction	Jung et al. (1992)
Turkey skin	Nisin	<1 Log reduction	Mahadeo and Tatini (1994)
RTE chicken	Nisin	4–5 Log reduction	Janes et al. (2002)
Frankfurters	Nisin	6 Log reduction	Lungu and Johnson (2005)
Fish fillets	Pediocin	Growth inhibition	Yin et al. (2007)
Turkey ham	Nisin	4.4 Log reduction	Ruiz et al. (2009)
Frankfurters	Pediocin PA-1	2 Log reduction	Nieto-Lozano et al. (2010)
Spanish dry-fermented sausages	*Pediococcus acidilactici*	2 Log reduction	Nieto-Lozano et al. (2010)

produce bacteriocins that exhibit potent antimicrobial activity. Bacteriocins are classified in several ways; for example, in one scheme, bacteriocins are named after the genus, species, or family of bacteria producing them, such as lantibiotics for the bacteriocins produced by *Lactobacillus* spp. and colicins for bacteriocins from *Escherichia coli* (Riley and Chavan 2006). Other classification schemes are based on the mechanism by which the bacteriocin is produced (ribosomal and nonribosomal) or on the mechanisms that bacteriocins use to kill microorganisms (pore formation, nuclease). One classification scheme divides bacteriocins into classes I–V (Tiwari et al. 2009). The class I bacteriocins are small peptides such as lantibiotics, and other modified bacteriocins, lanthionines, and β-lanthionines such as nisins A and Z, lacticin 481, lactocin S, and lacticin 3147 (Jack et al. 1995, Guder et al. 2000). The class II bacteriocins are small heat-stable peptides, which can be further divided into classes IIa, IIb, and IIc. Class IIa bacteriocins include the pediocins, which are well-known food preservatives because of their potent antilisterial activities (Nes and Holo 2000). Most of the bacteriocins of interest for food preservation are in class I or II, since class III bacteriocins are not heat-stable and class IV bacteriocins are large complex molecules requiring a carbohydrate or lipid moiety (Nes and Holo 2000). Bacteriocins alone are usually ineffective against Gram-negative bacteria because they cannot penetrate the outer membrane, but if the outer membrane is disrupted, the Gram-negative bacteria may then become sensitive to bacteriocins (Gillor et al. 2008). However, cyclic bacteriocins produced by *Carnobacterium maltaromaticum*, such as carnocyclin A (CclA) and carnobacteriocin BM1 (CbnBM1), can inhibit the growth of some Gram-negative bacteria, such as *E. coli*, *Pseudomonas aeruginosa*, and *Salmonella* Typhimurium (Martin-Visscher et al. 2011).

Lantibiotic or class 1 bacteriocins are peptides of 19 to more than 50 amino acids, many of which are unusual such as lanthionine, methyllanthionine, dehydrobutyrine, and dehydroalanine (Cleveland et al. 2001). Lantibiotics are additionally divided into subclasses 1a and 1b (Parada et al. 2007). Subclass 1a bacteriocins are composed of relatively elongated, flexible, and cationic peptides that form pores in the membranes of target organisms (Altena et al. 2000). Subclass 1b bacteriocins consist of rigid, globular peptides that are negatively charged or neutral and exert their action by disrupting enzymatic reactions essential to the bacteria they target (Deegan et al. 2006). One particular bacteriocin, nisin, has been widely used as a food preservative due to its nontoxic nature, stability in acidic conditions, and antimicrobial activity against a broad range of Gram-positive organisms, including *L. monocytogenes* (Rodriguez 1996).

15.3.1 Nisin

In the late 1920s and early 1930s, nisin was discovered and described as a *toxic* substance in milk that had adverse effects on the performance of cheese starter cultures (Rogers and Whittier 1928, Whitehead 1933). The first investigation into its potential as a food preservative was by Hirsch et al. (1951), and in the following year, McClintock et al. (1952) conducted research on the use of nisin as a preservative in processed cheese. Nisin is a ribosomally synthesized polycyclic peptide that contains unusual amino acids, including lanthionine, methyllanthionine, didehydroalanine, and didehydroaminobutyric acid (Hansen 1994). Commercial nisin preparations are made by fermentation of enzymatically digested skimmed milk with added yeast extract by nisin-producing strains of *L. lactis* ssp. *lactis* in continuous culture to prolong production (Kim 1997). Nisin is the most widely exploited and applied bacteriocin and is active against Gram-positive bacteria including highly pathogenic and food spoilage microorganisms including *L. monocytogenes* (Juneja et al. 2012). In the United States, nisin has been classified as GRAS since 1988 for use in cheese, heat-treated soups, and pasteurized cheese spreads, which are stored at refrigerated temperatures (Delves-Broughton et al. 1996).

The initial use of nisin for the preservation of foods was in processed cheese products, and this remains one of the major applications of nisin to date (McClintock et al. 1952, Delves-Broughton et al. 1996). The international unit (IU) for nisin activity is the amount of nisin required to inhibit one cell of *Streptococcus agalactiae* in 1 mL of broth and a standard preparation has been defined as 10^6 IU of nisin per gram of preparation (Tramer and Fowler 1964). The ability of nisin to inhibit *L. monocytogenes* in cheese products has been demonstrated with effective concentrations of 2000 IU/mL in cottage cheese (Ferreira and Lund 1996) and as low as 100 IU/mL in ricotta cheese (Cleveland et al. 2001).

It is important to note that the activity of nisin may be significantly impacted by the chemical composition and physical conditions of a food. A study by Davies et al. (1997) reported that addition of 2.5 mg/L nisin inhibited *L. monocytogenes* in ricotta cheese for up to 55 days at 6°C–8°C and that lowering the pH by adding acetic acid extended this inhibition for over 10 weeks. The solubility of nisin is reportedly 228 times greater at a pH of 2 than at a pH of 8 (Liu and Hansen 1990). The activity of nisin is also influenced by the fat content in food products. Jung et al. (1992) reported that a concentration of 1.25 mg/L nisin caused a four to six log reduction of *L. monocytogenes* in nonfat milk, while in half and half cream, less than one log reduction was achieved.

Due to a reduced solubility at high pH and interference by components such as phospholipids (De Vuyst and Vandamme 1994), some researchers have concluded that nisin is not ideal for use in meat products (Rayman et al. 1983). However, a study of nisin activity in sausage found that lower fat contents are associated with higher nisin activity (Davies et al. 1999) indicating that nisin has potential for use in lean meats. Several applications for the use of nisin in meat systems have been investigated including the addition of nisin-producing *Lactococcus lactis* ssp. *lactis* to meat systems in an effort to produce nisin *in situ* (Abee et al. 1994a), the direct incorporation of nisin into meat formulations (Gill and Holley 2000, Samelis et al. 2002) nisin dipping solutions (Ariyapitipun et al. 2000), and nisin-coated casings (Grower et al. 2004, Luchansky and Call 2004). Nisin has been documented to successfully inhibit *L. monocytogenes* in meat matrices including RTE turkey ham (Ruiz et al. 2009), RTE chicken (Janes et al. 2002), turkey breast, ham, and beef (Ming et al. 1997) and turkey franks (Lungu and Johnson 2005) and turkey skin (Mahadeo and Tatini 1994). Other bacteriocins have been studied for their effects on *L. monocytogenes* in a variety of food matrices.

15.3.2 Other Bacteriocins in Food

In refrigerated seafood, pediocin ACCEL has been found to be more effective than nisin in suppressing the growth of *L. monocytogenes* during 2 weeks of storage at 4°C (Yin et al. 2007). Nieto-Lozano et al. (2010) studied the effects of pediocin PA-1 and *Pediococcus acidilactici* MCH14 (a pediocin producing strain) on *L. monocytogenes* in frankfurters and in Spanish dry-fermented sausages, respectively. *L. monocytogenes* counts were reduced by two logs in sausages, compared to the control. In frankfurters, 5000 bacteriocin units/mL (BU/mL) of the pediocin PA-1 decreased *L. monocytogenes* by 2 and 0.6 log cycles during storage at 4°C for 60 days and 15°C for 30 days, respectively, compared to the control. Hartmann et al. (2011) tested cell-free culture supernatants (CFS) of eight bacterial strains of the order Lactobacillales for their efficacy to inhibit *L. monocytogenes* in different food matrices. Six of the CFS, leucocin A, leucocin B, mundticin L, pediocin PA-1, sakacin A, and sakacin X, were identified as the major antilisterial compounds. Each CFS was tested in culture broth, whole milk, and ground beef at 4°C for the ability to inhibit *L. monocytogenes*. While all bacteriocin-containing CFS were effective in broth at concentrations from 52 to 205 arbitrary unit (AU)/mL, significantly higher concentrations were needed when applied in food. An AU is defined as the reciprocal of the highest dilution factor that shows inhibition of an indicator strain (Rajaram et al. 2010). Best results were obtained using CFS containing pediocin PA-1, which required only 3- and 10-times greater amounts in milk and ground meat compared to broth, respectively. Sakacin A and sakacin X containing CFS gave anomalous results in that sakacin A was only effective in meat, while sakacin X was only effective in milk. However, in all cases inhibition of *L. monocytogenes* was only temporary and surviving or resistant bacteria grew during prolonged storage.

Experiments with nisin and other bacteriocins emphasize the importance of testing the effectiveness of bacteriocins against the selected target bacteria and in the food systems for which they are intended to be used. Furthermore, the observation of survivors or resistant bacteria outgrowth underlines the fact that bacteriocins alone may not be able to assure full inhibition of *L. monocytogenes* in a food product and may be better used in combination with other preservative measures.

15.3.3 Mode of Action of Nisin and Other Bacteriocins

Bruno and Montville (1993) proposed the hypothesis that bacteriocins produced by lactic acid bacteria have the same mode of action, which is to dissipate the proton motive force (PMF) of sensitive cells. The PMF consists of two components, the membrane potential and the pH gradient (the difference between the intracellular pH and the extracellular pH) (Kashket 1985). The PMF contributes to the

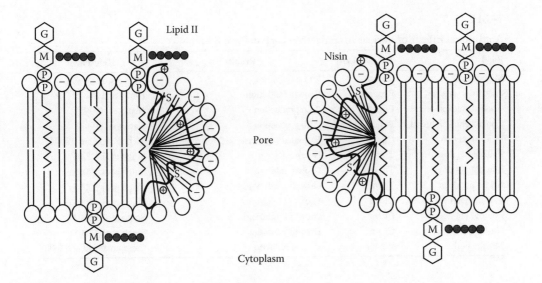

FIGURE 15.1   General mode of action of nisin: lipid II serves as a docking molecule that energetically facilitates the formation of pores by binding the molecule of nisin and allowing it to adopt the correct position for pore opening. (From Sobrino-López, A. and Martín-Belloso, O., *Int. Dairy J.*, 18, 329, 2008. With permission.)

generation of ATP by the movement of hydrogen ions across a membrane during cellular respiration (Mitchell 1966, Harold 1986). Nisin had been previously demonstrated to dissipate membrane potential in sensitive cells (Ruhr and Sahl 1985), and Bruno and Montville (1993) were able to show that pediocin PA-1 and leuconocin S produced PMF dissipation in *L. monocytogenes* Scott A. Nisin and pediocin PA-1 are known to completely dissipate both components of PMF in sensitive cells (Ruhr and Sahl 1985, Bruno and Montville 1993).

Several experiments have revealed that nisin forms poration complexes in target cell membranes through a multistep process that includes binding and insertion (Benz et al. 1991, Sahl 1991, Breukink and de Kruijff 1999). Linnett and Strominger (1973) studied isolated membranes in an *in vitro* system and found that nisin interfered with cell wall biosynthesis. Reisinger et al. (1980) identified the specific target as being the membrane-bound cell wall precursor lipid II (Figure 15.1). Nisin apparently uses all available lipid II molecules in the membrane to form stable pore complexes with a uniform structure consisting of eight nisin and four lipid II molecules (Hasper et al. 2004). Formation of pores by bacteriocins such as nisin and pediocin PA-1 allows efflux of any charged ions, but it has been shown that lactococcin G allows efflux of monovalent cations only (Montville and Chen 1998).

15.4 Essential Oils

Herbs and spices, historically used to add flavor and fragrance to foods, are also well known for their antimicrobial activities (Nychas et al. 2003). The active antimicrobial ingredients are mostly essential oils, most of which are classified as GRAS; however, their use as preservatives in foods is limited because of flavor and cost considerations (Kabara 1991, Klancnik et al. 2010).

Assessment of essential oils for antilisterial activities in food has a long history (Table 15.3). Aureli et al. (1992) examined the antimicrobial activity of several plant essential oils commonly used in food industry against *L. monocytogenes* and *L. innocua* and determined that cinnamon, clove, origanum, pimento, and thyme displayed activity against these organisms. When used in a minced pork product, thyme oil reduced *L. monocytogenes* by 2 log colony-forming units (CFU) during 1 week of storage (Aureli et al. 1992). Hao et al. (1998) found that pimento oil and eugenol were mildly effective against *L. monocytogenes* on refrigerated sliced beef. Menon and Garg (2001) tested the effects of clove oil when used at a final amount of 0.5% or 1.0% of substrate against *L. monocytogenes* inoculated into mozzarella cheese or fresh minced

TABLE 15.3

Antilisterial Effects of Various Essential Oils Used in Food Systems

Food	Essential Oil	Results	Reference
Minced pork	Thyme	2 Log reduction	Aureli et al. (1992)
Pork liver sausage	Rosemary	Growth inhibition	Pandit and Shelef (1994)
Cooked chicken breast	Eugenol	2 Log reduction	Hao et al. (1998)
Cooked chicken breast	Pimento	2 Log reduction	Hao et al. (1998)
Minced mutton	Clove	Extended lag phase, 1–3 log reduction	Menon and Garg (2001)
Full-fat hot dogs	Thyme	No effect	Singh et al. (2003)
Low-fat hot dogs	Thyme	Growth inhibition	Singh et al. (2003)
No-fat hot dogs	Thyme	Growth inhibition	Singh et al. (2003)
Full-fat hot dogs	Clove	Growth inhibition	Singh et al. (2003)
Low-fat hot dogs	Clove	Growth inhibition	Singh et al. (2003)
No-fat hot dogs	Clove	Growth inhibition	Singh et al. (2003)
Minced pork	Winter savory	<1 Log reduction	Carramiñana et al. (2008)

mutton. In this study, 1.0% clove oil was more effective, lowering *Listeria* counts by 1–3 log CFU/g compared to untreated controls during 5 days of storage at 30°C or 15 days of storage at 7°C.

Citrus oils have been a part of the human diet throughout history due to their abundance of flavor, vitamins, and antioxidants (Chalova et al. 2010, Hardin et al. 2010). Cold-pressed terpeneless Valencia oil (CPTVO) is present in the colored portion of orange peels, also known as the flavedo (Temelli et al. 1990), making it abundant in nature and relatively inexpensive (Crandall and Hendrix 2001). Citrus essential oils have shown widespread antimicrobial activity against a variety of foodborne pathogens (Nannapaneni et al. 2008, 2009a,b, O'Bryan et al. 2008, Pittman et al. 2011, Muthaiyan et al. 2012, Pendleton et al. 2012). *In vitro* testing has demonstrated that CPTVO has antimicrobial activities against *L. monocytogenes* (Friedly et al. 2009). The primary volatile components of CPTVO (linalool, decanal, and citral) may also be optimal candidates for natural alternatives to chemical-based antimicrobials (Kamdem et al. 2011, Liu et al. 2012, Klein et al. 2013). Gas chromatography–mass spectrometry (GC/MS) data found that CPTVO as a whole was composed of 20% linalool, 18% decanal, and 14% citral (Nannapaneni et al. 2009a). Results from disk diffusion assays have concluded that direct contact with 6%–10% linalool exhibits inhibitory effects on *L. monocytogenes* Scott A, producing a zone of inhibition of approximately 9 mm (Kim et al. 1995). Additional disk diffusion assays by Fisher and Phillips (2006) reported that contact with 10–100 μL of undiluted linalool produced zones of inhibition greater than 90 mm in *L. monocytogenes* ATCC 7644. Direct contact with citral has also been reported to inhibit *Listeria* spp. by drastically reducing the extent of a mild heating treatment required to have bactericidal effects on *L. innocua* in orange juice (Char et al. 2010). However, these citrus oils and components also have organoleptic impacts on foods, including flavor and aroma issues (Fisher and Phillips 2008). The combination of essential oils and bacteriocins together in a multiple hurdle approach might offer a natural way of improving food safety (Brul and Coote 1999, Mourey and Canillac 2002).

15.4.1 Mode of Action of Essential Oils

While there have been extensive investigations on the mode of action of bacteriocins, there are few studies that attempt such an insight for specific essential oils. Antimicrobial agents generally inhibit bacteria by one of three mechanisms: (1) inhibition of nucleic acid synthesis, (2) inhibition of cytoplasmic membrane function, or (3) inhibition of energy metabolism (Figure 15.2) (Tassou et al. 2000, Cox et al. 2001, Juglal et al. 2002, Ricke 2003, Naghmouchi et al. 2006, Rasooli et al. 2006). Essential oils are largely composed of terpenes and it has been postulated that their mode of action might be similar to that of other phenolic compounds (Tassou et al. 2000). Conner and Beuchat (1984a,b) postulated that essential oils acted to impair a variety of enzyme systems including those involved in energy production

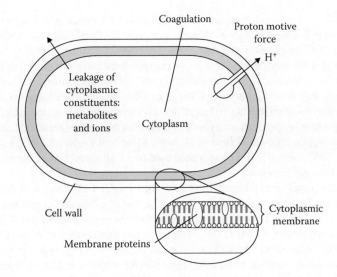

FIGURE 15.2 Locations and mechanisms in the bacterial cell thought to be sites of action for essential oils: degradation of the cell wall, damage to cytoplasmic membrane, damage to membrane proteins, leakage of cell contents, coagulation of cytoplasm, and depletion of the PMF. (From Burt, S., *Int. J. Food Microbiol.*, 94, 223, 2004.)

and structural component synthesis. Tassou et al. (2000) investigated the effects of mint essential oil on the consumption of energy sources by *Salmonella* Enteritidis and *Staphylococcus aureus*. They determined that glucose utilization was reduced drastically for both pathogens in the presence of essential oil, and as a consequence, the assimilation or formation of different compounds, such as lactate, formate, and acetate in the growth medium, was also affected. Cox et al. (2001) studied the effects of tea tree oil against *E. coli* and *S. aureus* and found that respiration was inhibited and permeability of the cytoplasmic membranes occurred, leading to potassium ion leakage, in both bacteria. Silva et al. (2011) found that Gram-positive bacteria exhibited lower susceptibility than the Gram-negative bacteria to coriander oil, although they determined that the mode of action is similar in both Gram-positive and Gram-negative bacteria. These researchers concluded that the primary mode of action of coriander oil is cell membrane permeabilization, resulting in membrane potential, respiratory activity, and efflux pump activity being interrupted. In a study of the essential oils from oregano and rosemary against *L. monocytogenes*, it was discovered that there was decreased glucose consumption and the loss of cellular material immediately after addition of the essential oils alone and in combination (De Azeredo et al. 2012). Electron microscopy revealed changes in cell wall structure, rupture of plasma membranes, shrinking of the cytoplasmic content, and leakage of intracellular material. Muthaiyan et al. (2012) found that cell lysis occurred in *S. aureus* treated with CPTVO when viewed by transmission electron microscopy. Microarray studies revealed that CPTVO induced the cell wall stress stimulon consistent with the inhibition of cell wall synthesis. Further studies are needed on the mode of action of certain essential oils alone and in combination with other antimicrobials such as bacteriocins against both pathogenic and spoilage microorganisms.

15.5 Combinations of Treatments

Multiple hurdle technology is a method for inhibiting microorganisms that involves the combination of more mild preservation techniques to prevent the survival and regrowth of foodborne pathogens (Leistner and Gorris 1995, Ricke et al. 2005). Synergistic inhibition of *Listeria* spp. associated with exposure to a combination of nisin and other antimicrobial treatments has been reported on multiple occasions including a combination treatment of nisin and thymol (Ettayebi et al. 2000) and nisin and garlic extract (Singh et al. 2001).

Ghalfi et al. (2007) studied a combination of cell-adsorbed bacteriocin (CAB) with oregano or savory essential oil to control *L. monocytogenes* in pork meat at 4°C. They determined that *Listeria* counts declined from 2 log CFU/g to below the detectable limit during 1 week of storage in samples treated with CAB combined with either essential oil. Counts increased after the third week of storage in all samples, except for those treated with the combination of CAB and oregano essential oil. A combination of CPTVO and nisin may be particularly well suited to a multiple hurdle design because they are both thought, through different mechanisms, to target the cell membrane (Abee et al. 1994b, Burt 2004).

Synergism using a hurdle approach may result in a lowered MIC of essential oils, therefore minimizing undesired organoleptic changes (Payne et al. 1989, Ming and Daeschel 1993, Leistner and Gorris 1995, Brewer et al. 2002, Burt 2004). The sequence used to introduce antimicrobial treatments may also have an impact on their effectiveness.

Shannon et al. (2011) evaluated the combination of nisin and CPTVO and found that the combination produced a zone of inhibition on disk diffusion assay that was significantly larger than zones of CPTVO or nisin alone. They also assayed the effects of this combination on the growth rate of *L. monocytogenes* and reported that exposure to CPTVO at 0 h followed by the introduction of nisin at 15 h resulted in a statistically significant reduction in growth as compared to the control (Shannon et al. 2011). This combination of treatments has potential as an all-natural, GRAS multiple hurdle intervention that may be applicable for RTE products in which *L. monocytogenes* is likely to be found.

15.6 Conclusions

Due to its high mortality rate in susceptible populations and the resistance of this pathogen to many traditional food preservation practices, *L. monocytogenes* is a pathogen of primary concern for the food industry. Advances in molecular subtyping methods, such as the development of pulsed field gel electrophoresis combined with resources such as PulseNet, have led to an increased ability to trace outbreaks of listeriosis back to sources of contamination (Williams et al. 2011). This surveillance of *Listeria* strains has implicated food handlers, food processing environments, and retail facilities with the transmission of and recontamination of RTE food products with *Listeria* spp. Properties of food products such as the presence of peptides or microflora may also contribute to the survival of *Listeria*. Consumers' desire for minimally processed foods with a reduced amount of chemical additives has increased interest in the use of natural antimicrobials for the inhibition of foodborne pathogens. Bacteriocins and essential oils are both well suited for this purpose because they are GRAS, natural, and have been previously reported to exhibit inhibitory effects on *L. monocytogenes*. Additionally, combining these antimicrobials for use as a hurdle technique may result in synergistic inhibition of *Listeria* and a reduction in undesired results associated with these treatments such as the development of bacteriocin-resistant strains or changes in the sensory qualities of food products.

ACKNOWLEDGMENTS

Preparation of this manuscript was supported partially by grants from the USDA Food Safety Consortium to authors Johnson and Ricke and by National Research Initiative grant 2007-35201-18380 to authors Crandall and Ricke.

REFERENCES

Abee, T., T. Klaenhammer, and L. Letellier. 1994a. Kinetic studies of the action of lactacin F, a bacteriocin produced by *Lactobacillus johnsonii* that forms poration complexes in the cytoplasmic membrane. *Applied and Environmental Microbiology* 60: 1006–1013.

Abee, T., F.M. Rombouts, J. Hugenholtz, G. Guihard, and L. Letellier. 1994b. Mode of action of nisin Z against *Listeria monocytogenes* Scott A grown at high and low temperatures. *Applied and Environmental Microbiology* 60: 1962–1968.

Altena, K., A. Guder, C. Cramer, and G. Bierbaum. 2000. Biosynthesis of the lantibiotic mersacidin: Organization of a type B lantibiotic gene cluster. *Applied and Environmental Microbiology* 66: 2565–2571.

Ariyapitipun, T., A. Mustapha, and A.D. Clarke. 2000. Survival of *Listeria monocytogenes* Scott A on vacuum-packaged raw beef treated with polylactic acid, lactic acid, and nisin. *Journal of Food Protection* 63: 131–136.

Aureli, P., A. Costantini, and S. Zolea. 1992. Antimicrobial activity of some plant essential oils against *Listeria monocytogenes*. *Journal of Food Protection* 55: 344–348.

Barnes, R., P. Archer, J. Strack, G.R. Istre, and Centers for Disease Control and Prevention. 1989. Epidemiologic notes and reports listeriosis associated with consumption of turkey franks. *Morbidity and Mortality Weekly Report* 38: 267–268.

Bell, C. and A. Kyriakides. 2005. *Listeria: A Practical Approach to the Organism and Its Control in Foods.* Wiley-Blackwell, Ames, IA, 336pp.

Benz, R., G. Jung, and H.G. Sahl. 1991. Mechanism of channel formation by lantibiotics in black lipid membranes. In: Jung, G. and H.G. Sahl (eds.), *Nisin and Novel Lantibiotics*. Escom, Leiden, the Netherlands, pp. 359–372.

Berghman, L.R., D. Abi-Ghanem, S.D. Waghela, and S.C. Ricke. 2005. Antibodies: An alternative for antibiotics? *Poultry Science* 84: 660–666.

Berry, E.D., R.W. Hutkins, and R.W. Mandigo. 1991. The use of bacteriocin-producing *Pediococcus acidilactici* to control postprocessing *Listeria monocytogenes* contamination of frankfurters. *Journal of Food Protection* 54: 681–686.

Berry, E.D., M.B. Liewen, R.W. Mandigo, and R.W. Hutkins. 1990. Inhibition of *Listeria monocytogenes* by bacteriocin-producing *Pediococcus* during the manufacture of fermented semidry sausage. *Journal of Food Protection* 53: 194–197.

Bille, J., D.S. Blanc, H. Schmid, K. Boubaker, A. Baumgartner, H.H. Siegrist, M.L. Tritten et al. 2006. Outbreak of human listeriosis associated with tomme cheese in northwest Switzerland, 2005. *Euro Surveillance* 11: 91–93.

Breukink, E. and B. de Kruijff. 1999. The lantibiotic nisin, a special case or not? *Biochimica et Biophysica Acta* 1462: 223–234.

Brewer, R., M. Adams, and S. Park. 2002. Enhanced inactivation of *Listeria monocytogenes* by nisin in the presence of ethanol. *Letters in Applied Microbiology* 34: 18–21.

Brul, S. and P. Coote. 1999. Preservative agents in foods: Mode of action and microbial resistance mechanisms. *International Journal of Food Microbiology* 50: 1–17.

Bruno, M.E.C. and T.J. Montville. 1993. Common mechanistic action of bacteriocins from lactic acid bacteria. *Applied and Environmental Microbiology* 59: 3003–3010.

Bula, C.J., J. Bille, and M.P. Glauser. 1995. An epidemic of foodborne listeriosis in western Switzerland: Description of 57 cases involving adults. *Clinical Infectious Diseases* 20: 66–72.

Burt, S. 2004. Essential oils: Their antibacterial properties and potential applications in foods—A review. *International Journal Food Microbiology* 94: 223–253.

Carramiñana, J.J., C. Rota, J. Burillo, and A. Herrera. 2008. Antibacterial efficiency of Spanish *Satureja montana* essential oil against *Listeria monocytogenes* among natural flora in minced pork. *Journal of Food Protection* 71: 502–508.

Cartwright, E.J., K.A. Jackson, S.D. Johnson, L.M. Graves, B.J. Silk, and B.E. Mahon. 2013. Listeriosis outbreaks and associated food vehicles, United States, 1998–2008. *Emerging Infectious Diseases* 19: 1–10. doi: 10.3201/eid1901.120393.

Castillo, A., L.M. Lucia, D.B. Roberson, T.H. Stevenson, I. Mercado, and G.R. Acuff. 2001. Lactic acid sprays reduce bacterial pathogens on cold beef carcass surfaces and in subsequently produced ground beef. *Journal of Food Protection* 64: 58–62.

CDC, Centers for Disease Control and Prevention. 1999. Update: Multistate outbreak of listeriosis—United States, 1998–1999. *Morbidity and Mortality Weekly Reports* 47: 1117–1118.

CDC, Centers for Disease Control and Prevention. 2001. Outbreak of listeriosis associated with homemade Mexican-style cheese-North Carolina, October 2000–January 2001. *Journal of the American Medical Association* 286: 664–665.

CDC, Centers for Disease Control and Prevention. 2007. Outbreak of *Listeria monocytogenes* infections associated with pasteurized milk from a local dairy—Massachusetts, 2007. *Morbidity and Mortality Weekly Report* 57: 1097–1100.

CDC, Centers for Disease Control and Prevention. 2012a. Multistate outbreak of listeriosis linked to whole cantaloupes from Jensen Farms, Colorado. Available at: http://www.cdc.gov/listeria/outbreaks/canta-loupes-jensen-farms/index.html. Accessed May 19, 2014.

CDC, Centers for Disease Control and Prevention. 2012b. Multistate outbreak of listeriosis linked to imported Frescolina Marte brand ricotta salata cheese. Available at: http://www.cdc.gov/listeria/outbreaks/cheese-09-12/index.html. Accessed May 19, 2014.

CDC, Centers for Disease Control and Prevention. 2013. Multistate outbreak of listeriosis linked to Crave Brothers Farmstead Cheeses. Available at: http://www.cdc.gov/listeria/outbreaks/cheese-07-13/index.html. Accessed May 19, 2014.

Chalova, V.I., P.G. Crandall, and S.C. Ricke. 2010. Microbial inhibitory and radical scavenging activities of cold-pressed terpeneless Valencia orange (*Citrus sinensis*) oil in different dispersing agents. *Journal of the Science of Food and Agriculture* 90: 870–876.

Char, C.D., S.N. Guerrero, and S.M. Alzamora. 2010. Mild thermal process combined with vanillin plus citral to help shorten the inactivation time for *Listeria innocua* in orange juice. *Food and Bioprocess Technology* 3: 752–761.

Cleveland, J., T.J. Montville, I.F. Nes, and M.L. Chikindas. 2001. Bacteriocins: Safe, natural antimicrobials for food preservation. *International Journal of Food Microbiology* 71: 1–20.

Collins, M.D., S. Wallbanks, D.J. Lane, J. Shah, R. Nietupskin, J. Smida, M. Dorsch, and E. Stackebrandt. 1991. Phylogenetic analysis of the genus *Listeria* based on reverse transcriptase sequencing of 16S rRNA. *International Journal of Systematic and Evolutionary Microbiology* 41: 240–246.

Conner, D.E. and L.R. Beuchat. 1984a. Sensitivity of heat-stressed yeasts to essential oils of plants. *Applied Environmental Microbiology* 47: 229–233.

Conner, D.E. and L.R. Beuchat. 1984b. Effects of essential oils from plants on growth of food spoilage yeasts. *Journal of Food Science* 49: 429–434.

Cox, S., C. Mann, J. Markham, H. Bell, J. Gustafson, J. Warmington, and S. Wyllie. 2001. The mode of antimicrobial action of the essential oil of *Melaleuca alternifolia* (tea tree oil). *Journal of Applied Microbiology* 88: 170–175.

Crandall, P. and C. Hendrix. 2001. Citrus processing. In: Schrobinger, U. (ed.), *Fruit and Vegetable Juices*, 3rd edn. Verlag Eugen Ulmer, Stuttgart, Germany, pp. 205–223.

Crandall, P.G., J.A. Neal Jr., C.A. O'Bryan, C.A. Murphy, B.P. Marks, and S.C. Ricke. 2011. Minimizing the risk of *Listeria monocytogenes* in retail delis by developing employee focused, cost effective training. *Agricultural Food and Analytical Bacteriology* 1: 159–174.

Daeschel, M.A. 1989. Antimicrobial substances from lactic acid bacteria for use as food preservatives. *Food Technology* 43: 164–167.

Davidson, P.M. and M.A. Harrison 2002. Resistance and adaptation to food antimicrobials, sanitizers, and other process controls. *Food Technology* 56: 69–78.

Davidson, P.M., T.M. Taylor, and S.E. Schmidt. 2013. Chemical preservatives and natural antimicrobial compounds. In: Doyle, M.P. and E.L. Buchanan (eds.), *Food Microbiology: Fundamentals and Frontiers*, 4th edn. ASM Press, Washington, DC, pp. 765–802.

Davies, E.A., H.E. Bevis, and J. Delves-Broughton. 1997. Use of the bacteriocin nisin as a preservative in ricotta-type cheeses to control the food-borne pathogen *Listeria monocytogenes*. *Letters in Applied Microbiology* 24: 343–346.

Davies, E.A., C.F. Milne, H.E. Bevis, R.W. Potter, J.M. Harris, G.C. Williams, L.V. Thomas, and J. Delves-Broughton. 1999. Effective use of nisin to control lactic acid bacterial spoilage in vacuum-packed bologna-type sausage. *Journal of Food Protection* 62: 1004–1010.

De Azeredo, G.A., R.C.B.Q. De Figueiredo, E.L. De Souza, and T.L.M. Stamford. 2012. Change in *Listeria monocytogenes* induced by *Origanum vulgare* L. and *Rosmarinus officinalis* L. essential oils alone and combined at subinhibitory amounts. *Journal of Food Safety* 32: 226–235.

Deegan, L.H., Cotter, P.D., Hill, C., and Ross, P. 2006. Bacteriocins: Biological tools for bio-preservation and shelf-life extension. *International Dairy Journal* 16: 1058–1071.

Delves-Broughton, J., P. Blackburn, R. Evans, and J. Hugenholtz. 1996. Applications of the bacteriocin, nisin. *Antonie Van Leeuwenhoek* 69: 193–202.

den Bakker, H.C., C.A. Cummings, V. Ferreira, P. Vatta, R.H. Orsi, L. Degoricija, M. Barker, O. Petrauskene, M.R. Furtado, and M. Wiedmann. 2010. Comparative genomics of the bacterial genus *Listeria*: Genome evolution is characterized by limited gene acquisition and limited gene loss. *BMC Genomics* 11: 688. doi: 10.1186/1471-2164-11-688.

De Vuyst, L. and E. Vandamme. 1994. Nisin, a lantibiotic produced by *Lactococcus lactis* subsp. *lactis*: Properties, biosynthesis, fermentation and applications. In: De Vuyst, L. and E, Vandamme (eds.), *Bacteriocins of Lactic Acid Bacteria: Microbiology, Genetics and Applications*. Chapman & Hall, London, U.K., pp. 151–221.

Doyle, M.P. 1988. Effect of environmental and processing conditions on *Listeria monocytogenes*. *Food Technology* 42: 169–171.

Ettayebi, K., J. El Yamani, and B.D. Rossi-Hassani. 2000. Synergistic effects of nisin and thymol on antimicrobial activities in *Listeria monocytogenes* and *Bacillus subtilis*. *FEMS Microbiology Letters* 183: 191–195.

Farber, J.M. and P.I. Peterkin. 1991. *Listeria monocytogenes*, a food-borne pathogen. *Microbiological Reviews* 55: 476–511.

Ferreira, M. and B. Lund. 1996. The effect of nisin on *Listeria monocytogenes* in culture medium and long-life cottage cheese. *Letters in Applied Microbiology* 22: 433–438.

Fisher, K. and C. Phillips. 2006. The effect of lemon, orange and bergamot essential oils and their components on the survival of *Campylobacter jejuni*, *Escherichia coli* O157, *Listeria monocytogenes*, *Bacillus cereus* and *Staphylococcus aureus in vitro* and in food systems. *Journal of Applied Microbiology* 101: 1232–1240.

Fisher, K. and C. Phillips. 2008. Potential antimicrobial uses of essential oils in food: Is citrus the answer? *Trends in Food Science and Technology* 19: 156–164.

Fleming, D.W., S.L. Cochi, K.L. MacDonald, J. Brondum, P.S. Hayes, B.D. Plikaytis, M.B. Holmes, A. Audurier, C.V. Broome, and S.L. Reingold. 1985. Pasteurized milk as a vehicle of infection in an outbreak of listeriosis. *New England Journal of Medicine* 312: 404–407.

Fretz, R., J. Pichler, U. Sagel, P. Much, W. Ruppitsch, A.T. Pietzka, A. Stoger et al. 2010. Update: Multinational listeriosis outbreak due to 'Quargel', a sour milk curd cheese, caused by two different *L. monocytogenes* serotype 1/2a strains, 2009–2010. *Eurosurveillance* 15: 2–3.

Friedly, E., P.G. Crandall, S.C. Ricke, M. Roman, C.A. O'Bryan, and V. Chalova. 2009. In vitro antilisterial effects of citrus oil fractions in combination with organic acids. *Journal of Food Science* 74: 67–72.

Gaulin, C., D. Ramsay, and S. Bekal. 2012. Widespread listeriosis outbreak attributable to pasteurized cheese, which led to extensive cross-contamination affecting cheese retailers, Quebec, Canada, 2008. *Journal of Food Protection* 75: 71–78.

Ghalfi, H., N. Benkerroum, D.D.K. Doguiet, M. Bensaid, and P. Thonart. 2007. Effectiveness of cell-adsorbed bacteriocin produced by *Lactobacillus curvatus* CWBI-B28 and selected essential oils to control *Listeria monocytogenes* in pork meat during cold storage. *Letters in Applied Microbiology* 44: 268–273.

Gill, A.O. and R.A. Holley. 2000. Inhibition of bacterial growth on ham and bologna by lysozyme, nisin and EDTA. *Food Research International* 33: 83–90.

Gillor, O., A. Etzion, and M.A. Riley. 2008. The dual role of bacteriocins as anti- and probiotics. *Applied Microbiology and Biotechnology* 81: 591–606.

Glass, K.A. and M.P. Doyle. 1989. Fate of *Listeria monocytogenes* in processed meat products during refrigerated storage. *Applied and Environmental Microbiology* 55: 1565–1569.

Gould, G.W. 1996. Industry perspectives on the use of natural antimicrobials and inhibitors for food applications. *Journal of Food Protection* 59(Suppl.): 82–86.

Goulet, V., J. Rocourt, I. Rebiere, C. Jacquet, C. Moyse, P. Dehaumont, G. Salvat, and P. Veit. 1998. Listeriosis outbreak associated with the consumption of rillettes in France in 1993. *Journal of Infectious Diseases* 177: 155–160.

Grower, J.L., K. Cooksey, and K.J.K. Getty. 2004. Development and characterization of an antimicrobial packaging film coating containing nisin for inhibition of *Listeria monocytogenes*. *Journal of Food Protection* 67: 475–479.

Guder, A., I. Wiedemann, and H.G. Sahl. 2000. Posttranslationally modified bacteriocins: The lantibiotics. *Biopolymers* 55: 62–73.

Hansen, J.N. 1994. Nisin as a model food preservative. *Critical Reviews in Food Science and Nutrition* 34: 69–93.

Hao, Y.Y., R.E. Brackett, and M.P. Doyle. 1998. Inhibition of *Listeria monocytogenes* and *Aeromonas hydrophila* by plant extracts in refrigerated cooked beef. *Journal of Food Protection* 61: 307–312.

Hardin, A., P.G. Crandall, and T. Stankus. 2010. Essential oils and antioxidants derived from citrus by-products in food protection and medicine: An introduction and review of recent literature. *Journal of Agriculture and Food Information* 11: 99–122.

Harold, F.M. 1986. *The Vital Force: A Study of Bioenergetics*. W.H. Freeman & Co., New York.

Hartmann, H.A., T. Wilke, and R. Erdmann. 2011. Efficacy of bacteriocin-containing cell-free culture supernatants from lactic acid bacteria to control *Listeria monocytogenes* in food. *International Journal of Food Microbiology* 146: 192–199.

Hasper, H.E., B. de Kruijff, and E. Breukink. 2004. Assembly and stability of nisin-lipid II pores. *Biochemistry* 43: 11567–11575.

Hirsch, A., E. Grinsted, H.R. Chapman, and A. Mattick. 1951. A note on the inhibition of an anaerobic sporeformer in Swiss-type cheese by a nisin-producing *Streptococcus*. *Journal of Dairy Research* 18: 205–206.

Hughey, V.L. and E.A. Johnson. 1987. Antimicrobial activity of lysozyme against bacteria involved in food spoilage and food-borne disease. *Applied and Environmental Microbiology* 53: 2165–2170.

Jack, R.W., J.R. Tagg, and B. Ray. 1995. Bacteriocins of gram-positive bacteria. *Microbiological Reviews* 59: 171–200.

Janes, M., S. Kooshesh, and M.G. Johnson. 2002. Control of *Listeria monocytogenes* on the surface of refrigerated, ready-to-eat chicken coated with edible zein film coatings containing nisin and/or calcium propionate. *Journal of Food Science* 67: 2754–2757.

Jenssen, H., P. Hamill, and R.E.W. Hancock. 2006. Peptide antimicrobial agents. *Clinical Microbiology Reviews* 19: 491–511.

Joerger, R.D. 2003. Alternatives to antibiotics: Bacteriocins, antimicrobial peptides and bacteriophages. *Poultry Science* 82: 640–647.

Johnston, M.A., H. Pivnick, and J.M. Samson. 1969. Inhibition of *Clostridium botulinum* by sodium nitrite in a bacteriological medium and in meat. *Canadian Institute of Food Technologists Journal* 2: 52–55.

Juglal, S., R. Govinden, and B. Odhav. 2002. Spice oils for the control of co-occurring mycotoxin-producing fungi. *Journal of Food Protection* 65: 683–687.

Juneja, V.K., H.P. Dwivedi, and X. Yan. 2012. Novel natural food antimicrobials. *Annual Reviews in Food Science and Technology* 3: 381–403.

Jung, D.S., F.W. Bodyfelt, and M.A. Daeschel. 1992. Influence of fat and emulsifiers on the efficacy of nisin in inhibiting *Listeria monocytogenes* in fluid milk. *Journal of Dairy Science* 75: 387–393.

Junttila, J.R., S.I. Niemelä, and J. Hirn. 1988. Minimum growth temperatures of *Listeria monocytogenes* and non-haemolytic *Listeria*. *Journal of Applied Bacteriology* 65: 321–327.

Kabara, J.J. 1991. Phenols and chelators. In: Russell, N.J. and Gould, G.W. (eds.), *Food Preservatives*. Blackie, London, pp. 200–214.

Kabuki, D.Y., A.Y. Kuaye, M. Wiedmann, and K.J. Boor. 2004. Molecular subtyping and tracking of *Listeria monocytogenes* in Latin-style fresh-cheese processing plants. *Journal of Dairy Science* 87: 2803–2812.

Kamdem, S.S., N. Belletti, R. Magnani, R. Lanciotti, and F. Gardini. 2011. Effects of carvacrol, (*E*)-2-hexenal, and citral on the thermal death kinetics of *Listeria monocytogenes*. *Journal of Food Protection* 74: 2070–2078.

Kashket, E.R. 1985. The proton motive force in bacteria: A critical assessment of methods. *Annual Reviews in Microbiology* 39: 212–242.

Kim, J., M.R. Marshall, and C. Wei. 1995. Antibacterial activity of some essential oil components against five foodborne pathogens. *Journal of Agriculture and Food Chemistry* 43: 2839–2845.

Kim, W.S. 1997. Nisin production by *Lactococcus lactis* using two phase batch culture. *Letters in Applied Microbiology* 25: 169–171.

Klancnik, A., S. Piskernik, B. Jersek, and S.S. Mozina. 2010. Evaluation of diffusion and dilution methods to determine the antibacterial activity of plant extracts. *Journal of Microbiological Methods* 81: 121–126.

Klein, G., C. Ruben, and M. Upmann. 2013. Antimicrobial activity of essential oil components against potential food spoilage microorganisms. *Current Microbiology* 67: 200–208.

Kornacki, J.L. and J.B. Gurtler. 2007. Incidence and control of *Listeria monocytogenes* in food processing facilities. In: Ryser, E.T. and E.H. Marth (eds.), *Listeria, Listeriosis and Food Safety*, 3rd edn. CRC Press/Taylor & Francis Group, Boca Raton, FL, pp. 681–786.

Lappi, V.R., J. Thimothe, K.K. Nightingale, K. Gall, M.W. Moody, and M. Wiedmann. 2004. Longitudinal studies on *Listeria* in smoked fish plants: Impact of intervention strategies on contamination patterns. *Journal of Food Protection* 67: 2500–2514.

Leistner, L. and L.G.M. Gorris. 1995. Food preservation by hurdle technology. *Trends in Food Science and Technology* 6: 41–46.

Li, M., A. Muthaiyan, C.A. O'Bryan, J.E. Gustafson, Y. Li, P.G. Crandall, and S.C. Ricke. 2011. Use of natural antimicrobials from a food safety perspective for control of *Staphylococcus aureus*. *Current Pharmaceutical Biotechnology* 12: 1240–1254.

Linnan, J.M., L. Mascola, X.D. Lou, V. Goulet, S. May, C. Salminen, D.W. Hird et al. 1988. Epidemic listeriosis associated with Mexican-style cheese. *New England Journal of Medicine* 319: 823–828.

Linnett, P.E. and J.L. Strominger. 1973. Additional antibiotic inhibitors of peptidoglycan synthesis. *Antimicrobial Agents and Chemotherapy* 4: 231–236.

Liu, K.H., Q.L. Chen, Y.J. Liu, X.Y. Zhou, and X.C. Wang. 2012. Isolation and biological activities of decanal, linalool, valencene, and octanal from sweet orange oil. *Journal of Food Science* 77: C1156–C1161.

Liu, W. and J.N. Hansen. 1990. Some chemical and physical properties of nisin, a small protein antibiotic produced by *Lactococcus lactis*. *Applied and Environmental Microbiology* 56: 2551–2558.

Lockie, S., K. Lyons, G. Lawrence, and K. Mummery. 2002. Eating 'green': Motivations behind organic food consumption in Australia. *Sociologia Ruralis* 42: 23–40.

Low, J. and W. Donachie. 1997. A review of *Listeria monocytogenes* and listeriosis. *The Veterinary Journal* 153: 9–29.

Luchansky, J.B. and J.E. Call. 2004. Evaluation of nisin-coated cellulose casings for the control of *Listeria monocytogenes* inoculated onto the surface of commercially prepared frankfurters. *Journal of Food Protection* 67: 1017–1021.

Lungu, B. and M.G. Johnson. 2005. Fate of *Listeria monocytogenes* inoculated onto the surface of model turkey frankfurter pieces treated with zein coatings containing nisin, sodium diacetate and sodium lactate at 4 degrees C. *Journal of Food Protection* 68: 855–859.

Lungu, B., S.C. Ricke, and M.G. Johnson. 2009. Growth, survival, proliferation and pathogenesis of *L. monocytogenes* under low oxygen or anaerobic conditions: A review. *Anaerobe* 15: 7–17.

Lyytikainen, O., T. Autio, R. Maijala, P. Ruutu, T. Honkanen-Buzalski, M. Miettinen, M. Hatakka et al. 2000. An outbreak of *Listeria monocytogenes* serotype 3a infections from butter in Finland. *The Journal of Infectious Diseases* 181: 1838–1841.

Mahadeo, M. and S. Tatini. 1994. The potential use of nisin to control *Listeria monocytogenes* in poultry. *Letters in Applied Microbiology* 18: 323–326.

Martin-Visscher, L.A., S. Yoganathan, C.C. Sit, C.T. Lohans, and J.C. Vederas. 2011. The activity of bacteriocins from *Carnobacterium maltaromaticum* UAL307 against gram-negative bacteria in combination with EDTA treatment. *FEMS Microbiological Letters* 317: 152–159.

McClintock, M., L. Serres, J. Marzolf, A. Hirsch, and G. Mocquot. 1952. Action inhibitrice des streptocoques producteurs de nisine sur le développement des sporulés anaérobies dans le fromage de Gruyère fondu. *Journal of Dairy Research* 19: 187–193.

Menon, K.V. and S.R. Garg. 2001. Inhibitory effect of clove oil on *Listeria monocytogenes* in meat and cheese. *Food Microbiology* 18: 647–650.

Ming, X. and M. Daeschel. 1993. Nisin resistance of foodborne bacteria and the specific resistance responses of *Listeria monocytogenes* Scott A. *Journal of Food Protection* 56: 944–948.

Ming, X., G.H. Weber, J.W. Ayres, and W.E. Sandine. 1997. Bacteriocins applied to food packaging materials to inhibit *Listeria monocytogenes* on meats. *Journal of Food Science* 62: 413–415.

Mitchell, P. 1966. Chemiosmotic coupling in oxidative and photosynthetic phosphorylation. *Biological Reviews of the Cambridge Philosophical Society* 41: 445–502.

Montville, T.J. and Y. Chen. 1998. Mechanistic action of pediocin and nisin: Recent progress and unresolved questions. *Applied Microbiology and Technology* 50: 511–519.

Montville, T.J. and M.L. Chikindas. 2013. Biological control of foodborne bacteria. In: Doyle, M.P. and E.L. Buchanan (eds.), *Food Microbiology: Fundamentals and Frontiers*, 4th edn. ASM Press, Washington, DC, pp. 803–822.

Mourey, A. and N. Canillac. 2002. Anti-*Listeria monocytogenes* activity of essential oils components of conifers. *Food Control* 13: 289–292.

Murray, E., R. Webb, and M. Swann. 1926. A disease of rabbits characterised by a large mononuclear leucocytosis, caused by a hitherto undescribed bacillus *Bacterium monocytogenes* (n. sp.). *Journal of Pathology and Bacteriology* 29: 407–439.

Murray, E.G.D. 1953. The story of *Listeria*. *Transactions of the Royal Society of Canada, Series 3, Section 5* 47: 15–21.

Muthaiyan, A., E. Martin, S. Natesan, P.G. Crandall, B.J. Wilkinson, and S.C. Ricke. 2012. Antimicrobial effect and mode of action of terpeneless cold pressed Valencia orange essential oil on methicillin-resistant *Staphylococcus aureus* cell lysis. *Journal of Applied Microbiology* 112: 1020–1033.

Naghmouchi, K., D. Drider, E. Kheadr, C. Lacroix, H. Prevost, and I. Fliss. 2006. Multiple characterizations of *Listeria monocytogenes* sensitive and insensitive variants to divergicin M35, a new pediocin-like bacteriocin. *Journal of Applied Microbiology* 100: 29–39.

Nannapaneni, R., V.I. Chalova, P.G. Crandall, S.C. Ricke, M.G. Johnson, and C.A. O'Bryan. 2009a. *Campylobacter* and *Arcobacter* species sensitivity to commercial orange oil fractions. *International Journal of Food Microbiology* 129: 43–49.

Nannapaneni, R., V.I. Chalova, R. Story, K.C. Wiggins, P.G. Crandall, S.C. Ricke, and M.G. Johnson. 2009b. Ciprofloxacin-sensitive and ciprofloxacin-resistant *Campylobacter jejuni* are equally sensitive to natural orange oil-based antimicrobials. *Journal of Environmental Science and Health, Part B* 44: 571–577.

Nannapaneni, R., A. Muthaiyan, P.G. Crandall, M.G. Johnson, C.A. O'Bryan, V.I. Chalova, T.R. Callaway et al. 2008. Antimicrobial activity of commercial citrus-based extracts against *Escherichia coli* O157:H7 isolates and mutant strains. *Foodborne Pathogens and Disease* 5: 695–699.

Nastro, L.J. and S.M. Finegold. 1972. Bactericidal activity of five antimicrobial agents against *Bacteroides fragilis*. *Journal of Infectious Diseases* 126: 104–107.

Nes, I.F. and H. Holo. 2000. Class II antimicrobial peptides from lactic acid bacteria. *Biopolymers* 55: 50–61.

Nielsen, J.W., J.S. Dickson, and J.S. Crouse. 1990. Use of a bacteriocin produced by *Pediococcus acidilactici* to inhibit *Listeria monocytogenes* associated with fresh meat. *Applied and Environmental Microbiology* 56: 2142–2145.

Nieto-Lozano, J.C., J.I. Reguera-Useros, M.D. Pelaez-Martinez, G. Sacristan-Perez-Minayo, A.J. Gutierrez-Fernandez, and A.H. de la Torre. 2010. The effect of the pediocin PA-1 produced by *Pediococcus acidilactici* against *Listeria monocytogenes* and *Clostridium perfringens* in Spanish dry-fermented sausages and frankfurters. *Food Control* 21: 679–685.

Nychas, G.J.E., C.C. Tassou, and P. Skandamis. 2003. Antimicrobials from herbs and spices. In: Roller, S. (ed.), *Natural Antimicrobials for the Minimal Processing of Foods*. Woodhead Publishers, Cambridge, U.K., pp. 176–200.

O'Bryan, C.A., P.G. Crandall, V.I. Chalova, and S.C. Ricke. 2008. Orange essential oils antimicrobial activities against *Salmonella* spp. *Journal of Food Science* 73: M264–M267.

Olsen, S.J., M. Patrick, S.B. Hunter, V. Reddy, L. Kornstein, W.R. MacKenzie, K. Lane et al. 2005. Multistate outbreak of *Listeria monocytogenes* infection linked to delicatessen turkey meat. *Clinical Infectious Diseases* 40: 962–967.

Pandit, V.A. and L.A. Shelef. 1994. Sensitivity of *Listeria monocytogenes* to rosemary (*Rosmarinus officinalis* L.). *Food Microbiology* 11: 57–63.

Parada, J.L., C.R. Caron, A.B.P. Medeiros, and C.R. Soccol. 2007. Bacteriocins from lactic acid bacteria: Purification, properties and use as biopreservatives. *Brazilian Archives of Biology and Technology* 50: 512–542.

Payne, K., E. Rico-Munoz, and P. Davidson. 1989. The antimicrobial activity of phenolic compounds against *Listeria monocytogenes* and their effectiveness in a model milk system. *Journal of Food Protection* 52: 151–153.

Peel, M., W. Donachie, and A. Shaw. 1988. Temperature-dependent expression of flagella of *Listeria monocytogenes* studied by electron microscopy, SDS-PAGE and western blotting. *Journal of General Microbiology* 134: 2171–2178.

Pendleton, S.J., P.G. Crandall, S.C. Ricke, L. Goodridge, and C.A. O'Bryan. 2012. Inhibition of *Escherichia coli* O157:H7 isolated from beef by cold pressed terpeneless Valencia orange oil at various temperatures. *Journal of Food Science* 77: M308–M311.

Perumalla, A.V.S., N.S. Hettiarachchy, K.F. Over, S.C. Ricke, E.E. Gbur, J. Zhang, and B. Davis. 2008. Effect of potassium lactate and sodium diacetate combination to inhibit *Listeria monocytogenes* in low and high fat chicken and turkey hotdog model systems. *The Open Food Science Journal* 6: 16–23.

PHAC (Public Health Agency of Canada). 2008. *Listeria monocytogenes* outbreak. Available at: http://www.phac-aspc.gc.ca/fs-sa/listeria/2008-lessons-lecons-eng.php. Accessed May 19, 2014.

Pirie, J.H.H. 1940a. The genus *Listerella* Pirie. *Science* 91: 383.

Pirie, J.H.H. 1940b. *Listeria*: Change of name for a genus bacteria. *Nature* 145: 264.

Pittman, C.I., S. Pendleton, B. Bisha, C. O'Bryan, L. Goodridge, P.G. Crandall, and S.C. Ricke. 2011. Validation of the use of citrus essential oils as a post-harvest intervention against *Escherichia coli* O157:H7 and *Salmonella* spp. on beef primal cuts. *Journal of Food Science* 76: M433–M438.

Rajaram, G., P. Manivasagan, B. Thilagavathi, and A. Saravanakumar. 2010. Purification and characterization of a bacteriocin produced by *Lactobacillus lactis* isolated from marine environment. *Advance Journal of Food Science and Technology* 2(2): 138–144.

Rasooli, I., M.B. Rezaei, and A. Allameh. 2006. Ultrastructural studies on antimicrobial efficacy of thyme essential oils on *Listeria monocytogenes*. *International Journal of Infectious Diseases* 10: 236–241.

Rayman, K., N. Malik, and A. Hurst. 1983. Failure of nisin to inhibit outgrowth of *Clostridium botulinum* in a model cured meat system. *Applied and Environmental Microbiology* 46: 1450–1452.

Reisinger, P., H. Seidel, H. Tschesche, and P. Hammes. 1980. The effect of nisin on murein synthesis. *Archives of Microbiology* 127: 187–193.

Ricke, S.C. 2003. Perspectives on the use of organic acids and short chain fatty acids as antimicrobials. *Poultry Science* 82: 632–639.

Ricke, S.C., M.M. Kundinger, D.R. Miller, and J.T. Keeton. 2005. Alternatives to antibiotics: Chemical and physical antimicrobial interventions and foodborne pathogen response. *Poultry Science* 84: 667–675.

Ricke, S.C. and M.E. Wideman. 2013. Cranberries and their potential application against foodborne pathogens. *OA Alternative Medicine* 1(2): 17.

Riedo, F.X., R.W. Pinner, M. Tosca, M.L. Cartter, L.M. Graves, M.W. Reeves, R.E. Weaver, B.D. Plikaytis, and C.V. Broome. 1994. A point-source foodborne listeriosis outbreak: documented incubation period and possible mild illness. *Journal of Infection and Disease* 170: 693–696.

Riley, M.A. and M.A. Chavan. 2006. *Bacteriocins, Ecology and Evolution*. Springer, Berlin, Germany.

Rodriguez, J. 1996. Review: Antimicrobial spectrum, structure, properties and mode of action of nisin, a bacteriocin produced by *Lactococcus lactis*. *Food Science and Technology International* 2: 61–68.

Rogers, L. and E. Whittier. 1928. Limiting factors in lactic acid fermentation. *Journal of Bacteriology* 16: 211–229.

Ruhr, E. and H.-G. Sahl. 1985. Mode of action of the peptide antibiotic nisin and influence on the membrane potential of whole cells and on cytoplasmic and artificial membrane vesicles. *Antimicrobial Agents and Chemotherapy* 27: 841–845.

Ruiz, A., S. Williams, N. Djeri, A. Hinton Jr., and G. Rodrick. 2009. Nisin, rosemary, and ethylenediamine-tetraacetic acid affect the growth of *Listeria monocytogenes* on ready-to-eat turkey ham stored at four degrees Celsius for sixty-three days. *Poultry Science* 88: 1765–1772.

Sahl, H.G. 1991. Pore-formation in bacterial membranes by cationic lantibiotics. In: Jung, G. and H.G. Sahl (eds.), *Nisin and Novel Lantibiotics*. Escom, Leiden, the Netherlands, pp. 347–358.

Samelis, J., G.K. Bedie, J.N. Sofos, K.E. Belk, J.A. Scanga, and G.C. Smith. 2002. Control of *Listeria monocytogenes* with combined antimicrobials after postprocess contamination and extended storage of frankfurters at 4°C in vacuum packages. *Journal of Food Protection* 65: 299–307.

Schlech, W.F., P.M. Lavigne, R.A. Bortolussi, A.C. Allen, E.V. Haldane, A.J. Wort, A.W. Hightower et al. 1983. Epidemic listeriosis: Evidence for transmission by food. *New England Journal of Medicine* 308: 203–206.

Shank, F.R., E.L. Elliot, I.K. Wachsmuth, and M.E. Losikoff. 1996. US position on *Listeria monocytogenes* in foods. *Food Control* 7: 229–234.

Shannon, E.M., S.R. Milillo, M.G. Johnson, and S.C. Ricke. 2011. Inhibition of *Listeria monocytogenes* by exposure to a combination of nisin and cold-pressed terpeneless Valencia oil. *Journal of Food Science* 76: M600–M604.

Silva, F., S. Ferreira, J.A. Queiroz, and F.C. Domingues. 2011. Coriander (*Coriandrum sativum* L.) essential oil: Its antibacterial activity and mode of action evaluated by flow cytometry. *Journal of Medical Microbiology* 60: 1479–1486.

Singh, A., R.K. Singh, A.K. Bhunia, and N. Singh. 2003. Efficacy of plant essential oils as antimicrobial agents against *Listeria monocytogenes* in hotdogs. *LWT Food Science and Technology* 36: 787–794.

Singh, B., M.B. Falahee, and M.R. Adams. 2001. Synergistic inhibition of *Listeria monocytogenes* by nisin and garlic extract. *Food Microbiology* 18: 133–139.

Sirsat, S.A., A. Muthaiyan, and S.C. Ricke. 2009. Antimicrobials for pathogen reduction in organic and natural poultry production. *Journal of Applied Poultry Research* 18: 379–388.

Sobrino-López, A. and O. Martín-Belloso. 2008. Use of nisin and other bacteriocins for preservation of dairy products. *International Dairy Journal* 18: 329–343.

Sofos, J.N., G. Flick, G.-J. Nychas, C.A. O'Bryan, S.C. Ricke, and P.G. Crandall. 2013. Meat, poultry, and seafood. In: Doyle, M.P. and R.L. Buchanan (eds.), *Food Microbiology—Fundamentals and Frontiers*, 4th edn. American Society for Microbiology, Washington, DC, pp. 111–167.

Tassou, C., K. Koutsoumanis, and G.J.E. Nychas. 2000. Inhibition of *Salmonella* Enteritidis and *Staphylococcus aureus* in nutrient broth by mint essential oil. *Food Research International* 33: 273–280.

Temelli, F., J.P. O'Connell, C.S. Chen, and R.J. Braddock. 1990. Thermodynamic analysis of supercritical carbon dioxide extraction of terpenes from cold-pressed orange oil. *Industrial Engineering and Chemical Research* 29: 618–624.

Tiwari, B.K., V.P. Valdramidis, C.P. O'Donnell, K. Muthukumarappan, P. Bourke, and P.J. Cullen. 2009. Application of natural antimicrobials for food preservation. *Journal of Agriculture and Food Chemistry* 57: 5987–6000.

Tompkin, B.A. 2002. Control of *Listeria* in the food-processing environment. *Journal of Food Protection* 65: 709–723.

Tramer, J. and G.G. Fowler. 1964. Estimation of nisin in foods. *Journal of the Science of Food and Agriculture* 15: 522–528.

Van Loo, E., V. Caputo, R.M. Nayga Jr., J.-F. Meullenet, P.G. Crandall, and S.C. Ricke. 2010. Effect of organic poultry purchase frequency on consumer attitudes toward organic poultry meat. *Journal of Food Science* 75: S384–S397.

Van Loo, E., V.V. Caputo, R.M. Nayga Jr., J.-F. Meullenet, and S.C. Ricke. 2011. Consumers' willingness to pay for organic chicken breast: Evidence from choice experiment. *Food Quality and Preference* 22: 603–613.

Van Loo, E.J., S.C. Ricke, C.A. O'Bryan, and M.G. Johnson. 2012. Historical and current perspectives in organic meat production. In: Ricke, S.C., E.J. Van Loo, M.G. Johnson, and C.A. O'Bryan (eds.), *Organic Meat Production and Processing*. Wiley Scientific/IFT, New York, pp. 1–9.

Vazquez-Boland, J.A., M. Kuhn, P. Berche, T. Chakraborty, G. Dominguez-Bernal, W. Goebel, B. Gonzalez-Zorn, J. Wehland, and J. Kreft. 2001. *Listeria* pathogenesis and molecular virulence determinants. *Clinical Microbiology Reviews* 14: 584–640.

Vit, M., R. Olejnik, J. Dlhý, R. Karpísková, J. Cástková, V. Príkazský, M. Príkazská, C. Benes, and P. Petrás. 2007. Outbreak of listeriosis in the Czech Republic, late 2006—Preliminary report. *Euro Surveillance* 12: E070208.1.

Walker, S., P. Archer, and J. Banks. 2008. Growth of *Listeria monocytogenes* at refrigeration temperatures. *Journal of Applied Microbiology* 68: 157–162.

Ward, T.J., L. Gorski, M.K. Borucki, R.E. Mandrell, J. Hutchins, and K. Pupedis. 2004. Intraspecific phylogeny and lineage group identification based on their *prfA* virulence gene cluster of *Listeria monocytogenes*. *Journal of Bacteriology* 186: 4994–5002.

Weststrate, J., G. Van Poppel, and P. Verschuren. 2002. Functional foods, trends and future. *British Journal of Nutrition* 88: 233–235.

Whitehead, H.R. 1933. A substance inhibiting bacterial growth, produced by certain strains of lactic *Streptococci*. *Biochemical Journal* 27: 1793.

Williams, S.K., S. Roof, E.A. Boyle, D. Burson, H. Thippareddi, I. Geornaras, J.N. Sofos, M. Wiedmann, and K. Nightingale. 2011. Molecular ecology of *Listeria monocytogenes* and other *Listeria* species in small and very small ready-to-eat meat processing plants. *Journal of Food Protection* 74: 63–77.

Winter, C.H., S.O. Brockmann, S.R. Sonnentag, T. Schaupp, R. Prager, H. Hof, B. Becker, T. Stegmanns, H.U. Roloff, G. Vollrath, A.E. Kuhm, B.B. Mezger, G.K. Schmolz, G.B. Klittich, G. Pfaff, and I. Piechotowski. 2009. Prolonged hospital and community-based listeriosis outbreak caused by ready-to-eat scalded sausages. *Journal of Hospital Infections* 73: 121–128.

Yin, L.J., C.W. Wu, and S.T. Jiang. 2007. Biopreservative effect of pediocin ACCEL on refrigerated seafood. *Fisheries Science* 73(4): 907–912.

16

Replacement of Conventional Antimicrobials and Preservatives in Food Production to Improve Consumer Safety and Enhance Health Benefits

Serajus Salaheen, Mengfei Peng, and Debabrata Biswas

CONTENTS

16.1 Introduction

According to the Center for Disease Control and Prevention, food-borne illnesses are a common, costly, but preventable public health problem. Food-borne pathogens remain a major cause of morbidity and mortality worldwide. In the United States, food-borne diseases cause 9.4 million illnesses, 55,961 hospitalizations, and 1,351 deaths each year (Scallan et al., 2011). The chief reservoir of common causative agents of food-borne illnesses are farm animals. *Salmonella*, *Campylobacter*, *Listeria*, and enterohemorrhagic *Escherichia coli* O157:H7 (EHEC) are the most common bacterial pathogens responsible for food-borne illness (Harrison et al., 2013). The Economic Research Service (ERS, 2000) reported that food-borne illnesses accounted for about 1 of every 100 US hospitalizations and 1 of every 500 US deaths. The ERS also estimated that, each year in the United States, five bacterial pathogens— *Campylobacter*, *Salmonella*, *E. coli* O157:H7, *Listeria monocytogenes*, and *Toxoplasma gondii*—were the source of more than 50% of food-borne illness and were responsible for $6.9 billion in medical expenditures, decreased productivity, and premature deaths. These numbers excluded any hidden costs that the victim or the family might have suffered, such as traveling cost, wasted working hours, lost time for looking after the sick family member, and obviously the pain and sufferings that accompany the illnesses. Not only that, sometimes the acute phase of food-borne illnesses may lead to more serious problems such as hemolytic uremic syndrome, irritable bowel syndrome, Guillain–Barré syndrome, cardiac problems, gut micro-biome imbalance, and mental disorders. According to the Food and Drug Administration (FDA), in 2%–3% of patients, food-borne illness leads to secondary long-term medical complications causing more economic burden.

To combat food-borne pathogens, antimicrobials have been used in both pre- and postharvest levels in agricultural animal production, in preserving foods, and thus improve consumer safety and health benefits. The use of antibiotics to improve microbiological quality of foods started in the mid-twentieth century. Since then, the use of antibiotics in agricultural farm animal production in various purposes including veterinary medicine as subtherapeutic agents and growth promoter to improve feed conversion efficiency caused the development of resistant strains of bacterial pathogens and led the human civilization toward the threat of mysterious postantibiotic era. As a result, for the control of food-borne bacterial pathogens in animal and plant products and to avoid chemical preservatives and antibiotics, the search for alternative natural and organic antimicrobials is more essential than ever.

Natural and organic antimicrobials have been used in different ethnic communities since ancient times when synthetic chemicals were not available. These include different phytochemicals present in plants and plant products. Bioactive compounds from plant origin have several benefits, such as effectiveness against broad spectrum of pathogenic microorganisms due to having more than one mechanism of action, obtained from natural sources so environmentally friendly, less toxic compared to synthetic chemicals, promotes health benefits, and availability from environmental sources. So the use of natural bioactive phytochemicals can reduce problems caused by antibiotics and synthetic chemicals.

The goal of this chapter is to focus on improving the safety of food products by the use of bioactive compounds of natural origin, specifically fruits and their by-products such as pomaces, to eliminate or reduce food-borne bacterial pathogens. In addition, we also explore the possible natural and organic alternatives for longer shelf lives and protection from microbial cross-contamination of the food products. To evaluate the scope of controlling food-borne bacterial infections caused by contaminated food products, we believe that the application of natural organic plant products and their by-products in food production (including pasture farming) in pre- and postharvest levels will control colonization of enteric bacterial pathogens in farm animal guts and prevent the cross-contamination of water, environment, and produce that will reduce the enteric infection and improve public health as well as nutritional values of the products and consumer confidence. In this chapter, we emphasize the published research findings in this area, the research needed to be done to replace common antimicrobial products with natural and organic components, and efforts to fill the gap for organic producers who cannot use synthetic chemicals to make their products safer. Replacing antimicrobial synthetic chemicals with natural organic compounds will also increase consumer confidence and improve public health.

16.2 Conventional Antimicrobials Used in Food Production

Antibiotics are used in both preharvest level farm animal production and postharvest level processing. In farm animals, antibiotics are used as veterinary drugs to reduce incidences of animal diseases, promote quick weight gain, and control the zoonotic pathogens in milk, eggs, meat, and meat products. In the postharvest level, these are also used to prevent cross-contamination and extend shelf life.

16.2.1 Antimicrobials Used in Farm Animal Production

Since the 1950s, large amounts of antibiotics have been used as growth promoters in agricultural animal production in the United States because of their effect on animal growth acceleration and their ability to enhance feed conversion efficiency. β-Lactam antibiotics (e.g., penicillin, lincosamide) and macrolids (e.g., erythromycin, tetracycline) are used mainly in pigs (Peter and John, 2004). Some other antimicrobials, such as arsenical compounds, bacitracin, flavophospholipol, pleuromutilins, quinoxalines, and

TABLE 16.1

FDA-Approved Antibiotics for Preharvest Subtherapeutic
Use in Cattle, Swine, and Poultry

Antibiotics	Cattle	Swine	Poultry
Apramycin		Yes	
Arsanilic acid		Yes	Yes
Avilamycin			Yes
Avoparcin			Yes
Bacitracin methylene disalicylate		Yes	
Bacitracin zinc	Yes	Yes	Yes
Bambermycins	Yes	Yes	Yes
Carbadox		Yes	
Chlortetracycline	Yes	Yes	Yes
Laidlomycin	Yes		
Lasalocid	Yes		
Lincomycin		Yes	Yes
Monensin	Yes	Yes	
Oxytetracycline	Yes	Yes	Yes
Penicillin		Yes	Yes
Roxarsone		Yes	Yes
Spiramycin			Yes
Tiamulin hydrogen fumarate		Yes	
Tilmicosin		Yes	
Tylosin	Yes	Yes	Yes
Virginiamycin	Yes	Yes	Yes

virginiamycin, are also used in pig production. In the cattle industry, flavophospholipol, monensin, and virginiamycin are generally used as growth promoters because these compounds play important roles in muscle formation and increase milk productivity (Peter and John, 2004). For poultry meat and egg production, flavophospholipol and virginiamycin are generally used in the United States (Peter and John, 2004). Some antibiotic candidates approved by FDA for preharvest use in cattle, swine, and poultry are listed in Table 16.1.

Preharvest use of antibiotic is common in farm animal production, but instead of specific targeted pathogens, antibiotics are used for diverse groups of pathogens. As a result, broad-spectrum antibiotics are used in preharvest level. Recent studies suggest that some antibiotic treatment can disrupt the dynamics of gut flora and therefore impair animal health and productivity, and even food safety (Croswell et al., 2009).

Despite these potential shortcomings of antibiotic used at the preharvest level, much recent research has found that at the farm animal level, most of the antibiotics have the potential effect in inhibiting foodborne pathogens and thus improving food safety. Most of the antibiotics are administered on a regular basis into animal diet to exert antimicrobial activities. For example, monensin is approved to be used in food animal production at the preharvest level to reduce the colonization of food-borne pathogenic microorganisms in cattle and pigs.

16.2.2 Antimicrobials Used in Veterinary Medicine

Therapeutic use of antimicrobials for animal diseases is essential for economical and sustainable farm animal production and also on ethical background. Most of the antimicrobials are administered orally through feed and water or by injection. In the case of large animals such as cattle, each of the sick animals may be treated individually, but in the case of small farm animals such as poultry, the whole herd/flock or portion of it needs to be treated. On the other hand, the spectrum of antibiotics to be used depends on the specific pathogen targets. Specific narrow-spectrum antibiotics could be used for treatment for known

TABLE 16.2

Therapeutic Use of Antibiotics in Cattle and Poultry

Animal Species	Disease	Causative Agent	Prescribed Antibiotics	References
Poultry	Necrotic enteritis	*Clostridium perfringens*	Tetracycline, streptomycin, neomycin, bacitracin, avilamycin	Dahiya et al. (2006), Wages (2001)
	Respiratory disease		Tylosin, tiamulin, tilmicosin, aivlosin, doxycycline, chlortetracycline, oxytetracycline, spiramycin, erythromycin, gentamicin, kitasamycin, neomycin, colistin	Loehren et al. (2008)
	Gangrenous dermatitis	*Clostridium septicum, Staphylococcus aureus, E. coli*	Erythromycin, penicillin, oxytetracycline	—
	Fowl cholera	*Pasteurella multocida*	Sulfaquinoxaline sodium, sulfamethazine, sulfadimethoxine, tetracycline, norfloxacin, streptomycin–dihydrostreptomycin combinational injection	Siti and Robert (2000)
	Gram-positive bacterial pathogens		Lincomycin, virginiamycin, spectinomycin, tylosin, erythromycin	Loehren et al. (2008)
Cattle	Respiratory tract disease	*Mannheimia, Pasteurella, Haemophilus*	Chlortetracycline, oxytetracycline, spectinomycin, neomycin, tilmicosin, erythromycin, tylosin, amoxicillin, ampicillin, cephalosporin, sulfamethazine, sulfadimethoxine, florfenicol, enrofloxacin	Barrett (2000), Rerat et al. (2012), Apley and Coetzee (2006), Apley (2001)
	Enteric disease	*E. coli, Salmonella*	Neomycin, chlortetracycline, oxytetracycline	—
	Mastitis	*Pseudomonas, Staphylococcus, E. coli, Mycoplasma, Pasteurella*	Novobiocin, pirlimycin, streptomycin, amoxicillin, cephapirin, cloxacillin, hetacillin, lincomycin, erythromycin	Kandasamy et al. (2011), Wagner and Erskine (2006)

causative agents. When the causative agent of the disease is unclear, a combination of several antibiotics or broad-spectrum antibiotics is commonly used. In general, the prominent animal diseases requiring therapeutic use of antibiotics are respiratory and enteric diseases in calves and pigs, necrotic enteritis in poultry, and mastitis in dairy cattle (Giguère et al., 2006; Radostits et al., 2007; Zimmerman et al., 2012). Antibiotics used against some prominent disease in cattle and poultry are summarized in Table 16.2.

The application/use of different antipathogenic strategies attacking either the specific pathogen of interest or a broad range of pathogens is the most straightforward or typical way to reduce food-borne pathogens at the preharvest levels. But the use of antibiotics in preharvest levels for reducing zoonotic pathogens has some limitations, and their use has decreased gradually due to growing worldwide concern about the transmission of antibiotic-resistant pathogens from farm animals to humans. Multiple antipathogenic strategies including bacteriocins, bacteriophages, enzymes, vaccines, organic acids, and bioactive phytochemicals are gradually replacing the conventional antibiotics. Promising antimicrobials, bacteriocins, are proteins or peptides produced by certain bacteria for inhibiting the growth of their competing bacterial strains in the environment. It is able to reduce the growth of several major food-borne pathogens including EHEC, *Salmonella*, and *Listeria* (Scham-berger and Diez-Gonzalez, 2002; Stahl et al., 2004; Patton et al., 2008). The application of bacteriocin isolated from *Lactobacillus salivarius* and *Paenibacillus polymyxa* in chicken intestinal tract has been documented to induce a significant reduction of *Campylobacter* in broiler chicken cecum (Svetoch and Stern, 2010). Bacteriophages were also found to be effective against EHEC and other food-borne pathogens in poultry and swine (Callaway et al., 2008). The main benefit of this strategy is that bacteriophages target specific pathogen, and other members of gut micro-biome remain unaltered. Also the consumption of large dose of bacteriophage does not negatively impact animal health. In addition to these strategies, vaccination and some natural products are also gaining attention for farm animal production.

16.2.3 Antimicrobials Used as Preservatives in Postharvest Levels

Because farm fresh food products are out of their reach, most of the world's population must rely on processed foods for their daily sustenance. As a result, external materials are required to preserve food quality and shelf life. External substances that are used in or sprayed on food products to preserve their appearance, odor, taste, and shelf life are called food preservatives. Preservatives play an important role during transportation and long-term storage of food products. The prominent mechanism of preservative action is the inhibition of those microorganisms responsible for food spoilage and degradation. The main property of an ideal preservative is its effectiveness against spoilage microorganism without any adverse effect on food and food products. But this is challenging because most preservatives show some detrimental effect on meat and other foods. Chemical preservatives are more problematic in this respect compared to their natural counterparts. Three classes of preservatives are commonly used in food preservation: natural, artificial, and microbial.

Natural preservatives have been used in food products since ancient times, and they are obtained from natural sources. Some examples of natural preservatives are sugar, salt, rosemary extracts, vinegar, oil herbs, and spices. These preservatives serve the preservation purpose in less toxic and environmentally friendly ways. But due to lack of knowledge, they are outcompeted by chemical preservatives, which pose significant health risks. The numbers of natural preservatives that are in practical use today are very few chiefly because of the lack of research in this area.

Artificial preservatives are substances of chemical origin that inhibit the growth of spoilage microorganisms. Some examples are benzoates, sorbates, EDTA, nitrites, calcium propionate, and sodium benzoate. Benzoates inhibit mold, yeasts, and bacteria in acidic drinks and liquid products like soft drinks. Sodium benzoate is used as an antimicrobial preservative agent in foods with a pH of 3.6 or lower, such as salad dressing, carbonated drinks, and fruit juices. Sorbates prevent the growth of mold, yeasts, and fungi in food products. Propionates are generally used in baked goods, and they are responsible for the inhibition of mold growth. Nitrites prevent the growth of bacteria, especially *C. botulinum* in meat and smoked fish. There are other preservatives that are used as antioxidants such as sulfites, vitamin E and C, and butylated hydroxytoluene.

Microbial preservatives can also inhibit bacterial or fungal growth and can act as oxygen absorbers in food constituents. Among preservatives of microbial origin, bacteriocins are the most important. The use of bacteriocin-producing strains of *Enterococcus* has been recorded, and their effects on pathogenic *Listeria, Bacillus, Staphylococcus, Vibrio,* and *Clostridium* have been documented (Giraffa, 2003). Microbial fermented tea has also shown promise as a food preservative (Mo et al., 2008).

16.3 Limitations of Conventional Antimicrobials Used in Food Production

Conventional antimicrobials, though served a lot in making foods safer, present certain limitations, which have led many people to rethink their application in food protection. There are many direct negative impacts as well as indirect effects of these antimicrobials toward life in general.

16.3.1 Antibiotic Resistance of Bacterial and Fungal Pathogens

Since the mid-twentieth century, when the application of antibiotics in human bacterial and fungal diseases was first introduced, it has been considered as the single most important medical event in human history. It has reduced morbidity and mortality in human and animal to a dramatic level. Since then, the use of antibiotics accounts for several million tons worldwide both for medication and in agricultural farm animal production (Wise, 2002; Andersson and Hughes, 2010). The intensive use of antibiotics caused a huge influence in the frequency of resistance among human bacterial and fungal pathogens. Enhanced rate of microbial resistance could be reduced or minimized if we could use the antibiotics effectively and properly in the treatment of human diseases and farm animal production. As the antibiotic-resistant microorganisms are more virulent and aggressive in respect to disease occurrence, this may increase the risk of complicated situations and fatality (Helms et al., 2002; Depuydt et al., 2008), may escalate the economic burden on health care systems (Cosgrove, 2006; Sipahi, 2008; Roberts et al., 2009), and may eventually introduce a dreadful postantibiotic era (Guay, 2008; Lew et al., 2008; Woodford and Livermore, 2009).

16.3.2 Antibiotic Residues in Food

A wide range of chemical and synthetic antimicrobials are used as conventional antimicrobials in farm animal production, and many of these persist in food and food products including meat, milk, and egg as residues (Nisha, 2008). Bioaccumulation might be an important way of increasing chemical residues in food, and consistent exposure to even lower concentrations of chemicals may cause harmful level of chemicals in food. As a result, these chemicals may cause detrimental effects on human health. Infants and children are more susceptible to these compounds because of their poor body protection. They are harmful to developing organs and bodily systems such as neurological and reproductive system. According to USDA Pesticide Data Program (http://www.ams.usda.gov/AMSv1.0/getfile?dDocName= STELPRDC5049946), conventionally grown produce contain three times more pesticide residues compared to organically grown produce because no chemical pesticides were used to grow them. The low level of residues are the probable reason of environmental factors because farmers are unable to control all the factors, and pesticides may remain in soil for several years even after the land has been switched to organic production (Doyle, 2006). Rawn et al. (2006) detected carbaryl and methomyl residues in conventionally produced meat, fruits, and deserts, but no residues were detected in organically produced food products.

16.4 Possible Options of Replacing Conventional Antibiotics with Organic and Natural Antimicrobials

16.4.1 Products of Plant Origin

Plants have been used as the major sources of natural and organic antimicrobials from the beginning of human race. In addition to medicinal plants, many common plants and plant products provide antibacterial, antifungal, or even antiviral activity. Antimicrobial properties of different plant products such as spices and herbs, pomaces from fruits, or vegetable extracts are well documented, which are described as follows.

16.4.1.1 Spices and Herbs

Since ancient times, spices and herbs have been used in foods not only as flavoring agents but also as folk medicine and food preservatives. A significant number of scientific research have been conducted on the antimicrobial properties of different spices and herbs such as mustard, garlic, cinnamon, cumin, bay, clove, thyme, pepper, rosemary, basil, and turmeric (Beuchat, 1994; Nakatani, 1994; Skrinjar et al., 2009). These spices, herbs, and their extracted products have been recommended to be used in food to reduce microbial contamination and increase overall shelf life of food products in a more natural and healthy way. Antimicrobials from spices and herbs are collected at different ways, from volatile or oily liquids, seeds, leaves, barks, and sometimes roots of plants (Tajkarimi et al., 2010). It has been concluded that the presence of alkaloids, glycosides, steroids, phenols, coumarins, and tannins are important antimicrobial elements but oily substances, especially essential oils (EOs), are the main factors responsible for their antimicrobial properties (Ebana et al., 1991).

EOs are aromatic and oily liquid substances present in particular regions of plants. They are generally secondary metabolites of plants and contain thymol, carvacrol, eugenol, and many other components that account for antimicrobial activity against broad spectrum of microorganisms including food-borne pathogens, spoilage bacteria, and fungus (Hammer et al., 1999; Dorman and Deans, 2000). Fungus, especially *Aspergillus* species, causes food and feed contamination and produces aflatoxin, which is a potent hepatocarcinogen in animal and human beings. Omidbeygi et al. (2007) showed different EOs having the inhibitory effect against *Aspergillus*, while Atanda et al. (2007) introduced other EOs causing reduction in aflatoxin production and mycelium growth. Numerous antibacterial and antifungal EOs have been documented (Skrinjar and Nemet, 2009), and their antimicrobial properties vary depending on surrounding conditions.

The antimicrobial property of EOs largely depends on the type of spice or herb and its composition, targeted microorganism, type of food to be used on, pH, and so on. Nedorostova et al. (2009) showed that minimum inhibitory concentration (MIC) of different EOs was higher in the case of Gram-positive bacteria than Gram-negative ones indicating that Gram-negative strains are generally more susceptible to EOs than their Gram-positive counterparts. Due to differences in composition, different herbs and spices show different levels of antimicrobial properties, some show higher and some might not show any effect against any specific target (Gutierrez et al., 2009). It has also been speculated that the use of certain EO can inhibit the growth of certain pathogen while it might stimulate the growth of others by providing favorable growth conditions (Davidson and Branen, 2005), though others provided information that this phenomenon will be less prone to happen because EOs have more than one mode of action. It has been suggested that a cocktail of EOs might be a better option to inhibit pathogenic and spoilage bacteria. Gutierrez et al. (2009) used fractional inhibitory concentration (FIC) index to check the synergy of EO combinations. The FIC was calculated as $FIC_A + FIC_B$, where $FIC_N = MIC_{N, \text{combination}}/MIC_{N, \text{alone}}$, and results interpreted as synergy ($FIC < 0.5$), addition ($0.5 < FIC < 1$), indifference ($1 < FIC < 4$), and antagonism ($FIC > 4$). They reported the presence of synergistic and addition effects but no antagonistic effect. Potential interactions between EO and food materials also play important role in the efficacies of EO. As a result, it is necessary to evaluate the effectiveness of EO within simulated food product and make sure that that specific EO is still active or its antimicrobial property has not been reduced while in food. Gutierrez et al. (2008) showed that oregano and thyme can be more effective at higher concentrations of hydrophobic proteins, which might help EOs to dissolute. On the other hand, its activity was reduced in the presence of high concentration of starch. It was also found that higher concentration of oil (e.g., sunflower oil) had a negative impact on the antibacterial activity of several EOs. The pH also plays a critical role in the activity of EOs, and EOs have higher hydrophobicity at lower-level pH. EOs in low pH dissolute the lipid portion of bacterial cell wall more efficiently (Juven et al., 1994), which indicates a lower pH as a better medium of EO action on target bacteria. All these modes of actions suggest that EOs confer complex effect on bacteria as well as surrounding food environment, which needs to be unearthed.

16.4.1.2 Pomace and By-Products of Fruits

Pomaces are the solid or semisolid remains of fruits after separating out the juice or oily portion. It consists of skins, stems, seeds, and small amount of pulp of the fruit. Juice and fruit processing industries produce a significant amount of pomaces. Pomace generally accounts for as much as 20%–30% of the weight of the whole fruit. According to National Agricultural Statistics Service (2009), a significant amount of pomaces were produced in the United States, which causes disposal problem as they cannot be used as animal feed due to their low protein content and acidic pH. As a result, an alternate use of these by-products will be highly appreciated. Different fruits are produced in different seasons, and they are stored by the juice factories to have a year-round supply of fruits confirming a year-round supply of pomaces. So these by-products can be used to extract polyphenolic compounds, which have been demonstrated to have numerous health benefits including anticarcinogenic, anti-atherosclerotic, anti-adhesive, anti-inflammatory, antiallergic, antihypertensive, antiarthritic, and antimicrobial properties (Boivin et al., 2007; Jepson and Craig, 2008). They also inhibit lipid peroxidation, acting as free radical scavengers and metal ion chelators.

The major bioactive components present in pomaces (a list of phytochemicals present in pomaces and EOs are summarized in Table 16.3) are polyphenols, phytosterols, tocopherols (mostly as α-tocopherol), fiber, protein, and biotic (Puupponen-Pimiä et al., 2001, 2005, 2008). There are different classes of polyphenols, and each class has unique characteristics that separate one from another. Phenolic compounds are one of the most diverse secondary metabolites found in edible plant. Major components of the polyphenols are anthocyanins (cyanidine-3-galactoside and cyanidine-3-glucoside), procyanidins, and hydroxycinnamic (α-cyano-4-hydroxycinnamic acid). Procyanidins, present in the by-products,

TABLE 16.3

Bioactive Phytochemicals Present in Pomaces and Essential Oils

Compound	Generalized Chemical Structure	Targeted Microorganism and Mechanism of Action	Reference
Procyanidin		Gram-positive and Gram-negative bacteria	Mayer et al. (2008)
Anthocyanin		Gram-positive bacteria and few Gram-negative bacteria Membrane damage and interaction with intracellular materials	Cisowska et al. (2011)
A-Tocopherol		Antioxidant, anti-inflammatory	Singh and Jialal (2004)
Phytosterol		*Staphylococcus, Streptococcus, E. coli, Pseudomonas, Klebsiella*	Sharma (1993)
Cholinergenic acid		*Staphylococcus, Streptococcus, Bacillus, E. coli, Shigella, Salmonella* Alteration of plasma membrane permeability	Lou et al. (2011)
Terpenes (monoterpenes)		Gram-positive and Gram-negative bacteria Biomembrane damage	Trombetta et al. (2005)

is a class of polymeric polyphenolics, and it contains catechin or epicatechin at a polymerized form (Djilas et al., 2009). Cyanidin-3-glucoside and cyanidin-3-galactoside are particular types of anthocyanin pigments found in many berries including blackberry, blueberry, cherry, cranberry, and raspberry (Sasaki et al., 2007). The highest concentrations of cyanidin are found in the skin and seed of these fruits because their biosynthesis is stimulated by sunlight (ultraviolet light), so higher concentrations are found in the most outer layer (skin) of the fruit. Different linkages between catechin and epicatechin units are found in procyanidins, the most common being B-type linkages. Oligomeric procyanidins demonstrated greater antimicrobial activity even at lower concentrations than catechin and epicatechin though the degree of polymerization on antimicrobial activity still remains unclear.

Pomaces are also found from citrus fruits, typically oranges, and the oil is found in oil glands in the colored portion of the peel and flavedo (Fisher and Phillips, 2008). Citrus/orange pomaces are complex mixture of compounds (around 400 compounds), depending on the citrus cultivar, extraction, and separation methods. There are two main divisions: the oxygenated compounds that constitute <5% of the volume and hydrocarbons that make up the majority, >95%, of most citrus oils. Hydrocarbons are not very water soluble, but can form emulsions with nonionic surfactants so they can be sprayed on food surfaces. The majority of the hydrocarbons are collectively termed as *terpenes*. Terpenes have the basic structure of $C_{10}H_{16}$ and are divided into cyclic and acyclic terpenes like mercene and ocimene. Cyclic terpene D-limonene, the principal citrus terpene, is a 10-carbon monoterpene that has been shown to be an antimicrobial for years (Mizrahi et al., 2006). It has a very high boiling point of 178°C, making it a likely candidate in food industry because it can withstand thermal processing. The most typical terpenes are limonene, with a slight citrus-like odor due to oxygenated impurities naturally present; γ-terpinene, waxy; terpinolene, green; and α-pinene and β-pinene, pine-like odors. These bioactive compounds play an essential role in exerting antimicrobial properties.

Bioactive compounds extracted from pomaces (especially berry and citrus fruits) have been shown to inhibit different food-borne pathogens as well as food-spoiling microorganisms. According to Puupponen-Pimiä et al. (2001), berry extracts inhibited the growth of *Salmonella*, EHEC, *Staphylococcus* but not *Lactobacillus* and *Listeria* species. Cavanagh et al. (2003) showed inhibitory effect of several berry extracts on the growth of wide range of Gram-positive and Gram-negative human pathogenic bacteria. Biswas et al. (2012) found that blueberry extract has a negative impact on the growth of *Salmonella*, EHEC, *Campylobacter*, and *Listeria*, whereas it stimulates the growth of probiotic *Lactobacillus*. In the other study, human intestinal bacteria such *as Bacteroides fragilis*, *C. perfringens*, *E. coli*, and *S.* Typhimurium were found to be inhibited by tannins while *Bifidobacterium infantis* and *L. acidophilus* were unaffected by it (Chung et al., 1998). The inhibitory effect of phenolic compounds, such as flavonoids, against multidrug-resistant (MDR) bacteria and methicillin-resistant *S. aureus* has been documented (Nanayakkara et al., 2002; Belofsky et al., 2004). Roccaro et al. (2004) found that catechin can increase tetracyclin activity against tetracyclin-resistant staphylococcal isolates. Like berry extracts, orange EOs also have been documented to have effect against various microorganisms. Muthaiyan et al. (2012) found that orange EOs can be used as anti-staphylococcal agent. Orange oil and terpineol were also found to extend the shelf life of milk to more than 56 days when stored at 4°C. According to Kim et al. (1995), carvacrol, citral, and geraniol had strong activity against *S.* Typhimurium and its rifampicin-resistant mutant in vitro but nerolidol, limonene, or β-ionone had no effect on similar kind of bacteria suggesting that different compounds have different target microorganism as well as different mechanisms of action.

Several mechanisms of action of pomace extracts have been proposed, which include cytoplasmic membrane destabilization, permeabilization of plasma membrane, extracellular microbial enzyme inactivation, direct effect on microbial metabolism, and deprivation of substrates mandatory for microbial growth (Puupponen-Pimiä et al., 2004). These extracts may also play a role in the alteration of host epithelial cell–pathogenic bacteria interactions, which is a prerequisite for colonization and infection of many pathogenic bacteria. Clifford (2004) documented that dietary phenolics are poorly absorbed in the small intestine, and 90%–95% accumulate in colon, which possess the capability to alter host cell–bacterial interactions. Cytoplasmic membrane, a semipermeable membrane in microbes, is a phospholipid bilayer, which contains embedded proteins (Maillard, 2002), that regulate the movements of solutes and different metabolites in and out of cell. Phenolics penetrate cytoplasmic membrane and interact with cellular proteins. Concentration-dependent mechanism is also proposed from several researches. Cellular enzyme activity is affected at lower concentration of phenolics, whereas higher concentration may cause protein denaturation. Walsh et al. (2003) reported membrane damage as a mode of action of phenolic EO components. They may interfere with electron transport, nutrient uptake, and nucleic acid synthesis. Carvacrol acts on cytoplasmic membrane, acts as a proton exchanger and hampers the pH gradient across the cytoplasmic membrane, disrupts proton motive force, and depletes intracellular ATP resulting in cellular death (Ultee et al., 2002). Hydroxycinnamic acids, due to having less polar side chain, can penetrate inside the cell very easily (Campos et al., 2003). Tannins inhibit oxidative phosphorylation by disrupting extracellular microbial enzymes or eliminating substrates for enzymatic reaction, which affect microbial metabolism (Scalbert, 1991). Tannins can also cause complexation of metal ions required for bacterial growth. The effect of phenolics on MDR bacteria is due to its ability to impair efflux pumps, which bacteria use to get rid of antibiotics and that cause an increase in the drug retention time inside bacterial cell. Yoda et al. (2004) concluded that green tea polyphenol, epigallocatechin gallate (EGCG), directly binds to cell wall components and causes inhibition, whereas Zhao et al. (2001) showed that EGCG and β-lactams attach to the same site and confer synergistic relationship to β-lactams. EGCG has also been shown to interfere with the transfer of conjugative R plasmid in *E. coli*, which is responsible for the conjugative plasmid-mediated antibiotic resistance in bacteria. Most of the acidic components of pomace extracts show similar mode of action; the pK_a of most phenolic acids ranges between pH 3 and 5, due to the higher pH of cytoplasmic membrane compared to surrounding medium. These compounds dissociate and release proton, which acidifies the cytoplasm (Cotter and Hill, 2003). Anionic portion of these acidic compounds, which cannot escape, accumulate within bacterial cell and impair metabolic function such as increasing the osmotic pressure, which causes incompatibility to bacterial cell. There are other hypothesized modes of actions that need to be evaluated.

16.4.1.3 Fruit and Vegetable Extracts

Different fruits and vegetables and their extracts are rich in bioactive compounds, which have antimicrobial activity. In plants, these compounds are naturally used against pathogenic microorganisms and other predators. These compounds are broadly grouped into phenolic compounds, EOs, terpenoids, alkaloids, polypeptides, lectins, etc. Different compounds have different roles and different modes of action on pathogenic and spoiling microorganisms. A few fruits and vegetables possessing these compounds are mentioned as follows.

Polyphenols are the main components of pomegranate. Clinical research reported that it takes up to 56 h for the ellagitannins, mostly punicalagins, to leave the colon, because they are absorbed very slowly and not completely (Cerda, 2004; Seeram et al., 2006). During this time, it is proved that pomegranate tannins inhibit a large number of pathogens in that section of the gut without affecting most beneficial bacteria (Bialonska et al., 2009). Pomegranate constituents seem to disturb both Gram-positive and Gram-negative intestinal pathogenic bacteria. According to previously published reports (Bialonska et al., 2009), ellagic acid inhibited the growth of *C. clostridioforme*, *C. perfringens*, and *C. ramosum*. Ellagic acid together with punicalagins is effective against the growth of *B. fragilis* and *S. aureus*. The effect on bifidobacteria is species specific. The number of *B. breve* and *B. infantis* increases in the presence of the pomegranate ellagitannins, which indicates that pomegranate products may help regulate pathogens without adverse effects on beneficial bacteria (Howell et al., 2013).

In red raspberries, the main phenolic compounds are ellagitannins, followed by flavonoids and anthocyanins. Antimicrobial activity of tannins against microorganisms is well documented. Phenolic compounds, including ellagitannins, anthocyanins, and flavonols (Nohynek et al., 2006), showed similar selective bactericidal effects on both Gram-positive and Gram-negative bacteria. Raspberries are effective inhibitors of *Staphylococcus* and *S.* Typhimurium (Puupponen-Pimiä et al., 2005). Other types of bacteria being affected by the phenolic compounds in raspberries include *E. coli* (Nohynek et al., 2006) and *S. enterica* (Puupponen-Pimiä et al., 2001). The growth of *Lactobacillus* strains showed no evidence to be inhibited by any of the raspberry extracts at low concentrations. However, in high concentrations, the growth of the probiotic was clearly disrupted (Puupponen-Pimiä et al., 2005).

The composition of the cranberry has been widely documented as a combination of flavonoids and phenolic acids such as catechin, myricetin, and benzoic acid, the latter of which is the most prominent one in a freshly squeezed sample of juice (Chen et al., 2001). Cranberries are also rich in a phenolic compound called proanthocyanidins. Proanthocyanidins have been cataloged as the ones responsible for the cranberry's great ability to disrupt bacterial adherence to human cells (O'May and Tufenkji, 2011). Howell et al. (2005) concluded that the proanthocyanidins in cranberry juice can prevent the adhesion of *E. coli* to the urinary tract thus preventing urinary infections. Furthermore, biofilm formation in uroepithelial cells can also be reduced by ingesting cranberry juice (Reid et al., 2001). The antimicrobial effects of cranberry juice have been recorded several times. Historically, women have been told to drink the juice in order to prevent and even cure urinary tract infections. In the early 1990s, researchers found that the monosaccharide fructose present in cranberry and blueberry juices competitively inhibited the adsorption of pathogenic *E. coli* to urinary tract epithelial cells, acting as an analogue for mannose (Zafriri et al., 1989). Different research studies have proven the antimicrobial properties of cranberries against both Gram-positive and Gram-negative bacteria. Cranberry extract has been demonstrated being effective toward *B. cereus*, *C. perfringens*, and *S. epidermis*; lyophilized cranberry had a bacterocidic impact on *S.* sv. Typhimurium, *S. aureus*, and *L. monocytogenes* (Puupponen-Pimiä et al., 2005). Also, coumaric acid, another phenolic acid present in cranberry juice, has confirmed efficacy against *L. plantarum* (Nualkaekul and Charalampopoulos, 2011). In a separate study, proanthocyanidins were proved effective in blocking part of the motile system of *P. aeruginosa*. In that study, results showed that the growth of *P. aeruginosa* was not inhibited; however, the characteristic migrating branching pattern of the bacteria was disrupted, which indicated a disruption in swarming motility in the presence of cranberry phenolics (O'May and Tufenkji, 2011)?

Grapes are rich in phenolic compounds, mostly in the skin and seeds. Grape skins are abundant in flavonols, including anthocyanins that are responsible for the color. Grape seeds are rich in monomeric phenolic compounds like catechins and procyanidins (Rodriguez et al., 2006). These perform well against pathogens in vitro, and they are also antiviral and anti-mutagenic agents. Other flavonols (quercetin, myricetin, kaempferol, isorhamnetin, and their glycosides) are potent antioxidants (Llobera and Canellas, 2007) as well. The level of anthocyanins is higher in the skin and lower in the seeds or the pulp, which is because these compounds are mostly stored in the vacuoles of the exocarp (peel) cell (Butkhup et al., 2010). Grape skins and seeds are more effective as bactericides than the whole fruit. Both skins and seeds have proven to inhibit Gram-positive bacteria such as *B. cereus*, *B. subtilis*, *S. aureus*, *S. faecalis*, and *S. cremoris*, and Gram-negative such as *E. coli*, *Shigella dysenteriae*, *S.* Typhimurium, and *V. cholerae*, although in a smaller scale. In addition, whole grapes are effective against the yeast *Kluyveromyces marxianus* (Butkhup et al., 2010).

Naturopaths and doctors with unconventional style prescribe grapefruit seed extract (GSE) for the treatment of gastrointestinal tract infections. In vitro experiments have shown its effectiveness on a huge number of microorganisms, including more than 800 strains of bacteria and virus, 100 fungal strains, and a significant number of single-celled parasites, and it works mainly by cell membrane disruption and releasing cytoplasmic content (Heggers et al., 2002). Because of its antioxidative and antifungal activity, GSE can be used for postharvest preservation of certain fruits (Xu et al., 2007). It can also be used as sanitizer, and for this purpose, GSE can be used either alone or in combination with nisin or citric acid. When used as a sanitizer, GSE can significantly reduce human bacterial pathogens from freshly cut and whole vegetables (Xu et al., 2006). Another added advantage is its effectiveness against diarrhea.

For hundreds of years, olive (*Olea europaea*) leaf has been widely used in folk medicine (Gucci and Tattini, 1997) and as antimalarial agent. The major bioactive compounds present in olive leaf are diosmetin-7-glucoside, caffeic acid, hydroxytyrosol, tyrosol, *p*-coumaric acid, vanillic acid, vanillin, oleuropein, luteolin, diosmetin, rutin, verbascoside, luteolin-7-glucoside, and apigenin-7-glucoside (Tasioula-Margari and Ologeri, 2001). Olive leaf can lower blood pressure and enhance blood circulation in the coronary arteries (Khayyal et al., 2002). It contains phenolic compounds that work as antioxidant (Skerget et al., 2005) as well as show antimicrobial properties against ulcer caused by *H. pylori*,

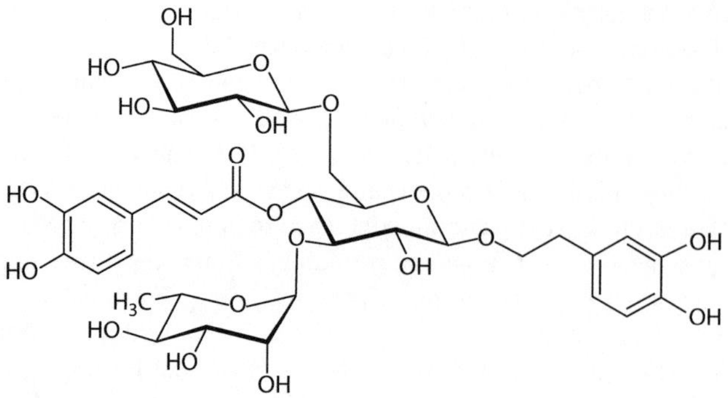

FIGURE 16.1   Chemical structure of echinacoside, the antibiotic compound in Echinacea.

C. jejuni, and *S. aureus*. Olive leaf extract has also been shown to stimulate the growth of probiotic *B. infantis* and *L. acidophilus*. Probiotic stimulation might be the result of having high fructooligosaccharide content, which is considered to be a prebiotic compound (Markin et al., 2003).

Traditionally, papaya has also been used as a therapeutic, immune-stimulating, and antioxidant agents. Almost every portion of papaya is important: green fruits and roots are used for their abortifacient properties (Cherian, 2000; Sharma and Mahanta, 2000); the seed is used as antifertility drug (Lohiya et al., 2000); the pulp is used for wound and burn healing; the seeds have anthelmintic property and are useful against gastrointestinal nematode infections (Stepek et al., 2004) and also show inhibitory effect against enteric pathogens (Krishna et al., 2008); leaves have been found to be effective against amoebic dysentery. Ethanol extract of papaya exhibited antimicrobial activity against *S. choleraesuis* and *S. aureus*, and it also induced significant wound healing (Nayak et al., 2012). Synergism between methanol extract and antibiotic has been shown to be effective against drug-resistant bacteria (Rakholiya and Chanda, 2012).

Echinacoside, an antibiotic compound from echinacea, has a wide array of activities, and for this reason, sometimes it is even compared with penicillin (Figure 16.1). Echinacea is used to treat bronchitis, tonsilitis, ear infections, wounds, abscesses, and other infectious diseases (Cowan, 1999). Active echinacea extract is thought to provide many beneficial effects to cold and flu patients, such as inactivation of the viral particles, inhibition of bacteria causing respiratory tract infections, and reversal of the pro-inflammatory responses induced by cold and flu agents without any cytopathic effects in tissue (Hudson, 2010). Echinacea root extract exerts antifungal activity at very low concentration (Dahui et al., 2011). Echinacea extract has also been suggested as an alternative to synthetic antioxidant and preservatives as it can inhibit mold growth and lipid oxidation during storage (Sabouri et al., 2012).

16.4.2 Products of Microbial Origin

Bacteriocins are the most important antimicrobials isolated from microbial origin, which are used in food processing. Many successful trials have been made to utilize these compounds in different foods. Nisin has been found to be effective to reduce spoilage bacteria in cheese and canned foods. Encapsulated nisin with low temperature was used to control *L. monocytogenes* (da Silva Malheiros et al., 2010). Similarly, nisin with high pressure can drastically (5-logs) reduce *E. coli* in apple and carrot juices (Pathanibul et al., 2009). Organic acids, such as citric, acetic, and lactic acids, are produced by microorganisms and have been used to preserve many foods and food products. Their use in bread (Pattison et al., 2004), juices (Corte et al., 2004), and meats (Drosinos et al., 2006) has been documented.

Reuterin, a broad-spectrum antimicrobial substance, is produced by *L. reuteri*. According to Arqués et al. (2004), reuterin can inhibit the growth of *L. monocytogenes*, *Aeromonas hydrophila*, *C. jejuni*, *Yersinia enterocolitica*, *S. aureus*, and *E. coli* O157:H7. Reuterin has been used as a biopreservative in different food products like beef sausage, milk, and cottage cheese. Combination of reuterin and nisin was tested against a wide array of Gram-positive and Gram-negative bacteria. Their combined effects were not significantly different from the effect of reuterin alone against Gram-negative bacteria (Arqués et al., 2004), but this combination showed better inhibition against Gram-positive bacteria including *L. monocytogenes* and *S. aureus* (Arqués et al., 2011).

Natamycin is an amphiphilic antifungal compound produced by *Streptomyces* sp. It is safe to consume, has no effect on food quality, and possesses prolonged antifungal activity especially on the surface of the food, and these properties make it an ideal broad-spectrum antifungal biopreservative to be used in foods and beverages (Juneja et al., 2012). It does not inhibit bacterial growth, as a result of which it does not affect fermentation bacteria.

Probiotics are defined as direct-fed mono- or mixed cultures of living microorganisms that can compete with undesired microbes and benefit the host by improving the properties of the indigenous microbiota (Fuller, 1992). These beneficial microbes are able to ameliorate the overall health of animal by improving the gut microbial balance; however, their exact mechanism is still under investigation. One major hypothesis for their actions could be due to their influence on intestinal metabolic activities, which include the improvement of bacteriocins, propionic acid, and vitamin B_{12} production, and increasing the villous length and nutrient absorption (Christina et al., 2009). Other possible mechanisms include competitive exclusion of pathogenic microorganisms and their immunostimulatory activities. Probiotics are also effective in boosting weight gain and feed conversion efficiency in newborn animals. However, several questions about the active strains, the maximum dosage, the effective delivery system, and the potential risk remain unanswered and need to be investigated in the future.

16.4.3 Products of Animal Origin

Several animal products have been documented so far:

Lysozyme is generally isolated from eggs and small peptides produced by protease digestion of lysozyme and have been documented to show direct or indirect antimicrobial effect against Gram-positive and Gram-negative bacteria. LzP, a commercially available lysozyme peptide, was found to be effective against bacterial spores (Abdou et al., 2007).

Lactoferrin, a glycoprotein isolated from milk, has been shown to inhibit the growth of pathogenic bacteria (Naidu et al., 2003; Taylor et al., 2004) when sprayed on beef. It binds to iron and thereby reduces or prevents pathogenic and spoilage bacterial growth.

Lactoperoxidase system (LPS) is useful to preserve the quality of raw milk when necessary refrigeration is not present. LPS can reduce pathogenic bacterial count on beef (Elliot et al., 2004). This enzyme oxidizes thiocyanate ions in the presence of reactive hydrogen peroxide whose products bind to sulfhydryl groups of bacterial enzyme and inhibit their effectiveness (Doyle, 2006). Monolaurin and LPS synergistically inhibited *E. coli* O157:H7 and *S. aureus* in broth and food products (McLay et al., 2002).

Ovotransferrin, also termed as conalbumin, is a glycoprotein isolated from eggs (Naidu, 2003). Being metalloproteinase in nature and having high affinity to iron, it reduces bacterial metabolism by making iron unavailable to bacteria. Its purified form shows bactericidal activity against both Gram-positive and Gram-negative bacteria. Pleurocidin, a heat-stable and salt-tolerant peptide, is present in myeloid cells and mucosal tissue of vertebrates and invertebrates. Inhibitory activities of pleurocidin against different food-borne pathogens such as *L. monocytogenes* and *E. coli* O157:H7 have been documented (Burrowes et al., 2004; Jung et al., 2007).

Defensins are cationic peptides having a broad spectrum of antibacterial, antiviral, and antifungal activities (Dhople et al., 2006; Jin et al., 2010). Protamine is also a cationic peptide found in Salmon and possesses natural antimicrobial properties against Gram-positive and Gram-negative bacteria as well as large number of fungi (Uyttendaele and Debevere, 1994).

Chitosan is a polysaccharide isolated from exoskeleton of crustaceans and arthropods (e.g., crabs and shrimps). It inhibits the growth of bacteria and mold and is used to enhance the quality as well as shelf life of foods. It can be used to reduce spoilage bacteria in cheese (Altieri et al., 2005) and inhibit *C. perfringens* spore germination in beef and turkey (Juneja et al., 2006). Chitosan coating has been shown to reduce trans-shell penetration of eggs by *S.* Enteritidis and *Pseudomonas*, *E. coli*, and *L. monocytogenes* (Leleu et al., 2011).

16.5 Advantages of Using Organic and Natural Antimicrobials

There are many ways in which natural and organic antimicrobials serve us, but the emerging issue of *antibiotic resistance* is an area where natural bioactive compounds can be most useful. Antibiotic resistance is a natural phenomenon adopted by bacteria for their survival in the presence of antibiotics and is part of microbial evolution and as a result is somewhat unavoidable. Due to natural selection, bacteria adopt certain mechanisms, such as (1) cross-resistance due to mutation in cellular genes (chromosomal mutation) and (2) gene transfer with the help of plasmids, transposons, integrons, and bacteriophages (Giedraitienė et al., 2011). These mechanisms help bacteria to inactivate antibiotics, target modifications, alter permeability, or bypass a metabolic pathway required for antibiotic action. These mechanisms are so diverse that they provide bacteria with different survival strategies that are responsible for the current condition of antibiotic resistance: resistance is observed against almost all of the antibiotics present. To deal with this serious problem, alternatives are essential. The first alternative is the use of natural and organic antimicrobials. Current antibiotics generally aim at one specific target such as cell wall synthesis, nucleic acid synthesis, or protein synthesis, and the required high dosage for effectivity very often causes bioavailability problems and the emergence of resistance. As a result, focusing on multiple targets through the use of antimicrobial adjuvants may solve this problem. Natural and organic antimicrobials generally have more than one target on a bacterial cell. Extracts of plant origin contain a combination of different phytochemicals that make even a complex mixture of antimicrobials against which resistance is less likely to be developed. Moreover, it has been suggested that synergism between traditional antibiotic treatment and supplemental phytochemicals may show better effect, and reversal of antibiotic sensitivity might occur (Simões et al., 2009).

The long-term use of prescribed antibiotics may result in several side effects such as feeling sick, bloating and indigestion, loss of appetite, allergic reactions, and most importantly antibiotic-associated diarrhea (AAD). High dosage and long-term use of antibiotics lead to an imbalance in colonic microbiota, which causes imbalance in carbohydrate metabolism with decreased absorption ability of short-chain fatty acid and imbalance of osmotic pressure, which ultimately cause diarrhea. *C. difficile* is responsible for 10%–20% of AAD cases. Because of gut flora imbalance and a lower number of *healthy* bacteria due to antibiotic treatment, *C. difficile* overgrow and produce enterotoxin. Prolonged use of antibiotics can cause oxidative stress. This is a condition when cells are programmed to produce chemically reactive oxygen molecules, which cause damage to DNA, enzymes, and cell membranes. Natural and organic antimicrobials can exert their role with mild effect on host cells and thus bypass side effect unlike antibiotics.

Improved consumer confidence is another beneficial impact of using natural agents as antimicrobials and preservatives in food production and processing. Consumers are today better educated and informed, and they tend to avoid chemicals in food and food products in part because they believe that organic food products provide healthier and safer food. As a result, foods with natural antimicrobials and natural preservatives are highly valued by most consumers and are thus in great demand.

16.6 Information Lacking for Introducing Organic and Natural Antimicrobials

Research is required to define the bioactive compounds within each of the natural agents and their mode of actions against microbial pathogens. For more extensive and effective use of natural and organic antimicrobials, elaborate research needs to be focused on (1) the effectiveness and functionality in food models and food systems; (2) their toxicology and their safety in in situ food formulations; (3) economic feasibility (e.g., extraction, isolation, and storage); (4) interactions with food ingredients and synergistic, antagonistic, or collaborative properties with other preservatives; (5) specific mode of actions against target microorganisms; and (6) the impact on food quality (e.g., nutritiousness and organoleptic) and flavor.

While the predictive effectiveness and some aspects of the mode of action of several phytochemicals have been studied, other factors such as the structure-activity relationship (SAR) are not well understood. Further research into the mechanisms by which phytochemicals cause inhibition of important cell functions and in some cases lysis of microbial cells is required in order to understand and hence exploit any mechanisms and apply them efficiently in new therapeutic or biocontrol strategies. The more structurally complex libraries inspired by natural products including phytochemicals might be tremendous sources of new antimicrobials. However, because of the relatively rapid mechanism of bacteria acquiring resistance, it must be taken into account that phytochemicals with antibacterial potential may have only a limited period of practical utility.

Antimicrobial compounds and herbal medicinal extracts have been used for hundreds of years for curing disease, preventing health problems, and improving the quality of life. However, methodical approaches for safety and effectiveness evaluation are not well documented. Recently, phytogenic compounds have gained more attention because of their potential role as an alternative to growth promoters. The holistic impact to health makes these compounds very attractive, but due to many factors, it makes scientific assessment very challenging. A few studies have shown their effectiveness in antimicrobial properties and animal nutrition, but these studies are difficult to compare as there are large variations of compositions in different sources, and the mechanisms behind are far from being elucidated (Jouany and Morgavi, 2007). Due to large variations in the composition of active compounds, their potential biological effects have not been standardized yet. More research is required to understand their mode of action and to determine circumstances where the use of these additives might be more beneficial. In complex mixtures, synergistic or antagonistic behavior of active compounds is also worth examination because these behaviors might be responsible for inconsistencies. Moreover, knowledge of their effect on complex gut flora, host histopathology, and the immune system is still rudimentary and must be investigated more in depth. Fruits, herbs, and vegetables also contain prebiotics like oligofructose, fructooligosaccharide, galactooligosaccharide, and inulin. But their use in food production and as feed additive does not show consistent results. Some show promising results in animal nutrition (Miguel et al., 2004), while others have no impact (Keegan et al., 2005). These inconsistencies need to be addressed.

Many people believe that plant-derived medicines are safe, but very often they are used in mixed forms and bioactive constituents. Therefore, quality control of bioactive compounds has a direct impact on their safety and efficiency (Ernst et al., 2005; Ribnicky et al., 2008). Very limited data on the quality, efficiency, and safety of these bioactive compounds are available, and further research on them is required. Moreover, some of the prominent issues that are also needed to be addressed include microbial resistance to natural antimicrobials. If this is the case, it will cause a similar problem to antibiotic resistance. The effects of homogeneously mixed antimicrobial compounds in food matrices or pilot-scale production of bioactive compounds from their natural sources without hampering efficacy and functionality are also important areas that warrant further study.

16.7 Conclusion

It is important to develop safe and effective disease prevention and treatments using natural, beneficial, and organic bioactive compounds to combat the infections caused by food-borne pathogens. We believe that the use of natural and organic bioactive compounds obtained from fruits, vegetables, and essentially their pomaces might be an effective and practical means of eliminating colonization and infection of food-borne pathogens in farm animal guts. The replacement of antibacterial synthetic chemicals used in farm animal production and food processing will help solve the emerging threat of antibiotic resistance and increase consumer confidence in the safety of food products. It will also reduce annual costs of medical care and production losses in the United States and around the world due to food-borne illness as well as improve public health. It will be applicable in both conventional and organic farming. However, there are many gaps and constraints in the practical application of natural antimicrobials in food processing because more research on their efficacy, consumer acceptability, and economic impact is needed. More research grants are required to carry out research and explore this promising field and its potential benefits.

REFERENCES

Abdou, A.M., Higashiguchi, S., Aboueleinin, A.M., Kim, M., and Ibrahim, H.R. 2007. Antimicrobial peptides derived from hen egg lysozyme with inhibitory effect against *Bacillus* species. *Food Control* 18:173–178.

Altieri, C., Scrocco, C., Sinigaglia, M., and Del Nobile, M.A. 2005. Use of chitosan to prolong mozzarella cheese shelf life. *Journal of Dairy Science* 88:2683–2688.

Andersson, D.I. and Hughes, D. 2010. Antibiotic resistance and its cost: Is it possible to reverse resistance? *Nature Reviews Microbiology* 8:260–271.

Apley, M. 2001. Animal husbandry and disease control: Cattle. In *FDA-CVM's Public Meeting: Use of Antimicrobial Drugs in Food Animals and the Establishment of Regulatory Thresholds on Antimicrobial Resistance*. January 22–24, 2001. Center for Veterinary Medicine, Food and Drug Administration, Rockville, MD.

Apley, M.D. and Coetzee, J.F. 2006. Antimicrobial drug use in cattle. In Giguère, S., Prescott, J.F., Baggot, J.D., Walker, R.D., and Dowling, P.M. (eds.), *Antimicrobial Therapy in Veterinary Medicine*, 4th edn. Blackwell, Oxford, U.K., pp. 485–506.

Arqués, J.L., Fernández, J., Gaya, P., Nuñez, M., Rodríguez, E., and Medina, M. 2004. Antimicrobial activity of reuterin in combination with nisin against food-borne pathogens. *International Journal of Food Microbiology* 95(2):225–229. doi:10.1016/j.ijfoodmicro.2004.03.009.

Arqués, J.L., Rodríguez, E., Nuñez, M., and Medina, M. 2011. Combined effect of reuterin and lactic acid bacteria bacteriocins on the inactivation of food-borne pathogens in milk. *Food Control* 22(3–4):457–461. doi:10.1016/j.foodcont.2010.09.027.

Atanda, O.O., Akpan, I., and Oluwafemi, F. 2007. The potential of some spice essential oils in the control of *A. parasiticus* CFR 223 and aflatoxin production. *Food Control* 18:601–607.

Barrett, D.C. 2000. Cost-effective antimicrobial drug selection for the management and control of respiratory disease in European cattle. *Veterinary Record* 146(19):545–550.

Belofsky, G., Percivill, D., Lewis, K., Tegos, G.P., and Ekart, J. 2004. Phenolic metabolites of Dalea versicolor that enhance antibiotic activity against model pathogenic bacteria. *Journal of Nature Products* 67:481–484.

Beuchat, L.R. 1994. Antimicrobial properties of spices and their essential oils. In Dillon, Y.M. and Board, R.G. (eds.), *Natural Antimicrobial Systems and Food Preservation*. CAB International, Wallingford, U.K., pp. 167–179.

Bialonska, D., Kasimsetty, S.G., Schrader, K.K., and Ferreira, D. 2009. The effect of pomegranate (*Punica granatum* L.) by-products and ellagitannins on the growth of human gut bacteria. *Journal of Agricultural and Food Chemistry* 57(18):8344–8349. ISSN 0021-8561.

Biswas, D., Wideman, N.E., O'Bryan, C.A., Muthaiyan, A., Lingbeck, J.M., Crandall, P.G., and Ricke, S.C. 2012. Pasteurized blueberry (*Vaccinium corymbosum*) juice inhibits growth of bacterial pathogens in milk but allows survival of probiotic bacteria. *Journal of Food Safety* 32(2):204–209. doi:10.1111/j.1745-4565.2012.00369.x.

Boivin, J., Bunting, L., Collins, J.A., and Nygren, K.G. 2007. International estimates of infertility prevalence and treatment-seeking: Potential need and demand for infertility medical care. *Human Reproduction* 22:1506–1512.

Burrowes, O.J., Hadjicharalambous, C., Diamond, G., and Lee, T.C. 2004. Evaluation of antimicrobial spectrum and cytotoxic activity of pleurocidin for food applications. *Journal of Food Science* 69(3):66–71.

Butkhup, L., Chowtivannaku, P.S., Gaensakoo, R., Prathepha, P., and Samappito, S. 2010. Study of the phenolic composition of shiraz red grape cultivar (*Vitis vinifera* L.) cultivated in north-eastern Thailand and its antioxidant and antimicrobial activity. *South African Journal of Enology & Viticulture* 31(2):89–98.

Callaway, T.R., T.S. Edrington, A.D. Brabban, R.C. Anderson, M.L. Rossman, M.J. Engler, M.A. Carr, K.J. Genovese, J.E. Keen, M.L. Looper, E.M. Kutter, and D.J. Nisbet. 2008. Bacteriophage isolated from feedlot cattle can reduce *Escherichia coli* O157:H7 populations in ruminant gastrointestinal tracts. *Foodborne Pathogens and Disease* 5:183–191.

Campos, F.M., Couto, J.A., and Hogg, T.A. 2003. Influence of phenolic acids on growth and inactivation of *Oenococcus oeni* and *Lactobacillus hilgardii*. *Journal of Applied Microbiology* 94:167–174.

Cavanagh, H.M., Hipwell, M., and Wilkinson, J.M. 2003. Antibacterial activity of berry fruits used for culinary purposes. *Journal of Medicinal Food* 1:57–61.

Cerda, B., Espin, J.C., Parra, S., Martinez, P., and Tomas-Barberan, F.A. 2004. The potent in vitro antioxidant ellagitannins from pomegranate juice are metabolised into bioavailable but poor antioxidant hydroxyl-6*H*-dibezopyran-6-one derivatives by the colonic microflora of healthy humans. *European Journal of Nutrition* 43:205–220.

Chen, H., Zuo, Y.G., and Deng, Y.W. 2001. Separation and determination of flavonoids and other phenolic compounds in cranberry juice by high-performance liquid chromatography. *Journal of Chromatography* 913:387–395.

Cherian, T. 2000. Effect of *papaya latex* extract on gravid and non-gravid rat uterine preparations in vitro. *Journal of Ethnopharmacology* 70:205–212.

Christina, E.W., Marie-Louise, H., and Olle, H. 2009. Probiotics during weaning reduce the incidence of eczema. *Pediatric Allergy & Immunology* 20(5):430–437.

Chung, K.T., Lu, Z., and Chou, M.W. 1998. Mechanism of inhibition of tannic acid and related compounds on the growth of intestinal bacteria. *Food and Chemical Toxicology* 36:1053–1060.

Cisowska, A., Wojnicz, D., Hendrich, A.B. 2011. Anthocyanins as antimicrobial agents of natural plant origin. *Natural Product Communications* 6:149–156.

Clifford, M.N. 2004. Diet-derived phenols in plasma and tissues and their implication for health. *Planta Medica* 70:1103–1114.

Corte, F.V., De Fabrizio, S.V., Salvatori, D.M., and Alzamora, S.M. 2004. Survival of *Listeria innocua* in apple juice as affected by vanillin or potassium sorbate. *Journal of Food Safety* 24:1–15.

Cosgrove, S.E. 2006. The relationship between antimicrobial resistance and patient outcomes: Mortality, length of hospital stay, and health care costs. *Clinical Infectious Diseases* 42:82–89.

Cotter, P.D. and Hill, C. 2003. Surviving the acid test: Responses of Gram-positive bacteria to low pH. *Microbiology and Molecular Biology Reviews* 67:429–453.

Cowan, M.M. 1999. Plant products as antimicrobial agents. *Clinical Microbiology Reviews* 12(4):564–582.

Croswell, A., Amir, E., Teggatz, P., Barman, M., and Salzman, N.H. 2009. Prolonged impact of antibiotics on intestinal microbial ecology and susceptibility to enteric Salmonella infection. *Infection and Immunity* 77(7):2741–2753. doi:10.1128/IAI.00006-09.

Dahiya, J.P., Wilkie, D.C., Van Kessel, A.G., and Drew, M.D. 2006. Potential strategies for controlling necrotic enteritis in broiler chickens in post-antibiotic era. *Animal Feed Science and Technology* 129(1–2):60–88.

Dahui, L., Zaigui, W., and Yunhua, Z. 2011. Antifungal activity of extracts by supercritical carbon dioxide extraction from roots of *Echinacea angustifolia* and analysis of their constituents using gas chromatography-mass spectrometry (GC-MS). *Journal of Medicinal Plants Research* 5(23):5605–5610.

da Silva Malheiros, P., Daroit, D.J., and Brandelli, A. 2010. Food applications of liposome encapsulated antimicrobial peptides. *Trends in Food Science & Technology* 21:284–292.

Davidson, P.M. and Branen, A.L. 2005. Food antimicrobials—An introduction. In Davidson, P.M., Sofos, J.N., and Branen, A.L. (eds.), *Antimicrobial in Food*. CRC Press/Taylor & Francis Group, Boca Raton, FL, pp. 1–9.

Depuydt, P.O., Vandijck, D.M., Bekaert, M.A., Decruyenaere, J.M., Blot, S.I., Vogelaers, D.P., and Benoit, D.D. 2008. Determinants and impact of multidrug antibiotic resistance in pathogens causing ventilator-associated-pneumonia. *Critical Care* 12:R142.

Dhople, V., Krukemeyer, A., and Ramamoorthy, A. 2006. The human β-defensin-3, an antibacterial peptide with multiple biological functions. *Biochimica et Biophysica Acta* 1758(9):1499–1512.

Djilas, S.M., Tumbas, V.T., Savatovic, S.S., Mandic, A.I., Markov, S.L., and Cvetkovic, D.D. 2009. Radical scavenging and antimicrobial activity of horsetail (*Equisetum arvense* L.) extracts. *Food Science and Technology* 44:269–278.

Dorman, H.J.D. and Deans, S.G. 2000. Antimicrobial agents from plants: Antibacterial activity of plant volatile oils. *Journal of Applied Microbiology* 88:308–316.

Doyle, M.E. 2006. Natural and organic foods: Safety considerations. A brief review of the literature (November). http://fri.wisc.edu/docs/pdf/FRIBrief_NaturalOrgFoods.pdf (accessed date May 16, 2014).

Drosinos, E.H., Mataragas, M., Kampani, A., Kritikos, D., and Metaxopoulos, I. 2006. Inhibitory effect of organic acid salts on spoilage flora in culture medium and cured cooked meat products under commercial manufacturing conditions. *Meat Science* 73:75–81.

Ebana, R.U.B., Madunagu, B.E., Ekpe, E.D., and Otung, I.N. 1991. Microbiological exploitation of cardiac glycosides and alkaloids from *Garcinia kola*, *Borreria ocymoides*, *Kola nitida* and *Citrus aurantiofolia*. *Journal of Applied Bacteriology* 71:398–401.

Elliot, R.M., McLay, J.C., Kennedy, M.J., and Simmonds, R.S. 2004. Inhibition of foodborne bacteria by the lactoperoxidase system in a beef cube system. *International Journal of Food Microbiology* 91:73–81.

Ernst, E., Schmidt, K., and Wider, B. 2005. CAM research in Britain: The last 10 years. *Complementary Therapies in Clinical Practice* 11:17–20.

Fisher, K. and Phillips, C. 2008. Potential antimicrobial uses of essential oils in food: Is citrus the answer? *Trends in Food Science & Technology* 19(3):156–164.

Fuller, R. 1992. *Probiotics: The Scientific Basis*. Chapman & Hall, London, U.K.

Giedraitienė, A., Vitkauskienė, A., Naginienė, R., and Pavilonis, A. 2011. Antibiotic resistance mechanisms of clinically important bacteria. *Medicina (Kaunas, Lithuania)* 47(3):137–146.

Giguère, S., Prescott, J.F., Baggot, J.D., Walker, R.D., and Dowling, P.M. 2006. *Antimicrobial Therapy in Veterinary Medicine*, 4th edn. Blackwell, Oxford, U.K.

Giraffa, G. 2003. Functionality of *enterococci* in dairy products. *International Journal of Food Microbiology* 88:215–222.

Guay, D.R. 2008. Contemporary management of uncomplicated urinary tract infections. *Drugs* 68:1169–1205.

Gucci, R. and Tattini, M. 1997. Salinity tolerance in olive. *Horticultural Reviews* 21:177–214.

Gutierrez, J., Barry-Ryan, C., and Bourke, P. 2008. The antimicrobial efficacy of plant essential oil combinations and interactions with food ingredients. *International Journal of Food Microbiology* 124:91–97.

Gutierrez, J., Barry-Ryan, C., and Bourke, P. 2009. Antimicrobial activity of plant essential oils using food model media: Efficacy, synergistic potential and interactions with food components. *Food Microbiology* 26:142–150.

Hammer, K.A., Carson, C.F., and Riley, T.V. 1999. Antimicrobial activity of essential oils and other plant extracts. *Journal of Applied Microbiology* 86:985–990.

Harrison, L.M., Balan, K.V., and Babu, U.S. 2013. Dietary fatty acids and immune response to food-borne bacterial infections. *Nutrients* 5(5):1801–1822. doi:10.3390/nu5051801.

Heggers, J.P., Cottingham, J., Gusman, J., Reagor, L., McCoy, L., Carino, E., Cox, R., Zhao, J.G., and Reagor, L. 2002. The effectiveness of processed grapefruit seed extract as an antibacterial agent. II. Mechanism of action and in-vitro toxicity. *Journal of Alternative and Complementary Medicine* 8:333–340.

Helms, M., Vastrup, P., Gerner-Smidt, P., and Molbak, K. 2002. Excess mortality associated with antimicrobial drug-resistant *Salmonella* Typhimurium. *Emerging Infectious Diseases* 8:490–495.

Howell, A.B., Reed, J.D., Krueger, C.G., Winterbottom, R., Cunningham, D.G., and Keahy, M. 2005. A-type cranberry proanthocyanidins and uropathogenic bacterial anti-adhesion activity. *Phytochemistry* 66:2281–2291.

Howell, A.B. and Souza, D.H.D. 2013. The pomegranate: Effects on bacteria and viruses that influence human health. *Evidence-Based Complementary and Alternative Medicine* 2013:606212.

Hudson, J.B. 2010. The multiple actions of the phytomedicine Echinacea in the treatment of colds and flu. *Journal of Medicinal Plants Research* 4(25):2746–2752.

Jepson, R.G. and Craig, J.C. 2008. Cranberries for preventing urinary tract infections. *Cochrane Database of Systematic Reviews* 1:Art. No.:CD001321. doi: 10.1002/14651858.CD001321.pub4.

Jin, J.Y., Zhou, L., Wang, Y., Li, Z., Zhao, J.G. et al. 2010. Antibacterial and antiviral roles of a fish β-defensin expressed both in pituitary and testis. *PLoS One* 5(12):e12883.

Jouany, J. P., and Morgavi. D. P. 2007. Use of 'natural' products as alternatives to antibiotic feed additives in ruminant production. *Animal* 1:1443–1466.

Juneja, V.K., Dwivedi, H.P., and Yan, X. 2012. Novel natural food antimicrobials. *Annual Review of Food Science and Technology* 3:381–403. doi:10.1146/annurev-food-022811-101241.

Juneja, V. K., Thippareddi, H., Bari, L., Inatsu, Y., Kawamoto, S., and Friedman, M. 2006. Chitosan protects cooked ground beef and turkey against *Clostridium perfringens* spores during chilling. *Journal of Food Science* 71:236–240.

Jung, H.J., Park, Y., Sung, W.S., Suh, B.K., Lee, J. et al. 2007. Fungicidal effect of pleurocidin by membrane-active mechanism and design of enantiomeric analogue for proteolytic resistance. *Biochimica et Biophysica Acta* 1768(6):1400–1405.

Juven, B.J., Kanner, J., Schved, F., and Weisslowicz, H. 1994. Factors that interact with the antibacterial action of thyme essential oil and its active constituents. *Journal of Applied Bacteriology* 76:626–631.

Kandasamy, S., Green, B.B., Benjamin, A.L., and Kerr, D.E. 2011. Between-cow variation in dermal fibroblast response to lipopolysaccharide reflected in resolution of inflammation during *Escherichia coli* mastitis. *Journal of Dairy Science* 94(12):5963–5975.

Keegan, T.P., Dritz, S.S., Nelssen, J.L., Derouchey, J.M., Tokach, M.D., and Robert, D. 2005. Peer reviewed effects of in-feed antimicrobial alternatives and antimicrobials on nursery pig performance and weight variation. *Journal of Swine Health and Production* 13:12–18.

Khayyal, M.T., El-Ghazaly, M.A., Abdallah, D.M., Nassar, N.N., Okpanyi, S.N., and Kreuter, M.H. 2002. Blood pressure lowering effect of an olive leaf extract (*Oleo europaea*) in L-NAME induced hypertension in rates. *Arzneimittel-Forschung/Drug Research* 52:797–802.

Kim, J.M., Marshall, M.R., Cornell, J.A., Preston, J.F., and Wei, C.I. 1995. Antibacterial activity of carvacrol, citral, and geraniol against *Salmonella* Typhimurium in culture medium and on fish cubes. *Journal of Food Science* 60:1364–1368.

Krishna, K.L., Paridhavi, M., and Jagruti, A.P. 2008. Review on nutritional, medicinal and pharmacological properties of Papaya (*Carica papaya* Linn.). *NPR* 7:364–373.

Leleu, S., Herman, L., Heyndrickx, M., De Reu, K., Michiels, C.W. et al. 2011. Effects on Salmonella shell contamination and trans-shell penetration of coating hens' eggs with chitosan. *International Journal of Food Microbiology* 145(1):43–48.

Lew, W., Pai, M., Oxlade, O., Martin, D., and Menzies, D. 2008. Initial drug resistance and tuberculosis treatment outcomes: Systematic review and meta-analysis. *Annals of Internal Medicine* 149:123–134.

Llobera, A. and Canellas, J. 2007. Dietary fiber content and antioxidant activity of Manto Negro red grape (*Vitis vinifera*): Pomace and stem. *Food Chemistry* 101:659–666.

Loehren, U., Ricci, A., and Cummings, T.S. 2008. Guidelines for antimicrobial use in poultry. In Guardabassi, L., Williamson, R., and Kruse, H. (eds.), *Guide to Antimicrobial Use in Animals*. Blackwell, Oxford, U.K., pp. 126–142.

Lohiya, N.K., Pathak, N., Mishra, P.K., and Manivannan, B. 2000. Contraceptive evaluation and toxicological study of aqueous extract of the seeds of *Carica papaya* in male rabbits. *Journal of Ethnopharmacology* 70:17–27.

Lou, Z., Wang, H., Zhu, S., Ma, C., and Wang, Z. 2011. Antibacterial activity and mechanism of action of chlorogenic acid. *Journal of Food Science* 76(6):M398–M403. doi:10.1111/j.1750-3841.2011.02213.x.

Maillard, J.Y. 2002. Bacterial target sites for biocide action. *Journal of Applied Microbiology (Symposium Supplement)* 90:16S–27S.

Markin, D., Duek, L., and Berdicevsky, I. 2003. In vitro antimicrobial activity of olive leaves. *Mycoses* 46(3–4):132–136.

Mayer, R., Stecher, G., Wuerzner, R., Silva, R. C., Sultana, T., Trojer, L., Feuerstein, I., Krieg, C., Abel, G., Popp, M., Bobleter, O., and Bonn, G. K. 2008. Proanthocyanidins: target compounds as antibacterial agents. *Journal of Agricultural and Food Chemistry* 56:6959–6966.

McLay, J.C., Kennedy, M.J., O'Rourke, A.L., Elliot, R.M., and Simmonds, R.S. 2002. Inhibition of bacterial foodborne pathogens by the lactoperoxidase system in combination with monolaurin. *International Journal of Food Microbiology* 73:1–9.

Miguel, J.C., Rodriguez-zas, S.L., and Pettigrew, J.E. 2004. Efficacy of a mannan oligosaccharide (Bio-Mos) for improving nursery pig performance. *Journal of Swine Health and Production* 12:296–307.

Mizrahi, B., Shapira, L., Domb, A.J., and Houri-Haddad, Y. 2006. Citrus oil and $MgCl_2$ as antibacterial and anti-inflammatory agents. *Journal of Periodontology* 77(6):963–968.

Mo, H., Zhu, Y., and Chen, Z. 2008. Microbial fermented tea—A potential source of natural food preservatives. *Trends in Food Science & Technology* 19:124–130.

Muthaiyan, A., Martin, E.M., Natesan, S., Crandall, P.G., Wilkinson, B.J., and Ricke, S.C. 2012. Antimicrobial effect and mode of action of terpeneless cold-pressed Valencia orange essential oil on methicillin-resistant *Staphylococcus aureus*. *Journal of Applied Microbiology* 112:1020–1033.

Naidu, A.S., Tulpinski, J., Gustilo, K., Nimmagudda, R., and Morgan, J.B. 2003. Activated lactoferrin. Part 2: Natural antimicrobial for food safety. *Agro Food Industry Hi-Tech* 14:27–31.

Nakatani, N. 1994. Antioxidative and antimicrobial constituents of herbs and spices. In Charalambous, G. (ed.), *Spices, Herbs and Edible Fungi*. Elsevier Science, New York, pp. 251–271.

Nanayakkara, N.P., Burandt, C.L., Jr., and Jacob, M.R. 2002. Flavonoids with activity against methicillin-resistant *Staphylococcus aureus* from *Dalea scandens* var. *paucifolia*. *Planta Medica* 68:519–522.

National Agricultural Statistics Service (NASS). 2010. Noncitrus fruits and nuts 2009 preliminary summary. US Department of Agriculture, Washington, DC, http://usda.mannilib.cornell.edu/usda/nass/noncFruiNu//2010s/2010/NoncFruiNu-01-22-2010_revision.pdf. NEWMOA. Metal painting and coating operations. Available at: http://www.istc.illinois.edu/main_sections/info_services/library_docs/manuals/coatings/toc.htm. Accessed: 6/9/2010.

Nayak, B.S., Ramdeen, R., Adogwa, A., Ramsubhag, A., and Marshall, J.R. 2012. Wound-healing potential of an ethanol extract of *Carica papaya* (Caricaceae) seeds. *International Wound Journal* 9(6):650–655.

Nedorostova, L., Kloucek, P., Kokoska, L., Stolcova, M., and Pulkrabek, J. 2009. Antimicrobial properties of selected essential oils in vapour phase against foodborne bacteria. *Food Control* 20:157–160.

Nisha, A.R. 2008. Antibiotic residues—A global health hazard. *Veterinary World* 1(12):375–377.

Nohynek, L.J., Alakomi, H., Kahkonen, M.P., Heinonen, M., Helander, I.M., Oksman-Caldentey, K., and Puupponen-Pimiä, R.H. 2006. Berry phenolics: Antimicrobial properties and mechanisms of action against severe human pathogens. *Nutrition and Cancer* 54(1):18–32.

Nualkaekul, S. and Charalampopoulos, D. 2011. Survival of *Lactobacillus plantarum* in model solutions and fruit juices. *International Journal of Food Microbiology* 146(2):111–117.

O'May, C. and Tufenkji, N. 2011. The swarming motility of *Pseudomonas aeruginosa* is blocked by cranberry proanthocyanidins and other tannin-containing materials. *Applied and Environmental Microbiology* 77:3061–3067.

Omidbeygi, M., Barzegar, M., Hamidi, Z., and Naghdibadi, H. 2007. Antifungal activity of thyme, summer savory and clove essential oils against *Aspergillus flavus* in liquid medium and tomato paste. *Food Control* 18:1518–1523.

Pathanibul, P., Taylor, T.M., Davidson, P.M., and Harte, F. 2009. Inactivation of *Escherichia coli* and *Listeria innocua* in apple and carrot juices using high pressure homogenization and nisin. *International Journal of Food Microbiology* 129:316–320.

Pattison, T.L., Lindsay, D., and Von Holy, A. 2004. Natural antimicrobials as potential replacements for calcium propionate in bread. *South African Journal of Science* 100:342–348.

Patton, B.S., Lonergan, S.M., Cutler, S.A., Stahl, C.H., and Dickson, J.S. 2008. Application of colicin E1 as a prefabrication intervention strategy. *Journal of Food Protection* 71:2519–2522.

Peter, H. and John, H. 2004. Antibiotic growth-promoters in food animals. Food and Agriculture Organization of the United Nations, Rome, Italy.

Puupponen-Pimiä, R., Aura, A.-M., Karppinen, S., Oksman-Caldentey, K.-M., and Poutanen, K. 2004. Interactions between plant bioactive food ingredients and intestinal flora—effects on human health. *Bioscience and Microflora* 23:67–80.

Puupponen-Pimiä, R., Nohynek, L., Alakomi, H.-L., and Oksman-Caldentey, K.-M. 2005. Bioactive berry compounds—Novel tools against human pathogens. *Applied Microbiology and Biotechnology* 67(1):8–18. doi:10.1007/s00253-004-1817-x.

Puupponen-Pimiä, R., Nohynek, L., Ammann, S., Oksman-Caldentey, K.M., and Buchert, J. 2008. Enzyme-assisted processing increases antimicrobial and antioxidant activity of bilberry. *Journal of Agricultural and Food Chemistry* 56:681–688.

Puupponen-Pimiä, R., Nohynek, L., Meier, C., Kahkonen, M., Heinonen, M., Hopia, A., and Oksman-Caldentey, K.M. 2001. Antimicrobial properties of phenolic compounds from berries. *Journal of Applied Microbiology* 90(4):494–507.

Radostits, O.M., Gay, C., Hinchcliff, K., and Constable, P. 2007. *Veterinary Medicine: A Textbook of the Diseases of Cattle, Horses, Sheep, Pigs, and Goats*, 10th edn. Elsevier, Edinburgh, U.K.

Rakholiya, K. and Chanda, S. 2012. In vitro interaction of certain antimicrobial agents in combination with plant extracts against some pathogenic bacterial strains. *Asian Pacific Journal of Tropical Biomedicine* 2(2):876–880.

Rawn, D.F., Quade, S.C., Shield, J.B., Conca, G., Sun, W.F., and Lacroix, G.M. 2006. Organophosphate levels in apple composites and individual apples from a treated Canadian orchard. *Journal of Agricultural and Food Chemistry* 54(5):1943–1948.

Reid, G., Hsiehl, J., Potter, P., Mighton, J., Lam, D., Warren, D., and Stephenson, J. 2001. Cranberry juice consumption may reduce biofilms on uroepithelial cells: Pilot study in spinal cord injured patients. *Spinal Cord* 39:26–30.

Rerat, M., Albini, S., Jaquier, V., and Hussy, D. 2012. Bovine respiratory disease: Efficacy of different prophylactic treatments in veal calves and antimicrobial resistance of isolated Pasteurellaceae. *Preventive Veterinary Medicine* 103(4):265–273.

Ribnicky, D.M., Poulev, A., Schmidt, B., Cefalu, W.T., and Raskin, I. 2008. The science of botanical supplements for human health: A view from the NIH botanical research centers. Evaluation of botanicals for improving human health. *American Journal of Clinical Nutrition* 87:472–475.

Roberts, R.R., Hota, B., and Ahmad, I. 2009. Hospital and societal costs of antimicrobial-resistant infections in a Chicago teaching hospital: Implications for antibiotic stewardship. *Clinical Infectious Diseases* 49:1175–1184.

Roccaro, A.S., Blanco, A.R., Giuliano, F., Rusciano, D., and Enea, V. 2004. Epigallocatechin-gallate enhances the activity of tetracycline in staphylococci by inhibiting its efflux from bacterial cells. *Antimicrobial Agents and Chemotherapy* 48:1968–1973.

Rodriguez, M.R., Romero, P.R., Chacon, V.J.L., Martinez, G.J., and Garcia, R.E. 2006. Phenolic compounds in skins and seeds of ten grape *Vitis vinifera* varieties grown in a warm climate. *Journal of Food Composition & Analysis* 19:687–693.

Sabouri, Z., Barzegar, M., Sahari, M.A., and Naghdi, B.H. 2012. Antioxidant and antimicrobial potential of *Echinacea purpurea* extract and its effect on extension of cake shelf life. *Journal of Medicinal Plants* 11:43.

Sasaki, R., Nishimura, N., Hoshino, H., Isa, Y., Kadowaki, M. et al. 2007. Cyanidin 3-glucoside ameliorates hyperglycemia and insulin sensitivity due to downregulation of retinol binding protein 4 expression in diabetic mice. *Biochemical Pharmacology* 74:1619–1627.

Scalbert, A. 1991. Antimicrobial properties of tannins. *Phytochemistry* 30:3875–3883.

Scallan, E., Hoekstra, R.M., Angulo, F.J., Tauxe, R.V., Widdowson, M.-A., Roy, S.L., Jones, J.L., and Griffin, P.M. 2011. Foodborne illness acquired in the United States—Major pathogens. *Emerging Infectious Diseases* 17(1):7–15. doi:10.3201/eid1701.P11101.

Schamberger, G.P. and Diez-Gonzalez, F. 2002. Selection of recently isolated colicinogenic *Escherichia coli* strains inhibitory to *Escherichia coli* O157:H7. *Journal of Food Protection* 65:1381–1387.

Seeram, N.P., Henning, S.M., Zhang, Y., Suchard, M., Li, Z., and Heber, D. 2006. Pomegranate juice ellagitannin metabolites are present in human plasma and some persist in urine for up to 48 h. *Journal of Nutrition* 136:2481–2485.

Sharma, H.N. and Mahanta, H.C. 2000. Modulation of morphological changes of endometrial surface epithelium by administration of composite root extract in albino rat. *The Journal of Contraception* 62:51–54.

Sharma, R.K. 1993. Phytosterols: Wide-spectrum antibacterial agents. *Bioorganic Chemistry* 21:49–60.

Simões, M., Bennett, R.N., and Rosa, E.S. 2009. Understanding antimicrobial activities of phytochemicals against multidrug resistant bacteria and biofilms. *Natural Product Reports* 26(6):746–757. doi:10.1039/b821648g.

Singh, U. and Jialal, I. 2004. Anti-inflammatory effects of alpha-tocopherol. *Annals of the New York Academy of Sciences* 1031(Cvd):195–203. doi:10.1196/annals.1331.019.

Sipahi, O.R. 2008. Economics of antibiotic resistance. *Expert Review of Anti Infective Therapy* 6:523–539.

Siti, M. and Robert, H. 2000. The immunogenicity and pathogenicity of *Pasteurella multocida* isolated from poultry in Indonesia. *Veterinary Microbiology* 72(1):27–36.

Skerget, M., Kotnik, P., Hadolin, M., Hras, A.R., Simonic, M., and Knez, Z. 2005. Phenols, proanthocyanidins, flavones and flavonols in some plant materials and their antioxidant activities. *Food Chemistry* 89:191–198.

Skrinjar, M., Mandic, A., Misan, A., Sakac, M., Saric, L., and Zec, M. 2009. Effect of mint (*Mentha piperita* L.) and caraway (*Carum carvi* L.) on the growth of some toxigenic *Aspergillus* species and aflatoxin B1 production. *Zbornik Matice Srpske Za Prirodne Nauke* 116:131–139.

Skrinjar, M. and Nemet, N. 2009. Antimicrobial effects of spices and herbs essential oils. *Acta Periodica Technologica* 220(40):195–209. doi:10.2298/APT0940195S.

Stahl, C.H., Callaway, T.R., Lincoln, L.M., Lonergan, S.M., and Genovese, K.J. 2004. Inhibitory activities of colicins against *Escherichia coli* strains responsible for postweaning diarrhea and edema disease in swine. *Antimicrobial Agents and Chemotherapy* 48:3119–3121.

Stepek, G., Behnke, J.M., Buttle, D.J., and Duce, I.R. 2004. Natural plant cysteine proteinases as anthelmintics? *Trends in Parasitology* 20:322–327.

Svetoch, E.A. and Stern, N.J. 2010. Bacteriocins to control *Campylobacter* spp. in poultry—A review. *Poultry Science* 89(8):1763–1768.

Tajkarimi, M.M., Ibrahim, S.A., and Cliver, D.O. 2010. Antimicrobial herb and spice compounds in food. *Food Control* 21(9):1199–1218.

Tasioula-Margari, M. and Ologeri, O. 2001. Isolation and characterization of virgin olive oil phenolic compounds by HPLC/UV and GC/MS. *Journal of Food Science* 66:530–534.

Taylor, S., Brock, J., Kruger, C., Berner, T., and Murphy, M. 2004. Safety determination for the use of bovine milk-derived lactoferrin as a component of an antimicrobial beef carcass spray. *Regulatory Toxicology and Pharmacology* 39:12–24.

Trombetta, D., Castelli, F., Sarpietro, M.G., Venuti, V., Cristani, M., Daniele, C., and Bisignano, G. 2005. Mechanisms of antibacterial action of three monoterpenes. *Antimicrobial Agents and Chemotherapy* 49(6):2474–2478. doi:10.1128/AAC.49.6.2474.

Ultee, A., Bennik, M.H.J., and Moezelaar, R. 2002. The phenolic hydroxyl group of carvacrol is essential for action against the food-borne pathogen *Bacillus cereus*. *Applied and Environmental Microbiology* 68:1561–1568.

US Department of Agriculture, Economic Research Service (ERS). 2000. Product liability and microbial food-borne illness, AER-799. Economic Research Service/USA, Washington, DC.

US Department of Agriculture, Pesticide Data Program Annual Summary 2005. http://www.ams.usda.gov/AMSv1.0/getfile?dDocName=STELPRDC5049946 (accessed date May 16, 2014).

Uyttendaele, M. and Debevere, J. 1994. Evaluation of the antimicrobial activity of protamine. *Food Microbiology* 11:417–427.

Wages, D. 2001. Animal husbandry and disease control: Poultry. In *FDA-CVM's Public Meeting: Use of Antimicrobial Drugs in Food Animals and the Establishment of Regulatory Thresholds on Antimicrobial Resistance*. January 22–24, 2001. Center for Veterinary Medicine, Food and Drug Administration, Rockville, MD.

Wagner, S. and Erskine, R. 2006. Antimicrobial drug use in bovine mastitis. In Giguère, S., Prescott, J.F., Baggot, J.D., Walker, R.D., and Dowling, P.M. (eds.), *Antimicrobial Therapy in Veterinary Medicine*, 4th edn. Blackwell, Oxford, U.K., pp. 507–517.

Walsh, S.E., Maillard, J.Y., Russell, A.D., Catrenich, C.E., Charbonneau, D.L., and Bartolo, R.G. 2003. Activity and mechanisms of action of selected biocidal agents on Gram-positive and -negative bacteria. *Journal of Applied Microbiology* 94:240–247.

Wise, R. 2002. Antimicrobial resistance: Priorities for action. *Journal of Antimicrobial Chemotherapy* 49:585–586.

Woodford, N. and Livermore, D.M. 2009. Infections caused by Gram-positive bacteria: A review of the global challenge. *Journal of Infection* 59(Suppl. 1):S4–S16.

Xu, W.T., Huang, K.L., Guo, F., Qu, W., Yang, J.J., Liang, Z.H., and Luo, Y.B. 2007. Postharvest grapefruit seed extract and chitosan treatments of table grapes to control *Botrytis cinerea*. *Postharvest Biology and Technology* 46:86–94.

Xu, W.T., Qu, W., Huang, K., Guo, F., Yang, J., Zhao, H., and Luo, Y. 2006. Antibacterial effect of grapefruit seed extract on foodborne pathogens and its application in the preservation of minimally processed vegetables. *Postharvest Biology and Technology* 45(1):126–133.

Yoda, Y., Hu, Z.-Q., and Zhao, W.-H. 2004. Different susceptibilities of *Staphylococcus* and Gram-negative rods to epigallocatechin gallate. *Journal of Infection and Chemotherapy* 10:55–58.

Zafriri, D., Ofek, I., Adar, R., Pocino, M., and Sharon, N. 1989. Inhibitory activity of cranberry juice on adherence of type 1 and type P fimbriated *Escherichia coli* to eucaryotic cells. *Antimicrobial Agents and Chemotherapy* 33:92–98.

Zhao, W.-H., Hu, Z.-Q., Okubo, S., Hara, Y., and Shimamura, T. 2001. Mechanism of synergy between epi-gallacatechin gallate and β-lactams against methicillin-resistant *Staphylococcus aureus*. *Antimicrobial Agents and Chemotherapy* 45:1737–1742.

Zimmerman, J., Karriker, L., Ramirez, A., Schwartz, K., and Stevenson, G. 2012. *Diseases of Swine*, 10th edn. Wiley-Blackwell, Ames, IA.

17

Control of Toxigenic Molds in Food Processing

Miguel A. Asensio, Félix Núñez, Josué Delgado, and Elena Bermúdez

CONTENTS

17.1 Introduction

Uncontrolled fungal growth in agricultural and food commodities causes various problems, from spoilage to health hazards due to mycotoxins. Thus, fighting fungal growth is a common task in the food industry. Mold growth can be efficiently controlled by various treatments such as chemical preservatives, physical treatments, or modified atmosphere packaging (MAP). However, the different treatments are adequate only for a particular application, such as for food contact surfaces and cooked or packaged foods. In addition, the main mycotoxins are not destroyed by processing treatments. For this, preventing fungal growth usually requires taking the advantages of an adequate combination of different treatments according to the various stages of food processing.

17.2　Synthetic Chemicals to Control Fungal Growth

Fungicides as benzimidazoles, aromatic hydrocarbons, and sterol biosynthesis inhibitors are being used especially in cereals, fruits, and vegetables during postharvest and storage. Compounds such as propiconazole decreased both *Fusarium* growth and mycotoxin accumulation in cereals (Liggitt et al. 1997), whereas other works found a reduction only in the incidence of the mold with no effect on the concentration of mycotoxins in stored grains (Ellis et al. 1996). On the contrary, treatment with thiabendazole in cereals resulted in the maximum reduction of mycotoxin level, but had no significant effect on *Fusarium* infection (Boyacioglu et al. 1992). The acquisition of resistance by some molds as a result of chemical overuse (Marín et al. 2008) has forced to increase the dose. Some of these compounds led to a strong induction of mycotoxin biosynthesis by toxigenic molds (Schmidt-Heydt et al. 2013), which implies that the effect on secondary metabolite biosynthesis should be tested. In addition, the hazard on human health due to their carcinogenicity, teratogenicity, high and acute residual toxicity, and long degradation period leads to strict restrictions in their use. Due to these problems, alternative treatments are being tested.

Salts of weak acids (sodium benzoate and potassium sorbate) require high concentrations to obtain optimum results leading to organoleptic changes in food. Other naturally occurring antimicrobials approved for foods, such as monolaurin and lactoperoxidase, show problems such as limited spectrum of activity, expensive application, development of fungal resistances, and organoleptic nondesirable changes on foods (da Cruz Cabral et al. 2013). Ozone is a potent disinfecting agent routinely used to control fungal alteration of cereals and fruits by applying dissolved ozone in water prior to food packaging and also by controlling the gaseous atmosphere of food storage areas. A significant effect on aflatoxin B_1 degradation was evident with gaseous ozone, whereas ozonated water was more effective on fungal counts (Zorlugenç et al. 2008). However, ozone is a strong oxidant, which is not adequate for nutritional value, color, and flavor in many foods.

Chemical compounds leave residues that can act for a while. On the other hand, they should meet some criteria to elude the formation of toxic residues, to negatively affect the nutritional and organoleptic quality of foods, and to be cost-effective, which means easy and cheap to apply. However, no treatment has been totally effective without negative impact on foods (Box 17.1).

17.3　Physical Methods

Physical methods leave no residues in foods, and some have been used to reduce mycotoxins to a safe level.

High temperatures are effective in fungal structures, but no damage is caused on mycotoxins (Betina 1989). Heat treatments can be applied as hot-water rinsing and brushing, hot-water dips, vapor heat,

---

BOX 17.1　ADVANTAGES AND LIMITATIONS OF SYNTHETIC CHEMICALS

Benzimidazoles, aromatic hydrocarbons, and sterol biosynthesis inhibitors

Advantages: effective on decreasing fungal growth and/or mycotoxin accumulation in cereals, fruits, and vegetables during postharvest and storage.
Limitations: strict restrictions in their use, and some of them can induce mycotoxin biosynthesis.

Salts of weak acids, monolaurin, and lactoperoxidase

Advantages: naturally occurring compounds approved for food use.
Limitations: low efficiency leading to organoleptic changes at active doses.

Ozone

Advantages: effective on fungal counts.
Limitations: not adequate for both the nutritional value and organoleptic characteristics of food.

and hot dry air (Fallik 2008). Mild treatments effectively reduce fungal growth in fruits and vegetables, but higher temperatures are responsible for deleterious effects on foods.

To increase the antifungal efficiency, high temperatures are combined with ethanol, 1-methylcyclopropene, sodium bicarbonate, or fungicides (Fallik 2008). A concentration of 10% of ethanol at 60°C inhibits fungal growth, but it is not adequate for many foods.

UV-C irradiation has been used to treat certain fruits and vegetables during storage to avoid microbial spoilage. Prestorage UV-C treatment of strawberry before mold inoculation results in lower diseases, which could be related to the increase in the transcription and activity of enzymes involved in the defense against the fungal pathogens (Pombo et al. 2011). On the contrary, this protective effect was not observed on papaya (Cia et al. 2007).

UV radiation has also been used on fruits to reduce mold growth, as well as to decrease aflatoxin concentration in 25% (Isman and Biyik 2009). To improve these results, UV radiation has been combined with other treatments such as infrared radiation (IR) or heating for surface decontamination of fruits (Hamanaka et al. 2011).

Given that all UV-C doses caused scald on the fruit, the use of gamma irradiation has been suggested as fungicidal treatment. However, many consumers reject irradiated foods, and other strategies should be used.

In vitro exposure of *Fusarium culmorum* to a *static magnetic field* affected calcium mobilization, which implies a reduction of the incidence and virulence of this mold (Albertini et al. 2003). More studies are necessary to translate in vitro observations to in vivo, because of the large number of cellular components, processes, and systems susceptible to this treatment.

Ultrasounds have been applied to reduce fungal growth on fruits because the cytolytic action is effective against microorganisms (Cao et al. 2010a). No significant texture changes or quality nutritional decay was observed on treated foods. Ultrasound treatment individually or combined with heating and pressure can be a choice technique to prevent microbiological spoilage in foods and extend shelf life during storage. On the other hand, these methods are not adequate for fermented foods where the microbial contribution is essential for the desired characteristics of the final product.

High pressure inactivates fungi even at a higher rate than bacteria (O'Reilly et al. 2000). Prestorage hypobaric treatments delay fungal decay in fruits, but no differences in spore viability and fungal growth were observed with a hypobaric treatment in vitro (Hashmi et al. 2013). This can be explained by the treatment effect activating the defense system of fruits, as it has also been described for UV-C (Pombo et al. 2011). On the other hand, due to the presence of sublethally injured cells after high pressure treatment (O'Reilly et al. 2000), this procedure cannot be adequate for long storing or processing times (Box 17.2).

Physical methods have instant effectiveness with no residual protection against future fungal growth. Thus, other means should be used when fungal recontamination is possible. In addition, all these methods can be used on foods where no mold is wanted. However, mold-ripened cheeses and meats are not suitable for these methods, because molds play a key role in developing the main characteristics of the final product. In most ripened cheeses, the raw material can be heated to eliminate the wild mold population, and fungal starter cultures may minimize the risk due to unwanted molds. On the contrary, heating is not adequate for mold-ripened raw meat products, such as dry-cured ham and sausages.

17.4 Modified Atmosphere and Active Packaging

MAP is an appropriate method to prevent fungal growth and mycotoxin production on foods. The gases used include carbon dioxide and monoxide, nitrogen, and sulfur dioxide, but the most effective one is carbon dioxide. Fungal growth and mycotoxin production in cheese were not totally reduced at 40% CO_2 (Taniwaki et al. 2001), whereas 80% CO_2 and 20% O_2 reduce fungal growth but not mycotoxin production (Taniwaki et al. 2010). For this, MAP combined with other treatments such as reduced water activity, low temperature, or antifungal products may be necessary to have a fungicidal effect in minimally processed foods. In bakery products, the combined use of the MAP with potassium sorbate made it possible to reduce the dose of the antifungal compound (Guynot et al. 2004).

BOX 17.2   ADVANTAGES AND LIMITATIONS OF PHYSICAL METHODS

Advantages common to all physical methods: instant effectiveness.

Limitations common to all physical methods: no residual protection against fungal growth and not adequate for mold-ripened cheeses and meats, where the natural mold population plays a key role in the final product.

Heat

Advantages: effective against fungal growth in fruits and vegetables.
Limitations: responsible for deleterious effects on foods.

Heat combined with ethanol

Advantages: effective against fungal growth.
Limitations: inadequate for many foods.

UV-C irradiation

Advantages: prevents mold growth in some fruits and vegetables.
Limitations: causes scald on fruits.

Static magnetic fields

Advantages: reduces the incidence of molds in vitro.
Limitations: more in vivo studies are needed.

Ultrasounds

Advantages: reduces fungal growth with little texture or quality decay in fruits.
Limitations: inadequate for microbial fermented foods.

High pressure

Advantages: inactivates fungi.
Limitations: inadequate for long storing time.

Growth and mycotoxin production by *Fusarium* species can be controlled with CO_2-enriched MAP combined with low water activity, but the highest effect was obtained in the absence of O_2 (Samapundo et al. 2007a,b). However, these environmental conditions are not adequate for ripened foods, such as cheeses or dry-cured meats, because of the negative effects on sensorial properties (Sánchez-Molinero and Arnau 2010). In addition, MAP may allow or even stimulate the growth of pathogen microorganisms (Farber 1991).

Fungicidal agents can be incorporated into the packaging structure itself, as in *active packaging*, to directly interact with molds or with the environment within the container. Active packaging can interact with the product or the headspace between the package and the food system to obtain a desired outcome. The aim of the active packaging is to increase the shelf life of foods maintaining their quality, safety, and sensory properties, without direct addition of the active agents to the product (Appendini and Hotchkiss 2002). Likewise, antimicrobial food packaging acts to reduce, inhibit, or retard the growth of microorganisms present in food. Inclusion of the active antimicrobial agents within the packaging material gives rise to active packaging. The gradual release of an antimicrobial from a packaging film to the food surface may have an advantage over dipping and spraying. Among these agents are oxygen absorbers, essential oils, and oleoresins from spices and herbs. Antimicrobial activity of essential oils or plant extracts may be rapidly lost due to interaction of these antimicrobials with food components as fat (Box 17.3).

Despite that satisfactory mold inhibition was reached combining MAP with some of these compounds, in some foods, off-flavor formation was observed, and the use of other preservation factors added to MAP has been suggested (Nielsen and Rios 2000).

> **BOX 17.3   ADVANTAGES AND LIMITATIONS OF MODIFIED**
> **ATMOSPHERE AND ACTIVE PACKAGING**
>
> *CO_2-enriched atmospheres*
>
> Advantages: delays growth of most molds.
> Limitations: additional hurdles may be necessary to prevent growth of the less sensitive pathogen
> microorganisms.
>
> *Active packaging*
>
> Advantages: facilitates the gradual release of an antimicrobial agent.
> Limitations: low efficiency in bulk packaging and lack of the same problems that the anti-
> microbial used.

17.5 Natural Products for Fungal Growth Control

Efforts in this field are aimed at finding effective and healthy products for the consumer and for the environment. Natural products are one of the fields that are being investigated, particularly those of plant origin (Tripathi and Dubey 2004; da Cruz Cabral et al. 2013). Many of the plants studied have been used traditionally in foods as flavorings (spices) and ingredients, which give them great security because they are considered *generally regarded as safe* (GRAS) products. Many compounds from plants have been tested for their effect on molds responsible for postharvest diseases in crops and plants.

17.5.1 Volatile Compounds from Plant Extracts

As they are aimed to be applied in warehouses and in large spaces, one of the most wanted features is that volatile compounds do not leave residues in foods. Some plant extracts contain antifungal volatile compounds, such as thymol, eugenol, or carvacrol, that can be concentrated in essential oils. The inhibitory effect seems to be due to the presence of OH groups that bind to intracellular enzymes, modifying some functions, membrane integrity, and permeability (da Cruz Cabral et al. 2013). Essential oils have high efficiency against fungi and other microorganisms in vitro, but this effect is achieved only in foods with higher concentrations of essential oils. This can be due to the fact that volatile compounds are bound to food components. The higher concentration of essential oils may cause negative organoleptic effects, by altering the natural taste of the food due to undesirable flavor and sensory changes (Hsieh et al. 2001). Most in vivo screenings of antifungal effect of essential oils and plant extracts are performed on cereals, vegetables, and fruits (da Cruz Cabral et al. 2013). In foods of animal origin, the activity of essential oils can be reduced by the high levels of fat, because of the solubility of essential oils in lipids (Lis-Balchin et al. 2003). Most studies in this type of foods are performed mainly with bacterial species (Holley and Patel 2005; Tajkarimi et al. 2010). Studies about the use of essential oils in foods of animal origin showed good results in the inhibition of *Aspergillus flavus* growth and aflatoxin production (Gandomi et al. 2009). However, adverse organoleptic effects were obtained because of the high doses of antifungal essential oil employed.

The effect of these compounds on mycotoxin production may be contradictory, since partial inhibition of fungal growth increases mycotoxin production due to the secondary metabolism activation as a response to stress. Another problem found is that the concentration of aromatic compounds used must be low to reduce any adverse sensorial impact on food. However, higher doses have to be applied in cheeses and meat products because the high levels of fat and carbohydrates greatly reduced the activity of the essential oils (Gutierrez et al. 2008). On the other hand, volatile compounds applied in fruits are more effective (Tripathi and Dubey 2004).

To avoid some of these problems, solutions such as edible films and encapsulation have been sought. Application of these compounds in the coating extends their antifungal activity, and the compounds are gradually released into the food to maintain a constant concentration during storage and preservation. Unfortunately, the impact of these aromatic compounds on sensory traits and the fact that coating is impracticable for some raw and ripened foods make it necessary to study alternative methods.

17.5.2 Resistance Inducers

Resistance inducers are compounds that increase plant defenses, sometimes exploiting their antimicrobial properties. Chitosan is one of the most important resistance inducers, and it has been checked as a fungicidal compound in crops and in several foods. It is active at very low concentration and can be used in solution, powder, or wettable coatings. Chitosan acts as a potent elicitor enhancing plant protection against pathogens (Amborabe et al. 2008). Commercial chitosan was more effective in reducing fungal disease than other resistance inducers such as benzothiadiazole, oligosaccharides, soybean lecithin, calcium, and organic acids (Romanazzi et al. 2013). The combined treatment of chitosan with ethanol improved the control of fungal growth on table grapes (Romanazzi et al. 2007). These treatments may serve for fresh vegetables, but no eliciting properties are to be expected in other foods.

17.5.3 Antimicrobial Peptides

Over the past few years, new antimicrobial peptides (AMPs) with promising characteristics to control food-unwanted molds are being discovered. AMPs are polypeptides of less than 100 amino acids that have antimicrobial activity at physiological concentrations under conditions prevailing in the tissues of origin (Ganz 2003). Short cationic amphipathic peptides with antimicrobial activities are naturally occurring in virtually every live form as a component of the innate immune defenses (Hancock and Sahl 2006).

Bacteriocins are a diverse group of AMPs produced by bacteria. The best-known prokaryotic AMP is nisin, produced by *Lactococcus lactis*. Nisin and other bacteriocins are active against many strains of Gram-positive bacteria and have been used as food preservatives for many years. However, bacteriocins have relatively narrow killing spectra and show no significant inhibition on molds.

Hundreds of different eukaryotic AMPs are produced by the different eukaryotic kingdoms. These peptides serve multicellular eukaryotes as defense mechanisms against invading pathogens, whereas they provide lower eukaryotes with the advantage to successfully compete with microorganisms that possess similar nutritional and ecological requirements (Hegedüs and Marx 2013). Most multicellular organisms secrete a cocktail of peptides from several structural classes with a high diversity of sequences, where single mutations can dramatically alter the biological activity of each peptide (Zasloff 2002).

AMPs are products of single genes and can swiftly be either constitutively expressed or rapidly transcribed upon induction with a minimal input of energy and biomass (Broekaert et al. 1995). They are derived from larger precursors that follow posttranslational modifications including proteolytic processing and, in some cases, glycosylation, amidation, isomerization, and halogenation (Zasloff 2002). However, those of most interest to fight molds in foods are highly amphipathic peptides of 1–6 kDa that usually possess a net positive charge at physiological pH.

The fundamental structural principle underlying AMPs is the ability of the molecule to adopt a shape with clusters of hydrophobic and cationic amino acids spatially organized in discrete sectors, in a sort of amphipathic design (Zasloff 2002). Antifungal proteins typically contain a high ratio of the cationic amino acids arginine and lysine.

Most AMPs from higher eukaryotes are broad-spectrum antimicrobials because they target a design feature of the microbial membrane that distinguishes microbes from multicellular eukaryotic organisms, whereas those from microorganisms are nontoxic to animals. Acquisition of resistance by a sensitive microbial strain against AMPs is improbable because the target of AMPs is the microbial membrane and the microbe would have to change the composition or the organization of its lipids (Zasloff 2002). For this, many AMPs are promising candidates to control unwanted fungi as relatively safe, because they target enzymes or structures essential for fungi but absent from mammals. However, destruction of AMPs by enzymatic proteolysis may occur, particularly when highly proteolytic microorganisms are present, as it is common in ripened cheeses and meat products (Box 17.4).

17.5.3.1 AMPs from Plants

Plants synthesize many AMPs, including defensins, thionins, chitinases, chitin-binding proteins, ribosome-inactivating proteins, thaumatin-like proteins, lectins, protease inhibitors, and lipid-transfer proteins (Selitrennikoff 2001; Wong et al. 2012).

Defensins are a large family of low-molecular-mass eukaryotic AMPs, isolated from plants, invertebrates, and mammals. Plant defensins are a large class of structurally similar peptides with 45–54 amino acids found throughout the plant kingdom that are currently classified in 18 groups (van der Weerden and Anderson 2013). The biological activities reported include antifungal activity, antibacterial activity, proteinase inhibitory activity, and insect amylase inhibitory activity (Stotz et al. 2009). Many are growth inhibitory toward fungi, including groups 3 from spruce and *Ginkgo biloba* trees, 5 from sugar beet, 7 from solanaceous plants, 8 from *Brassica* species, 9 from the Brassicaceae, Asteraceae, and Saxifragaceae families, 10 and 13 from the Fabaceae family, 14 from corn, and 16 from spinach (van der Weerden and Anderson 2013).

The two main defensin families, α- and β-defensins, are characterized by a triple-stranded β-sheet with a distinctive fold and a framework of six disulfide-linked cysteines (Ganz 2003). These disulfide bridges stabilize the antiparallel β-sheet conformation flanked by an α-helical segment (Hegedüs and Marx 2013). The α- and β-defensins differ in the length of the peptide segments between the six cysteines and the pairing of the cysteines connected by disulfide bonds (Ganz 2003). The amino acid sequences are highly variable except for the conservation of the cysteine framework in each defensin subfamily. Residues conserved in all sequences of plant defensins are restricted to the eight cysteines and four other amino acids (Broekaert et al. 1995). Some defensins are oligomeric, which is related to their membrane permeabilization mechanism of action (Hoover et al. 2001).

Thionins are another group of AMPs that protect plants against microbial infection. They occur in cereals and mistletoes and are toxic to bacteria, fungi, or yeasts, but also to various mammalian cell types (Thevissen et al. 1996). Thionin structure consists of a pair of antiparallel α-helices and a short antiparallel β-sheet.

17.5.3.2 AMPs from Invertebrates

Insect defensins, including sapesin from flesh fly *Sarcophaga peregrina*, drosomycin from the fruit fly *Drosophila melanogaster*, heliomicin from the tobacco budworm *Heliothis virescens*, termicin from the termite *Pseudocanthotermes spiniger*, and gallerimycin from the wax moth *Galleria mellonella*, are active against molds (*Aspergillus fumigatus*, *Aspergillus niger*, and *Metarhizium anisopliae*), but not against yeasts and bacteria (Hegedüs and Marx 2013). Psacotheasin from the yellow-spotted long-horned beetle *Psacothea hilaris* is active against both human bacterial and fungal pathogens, including *Candida albicans*, *Candida parapsilosis*, *Trichosporon beigelii*, and *Malassezia furfur* (Hwang et al. 2010).

Cecropins, after the cecropia moth *Hyalophora cecropia*, are cationic amphipathic molecules composed of 20–40 amino acid residues produced by many invertebrates in response to injury or infection. They contain no cysteine residues and form linear α-helical structures when bound to membranes. Cecropins display activity not only against bacteria, but also against *A. fumigatus* (Jenssen et al. 2006). Melittin, a component of the venom of the European honey bee (Habermann 1972), is similar to cecropins, being active against *C. albicans* (Jenssen et al. 2006). Cecropins display an amphipathic α-helical structure that can integrate in microbial membranes to form ion channels (Broekaert et al. 1995).

17.5.3.3 AMPs from Vertebrates

Vertebrates produce from the small (20–40 residues) basic linear AMPs like frog's magainins or single-looped by a disulfide bond also in frogs, to the antiparallel β-sheet stabilized by disulfide bonds bactenecins and defensins (Hancock and Sahl 2006). In humans and other mammals, defensins and cathelicidins are the main AMPs, whereas histatins, dermicidin, and anionic peptides are restricted to a few animal species (Ganz 2003).

Defensins have been found in mammals, birds, and snakes. Mammalian phagocytes and epithelial cells of the intestines and airways produce α- and β-defensins (25–45 residues), according to cysteine arrangement within the molecule. Cyclic θ-defensins of nine amino acids have also been identified in Rhesus monkeys as derived from α-defensins. In contrast to insect and plant defensins that mainly have antifungal activity, vertebrate α- and θ-defensins exhibit a wide antimicrobial spectrum (Wong et al. 2012). Vertebrate β-defensins show a narrower antimicrobial spectrum, except for some human β-defensins, such as HBD-3, that exhibit broad-spectrum antimicrobial activity against bacteria, fungi, and viruses (Weinberg et al. 2006; Parisien et al. 2007).

Cathelicidins display a remarkable variety of sizes, sequences, and structures characterized by an N-terminal cathelin domain. Human cathelicidin LL-37 exhibits a broad spectrum of antimicrobial activity against bacterial, fungal, and viral pathogens (Zanetti 2004).

Histatins are characterized by a high content of histidine and show antifungal activity against *C. albicans* and *Cryptococcus neoformans* (Tsai and Bobek 1998).

17.5.3.4 Fungal AMPs

AMPs from lower eukaryotes such as yeasts and molds will be discussed in Section 17.6.

Cordymin is a 10.9 kDa antifungal peptide from the medicinal mushroom *Cordyceps militaris*, which inhibited mycelial growth in *Bipolaris maydis*, *Mycosphaerella arachidicola*, *Rhizoctonia solani*, and *C. albicans*, but not in *A. fumigatus*, *Fusarium oxysporum*, and *Valsa mali* (Wong et al. 2010).

17.5.3.5 Phage-Encoded AMPs

The antifungal 10.4 kDa protein toxin zygocin is produced and secreted by a ZbV-M-infected killer strain of *Zygosaccharomyces bailii* (Weiler et al. 2002). Killer phenotype expression in this host is associated with the presence of an MZb-dsRNA-harboring killer virus that persists within the cytosol of the infected cell. Other killer toxins from yeasts may have a similar origin, as discussed later.

17.6 Biological Control

The limited efficacy of most physical and chemical procedures to control toxigenic molds in dry-cured foods and the increasing public demand for food free of additives have been requesting to use a more natural alternative approach. Biological control using antagonistic microorganisms is an emergent alternative to efficiently control fungi and mycotoxin production. There are a variety of microorganisms useful as biocontrol agents against toxigenic fungi that include bacteria, yeasts, and molds.

17.6.1 Lactic Acid Bacteria

Lactic acid bacteria (LAB) are commonly used for the production of fermented foods, and they are commonly found in dairy products, meat, meat-derived products, and cereal products. Several LAB, mainly from genera *Lactococcus* and *Lactobacillus* and, to a lesser extent, *Pediococcus* and *Leuconostoc*, are able to prevent toxigenic mold growth (Dalié et al. 2010), but their role is also related to inhibition of mycotoxin biosynthesis and binding of mycotoxins. *Enterococcus faecium*, *Streptococcus thermophilus*, and *Lactococcus casei* from cheese and yogurt showing strong antimicrobial effects on *Penicillium expansum*, *Botrytis cinerea*, and *Monilinia fructicola* have been proposed as biopreservatives in foods (Yang et al. 2012). *Pediococcus acidilactici* isolated from vacuum-packed fermented meat shows antifungal activity against *A. fumigatus*, *Aspergillus parasiticus*, *F. oxysporum*, and *Penicillium* spp. (Mandal et al. 2007). LAB may contribute to mycotoxin control in various fermented foods. The antifungal activity of LAB has been linked mainly to low-molecular-weight compounds, including organic acids, hydroxyl fatty acids, reuterin, hydrogen peroxide, phenolic compounds, lactones, cyclic dipeptides, and proteinaceous compounds (Dalié et al. 2010; Crowley et al. 2013).

LAB produce weak organic acids with antifungal properties such as lactic or acetic acid (Baek et al. 2012) or phenyllactic acid (Ryan et al. 2011).

Reuterin is a low-molecular-weight compound produced by *Lactobacillus reuteri*, which is able to inhibit the growth of several fungi (Axelsson et al. 1989), probably by causing oxidative stress to the cells (Crowley et al. 2013).

Cyclic dipeptides have been described from several LAB, and their antifungal mechanism of action remains unknown. Due to its high minimum inhibitory concentration for fungi, these cyclic dipeptides play a minimal role to inhibit mold growth (Ryan et al. 2009).

The antifungal abilities of hydroxyl fatty acids are attributed to a loss in the membrane integrity causing an uncontrolled release of electrolytes and proteins that leads to the disintegration of fungal cells (Avis and Belanger 2001). This activity is related to the structure of fatty acids, with a requirement of at least one hydroxyl and one double bond in its carbon backbone (Crowley et al. 2013).

The production of antifungal proteinaceous compounds has been reported in species of genera *Lactococcus*, *Streptococcus*, *Lactobacillus*, and *Pediococcus* (Crowley et al. 2013). A variety of structures have been described for these compounds: defensin-like proteins, peptides with homology with lacticin or peptides derived from proteolytic activity of LAB against caseins, or gliadin on sourdough fermented (Crowley et al. 2013). The mode of action of these compounds should be elucidated.

Some other compounds such as nucleosides (Ryan et al. 2011) or lactones (Yang et al. 2011) exhibit antagonistic activity against filamentous molds, but their effects on fungal cells remain unclear.

According to the consumers' demands toward natural food preservatives, LAB are good candidates for the biocontrol of toxigenic molds due to their GRAS and *qualified presumption of safety* statuses and their healthy probiotic properties. LAB could be applied in a variety of fermented foods such as fruits and vegetables, dairy and bakery products, or animal feed. However, the limited resistance of LAB to intermediate water activity values makes them unsuitable for long-ripening processing, such as dry-cured meat product, particularly when no sugar is added (Box 17.5).

17.6.2 Yeasts

Yeasts are considered one of the most potent biocontrol agents due to their biology and nontoxic properties (Pimenta et al. 2009). They have simple nutritional requirements, grow in fermenters on inexpensive media, are able to survive in a wide range of environmental conditions, and do not produce anthropotoxic compounds (Wilson and Wisniewski 1989). Several yeasts found naturally on foods have shown antifungal activities (Table 17.1) and/or the ability to decrease mycotoxin accumulation (Table 17.2). Therefore, these yeasts are considered promising ecological fungicides.

BOX 17.5   ADVANTAGES AND LIMITATIONS OF BIOLOGICAL CONTROL

Lactic acid bacteria

Advantages: active against some of the most concerning molds in dairy products and are good candidates for the biocontrol of toxigenic molds.

Limitations: the limited resistance to intermediate water activity values makes them unsuitable for dry-ripened foods, particularly when no sugar is added.

Yeasts

Advantages: provide various complementary antifungal mechanisms both to prevent mold growth and to reduce mycotoxin content.

Limitations: their use is limited to fermented and dry-ripened foods where yeast colonization is accepted.

Molds

Advantages: fight particular mycotoxigenic species and contribute to food ripening as well.

Limitations: narrow inhibition spectrum and potential use restricted to mold-ripened foods; a careful selection of safe strains is imposed to prevent the hazard due to toxic metabolites.

17.6.2.1 Mechanisms of Antifungal Activity

Mechanisms of action for mold growth inhibition by yeasts include competition for nutrient and space (Druvefors et al. 2003; Coelho et al. 2007), antibiosis by production of toxic volatile compounds (Masoud et al. 2005), killer toxins (Walker et al. 1995), or cell wall hydrolytic enzymes (Masih and Paul 2002) (Table 17.3).

Competition for essential factors, nutrients, and space could have a big impact over the metabolism of molds, and it has been proposed as the most common way involved in the biocontrol of filamentous fungi by yeast (Droby et al. 1989; Wilson and Wisniewski 1989). When antagonist yeasts are able to rapidly colonize and grow in a food surface, they can compete with the toxigenic mold for nutrients and space. Antagonistic yeasts should be able to rapidly consume the limiting nutrients for toxigenic molds, thus retarding fungal growth (Hernández-Montiel et al. 2010). *Candida guilliermondii* competes for nitrogen sources with *P. expansum* on apple (Ortu et al. 2003), and *P. anomala* for sugars with *P. roqueforti* (Druvefors et al. 2003). For this mechanism, there is a direct relationship between the population of the antagonist and its efficacy (Janisiewicz and Korsten 2002). Some yeasts, as fast-growing organisms, may colonize the outer layers of foods and suppress the adherence of filamentous fungi to surface, as it has been proposed for *P. anomala* in barley (Laitila et al. 2007). Importantly, a high yeast population (10^8 cfu/mL) may be necessary to restrict the availability of nutrients and sites for colonization, essential for the germination of mold spores (Björnberg and Schnürer 1993). As discussed later in this chapter, nutrient availability not only influences mold growth, but also directly affects secondary metabolism, including mycotoxin production (Luchese and Harrigan 1993). In some cases, the same mechanism of biocontrol can be effective against both mold growth and mycotoxin production. Nonetheless, biocontrol microorganisms may inhibit the growth of toxigenic fungi, but the metabolic activity of the existing hyphae may be not necessarily reduced. Given that secondary metabolites such as mycotoxins can be synthesized in response to stress, growth reduction triggered by other microorganisms can stimulate mycotoxin production, as shown for chemical fungicides (Gareis and Ceynowa 1994).

Volatile compounds from yeasts show antifungal properties (Masoud et al. 2005). The antifungal effect of some yeasts, such as *P. anomala*, has been related to synergistic influence of ethyl acetate and ethanol produced in an oxygen-limited environment (Druvefors et al. 2002). Yeasts isolated from coffee beans, *P. anomala*, *Pichia kluyveri*, and *Hanseniaspora uvarum* generate volatile alcohols and esters to inhibit the growth of *Aspergillus ochraceus* and ochratoxin production (Masoud et al. 2005). *P. anomala* generates phenethyl alcohol with high antifungal activity (Masoud et al. 2005). *Kluyveromyces lactis* produces

TABLE 17.1

Yeasts Isolated from Foods with Antifungal Activity in Different Substrates

Yeast	Target Mold	Substrate	Reference
Acremonium cephalosporium	*Aspergillus, Rhizopus, Botrytis*	Grapes	Zahavi et al. (2000)
Candida sake	*Penicillium expansum* and *B. cinerea*	Apples and pears	Nunes et al. (2002b)
Candida guilliermondii	*Aspergillus, Rhizopus, Botrytis*	Grapes	Zahavi et al. (2000)
Candida incommunis	*Aspergillus carbonarius, Aspergillus niger*	Grapes	Bleve et al. (2006)
Cryptococcus albidus	*P. expansum*	Apples and pears	Tian et al. (2002)
Cryptococcus laurentii	*Alternaria alternata, P. expansum, Botrytis cinerea, Rhizopus stolonifer*	Sweet cherry fruit	Qin et al. (2004)
Debaryomyces hansenii	*Penicillium digitatum*	Grape	Droby et al. (1989)
	Aspergillus sp., *Byssochlamys fulva, Byssochlamys nivea, Cladosporium* sp., *Eurotium chevalieri, Penicillium candidum, Penicillium roqueforti*	Yoghurt and cheese	Liu and Tsao (2009)
	P. roqueforti	Cheese agar	van den Tempel and Jakobsen (2000)
	Penicillium verrucosum	Meat products	Andrade et al. (2014)
Issatchenkia terricola	*A. carbonarius, A. niger*	Grapes	Bleve et al. (2006)
Issatchenkia orientalis	*A. carbonarius, A. niger*	Grapes	Bleve et al. (2006)
Lachancea thermotolerans	Ochratoxigenic fungi	Grapes	Ponsone et al. (2011)
M. pulcherrima	*P. expansum, Colletotrichum acutatum*	Apples	Conway et al. (2004)
Metschnikowia fructicola	*P. expansum*	Apple	Spadaro et al. (2013)
Metschnikowia pulcherrima	*A. carbonarius, A. niger*	Grapes	Bleve et al. (2006)
Pichia anomala	*Penicillium* spp., *Paecilomyces variotii, Aspergillus* spp., *Cladosporium cladosporioides, Fusarium* spp., *Botrytis cinerea, Mucor hiemalis, R. stolonifer, Eurotium amstelodami, Monascus ruber, Talaromyces flavus*	Wheat	Petersson and Schnürer (1995)
Pichia membranifaciens	*Rhizopus stolonifer*	Nectarine fruit	Fan and Tian (2000)
	A. alternata, P. expansum, B. cinerea, R. stolonifer	Sweet cherry fruit	Qin et al. (2004)
Rhodotorula glutinis	*A. alternata, P. expansum, B. cinerea, R. stolonifer*	Sweet cherry fruit	Qin et al. (2004)
Saccharomyces cerevisiae	*Aspergillus parasiticus*	Culture media	Armando et al. (2012a)
	A. carbonarius, Fusarium graminearum	Silage agar	Armando et al. (2013)
Torulaspora pullulans	*A. alternata, P. expansum, B. cinerea, R. stolonifer*	Sweet cherry fruit	Qin et al. (2004)

2-phenylethyl isobutyrate, ethyl acetate, isoamyl alcohol, and isobutyric and isovaleric acids that are considered responsible for the inhibition of *P. expansum* (Taczman-Brückner et al. 2005). *C. maltosa* isolated from cheese produces isoamyl alcohol and isoamyl acetate that inhibit spore germination of species from *Aspergillus, Penicillium, Rhizopus, Mucor*, and *Fusarium* genera (Ando et al. 2012).

The antimicrobial activity of alcohols is due to their absorption and accumulation in cell membrane affecting the organization and stability of the lipid bilayer, then increasing the permeability and accelerating the diffusion of essential ions and metabolites through the membrane (Ingram and Buttke 1984). Lipophilic alcohols such as 3-methyl-1-butanol and 2-methyl-1-butanol have high affinity for the membrane, producing higher toxicity than other less lipophilic alcohols as ethanol (Heipieper et al. 1994). The mode of action of other microbial volatiles on fungi is scarcely known. It has been proposed that volatiles may reduce the growth by changing protein expression (Humphris et al. 2002) and the activity of specific

TABLE 17.2

Yeasts Isolated from Foods Able to Reduce Mycotoxin Accumulation

Yeast	Target Mold	Mycotoxin	Substrate	Reference
Debaryomyces hansenii	*Penicillium verrucosum*	OTA	Dry-cured ham	Andrade et al. (2014)
Hanseniaspora uvarum	*Aspergillus ochraceus*	OTA	Culture media	Masoud et al. (2005)
Lachancea thermotolerans	Ochratoxigenic fungi	OTA	Grapes	Ponsone et al. (2011)
Metschnikowia fructicola	*Penicillium expansum*	Patulin	Apple	Spadaro et al. (2013)
Pichia kluyveri	*A. ochraceus*	OTA	Culture media	Masoud et al. (2005)
Pichia anomala	*P. verrucosum*	OTA	Culture media	Petersson et al. (1998)
	A. ochraceus	OTA	Culture media	Masoud et al. (2005)
Saccharomyces cerevisiae	*Aspergillus parasiticus*	Aflatoxin B1	Culture media	Armando et al. (2012a)
	Aspergillus carbonarius, Fusarium graminearum	TA, ZEA, DON	Silage agar	Armando et al. (2013)
	P. verrucosum	OTA	Culture media	Petersson et al. (1998)

enzymes (Wheatley 2002). Moreover, the metabolism of yeasts may significantly change the environmental gas composition increasing CO_2 and decreasing O_2, which may contribute to the inhibitory effect against molds (Taczman-Brückner et al. 2005). This effect could be of interest mainly in packed foods.

Killer proteins. Certain yeasts produce antimycotical low-molecular-mass proteins or glycoproteins, known as killer proteins that affect sensitive yeasts, bacteria, and filamentous fungi without cell–cell contact (Radler et al. 1993; Weiler and Schmitt 2003). Killer protein production is a widespread characteristic among yeast species of different genera including *Saccharomyces, Kluyveromyces, Zygosaccharomyces, Hanseniaspora, Pichia, Debaryomyces,* and *Schwanniomyces.* These proteins are associated with the presence of cytoplasmic dsRNA viruses, but they can also be encoded by linear dsDNA plasmids or chromosomally encoded (Schmitt and Breinig 2006), and they confer an ecological advantage on yeasts over their competitors (Magliani et al. 2008). Zygocin from *Z. bailii* is effective against *F. oxysporum* (Weiler and Schmitt 2003). *Candida guilliermondii, D. hansenii,* and *P. membranifaciens* showed antibiosis effect against *P. expansum* (Levy et al. 2002; Coelho et al. 2009). Cell-free filtrate from *D. hansenii* cultures was effective in inhibiting *Aspergillus* spp. (Liu and Tsao 2009). The inhibitory effect of killer yeasts against toxigenic molds has suggested the use of killer toxins as biocontrol agents.

The mechanism of action of killer toxins has been associated with the cell surface by interacting with components of the cell wall. The interactions might be dependent on protein signal recognition between yeast cells and hyphae, since protein-denaturing agents (SDS and mercaptoethanol) block the attachment, and other compounds (including mannose residues) may influence adhesion (Wisniewski et al. 1991; Chan and Tian 2005). Primary receptors of killer toxins are β-glucans, mannoproteins, or chitin. β-Glucans are primary receptors for K1 toxin of *Saccharomyces cerevisiae* (Al-Aidroos and Bussey 1978), HK toxin of *Hansenula mrakii* (Kasahara et al. 1994), wicaltin of *Williopsis californica* (Theisen et al. 2000), and for killer toxins produced by *H. uvarum* (Radler et al. 1990), *P. membranifaciens* (Santos et al. 2000), and *D. hansenii* (Santos et al. 2002). Mannoproteins are receptors for zigocin from *Z. bailii* (Radler et al. 1993) and KT28 of *S. cerevisiae* (Schmitt and Radler 1988). Chitin is the cell wall receptor for *Kluyveromyces lactis* killer toxin (Takita and Castilho-Valavicius 1993).

In sensitive cells, the adsorption of the killer toxin to cell wall is followed by disrupting the cytoplasmic membrane or by blocking DNA synthesis (Kasahara et al. 1994; Schmitt et al. 1996; Santos et al. 2000). Once linked to the cell-wall receptor, K1 toxin and zygocin produce a loss of cellular integrity through the disruption of cytoplasmic membrane by lethal ion channel formation (Martinac et al. 1990; Weiler and Schmitt 2003). Several yeasts, such as *Pichia guilliermondii, P. membranifaciens,* and *Candida saitoana* produced, in the presence of sensitive molds, lytic enzymes, such as β-(1–3) glucanase and exochitinase, which enhance the attaching ability of yeast to hyphae and cause the degradation of fungal hyphae (Wisniewski et al. 1991; Chan and Tian 2005). The glycosylated killer protein panomycocin produced by *P. anomala* belongs to the exo β-1,3 glucanase group (Izgu and Altinbay 2004) and hydrolyzes the β-1,3-glucans within the fungal cell wall, leading to the leakage of cytoplasmic components and

TABLE 17.3

Mechanisms of Action Described for Yeast with Antifungal and/or Antimycotoxigenic Activities

			Nitrogen Sources	
Competition	*Candida guilliermondii*	*P. expansum*	Nitrogen Sources	Ortu et al. (2003)
	Pichia anomala	*P. roqueforti*	Sugars	Druvefors et al. (2003)
	Saccharomyces cerevisiae	*A. parasiticus*	Carbohydrates	Weckbach and Marth (1977)
	P. anomala	Filamentous fungi	Space	Laitila et al. (2007)
Volatile compounds	*S. cerevisiae*	*Guignardia citricarpa*	Ethyl acetate, 3-methyl-1-butanol, 2-methyl-1-butanol, phenyl-ethyl alcohol, ethyl octanoate, and ethanol	Fialho et al. (2010)
	P. anomala	*Aspergillus ochraceus*	Ethyl acetate, isobutyl acetate, 2-phenyl ethyl acetate, ethyl propionate and isoamyl alcohol, phenyl ethyl alcohol	Masoud et al. (2005)
	Pichia kluyveri Hanseniaspora uvarum	*Aspergillus ochraceus*	Ethyl acetate, isobutyl acetate, 2-phenyl ethyl acetate, ethyl propionate, and isoamyl alcohol	Masoud et al. (2005)
	Kluyveromyces lactis	*Penicillium expansum*	2-Phenylethyl isobutyrate, ethyl acetate, isoamyl alcohol, isobutyric and isovaleric acids	Taczman-Brückner et al. (2005)
	C. maltosa	*Aspergillus, Penicillium, Rhizopus, Mucor,* and *Fusarium*	Isoamyl alcohol and isoamyl acetate	Ando et al. (2012)
Lytic enzymes	*Pichia guilliermondii*	*Botrytis cinerea*	β-(1-3) Glucanase	Wisniewski et al. (1991)
	P. membranifaciens	*Monilinia fructicola, Penicillium expansum, Rhizopus stolonifer*	β-(1-3) Glucanase, exo-chitinase	Chan and Tian (2005)
	Candida saitoana		β-(1-3) Glucanase, chitinase	El-Ghaouth et al. (1998)
Killer proteins	*Zygosaccharomyces bailii*	*Fusarium oxysporum*	Zygocin	Weiler and Schmitt (2003)
	C. guilliermondii	*P. expansum*	Killer toxin <3000 Da	Coelho et al. (2009)
	P. ohmeri	*P. expansum*	Killer toxin <3000 Da	Coelho et al. (2009)
	D. hansenii	*Aspergillus* sp.	Cell-free medium	Liu and Tsao (2009)
	Pichia anomala	*P. digitatum* and *P. italicum*	Panomycocin	Izgu et al. (2011)

(Continued)

TABLE 17.3 (Continued)

Mechanisms of Action Described for Yeast with Antifungal and/or Antimycotoxigenic Activities

Reduction of mycotoxins	S. cerevisiae	Patulin	Degradation	Moss and Long (2002); Coelho et al. (2007)
	Pichia ohmeri	Patulin	Degradation	Coelho et al. (2007)
	Trichosporon mycotoxinivorans	OTA, zearalenone	Degradation	Molnar et al. (2004)
	Saccharomyces cerevisiae	OTA	Degradation	Angioni et al. (2007)
	Kloeckera apiculata	OTA	Degradation	Angioni et al. (2007)
	Phaffia rhodozyma	OTA	Degradation	Péteri et al. (2007)
	Metschnikowia pulcherrima and Pichia guilliermondii	OTA	Degradation	Patharajan et al. (2011)
	S. cerevisiae	OTA	Adsorption	Bejaoui et al. (2004); Caridi et al. (2006); Caccioni et al. (1998); Garcia Moruno et al. (2005); Meca et al. (2010)
	Candida, Kloeckera, Rhodotorula	OTA	Adsorption	Var et al. (2009)
	S. cerevisiae	Zearalenone	Adsorption	Yiannikouris et al. (2004)
	D. hansenii	OTA	Adsorption	Gil-Serna et al. (2011)
	Candida guilliermondii, Cryptococcus laurentii, Candida krusei, Rhodotorula mucilaginosa, Pichia anomala, and Candida oleophila	Aflatoxins (A. flavus)	Block biosynthetic pathway	Hua et al. (1999)
	D. hansenii	OTA (A. westerdijkiae)		Gil-Serna et al. (2011)

cell death (Izgu et al. 2005). Similarly, killer toxins from *C. guilliermondii* and *Pichia ohmeri* inhibit conidial germination and hyphal growth of *P. expansum* probably due to a loss of cellular integrity (Coelho et al. 2009). On the other hand, HK toxin, wicaltin, and *D. hansenii* killer proteins inhibit cell wall synthesis (Yamamoto et al. 1986; Theisen et al. 2000; Santos et al. 2002). Panomycocin produces abnormal morphological changes in fungal hyphae and spores of *Penicillium digitatum* and *Penicillium italicum* isolated by citrus fruit (Izgu et al. 2011).

17.6.2.2 Mechanisms for Reduction of Mycotoxin Content

In addition to the indirect effect on mycotoxin reduction by inhibition of the producing mold, yeasts may exert a direct action to decrease mycotoxin content by adsorption, degradation, or blocking the biosynthetic pathway.

Adsorption is regulated by the chemical composition of the yeast cell wall (Caridi 2007), and it may be produced by a direct linkage between mycotoxins and cell wall molecules such as glucan/mannan. Strains of *S. cerevisiae* with high cell wall glucan content exhibited higher affinities for zearalenone (Yiannikouris et al. 2004). The cell wall thickness is also correlated with the ability of *S. cerevisiae* to bind aflatoxin B1, OTA, and zearalenone (Armando et al. 2011, 2012b). Adsorption of OTA to *S. cerevisiae* cell wall can occur during fermentation of must and wines (Meca et al. 2010). Adsorption of OTA has also been reported in yeast from other genera, such as *Phaffia*, *Candida*, *Kloeckera*, and *Rhodotorula* (Péteri et al. 2007; Var et al. 2009). Environmental factors can interfere with the adsorption process. The ionization state at low pH would improve OTA absorption by *D. hansenii*, while high pH values would reduce the ability of wall molecules to link toxins (Gil-Serna et al. 2011).

Degradation of mycotoxins can occur enzymatically during fermentation by an inducible process associated to mycotoxin presence, as it has been reported for patulin degradation by *S. cerevisiae* (Moss and Long 2002). *Pichia ohmeri* is also able to degrade patulin (Coelho et al. 2007). The yeasts *Trichosporon mycotoxinivorans* and *Phaffia rhodozyma* produce carboxypeptidases, which transform OTA to α-OTA, a less toxic derivative, and phenylalanine through the cleavage of the amide bond (Schatzmayr et al. 2003; Péteri et al. 2007). Strains of *S. cerevisiae*, *Kloeckera apiculata*, *M. pulcherrima*, and *P. guilliermondii* are able to degrade OTA by unknown ways (Angioni et al. 2007; Patharajan et al. 2011). *Trichosporon mycotoxinivorans* is also able to degrade zearalenone to carbon dioxide or other unidentified metabolites (Molnar et al. 2004).

Blocking the biosynthetic pathway of mycotoxins has been reported in aflatoxigenic *A. flavus* by saprophytic yeasts from nuts, including *C. guilliermondii*, *C. laurentii*, *Candida krusei*, *Rhodotorula mucilaginosa*, *P. anomala*, and *Candida oleophila* (Hua et al. 1999). Similarly, *D. hansenii* reduces the concentration of extracellular OTA in mixed cultures with *Aspergillus westerdijkiae* due to a reduction in the expression of OTA biosynthetic genes (Gil-Serna et al. 2009, 2011).

17.6.2.3 Yeasts for Biocontrol of Toxigenic Molds on Dry-Cured Foods

Yeasts are among the predominant microbial groups during the ripening of different intermediate moisture foods, such as fermented sausages (Encinas et al. 2000; Cocolin et al. 2006; Mendonça et al. 2013) or dry-cured ham (Comi and Cantoni 1983; Núñez et al. 1996; Simoncini et al. 2007). During manufacturing, the wild yeast population spontaneously grows on dry-cured products (Simoncini et al. 2007; Asefa et al. 2009; Mendonça et al. 2013). Several yeast species have been detected during sausage and ham maturation, such as *Candida famata*, *C. guilliermondii*, *Candida intermedia/curvata*, *Candida parapsilosis*, *Candida zeylanoides*, *Citeromyces matritensis*, *Cryptococcus albidus*, *D. hansenii*, *Endomycopsis fibuligera*, *Hansenula ciferri*, *Hansenula holstii*, *Hansenula sydowiorum*, *M. pulcherrima*, *Pichia triangularis*, *Rhodotorula glutinis*, *R. mucilaginosa*, *Rhodotorula rubra*, *Trichosporon ovoides*, and *Yarrowia lipolytica* (Comi and Cantoni 1983; Núñez et al. 1996; Encinas et al. 2000; Cocolin et al. 2006; Simoncini et al. 2007; Comi and Iacumin 2013; Mendonça et al. 2013). In the fully matured product, *D. hansenii* dominated on fermented sausages and dry-cured ham (Comi and Cantoni 1983; Núñez et al. 1996; Encinas et al. 2000; Cocolin et al. 2006). Some of the native yeasts have been proposed as starter cultures because they are involved in the generation of volatile compounds, contributing to the characteristic flavor of fermented sausages and dry-cured ham (Flores et al. 2004; Martín et al. 2006; Iucci et al. 2007). Moreover, growth of native yeasts in dry-cured hams and controlling air humidity of the room inhibit mold growth (Wang et al. 2006). Based on these characteristics, yeasts seem to be candidates for the biocontrol of toxigenic molds in dry-cured foods.

D. hansenii is widely used in food processes because of its characteristic metabolic properties (Breuer and Harms 2006). The high occurrence of *D. hansenii* in dry-cured meat products is most probably due to its moderately halophilic properties, accounting for its optimal growth at 3%–5% salt (Breuer and Harms 2006), being able to grow in up to 15% NaCl (Asefa et al. 2009). Native *D. hansenii* are nonpathogenic, and their tolerance to processing conditions, including temperature, pH, moisture, and water activity, enables *D. hansenii* to colonize fermented sausages and dry-cured ham reaching high population levels.

As it has been mentioned earlier in this chapter, *D. hansenii* inhibits growth and sporulation of several *Penicillium, Aspergillus, Eurotium,* and *Cladosporium* strains. Several strains of *D. hansenii* have been proposed as biocontrol agents against molds in fruits like grapes or citrus (Droby et al. 1989; Hernández-Montiel et al. 2010), yogurt, and cheese (Liu and Tsao 2009). Strains of *D. hansenii* isolated from cheese and about 60% of isolates from dry-cured ham showed killer activity against sensitive yeasts (Addis et al. 2001; Pérez-Nevado et al. 2006). Twenty-one *D. hansenii* isolated from dry-cured Iberian ham inhibited the growth of toxigenic molds commonly found in hams (Lara et al. 2012) (Figure 17.1). Some of these *D. hansenii* isolates grow fairly on dry-cured sausages and dry-cured ham, being able to control the growth of *P. verrucosum* (Lara et al. 2013; Andrade et al. 2014) (Figure 17.2). The antagonistic effect of these isolates of *D. hansenii* has been attributed to competition (Andrade et al. 2014) and to volatile compounds (unpublished data) (Figure 17.3), whereas the inhibitory activity of other strains of *D. hansenii* against the ochratoxigenic molds *A. ochraceus* or *Penicillium nordicum* has been linked to the production of killer toxins (Gil-Serna et al. 2009; Virgili et al. 2012). These differences can be explained by the fact that NaCl significantly enhanced the inhibitory activity of both *D. hansenii* (Gil-Serna et al. 2009; Virgili et al. 2012) and their killer proteins (Marquina et al. 2001). Moreover, *D. hansenii* isolated from ham is capable to reduce OTA accumulation by ochratoxigenic penicillia both in culture media (Virgili et al. 2012) and in ham (Andrade et al. 2014). This effect has also been reported against *A. westerdijkiae,* and it has been attributed to OTA adsorption to yeast cell wall and to a reduction in the expression of OTA biosynthetic genes (Gil-Serna et al. 2009, 2011).

Other species of yeasts isolated from dry-cured meat products have proved their potential as bioprotective cultures. *C. famata* and *E. fibuliger* inhibit the growth of the ochratoxigenic molds *A. niger, A. ochraceus, P. nordicum,* and *P. verrucosum,* and consequently OTA production on the surface of

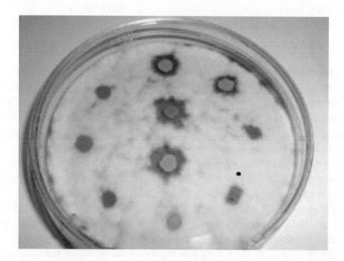

FIGURE 17.1   Inhibition of *Penicillium verrucosum* growth by *Debaryomyces hansenii* isolated from dry-cured hams.

FIGURE 17.2 Inhibition of *Penicillium verrucosum* on dry sausage by *Debaryomyces hansenii* after 10 days of incubation. Left: Inoculated solely with *P. verrucosum*; right: *P. verrucosum* coinoculated with *D. hansenii*.

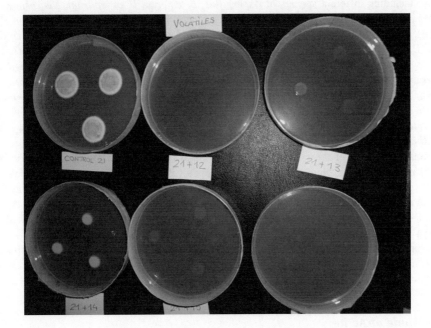

FIGURE 17.3 Inhibition of *Penicillium verrucosum* growth by volatile compounds produced by *Debaryomyces hansenii*. Upper left: *P. verrucosum* alone (control); upper center: *P. verrucosum* with volatile compounds from *D. hansenii* 12; upper right: *P. verrucosum* with volatile compounds from *D. hansenii* 13; lower left: *P. verrucosum* with volatile compounds from *D. hansenii* 14; lower center: *P. verrucosum* with volatile compounds from *D. hansenii* 15; and lower right: *P. verrucosum* with volatile compounds from *D. hansenii* 16.

San Daniele dry-cured ham (Comi and Iacumin 2013). *C. zeylanoides* was among the most effective yeasts in inhibiting *P. nordicum* in vitro (Virgili et al. 2012). However, *C. zeylanoides* decreases during processing steps in dry-cured hams and fails in maintaining high populations in fully matured dry-cured ham (Núñez et al. 1996; Simoncini et al. 2007; Asefa et al. 2009) having less ecological fitness than *D. hansenii*. Furthermore, *C. zeylanoides* should be considered unsuitable as biocontrol agent because it is an opportunistic pathogenic yeast (Levenson et al. 1991).

The concentration of toxigenic molds is a key factor for yeast effectiveness (Liu and Tsao 2009; Virgili et al. 2012); at lowest mold counts, yeast biocontrol is more effective. In applications of antagonistic yeasts for dry-cured ham biocontrol, the population of contaminating molds should be kept as low as possible, through HACCP procedures designed for this specific target (Asefa et al. 2010).

17.6.3 Molds

Several types of antifungal proteins are produced by molds. *Echinocandins* are cyclic lipo-hexapeptides synthesized by different ascomycetes that inhibit β-1,3-glucan synthesis in the fungal cell wall, resulting in being fungicidal to *Candida* spp. or fungistatic to *Aspergillus* spp. (Emri et al. 2013). *Aureobasidin A* is an antifungal cyclic depsipeptide antibiotic produced by *Aureobasidium pullulans* effective against protozoa and fungi, including *P. digitatum*, *P. italicum*, *P. expansum*, *B. cinerea*, and *Monilinia fructicola*. Given that this compound targets the inositol phosphorylceramide synthase, an enzyme essential for fungi but absent from mammals, its use has been proposed as relatively safe and promising fungicide candidates to control postharvest decays of fruits (Liu et al. 2013).

A group of defensin-like, cysteine-rich, highly basic, and low-molecular-weight (5.8–6.6 kDa) proteins from filamentous fungi has recently been described. *AFP* from *Aspergillus giganteus* (Nakaya et al. 1990) exhibited potent antifungal activity against certain phytopathogenic species of *Trichoderma*, *Fusarium*, *Penicillium*, and *Magnaporthe*, as well as the oomycete *Phytophthora infestans* (Lacadena et al. 1995; Vila et al. 2001). *PAF* from *Penicillium chrysogenum* (Marx et al. 1995) inhibited filamentous fungi, including different species of *Aspergillus*, *B. cinerea*, *F. oxysporum*, *Neurospora crassa*, *Trichoderma koningii*, *P. chrysogenum*, *Cochiobolus carbonum*, and *Gliocladium roseum* (Kaiserer et al. 2003). *Anafp* from *A. niger* inhibited the filamentous fungi *A. flavus*, *A. fumigatus*, *F. oxysporum*, *F. solani*, *T. beigelii*, as well as the yeasts *C. albicans* and *S. cerevisiae* (Lee et al. 1999), *AcAFP* from *Aspergillus clavatus* was active against mycelial growth of *F. oxysporum*, *F. solani*, *A. niger*, *B. cinerea*, and *Alternaria solani,* but did not affect yeast like *C. albicans*, *S. cerevisiae*, and *P. pastoris* (Skouri-Gargouri and Gargouri 2008), *PgAFP* from *P. chrysogenum* was active against *Penicillium echinulatum*, *Penicillium commune*, and *A. niger* (Acosta et al. 2009; Rodríguez-Martín et al. 2010a), and *NFAP* from *Neosartorya fischeri* inhibited two ascomycetes (*A. niger* and *Aspergillus nidulans*), as well as one zygomycete (*Rhizomucor miehei*) (Kovács et al. 2011).

The primary structure of these defensin-like antifungal proteins from molds differs from those of higher eukaryotes, such as plants, insects, or mammals (Marx et al. 2008). AFP, the defensin-like antifungal protein from *A. giganteus*, lacks the α-helix, but folds in five antiparallel β-strands, defining a small and compact β-barrel stabilized by internal disulfide bridges (Campos-Olivas et al. 1995). Similarly, PAF form *P. chrysogenum* comprises five β-strands forming two orthogonally packed β-sheets that share a common interface (Batta et al. 2009).

In addition, some fungal enzymes also show antifungal activity. *Glucose oxidase* from *P. chrysogenum* inhibits *A. niger*, *Aspergillus terreus*, and *Emericella nidulans* (Leiter et al. 2004) and the chitosanase *PgChP* from *P. chrysogenum* inhibits *P. commune* (Acosta et al. 2009; Rodríguez-Martín et al. 2010b).

17.6.3.1 Mode of Action

In the first stage, the cationic character of these antifungal proteins allows the interaction with negatively charged plasma membrane components of sensitive microorganisms. Then, two mechanisms can explain the antimicrobial action: insertion of the antifungal protein in the cell membrane to form a pore or interaction with specific receptors in the plasma membrane leading to selective alteration of membrane permeability. Pore formation leads to membrane depolarization, disintegration of plasma membrane, leakage of cell content, and death of the cell (Hwang et al. 2010). This has also been shown for AFP secreted by *A. giganteus* that accumulates within defined areas of the cell wall of the sensitive mold, but very little protein was bound to the cell wall of a resistant mold (Theis et al. 2005). On the contrary, most plant defensins and other defensin-like proteins from molds such as PAF do not form channels, but interact with phospholipids of the plasma membrane as a specific receptor, inducing membrane hyperpolarization and permeabilization with a rapid Ca^{2+} uptake (Thevissen et al. 1996, 2003; Leiter et al. 2005; Binder et al. 2010; Wong et al. 2012). The protein is internalized into the sensitive mold (Oberparleiter et al. 2003) to interact with intracellular targets activating signaling cascades, inducing an increase in the intracellular level of reactive oxygen species and apoptosis (Kaiserer et al. 2003; Leiter et al. 2005; De Coninck et al. 2013). The role of Ca^{2+} as a universal intracellular second messenger in eukaryotic cells seems to be linked to growth inhibition in molds (Binder et al. 2010).

FIGURE 17.4 Inefficient PgAFP inhibition of *Aspergillus flavus* on cheese. Left: Untreated uninoculated cheese; center: untreated cheese inoculated with *A. flavus*; right: PgAFP-treated cheese inoculated with *A. flavus*.

17.6.3.2 Influence of Environmental Factors

Transcription of AFP by *A. giganteus* and PAF by *P. chrysogenum* is enhanced by stress factors, such as heat shock, osmotic stress, carbon starvation, and alkaline pH (Marx 2004; Meyer et al. 2005). For this, production of these AMPs is expected to be high in ripened foods. In addition, the compact structure of AMPs from filamentous fungi is responsible for a remarkable resistance toward environmental factors, as it has been shown for PAF to pH (1.5–11), heat (80°C, 60 min), and proteases (pepsin, proteinase K, and pronase) (Skouri-Gargouri and Gargouri 2008; Batta et al. 2009; Kovács et al. 2011). Therefore, these AMPs will keep active for most food processes, including dry-ripening with highly proteolytic microorganisms.

Divalent cations and plasma proteins interfere with the antimicrobial activity of most defensins (Ganz 2003). Ca^{2+} inhibited the electrostatic interaction of thionins with membranes, which is the first step in the mode of action for some AMPs (Vernon and Rogers 1992). In addition, elevated $CaCl_2$ concentrations reduced the growth-inhibitory activity of PAF, because the high Ca^{2+} concentrations increase the activity of existing Ca^{2+} pumps/transporters to counteract the PAF-specific intracellular Ca^{2+} perturbation (Binder et al. 2010). The high Ca^{2+} normal concentration on cheese seems to be responsible for the lack of effect of PgAFP against sensitive molds (Figure 17.4).

17.6.4 Selection of Antifungal Agents

Biocontrol agents are usually identified through in vitro inhibition tests. However, there could be no correlation between the performance of biocontrol agents in vitro and in the food industry because laboratory conditions may favor the antagonist (Cotty and Mellon 2006). In addition to the aforementioned effect of Ca^{2+} and plasma proteins interfering with AMPs, changes in the environment can be decisive to determine the coexistence level or dominance of species in an ecological niche (Marín et al. 1998). In some cases, coculturing microorganisms can result in the stimulation of mycotoxin production. *Pichia burtonii* increased aflatoxin production by *A. flavus* in maize and rice substrates (Cuero et al. 1987).

The interspecific interactions between microbial species are profoundly influenced by abiotic factors such as water activity, temperature, and nutrient substrate (Magan and Lacey 1984). Extrinsic factors such as temperature and relative humidity, or intrinsic factors such as pH, nutrients, and water activity greatly change during processing of dry-cured foods. Given that biocontrol efficacy depends on survival and colonization of food surfaces, the selection should include testing the potential range of conditions under which the agent will be used.

Biocontrol agents should not affect the sensorial characteristics of foods. Moreover, biocontrol activity did not only depend on species or genus, being a strain-related characteristic (Bleve et al. 2006). When the strains intended to be used were not isolated from similar high-quality products, an in-depth evaluation of sensorial traits is required. In addition, the microbial population naturally present on dry-ripened foods includes highly proteolytic strains (Rodríguez et al. 1998; Sousa et al. 2001; Hughes et al. 2002). Selecting peptides that resist degradation by fungal proteases could increase the accumulation of active peptides to reach effective inhibitory concentrations (Oard et al. 2004). Thus, a careful selection of strains producing bioactive compounds resistant to the hydrolytic enzymes from the microbiota present on the food is a must.

The use of biocontrol agents to control mycotoxin contamination is a promising tool. Due to the variety of toxigenic molds potentially present, it is not feasible to use just one biological antagonist to control mycotoxins in ripened foods. Nevertheless, the complementary use of antagonists or combination with other means to control the fungal population could efficiently control this hazard. Several combinations of yeasts and fungicides have been proposed to control spoilage and toxigenic molds in plant foods. For example, ammonium molybdate improved the inhibitory effect of *P. membranifaciens*, *Candida sake*, and *C. laurentii* (Nunes et al. 2002a; Wan and Tian 2005; Cao et al. 2010b).

One of the most promising strategies to develop new antifungals relies on the small, cationic, cysteine-rich antifungal proteins produced commonly by eukaryotes. The use of small basic proteins secreted from filamentous fungi has been suggested for biopreservation purposes in food (Geisen 2000).

P. chrysogenum is a good option for ripened cheeses, dry-fermented sausages, and dry-cured meats. It is well adapted to the ripening process of cheeses and meats; the ripening conditions enhance PAF and PgAFP production; it also produces glucose oxidase and PgChP with antifungal activity; and the main antifungal proteins are adequate to withstand proteases. In addition, neither *P. chrysogenum* PAF nor *A. giganteus* AFP caused detrimental effects on mammalian including human cells and tissues when tested at effective doses in vitro (Szappanos et al. 2005).

Given that some of these molds, such as *P. chrysogenum*, produce compounds classified as GRAS by the US Food and Drug Administration and are commonly present in dry-cured foods, their use may prevent surface growth of toxigenic molds in mold-ripened foods.

A lot of research is still needed to fully understand the complex interactions among microorganisms and to obtain adequate strains to prevent mycotoxin production in dry-ripened foods. Nevertheless, the huge amount of evidence in favor of the effective and safe potential of biocontrol agents makes the use of protective cultures a valuable option to control toxigenic molds in food processing for the near future.

REFERENCES

Acosta, R., Rodríguez-Martín, A., Martín, A., Núñez, F., and Asensio, M.A. Selection of antifungal protein-producing molds from dry-cured meat products. *International Journal of Food Microbiology* 135 (2009): 39–46.

Addis, E., Fleet, G.H., Cox, J.M., Kolak, D., and Leung, T. The growth, properties and interactions of yeasts and bacteria associated with the maturation of Camembert and blue-veined cheeses. *International Journal of Food Microbiology* 69 (2001): 25–36.

Al-Aidroos, K. and Bussey, H. Chromosomal mutants of *Saccharomyces cerevisiae* affecting the cell wall binding site for killer factor. *Canadian Journal of Microbiology* 24 (1978): 228–237.

Albertini, M.C., Accorsi, A., Citterio, B. et al. Morphological and biochemical modifications induced by a static magnetic field on *Fusarium culmorum*. *Biochimie* 85 (2003): 963–970.

Amborabe, B.E., Bonmort, J., Fleurat-Lessart, P., and Roblin, G. Early events induced by chitosan on plant cells. *Journal of Experimental Botany* 59 (2008): 2317–2324.

Ando, H., Hatanaka, K., Ohata, I., Yamashita-Kitaguchi, Y., Kurata, A., and Kishimoto, N. Antifungal activities of volatile substances generated by yeast isolated from Iranian commercial cheese. *Food Control* 26 (2012): 472–478.

Andrade, M.J., Thorsen, L., Rodríguez, A., Córdoba, J.J., and Jespersen, L. Inhibition of ochratoxigenic moulds by *Debaryomyces hansenii* strains for biopreservation of dry-cured meat products. *International Journal of Food Microbiology* 170 (2014): 70–77.

Angioni, A., Caboni, P., Garau, A. et al. In vitro interaction between ochratoxin A and different strains of *Saccharomyces cerevisiae* and *Kloeckera apiculata*. *Journal of Agricultural and Food Chemistry* 55 (2007): 2043–2048.

Appendini, P. and Hotchkiss, J.H. Review of antimicrobial food packaging. *Innovative Food Science and Emerging Technologies* 3 (2002): 113–126.

Armando, M.R., Dogi, C.A., Poloni, V., Rosa, C.A.R., Dalcero, A.M., and Cavaglieri, L.R. In vitro study on the effect of *Saccharomyces cerevisiae* strains on growth and mycotoxin production by *Aspergillus carbonarius* and *Fusarium graminearum*. *International Journal of Food Microbiology* 161 (2013): 182–188.

Armando, M.R., Dogi, C.A., Rosa, C.A.R., Dalcero, A.M., and Cavaglieri, L.R. *Saccharomyces cerevisiae* strains and the reduction of *Aspergillus parasiticus* growth and aflatoxin B1 production at different interacting environmental conditions, in vitro. *Food Additives and Contaminants* 29 (2012a): 1443–1449.

Armando, M.R., Pizzolitto, R.P., Dogi, C.A. et al. Adsorption of ochratoxin A and zearalenone by potential probiotic *Saccharomyces cerevisiae* strains and its relation with cell wall thickness. *Journal of Applied Microbiology* 113 (2012b): 256–264.

Armando, M.R., Pizzolitto, R.P., Escobar, F. et al. *Saccharomyces cerevisiae* strains from animal environmental with aflatoxin B1 binding ability and anti-pathogenic bacteria influence in vitro. *World Mycotoxin Journal* 4 (2011): 59–68.

Asefa, D.T., Kure, C.F., Gjerde, R.O. et al. A HACCP plan for mycotoxigenic hazards associated with dry-cured meat production processes. *Food Control* 22 (2010): 831–837.

Asefa, D.T., Møretrø, T., Gjerde, R.O. et al. Yeast diversity and dynamics in the production processes of Norwegian dry-cured meat products. *International Journal of Food Microbiology* 133 (2009): 135–140.

Avis, T.J. and Belanger, R.R. Specificity and mode of action of the antifungal fatty acid *cis*-9-heptadecenoic acid produced by *Pseudozyma flocculosa*. *Applied and Environmental Microbiology* 67 (2001): 956–960.

Axelsson, L.T., Chung, T.C., Dobrogosz, W.J., and Lindgren, S.E. Production of a broad spectrum antimicrobial substance by *Lactobacillus reuteri*. *Microbial Ecology in Health and Disease* 2 (1989): 131–136.

Baek, E., Kim, H., Choi, H., Yoon, S., and Kim, J. Antifungal activity of *Leuconostoc citreum* and *Weissella confusa* in rice cakes. *Journal of Microbiology* 50 (2012): 842–848.

Batta, G., Barna, T., Gáspári, Z. et al. Functional aspects of the solution structure and dynamics of PAF—A highly-stable antifungal protein from *Penicillium chrysogenum*. *FEBS Journal* 276 (2009): 2875–2890.

Bejaoui, H., Mathieu, F., Taillandier, P., and Lebrihi, A. Ochratoxin A removal by synthetic and natural grape juices by selected oenological *Saccharomyces* strains. *Journal of Applied Microbiology* 97 (2004): 1038–1044.

Betina, V. *Mycotoxins: Chemical, biological and environmental aspects* (*Bioactive Molecules*. Vol. 9) (London, U.K.: Elsevier Applied Science, 1989), pp. 114–150.

Binder, U., Chu, M., Read, N.D., and Marx, F. The antifungal activity of the *Penicillium chrysogenum* protein PAF disrupts calcium homeostasis in *Neurospora crassa*. *Eukaryotic Cell* 9 (2010): 1374–1382.

Björnberg, A. and Schnürer, J. Inhibition of the growth of the grain-storage molds in vitro by the yeast *Pichia anomala* (Hansen) Kurtzman. *Canadian Journal of Microbiology* 39 (1993): 623–628.

Bleve, G., Grieco, F., Cozzi, G., Logrieco, A., and Visconti, A. Isolation of epiphytic yeasts with potential for biocontrol of *Aspergillus carbonarius* and *A. niger* on grape. *International Journal of Food Microbiology* 108 (2006): 204–209.

Boyacioglu, D., Hettiarachchy, N.S., and Stack, R.W. Effect of three systemic fungicides on deoxynivalenol (vomitoxin) production by *Fusarium graminearum* in wheat. *Canadian Journal of Plant Science* 72 (1992): 93–101.

Breuer, U. and Harms, H. *Debaryomyces hansenii*—An extremophilic yeast with biotechnological potential. *Yeast* 23 (2006): 415–437.

Broekaert, W.F., Terras, F.R.G., Cammue, B.P.A., and Osborn, R.W. Plant defensins: Novel antimicrobial peptides as components of the host defense system. *Plant Physiology* 108 (1995): 1353–1358.

Caccioni, D.R.L., Guizzardi, M., Biondi, D.M., Renda, A., and Ruberto, G. Relationship between volatile components of citrus fruit essential oils and antimicrobial action on *Penicillium digitatum* and *Penicillium italicum*. *International Journal of Food Microbiology* 43 (1998): 73–79.

Campos-Olivas, R., Bruix, M., Santoro, J. et al. NMR solution structure of the antifungal protein from *Aspergillus giganteus*: Evidence for cysteine pairing isomerism. *Biochemistry* 34 (1995): 3009–3021.

Cao, S., Hu, Z., Pang, B., Wang, H., Xie, H., and Wu, F. Effect of ultrasound treatment on fruit decay and quality maintenance in strawberry after harvest. *Food Control* 21 (2010a): 529–532.

Cao, S., Yuan, Y., Hu, Z., and Zheng, Y. Combination of *Pichia membranifaciens* and ammonium molybdate for controlling blue mould caused by *Penicillium expansum* in peach fruit. *International Journal of Food Microbiology* 141 (2010b): 173–176.

Caridi, A. New perspectives in safety and quality enhancement of wine through selection of yeasts based on the parietal adsorption activity. *International Journal of Food Microbiology* 120 (2007): 167–172.

Caridi, A., Galvano, F., Tafuri, A., Ritieni, A. Ochratoxin removal during winemaking. *Enzyme and Microbial Technology* 40 (2006): 122–126.

Chan, Z. and Tian, S. Interaction of antagonistic yeasts against postharvest pathogens of apple fruit and possible mode of action. *Postharvest Biology and Technology* 36 (2005): 215–223.

Cia, P., Pascholati, S.F., Benato, E.A., Camili, E.C., and Santos, C.A. Effects of gamma and UV-C irradiation on the postharvest control of papaya anthracnose. *Postharvest Biology and Technology* 43 (2007): 366–373.

Cocolin, L., Urso, R., Rantsiou, K., Cantoni, C., and Comi, G. Dynamics and characterization of yeasts during natural fermentation of Italian sausages. *FEMS Yeast Research* 6 (2006): 692–701.

Coelho, A.R., Celli, M.G., Ono, E.Y.S. et al. *Penicillium expansum* versus antagonist yeasts and patulin degradation in vitro. *Brazilian Archives of Biology and Technology* 50 (2007): 725–733.

Coelho, A.R., Tachib, M., Pagnoccac, F.C. et al. Purification of *Candida guilliermondii* and *Pichia ohmeri* killer toxin as an active agent against *Penicillium expansum*. *Food Additives and Contaminants* 26 (2009): 73–81.

Comi, G. and Cantoni, C. Yeasts in dry Parma hams. *Industrie Alimentari* 22 (1983): 102–104.

Comi, G. and Iacumin, L. Ecology of moulds during the pre-ripening and ripening of San Daniele dry cured ham. *Food Research International* 54 (2013): 1113–1119.

Conway, W.S., Leverentz, B., Janisiewicz, W.J., Blodgett, A.B., Saftner, R.A., and Camp, M.J. Integrating heat treatment, biocontrol and sodium bicarbonate to reduce post-harvest decay of apple caused by *Colletotrichum acutatum* and *Penicillium expansum*. *Post-Harvest Biology and Technology* 34 (2004): 11–20.

Cotty, P.J. and Mellon, J.E. Ecology of aflatoxin producing fungi and biocontrol of aflatoxin contamination. *Mycotoxin Research* 22 (2006): 110–117.

Crowley, S., Mahonya, J., and van Sinderen, D. Current perspectives on antifungal lactic acid bacteria as natural bio-preservatives. *Trends in Food Science and Technology* 33 (2013): 93–109.

Cuero, R.G., Smith, J., and Lacey, J. Increase of aflatoxin production by *A. flavus* in single or mixed culture either with *Pichia burtonii* or *Bacillus amyloliquefaciens* in gnotobiotic cracked maize and rice. *Applied Environmental Microbiology* 53 (1987): 1142–1146.

da Cruz Cabral, L., Fernández Pinto, V., and Patriarca, A. Application of plant derived compounds to control fungal spoilage and mycotoxin production in foods. *International Journal of Food Microbiology* 166 (2013): 1–14.

Dalié, D.K.D., Deschamps, A.M., and Richard-Forget, F. Lactic acid bacteria—Potential for control of mould growth and mycotoxins: A review. *Food Control* 2 (2010): 370–380.

De Coninck, B., Cammue, B.P.A., and Thevissen, K. Modes of antifungal action and *in planta* functions of plant defensins and defensin-like peptides. *Fungal Biology Reviews* 26 (2013): 109–120.

Droby, S., Chalutz, E., Wilson, C.L., and Wisniewski, M. Characterization of the biocontrol activity of *Debaryomyces hansenii* in the control of *Penicillium digitatum* on grape fruit. *Canadian Journal of Microbiology* 35 (1989): 794–800.

Druvefors, U., Jonsson, N., Boysen, M., and Schnürer, J. Efficacy of the biocontrol yeast *Pichia anomala* during long-term storage of moist feed grain under different oxygen and carbon dioxide regimes. *FEMS Yeast Research* 2 (2002): 389–394.

Druvefors, U.A., Passoth, V., and Schnürer, J. The role of nutrient competition and ethyl acetate formation in the mode of action of the biocontrol agent *Pichia anomala*. (Budapest, Hungary: *23rd International Specialised Symposium on Yeasts*, 2003), p. 115.

El-Ghaouth, A., Wilson, C.L., and Wisniewski, M. Ultrastructural and cytochemical aspects of the biological control of *Botrytis cinerea* by *Candida saitoana* in apple fruit. *Phytopathology* 88 (1998): 282–291.

Ellis, S.A., Gooding, M.J., and Thompson, A.J. Factors influencing the relative susceptibility of wheat cultivars (*Triticum aestivum* L.) to blackpoint. *Crop Protection* 15 (1996): 69–76.

Emri, T., Majoros, L., Tóth, V., and Póc, I. Echinocandins: Production and applications. *Applied Microbiology and Biotechnology* 97 (2013): 3267–3284.

Encinas, J.P., Lopez-Díaz, T.M., García-López, M.L., Otero, A., and Moreno, B. Yeast populations on Spanish fermented sausages. *Meat Science* 54 (2000): 203–208.

Fallik, E. Physical control of mycotoxigenic fungi. In *Mycotoxins in Fruits and Vegetables,* eds. R. Barkai-Golan and N. Paster (Amsterdam, the Netherlands: Academic Press, 2008), pp. 297–310.

Fan, Q. and Tian, S.P. Postharvest biological control of rhizopus rot on nectarine fruit by *Pichia membranefaciens* Hansen. *Plant Diseases* 84 (2000): 1212–1216.

Farber, J.M. Microbiological aspects of modified-atmosphere packaging technology: A review. *Journal of Food Protection* 54 (1991): 58–70.

Fialho, M.B., Toffano, L., Pedroso, M.P., Augusto, F., and Pascholati, S.F. Volatile organic compounds produced by *Saccharomyces cerevisiae* inhibit the in vitro development of *Guignardia citricarpa*, the causal agent of citrus black spot. *World Journal of Microbiology and Biotechnology* 26 (2010): 925–932.

Flores, M., Durá, M.A., Marco, A., and Toldrá, F. Effect of *Debaryomyces* spp. on aroma formation and sensory quality of dry-fermented sausages. *Meat Science* 68 (2004): 439–446.

Gandomi, H., Misaghi, A., Basti, A.A. et al. Effect of *Zataria multiflora* Boiss. essential oil on growth and aflatoxin formation by *Aspergillus flavus* in culture media and cheese. *Food and Chemical Toxicology* 47 (2009): 2397–2400.

Ganz, T. Defensins: Antimicrobial peptides of innate immunity. *Nature* 3 (2003): 710–720.

Garcia Moruno, E., Sanlorenzo, C., Boccaccino, B., and Di Stefano, R. Treatment with yeast to reduce the concentration of ochratoxin A in red wine. *American Journal of Enology and Viticulture* 56 (2005): 73–76.

Gareis, M. and Ceynowa, J. Influence of the fungicide Matador (tebuconazole/triadimenol) on mycotoxin production by *Fusarium culmorum*. *Zeitschrift für Lebensmittel-Untersuchung und -Forschung* 198 (1994): 244–248.

Geisen, R. *P. nalgiovense* carries a gene which is homologous to the *paf* gene of *P. chrysogenum* which codes for an antifungal peptide. *International Journal for Food Microbiology* 62 (2000): 95–101.

Gil-Serna, J., Patiño, B., Cortés, L., González-Jaén, M.T., and Vázquez, C. Mechanisms involved in reduction of ochratoxin A produced by *Aspergillus westerdijkiae* using *Debaryomyces hansenii* CYC 1244. *International Journal of Food Microbiology* 151 (2011): 113–118.

Gil-Serna, J., Patiño, B., González-Jaén, M.T., and Vázquez, C. Biocontrol of *Aspergillus ochraceus* by yeast. In *Current Research Topics in Applied Microbiology and Microbial Biotechnology*, ed. A. Mendez-Vilas (Singapore: World Scientific, 2009), pp. 368–372.

Gutierrez, J., Barry-Ryan, C., and Bourke, P. The antimicrobial efficacy of plant essential oil combinations and interactions with food ingredients. *International Journal of Food Microbiology* 124 (2008): 91–97.

Guynot, M.E., Marín, S., Sanchis, V., and Ramos, A.J. An attempt to minimize potassium sorbate concentration in sponge cakes by modified atmosphere packaging combination to prevent fungal spoilage. *Food Microbiology* 21 (2004): 449–457.

Habermann, E. Bee and wasp venoms. *Science* 177 (1972): 314–322.

Hamanaka, D., Norimura, N., Baba, N. et al. Surface decontamination of fig fruit by combination of infrared radiation heating with ultraviolet irradiation. *Food Control* 22 (2011): 375–380.

Hancock, R.E.W. and Sahl, H.G. Antimicrobial and host-defense peptides as new anti-infective therapeutic strategies. *Nature Biotechnology* 24 (2006): 1551–1557.

Hashmi, M.S., East, A.R., Palmer, J.S., and Heyes, J.A. Pre-storage hypobaric treatments delay fungal decay of strawberries. *Postharvest Biology and Technology* 77 (2013): 75–79.

Hegedüs, N. and Marx, F. Antifungal proteins: More than antimicrobials? *Fungal Biology Reviews* 26 (2013): 132–145.

Heipieper, H.J., Weber, F.J., Sikkema, J., Keweloh, H., and De Bont, J.A.M. Mechanisms of resistance of whole cells to toxic organic solvents. *Trends in Biotechnology* 12 (1994): 409–415.

Hernández-Montiel, L.G., Ochoa, J.L., Troyo-Diéguez, E., and Larralde-Corona, C.P. Biocontrol of postharvest blue mold (*Penicillium italicum* Wehmer) on Mexican lime by marine and citrus *Debaryomyces hansenii* isolates. *Postharvest Biology and Technology* 56 (2010): 181–187.

Holley, R.A. and Patel, D. Improvement in shelf-life and safety of perishable foods by plant essential oils and smoke antimicrobials. *Food Microbiology* 22 (2005): 273–292.

Hoover, D.M., Chertov, O., and Lubkowski, J. The structure of human β-defensin-1: New insights into structural properties of β-defensins. *The Journal of Biological Chemistry* 276 (2001): 39021–39026.

Hsieh, P.C., Mau, J.L., and Huang, S.H. Antimicrobial effect of various combinations of plant extracts. *Food Microbiology* 18 (2001): 35–43.

Hua, S.S.T., Baker, J.L., and Flores-Espiritu, M. Interaction of saprophytic yeast with a nor mutant of *Aspergillus flavus*. *Applied and Environmental Microbiology* 65 (1999): 2738–2740.

Hughes, M.C., Kerry, J.P., Arendt, E.K., Kenneally, P.M., McSweeney, P.L.H., and O'Neill, E.E. Characterization of proteolysis during the ripening of semi-dry fermented sausages. *Meat Science* 62 (2002): 205–216.

Humphris, S.N., Bruce, A., Buultjens, T.E.J., and Wheatley, R.E. The effects of volatile secondary metabolites on protein synthesis in *Serpula lacrymans*. *FEMS Microbiology Letters* 210 (2002): 215–219.

Hwang, B., Hwang, J.S., Lee, J., and Lee, D.G. Antifungal properties and mode of action of psacotheasin, a novel knottin-type peptide derived from *Psacothea hilaris*. *Biochemical and Biophysical Research Communications* 400 (2010): 352–357.

Ingram, L.O. and Buttke, T.M. Effect of alcohols on micro-organisms. *Advances in Microbial Physiology* 25 (1984): 253–300.

Isman, B. and Biyik, H. The aflatoxin contamination of fig fruits in Aydin city (Turkey). *Journal of Food Safety* 29 (2009): 318–330.

Iucci, L., Patrignani, F., Belletti, N. et al. Role of surface-inoculated *Debaryomyces hansenii* and *Yarrowia lipolytica* strains in dried fermented sausage manufacture. Part 2: Evaluation of their effects on sensory quality and biogenic amine content. *Meat Science* 75 (2007): 669–675.

Izgu, D.A., Kepekci, R.A., and Izgu, F. Inhibition of *Penicillium digitatum* and *Penicillium italicum* in vitro and in planta with Panomycocin, a novel exo-b-1,3-glucanase isolated from *Pichia anomala* NCYC 434. *Antonie van Leeuwenhoek* 99 (2011): 85–91.

Izgu, F. and Altinbay, D. Isolation and characterization of the K5-type yeast killer protein and its homology with an exo-β-1,3-glucanase. *Bioscience Biotechnology and Biochemistry* 68 (2004): 685–693.

Izgu, F., Altinbay, D., and Sertkaya, A. Enzymic activity of the K5-type yeast killer toxin and its characterization. *Bioscience, Biotechnology and Biochemistry* 69 (2005): 2200–2206.

Janisiewicz, W.J. and Korsten, L. Biological control of postharvest diseases of fruits. *Annual Review of Phytopathology* 40 (2002): 411–441.

Jenssen, H., Hamill, P., and Hancock, R.E.W. Peptide antimicrobial agents. *Clinical Microbiology Reviews* 19 (2006): 491–511.

Kaiserer, L., Oberparleiter, C., Weiler-Görz, R., Burgstaller, W., Leiter, E., and Marx, F. Characterization of the *Penicillium chrysogenum* antifungal protein PAF. *Archives of Microbiology* 180 (2003): 204–210.

Kasahara, S., Inoue, S.B., Mio, T. et al. Involvement of cell wall β-glucan in the action of HM-1 killer toxin. *FEBS Letters* 348 (1994): 27–32.

Kovács, L., Viragh, M., Tako, M., Papp, T., Vagvolgyi, C., and Galgoczy, L. Isolation and characterization of *Neosartorya fischeri* antifungal protein (NFAP). *Peptides* 32 (2011): 1724–1731.

Lacadena, J., Martínez del Pozo, A., Gasset, M. et al. Characterization of the antifungal protein secreted by the mould *Aspergillus giganteus*. *Archives of Biochemistry and Biophysics* 324 (1995): 273–281.

Laitila, A., Sarlin, T., Kotaviita, E., Huttunen, T., Home, S., and Wilhelmson, A. Yeasts isolated from industrial maltings can suppress *Fusarium* growth and formation of gushing factors. *Journal of Industrial Microbiology and Biotechnology* 34 (2007): 701–713.

Lara, M.S., Andrade, M.J., Gordillo, R., Sánchez-Montero, L., and Núñez, F. Application of yeasts isolated from dry-cured ham for biological control of toxigenic moulds in dry-cured meat products. (Ourique, Portugal: *VII World Congress of Dry-Cured Ham*, 2013), pp. 20–25.

Lara, M.S., Núñez, F., Asensio, M.A., Delgado, J., Sánchez-Montero, L., and Andrade, M.J. Selection of *Debaryomyces hansenii* isolated from dry-cured ham to control toxigenic molds. (Logroño, Spain: *XVIII Spanish Congress of Food Microbiology*, 2012), pp. 144–146.

Lee, D.G., Shin, S.Y., Maeng, C-Y., Jin, Z.Z., Kim, K.L., and Hahm, K.-S. Isolation and characterization of a novel antifungal peptide from *Aspergillus niger*. *Biochemical and Biophysical Research Communications* 263 (1999): 646–651.

Leiter, E., Marx, F., Pusztahelyi, T., Haas, H., and Pócsi, I. *Penicillium chrysogenum* glucose oxidase—A study on its antifungal effects. *Journal of Applied Microbiology* 97 (2004): 1201–1209.

Leiter, E., Szappanos, H., Oberparleiter, C. et al. Antifungal, protein PAF severely affects the integrity of the plasma membrane of *Aspergillus nidulans* and induces an apoptosis-like phenotype. *Antimicrobial Agents and Chemotherapy* 49 (2005): 2445–2453.

Levenson, D., Pfaller, M.A., Smith, M.A. et al. *Candida zeylanoides*: Another opportunistic yeast. *Journal of Clinical Microbiology* 29 (1991): 1689–1692.

Levy, R.M., Hayashi, L., Carreiro, S.C., Pagnocca, F.C., and Hirooka, E.Y. Inhibition of mycotoxigenic *Penicillium* sp. and patulin biodegradation by yeast strains. *Revista Brasileira de Armazenamento* 27 (2002): 41–47.

Liggitt, J., Jenkinson, P., and Parry, D.W. The role of saprophytic microflora in the development of *Fusarium* ear blight of winter wheat caused by *Fusarium culmorum*. *Crop Protection* 16 (1997): 679–685.

Lis-Balchin, M., Steyrl, H., and Krenn, E. The comparative effect of novel *Pelargonium* essential oils and their corresponding hydrosols as antimicrobial agents in a model food system. *Phytotherapy Research* 17 (2003): 60–65.

Liu, S.Q. and Tsao, M. Biocontrol of dairy moulds by antagonistic dairy yeast *Debaryomyces hansenii* in yoghurt and cheese at elevated temperatures. *Food Control* 20 (2009): 852–855.

Liu, X., Wang, J., Gou, P., Mao, C., Zhu, Z.-R., and Li, H. In vitro inhibition of postharvest pathogens of fruit and control of gray mold of strawberry and green mold of citrus by aureobasidin A. *International Journal of Food Microbiology* 119 (2013): 223–229.

Luchese, R.H. and Harrigan, W.F. Biosynthesis of aflatoxin—The role of nutrition factors. *Journal of Applied Microbiology* 74 (1993): 5–14.

Magan, N. and Lacey, J. Effects of temperature and pH on water relations of field and storage fungi. *Transactions of the British Mycological Society* 82 (1984): 71–81.

Magliani, W., Conti, S., Travassos, L.R., and Polonelli, L. From yeast killer toxins to antibodies and beyond. *FEMS Microbiology* 288 (2008): 1–8.

Mandal, V., Sen, S.K., and Mandal, N.C. Detection, isolation and partial characterization of antifungal compound(s) produced by *Pediococcus acidilactici* LAB 5. *Natural Product Communications* 2 (2007): 671–674.

Marín, S., Ramos, A.J., and Sanchís, V. Chemical control of mycotoxigenic fungi. In *Mycotoxins in Fruits and Vegetables*, eds. R. Barkai-Golan and N. Paster (Amsterdam, the Netherlands: Academic Press, 2008), pp. 279–296.

Marín, S., Sanchís, V., Ramos, A.J., Viñas, I., and Magan, N. Environmental factors, in vitro interactions, and niche overlap between *Fusarium moniliforme, F. proliferatum*, and *F. graminearum, Aspergillus* and *Penicillium* species from maize grain. *Mycological Research* 102 (1998): 831–837.

Marquina, D., Barroso, J., Santos, A., and Peinado, J.M. Production and characteristics of *Debaryomyces hansenii* killer toxin. *Microbiology Research* 156 (2001): 387–391.

Martín, M., Córdoba, J.J., Aranda, E., Córdoba, M.G., and Asensio, M.A. Contribution of a selected fungal population to the volatile compounds on dry-cured ham. *International Journal of Food Microbiology* 110 (2006): 8–13.

Martinac, B., Zhu, H., Kubalski, A., Zhou, X.L., Culbertson, M., Bussey, H., and Kung, C. Yeast K1 killer toxin forms ion channels in sensitive yeast spheroplasts and in artificial liposomes. *Proceedings of the National Academy of Sciences of the United States of America* 87 (1990): 6228–6232.

Marx, F. Small, basic antifungal proteins secreted from filamentous ascomycetes: A comparative study regarding expression, structure, function and potential application. *Applied Microbiology and Biotechnology* 65 (2004): 133–142.

Marx, F., Binder, U., Leiter, É., and Pócsi, I. The *Penicillium chrysogenum* antifungal protein PAF, a promising tool for the development of new antifungal therapies and fungal cell biology studies. *Cellular and Molecular Life Sciences* 65 (2008): 445–454.

Marx, F., Haas, H., Reindl, M., Stöffler, G., Lottpeich, F., and Redl, B. Cloning, structural organization and regulation of expression of the *Penicillium chrysogenum* paf gene encoding an abundantly secreted protein with antifungal activity. *Gene* 167 (1995): 167–171.

Masih, E.I. and Paul, B. Secretion of beta-1,3-glucanase by the yeast *Pichia membranifaciens* and its possible role in the biocontrol of *Botrytis cinerea* causing mold disease of the grapevine. *Current Microbiology* 44 (2002): 391–395.

Masoud, W., Poll, L., and Jakobsen, M. Influence of volatile compounds produced by yeasts predominant during processing of *Coffea arabica* in East Africa on growth and ochratoxin A (OTA) production by *Aspergillus ochraceus. Yeast* 22 (2005): 1133–1142.

Meca, G., Blaiotta, G., and Ritieni, A. Reduction of ochratoxin A during the fermentation of Italian red wine Moscato. *Food Control* 21 (2010): 579–583.

Mendonça, R.C.S., Gouvêa, D.M., Hungaro, H.M., Sodré, A.F., and Querol-Simon, A. Dynamics of the yeast flora in artisanal country style and industrial dry cured sausage (yeast in fermented sausage). *Food Control* 29 (2013): 143–148.

Meyer, V., Spielvogel, A., Funk, L., Tilburn, J., Arst Jr., H.N., and Stahl, U. Alkaline pH-induced up-regulation of the afp gene encoding the antifungal protein (AFP) of *Aspergillus giganteus* is not mediated by the transcription factor PacC: Possible involvement of calcineurin. *Molecular Genetics and Genomics* 274 (2005): 295–306.

Molnar, O., Schatzmayr, G., Fuchs, E., and Prillinger, H. *Trichosporon mycotoxinivorans* sp. nov., a new yeast species useful in biological detoxification of various mycotoxins. *Systematic and Applied Microbiology* 27 (2004): 661–671.

Moss, M.O. and Long, M.T. Fate of patulin in the presence of the yeast *Saccharomyces cerevisiae. Food Additives and Contaminants* 19 (2002): 387–399.

Nakaya, N., Omata, K., Okahashi, I., Nakamura, Y., Kolkenbrock, H.J., and Ulbrich, N. Amino-acid sequence and disulphide bridges of an antifungal-protein isolated from *Aspergillus giganteus*. *European Journal of Biochemistry* 193 (1990): 31–38.

Nielsen, P.V. and Ríos, R. Inhibition of fungal growth on bread by volatile components from spices and herbs, and the possible application in active packaging, with special emphasis on mustard essential oil. *International Journal of Food Microbiology* 60 (2000): 219–229.

Nunes, C., Usall, J., Teixido, N., and Vinas, I. Improvement of *Candida sake* biocontrol activity against post-harvest decay by the addition of ammonium molybdate. *Journal of Applied Microbiology* 92 (2002a): 927–935.

Nunes, C., Usall, J., Teixidó, N., Torres, R., and Viñas, I. Control of *Penicillium expansum* and *Botrytis cinerea* on apples and pears with the combination of *Candida sake* and *Pantoea agglomerans*. *Journal of Food Protection* 65 (2002b): 178–184.

Núñez, F., Rodríguez, M.M., Córdoba, J.J., Bermúdez, M.E., and Asensio, M.A. Yeast population during ripening of dry-cured Iberian ham. *International Journal of Food Microbiology* 29 (1996): 271–280.

Oard, S., Rush, M.C., and Oard, J.H. Characterization of antimicrobial peptides against a US strain of the rice pathogen *Rhizoctonia solani*. *Journal of Applied Microbiology* 97 (2004): 169–180.

Oberparleiter, C., Kaiserer, L., Haas, H., Ladurner, P., Andratsch, M., and Marx, F. Active internalization of the *Penicillium chrysogenum* antifungal protein PAF in sensitive Aspergilli. *Antimicrobial Agents and Chemotherapy* 47 (2003): 3598–3601.

O'Reilly, C.E., O'Connor, P.M., Kelly, A.L., Beresford, T.P., and Murphy, P.M. Use of hydrostatic pressure for inactivation of microbial contaminants in cheese. *Applied and Environmental Microbiology* 66 (2000): 4890–4896.

Ortu, G., Scherm, B., Muzzu, A., Budroni, M., and Migheli, Q. Biocontrol activity of antagonistic yeasts against *Penicillium expansum* on apple. (Budapest, Hungary: *23rd International Specialised Symposium on Yeasts*, 2003), p. 200.

Parisien, A., Allain, B., Zhang, J.R., Mandeville, R., and Lan, C.Q. Novel alternatives to antibiotics: Bacteriophages, bacterial cell wall hydrolases, and antimicrobial peptides. *Journal of Applied Microbiology* 104 (2007): 1–13.

Patharajan, S., Reddy, K.R.N., Spadaro, D. et al. Potential of yeast antagonists on in vitro biodegradation of ochratoxin A. *Food Control* 22 (2011): 290–296.

Pérez-Nevado, F., Córdoba, M.G., Aranda, A., Martín, A., Andrade, M.J., and Córdoba, J.J. Killer activity of yeasts isolated from Spanish dry-cured ham. In *Modern Multidisciplinary Applied Microbiology*, ed. A. Mendez-Vilas (Weinheim, Germany: Wiley-VCH, 2006), pp. 232–235.

Péteri, Z., Téren, J., Vágvölgyi, C., and Varga, J. Ochratoxin degradation and adsorption caused by astaxanthin-producing yeasts. *Food Microbiology* 24 (2007): 205–210.

Petersson, S., Hansen, M.W., Axberg, K., Hult, K., and Schnürer, J. Ochratoxin A accumulation in cultures of *Penicillium verrucosum* with the antagonist yeast *Picchia anomala* and *Saccharomyces cerevisiae*. *Mycological Research* 102 (1998): 1003–1008.

Petersson, S. and Schnürer, J. Biocontrol of mold in high moisture wheat stored airtight conditions by *Pichia anomala*, *Pichia guillermondii* and *Saccharomyces cerevisiae*. *Applied and Environmental Microbiology* 61 (1995): 1027–1032.

Pimenta, R.S., Morais, P.B., Rosa, C.A., and Correa, A. Utilization of yeast in biological control programs. In *Yeast Biotechnology: Diversity and Applications*, ed. T. Satyanarayana and G. Kunze (Berlin, Germany: Springer, 2009), pp. 199–214.

Pombo, M.A., Rosli, H.G., Martínez, G.A., and Civello, P.M. UV-C treatment affects the expression and activity of defense genes in strawberry fruit (*Fragaria × ananassa*, Duch.). *Postharvest Biology and Technology* 59 (2011): 94–102.

Ponsone, M.L., Chiotta, M.L., Combina, M., Dalcero, A., and Chulze, S. Biocontrol as a strategy to reduce the impact of ochratoxin A and *Aspergillus* section *Nigri* in grapes. *International Journal of Food Microbiology* 151 (2011): 70–77.

Qin, G., Tian, S., and Xu, Y. Biocontrol of post-harvest diseases on sweet cherries by four antagonistic yeasts in different storage conditions. *Post-Harvest Biology and Technology* 31 (2004): 51–58.

Radler, F., Herzberger, S., Schonig, I., and Schwarz, P. Investigation of a killer strain of *Zygosaccharomyces bailli*. *Journal of General Microbiology* 139 (1993): 495–500.

Radler, F., Schmitt, M.J., and Meyer, B. Killer toxin of *Hanseniaspora uvarum*. *Archives of Microbiology* 154 (1990): 175–178.

Rodríguez, M., Núñez, F., Córdoba, J.J., Bermúdez, M.E., and Asensio, M.A. Evaluation of proteolytic activity of micro-organisms isolated from dry cured ham. *Journal of Applied Microbiology* 85 (1998): 905–912.

Rodríguez-Martín, A., Acosta, R., Liddell, S., Núñez, F., Benito, M.J., and Asensio, M.A. Characterization of the novel antifungal protein PgAFP and the encoding gene of *Penicillium chrysogenum*. *Peptides* 31 (2010a): 541–547.

Rodríguez-Martín, A., Acosta, R., Liddell, S., Núñez, F., Benito, M.J., and Asensio, M.A. Characterization of the novel antifungal chitosanase PgChP and the encoding gene from *Penicillium chrysogenum*. *Applied Microbiology and Biotechnology* 88 (2010b): 519–528.

Romanazzi, G., Feliziani, E., Santini, M., and Landi, L. Effectiveness of postharvest treatment with chitosan and other resistance inducers in the control of storage decay of strawberry. *Postharvest Biology and Technology* 75 (2013): 24–27.

Romanazzi, G., Karabulut, O.A., and Smilanick, J.L. Combination of chitosan and ethanol to control postharvest gray mold of table grapes. *Postharvest Biology and Technology* 45 (2007): 134–140.

Ryan, L.A., Dal Bello, F., Arendt, E.K., and Koehler, P. Detection and quantitation of 2,5-diketopiperazines in wheat sourdough and bread. *Journal of Agriculture and Food Chemistry* 57 (2009): 9563–9568.

Ryan, L.A., Zannini, E., Dal Bello, F., Pawlowska, A., Koehler, P., and Arendt, E.K. *Lactobacillus amylovorus* DSM 19280 as a novel food-grade antifungal agent for bakery products. *International Journal of Food Microbiology* 146 (2011): 276–283.

Samapundo, S., De Meulenaer, B., Atukwase, A., Debevere, J., and Devlieghere, F. The influence of modified atmospheres and their interaction with water activity on the radial growth and fumonisin B1 production of *Fusarium verticillioides* and *F. proliferatum* on corn. Part I: The effect of initial headspace carbon dioxide concentration. *International Journal of Food Microbiology* 114 (2007a): 160–167.

Samapundo, S., De Meulenaer, B., Atukwase, A., Debevere, J., and Devlieghere, F. The influence of modified atmospheres and their interaction with water activity on the radial growth and fumonisin B1 production of *Fusarium verticillioides* and *F. proliferatum* on corn. Part II: The effect of initial headspace oxygen concentration. *International Journal of Food Microbiology* 113 (2007b): 339–345.

Sánchez-Molinero, F. and Arnau, J. Processing of dry-cured ham in a reduced-oxygen atmosphere: Effects on sensory traits. *Meat Science* 85 (2010): 420–427.

Santos, A., Marquina, D., Barroso, J., and Peinado, J.M. $(1 \rightarrow 6)$-β-D-glucan as the cell wall binding site for *Debaryomyces hansenii* killer toxin. *Letters in Applied Microbiology* 34 (2002): 95–99.

Santos, A., Marquina, D., Leal, J.A., and Peinado, J.M. $(1 \rightarrow 6)$-β-D-glucan as cell wall receptor for *Pichia membranifaciens* killer toxin. *Applied and Environmental Microbiology* 66 (2000): 1809–1813.

Schatzmayr, G., Heidler, D., Fuchs, E. et al. Investigation of different yeast strains for the detoxification of ochratoxin A. *Mycotoxin Research* 19 (2003): 124–128.

Schmidt-Heydt, M., Stoll, D., and Geisen, R. Fungicides effectively used for growth inhibition of several fungi could induce mycotoxin biosynthesis in toxigenic species. *International Journal of Food Microbiology* 166 (2013): 407–412.

Schmitt, M. and Radler, F. Molecular structure of the cell wall receptor for killer toxin KT28 in *Saccharomyces cerevisiae*. *Journal of Bacteriology* 170 (1988): 2192–2196.

Schmitt, M.J. and Breinig, F. Yeast viral killer toxins: Lethality and self-protection. *Nature Reviews Microbiology* 4 (2006): 212–221.

Schmitt, M.J., Klavehn, P., Wang, J., Schönig, I., and Tipper, D.J. Cell cycle studies on the mode of action of yeast K28 killer toxin. *Microbiology* 142 (1996): 2655–2662.

Selitrennikoff, C.P. Antifungal proteins. *Applied and Environmental Microbiology* 67 (2001): 2883–2894.

Simoncini, N., Rotelli, D., Virgili, R., and Quintavalla, S. Dynamics and characterization of yeasts during ripening of typical Italian dry-cured ham. *Food Microbiology* 24 (2007): 577–584.

Skouri-Gargouri, H. and Gargouri, A. First isolation of a novel thermostable antifungal peptide secreted by *Aspergillus clavatus*. *Peptides* 29 (2008): 1871–1877.

Sousa, M.J., Ardöb, Y., and McSweeney, P.L.H. Advances in the study of proteolysis during cheese ripening. *International Dairy Journal* 11 (2001): 327–346.

Spadaro, D., Lorè, A., Garibaldi, A., and Gullino, M.L. A new strain of *Metschnikowia fructicola* for postharvest control of *Penicillium expansum* and patulin accumulation on four cultivars of apple. *Postharvest Biology and Technology* 75 (2013): 1–8.

Stotz, H.U., Thomson, J.G., and Wang, Y. Plant defensins: Defense, development and application. *Plant Signaling & Behavior* 4 (2009): 1010–1012.

Szappanos, H., Szigeti, G.P., Pál, B. et al. The *Penicillium chrysogenum*-derived antifungal peptide shows no toxic effects on mammalian cells in the intended therapeutic concentration. *Naunyn-Schmiedeberg's Archives of Pharmacology* 371 (2005): 122–132.

Taczman-Brückner, A., Mohácsi-Farkas, C., Balla, C., and Kiskó, G. Mode of action of *Kluyveromyces lactis* in biocontrol of *Penicillium expansum*. *Acta Alimentaria* 34 (2005): 153–160.

Tajkarimi, M.M., Ibrahim, S.A., and Cliver, D.O. Antimicrobial herb and spice compounds in food. *Food Control* 21 (2010): 1199–1218.

Takita, M.A. and Castilho-Valavicius, B. Absence of cell wall chitin in *Saccharomyces cerevisiae* leads to resistance to *Kluyveromyces lactis* killer toxin. *Yeast* 9 (1993): 589–598.

Taniwaki, M.H., Hocking, A.D., Pitt, J.I., and Fleet, G.H. Growth of fungi and mycotoxin production on cheese under modified atmospheres. *International Journal of Food Microbiology* 68 (2001): 125–133.

Taniwaki, M.H., Hocking, A.D., Pitt, J.I., and Fleet, G.H. Growth and mycotoxin production by fungi in atmospheres containing 80% carbon dioxide and 20% oxygen. *International Journal of Food Microbiology* 143 (2010): 218–225.

Theis, T., Marx, F., Salvenmoser, W., Stahl, U., and Meyer, V. New insights into the target site and mode of action of the antifungal protein of *Aspergillus giganteus*. *Research in Microbiology* 156 (2005): 47–56.

Theisen, S., Molkenau, E., and Schmitt, M.J. Wicaltin, a new protein toxin secreted by the yeast *Williopsis californica* and its broad-spectrum antimycotic potential. *Journal of Microbiology and Biotechnology* 10 (2000): 547–550.

Thevissen, K., Ferket, K.K., Francois, I.E., and Cammue, B.P. Interactions of antifungal plant defensins with fungal membrane components. *Peptides* 24 (2003): 1705–1712.

Thevissen, K., Ghazi, A., De Samblanx, G.W., Brownlee, C., Osborn, R.W., and Broekaert, W.F. Fungal membrane responses induced by plant defensins and thionins. *The Journal of Biological Chemistry* 271 (1996): 15018–15025.

Tian, S.P., Fan, Q., Xu, Y., and Liu, H.B. Biocontrol efficacy of antagonist yeasts to grey mold and blue mold on apples and pears in controlled atmospheres. *Plant Diseases* 86 (2002): 848–853.

Tripathi, P. and Dubey, N.K. Exploitation of natural products as an alternative strategy to control postharvest fungal rotting of fruit and vegetables. *Postharvest Biology and Technology* 32 (2004): 235–245.

Tsai, H. and Bobek, L.A. Human salivary histatins: Promising anti-fungal therapeutic agents. *Critical Reviews in Oral Biology & Medicine* 4 (1998): 480–497.

van den Tempel, T. and Jakobsen, M. The technological characteristics of *Debaryomyces hansenii* and *Yarrowia lipolytica* and their potential as starter cultures for production of Danablu. *International Dairy Journal* 10 (2000): 263–270.

van der Weerden, N.L. and Anderson, M.A. Plant defensins: Common fold, multiple functions. *Fungal Biology Reviews* 26 (2013): 121–131.

Var, I., Erginkaya, Z., and Kabak, B. Reduction of ochratoxin A levels in white wine by yeast treatments. *Journal of the Institute of Brewing* 115 (2009): 30–34.

Vernon, L.P. and Rogers, A. Binding properties of *Pyrularia thionin* and *Naja naja* kaouthia cardiotoxin to human and animal erythrocytes and to murine P388 cells. *Toxicon* 30 (1992): 711–721.

Vila, L., Lacadena, V., Fontanet, P., Martínez del Pozo, A., and San Segundo, B. A protein from the mold *Aspergillus giganteus* is a potent inhibitor of fungal plant pathogens. *Molecular Plant-Microbe Interactions* 14 (2001): 1327–1331.

Virgili, R., Simoncini, N., Toscani, T., Leggieri, M.C., Formenti, S., and Battilani, P. Biocontrol of *Penicillium nordicum* growth and ochratoxin A production by native yeasts of dry cured ham. *Toxins* 4 (2012): 68–82.

Walker, G.M., McLeod, A.H., and Hodgson, V.J. Interactions between killer yeasts and pathogenic fungi. *FEMS Microbiology Letters* 127 (1995): 213–222.

Wan, Y.K. and Tian, S.P. Integrated control of postharvest diseases of pear fruits using antagonistic yeasts in combination with ammonium molybdate. *Journal of the Science of Food and Agriculture* 85 (2005): 2605–2610.

Wang, X., Ma, P., Jiang, D., Peng, Q., and Yang, H. The natural microflora of Xuanwei ham and the no-mouldy ham production. *Journal of Food Engineering* 77 (2006): 103–111.

Weckbach, L.S. and Marth, E.H. Aflatoxin production by *Aspergillus parasiticus* in a competitive environment. *Mycopathologia* 62 (1977): 39–45.

Weiler, F., Rehfeldt, K., Bautz, F., and Schmitt, M.J. The *Zygosaccharomyces bailii* antifungal virus toxin zygocin: Cloning and expression in a heterologous fungal host. *Molecular Microbiology* 46 (2002): 1095–1105.

Weiler, F. and Schmitt, M.J. Zygocin, a secreted antifungal toxin of the yeast *Zygosaccharomyces bailii*, and its effect on sensitive fungal cells. *FEMS Yeast Research* 3 (2003): 69–76.

Weinberg, A., Quinones-Mateu, M.E., and Lederman, M.M. Role of human beta-defensins in HIV infection. *Advances in Dentist Research* 19 (2006): 42–48.

Wheatley, R.E. The consequences of volatile organic compound mediated bacterial and fungal interactions. *Antonie van Leeuwenhoek* 81 (2002): 357–364.

Wilson, C.L. and Wisniewski, M.E. Biological control of postharvest diseases of fruit and vegetables: An emerging technology. *Annual Review of Phytopathology* 27 (1989): 425–441.

Wisniewski, M., Biles, C., Droby, S., McLaughlin, R., Wilson, C., and Chalutz, E. Mode of action of the postharvest biocontrol yeast, *Pichia guilliermondii*: I. Characterization of attachment to *Botrytis cinerea*. *Physiological and Molecular Plant Pathology* 39 (1991): 245–258.

Wong, J.H., Ip, D.C.W., Ng, T.B., Chan, Y.S., Fang, F., and Pan, W.L. A defensin-like peptide from *Phaseolus vulgaris* cv. 'King Pole Bean'. *Food Chemistry* 135 (2012): 408–414.

Wong, J.H., Ng, T.B., Wang, H. et al. Cordymin, an antifungal peptide from the medicinal fungus *Cordyceps militaris*. *Phytomedicine* 18 (2010): 387–392.

Yamamoto, T., Hiratani, T., Hirata, H., Imai, M., and Yamaguchi, H. Killer toxin from *Hansenula mrakii* selectively inhibits cell wall synthesis in a sensitive yeast. *FEBS Letters* 197 (1986): 50–54.

Yang, E., Fan, L., Jiang, Y., Doucette, C., and Fillmore, S. Antimicrobial activity of bacteriocin-producing lactic acid bacteria isolated from cheeses and yogurts. *AMB Express* 2 (2012): 1–12.

Yang, E.J., Kim, Y.S., and Chang, H.C. Purification and characterization of antifungal delta-dodecalactone from *Lactobacillus plantarum* AF1 isolated from kimchi. *Journal of Food Protection* 74 (2011): 651–657.

Yiannikouris, A., Francois, J., Poughon, L. et al. Alkali-extraction of β-D-glucans from cell wall of *Saccharomyces cerevisiae* and study of their adsorptive properties toward zearalenone. *Journal of Agriculture and Food Chemistry* 52 (2004): 3666–3673.

Zahavi, T., Cohen, L., Weiss, B. et al. Biological control of *Botrytis*, *Aspergillus* and *Rhizopus* rots on table and wine grapes in Israel. *Postharvest Biology and Technology* 20 (2000): 115–124.

Zanetti, M. Cathelicidins, multifunctional peptides of the innate immunity. *Journal of Leukocyte Biology* 75 (2004): 39–48.

Zasloff, M. Antimicrobial peptides of multicellular organisms. *Nature* 415 (2002): 389–395.

Zorlugenç, B., Zorlugenç, F.K., Öztekin, S., and Evliya, I.B. The influence of gaseous ozone and ozonated water on microbial flora and degradation of aflatoxin B_1 in dried figs. *Food and Chemical Toxicology* 46 (2008): 3593–3597.

18

Smart/Intelligent Nanopackaging Technologies for the Food Sector

Semih Otles and Buket Yalcin

CONTENTS

18.1 Introduction

A multidisciplinary approach of food engineering will involve certain components such as predictive microbiology as a tool to improve and evaluate food safety in traditional and new processing technologies, advanced methods for the detection of food contaminants, advanced processing technologies, advanced systems for controlling recontamination, and advanced systems for intelligent and active packaging (López-Gómez et al., 2009).

The main function of food packaging is to maintain the safety and quality of food products during transportation and storage and to extend the shelf life of food products by preventing unfavorable conditions or factors such as chemical contaminants, spoilage microorganisms, light, moisture, oxygen, and external force. In order to accomplish these functions, packaging materials should create proper physicochemical conditions and offer physical protection for products that are essential for maintaining food safety and quality and obtaining a satisfactory shelf life. The food package should hinder loss or gain of moisture, prevent microbial contamination, and act as a barrier against permeation of water vapor, carbon dioxide, oxygen, and other volatile compounds such as flavors, as well as guarding against defects of basic properties of packaging materials such as thermal, mechanical, and optical properties (Brody, 2003; Brown and Willims, 2003; Marsh and Bugusu, 2007; Mauriello et al., 2005; Rhim et al., 2013; Singh and Anderson, 2004; Singh and Singh, 2005; Suppakul et al., 2003).

The study of food safety engineering should include intervention technologies (traditional and novel chemical intervention technologies, nonthermal interventions, and hurdle approach), control/monitoring/identification techniques (biosensors), packaging applications in food safety (active packaging, smart or intelligent packaging, tamper-evident packaging), and traceability and tracking systems (López-Gómez et al., 2009; Ramaswamy et al., 2007).

Nanotechnology applications in the food industry range from intelligent packaging to the creation of on-demand interactive food that allows consumers to modify food, depending on the nutritional tastes and needs (Neethirajan and Jayas, 2011). Recent years have witnessed a great expansion of technology and research developments in the field of nanotechnology resulting in significant developments in the application of agricultural and food areas. This is particularly the case in the food-packaging field, where significant advances in the nanoreinforcement of biobased materials offer a more solid ground toward increasing the economical and technical competitiveness of renewable polymers for different applications. Besides, there is still a long way to go, not only in the energy consumption minimization parcels and materials development, but also for the full characterization of any particular potential environmental and toxicological impacts of the widespread commercialization of these novel nanostructured biopolymers (Lagaron and Lopez-Rubio, 2011).

18.2 Traditional Food Packaging

Packaging materials have traditionally been chosen to avoid unwanted interaction with food and for convenience (Kuswandi et al., 2011; Rooney, 1995). Nearly all of the drink and food that we buy and consume is packaged in some way. The main functions of food packaging are to preserve and protect the food, to maintain its safety and quality, and to reduce food waste. There is no doubt that modern food-packaging technologies and materials fulfill these packaging functions and play a key role in helping offer a nutritious and safe food supply. Besides, the beverage and food industry is always seeking new technologies to further improve the safety, quality, traceability, and shelf life of their products (Bradley et al., 2011, Chaudhry et al., 2008). The general properties that food-packaging materials are required to have could be classified as antimicrobial function, gas barrier, vapor barrier, aroma barrier, environment-friendly, thermal properties, optical properties, and mechanical properties (Rhim et al., 2013). Another definition of traditional food packaging is protecting food from external influences, such as off-odors, microorganisms, light, and oxygen, thus guaranteeing the convenience of food handling and preserving the food quality for an extended time period. The key point in the safety objective of traditional materials in contact with foods is to be as inert as possible, such that there should be a minimum of interaction between the food and the packaging materials (Dainelli et al., 2008).

According to another definition, food packaging should protect the food from physical damage as well as from insects and dirt. Food packages should also be used to dispense food, be easy to handle, and have many other attributes linked to the physical characteristics of the packaging material. Food packaging should help maintain freshness and protect the food against spoilage by humidity, oxygen ingress, light, odor, and contamination or the loss of flavor components. With the increasing move toward light-weighting of materials and providing extended shelf life to reduce food waste and increase convenience, materials that are thin but that have high barrier properties are in great demand. Conventional packaging is intended to be largely *passive* in that it serves a preservation and protection role as a barrier from and to the external environment. Active packaging concepts, however, are intended to change the composition or the nature of the food or of the atmosphere that surrounds the food in the pack. Surface biocides should not be confused with active packaging. The biocidal agent is intended to help maintain the hygienic condition of the food contact surface by reducing or preventing microbial growth and helping *cleanability* for surface biocides. There should be no preservative effect on the food. Surface biocides might have a useful function in food-processing equipment such as poultry lines and food-handling equipment, which might have conveyor belts that are difficult to clean. They might also have a role to play in reusable food containers such as crates and boxes and the inner lining of freezers and refrigerators. Their relevance to single-use disposable packaging is, however, questionable (Bradley et al., 2011).

18.3 Food-Packaging Improvement

Food packages are not only used as containers but also act as protective barriers with some innovative functions. In this sense, food packaging is quite different from other durable goods such as home appliances, furniture, and electronics due to its safety aspects and relatively very short shelf life. Basic packaging materials, such as glass, metal, paper and paper-board, plastic, and a combination of materials of various physical structures and chemical natures, are used to fulfill the requirements and functions of packaged foods depending on their type. Besides, there has been ever-increasing effort in the development of various packaging materials in order to enhance their effectiveness in maintaining the food quality with improved convenience for processing and final use (Rhim et al., 2013). Consumer trends have changed consciously in recent years. The demand for the substantial necessity of elevated safety, natural quality, ready-to-eat, longer shelf life, and minimally processed food has turned out to be a matter of major importance. The increasing rate of food-borne illnesses and the increasing awareness of environmental conservation have driven the food industry to create a more innovative solution such as bioactive packaging (Imran et al., 2010).

Nanotechnology is the creation and subsequent utilization of structures with at least one dimension on the nanometer scale of less than 100 nm, which creates novel phenomena and properties that are otherwise not displayed by either isolated bulk materials or molecules. Specifically in the food biopackaging sector, nanocomposites generally refer to materials containing, typically, 1–7 wt.% of modified nanoclays (Lagaron and Lopez-Rubio, 2011; Lagaron et al., 2005; Shonaike and Advani, 2003). The improvement of nanotechnology, which involves the use and manufacture of materials in the size range of up to about 100 nm in one or more dimensions, has opened up new opportunities for the development of new materials with improved properties for use as food contact materials. Therefore, it is not surprising that one of the sectors eager to embrace nanotechnology and realize the potential benefits is the beverage and food industry. Several of the world's largest food companies have been reported to be actively exploring the potential of nanomaterials for use in food or food packaging (Bradley et al., 2011; Chaudhry et al., 2008).

Three main categories of nanotechnology are used in food-packaging applications: nanoparticle (NP)-reinforced packaging, active and intelligent packaging, and biodegradable nanocomposite food packaging. In NP-reinforced packaging, the high surface area allows the NPs to dramatically improve the mechanical performance, such as reduced gaseous permeation, ultraviolet flame and light resistance, stability of humidity and temperature, and flexibility of packaging materials with a relatively low mass content (~5%, w/w). Active and intelligent food packaging is a novel type of packaging compared with traditional methods. In this category, the *active* refers to packaging that has the ability to remove undesirable flavors and tastes, and improve the smell or color of the packed food. Biodegradable nanocomposite food packaging is the last category. It involves new types of biodegradable materials, which in general could be made of montmorillonite (MMT) and polylactic acid (PLA). MMT is a widely available and relatively cheap natural clay derived from volcanic rocks and ashes (Avella et al., 2005; Chang et al., 2003; Chow and Lok, 2009; Han et al., 2011; McGlashan and Halley, 2003).

In the twentieth century, packaging developments have incorporated oxygen scavengers and antimicrobials in packages, thus establishing new customs for protecting food from environmental influences and prolonging shelf life. These new packaging systems are called active packaging (Kuswandi et al., 2011; Mahalik, 2009).

The main focus of new applications so far appears to be on health food products and food packaging, with only a few known examples in the mainstream beverage and food areas. According to market estimates, food-packaging applications make up the largest share of the short-term and current predicted market for nano-enabled products in the food sector. The most promising growth areas identified for the near future include functional food products, smart and active packaging, and health foods (Chaudhry and Castle, 2011).

Polymers have supplied most of the common packaging materials because they present several desired features such as lightness, transparency, and softness. Besides, increased use of synthetic packaging films has led to a serious ecological problem due to their total nonbiodegradability. Although their complete replacement with eco-friendly packaging films is just impossible to achieve, at least for specific applications such as food packaging, the use of bioplastics should be the future (Siracusa et al., 2008). New technological improvements show that nanotechnology-derived polymer composites provide new lightweight but stronger food-packaging materials that could keep food products safe from microbial pathogens, fresh for longer during storage, and secure during transportation (Chaudhry and Castle, 2011). Important issues associated with the use of bioplastics, such as the nonintended migration of plastic components to foods, could also be potentially reduced by the use of these nanoclays, and, recently, they have also presented great advantages in the formulation of active packaging technologies based on bioplastics such as more efficient oxygen scavenging, and antioxidant or antimicrobial biopackaging, which have direct implications on increasing packaged food safety and quality (Busolo et al., 2009; Lagaron et al., 2007; Lopez-Rubio et al., 2006; Sanchez-Garcia et al., 2008).

Bio-nanocomposites consist of a biopolymer matrix reinforced with nanoparticles having at least one dimension in the nanometer range of 1–100 nm. Bio-nanocomposites are a new class of materials presenting much improved properties as compared to the base biopolymers due to the high surface area and high aspect ratio of nanoparticles. Thus, much effort has been dedicated to develop biopolymer-based packaging materials as bio-nanocomposites for food-packaging films with improved rheological, mechanical, thermal, as well as barrier properties (Akbari et al., 2007; Arora and Padua, 2010; Bordes et al., 2009; Pandey et al., 2005; Pavlidou and Papaspyrides, 2008; Peterse et al., 1999; Rhim and Ng, 2007; Sinha Ray and Bousmina, 2005; Sorrentino et al., 2007; Yang et al., 2007).

In the food-manufacturing process, the functional nanoparticles of the additive surface that are characterized by their high surface reactive property, small size (≤100 nm), and large surface area significantly improve food texture, flavor, and color. They also increase food bioavailability and nutrition absorption and provide human health benefits (Aitken et al., 2007; Bouwmeester et al., 2007; Han et al., 2011; Moraru et al., 2003; Torchilin, 2007).

From another perspective, food packaging is nowadays the major area of application of metal and metal oxide nanomaterials. Nanomaterials finding use in packaging include nano-zinc and nano-silver oxide for antimicrobial action, plastic–polymer composites with nanoclay for gas barrier, nano-titanium nitride for mechanical strength and as a processing aid, nano-titanium dioxide for UV protection, nano-silica for hydrophobic surface coating among others (Chaudhry and Castle, 2011).

In the area of food packaging, the nanocomposite, which combines traditional food-packaging materials with the NMs, has been subjected to high-speed development in the food-packaging market with its excellent mechanical performance and good antibacterial and stronger resistant properties (Alexandra and Dubois, 2000; Han et al., 2011).

The high surface-to-volume ratio of many nanoscale structures, which favors improved performance of composite materials, is also ideal for applications that involve drug delivery and chemical reactions. Examples of their usefulness include the burst and controlled release of substances in functional and active energy storage applications and food-packaging technologies (e.g., intelligent food packaging) (Lopez-Rubio et al., 2006; Shonaike and Advani, 2003).

The development of chemical sensors and biosensors over several decades has been investigated, resulting in novel and interesting devices with great promise for many areas of applications including food technology. The incorporation of such kind of sensors into food-packaging technology has resulted in what we call intelligent or smart packaging (Kuswandi et al., 2011).

18.4 Smart/Intelligent Packaging

Intelligent packaging is a development technology that uses the communication function of the package to facilitate decision making in obtaining the benefits of enhanced food quality and safety. Thinking out of the box might have resulted in the development of intelligent packaging (Otles and Yalcin, 2008a). Intelligent food packaging seems to have more extensive application than active packaging. Generally, the intelligence aspect of food packaging indicates the concept of monitoring information about the quality of the packed food. For instance, packaging materials incorporating nanocapsules or nanosensors based on nanotechnology will be able to detect food spoilage organisms and trigger a color change to alert the consumer that the shelf life is ending or has ended. This kind of function may also incorporate a concept of *release-on-command*, which will offer a basis for intelligent preservative packaging technology that will release a preservative if food begins to spoil. Additionally, packaging containing sensor arrays is another novel technology based on the concept of an *electronic tongue*, which may signal and monitor the condition of the packed food (Han et al., 2011; Nachay, 2007).

Smart packaging might be defined as a new approach for the food-packaging system. This system can adjust its condition according to environmental changes such as moisture and temperature. It can also warn the consumer when food is contaminated (Otles and Yalcin, 2008b). A package could be made smart through its functional attributes, which adds benefits to the food and hence the consumers. It is originally an integrating method that deals with chemical, mechanical, electrical, and/or electronically driven functions, which enhance the effectiveness or usability of the food products in a proven way. Some aspects of smart packaging are time–temperature food quality labels, usage of self-cooling or self-heating containers with electronic displays indicating use-by information and dates regarding the origin and nutritional qualities of the product in numerous languages (Goddard et al., 1997). A conventional package may be made smart by a radio frequency identification (RFID) tag. The functionality is electronic and the major beneficiaries are the stakeholders along the entire supply chain (Mahalik and Nambiar, 2010). Smart packaging takes advantage of biosensor or chemical sensor to monitor the safety and quality of food from the producers to the costumers. This technology may result in a kind of sensor design that is suitable for monitoring food safety and quality, such as leakage, freshness, oxygen, pathogens, carbon dioxide, temperature, pH, or time. Therefore, this technology is needed as a safety and online quality control in terms of authorities, food producers, and consumers, and has great potential in the development of new sensing systems integrated in food packaging, which are beyond the existing conventional technologies, such as control of volume, color, appearance, and weight (Kuswandi et al., 2011). On the other hand, smart packaging devices can be defined as economical, small labels or tags that are attached onto packaging to facilitate communication throughout the supply chain so that appropriate actions might be taken to achieve the desired benefits in food safety and quality enhancement (Otles and Yalcin, 2008b).

According to another definition, intelligent or smart packaging was defined as an inherent property or integral part of a product, pack, or product/pack configuration that confirms intelligence appropriate for use and the function of the product itself (Summers, 1992). Other definitions state that intelligent packaging is the package function that switches on and off in response to changing internal/external conditions, and may include a communication to the end users or customers regarding the status of the product (Butler, 2001; Kuswandi et al., 2011). While active packaging incorporates compelling ways to control microbial growth, moisture, and oxidation, smart packaging designs facilitate the monitoring of food quality (Kerry et al., 2006). Additionally, smartness in packaging is a broad term that covers a number of functionalities, depending on the product being packaged, including beverage, food, pharmaceutical, and various types of household and health products (Kuswandi et al., 2011).

Early smartness in packaging covered a number of functionalities, including those from intelligent packaging and active packaging. However, eventually there was an important distinction between package functions that are smart/intelligent (microbial growth indicators, time–temperature indicators [TTIs], etc.) and those that become active (antimicrobial, oxygen scavenging, etc.) in response to a triggering event (Butler, 2001). In 2003, intelligent packaging was still represented as a system that monitors the condition of packaged foods to give information about the quality of the packaged food during storage and transport (Murphy et al., 2003). Then, a smart packaging definition is done as inexpensive, small tags or labels that are attached to primary or secondary packaging, to facilitate communication throughout the supply chain (Yam et al., 2005). Besides, the technologies and smart components that combine only the indicators and sensors, such as the TTIs and freshness indicators, and give the package the intelligent function of monitoring but do not allow data flow between the food chain agents, have limitations of use (Kerry et al., 2006). With the Information and Communication Technologies (ICT), such as RFID and Surface Acoustic Waves (SAW), it is possible to reach a complete traceability and data flow without line of sight between packages and the other supply chain agents (Berkenpas et al., 2006; Chang et al., 2007; Clarke et al., 2006; López-Gómez et al., 2009; Regattieri et al., 2007). On the other hand, there is another definition denoting that smart packaging development should be considered as a part of multidisciplinary food safety engineering, as it involves different active and intelligent packaging systems (oxygen scavengers, carbon dioxide scavengers/emitters, ethylene scavengers, preservative releasers, ethanol emitters, moisture absorbers, flavor/odor absorbers, temperature control packaging, temperature-compensating films, RFID and SAW tags, etc.) that must be engineered (different materials, biopolymers, nanomatters, and devices in contact with food) to achieve adequate food safety (Dainelli et al., 2008; Day, 2008; López-Gómez et al., 2009).

18.5 Application on Food Sector

While most nanotechnology-derived food products are still at the research and development stage, applications for food packaging are closer to becoming a commercial reality (Bradley et al., 2011; Smolander and Chaudhry, 2010). Recent developments in packaging implementation systems applying lean principles to food packaging and processing require accurate information regarding packaging operations. Packaging Execution System (PES) (Arc Advisory Group, 2007) is a derivative of its parent, Manufacturing Execution System (MES), which is specifically geared toward an instinctive aim. PES integrates packaging operations into the organization's overall operation management system, thus providing a holistic view of the entire organization. It explains how MES is tailored to meet the requirements of the packaging and assembly areas, thus improving efficiency and traceability (Baliga, 1998). Packaging systems are rarely integrated with the existing MES. It was therefore proposed that an Equipment Manager System be developed and integrated with the existing MES to control packaging machinery (Cheng et al., 2000; E. F. Scientist, 2008; Mahalik and Nambiar, 2010).

Food-packaging applications form the largest share of the short-term and current predicted market for nano-enabled products in the food sector. It has been estimated that nanotechnology-derived packaging, which includes food packaging, will make up to 19% of the share of nanotechnology-enabled applications and products in the global consumer goods industry by 2015 (Nanoposts Report, 2008). Nanomaterials' incorporation in plastic polymers has led to the development of novel or improved food-packaging materials (Chaudhry and Castle, 2011):

- Active packaging based on polymers incorporating nanomaterials with antimicrobial properties
- Nanobiosensors for smart packaging concepts
- Hydrophobic nano-coatings for self-cleaning surfaces and active nano-coatings for hygienic food contact materials and surfaces
- Polymer nanocomposites with improved properties in terms of durability, gas-barrier properties, moisture/temperature stability, flexibility

Nano-silica for surface coating, gas barrier, nano-titanium dioxide for UV protection, nano-zinc and nano-silver oxide for antimicrobial action, nano-titanium nitride for mechanical strength and as a processing aid are some examples of plastic polymers with nanoclays. The use of nanocomposites with biopolymers is expected to rise because they present the possibility of carbon-neutral biodegradable materials for packaging. This will present opportunities for developing countries to utilize their forestry and agricultural resources, wastes, and by-products for the development of biopolymer nanocomposites. Food-packaging applications also function at two levels (Chaudhry and Castle, 2011):

- Raw material manufacture, which will indicate a greater technical know-how, capacity, and infrastructure.
- Applications/uses of the material will not indicate large capacity or high technology. This will suit many developing countries, where new developments could be taken up by small and medium enterprises.

According to another definition, packaging applications of nanomaterials could be classified as intelligent, active packaging, nano-coatings, surface biocides, and nanocomposites.

Intelligent packaging: Incorporating nanosensors to report and monitor on the condition of the food.

Active packaging: Incorporating nanomaterials with antimicrobial or other properties such as antioxidant with intentional release into and consequent effect on the packaged food.

Nano-coatings: Incorporating nanomaterials into the packaging surface (either sandwiched as a layer in a laminate or the inside or outside surface) to especially improve barrier properties.

Surface biocides: Incorporating nanomaterials with antimicrobial properties acting on the packaging surface.

Nanocomposites: Incorporating nanomaterials into the packaging to improve barrier properties, physical performance, biodegradation, and durability (Bradley et al., 2011).

Concerns about these kinds of applications could also include the following:

- Possible migration into drinks and foods causing a toxicological risk
- Fate during recycling and recovery to make new packaging materials
- Fate in the environment after disposal of the packaging (Bradley et al., 2011)

18.5.1 Samples

Nanosensors can provide quality assurance by tracking contaminants, toxins, and microbes throughout the food-processing chain through data capture for automatic control of documentation and functions. Nanotechnology also allows the implementation of low-cost nanosensors in food packaging to monitor the quality of food during several stages of the logistic process to guarantee product quality up until consumption (Neethirajan and Jayas, 2011). An electronic tongue is developed for inclusion in food packaging that is composed of an array of nanosensors that are extremely sensitive to gases released by food when it spoils, causing the sensor strip to change color as a result, giving a visible signal of whether the food is fresh or not (Ruengruglikit et al., 2004). A freshness indicator material has been developed for the detection of *Escherichia coli* O157 enterotoxin with the possibility of applying it for the detection of other toxins as well. This indicator might be based on a color change of chromogenic substrates of enzymes produced by contaminating microbes or on the detection of microorganisms (Otles and Yalcin, 2008a).

PLA biocomposites, obtained by solvent casting, contain a novel silver-based antimicrobial-layered silicate additive for use in active food-packaging applications. The silver-based nanoclay demonstrated strong antimicrobial activity against gram-negative *Salmonella* spp. The films had strong biocidal properties and were highly transparent with enhanced water barrier (Busolo et al., 2010). Sliced meat, bakery, and cheese-like foods that are prone to spoiling on the surface could be protected by contact packaging replenished with antimicrobial nanoparticles. An antifungal active paper packaging has been developed that incorporates cinnamon oil with solid wax paraffin using nanotechnology as an active coating was shown to be an effective packaging material for bakery products (Rodriguez et al., 2008). Edible food films that are working with apple puree and oregano oil have been created that are able to kill *E. coli* bacteria (Rojas-Grau et al., 2006). Antimicrobial nanoparticles that have been tested and synthesized for applications in food storage boxes and antimicrobial packaging include nisin particles produced from the fermentation of bacteria (Gadang et al., 2008), silver oxide nanoparticles (Sondi and Salopek-Sondi, 2004), and magnesium oxide and zinc oxide nanoparticles (Jones et al., 2008).

Other examples include self-cooling beer that uses zeolite technology, making it possible to drink cool beer anywhere; a self-heating coffee container based on the CaO exothermic reaction is also becoming available. Another assessment that works for smart packaging is based on changing some characteristics of the materials coated inside the package (Goddard et al., 1997). Additionally, meat packaging includes new synthetic material that has been developed with silver nanoparticles. This material offers a longer shelf life and antibacterial properties for meat (Otles and Yalcin, 2008b).

Nanoparticles are ideal particles for active packaging as they cover a very large surface area and thus have huge potential for trapping or releasing chemicals. Examples include nanoparticles used for scavenging purposes, removing oxygen or odor and taint chemicals from within the pack. As an alternative, nano-encapsulates might be used to release additives like colors or preservatives onto the food surface where the additive is needed. This might reduce the amount of chemical additive needed, compared to having to distribute the additive throughout the whole food (Bradley et al., 2011).

Biosensors, chemical sensors, RFID, ripeness indicators, and TTIs are examples of components in smart packaging. Most of these smart devices have not had widespread commercial application, but the two that are gaining popularity are RFID and TTIs (Kuswandi et al., 2011).

The use of nano-silver is mainly for packaging and health food applications, but its use as an additive in antibacterial wheat flour is the subject of a recent patent application (Chaudhry and Castle, 2011; Park, 2005). Nano-silica is reported to be used in food-packaging applications and food contact surfaces, and as a free-flowing agent in powdered soups and some reports suggest its use in clarifying wines and beers (Chaudhry and Castle, 2011).

Generally, the occurrence of an increased CO_2 gas level is the prime indicator of food spoilage in packed foods; moreover, its maintenance at optimal levels is essential to avoid spoilage in food packaged under modified-atmosphere packaging (MAP) conditions. Therefore, a CO_2 sensor combined with the food package may efficiently monitor product quality till it reaches the consumer. Although much progress has been made in the development of sensors monitoring CO_2 so far, most of them are not versatile and suffer from limitations such as energy input requirement, bulkiness, high equipment cost, and safety concerns. Moreover, the development of efficient CO_2 sensors that might intelligently monitor the gas concentration changes inside a food package and specific to food-packaging applications is essential. Conventional and innovative optical CO_2 sensors (wet optical CO_2 indicators [pH-based], fluorescent CO_2, dry optical CO_2 sensors, sol–gel-based optical CO_2 sensor, photonic crystal sensors, polymer opal films, polymer hydrogel–based CO_2 sensor, color-changing metals) are used in intelligent food applications (Puligundla et al., 2012).

A UV-activated colorimetric oxygen indicator was developed that uses nanoparticles of TiO_2 to photosensitize the reduction of methylene blue by triethanolamine in a polymer encapsulation medium, using UVA light. Upon UV irradiation, the sensor bleaches and remains colorless, till it is exposed to oxygen, when it regains its original blue color. The rate of color recovery is proportional to the level of oxygen exposure (Lee et al., 2002). Nanocrystalline SnO_2 was used as a photosensitizer in a colorimetric O_2 indicator, with the color of the film varying depending on the O_2 exposure (Mills and Hazafy, 2009). Moreover, pH indicators based on organically modified silicate nanoparticles have been recently established (Jurmanović et al., 2010; Silvestre et al., 2011).

18.5.2 Benefits

Advantages and benefits of nanomaterials and nanotechnology in food packaging for the developed and the developing world are reflected in the near-market or actual applications in innovation, light-weighting, greater preservation and protection of food, and improved performance of biobased materials. These applications could be defined as follows:

Innovation: The main driver for applications of nanomaterials in food-packaging materials is new product development and innovation. New products can give greater consumer convenience and choice. New products can support lifestyles and social change. New products might create employment, open new markets, and drive economic growth.

Light-weighting: Using less packaging material but with the same technical performance presents lower material usage. This could give a lower carbon/environmental footprint from the transport and manufacture of the packaged food and the packaging.

Greater preservation and protection of food: Better barrier properties may help increase shelf life and maintain food quality without additional chemical preservatives. This might provide potentially more reliable and cheaper food supply, better nutrition, and less food waste.

Improved performance of biobased materials: The performance of biobased materials could be inferior to synthetic polymers based on gas or oil coal. If the technical performance can be boosted by using nanotechnology tools to improve production methods and study material properties or by incorporating nanomaterials, this might enable synthetic polymers to be substituted by locally sourced biobased materials (Bradley et al., 2011).

18.5.3 Safety and Risk Assessments

The main risk of consumer exposure to nanoparticles from food packaging is likely to be through the potential migration of nanoparticles into drink and food. Besides, migration experimental data are not available, despite the fact that some food-packaging types containing nanoparticles are already in commercial use and available in some countries. Migration studies of nanoparticles in food contact materials are relative rare. The difficulties in characterizing nanoparticles in composites and the lack of methods for quantitative and qualitative analysis are responsible for this limited research on nanoparticle migration in nanocomposites. On the other hand, according to the scientific report from the European Centre for Ecotoxicology and Toxicology of Chemicals, there are three main routes for human exposure to nanoparticles: dermal contact, oral ingestion, and inhalation (Paul et al., 2006). In general terms, smaller nanoparticles have a higher possibility of affecting the epithelium of the lung structure. Besides, studies are increasingly confirming that the nanoparticles taken up by the skin remain on the surface of the stratum corneum or accumulate in the follicle orifices; whichever penetration pathway they take, they do not penetrate the follicle or the skin (Alvarez-Roman et al., 2004; Avella et al., 2005; Han et al., 2011; Pflücker et al., 2001). The migration of nanoparticles from packaging to packed food raises public concern and has been corroborated by in vitro cell experiments and animal oral administration. Generally, two kinds of mechanism could be adopted to explain the toxicity effects on humans. One is that the toxicity is independent of the nanoparticles, and might be realized by generating reactive oxygen species (ROS) within the cells. The other is that the toxicity has a strong relationship with the chemical component of the nanoparticles (Han et al., 2011).

18.6 Conclusion

Intelligent and smart food-packaging applications in food-packaging systems are innovative technologies that have been developed in recent years. These kinds of development are becoming an alternative to enhance the performance of packaging materials and systems. In intelligent and smart packaging, there are two points to be considered: the first is a system outside of the package and the second includes nanomaterials incorporated within the packages. The aim of these new technologies could be briefly classified as follows: extend shelf life, enhance safety, improve quality and packaging, enhance absorption of nutrients, warn consumers about problems, and control external and internal conditions of food products that are packed. Food-packaging applications form the largest share of the short-term and current predicted market for nano-enabled products. The safety assessment of nanomaterials and nanoparticles should be considered before they become a part of the daily life of people. It is believed that the invention of novel detection techniques will give nanomaterials a justified evaluation in the future by developing high-tech products in the food-packaging applications. Talking about nanocomposite food contact materials, nanotechnology is already one of the most powerful forces for innovations with this kind of materials. The reduction of weight and the enhancement of the barrier and mechanical properties of polymers and more attractive materials are some benefits of these applications.

REFERENCES

Aitken, R.J., Hankin, S.M., Tran, C.L., Donaldson, K., Stone, V., Cumpson, C., Johnstone, J., Chaudhry, Q., Cash, S. 2007. REFNANO: Reference materials for engineered nanoparticle toxicology and metrology, Final report on Project CB01099, www.iom-world.org.

Akbari, Z., Ghomashchi, T., Moghadam, S. 2007. Improvement in foodpackaging industry with biobased nanocomposites. *International Journal of Food Engineering*, 3(4): 1–24 (article 3).

Alexandra, M., Dubois, P. 2000. Polymer-layered silicate nanocomposites: Preparation, properties and uses of a new class of materials. *Material Science and Engineering: R*, 28: 1–63.

Alvarez-Roman, R., Naik, A., Kalia, Y.N., Guya, R.H., Fessi, H. 2004. Skin penetration and distribution of polymeric nanoparticles. *Journal of Controlled Release*, 99: 53–62.

Arc Advisory Group. 2007. Packaging execution systems benefit lean manufacturing initiatives. http://www.systech-tips.com/pes/pdfs/pes-english.pdf. Accessed August 2013.

Arora, A., Padua, G.W. 2010. Review: Nanocomposite in food packaging. *Journal of Food Science*, 75(1): 43–49.

Avella, M., De Vlieger, J.J., Errico, M.E., Fischer, S., Vacca, P., Volpe, M.G. 2005. Biodegradable starch/clay nanocomposite films for food packaging applications. *Food Chemistry*, 93: 467–474.

Baliga, J. 1998. Packaging foundry employes MES to manage production. *Semiconductor International*, 21(2): 46.

Berkenpas, E., Millard, P., Pereira, D.C. 2006. Detection of *Escherichia coli* O157:H7 with langasite pure shear horizontal surface acoustic wave sensors. *Biosensors and Bioelectronics*, 21: 2255–2262.

Bordes, P., Pollet, E., Avérous, L. 2009. Nano-biocomposites: Biodegradable polyester/nanoclay systems. *Progress in Polymer Science*, 20: 125–155.

Bouwmeester, H., Dekkers, S., Noordam, M., Hagens, W., Bulder, A., de Heer, C., ten Voorde, S., Wijnhoven, S., Sips, A. 2007. Health impact of nanotechnologies in food production. Technical Report, RIKILT— Institute of Food Safety, Wageningen University and Research Centre, Wageningen, the Netherlands.

Bradley, E.L., Castle, L., Chaudhry, Q. 2011. Applications of nanomaterials in food packaging with a consideration of opportunities for developing countries. *Trends in Food Science & Technology*, 22: 604–610.

Brody, A. 2003. Nanocomposite technology in food packaging. *Food Technology*, 61(10): 80–83.

Brown, H., Willims, J. 2003. Packaged product quality and shelf life. In: Coles, R., Mcdowell, D., Kirwan, M.J. (eds.), *Food Packaging Technology*. Oxford, U.K.: Blackwell Publishing Ltd., pp. 65–94.

Busolo, M.A., Fernandez, P., Ocio, M.J., Lagaron, J.M. 2010. Novel silver-based nanoclay as an antimicrobial in polylactic acid food packaging coatings. *Food Additives and Contaminants*, 27(11): 1617–1626 (Taylor & Francis Group).

Busolo, M.A., Ocio, M.J., Lagaron, J.M. 2009. Development of antimicrobial PLA nanocomposites with silver containing layered nanoclays for packaging and coating applications. *Conference Proceedings: Annual Technical Conference (ANTEC)*, Chicago, IL, pp. 236–239.

Butler, P. 2001. Smart packaging—Intelligent packaging for food, beverages, pharmaceuticals and household products. *Materials World*, 9(3): 11–13.

Chang, J.H., An, Y.U., Cho, D., Giannelis, E.P. 2003. Poly(lactic acid) nanocomposites: Comparison of their properties with montmorillonite and synthetic mica (II). *Polymer*, 44: 3715–3720.

Chang, K.S., Chang, C.K., Chen, C.Y. 2007. A surface acoustic wave sensor modified from a wireless transmitter for the monitoring of the growth of bacteria. *Sensors and Actuators B: Chemical*, 125: 207–213.

Chaudhry, Q., Castle, L. 2011. Food applications of nanotechnologies: An overview of opportunities and challenges for developing countries. *Trends in Food Science & Technology*, 22: 595–603.

Chaudhry, Q., Scotter, M., Blackburn, J., Ross, B., Boxall, A., Castle, L., Aitken, R., Watkins, R. 2008. Applications and implications of nanotechnologies for the food sector. *Food Additives and Contaminants*, 25: 241–258.

Cheng, F.T., Yang, H.C., Kuo, T.L., Feng, C., Jeng, M. 2000. Modeling and analysis of equipment managers in manufacturing execution systems for semiconductor packaging. *IEEE Transactions on Systems, Man, and Cybernetics, Part B: Cybernetics*, 30(5): 772–782.

Chow, W.S., Lok, S.K. 2009. Thermal properties of poly(lactic acid)/organo-montmorillonite nanocomposites. *Journal of Thermal Analysis and Calorimetry*, 95: 627–632.

Clarke, R.H., Twede, D., Tazelaar, J.R., Boyer, K.K. 2006. Radio Frequency Identification (RFID) performance: The effect of tag orientation and package contents. *Packaging Technology and Science*, 19: 45–54.

Dainelli, D., Gontard, N., Spyropoulos, D., Zondervan-van den Beukend, E., Tobback, P. 2008. Active and intelligent food packaging: Legal aspects and safety concerns. *Trends in Food Science & Technology*, 19: 103–112.

Day, B.P.F. 2008. Active packaging of food. In: Kerry, J., Butler, P. (eds.) *Smart Packaging Technologies*. West Sussex, England: John Wiley & Sons Ltd, pp. 1–325.

E.F. Scientist. 2008. Manufacturing execution system boosts accuracy and transparency. *E.F. Scientist*.

Gadang, V.P., Hettiarachchy, N.S., Johnson, M.G., Owens, C. 2008. Evaluation of antibacterial activity of whey protein isolate coating incorporated with Nisin, grape seed extract, malic acid, and EDTA on a turkey frankfurter system. *Journal of Food Science*, 73(8): 389–394.

Goddard, N.D.R., Kemp, R.M., Lane, R. 1997. An overview of smart technology. *Packaging Science and Technology*, 10: 129–143.

Han, W., Yu, Y.J., Li, N.T., Wang, L.B. 2011. Application and safety assessment for nano-composite materials in food packaging. *Chinese Science Bulletin*, 56: 1216–1225. doi: 10.1007/s11434-010-4326-6.

Imran, M., Revol-Junelles, A.M., Martyn, A., Tehrany, E.A., Jacquot, M., Linder, M., Desobry, S. 2010. Active food packaging evolution: Transformation from micro- to nanotechnology. *Critical Reviews in Food Science and Nutrition*, 50: 799–821 (Copyright C@ Taylor & Francis Group, LLC).

Jones, N., Ray, B., Ranjit, K.T., Manna, A.C. 2008. Antibacterial activity of ZnO nanoparticle suspensions on a broad spectrum of microorganisms. *FEMS Microbiology Letters*, 279: 71–76.

Jurmanović, S., Kordić, S., Steinberg, M.D., Murković Steinberg, I. 2010. Organically modified silicate thin films doped with colorimetric pH indicators methyl red and bromocresol green as pH responsive sol–gel hybrid materials. *Thin Solid Films*, 518: 2234–2240.

Kerry, J.P., O'Grady, M.N., Hogan, S.A. 2006. Past, current and potential utilization of active and intelligent packaging systems for meat and muscle-based products: A review. *Meat Science*, 74: 113–130.

Kuswandi, B., Wicaksono, Y., Jayus, J., Abdullah, A., Heng, L.Y., Ahmad, M. 2011. Smart packaging: Sensors for monitoring of food quality and safety. *Sensing and Instrumentation for Food Quality and Safety*, 5: 137–146.

Lagaron, J.M., Cabedo, L., Cava, D., Feijoo, J.L., Gavara, R., Gimenez, E. 2005. Improving packaged food quality and safety. Part 2: Nanocomposites. *Food Additives and Contaminants*, 22: 994–998.

Lagaron, J.M., Gimenez, E., Cabedo, L. 2007. Preparing laminar nanocomposites that contain intercalated organic materials, useful e.g. for reinforcing plastic packaging and for controlled release of pharmaceuticals. Patent number WO2007074184-A1.

Lagaron, J.M., Lopez-Rubio, A. 2011. Nanotechnology for bioplastics: Opportunities, challenges and strategies. *Trends in Food Science & Technology*, 22: 611–617.

Lee, S.W., Mao, C., Flynn, C.E., Belcher, A.M. 2002. Ordering of quantum dots using genetically engineered viruses. *Science*, 296: 892–895.

López-Gómez, A., Fernández, P.S., Palop, A., Periago, P.M., Martinez-López, A., Marin-Iniesta, F., Barbosa-Cánovas, G.V. 2009. Food safety engineering: An emergent perspective. *Food Engineering Reviews*, 1: 84–104. doi: 10.1007/s12393-009-9005-5.

Lopez-Rubio, A., Gavara, R., Lagaron, J.M. 2006. Bioactive packaging: Turning foods into healthier foods through biomaterials. *Trends in Food Science & Technology*, 17(10): 567–575.

Mahalik, N.P. 2009. Processing and packaging automation systems: A review. *Sensing and Instrumentation for Food Quality and Safety*, 3: 12–25.

Mahalik, N.P., Nambiar, A.N. 2010. Trends in food packaging and manufacturing systems and technology. *Trends in Food Science & Technology*, 21: 117–128.

Marsh, K., Bugusu, B. 2007. Food packaging—Roles, materials, and environ-mental issues. *Journal of Food Science*, 72(3): 38–55.

Mauriello, G., De Luca, E., La Satoria, A., Villani, F., Ercolini, D. 2005. Antimicrobial activity of a nisin-activated plastic film for food packaging. *Letters in Applied Microbiology*, 41: 464–469.

McGlashan, S.A., Halley, P.J. 2003. Preparation and characterisation of biodegradable starch-based nanocomposite materials. *Polymer International*, 52: 1767–1773.

Mills, A., Hazafy, D. 2009. Nanocrystalline SnO_2-based, UVB-activated, colourimetric oxygen indicator. *Sensors and Actuators B*, 136: 344–349.

Moraru, C.I., Panchapakesan, C.P., Huang, Q., Takhistov, P., Liu, S., Kokini, J.L. 2003. Nanotechnology: A new frontier in food science. *Food Technology*, 57: 24–29.

Murphy, A., Millar, N., Cuney, S. 2003. Active and intelligent packaging. The kitchen of the future. Presentation to innovation day, Cambridge Consultants Ltd. www.CambridgeConsultants.com. Accessed November 4, 2003.

Nachay, K. 2007. Analyzing nanotechnology. *Food Technology*, 61: 34–36.

Nanoposts Report. 2008. Nanotechnology and consumer goods e Market and applications to 2015, 2008. Nanoposts.com.

Neethirajan, S., Jayas, D.S. 2011. Nanotechnology for the food and bioprocessing industries. *Food and Bioprocess Technology*, 4: 39–47.

Otles, S., Yalcin, B. 2008a. Intelligent food packaging. *LogForum*, 4(Issue 4, No. 3): 1–9.

Otles, S., Yalcin, B. 2008b. Smart food packaging. *LogForum*, 4(Issue 3, No. 4): 1–7.

Pandey, J.K., Kumar, A.P., Misra, M., Mohanty, A.K., Drzal, L.T., Singh, R.P. 2005. Recent advances in biodegradable nanocomposites. *Journal for Nanoscience and Nanotechnology*, 5: 497–526.

Park, K.H. 2005. Preparation method antibacterial wheat flour by using silver nanoparticles. Korean Intellectual Property Office (KIPO). Publication number/date 1020050101529A/24.10.2005.

Paul, J.A.B., David, R., Stephan, H., Kuhlbusch, T., Fissan, H., Donaldson, K., Schins, R. et al. 2006. The potential risks of nanomaterials: A review carried out for ECETOC. *Particle and Fibre Toxicology*, 3: 1–35.

Pavlidou, S., Papaspyrides, C.D. 2008. A review on polymer-layered silicate nanocomposites. *Progress in Polymer Science*, 33: 1119–1198.

Peterse, K., Nielsen, P.V., Bertelsen, G., Lawther, M., Olsen, M.B., Nilsson, N.H., Mortensen, G. 1999. Potential of biobased materials for food packaging. *Trends in Food Science & Technology*, 10: 52–68.

Pflücker, F., Wendel, V., Hohenberg, H., Gärtner, E., Will, T., Pfeiffer, S., Wepf, R., GersBarlag, H. 2001. The human stratum corneum layer: An effective barrier against dermal uptake of different forms of topically applied micronised titanium dioxide. *Skin Pharmacology and Applied Skin Physiology*, 14(Suppl. 1): 92–97.

Puligundla, P., Jung, J., Ko, S. 2012. Carbon dioxide sensors for intelligent food packaging applications. *Food Control*, 25: 328–333.

Ramaswamy, R., Ahn, J., Balasubramaniam, V.M., Rodríguez-Saona, L., Yousef, A.E. 2007. Food safety engineering. In: Kutz, M. (ed.), *Handbook of Farm, Dairy and Food Machinery*. New York: William Andrew Publishing Inc, pp. 45–70.

Regattieri, A., Gamberi, M., Manzini, R. 2007. Traceability of food products: General framework and experimental evidence. *Journal of Food Engineering*, 81: 347–356.

Rhim, J.W., Ng, P.K. 2007. Natural biopolymer-based nanocomposite films for packaging applications. *Critical Reviews in Food Science and Nutrition*, 47: 411–433.

Rhim, J.W., Park, H.M., Ha, S.C. 2013. Bio-nanocomposites for food packaging applications. *Progress in Polymer Science* 38, 10–11, 1629–1652.

Rodriguez, A., Nerin, C., Batlle, R. 2008. New cinnamon-based active paper packaging against Rhizopusstolonifer food spoilage. *Journal of Agricultural and Food Chemistry*, 56(15): 6364–6369.

Rojas-Grau, M.A, Bustillos, A.R.D, Friedman, M., Henika, P.R., Martin-Belloso, O., Mc Hugh, T.H. 2006. Mechanical, barrier and antimicrobial properties of apple puree edible films containing plant essential oils. *Journal of Agricultural and Food Chemistry*, 54: 9262–9267.

Rooney, M.L. 1995. Overview of active food packaging. In: Rooney, M.L. (ed.), *Active Food Packaging*. New York: Chapman & Hall, pp. 1–33.

Ruengruglikit, C., Kim, H., Miller, R.D., Huang, Q. 2004. Fabrication of nanoporous oligonucleotide microarrays for pathogen detection and identification. *Polymer Preprints*, 45: 526.

Sanchez-Garcia, M.D., Gimenez, E., Lagaron, J.M. 2008. Morphology and barrier properties of solvent cast composites of thermoplastic biopolymers and purified cellulose fibers. *Carbohydrate Polymers*, 71(2): 235–244.

Shonaike, G.O., Advani, S.G. (eds.). 2003. *Advanced Polymeric Materials: Structure Property Relationships*. Boca Raton, FL: CRC Press.

Silvestre, C., Duraccio, D., Cimmino, S. 2011. Food packaging based on polymer nanomaterials. *Progress in Polymer Science*, 36: 1766–1782.

Singh, R.K., Singh, N. 2005. Quality of packaged food. In: Han, J.H. (ed.) *Innovations in Food Packaging*. San Diego, CA: Elsevier Academic Press, pp. 24–44.

Singh, R.P., Anderson, B.A. 2004. The major types of food spoilage: An overview. In: Woodhead, S.R. (ed.), *Understanding and Measuring the Shelf-Life of Food*. Cambridge, U.K.: Woodhead Publishing Ltd., pp. 3–23.

Sinha Ray, S., Bousmina, M. 2005. Biodegradable polymers and their layered silicate nanocomposites: In greening the 21st century materials world. *Progress in Materials Science*, 50: 962–1079.

Siracusa, V., Rocculi, P., Romani, S., Rossa, M.D. 2008. Biodegradable polymers for food packaging: A review. *Trends in Food Science & Technology*, 19: 634–643.

Smolander, M., Chaudhry, Q. 2010. Nanotechnologies in food packaging. In: Chaudhry, Q., Castle, L., Watkins, R. (eds.), *Nanotechnologies in Food*. London, U.K.: Royal Society of Chemistry Publishers, 101pp., ISBN 978-0-85404-169-5, 86e.

Sondi, I., Salopek-Sondi, B. 2004. Silver nanoparticles as antimicrobial agent: A case study on *E. coli* as a model for Gram-negative bacteria. *Journal of Colloid and Interface Science*, 275: 177–182.

Sorrentino, A., Gorrasi, G., Vittoria, V. 2007. Potential perspectives of bio-nanocomposites for food packaging applications. *Trends in Food Science & Technology*, 18: 84–95.

Summers, L. 1992. *Intelligent Packaging*. London, U.K.: Centre for Exploitation of Science & Technology.

Suppakul, P., Miltz, J., Sonnenveld, K., Bigger, S.W. 2003. Active packaging technologies with an emphasis on antimicrobial packaging and its applications. *Journal of Food Science*, 68: 408–420.

Torchilin, V.P. 2007. Targeted pharmaceutical nanocarriers for cancer therapy and imaging. *AAPS Journal*, 9: E128–E147.

Yam, K.L., Takhistov, P.T., Miltz, J. 2005. Intelligent packaging: Concept and applications. *Journal of Food Science*, 70(1): 1–10.

Yang, K.K., Wang, X.L., Wang, Y.Z. 2007. Progress in nanocomposite of biodegradable polymer. *Journal of Industrial and Engineering Chemistry*, 13: 485–500.

19

Plant Extracts as Natural Antimicrobials in Food Preservation

Yasmina Sultanbawa

CONTENTS

19.1 Introduction: Background and Consumer Needs

Microbial safety of food and preservation by reduction of food spoilage throughout the supply chain still remains a challenge to food industries, regulatory agencies, and consumers (Negi, 2012). The concern of the consumer regarding synthetic chemical additives in food, awareness of clean labels and preservation using natural additives, underpins the importance of finding alternative sources of safe and effective preservatives (Burt, 2004). Plants fulfill this requirement as herbs, and spices have been used traditionally to preserve food for long periods of time in different cultures (Tajkarimi et al., 2010). Plant extracts with known antimicrobial activity alone will not extend the shelf life of food, but they in combination with other technologies such as processing and packaging can preserve food for longer periods of time. The application of plant extracts as natural preservatives has the greatest potential in fresh food such as fresh-cut salads, marinated chilled fish and meat products, and ready-to-eat or ready-to-cook food as the consumer is looking for convenient food and has the desire to lead a healthy lifestyle. An extended storage life for fresh food, which is highly perishable, is also beneficial to the food retailer as it gives a longer period on the shelf. However, growth of the fresh food market adds new risks due to the changes in food production practices and distribution chains (Havelaar et al., 2010). Plant extracts as an emerging technology to extend storage life of food can be used by the industry to address the needs of today's consumer looking for clean, green technologies to preserve food and to overcome some of the food safety issues associated with fresh food.

19.2 Plant Extracts as Natural Antimicrobials

Antimicrobials are chemical substances that delay microbial growth or cause microbial cell death on entering a food matrix. Antimicrobials can be classified as traditional (chemical preservatives) and novel substances called *natural*. Natural antimicrobials can be obtained from animal, plant, microbial, and mineral sources. Antimicrobials obtained from plants such as vegetables, fruits, and herbs/spices are termed as plant extracts or plant antimicrobials, which also provide other functional properties like antioxidant activity and novel flavors to the food (Sultanbawa, 2011). In this chapter, plant extracts refer to the whole plant material or extracts. Plant extracts contain phytochemicals that are important for the functioning of the plant, and plants use these chemicals as defense against microorganisms and other predators (Cowan, 1999, Molyneux et al., 2007). Phytochemicals in plants are broadly classified into phenolic compounds, terpenoids and essential oils, alkaloids, lectins, and polypeptides (Balasundram et al., 2006, Cowan, 1999). Phenolic compounds are important contributors to the sensory and nutritional quality of plants, which include fruits and vegetables. These compounds are known as the most widely occurring group of phytochemicals with bioactive properties (Ignat et al., 2011). In particular, phenolic compounds occurring naturally have been reported to have excellent antimicrobial properties as food preservatives (Brul and Coote, 1999, Tajkarimi et al., 2010). Phenolic compounds or polyphenols are divided into several classes according to the number of phenol rings that they contain and the structural elements that bind to these rings. The main groups of polyphenols are flavonoids, phenolic acids, tannins, stilbenes, and lignans (Ignat et al., 2011). Other classes of plant-derived compounds such as polyamines, glucosinolates, and glucosides have been assessed for their antimicrobial properties, and glucosinolates in particular indicate antifungal and antibacterial properties (Vig et al., 2009). In mustard and horseradish, glucosinolates are converted by the enzyme myrosinase to yield a variety of isothiocyanates including the allyl form, which has strong antimicrobial properties (Delaquis and Mazza, 1995). Plants with known antimicrobial properties or other edible plant species or waste streams from processing of plant material can be assessed for antimicrobial efficacy and their potential to be used as natural preservatives in food systems. Figure 19.1 refers to a flow chart detailing a procedure for the testing of a plant source for antimicrobial activity: screening against a range of pathogenic and food spoilage organisms, optimizing the efficacy by studying in vitro synergies with other plant antimicrobials and known natural antimicrobials like organic acids, in vivo application to a food system in combination with appropriate processing, packaging, and storage. Once the efficacy of the plant extract is proven, the food product is evaluated for sensory and physicochemical characteristics and challenge tested against relevant food pathogens depending on the food system and the processing, packaging, and storage conditions. Table 19.1 gives examples of phytochemicals from some plant sources indicating broad-spectrum antimicrobial activity against food spoilage and pathogenic bacteria, yeasts, and fungi. This demonstrates the potential of using plant extracts as natural antimicrobials.

19.3 Synergistic Effects with Other Natural Antimicrobials

Many studies have shown the efficacy of plant extracts as natural antimicrobials; however, there are limitations in their usage due to the limited spectrum of activity of individual extracts, application costs, and effects on the organoleptic quality of foods (Dufour et al., 2003). Methods used to determine antimicrobial interactions (phytosynergy) include basic combination effects, the sum of fractional inhibitory concentration index (FIC), isobole interpretations, and time-kill assays (vanVuuren and Viljoen, 2011). The microdilution checkerboard method is one of the techniques most frequently used to assess the in vitro antimicrobial combinations (Schelz et al., 2006). This 96-well microtiter plate assay is used to quantify the synergies between compounds by calculating the FIC indices using the checkerboard method. For example, for two samples A and B, the FIC indices for the combinations were calculated as $FIC_A + FIC_B$, which are the minimum inhibitory concentrations that inhibited the bacterial growth for samples A and B. $FIC_A = (MIC_A$ in combination with $MIC_B)/MIC_A$ alone; similarly $FIC_B = (MIC_B$ in combination with $MIC_A)/MIC_B$ alone. The mean FIC index $= FIC_A + FIC_B$, and the results are interpreted as follows: synergistic (<0.5), additive (0.5–1.0), indifferent (>1.0), or antagonistic (>4.0) (Gutierrez et al., 2008, Schelz et al., 2006). Table 19.2 gives examples of possible combination effects of plant extracts and other natural antimicrobials. The advantage of synergistic combinations is the enhanced efficacy and cost benefit as lower levels of the plant extract are used. It also gives a wider range of phytochemicals that a microorganism needs to build resistance to and this reduces the likelihood of developing pathogens resistant to preservatives.

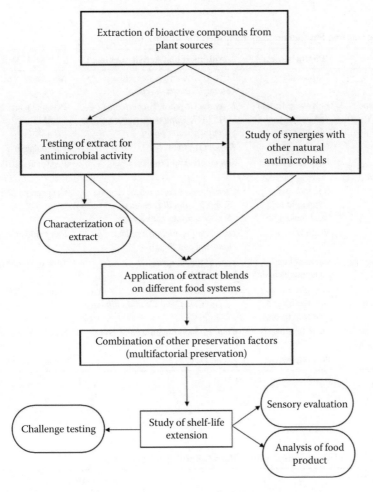

FIGURE 19.1 Flow chart detailing the selection and application of plant extracts as natural antimicrobials in a food system.

19.4 Application to Different Food Systems

Food preservation during processing is carried out to maintain the quality of raw materials, physicochemical properties, and microbial safety of products and minimize spoilage potential. This is achieved through processing and varies from one product to another. Therefore, preservation processes, which are about a third of the unit operations in food processing, employ combination processes such as a mild heat treatment and low concentration of preservatives to achieve a high-quality, microbially safe, and shelf-stable product (Brul and Coote, 1999). The antimicrobial activity of plant extracts that is observed in laboratory screening assays is quite different from that of food systems. Depending on the composition of the food, whether it is high in fat, protein, or carbohydrate, the interaction of plant extracts with the microorganism is different and in most cases the antimicrobial activity will be decreased. Therefore, a higher concentration of plant extract is required for sufficient efficacy in food systems in comparison to the minimum inhibitory concentration that is determined in in vitro assays (Burt, 2004). In addition to the composition of the food, other factors such as processing will affect the efficacy of the plant extract. Heat treatment is the main cause for the degradation of tea catechins during processing of tea leaves, tea beverages, and bakery products. The degradation is also

TABLE 19.1

Some Plant Sources and Phytochemicals Identified for Antimicrobial Activity

Plant Source	Phytochemicals	Antimicrobial Activity against	Reference
Almond skins	Flavonoids	*Listeria monocytogenes* and *Staphylococcus aureus*	Mandalari et al. (2010)
Green tea and grape seed extracts	Polyphenolic and proanthocyanidin compounds	*L. monocytogenes, Escherichia coli* O157:H7, *Salmonella typhimurium, Campylobacter jejuni*	Perumalla and Hettiarachchy (2011)
Mustard flour	Glucosinolates	*E. coli* O157:H7	Nadarajah et al. (2005)
Blueberry and muscadine grape	Phenolic extracts	*Salmonella* and *Listeria* strains	Park et al. (2011)
Horsetail (culinary herb)	Benzoic and cinnamic acid and flavonoids	*Pseudomonas aeruginosa, E. coli, S. aureus, Bacillus cereus, Saccharomyces cerevisiae*	Čanadanović-Brunet et al. (2009)
Vanilla beans	Vanillin	*Lactobacillus plantarum, Listeria innocua,* and *E. coli*	Fitzgerald et al. (2004)
Oregano and cinnamon essential oils	Carvacrol and cinnamaldehyde	*Penicillium notatum*	Tunc et al. (2007)
Waste streams from olive oil and wine production	Phenolic extracts, quercetin, and hydroxytyrosol	*E. coli, Salmonella poona, B. cereus, S. cerevisiae,* and *Candida albicans*	Serra et al. (2008)
Sorghum syrup from the biofuel industry	Phenolic extracts	*Campylobacter jejuni* and *E. coli*	Navarro et al. (2013)
Cranberry	Phenolic extracts	*E. coli* O157:H7	Wu et al. (2009)

TABLE 19.2

Combination Effects of Plant Extracts and Other Natural Antimicrobials

Plant Extracts+Other Antimicrobials	Microorganism	Combination Effect+Model System	Reference
Thymol, grapefruit seed extract, lemon extract	*Pseudomonas fluorescens, Photobacterium phosphoreum,* and *Shewanella putrefaciens*	Synergistic in fish hamburger	Corbo et al. (2008)
Oregano + lemon balm	*Listeria monocytogenes*	Indifference in lettuce leaf model media	Gutierrez et al. (2009)
Oregano + thyme	*Pseudomonas fluorescens*	Additive in lettuce leaf model media	Gutierrez et al. (2009)
Corni fructus, cinnamon, Chinese chive	Total bacterial count	Synergistic in dumplings, guava juice, and tea	Hsieh et al. (2001)
Grape seed, green tea, organic acids	*L. monocytogenes, Escherichia coli* O157:H7, *Salmonella* Typhimurium	Synergistic in chicken meat system	Over et al. (2009)
Chitosan, rosemary essential oil	Lactic acid bacteria, *Brochothrix thermosphacta,* Enterobacteriaceae, *Pseudomonas* spp.	Synergistic in chilled turkey meat	Vasilatos and Savvaidis (2013)
Chitosan, ginger, onion, garlic	Total bacterial count	Synergistic in chilled stewed pork	Cao et al. (2013)
Chitosan, oregano essential oil	*L. monocytogenes,* total bacterial count	Synergistic in chicken meat fillets	Khanjari et al. (2013)
Cilantro and *Eucalyptus dives* essential oils	*Yersinia enterocolitica*	Synergistic in trypticase soy yeast extract (TSYE) broth	Delaquis et al. (2002)
Cilantro and *Eucalyptus dives* essential oils	*Pseudomonas fragi, Enterobacter agglomerans*	Antagonistic in TSYE broth	Delaquis et al. (2002)

dependent on the initial concentration of the tea catechins and other ingredients including the pH of the system and processing time and temperature (Ananingsih et al., 2013).

19.4.1 Challenges and Benefits in Applying Plant Extracts as Preservatives

The challenge of using plant extracts in food systems is the lack of reproducibility of their activity due to the diversity of compounds that are present in plants (Negi, 2012). These plant-derived compounds have variations due to different cultivars, growing conditions, harvesting times, and environmental factors. After harvesting, the stability of phytochemicals must be determined during all stages of processing required for their inclusion in food including packaging and storage (Harbourne et al., 2013). Volatile bioactive compounds in three Australian native herbs known for their antimicrobial properties were retained by using a high-barrier packaging material in comparison to high- or low-density polyethylene. The high-barrier packaging retained in excess of 80% of volatiles in lemon myrtle, anise myrtle, and Tasmanian pepper leaf during 6 months of storage at ambient temperature, demonstrating the importance of packaging in retaining the bioactivity of plant-derived compounds (Chaliha et al., 2013). Extraction methods used to concentrate phytochemicals and delivery methods will also affect the efficacy of the plant extract. Conventional extraction methods include infusions, maceration, and percolation and nonconventional extraction methods are supercritical fluid extraction, microwave extraction, pressurized liquid extraction, and ultrasonic extraction. The most suitable solvent to extract plant bioactive compounds is water as most of these compounds are water soluble (Harbourne et al., 2013, Peschel et al., 2006). Plants are a rich source of antioxidants evident from their diverse range of phytochemicals; antioxidants in plants are not only beneficial for their health properties but also retain the freshness in food by absorbing the off flavors, and the formation of stale or rancid odors is prevented due to the absorption of these compounds by phenolic compounds. Microencapsulation and addition of the phenolic antioxidant caffeic acid to olive oil improved its storage stability by slowing down oxidative changes and preserving unsaturated fatty acids (Sun-Waterhouse et al., 2011). Plant extracts also contribute to the flavor of a food product, and this can be a negative if it imparts a strong flavor when applied as a natural preservative, however, beneficial if the flavor is complementary to the food product. Even though there are issues of stability of the bioactive compounds in plant extracts during processing, packaging, and storage, the benefits of retaining the freshness of the product and prevention of oxidation and off-flavor formation in addition to extending the shelf life far outweigh the challenges (Table 19.3).

19.4.2 Multifactorial Preservation

The application of the hurdle technology concept includes the combination of multiple preservative factors (hurdles); these can be used to improve the microbial stability and the sensory acceptability of foods in addition to retaining their quality and nutritional properties during storage (Leistner, 2000, Mansour and Millière, 2001). This multifactorial preservation model is suitable for plant extracts as lower levels of plant extracts can be used, thereby overcoming the challenge of strong flavors imparted to the food system especially in the case of essential oils.

19.5 Innovative Delivery Systems

For plant extracts to be effective in a food system, bioactive compounds responsible for antimicrobial activity should be available to react with the microorganism. This is a challenge in a complex food matrix, and this has led researchers to look for ways of concentrating and protecting these compounds through novel delivery systems. This ensures the release of bioactive compounds once they are inside and allows their slow release over time, thereby giving a longer storage life. The challenges in the introduction of bioactive compounds into foods include incorporation into free-flowing powders or solutions, degradation during processing and storage, and insoluble compounds requiring a delivery method to increase solubility and dispersability in a hydrophilic food matrix or liquid. Some of these challenges can be overcome by the use of nanotechnology (Shimoni, 2009). Nanotechnology involves the application,

TABLE 19.3

Plant Extracts in Combination with Other Preservation Methods in Extending Shelf Life of Fresh Food

Plant Extracts+Other Natural Antimicrobials	Other Preservation Methods	Type of Food and Shelf Life Extension	Reference
Thymol, lemon extract, chitosan, and grapefruit seed extract	High-barrier packaging material and chilled temperature (4°C)	Amaranth-based fresh pasta, 25 days	Del Nobile et al. (2009b)
Thyme, oregano essential oils	Modified atmosphere packaging (MAP) and chilled temperature (4°C)	Lamb meat chunks, 21–22 days	Karabagias et al. (2011)
Lemongrass essential oil	Gelatin film incorporated with lemongrass essential oil and chilled temperature (4°C)	Sea bass fish fillets, 12 days	Ahmad et al. (2012)
Extract of ginger, garlic, onion, chitosan	Heat treated and chilled temperature (4°C)	Stewed pork, 12 days	Cao et al. (2013)
Horseradish root powder (isothiocyanates)	Packaging and chilled temperature (10°C)	Tofu (bean curd), 10 days	Shin et al. (2010)
Green tea extract, chitosan	Active packaging film and chilled temperature (4°C)	Pork sausages, 20 days	Siripatrawan and Noipha (2012)
Oregano essential oil, salt	Vacuum packaging and chilled temperature (4°C)	Rainbow trout fillets, 11–12 days	Frangos et al. (2010)
Extract of green leafy vegetable fatsia (*Aralia elata*)	Packaging and chilled temperature (4°C)	Raw beef patties, 12 days	Kim et al. (2013)
Citrus green extract	Vacuum packaging and chilled temperature (4°C)	*Tzatziki* traditional Greek salad, in excess of 10 days compared to the control	Tsiraki and Savvaidis (2014)
Thymol, lemon extract, grapefruit seed extract	MAP and chilled temperature (4°C)	Fish burgers with hake and mackerel, 23 days	Del Nobile et al. (2009a)
Grape seed extracts, malic and lactic acid	Electrostatic spraying and chilled temperature (4°C)	Spinach leaves, reduction in microbial load	Ganesh et al. (2010)

production, and processing of materials at the nanometer scale (Augustin and Sanguansri, 2009). Novel nanoencapsulation has demonstrated potential to protect functional ingredients and control their delivery to sites of action. Microencapsulation is defined as a technology of packaging solids, liquids, or gaseous materials in miniature, sealed capsules that can release their contents at controlled rates under specific conditions (Fang and Bhandari, 2010). In comparison to microcapsules, nanoencapsulated materials have superior physical and chemical stability, better compatibility with food matrices, the ability to target bacterial surfaces through tailoring of interfacial properties, and the ability to deliver high concentrations of lipophilic bioactive compounds that can be evenly distributed in the water phase of the food. There are four nanoencapsulation systems that can be used for the incorporation of plant extracts, namely, nanoemulsions, solid lipid nanoparticles, microemulsions, and liposomes (Weiss et al., 2009). Examples of microencapsulated plant extracts in food systems to improve safety and quality are given in Table 19.4.

19.6 Packaging Solutions for Shelf Life Extension

Bioactive packaging incorporating natural antimicrobials in films allows for the functional effect at the food surface, where most of the microbial growth is localized. Examples of antimicrobial packaging would include adding a sachet into the package, dispersing bioactive compounds in the packaging material or coating on the surface of the packaging material, and using antimicrobial macromolecules with film-forming properties as edible packaging material (Coma, 2008). The application of antimicrobial

TABLE 19.4

Innovative Delivery Methods Incorporating Plant Extracts

Plant Extracts + Other Natural Antimicrobials	Delivery Method	Type of Food, Effect on Safety and Quality	Reference
Caffeic acid	Microencapsulated in alginate microspheres	Olive oil, reduction in oxidation	Sun-Waterhouse et al. (2011)
Allyl isothiocyanate	Microencapsulated in gum acacia	Chopped beef, reduction or elimination of *Escherichia coli* O157:H7	Chacon et al. (2006)
Oregano, thyme essential oils	Incorporated into soy protein edible films	Ground beef patties, reduction in coliform and *Pseudomonas* spp.	Emiroglu et al. (2010)
Catechin, cinnamaldehyde	Incorporated into *Gelidium corneum* film	Sausage, reduction in lipid oxidation and *E. coli* O157:H7, *Listeria monocytogenes*	Ku et al. (2008)
Green tea extracts	Edible coatings with tapioca starch and hsian-tsao leaf gum	Fruit-based salads, romaine hearts, pork slices. Reduction in total bacteria, yeast, and molds, *Staphylococcus aureus*, *L. monocytogenes, Bacillus cereus*	Chiu and Lai (2010)
Oregano, rosemary essential oils and chitosan	Incorporated into gelatin film	Cold smoked sardine, reduction in lipid oxidation and total bacteria	Gómez-Estaca et al. (2007)
Limonene, peppermint, oregano extract, red thyme	Incorporated into chitosan coating	Strawberries reduction in molds and total bacteria	Vu et al. (2011)
Green tea extracts	Incorporated into chitosan film	Pork sausages, inhibition of lipid oxidation and microbial growth	Siripatrawan and Noipha (2012)

edible films and coating on fresh-cut fruits has provided inhibitory effects against spoilage and pathogenic bacteria by maintaining effective concentrations of the bioactive compounds on the surface. Lemongrass essential oil, oregano essential oil, and vanillin in alginate and apple puree as a coating material on fresh-cut apple slices have shown a reduction in psychrophilic aerobe, mold, and yeast counts (Rojas-Graue et al., 2007). Other examples of using bioactive packaging to extend storage life are given in Table 19.4. Antimicrobial packaging is important in reducing the risk of pathogen contamination and in extending the storage life of foods; however, it should not substitute for good quality raw materials, properly processed foods, and good manufacturing practices. It should be considered a hurdle technology that is used in food preservation (Appendini and Hotchkiss, 2002).

19.7 Conclusion and Future Directions

Natural antimicrobials from plants are sourced from a variety of plant sources, and they have broad-spectrum antimicrobial activity against food-related organisms. They have shown promise as natural preservatives and the benefit of having antioxidant properties and unique flavors. Plant extracts can be successfully used as one of the hurdles in a multifactorial preservation model to extend the shelf life of food products. The potential of using a wider variety of plants as sources of antimicrobial agents is still unexploited, and there is a huge opportunity to identify and expand this list of plant antimicrobials for food preservation. Research should focus on understanding the mechanism of antimicrobial inhibition of plant extracts within complex food systems and increasing microbial efficacy by studying synergies between different phytochemicals. Cost reduction and improvement in efficacy can be achieved by synergistic combinations, and this could overcome the barrier of replacing chemical preservatives that are favored due to the low cost of production. Tailored delivery methods should be developed to enhance the efficacy of bioactive compounds. The challenge of retaining the stability of bioactive compounds during extraction, processing, packaging, and storage of the food product should also be addressed. It is clear from the literature that natural antimicrobials from plants can be effectively used to reduce food spoilage and pathogenic microorganisms, indicating their potential use as a natural preservative in food.

REFERENCES

Ahmad, M., S. Benjakul, P. Sumpavapol, and N. P. Nirmal. Quality changes of sea bass slices wrapped with gelatin film incorporated with lemongrass essential oil. *International Journal of Food Microbiology* 2012;155(3): 171–178.

Ananingsih, V. K., A. Sharma, and W. Zhou. Green tea catechins during food processing and storage: A review on stability and detection. *Food Research International* 2013;50(2): 469–479.

Appendini, P. and J. H. Hotchkiss. Review of antimicrobial food packaging. *Innovative Food Science & Emerging Technologies* 2002;3(2): 113–126.

Augustin, M. A. and P. Sanguansri. Nanostructured materials in the food industry. In *Advances in Food and Nutrition Research*, ed. Taylor, S. L., pp. 183–213. San Diego, CA: Academic Press, 2009.

Balasundram, N., K. Sundram, and S. Samman. Phenolic compounds in plants and agri-industrial by-products: Antioxidant activity, occurrence, and potential uses. *Food Chemistry* 2006;99(1): 191–203.

Brul, S. and P. Coote. Preservative agents in foods: Mode of action and microbial resistance mechanisms. *International Journal of Food Microbiology* 1999;50(1–2): 1–17.

Burt, S. Essential oils: Their antibacterial properties and potential applications in foods—A review. *International Journal of Food Microbiology* 2004;94(3): 223–253.

Čanadanović-Brunet, J. M., G. S. Ćetković, S. M. Djilas et al. Radical scavenging and antimicrobial activity of horsetail (*Equisetum arvense* L.) extracts. *International Journal of Food Science & Technology* 2009;44(2): 269–278.

Cao, Y., W. Gu, J. Zhang et al. Effects of chitosan, aqueous extract of ginger, onion and garlic on quality and shelf life of stewed-pork during refrigerated storage. *Food Chemistry* 2013;141(3): 1655–1660.

Chacon, P. A., R. A. Buffo, and R. A. Holley. Inhibitory effects of microencapsulated allyl isothiocyanate (AIT) against *Escherichia coli* O157:H7 in refrigerated, nitrogen packed, finely chopped beef. *International Journal of Food Microbiology* 2006;107(3): 231–237.

Chaliha, M., A. Cusack, M. Currie, Y. Sultanbawa, and H. Smyth. Effect of packaging materials and storage on major volatile compounds in three Australian native herbs. *Journal of Agricultural and Food Chemistry* 2013;61(24): 5738–5745.

Chiu, P.-E. and L.-S. Lai. Antimicrobial activities of tapioca starch/decolorized hsian-tsao leaf gum coatings containing green tea extracts in fruit-based salads, romaine hearts and pork slices. *International Journal of Food Microbiology* 2010;139(1–2): 23–30.

Coma, V. Bioactive packaging technologies for extended shelf life of meat-based products. *Meat Science* 2008;78(1–2): 90–103.

Corbo, M. R., B. Speranza, A. Filippone et al. Study on the synergic effect of natural compounds on the microbial quality decay of packed fish hamburger. *International Journal of Food Microbiology* 2008;127(3): 261–267.

Cowan, M. M. Plant products as antimicrobial agents. *Clinical Microbiology Review* 1999;12(4): 564–582.

Del Nobile, M. A., M. R. Corbo, B. Speranza et al. Combined effect of map and active compounds on fresh blue fish burger. *International Journal of Food Microbiology* 2009a;135(3): 281–287.

Del Nobile, M. A., N. Di Benedetto, N. Suriano et al. Use of natural compounds to improve the microbial stability of amaranth-based homemade fresh pasta. *Food Microbiology* 2009b;26(2): 151–156.

Delaquis, P. J. and G. Mazza. Antimicrobial properties of isothiocyanates in food preservation. *Food Technology* 1995;49(11): 73–84.

Delaquis, P. J., K. Stanich, B. Girard, and G. Mazza. Antimicrobial activity of individual and mixed fractions of dill, cilantro, coriander and eucalyptus essential oils. *International Journal of Food Microbiology* 2002;74(1–2): 101–109.

Dufour, M., R. S. Simmonds, and P. J. Bremer. Development of a method to quantify *in vitro* the synergistic activity of "natural" antimicrobials. *International Journal of Food Microbiology* 2003;85(3): 249–258.

Emiroglu, Z. K., G. P. Yemis, B. K. Coskun, and K. Candogan. Antimicrobial activity of soy edible films incorporated with thyme and oregano essential oils on fresh ground beef patties. *Meat Science* 2010;86(2): 283–288.

Fang, Z. and B. Bhandari. Encapsulation of polyphenols—A review. *Trends in Food Science & Technology* 2010;21(10): 510–523.

Fitzgerald, D. J., M. Stratford, M. J. Gasson et al. Mode of antimicrobial action of vanillin against *Escherichia coli*, *Lactobacillus plantarum* and *Listeria innocua*. *Journal of Applied Microbiology* 2004;97(1): 104–113.

Frangos, L., N. Pyrgotou, V. Giatrakou, A. Ntzimani, and I. N. Savvaidis. Combined effects of salting, oregano oil and vacuum-packaging on the shelf-life of refrigerated trout fillets. *Food Microbiology* 2010;27(1): 115–121.

Ganesh, V., N. S. Hettiarachchy, M. Ravichandran et al. Electrostatic sprays of food-grade acids and plant extracts are more effective than conventional sprays in decontaminating *Salmonella* Typhimurium on spinach. *Journal of Food Science* 2010;75(9): M574–M579.

Gómez-Estaca, J., P. Montero, B. Giménez, and M. C. Gómez-Guillén. Effect of functional edible films and high pressure processing on microbial and oxidative spoilage in cold-smoked sardine (*Sardina pilchardus*). *Food Chemistry* 2007;105(2): 511–520.

Gutierrez, J., C. Barry-Ryan, and P. Bourke. The antimicrobial efficacy of plant essential oil combinations and interactions with food ingredients. *International Journal of Food Microbiology* 2008;124(1): 91–97.

Gutierrez, J., C. Barry-Ryan, and P. Bourke. Antimicrobial activity of plant essential oils using food model media: Efficacy, synergistic potential and interactions with food components. *Food Microbiology* 2009;26(2): 142–150.

Harbourne, N., E. Marete, J. C. Jacquier, and D. O'Riordan. Stability of phytochemicals as sources of anti-inflammatory nutraceuticals in beverages—A review. *Food Research International* 2013;50(2): 480–486.

Havelaar, A. H., S. Brul, A. de Jong et al. Future challenges to microbial food safety. *International Journal of Food Microbiology* 2010;139(Supplement 1): S79–S94.

Hsieh, P.-C., J.-L. Mau, and S.-H. Huang. Antimicrobial effect of various combinations of plant extracts. *Food Microbiology* 2001;18(1): 35–43.

Ignat, I., I. Volf, and V. I. Popa. A critical review of methods for characterisation of polyphenolic compounds in fruits and vegetables. *Food Chemistry* 2011;126(4): 1821–1835.

Karabagias, I., A. Badeka, and M. G. Kontominas. Shelf life extension of lamb meat using thyme or oregano essential oils and modified atmosphere packaging. *Meat Science* 2011;88(1): 109–116.

Khanjari, A., I. K. Karabagias, and M. G. Kontominas. Combined effect of *N,O*-carboxymethyl chitosan and oregano essential oil to extend shelf life and control *Listeria monocytogenes* in raw chicken meat fillets. *LWT—Food Science and Technology* 2013;53(1): 94–99.

Kim, S.-J., A. R. Cho, and J. Han. Antioxidant and antimicrobial activities of leafy green vegetable extracts and their applications to meat product preservation. *Food Control* 2013;29(1): 112–120.

Ku, K. J., Y. H. Hong, and K. B. Song. Mechanical properties of a *Gelidium corneum* edible film containing catechin and its application in sausages. *Journal of Food Science* 2008;73(3): C217–C221.

Leistner, L. Basic aspects of food preservation by hurdle technology. *International Journal of Food Microbiology* 2000;55(1–3): 181–186.

Mandalari, G., C. Bisignano, M. D'Arrigo et al. Antimicrobial potential of polyphenols extracted from almond skins. *Letters in Applied Microbiology* 2010;51(1): 83–89.

Mansour, M. and J.-B. Millière. An inhibitory synergistic effect of a nisin–monolaurin combination on *Bacillus* sp. vegetative cells in milk. *Food Microbiology* 2001;18(1): 87–94.

Molyneux, R. J., S. T. Lee, D. R. Gardner, K. E. Panter, and L. F. James. Phytochemicals: The good, the bad and the ugly? *Phytochemistry* 2007;68(22–24): 2973–2985.

Nadarajah, D., J. H. Han, and R. A. Holley. Use of mustard flour to inactivate *Escherichia coli* O157:H7 in ground beef under nitrogen flushed packaging. *International Journal of Food Microbiology* 2005;99(3): 257–267.

Navarro, M., S. Roger, A. Cusack, and Y. Sultanbawa. Anti-*Campylobacter* activity of food industry by-products. In *Proceedings of the 46th Annual Australian Institute of Food Science and Technology*, p. 43. Brisbane, Queensland, Australia: AIFST, 2013.

Negi, P. S. Plant extracts for the control of bacterial growth: Efficacy, stability and safety issues for food application. *International Journal of Food Microbiology* 2012;156(1): 7–17.

Over, K. F., N. Hettiarachchy, M. G. Johnson, and B. Davis. Effect of organic acids and plant extracts on *Escherichia coli* O157:H7, *Listeria monocytogenes*, and *Salmonella* Typhimurium in broth culture model and chicken meat systems. *Journal of Food Science* 2009;74(9): M515–M521.

Park, Y. J., R. Biswas, R. D. Phillips, and J. Chen. Antibacterial activities of blueberry and muscadine phenolic extracts. *Journal of Food Science* 2011;76(2): M101–M105.

Perumalla, A. V. S. and N. S. Hettiarachchy. Green tea and grape seed extracts—Potential applications in food safety and quality. *Food Research International* 2011;44(4): 827–839.

Peschel, W., F. Sánchez-Rabaneda, W. Diekmann et al. An industrial approach in the search of natural antioxidants from vegetable and fruit wastes. *Food Chemistry* 2006;97(1): 137–150.

Rojas-Graue, M. A., R. M. Raybaudi-Massilia, R. C. Soliva-Fortuny et al. Apple puree-alginate edible coating as carrier of antimicrobial agents to prolong shelf-life of fresh-cut apples. *Postharvest Biology and Technology* 2007;45(2): 254–264.

Schelz, Z., J. Molnar, and J. Hohmann. Antimicrobial and antiplasmid activities of essential oils. *Fitoterapia* 2006;77(4): 279–285.

Serra, A. T., A. A. Matias, A. V. M. Nunes et al. In vitro evaluation of olive- and grape-based natural extracts as potential preservatives for food. *Innovative Food Science & Emerging Technologies* 2008;9(3): 311–319.

Shimoni, E. Nanotechnology for foods: Delivery systems. In *Global Issues in Food Science and Technology*, eds. Barbosa-Cánovas, G., Mortimer, A., Lineback, D., Spiess, W., Buckle, K., and Colonna, P., pp. 411–424. San Diego, CA: Academic Press, 2009.

Shin, I.-S., J.-S. Han, K.-D. Choi et al. Effect of isothiocyanates from horseradish (*Armoracia rusticana*) on the quality and shelf life of tofu. *Food Control* 2010;21(8): 1081–1086.

Siripatrawan, U. and S. Noipha. Active film from chitosan incorporating green tea extract for shelf life extension of pork sausages. *Food Hydrocolloids* 2012;27(1): 102–108.

Sultanbawa, Y. Plant antimicrobials in food applications: Minireview. In *Science against Microbial Pathogens: Communicating Current Research and Technological Advances*, ed. Méndez-Vilas, A., Vol. 2, pp. 1084–1093. Badajoz, Spain: Formatex Research Center, 2011.

Sun-Waterhouse, D., J. Zhou, G. M. Miskelly, R. Wibisono, and S. S. Wadhwa. Stability of encapsulated olive oil in the presence of caffeic acid. *Food Chemistry* 2011;126(3): 1049–1056.

Tajkarimi, M. M., S. A. Ibrahim, and D. O. Cliver. Antimicrobial herb and spice compounds in food. *Food Control* 2010;21(9): 1199–1218.

Tsiraki, M. I. and I. N. Savvaidis. Citrus extract or natamycin treatments on "Tzatziki"—A traditional Greek salad. *Food Chemistry* 2014;142: 416–422.

Tunc, S., E. Chollet, P. Chalier, L. Preziosi-Belloy, and N. Gontard. Combined effect of volatile antimicrobial agents on the growth of *Penicillium notatum*. *International Journal of Food Microbiology* 2007;113(3): 263–270.

van Vuuren, S. and A. Viljoen. Plant-based antimicrobial studies—Methods and approaches to study the interaction between natural products. *Planta Medica* 2011;77(11): 1168–1182.

Vasilatos, G. C. and I. N. Savvaidis. Chitosan or rosemary oil treatments, singly or combined to increase turkey meat shelf-life. *International Journal of Food Microbiology* 2013;166(1): 54–58.

Vig, A. P., G. Rampal, T. S. Thind, and S. Arora. Bio-protective effects of glucosinolates—A review. *LWT—Food Science and Technology* 2009;42(10): 1561–1572.

Vu, K. D., R. G. Hollingsworth, E. Leroux, S. Salmieri, and M. Lacroix. Development of edible bioactive coating based on modified chitosan for increasing the shelf life of strawberries. *Food Research International* 2011;44(1): 198–203.

Weiss, J., S. Gaysinsky, M. Davidson, and J. McClements. Nanostructured encapsulation systems: Food antimicrobials. In *Global Issues in Food Science and Technology*, eds. Barbosa-Cánovas, G., Mortimer, A., Lineback, D., Spiess, W., Buckle, K., and Colonna, P., pp. 425–479. San Diego, CA: Academic Press, 2009.

Wu, V. C. H., X. Qiu, B. G. de los Reyes, C.-S. Lin, and Y. Pan. Application of cranberry concentrate (*Vaccinium macrocarpon*) to control *Escherichia coli* O157:H7 in ground beef and its antimicrobial mechanism related to the downregulated Slp, Hdea and Cfa. *Food Microbiology* 2009;26(1): 32–38.

20

Hurdle Technology

Nada Smigic and Andreja Rajkovic

CONTENTS

20.1 Introduction

Food-borne illness as an outcome after the ingestion of contaminated food products indicates a broad group of illnesses caused by pathogenic microorganisms, chemical and physical contaminants that can contaminate food at several points during production and preparation process. Although the research in this field has been very intensive during the last decades with the same trend that will continue in future, and many preventive and control measures that have already been applied in the food industry, the number of food-borne illnesses stays at unacceptably high level (Havelaar et al., 2010). There are several reasons for this. First, advances in the food microbiology allowed more food-borne pathogens to be identified (e.g., *Escherichia coli* O157:H7, *Listeria monocytogenes*, and *Cronobacter sakazakii*). Some known pathogens have expressed unexpected characteristics regarding survival/growth and occurrence in food not commonly associated with the specific pathogen (e.g., *E. coli* O157:H7 was found in fresh produce, apple cider, and cookie dough). Additionally, consumers' demands have changed; nowadays, consumers prefer more fresh-like food with unchanged natural properties with long shelf life, and demographic characteristics including age, gender, education, and income have also changed. All these factors create the environment where food producers and scientists are facing new challenges and constantly search for new and enhanced preservation treatments to improve microbial safety of food products.

    Different food preservation treatments are used to maintain edible food products for human consumption. Preservation treatments are meant to slow down or inhibit chemical deterioration and microbial multiplication in food, including both pathogenic and spoilage microorganisms. The growth of microorganisms is affected by intrinsic (pH, oxidation-reduction potential, water activity, natural antimicrobials, and barriers) and extrinsic (atmosphere, temperature, and humidity) factors in food. The traditional preservation treatments include temperature reduction (chill or frozen storage), water activity reduction (drying and curing), reduction in pH values (fermentation or addition of acids), removal of available oxygen (vacuum or modified atmosphere), changing the atmosphere in packaging, or addition of different preservatives. Along with these inhibitory treatments, the food industry successfully applies treatments with the aim to reduce or completely inactivate present microorganisms in food. Although the traditional and most utilized heat treatment, being pasteurization and sterilization, is effective in obtaining and keeping the safety of food products, often the quality of these foods is shifted from natural fresh-like taste and natural nutritional values. Heat treatment may result in thermal inactivation of some food enzyme and destruction of food constituents, resulting in compromised food sensorial and nutritional characteristics of final products. Therefore, heat treatment not only is energy intensive but also unfavorably affects flavor, chemical composition, and nutritional quality of the treated food.

Consumers' demand for less processed, fresher-tasting nutritive foods without the use of heat or chemical preservatives and safe food products has presented particular challenges to food processors. In order to adapt to consumers' changed requirements, food preservation treatments are constantly developing, and nowadays alternative, nonthermal, mild, novel food processing treatments are being explored, and they have attracted the attention of many food manufacturers and research institutions.

Some nonthermal processing developed in last decades include physical treatments such as irradiation, high pressure, pulsed electric fields (PEFs), intense light pulses (ILP), ultrasound, and ultraviolet radiation; and chemical and biochemical treatments such as the application of organic acids, antimicrobials, ozone, electrolyzed oxidizing water, and chlorine dioxine. It is important to note that each of these treatments has its own advantages and disadvantages, and their practical application in the food industry will depend not only on their own characteristics and compositions of the food products but also on the pathogens related to the food product. In some cases, the successful application of these nonthermal food processing treatments is connected with its high intensities to inactivate sufficient number of microorganisms and/or spores to reach the required level of microbial safety. These high intensities can still provoke some unacceptable changes in food properties. For example, high hydrostatic pressure can cause the alteration of protein and polysaccharide structure, and therefore causing texture changes, sensorial appearance, and functionality (Considine et al., 2008). Therefore, the possible solution to minimize limitations and drawbacks of nonthermal treatments is to combine lower intensities of nonthermal treatments with low temperature, mild heat, modified atmosphere packaging, antimicrobial agents, or other preservation treatments, in order to obtain sufficient shelf life and product safety with minimal loss of food quality. A great number of scientific literatures have been published lately covering the subject of combined technology, which indicated the importance, popularity, and potential of this concept to be applied in the food industry.

20.2 Combined Treatments

The usage of multiple barriers such as traditional and/or nonthermal preservation treatments to preserve food and to form food environment that microorganisms should not be able to overcome is known as a hurdle technology. In literature, the term hurdle technology is used in parallel with the terms such as combined methods, combined processes, combined preservation, combination techniques, barrier technology, and intervention technology (Raso and Barbosa-Cánovas, 2003; Ross et al., 2003).

This concept of intelligent technology was launched in 1978 by Lothar Leistner at the Federal Centre for Meat Research in Kulmbach, Germany, initially designed to improve the safety of meat products, such as mildly heated fresh-like meat stored without refrigeration. As a result of applied intelligent combination of hurdles, shelf-stable sausage was produced (Leistner and Gorris, 1995). Several categories of these shelf-stable meat products have evolved, which are present in large quantities on the German market. As a result of applied hurdle technology, a shelf-stable pork sausage was produced to be stable during storage at an ambient temperature of 37°C. The applied hurdles included low pH, low water activity, vacuum packaging, and postpackage reheating (Thomas et al., 2008). Chawla and Chander (2004) also reported a successful combination of hurdles for the production of ready-to-use shelf-stable meat products, mutton kebabs. The combination consisted of water activity lower than 0.85, which not only affected the growth of challenge organisms such as *Clostridium sporogenes* and *Staphylococcus aureus*, but also reduced *Bacillus cereus* counts during storage, together with irradiation, which is used to eliminate pathogenic microorganisms, yeast and molds in meat product. This concept is limited not only to meat products but also to a wider range of food such as fruits and vegetables, and the products thereof, dairy, poultry, fish, bakery products, etc.

Even before this concept was first released by Leistner using this particular term, food preservation by combining several preservation factors had been used in food processing. One simple example is the preservation of fruit jams, in which high temperature, low pH of fruits, low water activity obtained by initial sugar from fruits and added sugar during processing, and anaerobic packaging are used to reduce currently present microorganisms as well as to suppress the growth of survivors during shelf life at room temperature. Nevertheless, the long-term preservation of meat, dairy, or fish products at refrigerated temperature or even at ambient temperature is more challenging.

Chilled food products are foods that have received minimal processing with an improved but still limited shelf life in which refrigeration is a key preservation measure. These foods include conventional products, such as luncheon meats and cured meats, seafood, egg, and vegetable salads, fresh pasta and pasta sauces, other sauces, soups, entrées, complete meals, and uncured meat and poultry items. In the case of chilled foods that are initially heat treated, the heat treatment is less than required for the commercial sterility. The combined preservation factors might be used as a preventive measure for the storage of these foods, which introduces additional and preventive hurdles into the production of chilled food in order to avoid worst-case scenarios—temperature abuse—that might occur during cold chain. Additional hurdles, such as modified atmosphere packaging, might back up the situation where temperature abuse occurs and therefore play the role of safeguards in chilled foods, ensuring that they remain microbiologically stable and safe during storage in retail as well as at home storage (Leistner and Gorris, 1995). The effectiveness of the combination of hurdles used for chilled food products needs to be verified within the appropriate challenge studies.

As mentioned earlier, the individual hurdle applied at specific intensity might be ineffective, but when using several hurdles together, their effect can be additive or even synergistic and therefore the target organisms fail to sustain during shelf life. It is assumed that the hurdles that will affect the same elements within the cell have only an additive inhibitory effect (Leistner, 1992, 1994). When different cellular targets are attacked, the synergistic effect can be obtained as the organisms will not be able to repair simultaneously each damage that occur in the microbial cell, and also the activation of stress shock proteins will be more complicated (Leistner, 2000). If the initial treatment induces cellular damage that will make cells more sensitive to subsequent or simultaneous treatments, the lethal effect will be greater than the simple addition of individual effect of each hurdle, and synergistic effect will be obtained. For example, PEF treatment and acidification with organic acids may result in synergistic effect for *E. coli* (Liu et al., 1997), as PEF might increase the permeability of cell wall and membrane and, in this way, allow easier penetration of undissociated acids into bacterial cells. However, the opposite effect may also occur when one treatment induced increased resistance of bacterial cells for the subsequent treatments, and the lethal effect of combined treatment is lower than individual effects. This is called an antagonistic effect. The low water activity sometimes may result in increased resistance against heat treatment. This effect is also seen through adaptation or cross-protection, when the part of bacterial population survive exposure to stress, which results in a protection against the subsequent lethal treatments, including stresses to which bacteria have not been previously exposed. The acid adaptation of *E. coli* O157:H7 cells resulted in bacterial protection against heat and osmotic stress. High hydrostatic-resistant mutants of *L. monocytogenes* showed more resistance against heat, acid, and H_2O_2 treatments compared to wild-type strains (Karatzas and Bennik, 2002). Sometimes applying a specific hurdle at lower intensities can induce cellular damages, which trigger the metabolic pathways that provide the better survival under following conditions and treatments.

Recently, Lee and Kang (2009) reported that various combinations of salt, heat, and acid resulted in completely different results. The outcome of combined salt and heat was seen as an additive effect, combination of acid and heat resulted in synergistic effect, while in the combination of salt and acid, salt gave protection against acid treatment and thus resulted in antagonistic effect, giving lower reduction of *E. coli* O157:H7 in the combined treatment than in an individual treatment. Under combined stress-ful conditions applied during food preservation, pathogenic cells might show various responses, and sometimes instead of being advantageous, the approach becomes problematic for the application in the food industry.

20.3 Microbial Stress Response

Bacterial cells present in food are subjected to various stresses of physical, chemical, and biological origin, which might occur throughout the food chain and consequently lead to different levels of bac-terial damage/injury. In order to survive sudden potentially lethal changes in the environments that they encounter, bacteria must be able to sense and respond rapidly and appropriately to a vast array of stresses. Response to stress and maintenance of the constant internal environment in bacterial cells is an active process and requires the expenditure of energy. Due to stress, induced via environmental and processing conditions, different modifications may occur in cells to sustain stress, such as maintenance of intracellular pH (e.g., acid stress or lowered pH), modified cell membrane to maintain satisfactory fluidity (e.g., low temperature), enzymatic protection from oxygen-derived free radicals (e.g., high-oxygen atmosphere), and repair of single-strand breaks in DNA (e.g., ionizing radiation). However, homeostasis might be disturbed when several stresses/hurdles are applied, and cells will have less available energy for functioning and repairing disturbed homeostasis. As a consequence, lag phase will be prolonged, and when homeostasis is not reestablished, the bacterial cell might die. Therefore, the successful multitarget combination of preservation treatments (nonthermal and traditional) should be the most efficient, with the aim to attack and disturb cells in several ways, and different sites of the cells to be perturbed, such as cell wall, membrane transport, signal transduction, enzyme system, and control of gene expression.

It is the aim to prevent the possibility of bacterial cells to adapt to new environment to make resis-tance or cross-protection against antimicrobial agents and different hurdles. Therefore, it is not an easy job to create a sequence of hurdles that will answer and fulfill all requirements for producing safe and nutritional food.

20.4 Available Hurdles

The most important hurdles used in the food industry are temperature (high and low), water activ-ity, acidity, redox potential, selected competitors, and usage of preservatives. Nevertheless, along with these traditional and commonly used hurdles/preservative factors, the list of hurdles is much longer with more than 60 hurdles that were identified to be appropriate for foods of animal or plant origin. Leistner and Gorris (1995) divided hurdles to be used in the food industry into four different categories, namely, the following:

- *Physical hurdles* (high and low temperatures, ionizing radiation, high pressure, ultrasonication, ultraviolet radiation, modified atmosphere packaging, packaging films, aseptic packaging, etc.)
- *Physicochemical hurdles* (low pH, low water activity, organic acids, phosphates, salts, smok-ing, sodium nitrate/nitrite, sodium sulfite, potassium sulfite, oxygen, ozone, species, herbs, sur-face treatment agents, etc.)
- *Microbial-derived hurdles* (competitive flora, protective cultures, antibiotics, and bacteriocins)
- *Miscellaneous hurdles* (free fatty acids, chlorine, chitosan, etc.)

Each food product requires specific combination of hurdles, depending on various factors, such as the initial microbial load of the product that will be preserved, intrinsic factors of the food products, and favorable conditions within the food for the growth/survival of microorganisms, but the targeted shelf life of the food products also will influence the possibility to combine different hurdles in great deal. Hurdles may have multiple functions in food; in some cases, hurdles may play a role of preservative factor at the same time influencing the quality of food, such as products of Maillard reaction (Leistner and Gorris, 1995). One hurdle may have positive or negative effect depending on the level of intensity applied. For example, low pH is favorable to prevent pathogen growth and/or survival, but at the same time, food with very low pH might be inappropriate for human consumption.

Hurdles may be applied in a sequence, such as a series of different processing steps/hurdles, or simultaneously one hurdle at a time. Some studies showed that the order of applied hurdles may have an influence on the effectiveness of the combination. In a study of the inactivation efficiency of combined PEF and manothermosonication (MTS) for the control of *Listeria innocua* in a smoothie-type beverage, the sequence in which these two treatments were applied was found to have a significant impact on the level of microbial reduction. When MTS was followed by PEF, the log reduction of *L. innocua* was 5.6 CFU/mL compared to 4.2 CFU/mL for the reverse sequence (Palgan et al., 2012). The obtained results may be due to greater cell damage induced by MTS when applied as a first hurdle, through cavitation, possibly causing increased susceptibility to the subsequent PEF treatment.

Although hurdle technology with traditional preservation treatments is often found in the food industry, this hurdle concept gained more attention after the novel nonthermal preservation treatments were introduced in the food industry and their benefits attracted the producers. As for all combined treatments, a good understanding of the mechanism of inactivation of each of the treatments is very important and seems to be crucial in order to achieve additive or synergistic effect (Leistner, 2000). The exact mechanism of inactivation for most of novel nonthermal preservation treatments is still not completely understood, while assumptions are made for most of them (Rajkovic et al., 2010). Regarding the understanding of emerging food preservation treatments, the unknown field still remains the microbial cell physiological responses. Microbial response is very complex and more difficult compared to traditional preservation factors. In addition, the characteristics of individual hurdle applied can change over time in both directions, either to increase or decrease in its values. Even though mechanisms of microbial inactivation are still missing, nonthermal preservation treatments are successfully combined and results are promising. Some of the recent studies will be presented in this chapter.

Nonthermal treatment might result in incomplete inactivation of microorganisms and remaining population of sublethally injured cells. The ability of injured bacterial cells to resuscitate and resume growth under favorable conditions, when and if they occurred, depends on the type and level of the injury, type of the microorganism, as well as the conditions in the surrounding environment (Rajkovic et al., 2009; Smigic et al., 2009, 2010; Tiganitas et al., 2009; Van Houteghem et al., 2008). In addition to the inactivation technologies applied to foods, both microbial growth and survival can be influenced by different intrinsic factors of the food. This further means that intrinsic factors (water activity, pH, and nutrients), alone or combined with the extrinsic factors (modified atmosphere, temperature, and humidity), can enhance or inhibit recovery and growth of microbial cells. Therefore, the safety and stability of food can be improved using an appropriate combination of several factors that will prevent survival and proliferation of sublethally injured cells.

20.5 Nonthermal Preservation Treatments and Hurdle Technology

The commercial application of novel nonthermal preservation treatments can be improved by improving their effectiveness in obtaining safe food. This can be improved by combining with other processing. Due to the fact that each nonthermal process has different mechanisms of microbial inactivation, the appropriate combinations with other treatments will also be different, without general rules and application.

20.5.1 Combined Treatments: High-Pressure Processing

High-pressure processing (HPP) has a considerable potential to be an alternative to thermal food preservation, due to its ability to reduce microbial load in food products, without altering the organoleptic characteristics of food. It is energy efficient, and is employed to destroy contaminants using an isostatic pressure between 100 and 600 MPa. The inactivation mechanism of HPP proceeds through low energy and does not promote the formation of unwanted chemical compounds or free radicals. The microbial inactivation is mainly related to the inactivation of cell membrane, such as modification in permeability and ion exchange. Some changes in cell morphology and biochemical reactions, protein denaturation, and inhibition of genetic mechanism may occur in microbial cells as a result of this treatment. Different microorganisms show different levels of resistance against HPP (Mújica-Paz et al., 2011; Torres and Velazquez, 2005).

The combination of high-pressure treatment with other physical and chemical treatments was often investigated and tested in different food products, with the aim to reach lethality effect, which is greater than of either treatment alone. As the population of sublethally injured cells may occur after the application of high pressure (Rajkovic et al., 2010; Wesche et al., 2009), these sublethally injured cells may be more sensitive to other treatments (Hauben et al., 1997), and synergistic effects have been determined when HPP was combined with mild heat, antimicrobials, essential oils, PEFs, irradiation, or ultrasonication.

20.5.1.1 High Pressure and Mild Temperature

Although many HPP treatments are usually performed at ambient temperature, a study of Kalchayanand et al. (1998) showed that an increase or decrease in temperature resulted in greater inactivation level of different pathogens. The bacterial cells are less sensitive to the pressure treatment combined with a temperature between 20°C and 30°C, but above 30°C, they become more sensitive, probably due to phase transition of membrane lipids (Kalchayanand et al., 1998). Combining pressure treatment with 30°C resulted in 6 log CFU/mL reduction of *E. coli* O157:H7 (Linton et al., 1999a,b). The combination of HPP and higher temperature of 50°C during 5 min resulted in greater reduction (>8 log) for *Bacillus* spp., *L. monocytogenes*, *E. coli* O157:H7, *S. aureus*, and *Salmonella* in orange juice and pasteurized milk, with an exception of one *S. aureus* strain reduced only by 5.5 log in pasteurized milk (Alpas and Bozoglu, 2000). Recently, Neetoo et al. (2009) investigated the effectiveness of HPP treatment combined with mild temperature to eliminate the population of *E. coli* O157:H7 inoculated on alfalfa seeds and to determine the effect of these processes on the seeds' viability. They found that the temperature plays a significant role in the inactivation level, as 550 MPa for 2 min combined with 40°C eliminated population *E. coli* O157:H7 for 5 log units, without altering alfalfa seeds' viability.

Regarding the bacterial spores, the greater level of lethality is attributed to HPP treatments at higher temperatures mainly through the germination and later loss of heat resistance (Raso and Barbosa-Cánovas, 2003). It has been reported that the pressure between 50 and 250 MPa triggered the germination of *B. subtilis* and *B. cereus* spores (Murrell and Wills, 1977; Paidhungat et al., 2002). The subsequent heat treatment using temperatures as low as 80°C could inactivate bacterial spores efficiently without intensive heating. According to the results presented in the report of Gao et al. (2006), reduction level of 6 log could be obtained for a combination of a temperature of 87°C, 576 MPa during 13 min, for *B. subtilis* spores. The study of Bull et al. (2009) reported that the synergistic effect of the inactivation of *Clostridium botulinum* spores is strain and product type dependent.

20.5.1.2 High Pressure and Antimicrobials

The combination of HPP treatment with antimicrobial agents can improve the effectiveness of the pressure treatment without altering food quality (Raso and Barbosa-Cánovas, 2003). The synergistic effect was observed when 250 MPa was combined with lacticin 3147 against *S. aureus* and *L. innocua* in milk (Morgan et al., 2000). Single treatments resulted in 2.2 and 1 log reduction for pressure and bacteriocin treatment, respectively, while combination resulted in >6 log reductions. Combining HPP and nisin gave a greater inactivation of *E. coli*, *Pseudomonas fluorescens*, *L. innocua*, and *Lactobacillus viridescens* in milk than when either of the treatment was applied individually (Black et al., 2005). Several studies reported the effectiveness of PEF treatment and antimicrobials (Table 20.1) (Alpas and Bozoglu, 2000; Monfort et al., 2012; Ponce et al., 1998; Zhao et al., 2013). The synergistic effects observed between HHP

TABLE 20.1

Level of Reduction for Combined High-Pressure Treatment

Combined High-Pressure Treatment	Microorganism	Food	Log Reduction	Reference
500 MPa, 5 min, 20°C + nisin (500 IU/mL)	*L. innocua*	Milk	>8 log	Black et al. (2005)
450 MPa, 5 min, 20°C + nisin (500 IU/mL)	*E. coli*	Milk	>8 log	
345 MPa, 5 min, 50°C + bacteriocin (5000 AU/mL)	*S. aureus* *L. monocytogenes*	Milk	>8 log	Alpas and Bozoglu (2000)
300 MPa, for 3 min, 20°C + triethyl citrate (2%)	*E. coli*	Liquid whole eggs	4.4 log	Monfort et al. (2012)
300 MPa, for 3 min, 20°C + triethyl citrate (2%)	*L. innocua*		1.0 log	
450 MPa, 10 min, 20°C + nisin (5 mg/L)	*L. innocua*	Liquid whole eggs	>6 log	Ponce et al. (1998)
450 MPa, 10 min, 20°C + nisin (5 mg/L)	*E. coli*		>5 log	
400 MPa, 4 min, 25°C + nisin (100 IU/mL)	Total aerobic count	Cucumber juice drink	≈4 log	Zhao et al. (2013)
300 MPa, for 20 min + (+)-limonene (200 µL/L)	*E. coli* O157:H7	Apple or orange juice	5 log	Espina et al. (2013)

and antimicrobials might be due to the damage that high pressure induces to cells and easier entrance of antimicrobial molecules into the pressure-treated cells.

No viable cells were determined when pediocin AcH (3000 AU/mL) was combined with high-pressure treatment (345 MPa) at 50°C for 5 min (Kalchayanand et al., 1998). When lower temperature of 25°C was applied, the results indicated an additional reduction of 0.6–2.1 log CFU/mL above the effect of bacteriocins alone or HHP treatment alone in *S. aureus*, *L. monocytogenes*, and *E. coli* O157:H7.

The effect of combined HHP (400 MPa) and enterocin LM-2 (2560 AU/g) on the refrigerated shelf life of sliced cooked ham was evaluated during storage of 90 days at 4°C (Liu et al., 2012). The results obtained in this study indicated that this combination inactivated *L. monocytogenes* and *S.* Enteritidis and completely inhibited the growth of surviving bacterial cells of sliced cooked ham and extended the shelf life of refrigerated sliced ham, without affecting sensorial properties of food.

Similarly, the synergistic effect can be obtained when high-pressure treatment was combined with essential oils and their chemical constitutes due to similar and joint effect on the microbial structure (Espina et al., 2013; Gayán et al., 2012). The possible mechanism of inactivation by this combined treatment suggested that the HPP treatment leaves the population of sublethally injured cells due to temporary disturbed membrane, while essential oils might also disturb cytoplasmic membrane, which resulted in increased membrane permeability, pH gradient in the cells that might decrease, and modification of osmoregulatory functions (Gayán et al., 2012).

20.5.2 Combined Treatments: Pulsed Electric Field

PEF is considered as a potential and promising nonthermal treatment for food preservation. Microbial inactivation is mainly connected to the electromechanical instability and irreversible electroporation of the bacterial cell membrane, which occurred after microbial exposure to high-voltage PEFs and leads to the leakage of intracellular content and eventually cell lyses (Jeyamkondan et al., 1999). The application of PEF treatment is restricted only to the foods that can sustain high electric fields, have low electrical conductivity, and have no bubbles. These are the reasons for quite a limited number of foods in which PEF was successfully applied, and they are mainly fruit juices, liquid eggs, fruit smoothies, fermented beverages, and milk and dairy products (Martín-Belloso and Sobrino-López, 2011).

Along with many studies that investigated the type of PEF equipment used, treatment parameters, media/food processed, and target microorganism, lately the research is mainly focused on the possibility to obtain greater reduction level when PEF is combined with other preservation. The combination of PEF with other treatments may be more effective than individual treatments, especially with those treatments that affect the cell membrane integrity. Synergistic and/or additive effects were determined when PEF treatment was combined with mild heat treatment, different antimicrobials, ultraviolet light, etc.

20.5.2.1 *PEF and Mild Temperature*

PEF treatment can inactivate microorganisms at nonlethal temperatures. The possible explanation for this is that the mild heat treatment causes loss of cell membrane integrity and elasticity as a result of the transition of membrane phospholipids from gel to liquid crystalline structure and therefore become more sensitive to PEF treatment. Several studies reported the effectiveness of PEF treatment at nonlethal temperature levels (Table 20.2) (Amiali et al., 2007; Bazhal et al., 2006; Fleischman et al., 2004; Jin et al., 2009; Ravishankar et al., 2002; Reina et al., 1998; Walkling-Ribeiro et al., 2008, 2009a). However, the underlying mechanism by which PEF sensitivity increases with temperature is not completely understood.

When native microbiota in whole raw milk was treated with the combined effect of heat (50°C) and PEF (40 kV/cm, 33 μs), similar reduction level up till 6.0 log CFU/mL was obtained, being almost the same to the reduction obtained by thermal pasteurization only (72°C during 26 s, 6.7 log CFU/mL). Nevertheless, this combined treatment resulted in longer microbiological shelf life of 21 days at 4°C than the shelf life of thermally treated milk of only 14 days at 4°C (Walkling-Ribeiro et al., 2009a).

Due to the aggregation of egg protein at a temperature of 56°C, the mild heat treatment combined with PEF can be effectively used for the production of safe liquid eggs. Bazhal et al. (2006) investigated the combined PEF treatment and mild temperature in whole liquid eggs, and they determined 2.5 log CFU/mL reduction of *E. coli* O157:H7 when temperature of 55°C was applied with PEF of 15, 11, and 9 kV/cm. With a temperature increase up till 60°C, reduction level increased up till 3.5–4 log CFU/mL (Bazhal et al., 2006). The reduction level of 3 log CFU/mL of *S.* Typhimurium in liquid whole egg was observed after the application of PEF (25 kV/cm, 250 μs) combined with a mild temperature of 55°C. This obtained reduction level was comparable with the one obtained solely by heat treatment of 60°C during 3.5 min (Jin et al., 2009). Therefore, the combination of PEF treatment with mild heat resulted in reduced liquid egg pasteurization temperature while achieving similar reduction in the observed pathogen, making this hurdle technology applicable in the food industry.

20.5.2.2 *PEF and Antimicrobials*

Combinations of PEF and antimicrobials have been reported to be effective in inactivating different pathogens and spoilage microorganisms. The mechanism of inactivation is still not completely clear, but it is assumed that PEF treatments allow antimicrobials to enter easier through the cell membrane and

TABLE 20.2

Level of Reduction for Combined PEF Treatments and Mild Heat

Combined PEF Treatment	Microorganism	Substrate	Reduction (log CFU/mL)	Reference
30 kV/cm + 50°C	*L. monocytogenes*	Milk	4 log	Reina et al. (1998)
20 kV/cm + 55°C	*L. monocytogenes*	Water	>9 log	Fleischman et al. (2004)
30 kV/cm + 55°C	*E. coli* O157:H7	Water	4 log	Ravishankar et al. (2002)
9, 11, 15 kV/cm + 50°C	*E. coli* O157:H7	Liquid whole eggs	2.5 log	Bazhal et al. (2006)
9, 11, 15 kV/cm + 60°C	*E. coli* O157:H7	Liquid whole eggs	3.5–4 log	
25 kV + 55°C	*S.* Typhimurium	Liquid whole eggs	3 log	Jin et al. (2009)
40 kV/cm + 50°C	Natural microbiota	Raw milk	6 log	Walkling-Ribeiro et al. 2009a)

react inside the cell interior (Calderón-Miranda et al., 1999; Dutreux et al., 2000). Although antimicrobials are mainly applied against Gram-positive bacteria, due to the outer lipopolysaccharide component of the membrane of Gram-negative bacteria that prevents antimicrobial components to access the cytoplasmic membrane or peptidoglycan layer (Helander and Mattila-Sandholm, 2000), they might be also applied against Gram-negative bacteria using hurdle concept and combination with PEF treatment. It is assumed that the effectiveness of these antimicrobials is improved due to initially disturbing the barrier properties of the outer membrane by PEF treatment (Smith et al., 2002).

McNamee et al. (2010) reported the synergistic effect of PEF treatment (40 kV/cm, 100 μs, max temperature 56°C) combined with 2.5 ppm of nisin in orange juice. This combination reduced a population of *L. innocua* and *E. coli* K12 for 5.6 and 7.9 log CFU/mL, respectively, which was 1.7 and 0.8 log CFU/mL greater than the sum of the individual treatments (McNamee et al., 2010). Similarly, the synergistic effect was determined when PEF treatment, which consisted of 80 kV/cm and 20 pulses, was combined with 100 IU/mL nisin in tomato juice. The obtained reductions in natural microbiota were greater than the sum of individual treatments (Nguyen and Mittal, 2007).

Additionally, the effect of combined PEF treatment and nisin against Gram-positive bacteria seems to be dependent on the temperature. The additive or slightly synergistic effect was observed for *L. monocytogenes* at all tested temperatures (Saldana et al., 2011). However, this effect was synergistic for *S. aureus* only at low temperatures and tended to disappear with a temperature increase. The authors reported 4.5 and 5.5 log CFU/mL reduction in the populations of *S. aureus* and *L. monocytogenes*, respectively, in a laboratory medium of pH 3.5 in the presence of 200 μg/mL of nisin at 50°C (Saldana et al., 2011).

Some researchers also investigated if the sequence of the preservation factors may induce different results, and it seems that for some preservation factors, the order of application is very important. Gallo et al. (2007) showed in their work that the initial exposure of *L. innocua* to PEF treatment and subsequent exposure to nisin resulted in an antagonistic effect on bacterial inactivation. It is supposed that this might be due to the changes in the cell envelope and modifications of the medium caused by the application of PEF treatment. Therefore, the inactivation effect of nisin was reduced, showing an increase in the resistance of *L. innocua* to nisin. The opposite order of preservation factors, the addition of nisin prior to PEF treatment, exhibited an additive and slightly synergistic effect, indicating the possibility of binding nisin to the cell membrane, which resulted in greater susceptibility of the microorganism to PEF treatment (Gallo et al., 2007). However, Calderón-Miranda et al. (1999) reported no effect of the sequence of order of nisin and PEF treatment in milk, while many researchers have not included in their reports the investigation of the effect of sequence of order.

The combined effect of lactic acid (500 ppm) and PEF (40 kV/cm, 100 μs, 56°C) was found to have synergistic effect against *Pichia fermentans*, a spoilage yeast in orange juice (McNamee et al., 2010). The authors reported 7.8 log CFU/mL reduction, which is 3.1 log CFU/mL greater than the sum of individual treatments. The possible explanation for this synergy can be found in the simultaneous influx of lactic acid molecules and hydrogen ions into the microbial cytoplasm and the increased permeability of membrane due to PEF treatment. The combination of benzoic acid (100 ppm) and PEF treatment inactivated *P. fermentans* in an additive manner, since 5.5 log CFU/mL reduction was not significantly different from the sum of the effects of individual treatments (5.6 log CFU/mL) (McNamee et al., 2010).

Similar to the synergistic effect of essential oils and HPP, these oils have also shown synergistic effect when combined with PEF treatment. Outstanding synergistic lethal effects were determined using mild heat (54°C, 10 min) or PEF (30 kV/cm, 25 pulses) combined with 0.2 μL/mL of most essential oils tested, against *E. coli* O157:H7 and *L. monocytogenes* in apple and orange juices (Ait-Ouazzou et al., 2011, 2012).

20.5.3 Combined Treatments: Ultrasound

Ultrasound treatment is one of the emerging nonthermal methods used in the food industry. It involves the application of high-intensity ultrasonic waves, which causes cavitation of the cells, and even low intensity ultrasound may result in the modification of cells' metabolism, thinning the membrane, localized heating, and production of free radicals. Cavitation is the formation of vapor bubbles of a flowing liquid in a region where the pressure of the liquid falls below its vapor pressure. Ultrasound obtained special attention when combined with external hydrostatic pressure within the name manosonication (MS), which resulted in greater level of lethality. Also, the successful combination of ultrasound and heat was determined, within the name thermosonicaiton (TS), or a combination of ultrasound, pressure and heat, and MTS. The enhanced mechanical disruption of cells is the reason for the enhanced lethality when ultrasound is combined with heat or pressure.

20.5.3.1 Ultrasound and Mild Temperature

The inactivation effect of ultrasound increased when applied with mild temperatures (40°C–70°C) (Raso et al., 1998). A study of Czank et al. (2010) reported that the TS (150 W, 20 kHz at 45°C and 50°C) was more effective in inactivating *E. coli* and *Staphylococcus epidermidis* compared to ultrasound treatment alone in human milk. The combination of sublethal temperatures of 40°C, 45°C, and 50°C and ultrasound (20 kHz, 0.46 W/mL up to 20 min) was more effective in inactivating *E. coli* in apple cider, than the thermal treatment alone (Ugarte-Romero et al., 2006). Even more, the quality characteristics of cider were not significantly affected. The effect of ultrasound and mild temperature of 50°C and 60°C on *E. coli* O157:H7 and *S.* Enteritidis were investigated in mango juice (Kiang et al., 2013). Their results showed that mango juice samples treated with sonication (25 kHz and 200 W) at 60°C during 7 min resulted in approx. 5 log CFU/mL reduction of *E. coli* O157:H7, whereas the reduction was only 4.4 log CFU/mL when thermal treatment alone was applied. Greater reduction level of approx. 9 log CFU/mL was determined for *S.* Enteritidis after 3 and 7 min with sonication at 60°C and 50°C, respectively, indicating that *S.* Enteritidis is more susceptible to this combination than *E. coli* O157:H7. Synergistic effect between ultrasound and mild heat was also confirmed in the study of Wordon et al. (2012).

It has been shown that the combined use of ultrasound, moderate heat, and pressure may improve the bactericidal effect in either an additive or synergistic way (Lee et al., 2009). These authors investigated the effect of ultrasound (20 kHz, 124 μm amplitude), at 40°C, 47°C, 54°C, and 61°C and 100, 300, 400, and 500 kPa on the inactivation level of *E. coli* in phosphate buffer. The combination of lethal effects significantly shortened the time that was needed to obtain 5 log CFU/mL reduction. More efficient was TS and MTS compared to MS and ultrasound. The effect of simultaneous application of heat and ultrasonic waves under pressure (MTS) on the inactivation of *C. sakazakii* in apple juice was studied by Arroyo et al. (2012). Their results indicated that temperature below 45°C showed no effect on the inactivation level of pathogen and that the synergistic effect was observed in the temperature range from 45°C till 64°C, with the maximum obtained at 54°C. They also determined that the cells that survived MTS treatment showed decrease in number during subsequent storage of apple juice under refrigeration.

20.5.3.2 Ultrasound and Antimicrobials

The combination of naturally occurring and synthetic antimicrobials with nonthermal processing techniques has been proven to be effective in microbial inactivation, although the limited number of publication is available regarding the combination of ultrasound and antimicrobial agents. High-intensity ultrasound combined with mild heat treatment and natural antimicrobials may be effective in inactivating *L. monocytogenes* in orange juice as reported by Ferrante et al. (2007). The treatment involved the combination of moderate temperature (45°C), high-intensity ultrasound (600 W, 20 kHz), and the addition of different levels of vanillin (0, 1000, 1500, and 2000 ppm), citral (0, 75, and 100 ppm), or both. The presence of antimicrobial agents increased the bactericidal effect of ultrasound and when both antimicrobials were applied with ultrasound. Also, the greatest multitarget inactivation effect was observed between 45°C and 55°C, when a combination of ultrasound, vanillin, and temperature was used against *L. innocua* (Gastélum et al., 2012).

Low-weight chitosan (1000 ppm) was used in combination with high-intensity ultrasound (20 kHz, 45°C) against *Saccharomyces cerevisiae*. The obtained results indicated that the addition of chitosan enhanced the inactivation by ultrasound (Guerrero et al., 2005). This is especially important having in mind the fact that yeast are resistant to ultrasound and on the other have low-acid fruit juices are favorable for molds and yeast. The obtained combination can be promising for minimally processed fruit beverages.

The combined treatment of ultrasound with organic acids (malic acid, lactic acid, and citric acid) was effective at increasing the reduction of *E. coli* O157:H7, *S.* Typhimurium, and *L. monocytogenes* on the surface of lettuce compared to the treatment with organic acids alone without significantly affecting the quality of organic fresh lettuce (Sagong et al., 2011). They suggested that the obtained results are mainly attributed to the cavitation caused by ultrasound-removed pathogens from lettuce leaves and simultaneously enhanced the access of organic acids to the sites of the leaves not easily accessible such as cut surfaces, punctures, and cracks in produce surfaces. Therefore, this combination can be used in the food industry to increase the microbial safety of fresh produce.

20.5.4 Combined Treatments: Intense Light Pulses

ILP is one of the emerging nonthermal treatments investigated to be an alternative to the traditional thermal treatments. The treatment consisted of short high-intensity pulses of broad-spectrum light (rich in UV-C light, the portion of electromagnetic spectrum corresponding to the band between 200 and 280 nm) that is used to inactivate microorganisms. The mechanism of inactivation is mainly related to its photochemical effect, which includes the chemical modification and cleavage of DNA and denaturation of proteins.

Only limited research is available regarding the combined treatments with ILP. Uesugi and Moraru (2009) have recently evaluated the effect of combined pulsed light and nisin on the inactivation of *L. innocua* in ready-to-eat canned sausages. Nisin at a concentration of 5000 IU/mL alone resulted in 2.35 log CFU, and pulsed light (9.4 J/cm^2) alone reduced *L. innocua* by 1.37 log CFU. The combined pulsed light and nisin resulted in a significantly greater reduction compared with that achieved with individual treatments. This indicates that pulsed light and nisin could have an additive effect on ready-to-eat sausages. The additive effect of these two treatments was pronounced during further storage at 4°C and resulted in a total reduction of 4.03 log CFU after 48 h storage at 4°C. The authors suggested that this combination of nonthermal treatments might be used as an effective step in the production of ready-to-use foods (Uesugi and Moraru, 2009).

20.5.5 Combinations of Nonthermal Treatments

Nonthermal treatments are often combined with antimicrobials, mild heat, or low pH, but often they are combined with each other. These combinations may be successful only if the adequate level of inactivation is obtained and if the treatments are compatible in adequate technical ways.

Additive effect of PEF (11.3 pulses, 60 kV/cm, specific energy 162 J/mL) and UV radiation (length of 30 cm, treatment time of 1.8 s, and flow rate of 8 mL/min) was determined in apple juice (Gachovska et al., 2008). The observed reduction level of *E. coli* was 5.3 log CFU/mL. The same authors have not reported any synergistic effect of combined PEF and UV treatments in apple juice. Nevertheless, PEF applied with UV treatment in poultry chiller water resulted in >6 log CFU/mL of *E. coli* O157:H7 in poultry chiller water (Ngadi et al., 2004). The authors reported synergistic antimicrobial effect between PEF and UV treatments when PEF treatment with <50 pulses was applied, while additive effect was seen only with larger pulse number. The authors characterized this synergism as a result of complete inactivation of the injured cells resulting from the UV treatment up till 50 pulses (Ngadi et al., 2004).

When combination of UV pretreatment was followed by PEF at 40 kV/cm, 100 pulses (1 µs, 15 Hz) in freshly squeezed apple juice, 7.1 log CFU/mL reduction of natural microbiota was observed (Noci et al., 2008). The opposite order of applied hurdles (PEF followed by UV) gave 6.2 log CFU/mL reduction level in apple juice. Although no synergistic effect was obtained, the quality attributes of apple juice after combined treatments applied were similar to those observed in the milder heat process (72°C) and consistently superior when compared to the severe heat treatment (94°C), with the exception of enzyme inactivation. This indicates the potential for the usage of combined treatments in the processing of freshly squeezed apple juice (Noci et al., 2008).

The enhanced synergistic effect observed with the PEF (24 kV/cm with 89 pulses) and high-intense light pulses (HILP with pulse length 360 µs, 3 Hz, 4 J/cm^2) sequence was applied against *E. coli* K12 in apple juice. This may be due to the greater damage to the cell membrane induced by PEF as a first hurdle, possibly causing increased susceptibility to the UV component of the subsequent HILP treatment. Nevertheless, the opposite order of treatment, meaning HILP and subsequent PEF, resulted in only additive effect, being 4.95 log CHU/mL compared to 5.1 log CFU/mL summed individual effects of HILP and PEF (Caminiti et al., 2011).

It is very important to evaluate the right combination of different hurdles in specific food products, against pathogenic and spoilage bacteria, which might be commonly found in that food product and might cause food safety- or quality-related issues. Therefore, the studies investigating the effectiveness of the combined treatments on the microbiological safety should also incorporate additional studies where adequate quality parameters and sensorial characteristics are tested (Mosqueda-Melgar et al., 2012).

Recently, the improved microbial inactivation was reported when ultrasound was combined with other nonlethal emerging treatments, such as PEF. The study of Noci et al. (2009) showed that the pretreatment of milk at 55°C followed by thermosonication (24 kHz, 400 W) and PEF (40 kV/cm) resulted in the reduction level of *L. innocua*, which is comparable with thermal pasteurization. The thermosonication treatment decreased the severity of the temperature/time exposure over thermal treatment alone. Similar observation was found for thermosonication treatment combined with PEF for orange juice (Walkling-Ribeiro et al., 2009b).

When MTS (100%, 160 mL/min, 200 kPa) was followed by PEF (34 kV/cm, 32 ms), the log reduction of *L. innocua* in milk-based smoothie was 5.6 CFU/mL compared to 4.2 CFU/mL for the reverse sequence (Palgan et al., 2012). When comparing the results obtained for MTS followed by PEF or the reverse sequence with the theoretical sum of reductions for each of the individual hurdles, the effect was found to be additive as the inactivation obtained by these hurdle combinations was not significantly different from the sum of the two hurdles used individually (5.7 CFU/mL) under the same conditions. The greater inactivation observed with the MTS + PEF sequence may be due to greater cell damage induced by MTS (as a first hurdle) through cavitation, possibly causing increased susceptibility to the subsequent PEF treatment (Palgan et al., 2012).

20.6 Final Remarks

Hurdle technology implies the combination of existing and novel nonthermal preservation treatments to establish a series of various hurdles that microorganism should not be able to overcome (Leistner, 1992; Raso and Barbosa-Cánovas, 2003). This is particularly important for the microorganisms and bacterial spores that are very resistant to applied inactivation treatments and are a drawback for the application of nonthermal treatments. To apply principles of food preservation by combined processes correctly, an understanding of the mechanisms of action of the individual treatments alone and in its combination is needed. This will further allow justified and well-balanced combination of hurdles in order to achieve the desired level of safety and quality (Raso and Barbosa-Cánovas, 2003). This approach of combining hurdles will allow the application of factors at lower intensities, instead of applying only one factor at such high intensity that causes severe changes in the quality of food. Selection of adequate nonthermal treatments is difficult and requires extensive evaluation. Although these treatments are useful in reducing pathogen prevalence and controlling microbial growth, they may not deliver complete inactivation of present pathogenic microorganisms. The advantage of using combined preservation treatments can be seen only in the case the multihurdles are carefully selected and applied in correct order, in such a way to cause metabolic exhaustion of surviving pathogenic cells, as a result of great expenditure of energy. Combination of treatments should be selected in a way that will not allow survivors from the first hurdle to successfully cope with a subsequent hurdle or with the final acid stress in stomach. Nowadays, combinations should be made based on the available information regarding the mode of action of specific treatment, in order to obtain the most successful combination. Nevertheless, more research and clarification in the field of resistance, cross-protection, damage, and recovery are needed, as the effectiveness of the treatment combination can be overestimated, while contamination level underestimated in the case that injured bacteria are able to resuscitate into a viable state (Jasson et al., 2009).

REFERENCES

Ait-Ouazzou, A., Cherrat, L., Espina, L., Lorán, S., Rota, C., and Pagán, R. 2011. The antimicrobial activity of hydrophobic essential oil constituents acting alone or in combined processes of food preservation. *Innovative Food Science and Emerging Technologies*, 12(3), 320–329.

Ait-Ouazzou, A., Espina, L., Cherrat, L., Hassani, M., Laglaoui, A., Conchello, P., and Pagán, R. 2012. Synergistic combination of essential oils from Morocco and physical treatments for microbial inactivation. *Innovative Food Science and Emerging Technologies*, 16, 283–290.

Alpas, H. and Bozoglu, F. 2000. The combined effect of high hydrostatic pressure, heat and bacteriocins on inactivation of foodborne pathogens in milk and orange juice. *World Journal of Microbiology and Biotechnology*, 16(4), 387–392.

Amiali, M., Ngadi, M. O., Smith, J. P., and Raghavan, G. S. V. 2007. Synergistic effect of temperature and pulsed electric field on inactivation of *Escherichia coli* O157:H7 and *Salmonella* Enteritidis in liquid egg yolk. *Journal of Food Engineering*, 79(2), 689–694.

Arroyo, C., Cebrián, G., Pagán, R., and Condón, S. 2012. Synergistic combination of heat and ultrasonic waves under pressure for *Cronobacter sakazakii* inactivation in apple juice. *Food Control*, 25(1), 342–348.

Bazhal, M. I., Ngadi, M. O., Raghavan, G. S. V., and Smith, J. P. 2006. Inactivation of *Escherichia coli* O157:H7 in liquid whole egg using combined pulsed electric field and thermal treatments. *LWT—Food Science and Technology*, 39(4), 420–426.

Black, E. P., Kelly, A. L., and Fitzgerald, G. F. 2005. The combined effect of high pressure and nisin on inactivation of microorganisms in milk. *Innovative Food Science and Emerging Technologies*, 6(3), 286–292.

Bull, M. K., Oliver, S. A., van Diepenbeek, R. J., Kormelink, F., and Chapman, B. 2009. Synergistic inactivation of spores of proteolytic *Clostridium botulinum* strains by high pressure and heat is strain and product dependent. *Applied and Environmental Microbiology*, 75(2), 434–445.

Calderón-Miranda, M. L., Barbosa-Cánovas, G. V., and Swanson, B. G. 1999. Inactivation of *Listeria innocua* in liquid whole egg by pulsed electric fields and nisin. *International Journal of Food Microbiology*, 51(1), 7–17.

Caminiti, I. M., Palgan, I., Noci, F., Muñoz, A., Whyte, P., Cronin, D. A., Morgan, D. J., and Lyng, J. G. 2011. The effect of pulsed electric fields (PEF) in combination with high intensity light pulses (HILP) on *Escherichia coli* inactivation and quality attributes in apple juice. *Innovative Food Science and Emerging Technologies*, 12(2), 118–123.

Chawla, S. P. and Chander, R. 2004. Microbiological safety of shelf-stable meat products prepared by employing hurdle technology. *Food Control*, 15(7), 559–563.

Considine, K. M., Kelly, A. L., Fitzgerald, G. F., Hill, C., and Sleator, R. D. 2008. High-pressure processing—Effects on microbial food safety and food quality. *FEMS Microbiology Letters*, 281(1), 1–9.

Czank, C., Simmer, K., and Hartmann, P. E. 2010. Simultaneous pasteurization and homogenization of human milk by combining heat and ultrasound: Effect on milk quality. *Journal of Dairy Research*, 77(02), 183–189.

Dutreux, N., Notermans, S., Wijtzes, T., Góngora-Nieto, M. M., Barbosa-Cánovas, G. V., and Swanson, B. G. 2000. Pulsed electric fields inactivation of attached and free-living *Escherichia coli* and *Listeria innocua* under several conditions. *International Journal of Food Microbiology*, 54(1–2), 91–98.

Espina, L., García-Gonzalo, D., Laglaoui, A., Mackey, B. M., and Pagán, R. 2013. Synergistic combinations of high hydrostatic pressure and essential oils or their constituents and their use in preservation of fruit juices. *International Journal of Food Microbiology*, 161(1), 23–30.

Ferrante, S., Guerrero, S., and Alzamora, S. M. 2007. Combined use of ultrasound and natural antimicrobials to inactivate *Listeria monocytogenes* in orange juice. *Journal of Food Protection*, 70(8), 1850–1856.

Fleischman, G. J., Ravishankar, S., and Balasubramaniam, V. M. 2004. The inactivation of *Listeria monocytogenes* by pulsed electric field (PEF) treatment in a static chamber. *Food Microbiology*, 21(1), 91–95.

Gachovska, T. K., Kumar, S., Thippareddi, H., Subbiah, J., and Williams, F. 2008. Ultraviolet and pulsed electric field treatments have additive effect on inactivation of *E. coli* in apple juice. *Journal of Food Science*, 73(9), M412–M417.

Gallo, L. I., Pilosof, A. M. R., and Jagus, R. J. 2007. Effect of the sequence of nisin and pulsed electric fields treatments and mechanisms involved in the inactivation of *Listeria innocua* in whey. *Journal of Food Engineering*, 79(1), 188–193.

Gao, Y.-L., Ju, X.-R., and Jiang, H.-H. 2006. Studies on inactivation of *Bacillus subtilis* spores by high hydrostatic pressure and heat using design of experiments. *Journal of Food Engineering*, 77(3), 672–679.

Gastélum, G., Avila-Sosa, R., López-Malo, A., and Palou, E. 2012. *Listeria innocua* multi-target inactivation by thermo-sonication and vanillin. *Food and Bioprocess Technology*, 5(2), 665–671.

Gayán, E., Torres, J. A., and Paredes-Sabja, D. 2012. Hurdle approach to increase the microbial inactivation by high pressure processing: Effect of essential oils. *Food Engineering Reviews*, 4(3), 141–148.

Guerrero, S., Tognon, M., and Alzamora, S. M. 2005. Response of *Saccharomyces cerevisiae* to the combined action of ultrasound and low weight chitosan. *Food Control*, 16(2), 131–139.

Hauben, K. J. A., Bartlett, D. H., Soontjens, C. C. F., Cornelis, K., Wuytack, E. Y., and Michiels, C. W. 1997. *Escherichia coli* mutants resistant to inactivation by high hydrostatic pressure. *Applied and Environmental Microbiology*, 63(3), 945–950.

Havelaar, A. H., Brul, S., de Jong, A., de Jonge, R., Zwietering, M. H., and ter Kuile, B. H. 2010. Future challenges to microbial food safety. *International Journal of Food Microbiology*, 139(Suppl. 1), S79–S94.

Helander, I. M. and Mattila-Sandholm, T. 2000. Permeability barrier of the Gram-negative bacterial outer membrane with special reference to nisin. *International Journal of Food Microbiology*, 60(2–3), 153–161.

Jasson, V., Rajkovic, A., Baert, L., Debevere, J., and Uyttendaele, M. 2009. Comparison of enrichment conditions for rapid detection of low numbers of sublethally injured *Escherichia coli* O157 in food. *Journal of Food Protection*, 72(9), 1862–1868.

Jeyamkondan, S., Jayas, D. S., and Holley, R. A. 1999. Pulsed electric field processing of foods: A review. *Journal of Food Protection*, 62(9), 1088–1096.

Jin, T., Zhang, H., Hermawan, N., and Dantzer, W. 2009. Effects of pH and temperature on inactivation of *Salmonella* Typhimurium DT104 in liquid whole egg by pulsed electric fields. *International Journal of Food Science and Technology*, 44(2), 367–372.

Kalchayanand, N., Sikes, A., Dunne, C. P., and Ray, B. 1998. Factors influencing death and injury of foodborne pathogens by hydrostatic pressure-pasteurization. *Food Microbiology*, 15(2), 207–214.

Karatzas, K. A. G. and Bennik, M. H. 2002. Characterization of a *Listeria monocytogenes* Scott A isolate with high tolerance towards high hydrostatic pressure. *Applied and Environmental Microbiology*, 68(7), 3183–3189.

Kiang, W. S., Bhat, R., Rosma, A., and Cheng, L. H. 2013. Effects of thermosonication on the fate of *Escherichia coli* O157:H7 and *Salmonella* Enteritidis in mango juice. *Letters in Applied Microbiology*, 56(4), 251–257.

Lee, H., Zhou, B., Liang, W., Feng, H., and Martin, S. E. 2009. Inactivation of *Escherichia coli* cells with sonication, manosonication, thermosonication, and manothermosonication: Microbial responses and kinetics modeling. *Journal of Food Engineering*, 93(3), 354–364.

Lee, S.-Y. and Kang, D.-H. 2009. Combined effects of heat, acetic acid, and salt for inactivating *Escherichia coli* O157:H7 in laboratory media. *Food Control*, 20(11), 1006–1012.

Leistner, L. 1992. Food preservation by combined methods. *Food Research International*, 25(2), 151–158.

Leistner, L. 1994. Further developments in the utilization of hurdle technology for food preservation. *Journal of Food Engineering*, 22(1–4), 421–432.

Leistner, L. 2000. Basic aspects of food preservation by hurdle technology. *International Journal of Food Microbiology*, 55(1–3), 181–186.

Leistner, L. and Gorris, L. G. M. 1995. Food preservation by hurdle technology. *Trends in Food Science and Technology*, 6(2), 41–46.

Linton, M., McClements, J. M. J., and Patterson, M. F. 1999a. Inactivation of *Escherichia coli* O157:H7 in orange juice using a combination of high pressure and mild heat. *Journal of Food Protection*, 62(3), 277–279.

Linton, M., McClements, J. M. J., and Patterson, M. F. 1999b. Survival of *Escherichia coli* O157:H7 during storage in pressure-treated orange juice. *Journal of Food Protection*, 62(9), 1038–1040.

Liu, G., Wang, Y., Gui, M., Zheng, H., Dai, R., and Li, P. 2012. Combined effect of high hydrostatic pressure and enterocin LM-2 on the refrigerated shelf life of ready-to-eat sliced vacuum-packed cooked ham. *Food Control*, 24(1–2), 64–71.

Liu, X. I. A., Yousef, A. E., and Chism, G. W. 1997. Inactivation of *Escherichia coli* O157:H7 by the combination of organic acids and pulsed electric field. *Journal of Food Safety*, 16(4), 287–299.

Martín-Belloso, O. and Sobrino-López, A. 2011. Combination of pulsed electric fields with other preservation techniques. *Food and Bioprocess Technology*, 4(6), 954–968.

McNamee, C., Noci, F., Cronin, D. A., Lyng, J. G., Morgan, D. J., and Scannell, A. G. M. 2010. PEF based hurdle strategy to control *Pichia fermentans*, *Listeria innocua* and *Escherichia coli* k12 in orange juice. *International Journal of Food Microbiology*, 138(1–2), 13–18.

Monfort, S., Ramos, S., Meneses, N., Knorr, D., Raso, J., and Álvarez, I. 2012. Design and evaluation of a high hydrostatic pressure combined process for pasteurization of liquid whole egg. *Innovative Food Science and Emerging Technologies*, 14, 1–10.

Morgan, S. M., Ross, R. P., Beresford, T., and Hill, C. 2000. Combination of hydrostatic pressure and lacticin 3147 causes increased killing of *Staphylococcus* and *Listeria*. *Journal of Applied Microbiology*, 88(3), 414–420.

Mosqueda-Melgar, J., Raybaudi-Massilia, R. M., and Martín-Belloso, O. 2012. Microbiological shelf life and sensory evaluation of fruit juices treated by high-intensity pulsed electric fields and antimicrobials. *Food and Bioproducts Processing*, 90(2), 205–214.

Mújica-Paz, H., Valdez-Fragoso, A., Samson, C., Welti-Chanes, J., and Torres, J. A. 2011. High-pressure processing technologies for the pasteurization and sterilization of foods. *Food and Bioprocess Technology*, 4(6), 969–985.

Murrell, W. G. and Wills, P. A. 1977. Initiation of *Bacillus* spore germination by hydrostatic pressure: Effect of temperature. *Journal of Bacteriology*, 129(3), 1272–1280.

Neetoo, H., Pizzolato, T., and Chen, H. 2009. Elimination of *Escherichia coli* O157:H7 from Alfalfa seeds through a combination of high hydrostatic pressure and mild heat. *Applied and Environmental Microbiology*, 75(7), 1901–1907.

Ngadi, M., Jun, X., Smith, J., and Raghavan, G. S. V. 2004. Inactivation of *Escherichia coli* O157:H7 in poultry chiller water using combined ultraviolet light, pulsed electric field and ozone treatments. *International Journal of Poultry Science*, 3(11), 733–737.

Nguyen, P. and Mittal, G. S. 2007. Inactivation of naturally occurring microorganisms in tomato juice using pulsed electric field (PEF) with and without antimicrobials. *Chemical Engineering and Processing: Process Intensification*, 46(4), 360–365.

Noci, F., Riener, J., Walkling-Ribeiro, M., Cronin, D. A., Morgan, D. J., and Lyng, J. G. 2008. Ultraviolet irradiation and pulsed electric fields (PEF) in a hurdle strategy for the preservation of fresh apple juice. *Journal of Food Engineering*, 85(1), 141–146.

Noci, F., Walkling-Ribeiro, M., Cronin, D. A., Morgan, D. J., and Lyng, J. G. 2009. Effect of thermosonication, pulsed electric field and their combination on inactivation of *Listeria innocua* in milk. *International Dairy Journal*, 19(1), 30–35.

Paidhungat, M., Setlow, B., Daniels, W. B., Hoover, D., Papafragkou, E., and Setlow, P. 2002. Mechanisms of induction of germination of *Bacillus subtilis* spores by high pressure. *Applied and Environmental Microbiology*, 68(6), 3172–3175.

Palgan, I., Muñoz, A., Noci, F., Whyte, P., Morgan, D. J., Cronin, D. A., and Lyng, J. G. 2012. Effectiveness of combined Pulsed Electric Field (PEF) and Manothermosonication (MTS) for the control of *Listeria innocua* in a smoothie type beverage. *Food Control*, 25(2), 621–625.

Ponce, E., Pla, R., Sendra, E., Guamis, B., and Mor-Mur, M. 1998. Combined effect of nisin and high hydrostatic pressure on destruction of *Listeria innocua* and *Escherichia coli* in liquid whole egg. *International Journal of Food Microbiology*, 43(1–2), 15–19.

Rajkovic, A., Smigic, N., and Devlieghere, F. 2010. Contemporary strategies in combating microbial contamination in food chain. *International Journal of Food Microbiology*, 141(Suppl. 1), S29–S42.

Rajkovic, A., Uyttendaele, M., Van Houteghem, N., Gómez, S. M. O., Debevere, J., and Devlieghere, F. 2009. Influence of partial inactivation on growth of *Listeria monocytogenes* under sub-optimal conditions of increased NaCl concentration or increased acidity. *Innovative Food Science and Emerging Technologies*, 10(2), 267–271.

Raso, J. and Barbosa-Cánovas, G. V. 2003. Nonthermal preservation of foods using combined processing techniques. *Critical Reviews in Food Science and Nutrition*, 43(3), 265–285.

Raso, J., Pagán, R., Condón, S., and Sala, F. J. 1998. Influence of temperature and pressure on the lethality of ultrasound. *Applied and Environmental Microbiology*, 64(2), 465–471.

Ravishankar, S., Fleischman, G. J., and Balasubramaniam, V. M. 2002. The inactivation of *Escherichia coli* O157:H7 during pulsed electric field (PEF) treatment in a static chamber. *Food Microbiology*, 19(4), 351–361.

Reina, L. D., Jin, Z. T., Zhang, Q. H., and Yousef, A. E. 1998. Inactivation of *Listeria monocytogenes* in milk by pulsed electric field. *Journal of Food Protection*, 61(9), 1203–1206.

Ross, A. I. V., Griffiths, M. W., Mittal, G. S., and Deeth, H. C. 2003. Combining nonthermal technologies to control foodborne microorganisms. *International Journal of Food Microbiology*, 89(2–3), 125–138.

Sagong, H.-G., Lee, S.-Y., Chang, P.-S., Heu, S., Ryu, S., Choi, Y.-J., and Kang, D.-H. 2011. Combined effect of ultrasound and organic acids to reduce *Escherichia coli* O157:H7, *Salmonella* Typhimurium, and *Listeria monocytogenes* on organic fresh lettuce. *International Journal of Food Microbiology*, 145(1), 287–292.

Saldana, G., Minor-Perez, H., Raso, J., and Alvarez, I. 2011. Combined effect of temperature, pH, and presence of nisin on inactivation of *Staphylococcus aureus* and *Listeria monocytogenes* by pulsed electric fields. *Foodborne Pathogens and Disease*, 8(7), 797–802.

Smigic, N., Rajkovic, A., Antal, E., Medic, H., Lipnicka, B., Uyttendaele, M., and Devlieghere, F. 2009. Treatment of *Escherichia coli* O157:H7 with lactic acid, neutralized electrolyzed oxidizing water and chlorine dioxide followed by growth under sub-optimal conditions of temperature, pH and modified atmosphere. *Food Microbiology*, 26(6), 629–637.

Smigic, N., Rajkovic, A., Nielsen, D. S., Arneborg, N., Siegumfeldt, H., and Devlieghere, F. 2010. Survival of lactic acid and chlorine dioxide treated *Campylobacter jejuni* under suboptimal conditions of pH, temperature and modified atmosphere. *International Journal of Food Microbiology*, 141(Suppl. 1), S140–S146.

Smith, K., Mittal, G. S., and Griffiths, M. W. 2002. Pasteurization of milk using pulsed electrical field and antimicrobials. *Journal of Food Science*, 67(6), 2304–2308.

Thomas, R., Anjaneyulu, A. S. R., and Kondaiah, N. 2008. Development of shelf stable pork sausages using hurdle technology and their quality at ambient temperature (37 ± 1°C) storage. *Meat Science*, 79(1), 1–12.

Tiganitas, A., Zeaki, N., Gounadaki, A. S., Drosinos, E. H., and Skandamis, P. N. 2009. Study of the effect of lethal and sublethal pH and aw stresses on the inactivation or growth of *Listeria monocytogenes* and *Salmonella* Typhimurium. *International Journal of Food Microbiology*, 134(1–2), 104–112.

Torres, J. A. and Velazquez, G. 2005. Commercial opportunities and research challenges in the high pressure processing of foods. *Journal of Food Engineering*, 67(1–2), 95–112.

Uesugi, A. R. and Moraru, C. I. 2009. Reduction of *Listeria* on ready-to-eat sausages after exposure to a combination of pulsed light and nisin. *Journal of Food Protection*, 72(2), 347–353.

Ugarte-Romero, E., Feng, H., Martin, S. E., Cadwallader, K. R., and Robinson, S. J. 2006. Inactivation of *Escherichia coli* with power ultrasound in apple cider. *Journal of Food Science*, 71(2), E102–E108.

Van Houteghem, N., Devlieghere, F., Rajkovic, A., Gómez, S. M. O., Uyttendaele, M., and Debevere, J. 2008. Effects of CO_2 on the resuscitation of *Listeria monocytogenes* injured by various bactericidal treatments. *International Journal of Food Microbiology*, 123(1–2), 67–73.

Walkling-Ribeiro, M., Noci, F., Cronin, D. A., Lyng, J. G., and Morgan, D. J. 2009a. Antimicrobial effect and shelf-life extension by combined thermal and pulsed electric field treatment of milk. *Journal of Applied Microbiology*, 106(1), 241–248.

Walkling-Ribeiro, M., Noci, F., Cronin, D. A., Lyng, J. G., and Morgan, D. J. 2009b. Shelf life and sensory evaluation of orange juice after exposure to thermosonication and pulsed electric fields. *Food and Bioproducts Processing*, 87(2), 102–107.

Walkling-Ribeiro, M., Noci, F., Cronin, D. A., Riener, J., Lyng, J. G., and Morgan, D. J. 2008. Reduction of *Staphylococcus aureus* and quality changes in apple juice processed by ultraviolet irradiation, preheating and pulsed electric fields. *Journal of Food Engineering*, 89(3), 267–273.

Wesche, A. M., Gurtler, J. B., Marks, B. P., and Ryser, E. T. 2009. Stress, sublethal injury, resuscitation, and virulence of bacterial foodborne pathogens. *Journal of Food Protection*, 72(5), 1121–1138.

Wordon, B. A., Mortimer, B., and McMaster, L. D. 2012. Comparative real-time analysis of *Saccharomyces cerevisiae* cell viability, injury and death induced by ultrasound (20 kHz) and heat for the application of hurdle technology. *Food Research International*, 47(2), 134–139.

Zhao, L., Wang, S., Liu, F., Dong, P., Huang, W., Xiong, L., and Liao, X. 2013. Comparing the effects of high hydrostatic pressure and thermal pasteurization combined with nisin on the quality of cucumber juice drinks. *Innovative Food Science and Emerging Technologies*, 17, 27–36.

21

Quorum Sensing in Food-Borne Bacteria and Use of Quorum Sensing Inhibitors as Food Intervention Techniques

Jamuna A. Bai and V. Ravishankar Rai

CONTENTS

21.1 Introduction

In the last few years, significant progress has been made in understanding the virulence, transmission, survival, stress response, and interactions of food-borne bacteria with other microbiota. However, the knowledge we have about the ecology of food spoilage bacteria and the biochemical mechanisms behind spoilage at the molecular level is considerably less. A number of studies imply that virulence- and spoilage-regulated phenotypes in food-borne bacteria are cell density–dependent phenomenon regulated at the genetic level by the mechanism of quorum sensing. In quorum sensing (QS) or cell-to-cell communication, bacteria produce, detect, and respond to the signaling molecules, also known as autoinducers. Bacteria sense their population density by monitoring the signal molecules in their environment and respond to it by activating or repressing target genes and the expression of certain phenotypes. In this chapter, various quorum sensing systems and signaling molecules used by bacteria are reviewed. The quorum sensing mechanisms employed by food-borne pathogens to express virulence factors and the food spoilage bacteria to carry out spoilage activities in foods are summarized. The antiquorum sensing strategies that can attenuate coordinated activity in bacteria are briefed. The application of quorum sensing inhibitors that are commercially available and those derived from foods and phytochemicals as potential food intervention techniques to increase the safety and quality of foods will be discussed.

21.2 Quorum Sensing Systems in Bacteria

21.2.1 Acylated Homoserine Lactones

Acylated homoserine lactones (AHLs) are the signaling molecules used by the gram-negative bacteria to carry out cell-to-cell communication. AHLs have a core homoserine lactone ring that is N-acylated with a fatty acyl group at the C-1 position. The N-acyl side chain varies in length (from 4 to 18 carbons), saturation level (may contain double bonds), and oxidation state (an oxo or hydroxyl substituent is present at the C-3 position). An enzyme of the LuxI family AHL synthases is involved in the synthesis of AHLs. It utilizes *S*-adenosylmethionine (SAM) and an acyl-acyl carrier protein for AHL synthesis (Figure 21.1). The LuxR family of the transcriptional regulators detects and responds to these signaling molecules. The LuxR on binding to AHLs forms a LuxR/AHL complex and undergoes conformational changes. This complex is responsible for the regulation and expression of various target genes as shown in Figure 21.2 (Whitehead et al., 2001). Bacterial species are capable of synthesizing more than one type of AHL, while the same type of AHL may be produced by representatives of different bacterial genera. The short-chain AHLs diffuse through the bacterial membrane and long-chain AHLs are actively transported in and out of the cells via efflux and influx systems. The factors that influence the concentration and type of AHL production are temperature, pH, NaCl, growth medium, inoculum size, and bacterial growth phase (Skandamis and Nychas, 2012).

21.2.2 Autoinducer 2

Autoinducer-2 (AI-2) is the *universal* signal molecule utilized by both gram-positive and gram-negative bacteria for inter- and intracellular communication. AI-2 is a species-nonspecific signaling molecule, and different bacteria respond to it in different ways (Roy et al., 2011). AI-2 is synthesized by the enzymeAI-2 synthase (LuxS) encoded by the luxS gene from SAM until the formation of the AI-2

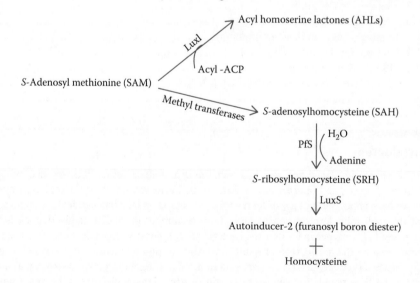

FIGURE 21.1 Pathway for the synthesis of quorum sensing signaling molecules: acylated homoserine lactones and autoinducer-2.

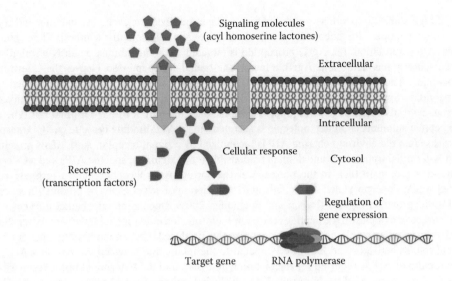

FIGURE 21.2 Overview of acyl homoserine lactone–mediated quorum sensing in bacteria.

precursor, 4,5-dihydroxy-2,3-pentanedione (DPD). DPD is unstable and spontaneously cyclizes into a mixture of compounds of which some function as signaling molecules (Figure 21.1). AI-2 differentiation occurs mainly at the level of transduction, and different bacterial species recognize chemically distinct forms of AI-2 (Miller et al., 2004). The LuxP receptor protein of *Vibrio harveyi* binds to the 2,3-borate diester of the hydrated α-anomer, whereas in *Salmonella enterica* serovar Typhimurium, the receptor protein LsrB recognizes the hydrated β-anomer (Gori et al., 2011). AI-2 has been found to control biofilm formation and virulence in the food-borne pathogens *Listeria monocytogenes*, *Staphylococcus aureus*, and *Escherichia coli* (Barrios et al., 2006). It is also known to control acid stress regulation in dairy-relevant starter cultures and probiotic bacteria such as in *Lactococcus lactis* and *Lactobacillus* spp., respectively (Moslehi-Jenabian et al., 2009).

21.2.3 Autoinducer-3/QseC System

The autoinducer-3(AI-3) or QseC cell-to-cell signaling system is the least studied quorum sensing system with the structure and synthesis of the signal molecule unclear (Walters et al., 2006). A number of commensal bacteria such as nonpathogenic *E. coli* and *Enterobacter cloacae*, as well as pathogenic *Shigella*, *Salmonella*, and *Klebsiella* species are known to produce AI-3. The detection of AI-3 is through a two-component system comprising the sensor kinase QseC and response regulator QseB. In the presence of periplasmic AI-3, QseC first undergoes autophosphorylation and then transfers this phosphate to QseB, which activates the genes responsible for flagella biosynthesis and motility by upregulating the master flagellar regulator gene *flhDC* (Clarke et al., 2006). In enterohemorrhagic *E. coli* (EHEC), AI-3 plays a role in the formation of attaching and effacing lesions through upregulation of five separate loci of enterocyte effacement (LEE) operons located within the EHEC chromosome (Sperandio et al., 2003). The AI-3 produced by the host's flora present in the GI tract is used by the enteric pathogens for upregulation of flagella and motility needed to penetrate the mucosal lining of the colon to reach epithelial cells and to colonize host cells. The QseBC cascade is also known to respond to the adrenergic signals, adrenaline and noradrenaline, found in the GI tract, implying that QseC responds to bacterial and host signals simultaneously (Parker and Sperandio, 2009).

21.2.4 AIP/Agr System

Autoinducing peptides (AIPs) are used by gram-positive bacteria for cell-to-cell communication. AIPs are peptides or modified peptides characterized by a small size (i.e., ranging from 5 to 26 amino acid residues), high stability, specificity, and diversity and can be linear or cyclic (Skandamis and Nychas, 2012). These peptides are encoded by the *agrD* gene and ribosomally synthesized as precursor peptides. After translation, the AgrD propeptide is targeted to the membrane by an N-terminal signal sequence. Once at the membrane, AgrB, a membrane-bound endopeptidase, cleaves the C-terminus of the propeptide. The N-terminus of the propeptide, including the signal sequence, is removed by the signal peptidase SpsB. Finally, the C-terminus of this processed polypeptide is covalently linked to a conserved, centrally located cysteine to form a thiolactone ring with a free N-terminal tail. The active mature peptide autoinducer signal molecule is secreted via an ATP-binding cassette (ABC) transporter. After release into the environment, the AIP is recognized by a signal receptor, AgrC. This protein contains an N-terminal transmembrane domain responsible for recognizing specific AIPs and a C-terminal histidine kinase domain that, in the presence of the correct AIP, phosphorylates a response regulator called AgrA. Phosphorylated AgrA activates transcription of select genes by binding direct repeats located in the promoter regions (Parker and Sperandio, 2009). Depending on whether the sensor is on the cell surface or cytoplasm, the peptides can exert their function either intercellularly or extracellularly. The AIPs can act as an autoinducer for the organism that produced it but an inhibitor to other organisms. This dual role as activator and inhibitor is related to the interaction between the AIP and ArgC. The cyclic structure of AIP is required for interaction with AgrC, and the N-terminal tail is responsible for AgrC activation. Removal of the N-terminal tail results in a universal inhibitor that binds to AgrC but is unable to activate the Agr system (Lyon et al., 2000).

21.3 Mechanism of Quorum Sensing in Food-Borne Pathogens

21.3.1 *E. coli* and *Salmonella* Typhimurium

Escherichia and *Salmonella* do not synthesize AHLs; however, they have a functional AHL receptor encoded by SdiA and are capable of detecting AHLs. *Escherichia*, *Salmonella*, and related bacteria such as *Klebsiella*, *Enterobacter*, and *Citrobacter* encode a single LuxR homolog named SdiA, but lack a corresponding LuxI homolog and, therefore, do not produce AHLs (Ahmer et al., 1998). SdiA of *Salmonella* can detect a wide range of AHL structures, but preferentially binds to oxoC8. SdiA can also detect AHLs that are not modified at position 3 and can detect chain lengths of 4–12. The detection sensitivity of SdiA for the various AHLs is in the range of 1 nM to 1 µM. Even the detection of nonideal AHLs with micromolar sensitivities may be physiologically relevant since *Salmonella* can detect C4 and oxoC12 AHLs produced by *Pseudomonas aeruginosa* (Michael et al., 2001). SdiA is hypothesized to function in the interspecies cross talk with AHL-producing bacteria (Ahmer, 2004; Smith et al., 2008). In response to AHL under laboratory conditions, *Salmonella* SdiA activates the expression of the chromosomally encoded *srgE* gene and the *rck* operon, which is borne on the *S. enterica* sv. Typhimurium virulence plasmid pSLT (Ahmer et al., 1998; Michael et al., 2001). The *rck* operon contains six genes: *pefI*, *srgD*, *srgA*, *srgB*, *rck*, and *srgC* (Ahmer, 2004). PefI is a regulator of the upstream *pef* operon, which encodes fimbriae that bind to blood antigens (Chessa et al., 2008). Rck is an outer membrane protein with functions in resistance to complement killing and adhesion to fibronectin and laminin. In *E. coli*, the SdiA regulon is entirely different than the regulon in *S.* Typhimurium (Soares and Ahmer, 2011). SdiA activates the *gad* genes of the acid fitness island of *E. coli* K-12 and EHEC. SdiA also represses flagellar genes and the genes of the LEE pathogenicity island (LEE) of EHEC. Thus, SdiA controls the expression of virulence factors in EHEC (Kanamaru et al., 2000).

In *E. coli* and *S.* Typhimurium, AI-2 synthesis, secretion, uptake, and signal transduction are regulated by the luxS-regulated operon (lsr) (Taga et al., 2003). In *E. coli*, the lsr operon is largely homologous to the *S.* Typhimurium lsr operon, although some gene functions and their regulation are distinct (Roy et al., 2011). The enzymes Pfs and LuxS convert S-adenosylhomocysteine (SAH) to DPD inside the cell, releasing adenine and homocysteine as intermediates. DPD cyclizes to form AI-2 and is exported into the extracellular medium by a membrane spanning protein, YdgG. When the AI-2 concentration in the extracellular medium exceeds a threshold, the signal transduction cascade for the uptake and processing of AI-2 is triggered and it is imported into the cell via an Lsr transporter. AI-2 gets phosphorylated by kinase, LsrK, to form phospho-AI-2. Phospho-AI-2 activates the quorum sensing (QS) circuit by binding to the repressor, LsrR, causing its release. In the absence of phospho-AI-2, the repressor is bound to the lsr promoter region and prevents operon expression. Once phopsho-AI-2 induces the lsr operon, the circuit uses a positive feedback mechanism, causing increased expression of the transporter on the cell surface, and results in increased internalization of AI-2. Also upregulated is LsrG, which is involved in the degradation of phospho-AI-2 (Xavier and Bassler, 2005; Xavier et al., 2007).

21.3.2 *V. harveyi*

The AHLs of *V. harveyi* are known to regulate density-dependent bioluminescence production through the operon luxCDABE in *V. harveyi*. The AI-2, a furanosyl borate diester, was first discovered and characterized in *V. harveyi* (Miller et al., 2004). The third autoinducer termed CAI-1 has been identified and is (*S*)-3-hydroxytridecan-4-one (Henke and Bassler, 2004). The three autoinducers and the three cognate receptors function in parallel and channel the information into a single, shared regulatory pathway. In this circuit, AI-1 is produced by LuxM and binds to the membrane-bound protein, LuxN. LuxS produces AI-2, which binds to LuxP; the LuxP-AI-2 complex then interacts with LuxQ in the membrane. CAI-1 is produced by the CqsA enzyme and interacts with CqsS. The cognate receptors, CqsS, LuxN, and LuxQ, are all membrane-bound histidine sensor kinases. The sensory information feeds into a two-component response regulatory pathway, which ends at the production of LuxR. The LuxR protein controls the expression of bioluminescence-related genes. At low cell densities, the cognate receptors function as kinases and lead to the synthesis of phospho-LuxO via LuxU. The phosphorylated LuxO, in conjunction with σ^{54}, synthesizes sRNAs that interact with the RNA chaperone Hfq to degrade luxR mRNA. At high cell densities, the membrane sensors bind to their respective autoinducers and switch from kinase to phosphatase activity and remove phosphates from LuxO to LuxU. This allows translation of luxR mRNA to LuxR and expression of bioluminescence (Waters and Bassler, 2005). Phenotypes that are controlled by the *V. harveyi* quorum sensing system in vitro include bioluminescence and the production of several virulence factors such as a type III secretion system, extracellular toxin, metalloprotease, and siderophore (Defoirdt et al., 2007).

21.3.3 *Vibrio cholerae*

In *V. cholerae*, production and expression of the enterotoxin called cholera toxin are controlled by QS (Zhu et al., 2002). *V. cholerae* produces and responds to two AIs using two parallel QS circuits. One AI is (*S*)-3-hyroxytridecan-4-one (CAI-1), which is synthesized by the CqsA enzyme using SAM and decanoyl-coenzyme A as substrates (Wei et al., 2011). The second AI, AI-2, is synthesized by LuxS. LuxS converts the SAM cycle intermediate *S*-ribosylhomocysteine to DPD and homocysteine. DPD spontaneously converts into AI-2 (Chen et al., 2002). Thus, using two different AIs presumably allows *V. cholerae* to detect both the number of other vibrios and the total number of bacteria in the environment. *V. cholerae* detects CAI-1 and AI-2 using two parallel membrane-bound two-component receptors. CqsS detects CAI-1 and LuxPQ detects AI-2 (Ng et al., 2011). Along with other regulators, QS also controls biofilm formation in *V. cholerae*. Biofilm formation is activated at low cell density and repressed at high cell density. HapR, expressed at high cell density directly represses genes encoding components of the biofilm, represses expression of two transcription factors that activate biofilm-formation genes, VpsR and VpsT, and represses synthesis of the second messenger cyclic-di-GMP, which is sensed through the transcription factor VpsT to activate biofilm formation (Krasteva et al., 2010). At low cell density, biofilm lifestyle allows *V. cholerae* to remain attached to the host tissue when virulence factors are expressed. At high cell density, the QS repression of biofilms and virulence factor production facilitates dispersal of *V. cholerae* back into the environment. This strategy allows *V. cholerae* to maximally compete for nutrients in the host at low cell density (LCD) and maximize its ability to exit and spread to other hosts once it reaches high cell density (HCD) (Nadell and Bassler, 2011; Rutherford and Bassler, 2012).

21.3.4 *Bacillus cereus*

Quorum sensing in *B. cereus* requires the transcription factor PlcR, which controls the expression of most *B. cereus* virulence factors following binding to the intracellular AIP derived from the PapR protein (Slamti and Lereclus, 2002). *papR* is encoded 70 basepairs downstream of *plcR*. PapR is 48 amino acids long and contains an amino-terminal signal peptide that targets it for the secretory pathway (Okstad et al., 1999). Once outside the cell, the PapR pro-AIP is processed by the secreted neutral protease B (NprB) to form the active AIP (Pomerantsev et al., 2009). *nprB* is encoded divergently from *plcR*, and *nprB* expression is activated by AIP-bound PlcR (Okstad et al., 1999). NprB cleaves the pro-AIP PapR into peptides of 5, 7, 8, and 11 amino acids in length, all of which are derived from the carboxyl terminus of full-length PapR (Pomerantsev et al., 2009). Only the pentapeptide and heptapeptide activate PlcR activity; however, the heptapeptide causes maximal activation and is more prevalent in vivo (Bouillaut et al., 2008; Pomerantsev et al., 2009). Once inside the cell, the AIP binds to the transcription factor PlcR, and this causes conformational changes in the DNA-binding domain of PlcR and facilitates PlcR oligomerization, DNA binding, and regulation of transcription (Declerck et al., 2007). When PlcR interacts with the PapR AIP and oligomerizes, it binds to *PlcR boxes* to regulate the transcription of target genes (Agaisse et al., 1999). PlcR controls the expression of 45 genes, many of which encode extracellular proteins including several enterotoxins, hemolysins, phospholipases, and proteases (Gohar et al., 2008). As is the case in all QS systems, AIP-bound PlcR also feedback activates the expression of *papR* (Rutherford and Bassler, 2012).

21.4 Quorum Sensing in Food Spoilage Bacteria

Interestingly, several bacteria that are typically associated with food spoilage produce *N*-acyl-homoserine lactones (AHLs), the major type of quorum sensing signal molecules (Bruhn et al., 2004; Flodgaard et al., 2005; Gram et al., 1999, 2002; Liu et al., 2006; Ravn et al., 2003), and AHLs can be detected in a number of food products that carry large bacterial populations such as meat and bean sprouts (Bruhn et al., 2004; Gram et al., 2002; Rasch et al., 2005). Furthermore, several enzymes believed to play a role in spoilage, such as protease, lipase, cellulase, and pectinase, are produced under the control of quorum sensing systems in a variety of bacteria (Christensen et al., 2003; Liu et al., 2007). More direct evidence came from the demonstration that spoilage of bean sprouts by a *Pectobacterium* strain or of milk by *Serratia proteamaculans* was delayed or reduced when bacterial AHL production was knocked out by mutation (Christensen et al., 2003; Rasch et al., 2005). However, not all studies support a role for quorum sensing in spoilage. For example, vacuum-packed meat inoculated with *Hafnia alvei* wild type spoiled at the same rate as with its AHL-deficient mutant (Bruhn et al., 2004). Also, the use of quorum sensing inhibitors, which interfere with the quorum signaling cascade, did not convincingly reduce spoilage of bean sprouts by *Pectobacterium* (Rasch et al., 2005). More work is clearly needed to determine the role of quorum sensing in various food spoilage systems. In addition, the precise role of quorum sensing, that is, which quorum sensing–regulated functions or pathways actually contribute to spoilage, remains largely unknown (Bai and Rai, 2011).

21.5 Biofilms in Food Systems

Biofilms are a major problem in brewing, dairy processing, fresh produce, poultry processing, and red meat processing food industries (Chen et al., 2007). Biofilm formation in dairy processing plants can lead to serious hygiene problems and economic losses due to food spoilage and equipment impairment. Microorganisms in biofilms cause metal corrosion in pipelines and tanks, and when the biofilms become sufficiently thick at plate heat exchangers and pipelines, they can reduce the heat transfer efficacy (Bremer et al., 2006; Gram et al., 2007). Moreover, the persistence of food-borne pathogens on food contact surfaces and biofilms can affect the quality and safety of food products. The presence of pathogens such as *L. monocytogenes*, *Yersinia enterocolitica*, *Campylobacter jejuni*, *Salmonella* spp.,

Staphylococcus spp., and *E. coli* O157:H7 as biofilms in foods and food processing units has led to outbreaks of food-borne diseases. Biofilm formation enhances the capacity of food-borne bacteria to survive stresses that are commonly encountered within food processing plants such as refrigeration, acidity, salinity, and disinfection (Giaouris et al., 2012). Biofilms are difficult to eradicate due to their resistance to antimicrobials, and these can harbor other microorganisms less prone to biofilm formation, increasing the probability of pathogen survival and further dissemination during food processing. In the meat industry, biofilms formed by pathogenic bacteria, such as *S. enterica*, *L. monocytogenes*, and *E. coli*, together with common meat spoilage bacteria, such as *Pseudomonas* spp., *Brochothrix thermosphacta*, and *Lactobacillus* spp., create a persistent source of product contamination, resulting in serious hygiene problems and also economic losses due to food spoilage (Jessen and Lammert, 2003; Sofos and Geornaras, 2010).

There are a number of mechanisms by which many microbial species are able to come into closer contact with a surface, attach firmly to it, promote cell–cell interactions, and grow as a complex structure. Biofilm formation in food processing plants comprises a sequence of steps: (1) preconditioning of the adhesion surface either by macromolecules present in the bulk liquid or intentionally coated on the surface, (2) transport of planktonic cells from the bulk liquid to the surface, (3) adsorption of cells at the surface, (4) desorption of reversibly adsorbed cells, (5) irreversible adsorption of bacterial cells at a surface, (6) production of cell–cell signaling molecules, (7) transport of substrates to and within the biofilm, (8) substrate metabolism by the biofilm-bound cells and transport of products out of the biofilm; these processes are accompanied by cell growth, replication, and EPS production, and (9) biofilm removal by detachment or sloughing (Simoes et al., 2010).

As biofilms typically contain high concentration of cells, quorum sensing regulation of gene expression has been proposed as an essential component of biofilm physiology (Kjelleberg and Molin, 2002; Parsek and Greenberg, 2005). It is believed that quorum sensing inhibition may represent a natural, widespread, antibiofilm strategy (Chorianopoulos et al., 2010; Simoes et al., 2010). Unambiguously, a deep understanding of the quorum sensing phenomenon in bacteria relevant to food processing may be used to control their biofilm formation through the identification of products that could affect QS and thus biofilm formation (Giaouris et al., 2013; Lazar, 2011).

21.6 Quorum Sensing Inhibitory Mechanisms

Cell-to-cell communication in bacteria can be disrupted by the following mechanisms: reducing the activity of signal cognate receptor protein or signal synthase, inhibiting the production of the signaling molecules, degradation of the signaling molecule, and mimicking the signal molecules primarily by using synthetic compounds as analogs of signal molecules (Figure 21.3) (Kalia, 2013).

21.6.1 Inhibition of Signal Generation/Degradation

The AHLs and AI-2 signaling molecules are derived from SAM as part of the bacterial 1-carbon metabolism. Methylthioadenosine/*S*-adenosylhomocysteine nucleosidase (MTAN/Pfs) catalyzes the hydrolytic deadenylation of MTA or SAH, and accumulation of either product inhibits the AI-1 or AI-2 pathways, respectively. MTAN inhibitors can potentially alter both AI-1 and AI-2 signaling by disrupting signal generation (Guttierez et al., 2009).

QS inhibition can be achieved by AHL inactivation or complete degeneration of the signal molecules themselves (Dong et al., 2001). Although AHL degradation techniques have been the focus of many investigations, the degradation of AI-2 ex vivo to attenuate the QS response is also possible. On exogenous addition of LsrK (AI-2 kinase) and ATP to a bacterial culture, AI-2 is phosphorylated extracellularly and AI-2-controlled bacterial cross talk gets significantly reduced. In vitro synthesized phospho-AI-2 quenched the QS response in *E. coli*, *S.* Typhimurium, and *V. harveyi*, as both pure cultures and a synthetic ecosystem—a coculture of all three organisms. The phosphorylated AI-2 is more hydrophilic and unable to cross the cell membrane and serve as a signal molecule resulting in reduced QS response (Roy et al., 2010a).

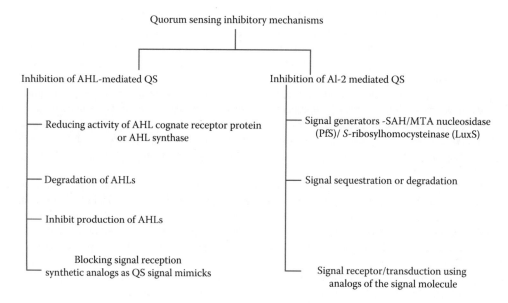

FIGURE 21.3   Quorum sensing inhibitory strategies in bacteria: signal degradation, interception of signal synthesis, and reception.

21.6.2 Inhibition of Signal Reception/Transduction

Probably the most intensively investigated strategy for the inhibition of QS is the use of small molecules to block the activation of LuxR homologs. Blocking the AHL receptor site with an AHL analog is a classical pharmacological approach to receptor antagonism. Most compounds have substitutions, such as a keto-oxygen or an extra carbon atom in the HSL ring. The compounds able to bind to LuxR were not only able to displace the AHL but were also able to activate the LuxR protein. Hence, such compounds are not true inhibitors of QS but are competitive agonists (Rasmussen and Givskov, 2006).

The structural mimics of quorum-sensing signals, such as the halogenated furanones and the synthetic AIPs are similar to AHL and AIP signals, respectively. Evidence shows that these inhibitors act by interfering with the corresponding signal binding to the receptor or decreasing the receptor concentration (Dong et al., 2007).

DPD is a linear precursor of AI-2, which cyclizes to form various isomers in solutions. On screening DPD analogs consisting of linear (C1–C7), branched (isobutyl, isopropyl, neopentyl, 2-methylpropyl), cyclic (cyclopropyl), and deoxy versions (methyl and isobutyl) for quorum sensing inhibitory activity, it was found that some of them were potent quorum quenchers in *E. coli* and were phosphorylated in vitro by the *E. coli* AI-2 kinase, LsrK. The data also showed that LsrR is required for analog-mediated repression of *lsr*-based QS response. Interestingly, the addition of a single carbon to the C1-analog alkyl chain plays a critical role in determining the effect of the analog on the QS response. While an ethyl modified analog is an agonist, propyl is an antagonist of the *E. coli* QS circuit. To determine the effectiveness of these analogs in a polymicrobial infection, the analogs were tested in a self-assembled trispecies synthetic ecosystem comprised of *E. coli*, *S.* Typhimurium, and *V. harveyi*. They showed both cross-species and species-specific anti-AI-2 QS activities by the analogs (Roy et al., 2010b).

21.6.3 Nonspecific QS Inhibition

The natural *Dalea pulchra* furanone, (5Z)-4-bromo-5-(bromomethylene)-3-butyl-2(5H)-furanone, inhibits swarming and biofilm formation in *E. coli*. The furanone inhibited AI-1-based QS in *V. harveyi* 3300-fold and AI-2-based QS 5500-fold; it also inhibited AI-2 QS in *E. coli* (Ren et al., 2002). DNA microarray analysis showed that 90 genes were differentially expressed by furanone-treated *E. coli*

(34 were induced and 56 genes were repressed). Most of the repressed genes were involved in chemotaxis, motility, and flagella synthesis. Furthermore, 80% of the genes repressed were known to be induced by AI-2. This suggested that the furanone may interfere with the AI-2 regulon (Ren et al., 2004). An investigation of a synthetic furanone with two to six carbon atom side chains in *S.* Typhimurium was found to significantly decrease biofilm formation at non-growth-inhibiting concentrations (Janssens et al., 2008).

Defoirdt et al. (2008) employed mutant strains of *V. harveyi* to identify various furanone targets. They found that all three QS systems were blocked by furanone addition, suggesting altered LuxR binding was altering QS gene regulation.

21.7 Quorum Sensing Inhibitors as Food Preservatives

The use of antiquorum sensing agents for inhibiting growth of pathogenic/spoilage bacteria can lead to the development of novel preservative or intervention techniques (Table 21.1).

TABLE 21.1

Overview of Quorum Sensing Inhibitors and Their Target Food-Borne Bacteria

Quorum Sensing Inhibitors	Target Organism	Mode of Action	References
2(5*H*)-furanones	*Pseudomonas* spp.	Inhibits AHL (hexanoyl homoserine lactone [HHSL] and butryl homoserine lactone [BHSL]) production	Shobharani and Agarwal (2010)
Synthetic MTAN inhibitors	*V. cholerae* and EHEC	Blocks AI-2 production	Schramm et al. (2008), Brackman et al. (2009)
Culture extracts of *B. longum* ATCC 15707	EHEC O157:H7 and *V. harveyi*	Inhibit AI-2-regulated activity	Kim et al. (2012)
Culture extracts of *L. acidophilus* A4	EHEC O157:H7	Degradation or inactivation of AI-2 molecules	Kim et al. (2008)
Fatty acids from ground beef and poultry meat	*V. harveyi* and *E. coli* K-12	Inhibit AI-2 activity and biofilm formation	Lu et al. (2004), Soni et al. (2008a,b), Widmer et al. (2007)
Furocoumarins	*V. harveyi* reporter strains BB886 and BB170	Inhibit AI-1- and AI-2-regulated activities	Girennavar et al. (2008)
Purified furocoumarins—dihydroxybergamottin and bergamottin	*E. coli* O157:H7, *S.* Typhimurium, and *P. aeruginosa*	Inhibit biofilm formation	Girennavar et al. (2008)
Limonoids—isolimonic acid, ichangin, and deacetyl nomilinic acid 17 β-ᴅ-glucopyranoside	*V. harveyi*	Inhibit AI-2 activity	Vikram et al. (2011)
Flavonoids—naringenin, kaempferol, quercetin, and apigenin	*V. harveyi* BB886 and MM32	Inhibited harveyi autoinducer (HAI)-1- or AI-2-mediated bioluminescence	Vikram et al. (2010)
Naringenin, quercetin	*V. harveyi* BB120 and *E. coli* O157:H7	Inhibit biofilm formation	Vikram et al. (2010)
Essential oil and hydrosol of *S. thymbra*	*Salmonella*, *Listeria*, *Pseudomonas*, *Staphylococcus*, and *Lactobacillus* spp.	Inhibit biofilm formation	Chorianopoulos et al. (2008)
Cinnamaldehyde	Vibrios	Inactivate HapR and its homologs	Niu et al. (2006), Brackman et al. (2008, 2009)
Broccoli extract (BE) and its flavonoid constituents	*E. coli* O157:H7	AI-2 synthesis and AI-2-mediated bacterial motility	Lee et al. (2011)

21.7.1 Synthetic Quorum Sensing Inhibitors

A synthetic quorum sensing inhibitor, 2(5*H*)-furanone, when added to fermented milk, was capable of extending the shelf life of milk by 9 days. Furanones have structural similarity to that of AHL, and by inhibiting the HHSL and BHSL production and reducing the expression of rhamnolipid, exoprotease, and motility, they could prevent spoilage of fermented milk by *Pseudomonas* spp. (Shobharani and Agarwal, 2010).

In the SAM cycle, MTAN is a hydrolase that turns over methylthioadenosine to maintain the homeo-static SAM pool. This step is critical for the synthesis of both CAI-1 and AI-2 (Schauder et al., 2001; Wei et al., 2011). A synthetic inhibitor of MTAN blocks QS in *V. cholerae* without affecting growth (Schramm et al., 2008). This MTAN inhibitor blocks AI-2 production in EHEC as well. In a screen for nucleoside analogs able to disrupt AI-2 QS, another molecule was identified that disrupts AI-2-based QS without affecting growth in *V. harveyi*, a close relative of *V. cholerae* (Brackman et al., 2009). This compound likely targets the signaling pathway for the AI-2 receptor, LuxPQ, which is present in numerous vibrio species including *V. cholerae* (Rutherford and Bassler, 2012).

21.7.2 Quorum Quenching by Bacteria

The cell extracts of *Bifidobacterium longum* ATCC 15707 resulted in a 98-fold reduction in AI-2 activity in EHEC O157:H7 as well as in the *V. harveyi* reporter strain, even though they did not inhibit the growth of EHEC O157:H7. In addition, they resulted in a 36% reduction in biofilm formation by the organism. Consistently, the virulence of EHEC O157:H7 was significantly attenuated by the presence of cell extracts of *B. longum* ATCC 15707 in the *Caenorhabditis elegans* nematode in vivo model. By a proteome analysis using two-dimensional electrophoresis (2-DE), it was shown that seven proteins including the formation of iron-sulfur protein (NifU), thiol:disulfide interchange protein (DsbA), and flagellar P-ring protein (FlgI) were differentially regulated in the EHEC O157:H7 when supplemented with cell extracts of *B. longum* ATCC 15707. These findings propose a novel function of a dairy adjunct in repressing the virulence of EHEC O157:H7 by inhibiting AI-2 activity (Kim et al., 2012).

The cell extracts isolated from *Lactobacillus* interfered with autoinducer (AI)-2-like activity and affected the virulence of EHEC O157:H7, including attachment, biofilm formation, and the killing of a surrogate host, the nematode *C. elegans*. A number of active components from *Lactobacillus acidophilus* A4 were involved in the inhibition of virulence development involved in attachment and biofilm formation, which is controlled by AI-2-like activity. These cell extracts that are associated with virulence control in EHEC O157:H7 can be used as quenching agents (degradation or inactivation) of AI-2-like activity for the novel development of antimicrobials (Kim et al., 2008).

21.7.3 QS Inhibitors from Food

Certain food matrices, including ground beef extracts, possess compounds capable of inhibiting AI-2 activity. Foods such as turkey patties, chicken breast, homemade cheeses, beef steak, and beef patties showed 84.4%–99.8% inhibition of AI-2 activity (Lu et al., 2004; Soni et al., 2008a). Long chain fatty acids derived from poultry meat had strong inhibitory effect (25%–99%) on AI-2 activity of *V. harveyi* BB170. Further studies revealed that these inhibitors resulted in differential expression of virulence-related genes such that 87.5% reduction in bioluminescence occurred in comparison to that caused by cell free supernatant and 60% in response to in vitro synthesized AI-2 (Widmer et al., 2007). It was also shown that ground beef contains compounds such as palmitic acid ($C_{16:0}$), stearic acid ($C_{18:0}$), oleic acid ($C_{18:1}\omega9$), and linoleic acid ($C_{18:2}\omega6$), which can inhibit AI-2-mediated bioluminescence in *V. harveyi* by >90%. Ground beef extract reversed the AI-2-upregulated virulence genes such as *yadK* and *hha* (Soni et al., 2008b). These fatty acids were tested (using *V. harveyi* BB170 and MM32 reporter strains) at different concentrations (1, 5, and 10 mM) to identify differences in the level of AI-2 activity inhibition. AI-2 inhibition ranged from 25% to 90%. A mixture of these fatty acids (prepared at concentrations equivalent to those present in the ground beef extract) produced 52%–65% inhibition of AI-2 activity.

The fatty acid mixture also negatively influenced *E. coli* K-12 biofilm formation. These results demonstrate that both medium- and long-chain fatty acids in ground beef have the ability to interfere with AI-2-based cell signaling. The mode of action for these compounds was not established, however, and currently remains unknown. These molecules represent a distinct chemical class of natural QS inhibitors (Roy et al., 2011; Skandamis and Nychas, 2012).

21.7.4 Phytochemicals as Quorum Sensing Inhibitors

Plant extracts or phytochemicals having chemical structure similar to those of QS signals (AHL) can be used as quorum sensing inhibitors (Teplitski et al., 2011; Vattem et al., 2007). Furocoumarins from grapefruit are capable of inhibiting AI-1- and AI-2-regulated activities in *V. harveyi* reporter strains BB886 and BB170. These phytochemicals could also inhibit biofilm formation by pathogens such as *E. coli* O157:H7, *S.* Typhimurium, and *P. aeruginosa*. Purified furocoumarins—dihydroxybergamottin and bergamottin—caused AI inhibitions in the range of 94.6%–97.7% (Girennavar et al., 2008). Limonoids present in orange seeds such as isolimonic acid, ichangin, and deacetyl nomilinic acid 17 β-D-glucopyranoside at 100 μg/mL inhibited AI-2 activity in *V. harveyi* by more than 90% (Vikram et al., 2011). The QSI abilities of furocoumarins and limonoids have been attributed to the furan moiety they share with the synthetic furanones (Lönn-Stensrud et al., 2007).

Flavonoids have been the focus of research for their roles as antioxidant, anti-inflammatory, and anticancer agents. Keeping in view these health benefits, flavonoids such as naringenin, kaempferol, quercetin, and apigenin were evaluated for their QSI activities. All these flavonoids inhibited HAI-1- or AI-2-mediated bioluminescence in *V. harveyi* BB886 and MM32. Quercetin and naringenin were found to inhibit biofilm formation by *V. harveyi* BB120 and *E. coli* O157:H7 (Vikram et al., 2010). Flavan-3-ol catechin, one of the flavonoids from the bark of *Combretum albiflorum*, reduces the production of QS-mediated virulence factors—pyocyanin and elastase—and biofilm formation by *P. aeruginosa* PAO1 (Vandeputte et al., 2010).

Herbal plants like tea also produce catechins that affect transfer of conjugative R plasmid in *E. coli* (Zhao et al., 2001). Rosmarinic acid produced by the roots of *Ocimum basilicum* (sweet basil) could inhibit QS-regulated activities in *P. aeruginosa* (Walker et al., 2004). Ursolic acid purified from plant extracts at a concentration of 10 μg/mL inhibited biofilm formation by 79% in *E. coli* and 57%–95% in *V. harveyi* and *P. aeruginosa* PAO1 (Ren et al., 2005). Essential oil and hydrosol of *Satureja thymbra* and polytoxinol were effective against biofilms formed by *Salmonella*, *Listeria*, *Pseudomonas*, *Staphylococcus*, and *Lactobacillus* spp. (Chorianopoulos et al., 2008).

Cinnamaldehyde is a natural product that has the ability to inhibit QS in vibrios (Niu et al., 2006). Cinnamaldehyde and its derivatives are proposed to target HapR and its homologs, thus inhibiting virulence factor production and biofilm formation without dramatically affecting growth (Brackman et al., 2008, 2009).

Broccoli extract (BE) has numerous beneficial effects on human health including anticancer activity. BE suppressed AI-2 synthesis and AI-2-mediated bacterial motility in a dose-dependent manner in *E. coli* O157:H7. In addition, the expression of the *ler* gene that regulates AI-3 QS system was also diminished in response to the treatment with BE. Furthermore, in an in vivo efficacy test using *C. elegans* as a host organism, *C. elegans* fed on *E. coli* O157:H7 in the presence of BE survived longer than those fed solely on the pathogenic bacteria. Quantitative real-time PCR analysis indicated that quercetin was the most active among the tested broccoli-derived compounds in downregulating virulence gene expression, while treatment with myricetin significantly suppressed the expression of the *eae* gene involved in type III secretion system. BE and its flavonoid constituents can inhibit expression of QS-associated genes, thereby downregulating the virulence attributes of *E. coli* O157:H7 both in vitro and in vivo (Lee et al., 2011).

QS in food-borne pathogens and food spoilers regulates virulence factor expression and spoilage phenotype production, respectively. As QS circuits offer several possible targets for the discovery or design of novel antimicrobial agents, safe QS inhibitors could be developed and used as food intervention or preservative techniques to prolong shelf life and safety of foods.

REFERENCES

Agaisse, H., Gominet, M., Okstad, O.A., Kolsto, A.B., Lereclus, D. 1999. PlcR is a pleiotropic regulator of extracellular virulence factor gene expression in *Bacillus thuringiensis*. *Molecular Microbiology* 32:1043–1053.

Ahmer, B.M.M. 2004. Cell-to-cell signalling in *Escherichia coli* and *Salmonella enterica*. *Molecular Microbiology* 52:933–945.

Ahmer, B.M.M., Reeuwijk, J., Timmers, C.D., Valentine, P.J., Heffron, F. 1998. *Salmonella typhimurium* encodes an sdia homolog, a putative quorum sensor of the luxr family, that regulates genes on the virulence plasmid. *Journal of Bacteriology* 180(5):1185–1193.

Bai, A.J., Rai, V.R. 2011. Bacterial quorum sensing and food industry. *Comprehensive Reviews in Food Science and Food Safety* 10:184–194.

Barrios, A.F.G., Zuo, R.J., Hashimoto, Y., Yang, L., Bentley, W.E., Wood, T.K. 2006. Autoinducer 2 controls biofilm formation in *Escherichia coli* through a novel motility quorum-sensing regulator (MqsR, B3022). *Journal of Bacteriology* 188:305–316.

Bouillaut, L., Perchat, S., Arold, S., Zorrilla, S., Slamti, L., Henry, C., Gohar, M., Declerck, N., Lereclus, D. 2008. Molecular basis for group-specific activation of the virulence regulator PlcR by PapR heptapeptides. *Nucleic Acids Research* 36:3791–3801.

Brackman, G., Celen, S., Baruah, K., Bossier, P., Van Calenbergh, S., Nelis, H.J., Coenye, T. 2009. AI-2 quorum-sensing inhibitors affect the starvation response and reduce virulence in several *Vibrio* species, most likely by interfering with LuxPQ. *Microbiology* 155:4114–4122.

Brackman, G., Defoirdt, T., Miyamoto, C., Bossier, P., Van Calenbergh, S., Nelis, H. 2008. Cinnamaldehyde and cinnamaldehyde derivatives reduce virulence in *Vibrio* spp. by decreasing the DNA-binding activity of the quorum sensing response regulator LuxR. *BMC Microbiology* 8:149.

Bremer, P.J., Fillery, S., McQuillan, A.J. 2006. Laboratory scale clean-in-place (CIP) studies on the effectiveness of different caustic and acid wash steps on the removal of dairy biofilms. *International Journal of Food Microbiology* 106:254–262.

Bruhn, J.B., Christensen, A.B., Flodgaard, L.R., Nielsen, K.F., Larsen, T.O., Givskov, M., Gram, L. 2004. Presence of acylated homoserine lactones (AHLs) and AHL-producing bacteria in meat and potential role of AHL in spoilage of meat. *Applied and Environmental Microbiology* 70:4293–4302.

Chen, J., Rossman, M.L., Pawar, D.M. 2007. Attachment of enterohemorrhagic *Escherichia coli* to the surface of beef and a culture medium. *LWT—Food Science and Technology* 40:249–254.

Chen, X., Schauder, S., Potier, N., Van Dorsselaer, A., Pelczer, I., Bassler, B.L., Hughson, F.M. 2002. Structural identification of a bacterial quorum-sensing signal containing boron. *Nature* 415:545–549.

Chessa, D., Dorsey, C.W., Winter, M., Baumler, A.J. 2008. Binding specificity of *Salmonella* plasmid-encoded fimbriae assessed by glycomics. *Journal of Biological Chemistry* 283:8118–8124.

Chorianopoulos, N.G., Giaouris, E.D., Kourkoutas, Y., Nychas, G.-J. 2010. Inhibition of the early stage of *Salmonella enterica* serovar Enteritidis biofilm development on stainless steel by cell-free supernatant of a *Hafnia alvei* culture. *Applied and Environmental Microbiology* 76(6):2018–2022.

Chorianopoulos, N.G., Giaouris, F.D., Skendamis, P.N., Haroutounian, S.A., Nychas, G.-J.E. 2008. Disinfectant test against monoculture and mixed-culture biofilms composed of technological, spoilage and pathogenic bacteria: Bactericidal effect of essential oil and hydrosol of *Satureja thymbra* and comparison with standard acid–base sanitizers. *Journal of Applied Microbiology* 104:1586–1596.

Christensen, A.B., Riedel, K., Eberl, L., Flodgaard, L.R., Molin, S., Gram, L., Givskov, M. 2003. Quorum sensing-directed protein expression in *Serratia proteamaculans* B5a. *Microbiology* 149:471–483.

Clarke, M.B., Hughes, D.T., Zhu, C., Boedeker, E.C., Sperandio, V. 2006. The QseC sensor kinase: A bacterial adrenergic receptor. *Proceedings of the National Academy of Sciences of the United States of America* 103:10420–10425.

Declerck, N., Bouillaut, L., Chaix, D., Rugani, N., Slamti, L., Hoh, F., Lereclus, D., Arold, S.T. 2007. Structure of PlcR: Insights into virulence regulation and evolution of quorum sensing in Gram-positive bacteria. *Proceedings of the National Academy of Sciences of the United States of America* 104:18490–18495.

Defoirdt, T., Boon, N., Sorgeloss, P., Verstraete, W., Bossier, P. 2008. Quorum sensing and quorum quenching in *Vibrio harveyi*: Lessons learned from in vivo work. *The ISME Journal* 2:19–26.

Defoirdt, T., Miyamoto, C.M., Wood, T.K., Meighen, E.A., Sorgeloos, P., Verstraete, W. 2007. The natural furanone (5Z)-4-bromo-5-(bromomethylene)-3-butyl-2(5H)-furanone disrupts quorum sensing-regulated gene expression in *Vibrio harveyi* by decreasing the DNA-binding activity of the transcriptional regulator protein luxR. *Environmental Microbiology* 9:2486–2495.

Dong, Y.H., Wang, L.H., Xu, J.L., Zhang, H.B., Zhang, X.F., Zhang, L.H. 2001. Quenching quorum sensing-dependent bacterial infection by an *N*-acyl homoserine lactonase. *Nature* 411:813–817.

Dong, Y.H., Wang, L.H., Zhang, L.H. 2007. Quorum-quenching microbial infections: Mechanisms and implications. *Philosophical Transactions of the Royal Society B: Biological Sciences* 362:1201–1211.

Flodgaard, L.R., Dalgaard, P., Andersen, J.B., Nielsen, K.F., Givskov, M., Gram, L. 2005. Nonbioluminescent strains of *Photobacterium phosphoreum* produce the cell-to-cell communication signal *N*-(3-hydroxyoctanoyl)homoserine lactone. *Applied and Environmental Microbiology* 71:2113–2120.

Giaouris, E., Chorianopoulos, N., Skandamis, P., Nychas, G.-J. 2012. Attachment and biofilm formation by *Salmonella* in food processing environments. In B.S.M. Mahmoud (Ed.), *Salmonella: A Dangerous Foodborne Pathogen* (pp. 157–180). Intech Open Access Publisher: Rijeka, Croatia.

Giaouris, E., Heir, E., Hebraud, M., Chorianopoulos, N., Langsrud, S., Moretro, T., Habimana, O., Desvaux, M., Renier, S., Nychas, G.-J. 2013. Attachment and biofilm formation by foodborne bacteria in meat processing environments: Causes, implications, role of bacterial interactions and control by alternative novel methods. *Meat Science* 97:298–309.

Girennavar, B., Cepeda, M.L., Soni, K.A., Vikram, A., Jesudhasan, P., Jayaprakasha, G.K. 2008. Grapefruit juice and its furocoumarins inhibit autoinducer signaling and biofilm formation in bacteria. *International Journal of Food Microbiology* 125:204–208.

Gohar, M., Faegri, K., Perchat, S., Ravnum, S., Okstad, O.A., Gominet, M., Kolsto, A.B., Lereclus, D. 2008. The PlcR virulence regulon of *Bacillus cereus*. *PLoS One* 3:e2793.

Gori, K., Moslehi-Jenabian, S., Purrotti, M., Jespersen, L. 2011. Autoinducer-2 activity produced by bacteria found in smear of surface ripened cheeses. *International Dairy Journal* 21:48–53.

Gram, L., Bagge-Ravn, D., Yin Ng, Y., Gymoese, P., Vogel, B.F. 2007. Influence of food soiling matrix on cleaning and disinfection efficiency on surface attached *Listeria monocytogenes*. *Food Control* 18:1165–1171.

Gram, L., Christensen, A.B., Ravn, L., Molin, S., Givskov, M. 1999. Production of acylated homoserine lactones by psychrotrophic members of the Enterobacteriaceae isolated from foods. *Applied and Environmental Microbiology* 65:3458–3463.

Gram, L., Ravn, L., Rasch, M., Bruhn, J.B., Christensen, A.B., Givskov, M. 2002. Food spoilage—Interactions between food spoilage bacteria. *International Journal of Food Microbiology* 78:79–97.

Gutierrez, J.A., Crowder, T., Rinaldo-Matthis, A., Ho, M.C., Almo, S.C., Schramm, V.L. 2009. Transition state analogs of 5'-methylthioadenosine nucleosidase disrupt quorum sensing. *Nature Chemical Biology* 5:251–257.

Henke, J.M., Bassler, B.L. 2004. Three parallel quorum-sensing systems regulate gene expression in *Vibrio harveyi*. *Journal of Bacteriology* 186:6902–6914.

Janssens, J.C., Steenackers, H., Robijns, S., Gellens, E., Levin, J., Zhao, H. 2008. Brominated furanones inhibit biofilm formation by *Salmonella enterica* serovar Typhimurium. *Applied and Environmental Microbiology* 74:6639–6648.

Jessen, B., Lammert, L. 2003. Biofilm and disinfection in meat processing plants. *International Biodeterioration and Biodegradation* 51:265–279.

Kalia, V.C. 2013. Quorum sensing inhibitors: An overview. *Biotechnology Advances* 31:224–245.

Kanamaru, K., Kanamaru, K., Tatsuno, I., Tobe, T., Sasakawa, C. 2000. Sdia, an *Escherichia coli* homologue of quorum-sensing regulators, controls the expression of virulence factors in enterohaemorrhagic *Escherichia coli* O157:H7. *Molecular Microbiology* 38:805–816.

Kim, Y., Lee, J.W., Kang, S.-G., Oh, S., Griffith, M.W. 2012. *Bifidobacterium* spp. influences the production of autoinducer-2 and biofilm formation by *Escherichia coli* O157:H7. *Anaerobe* 18:539–545.

Kim, Y., Oh, S., Park, S., Seo, J.B., Kim, S.-H. 2008. *Lactobacillus acidophilus* reduces expression of enterohemorrhagic *Escherichia coli* O157:H7 virulence factors by inhibiting autoinducer-2-like activity. *Food Control* 19:1042–1050.

Kjelleberg, S., Molin, S. 2002. Is there a role for quorum sensing signals in bacterial biofilms? *Current Opinion in Microbiology* 5(3):254–258.

Krasteva, P.V., Fong, J.C., Shikuma, N.J., Beyhan, S., Navarro, M.V., Yildiz, F.H., Sondermann, H. 2010. *Vibrio cholerae* VpsT regulates matrix production and motility by directly ensing cyclic di-GMP. *Science* 327(5967):866–868.

Lazar, V. 2011. Quorum sensing in biofilms—How to destroy the bacterial citadels or their cohesion/power? *Anaerobe* 17(6):280–285.

Lee, K.-M., Lim, J., Nam, S., Yoon, M.Y., Kwon, Y.-K., Jung, B.Y., Park, Y., Park, S., Yoon, S.S. 2011. Inhibitory effects of broccoli extract on *Escherichia coli* O157:H7 quorum sensing and in vivo virulence. *FEMS Microbiology Letters* 321:67–74.

Liu, M., Gray, J.M., Griffiths, M.W. 2006. Occurrence of proteolytic activity and *N*-acyl-homoserine lactone signals in the spoilage of aerobically chill-stored proteinaceous raw foods. *Journal of Food Protection* 69:2729–2737.

Liu, M., Wang, H., Griffiths, M.W. 2007. Regulation of alkaline metalloprotease promoter by *N*-acyl homoserine lactone quorum sensing in *Pseudomonas fluorescens*. *Journal of Applied Microbiology* 103:2174–2184.

Lönn-Stensrud, J., Petersen, F.C., Benneche, T., Scheie, A.A. 2007. Synthetic bromated furanone inhibits autoinducer-2-mediated communication and biofilm formation in oral streptococci. *Oral Microbiology and Immunology* 22:340–346.

Lu, L., Hume, M.E., Pillai, S.D. 2004. Autoinducer-2-like activity associated with foods and its interaction with food additives. *Journal of Food Protection* 67:1457–1462.

Lyon, G.J., Mayville, P., Muir, T.W., Novick, R.P. 2000. Rational design of a global inhibitor of the virulence response in *Staphylococcus aureus*, based in part on localization of the site of inhibition to the receptor histidine kinase, AgrC. *Proceedings of the National Academy of Sciences of the United States of America* 97:13330–13335.

Michael, B., Smith, J.N., Swift, S., Heffron, F., Ahmer, B.M. 2001. Sdia of *Salmonella enterica* is a luxr homolog that detects mixed microbial communities. *Journal of Bacteriology* 183(19):5733–5742.

Miller, S.T., Xavier, K.B., Campagna, S.R., Taga, M.E., Semmelhack, M.F., Bassler, B.L. 2004. *Salmonella* Typhimurium recognizes a chemically distinct form of the bacterial quorum-sensing signal AI-2. *Molecular Cell* 15:677–687.

Moslehi-Jenabian, S., Gori, K., Jespersen, L. 2009. AI-2 signalling is induced by acidic shock in probiotic strains of *Lactobacillus* spp. *International Journal of Food Microbiology* 135:295–302.

Nadell, C.D., Bassler, B.L. 2011. A fitness trade-off between local competition and dispersal in *Vibrio cholerae* biofilms. *Proceedings of the National Academy of Sciences of the United States of America* 108:14181–14185.

Ng, W.L., Perez, L.J., Wei, Y., Kraml, C., Semmelhack, M.F., Bassler, B.L. 2011. Signal production and detection specificity in *Vibrio* CqsA/CqsS quorum-sensing systems. *Molecular Microbiology* 79:1407–1417.

Niu, C., Afre, S., Gilbert, E.S. 2006. Subinhibitory concentrations of cinnamaldehyde interfere with quorum sensing. *Letters in Applied Microbiology* 43:489–494.

Okstad, O.A., Gominet, M., Purnelle, B., Rose, M., Lereclus, D., Kolsto, A.B. 1999. Sequence analysis of three *Bacillus cereus* loci carrying PIcR-regulated genes encoding degradative enzymes and enterotoxin. *Microbiology* 145:3129–3138.

Parker, T.C., Sperandio, V. 2009. Cell-to-cell signalling during pathogenesis. *Cell Microbiology* 11(3):363–369.

Parsek, M.R., Greenberg, E.P. 2005. Sociomicrobiology: The connections between quorum sensing and biofilms. *Trends in Microbiology* 13(1):27–33.

Pomerantsev, A.P., Pomerantseva, O.M., Camp, A.S., Mukkamala, R., Goldman, S., Leppla, S.H. 2009. PapR peptide maturation: Role of the NprB protease in *Bacillus cereus* 569 PlcR/PapR global gene regulation. *FEMS Immunology and Medical Microbiology* 55:361–377.

Rasch, M., Andersen, J.B., Nielsen, K.F., Flodgaard, L.R., Christensen, H., Givskov, M., Gram, L. 2005. Involvement of bacterial quorum-sensing signals in spoilage of bean sprouts. *Applied and Environmental Microbiology* 71:3321–3330.

Rasmussen, T.B., Givskov, M. 2006. Quorum-sensing inhibitors as anti-pathogenic drugs. *International Journal of Medical Microbiology* 296:149–161.

Ravn, L., Christensen, A.B., Molin, S., Givskov, M., Gram, L. 2003. Influence of food preservation parameters and associated microbiota on production rate, profile and stability of acylated homoserine lactones from food-derived Enterobacteriaceae. *International Journal of Food Microbiology* 84:145–156.

Ren, D., Bedzyk, L.A., Ye, R.W., Thomas, S.M., Wood, T.K. 2004. Differential gene expression shows natural brominated furanones interfere with the autoinducer-2 bacterial signaling system of *Escherichia coli*. *Biotechnology and Bioengineering* 88:630–642.

Ren, D., Sims, J.J., Wood, T.K. 2002. Inhibition of biofilm formation and swarming of *Bacillus subtilis* by (5*Z*)-4-bromo-5-(bromomethylene)-3-butyl-2(5*H*)-furanone. *Letters of Applied Microbiology* 34:293–299.

Ren, D., Zuo, R., González-Barrios, A.F., Bedzyk, L.A., Eldridge, G.R., Pasmore, M.E. 2005. Differential gene expression for investigation of *Escherichia coli* biofilm inhibition by plant extract ursolic acid. *Applied and Environmental Microbiology* 71:4022–4034.

Roy, V., Adams, B.L., Bentely, W.E. 2011. Developing next generation antimicrobials by intercepting AI-2 mediated quorum sensing. *Enzyme and Microbial Technology* 49:113–123.

Roy, V., Fernandes, R., Tsao, C.Y., Bentley, W.E. 2010a. Cross species quorum quenching using a native AI-2 processing enzyme. *ACS Chemical Biology* 5:223–232.

Roy, V., Smith, J.A., Wang, J., Stewart, J.E., Bentley, W.E., Sintim, H.O. 2010b. Synthetic analogs tailor native AI-2 signaling across bacterial species. *Journal of American Chemical Society* 132:11141–11150.

Rutherford, S.T., Bassler, B.L. 2012. Bacterial quorum sensing: Its role in virulence and possibilities for its control. *Cold Spring Harbor Perspectives in Medicine* 2:a012427.

Schauder, S., Shokat, K., Surette, M.G., Bassler, B.L. 2001. The LuxS family of bacterial autoinducers: Biosynthesis of a novel quorum-sensing signal molecule. *Molecular Microbiology* 41:463–476.

Schramm, V.L., Gutierrez, J.A., Cordovano, G., Basu, I., Guha, C., Belbin, T.J., Evans, G.B., Tyler, P.C., Furneaux, R.H. 2008. Transition state analogues in quorum sensing and SAM recycling. *Nucleic Acids Symposium Series (Oxford)* (52):75–76.

Shobharani, P., Agarwal, R. 2010. Interception of quorum sensing signal molecule by furanone to enhance shelf life of fermented milk. *Food Control* 21:61–69.

Simoes, M., Simoes, L., Vieira, M.J. 2010. A review of current and emergent biofilm control strategies. *LWT— Food Science and Technology* 43:573–583.

Skandamis, P.N., Nychas, G.-J.E. 2012. Quorum sensing in the context of food microbiology. *Applied and Environmental Microbiology* 78(16):5473–5482.

Slamti, L., Lereclus, D. 2002. A cell–cell signaling peptide activates the PlcR virulence regulon in bacteria of the *Bacillus cereus* group. *The EMBO Journal* 21:4550–4559.

Smith, J.N., Dyszel, J.L., Soares, J.A., Ellermeier, C.D., Altier, C., Lawhon, S.D., Adams, L.G. et al. 2008. Sdia, an *n*-acyl homoserine lactone receptor, becomes active during the transit of *Salmonella enterica* through the gastrointestinal tract of turtles. *PLoS One* 3(7):e2826.

Soares, J.A., Ahmer, B.M.M. 2011. Detection of acyl-homoserine lactones by *Escherichia* and *Salmonella*. *Current Opinion in Microbiology* 14(2):188–193.

Sofos, J.N., Geornaras, I. 2010. Overview of current meat hygiene and safety risks and summary of recent studies on biofilms, and control of *Escherichia coli* O157:H7 in nonintact, and *Listeria monocytogenes* in ready-to-eat, meat products. *Meat Science* 86:2–14.

Soni, K.A., Lu, L., Jesudhasan, P.R., Hume, M.E., Pillai, S.D. 2008b. Influence of autoinducer-2 (AI-2) and beef sample extracts on *E. coli* O157:H7 survival and gene expression of virulence genes yadK and hhA. *Journal of Food Science* 73:135–139.

Soni, K.A., Palmy, J., Martha, C., Kenneth, W., Jayaprakasha, G.K., Patil, B.S., Hume, M.E., Pillai, S.D. 2008a. Identification of ground beef-derived fatty acid inhibitors of autoinducer-2-based cell signaling. *Journal of Food Protection* 71:134–138.

Sperandio, V., Torres, A.G., Jarvis, B., Nataro, J.P., Kaper, J.B. 2003. Bacteria-host communication: The language of hormones. *Proceedings of the National Academy of Sciences of the United States of America* 100:8951–8956.

Taga, M.E., Miller, S.T., Bassler, B.L. 2003. Lsr-mediated transport and processing of AI-2 in *Salmonella* Typhimurium. *Molecular Microbiology* 50:1411–1427.

Teplitski, M., Mathesius, U., Rambaugh, K.P. 2011. Perception and degradation of *N*-acyl homoserine lactone quorum sensing signals by mammalian and plant cells. *Chemical Reviews* 111:100–116.

Vandeputte, O.M., Kiendrebeogo, M., Rajaonson, S., Diallo, B., Mol, A., Jaziri, M.E. 2010. Identification of catechin as one of the flavonoids from *Combretum albiflorum* bark extract that reduces the production of quorum-sensing-controlled virulence factors in *Pseudomonas aeruginosa* PAO1. *Applied Environmental Microbiology* 71:243–253.

Vattem, D.A., Mihalik, K., Crixell, S.H., McLean, R.J.C. 2007. Dietary phytochemicals as quorum sensing inhibitors. *Fitoterapia* 78:302–310.

Vikram, A., Jayaprakasha, G.K., Jesudhasan, P.R., Pillai, S.D., Patil, B.S. 2010. Suppression of bacterial cell-cell signaling, biofilm formation and type III secretion system by citrus flavonoids. *Journal of Applied Microbiology* 109:515–527.

Vikram, A., Jesudhasan, P.R., Jayaprakasha, G.K., Pillai, S.D., Patil, B.S. 2011. Citrus limonoids interfere with *Vibrio harveyi* cell-cell signaling and biofilm formation by modulating the response regulator LuxO. *Microbiology* 157:99–110.

Walker, T.S., Bais, H.P., Deziel, E., Schweizer, H.P., Rahme, L.G., Fall, R. 2004. *Pseudomonas aeruginosa*-plant root interactions, pathogenicity, biofilm formation, and root exudation. *Plant Physiology* 34:320–331.

Walters, M., Sircili, M.P., Sperandio, V. 2006. AI-3 synthesis is not dependent on *luxS* in *Escherichia coli*. *Journal of Bacteriology* 188:5668–5681.

Waters, C.M., Bassler, B.L. 2005. Quorum sensing: Cell-to-cell communication in bacteria. *Annual Review of Cell and Developmental Biology* 21:319–346.

Wei, Y., Perez, L.J., Ng, W.L., Semmelhack, M.F., Bassler, B.L. 2011. Mechanism of *Vibrio cholerae* autoinducer-1 biosynthesis. *ACS Chemical Biology* 6:356–365.

Whitehead, N.A., Barnard, A.M., Slater, H., Simpson, N.J., Salmond, G.P. 2001. Quorum-sensing in gram-negative bacteria. *FEMS Microbiology Review* 25:365–404.

Widmer, K.W., Soni, K.A., Hume, M.E., Beier, R.C., Jesudhasan, P., Pillai, S.D. 2007. Identification of poultry meat-derived fatty acids functioning as quorum sensing signal inhibitors to autoinducer-2 (AI-2). *Journal of Food Science* 72:363–368.

Xavier, K.B., Bassler, B.L. 2005. Regulation of uptake and processing of the quorumsensing autoinducer AI-2 in *Escherichia coli*. *Journal of Bacteriology* 187:238–248.

Xavier, K.B., Miller, S.T., Lu, W., Kim, J.H., Rabinowitz, J., Pelczer, I. 2007. Phosphorylation and processing of the quorum-sensing molecule autoinducer-2 in enteric bacteria. *ACS Chemical Biology* 2:128–136.

Zhao, W.H., Hu, Z.Q., Hara, Y., Shimamura, T. 2001. Inhibition by epigallocatechin gallate (EGCg) of conjugative R plasmid transfer in *Escherichia coli*. *Journal of Infection and Chemotherapy* 7:195–197.

Zhu, J., Miller, M.B., Vance, R.E., Dziejman, M., Bassler, B.L., Mekalanos, J.J. 2002. Quorum-sensing regulators control virulence gene expression in *Vibrio cholerae*. *Proceedings of the National Academy of Sciences of the United States of America* 99:3129–3134.

22

Plasma Technology as a New Food Preservation Technique

Rocío Rincón, Cristina Yubero, M.D. Calzada, Lourdes Moyano, and Luis Zea

CONTENTS

22.1 Introduction

The food unhealthiness due to food deterioration during storage has been a health problem for humans since the dawn of history. While governments around the world are doing their best to improve the safety of the food supply, the existence of food-borne diseases (FBDs) remains a significant health problem in all countries, from the most to the least developed ones. According to the World Health Organization (WHO), FBDs remain responsible for high levels of mortality in the general population, which points out the necessity of developing and applying food preservation techniques.

Besides the beneficial health implication, from a medical standpoint, these techniques take on account a new perspective. Not only is it important to keep the food away from any microorganisms that may lead to FBDs, but it is also important to try to preserve the food in the best healthy, nutritional, and organoleptic condition.

According to Raso et al. (Raso 2003), an *ideal* method of food preservation has to meet the following characteristics: (1) it improves shelf life and safety by inactivating spoilage and pathogenic microorganisms, (2) it does not change organoleptic and nutritional attributes, or (3) it does not leave residues. The food industry is devoting considerable resource and expertise to find a replacement for traditional food preservation techniques such as intense heat treatments, salting, acidification, dying, or chemical preservation by new preservation techniques (Devlieghere et al. 2004) due to the increased consumer demand for tasty, nutritious, natural, and easy-to-handle food products. The most investigated new preservation techniques are nonthermal inactivation technologies such as high hydrostatic pressure (HHP) (Capellas et al. 1996, de Daucos et al. 2000, Yuste et al. 2002), pulsed electric field (PEF) (Gould 1995, Góngora-Nieto et al. 2002, Raso 2003, Devlieghere et al. 2004), ultrasound (Raso 2003), or the use of ultraviolet (UV) radiation (Noble 2002, Artés et al. 2009).

In Noble (2002), the action of UV-B radiation from a 40 W fluorescent bulb on seed germination was assessed. This study shows that UV radiation (280–320 nm) causes a more rapid germination of seeds, while, once germinated, the effects of continued UV radiation are translated into a markedly retarded growth. Notwithstanding that the source of UV radiation is taken frequently from spectral lamps, nonequilibrium atmospheric plasmas have recently been playing the role of UV and vacuum ultra violet (VUV) sources (Eliasson and Koglschatz 1991).

Plasma is a partially ionized gas that is considered the fourth state of matter. It consists of electrons, ions, and neutrals in fundamental and excited states, which make it a chemically active media. Within plasma, reactions (which are not allowed to take place in other media) can be achieved. In addition, depending on the way plasmas are created and their working power, low or very high temperatures can be generated. This wide temperature range enables many applications for plasma technologies such as deposition of surface coatings (Feugeas et al. 2003), chemical analysis (Calzada et al. 1992), nanomaterial synthesis (Zajickova et al. 2006), hydrogen production (Jiménez et al. 2013), or sterilization (Park et al. 2003). In fact, as far as sterilization is concerned, the role of various agents emanating from the plasma in the sterilization process has been widely studied (Boudam et al. 2006). These agents are mainly: (i) UV radiation, (ii) reactive species, and (iii) charged particles (electrons and ions), which are produced in plasmas generated using a mixture of gases, particularly oxygen and nitrogen. Taking advantage of the plasma's capabilities, several researches applying plasma in food preservation have been carried out.

Song et al. (2009) assessed the use of cold atmospheric pressure plasma to improve the safety of sliced cheese and ham inoculated by three-strain cocktail *Listeria monocytogenes*. The process parameters considered were input power and plasma exposure time. The microbial reduction increased with the increase of both parameters; however, the effect of plasma depended strongly on the type of food. On the other hand, Basaran et al. (2008) tested a low-pressure cold plasma for antifungal efficacy against *Aspergillus parasiticus* on various nut samples. In the study, a rapid functional clean-up method for the elimination of *aflatoxin*-producing fungus from shelled and unshelled nuts was researched as a suitable fungal decontamination method. In Selcuk et al. (2008), the same kind of plasma was used to eliminate and/or inactivate two pathogenic fungi (*Aspergillus* spp. and *Penicillium* spp.) artificially contaminated on seed surfaces. Plasma treatment reduced the fungal-attachment to seeds preventing, at the same time seed, quality during germination process. In the research carried out by Perni et al. (2008), the inactivation of one pathogenic and three spoilage microorganisms on the pericarps of mangoes and melons by cold atmospheric plasmas is described. A relationship between reactive plasma species (particularly oxygen atoms) and the level of bacterial inactivation was pointed out—the higher the oxygen atoms, the more successful the inactivation. In Niemira and Sites (2008), a cold plasma generated using a gliding arc was applied to outbreak strains of *Escherichia coli* O157:H7 and *Salmonella* Stanley on agar plates and inoculated onto the surfaces of Golden Delicious apples. They found that plasma is a nonthermal process that can effectively reduce human pathogens inoculated onto fresh produce. Volin et al. (2000) carried out a study concerning the modification of seed germination performance through cold plasma chemistry technology where the survival of seeds after plasma treatment was found. Additionally, in the light of the results, the authors concluded that plasma technique can be considered an important technology to modify seed germination characteristics in agricultural plant species.

In the present research, the capability of surface wave discharges (SWDs), a special type of microwave atmospheric pressure plasmas, in reducing lentils' germination and avoiding browning of Sherry Fino wine is assessed. On the one hand, Ar, Ar-N$_2$, and Ar-N$_2$O plasmas were applied to different groups of lentils in order to figure out the respective role of UV radiation and plasma species in the germination process. Besides, to evaluate the action of UV radiation on the growth rate of lentils, a UV lamp and an array of UV light emitting diodes (LEDs) were applied. On the other hand, Ar-N$_2$ gas, Ar, and Ar-N$_2$ plasmas were applied to Fino samples in order to study the action of different plasma species on browning of Sherry Fino wine. Emission spectroscopy techniques were used to identify radiation from the plasma and to control the different species found in the discharge.

22.2 Experimental Set-Up

22.2.1 Plasma Devices

The plasma was created in a quartz tube with one of its end opened to the atmosphere. The inner and outer diameters of the discharge tube were 4 and 6 mm, respectively. In order to generate both UV radiation within the range 200–400 nm and active species, different gas mixtures of argon (Ar), nitrogen (N_2), nitrogen protoxide (N_2O), hydrogen peroxide (H_2O_2), and air were utilized as plasma gases. The plasma was always ignited using an Ar flow, and then other gases (N_2, N_2O, and air) were introduced. As for the introduction of hydrogen peroxide, a method called bubbling was employed to introduce it (Jiménez et al. 2008) into the discharge tube. The total Ar gas flow was divided into two parts: one part was used as the carrier gas and was bubbled into the recipient containing the hydrogen peroxide to drag its molecules. In the case of Ar–H_2O_2 mixtures, the resulting mixture (Ar and hydrogen peroxide) was then united with the rest of the argon flow before introducing it into the quartz tube that contained the discharge. In all cases, the different flows were controlled by hi-tech flow controllers (IB 31) with different maximum flow limits (0.25 and 5.00 standard liters per minute, slm).

The microwave (2.45 GHz) power used to create and maintain the discharge was supplied in a continuous mode by an SAIREM generator and coupled to the plasma by means of a surfatron device (Moisan et al. 1974; Figure 22.1a) and a surfaguide device (Moisan et al. 1996; Figure 22.1b). A stub system was also used in order to ensure that the reflected power was less than 5% of the incident power. The power value was kept equal to 200 W.

The radiation emitted by the plasma was recorded by an optical fiber and directed to the entrance of a Jobin Yvon monochromator (1000M, Czerny-Turner type) with 1 m of focal length and a holographic diffraction grating of 2400 lines/mm. A CCD camera was used as a radiation detector (Symphony CCD-1024 × 256-OPEN-STE).

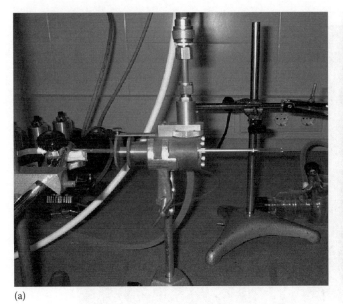

(a)   (b)

FIGURE 22.1 Experimental devices used to create and maintain an SWD: (a) surfatron and (b) surfaguide.

22.2.2 Lentils and Sherry Fino Wine Treatments

Lentils were bought in a local supermarket; lentils with best visual aspect (not stained or broken) were chosen. Different groups of 20 lentils were deposited in petri dishes, at least one group was considered as the control one to compare the effect of different treatments. All treatments were done twice in order to ensure their reproducibility. After the treatment, the treated seeds were placed on wet cottons (Peñuelas 2004, Sangronis and Machado 2007, Tkalec et al. 2009) in a sufficiently ventilated and dark ambience at about 25°C. In order to keep hydrated the samples, water was sprayed on the samples (Sangronis and Machado 2007, Tkalec et al. 2009, www.anar_ae.org 2009). Once germinated, the seeds were planted using special soil for plants.

As far as the treatment of Sherry Fino wine is concerned, three samples of 23 mL each of the Sherry Fino wine from *Los Raigones* winery, Córdoba, Spain, were treated. One sample was kept untreated in order to be able to carry out some comparisons. After the treatment, the samples were kept closed and stored in a dark environment for 2 years.

22.3 Active Species and UV Photons in Plasmas at Atmospheric Pressure

22.3.1 Active Species in Plasmas at Atmospheric Pressure

The appearance of an SWD generated using a power of 200 W and a flow of Ar and N_2 gases of 4.90 and 0.10 slm ($\approx$2% N_2), respectively, is shown in Figure 22.2. There are two easily distinguishable areas: (1) the discharge, which is the brightest zone and contains charged particles, and (2) the postdischarge area characterized by the absence of charged particles where the emitted light diffuses the most. The spectrum registered at the postdischarge is shown in Figure 22.3a. Thus, the radiation produced due to the transition of excited nitrogen molecule from the state $N_2(B^3\Sigma_g)$ to $N_2(A^3\Sigma_u^+)$ can be observed with band heads at 580.4 nm and 654.6 nm. These metastable molecules are responsible for maintaining postdischarge due to their long natural life span.

FIGURE 22.2 Image of an Ar-N_2 SWD generated with a flow of 4.90 and 0.10 slm, respectively, using a surfatron as a coupler device.

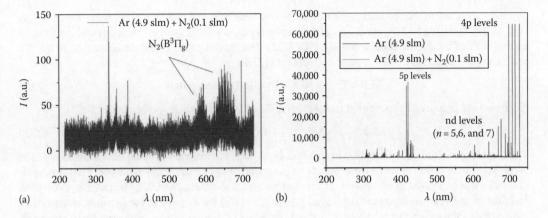

FIGURE 22.3 (a) Spectrum of an Ar-N2 (4.90 and 0.10 slm) postdischarge; (b) spectra of an Ar (4.90 slm) and an Ar-N$_2$ (4.90–0.10 slm) discharge.

Focusing on the kinetics to generate $N_2(A^3\Sigma_u^+)$ molecules in the discharge, which are transported by the gas flow toward the postdischarge, they are obtained from the de-excitation of the $N_2(B^3\Pi_g)$ state by the following reaction (Callade 1991, Ricard 1994, 1997):

$$N_2(B^3\Pi_g) \rightarrow N_2(A^3\Sigma_u^+) + h\nu \text{ (first positive system)} \quad (22.1)$$

Thus, knowing the channel to produce $N_2(B^3\Pi_g)$ molecules is essential to study postdischarge behavior during its application. Two channels are possible for this purpose.

The first corresponds to the Penning excitation Reaction 22.2 followed by spontaneous de-excitation to the $N_2(B^3\Pi_g)$ state (Reaction 22.3). Reactions 22.2 and 22.3 can be expressed as (Tatarova et al. 2006)

$$Ar^m + N_2(X^1\Sigma_g^+) \rightarrow N_2(C^3\Pi_u) + Ar^0 \quad (22.2)$$

$$N_2(C^3\Pi_u) \rightarrow N_2(B^3\Pi_g) + h\nu \text{ (second positive system)} \quad (22.3)$$

Another possibility is to start from the charge transfer Reaction 22.4, followed by collisions of $N_2^+(X^1\Sigma_g^+)$ with electrons, leading to the production of nitrogen atoms (22.5). These atoms participate in processes of three-body collisions, giving place to $N_2(B^3\Pi_g)$ molecules:

$$Ar^+ + N_2(X^1\Sigma_g^+) \rightarrow Ar^0 + N_2 + (X^1\Sigma_g^+) \quad (22.4)$$

$$N_2 + (X^2\Sigma_g^+) + e^- \rightarrow N + N \quad (22.5)$$

$$N + N + N_2(X^1\Sigma_g^+) \rightarrow N_2(B^3\Pi_g) + N_2(X^1\Sigma_g^+) \quad (22.6)$$

Finally, $N_2(A^3\Sigma_u^+)$ molecules are generated from (22.1).

The analysis of the spectra of Ar (4.9 slm) and Ar-N$_2$ (4.9-0.1 slm) discharges (Figure 22.3b) shows that the intensities of 4p levels in Ar-N$_2$ discharges are lower than that of 4p levels in Ar discharges, while 5p and higher levels disappear when N$_2$ is added. Thus, Reaction 22.2 leads to decreased metastable Ar density. Observing the spectra in Figures 22.3a and 22.3b, the kinetics in an Ar-N$_2$ discharge to generate a postdischarge where exist metastables $N_2(A^3\Sigma_u^+)$ particles is controlled by Reactions 22.2, 22.3, and 22.1.

22.3.2 UV Photons in Plasmas at Atmospheric Pressure

Moving to the generation of UV photons in the discharge of an SWD at atmospheric pressure, it is possible to obtain UV emission from excited NO molecules generated in discharges achieved using a mixture of N$_2$ with an oxygen-containing molecule. At atmospheric pressure, only the NO$_\gamma$ band is emitted in contrast to the reduced-pressure case where both NO_β and NO_γ bands are observed (Boudam et al. 2006). Focusing on NO_γ emission, gamma band is produced due to an electronic transition from an excited state of the NO molecule, namely, $NO(A^2\Sigma^+)$ state, to the ground state designated as $NO(X^2\Pi)$, producing UV emission in the range of 220–260 nm (22.7):

$$NO(A^2\Sigma^+) \rightarrow NO(X^2\Pi) + h\nu \text{ (gamma band)} \tag{22.7}$$

NO bands are produced by excitation transfer from $N_2(A^3\Sigma^+_u)$ metastable molecules (22.8):

$$NO(X) + N_2(A^3\Sigma^+_u) \rightarrow NO(A^2\Sigma^+) + N^+_2 (X^1\Sigma^+_g) \tag{22.8}$$

Thus, in order to obtain UV from NO_γ in an SWD, $N_2(A^3\Sigma^+_u)$ active species should be first created. In the current research, different set of experiments were carried out to generate UV emission. First, an Ar-N_2 discharge was generated in conditions similar to those used to obtain $N_2(A^3\Sigma^+_u)$ metastable molecules, and then oxygen-containing gases (hydrogen peroxide and air) were added to the plasmas, as described previously in the experimental section. Second, an Ar-N_2O discharge was evaluated to produce UV emission. Radiation from these three discharges was taken and is shown in Figure 22.4 (Ar-N_2-H_2O_2 plasma), Figure 22.5 (Ar-N_2-air), and Figure 22.6 (Ar-N_2O plasma).

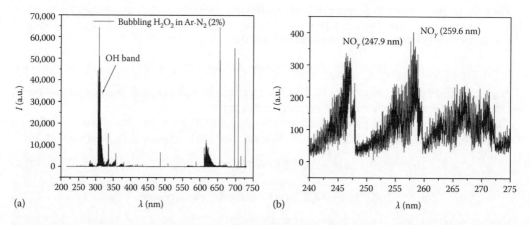

FIGURE 22.4 Spectra emitted by plasma generated by Ar-N_2-H_2O_2 mixture: (a) total spectrum and (b) amplified 220–290 nm UV region.

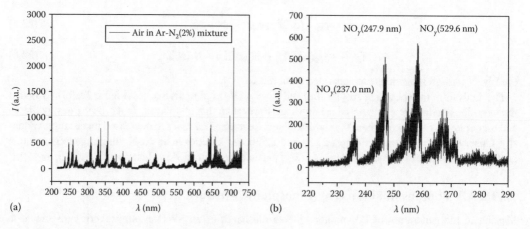

FIGURE 22.5 Spectra emitted by plasma generated by Ar-N_2-air mixture: (a) total spectrum and (b) amplified 220–290 nm UV region.

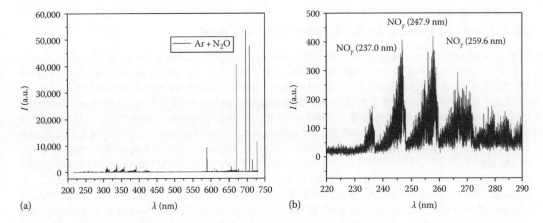

FIGURE 22.6 Spectra emitted by plasma generated with Ar-N$_2$O mixture: (a) total spectrum and (b) amplified 220–290 nm UV region.

The mixture first used as plasma gas to generate NO(A$^2\Sigma^+$) species was Ar-N$_2$-H$_2$O$_2$. The dissociation of H$_2$O$_2$ molecules can give oxygen atoms and water according to (22.9), and the association of this oxygen with the nitrogen atoms produces the NO radical at an excited level.

The bubbling method was employed to introduce H$_2$O$_2$ into the plasma at room temperature. In this sense, the total gas flow of Ar (0.56 slm) was divided into two parts: one (flow 1), with a flow of 0.07 slm, was bubbled into the recipient containing H$_2$O$_2$ to drag its molecules. The resulting mixture (gas and H$_2$O$_2$) was then added to a mixture containing the rest of argon (0.49 mL/min) (flow 2) and a flow of nitrogen (0.12 slm) ($\approx$2% N$_2$) (flow 3) before introducing it into the quartz tube that contained the discharge. The spectrum registered from the plasma emission is shown in Figure 22.4, observing molecular bands of OH radical with the highest intensity. Only if the interval between 220 and 275 nm is amplified, the molecular bands corresponding to NO$_\gamma$ species whose band heads correspond to 247.9 and 259.6 nm can be observed.

Two different processes for the decomposition of H$_2$O$_2$ molecules by the plasma particles can take place in the discharge:

$$H_2O_2 \rightarrow \tfrac{1}{2}O_2 + H_2O \tag{22.9}$$

$$H_2O_2 \rightarrow 2OH \tag{22.10}$$

From Figure 22.4, it can be observed that the intensity of OH band is stronger than NO$_\gamma$ one, hence, the dissociation of H$_2$O$_2$ molecules can be pointed out to be produced mainly by Reaction 22.10.

Another mixture, Ar-N$_2$-air, was studied, using flows of 5.00, 0.12, and 0.18 slm for Ar, nitrogen ($\approx$2%), and air, respectively. Figure 22.5 shows the spectrum obtained from this mixture, and in the amplified 220–290 nm region, NO$_\gamma$ can be observed. Besides, a strong emission of OH band can be noticed. Other molecular bands are observed that belong to different species that are present in the air added to Ar-N$_2$ mixtures.

The possibility of obtaining UV photons by a plasma generated by the addition of N$_2$O to Ar gas with flows of 0.02 and 0.56 slm, respectively, was also considered. Figure 22.6 shows the spectra emitted by this discharge. In Figure 22.6a, the intensities of NO$_\gamma$ bands are observed when the region in the 200–290 nm interval is amplified (see Figure 22.6b). This result is similar to what was found using the Ar-N$_2$-H$_2$O$_2$ plasma; however, the emission of UV photons coming from the OH band is smaller in comparison to the initially studied mixture. Thus, the use of Ar-N$_2$O plasmas can be considered the most suitable since UV emission comes only from NO$_\gamma$ gamma emission and not form the OH band.

22.4 Action of Plasma on the Growth Rate of Lentils

The experimental device used in the treatment of various groups of lentils is shown in Figure 22.1b.

Four groups of 20 lentils were considered. The first group was considered as the control to compare the effect of different treatments. The second group of lentils was treated with a postdischarge of an Ar (0.49 slm) and N_2 (0.05 slm $\approx$ 9%) SWD where the postdischarge cannot be noticed and thus active species are not created due to a low argon flow. The petri with the lentil population stood 7 cm from the end of the quartz tube to ensure that the lentils do not suffer a temperature higher than 55°C; besides, treatment exposure time to this plasma was 15 min (Figure 22.7a). As for the second treatment, the group of lentils was also treated with a mixture of Ar (4.90 slm) and N_2 (0.10 slm $\approx$ 2%) SWD but with a different percentage of nitrogen, which enables the production of active species as discussed previously in Section 22.3.1. To ensure that the temperature did not exceed 55°C, the dish was placed 8 cm from the end of the tube and was placed on a container filled with water to keep the sample cool. The postdischarge was also applied for 15 min. The last group of lentils was treated with an Ar (4.9 slm) and N_2O (0.05 slm) SWD. As in the second treatment, a water-filled container was used as a coolant and the samples were placed at 8 cm from the end of the tube; the exposure time was the same as in the other treatments. After each treatments, the different groups of lentils were kept in a damp environment in order to favor the lentils' germination.

The aspect of the groups of lentils after 72 h and 6 days of treatment is shown in Figures 22.7 and 22.8, respectively. As it can be seen in Figure 22.7, the group of lentils treated with an Ar-N_2O plasma is the least germinated group followed by a group treated with a postdischarge of an Ar-N_2 plasma with a large flow of Ar. According to Callede et al. (1991) and Ricard et al. (1991) and Figures 22.3 and 22.4, it has been found that the most successful treatment is when they are both combined, active species and UV photons, followed by only using active species, and finally when there are none. As for the growth rate in lentils, it is found that the behavior is the same as in germination rate, as shown in Figure 22.8. The growth rate in lentils is affected by the treatment that uses both active species and UV photons.

In order to check the influence of UV radiation on the growth rate of lentils, the action of different UV emission on seed germination has been assessed by means of using UV lamps (at 254 and 366 nm) and an array of UV LEDs. The UV lamp was a mercury lamp (DUO-UV-source for thin-layer and column chromatography, MinUVIS, DESAGA), which works at 220 V and 30 W. As for the array of 90 UV LEDs, the working voltage was 12 V.

Three different groups of 20 lentils were each treated with different UV radiation sources. The difference between each treatment was the time for which the lentils were exposed to UV radiation: 15, 60, and 120 min. Another group was considered as the control with no treatment for comparison purposes. In Figure 22.9, the average lengths of the germinated seeds after 72 h of treatment are shown. As it can be seen, treatment of 15 min retards the growth of lentils. In other treatments, the germination was favored, which is in agreement with what was found in Noble (2002). Thus, the action of plasma in seed germination can be pointed out as crucial in the growth rate of lentils.

22.5 Action of Plasma on Browning of Sherry Fino Wine

The experimental device used in the treatment of the Sherry Fino wine samples was the one shown in Figure 22.1b but arranged vertically. The quartz tube protruded 11.5 cm from the surfatron and the distance between the end of the tube and the sample of wine was approximately 1 cm. Four samples of 23 mL each of Sherry Fino wine from *Los Raigones* winery, Córdoba, Spain, were treated. The first sample was considered as the control just to be able to compare and contrast the effect of different treatments (gas and plasma). A flowing of gas mixture with Ar (4.90 slm) and N_2 (0.10 slm) was applied to the second sample. The postdischarge of pure Ar (4.90 slm) and Ar-N_2 (4.90–0.10 slm, 2%) plasmas were applied to the third and fourth samples, respectively. In the last two cases, the wine samples were refrigerated by placing them in an ice-water container in order to avoid their heating during the treatment. In all cases, the treatment time was over 5 s. Next, the samples were kept for 2 years in an airtight tube and stored in a dark environment where the temperature was lower than 20°C.

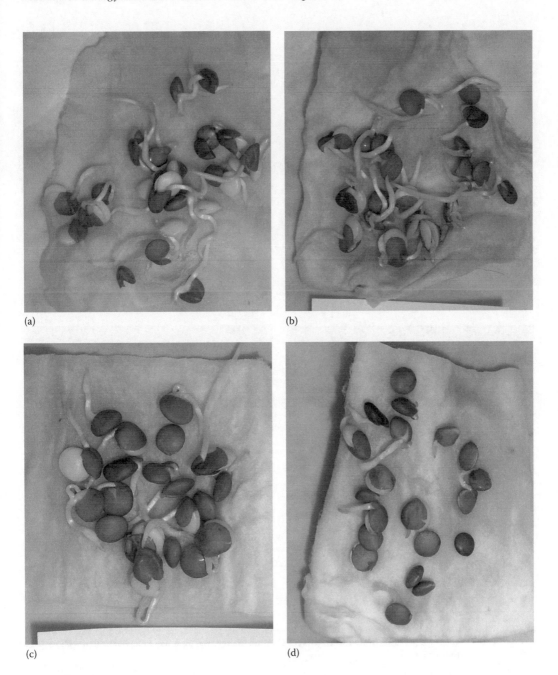

FIGURE 22.7 Germination rate after 72 h of treatment: (a) control group of lentils, (b) group of lentils treated with an Ar-N_2 (9%) SWD, (c) group of lentils treated with Ar-N_2 (2%) postdischarge, and (d) group of lentils treated with Ar-N_2O (4.90–0.05 slm) postdischarge.

One important property of Fino wines, which are obtained by biological aging, is their pale yellow color. It should be pointed out that this type of wine does not brown under yeast, preserving its pale color with slight oscillations for years. This has traditionally been ascribed to flor yeasts that grow on its surface making the diffusion of atmospheric oxygen through the flor film difficult and protecting the wine from oxidation. This type of wine is a delicate product because it tends to brown after bottling. As far as other organoleptic qualities such as smell or flavor are concerned, in the current research, no analysis has been

FIGURE 22.8 Growth rate after 6 days of treatment: (a) control group of lentils, (b) group of lentils treated with an Ar-N$_2$ (9%) SWD, (c) group of lentils treated with Ar-N$_2$ (2%) postdischarge, and (d) group of lentils treated with Ar-N$_2$O (4.90–0.05 slm) postdischarge.

carried out since the main objective was focused on the study of browning of sherry Fino wine which is color quality. The external aspect of the samples is shown in Figure 22.10. As it can be observed, the sample that shows less browning is the one that was treated with a postdischarge of Ar-N$_2$ (2%).

In addition to a direct observation of the sample, other analyses were carried out in order to know which of the treatments used, gas or plasma, increased the resistance of Fino wines to browning.

Particular chromatic parameters for the samples were obtained from direct spectrophotometric measurements. These were carried out on a path length of 10 mm using a Perkin Elmer Lambda 25 spectrophotometer. Color analyses were carried out following International Commission on Illumination recommendations (C.I.E. 2004) and using the visible spectrum obtained from 380 to 780 nm. In this work, the following CIELab uniform space colorimetric parameters have been considered: $a*$ and $b*$

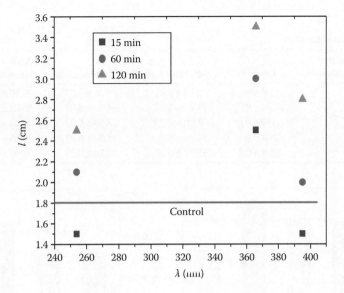

FIGURE 22.9 Average length of roots of germinated lentils versus UV radiation used in the treatments.

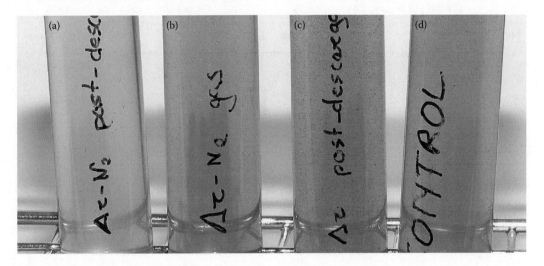

FIGURE 22.10 Aspect of the samples after 2 years of storage treated with (a) Ar-N$_2$ (2%) postdischarge, (b) Ar-N$_2$ (2%) gas, (c) Ar postdischarge, and (d) control.

(chromatic coordinates representing red-green and yellow-blue axes, respectively), rectangular coordinates L_{ab}^* (black-white component, lightness), and the cylindrical coordinates C_{ab}^* (chroma or saturation) and h_{ab} (hue angle). These parameters were measured using as references the C.I.E. 1964 standard observer (10 visual field) and the CIE standard illuminant D65. Besides, the absorbance at 420 nm can be also used as a measure of wine browning (Fabios et al. 2000, Moreno 2005).

Table 22.1 shows the CIELab parameters a^* (red/green coordinate), b^* (yellow/blue coordinate), lightness (L_{ab}^*), chroma (C_{ab}^*), and hue (h_{ab}), for the control Fino wine and for the Fino wines subjected to Ar-N$_2$ (2%) gas and postdischarges of Ar and Ar-N$_2$ (2%) plasmas.

As can be seen from Table 22.1, CIELab coordinates a^* and b^* decreased in the Fino wine with plasma-Ar-N$_2$ (2%) treatment. Chroma (C_{ab}^*) measures the contribution of a^* and b^* to the color of wine, indicating values close to 50 saturated colors (Gil-Muñoz et al. 1997). C_{ab}^* was also lower in the wine subjected to the aforementioned treatment, and with substantial differences with respect to the control wine

TABLE 22.1

CIElab Parameters

Chromatic Characteristics	Control	Ar-N₂ (2%) Gas	Ar Postdischarge	Ar-N₂ (2%) Postdischarge
a^*	1.96 ± 0.02	2.30 ± 0.01	3.26 ± 0.02	-0.36 ± 0.01
b^*	31.30 ± 0.02	31.20 ± 0.02	34.90 ± 0.01	24.10 ± 0.02
L_{ab}^*	82.40 ± 0.03	81.90 ± 0.04	78.60 ± 0.02	90.10 ± 0.04
C_{ab}^*	31.4 ± 0.1	31.30 ± 0.02	35.00 ± 0.03	24.10 ± 0.02
h_{ab}	86.4 ± 0.4	85.80 ± 0.03	84.70 ± 0.03	90.80 ± 0.04

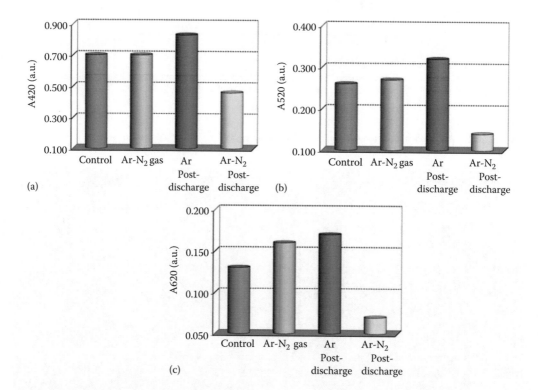

FIGURE 22.11   Measure of absorbance for each sample (a) at 420 nm, (b) at 520 nm, and (c) at 620 nm.

and the rest of the Fino wines. Both, lightness (L_{ab}^*) range from 0 (black) to 100 (white), and hue (h_{ab}), dened as the arctan (b^*/a^*), slightly increased in the Fino wine subjected to postdischarge Ar-N₂ plasma.

As far as the measure of absorbance is concerned, Figure 22.11 shows the absorbance at 420 nm (Figure 22.11a), at 520 nm (Figure 22.11b), and at 620 nm (Figure 22.11c) for each of the wine sample. In the light of the graphs, the results show high values in A420 for the control Fino wine and for the wines subjected to Ar-N₂ (2%) gas and Ar postdischarge, which reached values above 0.70 a.u. However, the Fino wine with Ar-N₂ (2%) postdischarge treatment showed the lowest value near to 0.46 a.u. It is corresponding to a decrease of 65% with respect to the control wine, which points out that the plasma treatment of a postdischarge, where active species can be found, keeps the characteristic pale color, avoiding browning of the wine. The absorbance at 520 nm, which is a measure of brown-red color, and at 620 nm (blue), exhibited the same trend that was observed for 420 nm.

From these preliminary results, the mechanisms induced by the active species and radiation of the postdischarge in the resistance increase of the Fino wine to browning cannot be known. However, in a first approximation, it is possible to know that are the active species in the Ar-N₂ (2%) postdischarge which could be capable of acting on the samples according to the discussion carried out in Section 22.3.1.

22.6 Conclusions

One serious problem for the food industry is the deterioration of food products during their storage. Besides, the consumers are increasing their demands related to the use of the preservation process, which should be nondetrimental for human health. In order to reconcile both consumers' demands and the decrease of economic losses by food companies, the industry involved in this matter is devoting important economic resources and developing new techniques for the production and the storage of their products. In this research, a study of the capability of SWDs at atmospheric pressure for food preservation has been carried out.

Firstly, in the current research active species and UV radiation have been generated following previous results in the sterilization plasma-process. In this regard, it has been demonstrated that the joint effort of both active species and UV radiation from plasma has inhibited the germination of lentils. Furthermore, the application of an Ar-N$_2$ postdischarge to a sample of Sherry Fino wine manages to keep the characteristic pale yellow color of this wine avoiding its browning.

According to this preliminary research, there seems to be enough evidence to suggest that the use of plasma as a food preservation technique is a suitable option, so this work could be considered as the first step in this field.

Plasma technology not only improves the shelf life and safety of food by inactivating spoilage and pathogenic microorganisms by means of the well-studied sterilization mechanisms taking place in the plasma (both UV radiation and active species), but its use also does not change organoleptic attributes as pointed out by the action of plasma on the appearance of lentil seeds and the characteristic yellow pale color of the Sherry Fino wine and its use does not leave behind any residues. In spite of the noteworthy advantages on the use of plasma technology as a new food preservation technique, this technique involves several fields of knowledge which suggests that expertise from plasma physics should be combined with others thus implicating a wider dimension of the research.

ACKNOWLEDGMENT

This research has been supported by the Ministry of Science and Innovation (Spain) and FEDER Funds within the framework of the project No. ENE2008-01015/FTN.

REFERENCES

Artés, F., P. Gómez, E. Aguayo, V. Escalona, and F. Artés-Hernández. 2009. Sustainable sanitation techniques for keeping quality and safety of fresh-cut plant commodities. *Postharvest Biology and Technology* 51: 287–296.

Basaran, P., N. Basaran-Akgul, and L. Oksuz. 2008. Elimination of *Aspergillus parasiticus* from nut surface with low pressure cold plasma (LPCP) treatment. *Food Microbiology* 25: 626–632.

Boudam, M.K., M. Moisan, B. Saoudi, C. Popovici, N. Gherrdi, and F. Massines. 2006. Bacterial spore inactivation by atmospheric-pressure plasmas in the presence or absence of UV photons as obtained with the same gas mixture. *Journal of Physics D: Applied Physics* 39: 3949–3507.

Callede, G., J. Deschanps, J.L. Godart, and A. Ricard. 2001. Active nitrogen atoms in an atmospheric pressure flowing Ar-N$_2$ microwave discharge. *Journal of Physics D: Applied Physics* 24: 909–914.

Callede, G., J. Deschanps, J.L. Godart, and A. Ricard. 1991. Active nitrogen atoms in atmospheric pressure flowing Ar-N$_2$ microwave discharge. *Journal of Physics D* 24: 909–914.

Calzada, M.D., M.C. Quintero, A. Gamero, and M. Gallego. 1992. Chemical generation of chlorine, bromine and iodine for sample introduction into a surfatron-generated argon microwave-induced plasma. *Analytical Chemistry* 64: 1374–1378.

Capellas, M., M. Mon-Mur, E. Sendra, R. Pla, and B. Guamis. 1996. Populations of aerobic mesophiles and inoculated *E. coli* during storage of fresh goat's milk cheese treated with high pressure. *Journal of Food Protection* 59: 582–587.

de Daucos, B., M.P. Cano, and R. Gómez. 2000. Characteristics of stirred low-fat yogurt as affected by high pressure. *International Dairy Journal* 10: 105–111.

Devlieghere, F., L. Vermeiren, and J. Debevere. 2004. New preservation technologies: Possibilities and limitations. *International Dairy Journal* 14: 273–285.

Eliasson, B. and V. Koglschatz. 1991. Nonequilibrium volume plasma processing. *IEEE Transactions on Plasma Science* 19: 1063–1077.

Fabios, M., A. Lopez-Toledano, M. Mayen, J. Mérida, and M. Medina. 2000. Phenolic compounds and browning in sherry wines subjected to oxidative and biological ageing. *Journal of Agricultural and Food Chemistry* 48: 2155–2159.

Feugeas, J.N., B.J. Gomez, L. Nachez, and J. Lesage. 2003. Steel surface treatment by a dual process of ion nitriding and thermal shock. *Thin Solid Film* 424: 125–129.

Gil-Muñoz, R., E. Gomez-Plata, A. Martinez, and J. Lopez-Roca. 1997. Evolution of the CIELAB and other spectrophotometric parameters during wine fermentation. Influence of some pre and postfermentation factors. *Food Research International* 30: 699–670.

Góngora-Nieto, M.M., D.R. Sepúlveda, P. Pedrow, G.V. Barbarosa-Cánovas, and B.G. Swanson. 2002. Food processing by pulsed electric fields treatment delivery inactivation level, and regulation aspects. *Lebensmittel-Wissenschaft und Technologie* 35: 375–388.

Gould, G.W. 1995. Biodeterioration of foods and overview of preservation in the food and dairy industries. *International Biodeterioration & Biodegradation* 36: 267–277.

Jiménez, M., R. Rincón, A. Marinas, and M.D. Calzada. 2013. Hydrogen production from ethanol decomposition by a microwave plasma: Influence of the plasma gas flow. *International Journal of Hydrogen Energy* 38: 8708–8719.

Jiménez, M., C. Yubero, and M.D. Calzada. 2008. Study on the reforming of alcohols in a surface wave discharge at atmospheric pressure. *Journal of Physics D: Applied Physics* 41: 175201.

Moisan, M., E. Etemadi, and J.C. Rostaing. 1998. French Patent No. 2 762 748, European Patent No. EP 0874 537 A1.

Moisan, M., P. Leprince, C. Beaudry, and E. Bloyet. 1974. Perfectionnements apportés aux dispositifs d'excitation par des ondes HF d'une colonne de gaz enfermée dans un enveloppe. Brevet, France.

Moreno, A. 2005. Influencia del tipo de envejecimiento sobre el perfil aromático de vinos generosos andaluces. PhD thesis, Universidad de Córdoba, Córdoba, Spain.

Niemira, B.A. and J. Sites. 2008. Cold plasma inactivates *Salmonella* Stanley and *Escherichia coli* O157:H7 inoculated on Golden Delicious apples. *Journal of Food Protection* 71: 1357–1365.

Noble, R.E. 2002. Effects of UV-irradiation on seed germination. *Science of the Total Environment* 299: 173–176.

Park, B.J., D.H. Lee, J.C. Park, I.S. Lee, K.Y. Lee, S.O. Hyun, M.S. Chun, and K.H. Chung. 2003. Sterilisation using a microwave-induced argon plasma system at atmospheric pressure. *Physics of Plasmas* 10: 4539–4544.

Peñuelas, J., J. Llusià, B. Martínez, and J. Fontcuberta. 2004. Diamagnetic susceptivility and root growth responses to magnetic fields in Lens culinaris, Glycine soja, and Triticum aestivum. *Electromagnetic Biology and Medicine* 23: 97–122.

Perni, S., D.W. Liu, G. Shama, and M.G. Kong. 2008. Cold atmospheric plasma decontamination of the pericarps of fruit. *Journal of Food Protection* 71: 302–308.

Raso, J., and G.V. Barbosa-Cánovas. 2003. Nonthermal Preservation of foods using combined processing techniques. *Critical Reviews in Food Science and Nutrition* 43: 265–285.

Ricard, A. 1997. The production of active plasma species for surface treatment. *Journal of Physics D* 30: 2261–2269.

Ricard, A., H. Malvos, S. Bordeleau, and J. Hubert. 1994. Production of active species in common flowing post-discharge of Ar-N_2 plasma and an A_r-H_2-CH_4 plasma. *Journal of Physics D: Applied Physics* 27: 504–508.

Ricard, A., J. Tétreault, and J. Hubert. 1991. Nitrogen atom recombination in high pressure Ar-N_2 flowing post-discharges. *Journal of Physics B: Atom, Molecular and Optical Physics* 24: 1115–1123.

Sangronis, E. and C.J. Machado. 2007. Influence of germination on the nutritional quality of *Phaseous vulgaris* and *Cajanus cajan*. *LWT-Food Science and Technology* 40: 116–120.

Selcuk, M., L. Oksuz, and P. Basaran. 2008. Decontamination of grains and legumes infected with *Aspergillus* spp. and *Penicillium* spp. by cold plasma treatment. *Bioresource Technology* 99: 5104–5109.

Song, Pa., B. Kim, J. Ho Choe, S. Jung, S.Y. Moon, W. Choe, and C. Jo. 2009. Evaluation of atmospheric pressure plasma to improve the safety of sliced cheese and from inoculated by 3-strain cocktail *Listeria monocytogenes*. *Food Microbiology* 26: 432–436.

Tatarova, E., F.M. Dias, J. Henriques, and C.M. Ferreira. 2006. Large-scale Ar and N_2-Ar microwave plasma sources. *Journal of Physics D* 39: 2747–2753.

Tkalec, M., K. Malavic, M. Pavliva, B. Pevalek-Kozlina, and Z. Vidakovic-Cifrek. 2009. Effects of radiofrequency electromagnetic fields on seed germination and root meristematic cells of *Allium cepa* L. *Mutation Research* 672: 76–81.

Voli, J.C., F.S. Denes, R.A. Young, and S.M. Park. 2000. Modification of seed germination performance through cold plasma chemistry technology. *Crop Science* 40: 1706–1718.

Yuste, J., M. Capellas, R. Pla, D.Y.C. Fung, and M. Mon-Mur. 2002. High pressure processing for food safety and preservation: A review. *Journal of Rapid Methods and Automatization in Microbiology* 9: 1–10.

Zajickova, L., M. Elias, O. Jasek, V. Kurdle, Z. Frgala, J. Matecjková, A. Rek, J. Bursik, and M. Kadleciková. 2006. Atmospheric pressure microwave torch at atmospheric pressure. *Materials Science and Engineering C* 26: 1189.

23

Broad-Spectrum Hybrid and Engineered Bacteriocins for Food Biopreservation: What Will Be the Future of Bacteriocins?

Augusto Bellomio, Carlos Javier Minahk, and Fernando Dupuy

CONTENTS

23.1 Introduction: Lactic Acid Bacteria and Food Biopreservation

Fermented foods were discovered a long time ago, when humans began storing raw food for use in times of scarcity. If we could figure out how our ancestors lived at that time and the containers they used for this purpose, we would easily notice that the living and sanitary conditions differed diametrically from those existing in our current food stores or kitchens. As microorganisms growing in provisions cause fermentation, raw food was *naturally processed*, acquiring unique organoleptic characteristics. In many circumstances, perhaps ravaged by hunger, our ancestors were *forced* to eat those fermented foods and some of them may have found them tasty. They learned how to treat the raw material to obtain fermented foods and also as a strategy to extend the storage period. In that way, they could dispose of safe food to be consumed at times of the year when fresh foods were unavailable. For instance, products such as curd or cheese would have been obtained for the first time after storing fresh milk for long periods of time. Without any doubt, the mastering of each method of fermentation must have taken many years. The people who populated the earth developed various cultures and fermented foods, using different raw materials and ferments. Most fermented foods are still produced using traditional methods in small farms. However, the production of many of them has been escalated to the industrial level using advanced technology, making possible the globalization of many fermented foods (Hutkins, 2006). Ancient techniques that allowed the preparation of many fermented foods have been greatly improved at present. Now, we know that lactic acid bacteria (LAB) are the major contributors to the desired taste properties, improved nutritional qualities, functionality, and increased shelf life, which many fermented foods have. LAB are nonsporulating gram-positive bacteria that produce lactic acid from available sugars. They can also produce other organic acids, ethanol, carbon dioxide, and even aromatic molecules, which contribute to the flavor and taste of fermented foods.

Fermented foods obtained under controlled conditions using modern industrial processes have undergone significant improvements in terms of nutritional quality, flavor, aroma, and most of all, high microbiological quality. Rigorous quality control must be followed during the different elaboration stages—handling of the raw material, processing, packaging, and distribution—in order to eliminate the risk of pathogens or microorganisms producing unwanted metabolites that could be toxic for the consumer.

Certainly, the quality control process in the food chain industry is the one which raises concern; hence, it requires closer attention by health authorities at the level of government and international health organizations. Manufacturers want to sell products with high demand satisfying the needs of the distributors, retailers, and consumers. Everyone wants a product suitable for consumption with the longest shelf life. At the same time, consumers require foodstuff as tasty and healthy as possible. That is, they want food with the best organoleptic and nutritional properties but also not containing chemical additives, which could potentially produce harmful health effects. The mere fact that foodstuff additives are *nonnatural* raises concern among consumers about what they are really eating. Actually, it is not really tested how damaging this could be for human health.

The growth of LAB in fermented foods or even on the surface of unfermented food produces depletion of the available nutrients for other undesirable contaminant bacteria. Moreover, their metabolic products are inhibitory for bacterial growth; hence, an increased shelf life of food is achieved.

Additionally, many LAB strains produce compounds with powerful antibacterial activity called bacteriocins. In fact, many strains that are used as starter cultures produce bacteriocins and thereby eliminate undesirable bacteria. Among the genera of pathogenic bacteria that can cause food-borne diseases, the gram-negative bacteria *Campylobacter, Salmonella, Escherichia, Shigella, Brucella, Yersinia,* and *Vibrio* and the gram-positive bacteria *Listeria, Bacillus, Clostridium,* and *Staphylococcus* can be highlighted.

23.2 Food-Borne Diseases and Bacterial Pathogens

Food-borne diseases are illnesses derived from the ingestion of contaminated food and tap water. Even though they are mainly represented by gastrointestinal symptoms, complications such as neurological, gynecological, and immunological problems or even death may occur. They represent a huge burden to the public health system as millions are affected every year worldwide and encompass diseases caused by microorganisms as well as chemical hazards (Brown, 2011). This section focuses on bacteria associated with food-borne diseases and the perspective of using natural bacteriocins and hybrid bacteriocins for the prevention of food-related illnesses.

Both gram-positive and gram-negative bacteria can be associated with food-borne disease. Among the well-known gram-negative bacteria, *Campylobacter* spp., *Salmonella* spp., *Escherichia coli* O157:H7, *Shigella* spp., and *Vibrio cholerae* are the most representative ones. On the other hand, *L. monocytogenes, S. aureus, B. cereus, C. perfringens,* and *C. botulinum* are the more frequently cited gram-positive bacteria associated with food poisoning.

23.2.1 Campylobacter

Campylobacter spp. is the most frequent microorganism isolated from people with gastrointestinal symptoms in Europe (Eurosurveillance Editorial Team, 2013; Havelaar et al., 2012) and the third bacterium in the United States only behind *Salmonella* spp. and *C. perfringens* (Scallan et al., 2011). On the other hand, developing countries do not have national surveillance programs for *Campylobacter* infections; thus, the real incidence of campylobacteriosis is surely underestimated (Coker et al., 2002). Moreover, as *Campylobacter* is fastidious to culture, it generally cannot be accurately assessed, unless modern techniques such as real-time quantitative polymerase chain reaction (PCR) are used on a regular basis in clinical laboratories (Inglis and Kalischuk, 2004). Therefore, there is no solid statistics on the real incidence of campylobacteriosis in developing countries but a significant impact should be assumed. Indeed, *Campylobacter* spp. are thought to be the most prevalent bacteria associated with gastrointestinal diseases in the world (Roux et al., 2013).

Campylobacter coli and particularly *Campylobacter jejuni* are the most common species that cause human infections. The symptoms span from diarrhea to fever and abdominal pain (Peterson, 1994). The Guillain–Barré syndrome, the most common cause of acute flaccid paralysis, may follow a *Campylobacter* infection due to an autoimmune response triggered by the bacterium (Ang et al., 2004; Louwen et al., 2013). In addition, rare gastrointestinal perforation and occasionally death may also take place (Elshafie et al., 2010).

Bacteria from the *Campylobacter* genus are microaerophilic and thermophilic. In fact, *C. jejuni* has an optimal growth at 42°C with very low oxygen concentrations (Penner, 1988); hence, it is well adapted for growing in the intestines of warm-blooded birds and mammals. In fact, *C. jejuni* and *C. coli* are frequently present in a wide range of asymptomatic domesticated farm animals. The main vehicles of infection include contaminated water, unpasteurized milk, and especially chicken (Altekruse, 1999).

Although campylobacteriosis is a self-limited disease, some severe cases must be treated with antibiotics. In this regard, antibiotic resistance of *C. jejuni* is a growing problem worldwide, and in particular resistance to quinolones, likely due to their extensive use in poultry feed (Nelson et al., 2007). Even though macrolides are the usual and more effective antibiotics prescribed for treating campylobacteriosis, resistance toward this group of antibiotics is rising as well (Belanger and Shryock, 2007). Therefore, newer approaches must be adopted in order to treat and prevent infections by *Campylobacter* spp., such as bacteriocins. In fact, a number of anti-*Campylobacter* peptides have been recently described (Messaoudi et al., 2012; Svetoch and Stern, 2010).

23.2.2 Salmonella

Salmonella spp. is a worldwide enteropathogenic bacterium (Aarts et al., 2011; Thiele et al., 2011). Although it is not as prevalent as *Campylobacter* spp., it is believed to be the deadliest gram-negative bacterium associated with food poisoning (Majowicz et al., 2010). In fact, *Salmonella enterica* subspecies *enterica* is a widespread pathogen that mainly presents as a common inhabitant in humans as well as animals (Stevens et al., 2009). Interestingly, *S. enterica* has six subspecies that are divided into more than 2500 serovars (Sánchez-Jiménez et al., 2010), which are classified by the antigen type present in the cell membrane lipopolysaccharide (LPS) like the somatic O, flagellar H, or Vi antigen, as well as by proteomic analysis (de Jong and Ekdahl, 2006; Dieckmann et al., 2008). Even though these serovars are genetically close, there are wide variations in the host-specificity, virulence, and disease manifestations.

Salmonellosis has two different clinical outcomes: typhoid-like disease or nontyphoid disease. *S.* Typhi and *S.* Paratyphi are involved in typhoid-like cases and may cause death because of a number of complications such as dehydration, encephalitis, hemorrhage, and peritonitis, among others. On the other hand, nontyphoid disease represents a mild to severe gastroenteritis caused mainly by *S.* Enteritidis and *S.* Typhimurium serovars. It is interesting to note that *S.* Typhimurium causes typhoid fever in mice (Raffatellu et al., 2006; Sánchez-Vargas et al., 2011).

Regardless of the salmonellosis type, the classic sources of infection are either contaminated water or raw and poorly cooked food, especially egg-containing food (Betancor et al., 2010; Ricke et al., 2013).

Salmonella infection implies an intracellular cycle of the pathogen. However, nontyphoid processes only affect the intestinal mucosa, unless some sort of immunodeficiency occurs; in these cases, an invasive infection may take place (Gordon et al., 2010). On the other hand, typhoid-like diseases are classically associated with the invasion and spreading of pathogens (de Jong et al., 2012b). *Salmonella*–eukaryotic cell interaction involves the following steps: bacterial adhesion, cell invasion, *Salmonella*-containing vacuoles maturation, and replication (Ibarra and Steele-Mortimer, 2009; Lahiri et al., 2010). So far, a number of genes encoding virulence factors have been identified as key players for *Salmonella* pathogenesis. Interestingly, most of these genes are located close to each other in the bacterial genome, in groups denominated "*Salmonella* pathogenicity islands" (SPI) (Dieye et al., 2009).

As for *Campylobacter* spp., the rise of *Salmonella* resistance toward clinical antibiotics is a serious concern. *Salmonella* is also zoonotic; hence, the observed resistance is again mainly due to the use of antibiotics in farm animals (de Jong et al., 2012a; Fluit, 2005; Koluman and Dikici, 2013). In this regard, integrons class I and the *Salmonella* genomic island 1 play a crucial role in classic antibiotic resistance (Fluit and Schmitz, 2004). For invasive salmonellosis, fluoroquinolones used to be the antibiotics of choice. However, there is a marked decrease in susceptibility nowadays (Cooke et al., 2006; Medalla et al., 2011). Therefore, the use of third-generation cephalosporins, such as ceftriaxone, has gained acceptance in the past years. However, resistance to third-generation cephalosporins was recently described in *Salmonella* as well (Majtan et al., 2010; Miriagou et al., 2004). Moreover, there is no approved vaccine against this food-borne pathogen; thus, there are several approaches that are currently being developed in order to tackle the *Salmonella* problem. On the one hand, there are some vaccines that are being tested (Chalón et al., 2012; Martin, 2012). On the other hand, hurdle technology received much attention because of its promising use in the preservation of food products (Leistner, 2000). One of the barriers being considered is represented by antimicrobial peptides, the so-called bacteriocins, in particular, those produced by LAB (Gálvez et al., 2007). As mentioned earlier for *Campylobacter*, there are a number of peptides that proved to be efficient in reducing *Salmonella* counts in food (Chalón et al., 2012).

23.2.3 *E. coli* and *Shigella*

Two closely related gram-negative food-borne bacteria can be mentioned as well: the enterohemorragic *E. coli* O157:H7 and *Shigella* spp. The latter genus has two main species generally involved in food poisoning: *Shigella sonnei* and *Shigella flexneri* (Paula et al., 2010). As opposed to *Campylobacter* and *Salmonella*, humans and a few primates are the only hosts for *Shigella* (Paula et al., 2010). However, like *Salmonella*, *Shigella* spp. are also bacteria that can acquire an intracellular cycle (Ashida et al., 2011). Moreover, *Shigella* can infect epithelial as well as M cells, and after being enclosed by membrane vacuoles, they are able to disrupt these structures and replicate in the cytoplasm in a similar way *Listeria* does (Ashida et al., 2009). Even though the impact of *Shigella* on the healthcare system of developed countries is not as important as *Salmonella*, the impact of shigellosis is still very important in poor and developing countries (Chang et al., 2012). Although many attempts have been made so far, vaccines are not licensed worldwide yet (Kweon, 2008), but some of the vaccines that are being developed seem to be very promising (Martinez-Becerra et al., 2012).

Verotoxigenic *E. coli* strains, on the other hand, appear to be other food-related threats, closely following *Salmonella* spp. (Paton and Paton, 1998). They cause diarrhea, generally associated with bloody stools and abdominal cramps. Furthermore, hemolytic–uremic syndrome, which is a major cause of acute renal failure in children, may follow verotoxigenic *E. coli* infection. This complication can cause death in up to 10% of affected patients (Scheiring et al., 2008; Zoja et al., 2010). The main pathogenic factor is represented by toxins that are collectively known as verotoxins or Shiga-like toxins because they resemble the toxin produced by *Shigella dysenteriae* type 1. Although there are more than a hundred verotoxigenic strains, the one the most often reported and best characterized is the *E. coli* O157:H7 strain, designated by its somatic (O) and flagellar (H) antigens. This strain was first described more than 30 years ago by Riley et al. (1983). Since then, most infections have been linked to undercooked ground beef (Taylor et al., 2012). However, this is not the only mode of transmission; in fact, unchlorinated water supply can be another source (Olsen et al., 2002; Swerdlow et al., 1992).

A key step in *E. coli* O157:H7 infection is the adhesion to the host mucosa during the large intestine colonization by means of fimbrial and afimbrial adhesins (Lloyd et al., 2012). In this regard, two adhesins have been proved to be crucial for colon colonization: intimin, an outer membrane protein, and the long polar fimbriae (Fitzhenry et al., 2002; Mundy et al., 2007). A direct consequence of the tight and efficient adhesion is the fact that a relatively small inoculum is sufficient for a successful infection. Indeed, fewer than 100 colony-forming units are more than enough to colonize the human intestine (Mead and Griffin, 1998). Once *E. coli* O157:H7 is firmly attached to the large intestine, Shiga-toxin(s) secretion takes place. These toxins are composed of two subunits designated as subunits A and B (Paton and Paton, 1998). Subunit A displays an RNA *N*-glycosidase activity that inhibits protein synthesis, thus inducing apotosis (Karmali et al., 2010). On the other hand, subunit B forms a pentamer that binds to globotriaosylceramide-3 (Lingwood et al., 1987), which represents the host receptor to the toxin. Once the toxins get absorbed, they enter the bloodstream and target cells expressing globotriaosylceramide-3. In fact, the renal glomerular endothelium expresses this receptor in high levels. That is why hemolytic–uremic syndrome may occur (Nguyen and Sperandio, 2012). It is important to note that Shiga toxins may not be necessary for triggering diarrhea. Toxins are encoded by the *stx* genes, which are located on lambdoid prophages (Loś et al., 2012). Therefore, the production of Shiga toxins occurs only after prophage induction. As toxin-producing cells are lysed, the efficiency of Shiga toxin–converting prophage is believed to be low. Therefore, a large proportion of the bacterial population can survive (Loś et al., 2012).

The use of antibiotics is not recommended because they can exacerbate Shiga toxin–mediated cytotoxicity by enhancing the replication and expression of *stx* genes (Nguyen and Sperandio, 2012). Moreover, no vaccine for *Shigella* is available either. Hurdle technology once again appears to be a good solution for combating enterobacteria without the use of chemical food preservatives. There are a growing number of bacteriocins that seem to be useful for reducing *E. coli* as well as *Shigella* spp. counts (Chalón et al., 2012; Patton et al., 2008; Pomares et al., 2009).

23.2.4 Other Gram-Negative Bacteria

Other gram-negative bacteria that may be involved in food poisoning are *Aeromonas* spp. and *V. cholerae*, which are mainly associated with water-borne diseases. Interestingly, humans are the only known hosts for *V. cholerae* and it used to be a major pathogen worldwide. In fact, there are seven reported pandemic waves due to *V. cholerae* (Bishop and Camilli, 2011). Notwithstanding, it remains endemic in poor countries (Bishop and Camilli, 2011). *Vibrio cholerae* infection follows a similar pattern as *E. coli* O157:H7 in the sense that it first get firmly attached to the mucosa (Taguchi et al., 1997) and then it secretes a powerful toxin, the so-called CT toxin, that indirectly disrupts the activity of ion channels in epithelial cells (Bharati and Ganguly, 2011). This toxin is also encoded by a prophage (Safa et al., 2010). However, *V. cholerae* colonizes mainly the small intestine as opposed to *E. coli* O157:H7 (Peterson, 2002). As stated for *Shigella* and *V. cholera*, *Aeromonas* spp. also lead to food-borne and especially water-borne gastroenteritis in developing countries (Ghenghesh et al., 2008). Even though several virulence factors have been described so far, a hemolysin known as aerolysin is the better studied one. This protein has both hemolytic and enterotoxic activities (Bücker et al., 2011; Fivaz et al., 2001). While watery and self-limited diarrhea is the most common outcome, cholera-like diarrhea has been reported as well in some cases (Ghenghesh et al., 2008).

So far, there is little information regarding bacteriocins against these latter microorganisms; some gram-positive bacteriocins turned out to be effective against them but in combination with other treatments (Cobo Molinos et al., 2008). However, the impact of *V. cholerae* and *Aeromonas* spp. is quantitatively much lower as compared to the global impact of *E. coli* O147:H7, *Shigella* spp., and especially *Campylobacter* spp. and *Salmonella* (Centers for Disease Control and Prevention [CDC], 2013).

23.2.5 Gram-Positive Bacteria

As noted earlier, there are gram-positive bacteria that also cause great concern in food poisoning. It can be mentioned first that the toxin-secreting microorganisms such as *B. cereus*, the anaerobic *C. perfringens*, and the facultative anaerobic *S. aureus*. *Bacillus cereus* and *C. perfringens* are spore-forming microorganisms that have the ability to survive harsh conditions (Abee et al., 2011; Márquez-González et al., 2012). Therefore, it is extremely difficult to eradicate them by an easy protocol. Ultimately, toxins are responsible for the gastroenteritis related to these pathogens. This is the reason why the symptoms begin much quicker than in the other cases of food poisoning. Actually, symptoms are evident less than 24 h after food ingestion (McMullan et al., 2007). *Clostridium perfringens* is the second most common agent associated with bacterial food-borne diseases in the United States, only behind *Salmonella* (Le Loir et al., 2003). In some rare cases, a more serious disease can be presented, such as the so-called necrotic enteritis or pig-bel disease, where necrosis of the intestines and septicemia take place (Borriello, 1995). It is important to note that a relatively high infective dose is needed (Le Loir et al., 2003). Even though more than 10 toxins have been described so far, the *C. perfringens* enterotoxin (CPE) is the most well characterized and it is believed to play the principal role in the human intestine in *C. perfringens* food poisoning cases (Lindsay, 1996). It is synthesized during sporulation and released when sporulation is completed. CPE has two domains: the N-terminal domain is able to form pores at the plasma membrane of target cells, while the C-terminal domain binds to the toxin receptors, that is, claudins 3 and 4 (Mitchell and Koval, 2010).

A closely related bacterium, *C. botulinum*, can also be found in food, especially in incorrectly or minimally processed food. This is another anaerobic spore-forming microorganism that produces a toxin (Peck et al., 2011). However, in this case, it synthesizes a powerful neurotoxin, which is a heat-labile protein that can cause muscular paralysis and eventually asphyxia and death (Wheeler and Smith, 2013).

Bacillus cereus is present mainly in decaying organic matter, water, vegetables, and also in the intestinal tract of invertebrates (Jensen et al., 2003). Once food products become contaminated from these reservoirs, *B. cereus* can transiently colonize the human intestine. The detrimental effect on the intestines can be explained by the concerted action of several factors such as three pore-forming enterotoxins: cytotoxin K, hemolysin BL, and nonhemolytic enterotoxin. Besides, *B. cereus* can produce several phospholipases as well as other hemolysins (Bottone, 2010; Granum, 1994). It also secretes an emetic toxin called cereulide (Toh et al., 2004). Like *C. perfringens*, *B. cereus* can also be linked to gas gangrene–like infections (Bottone, 2010). What is worrying is that most *B. cereus* isolates are resistant to penicillins and cephalosporins because they generally produce beta-lactamases. Resistance to other antibiotics has been reported as well (Savini et al., 2009).

Last but not least, *L. monocytogenes* is the other important food gram-positive pathogen. *Listeria monocytogenes* is a motile, rod-shaped, facultative-anaerobic, and non-spore-forming bacterium (Gray and Killinger, 1966). It is a ubiquitous microorganism that can be isolated from polluted water, sludge, soil, plants, and even from different farm and domestic animals, either in planktonic form or as a biofilm (da Silva and De Martinis, 2013). The main problem with *L. monocytogenes* is that it can grow at low temperatures and also live in a wide range of pH, with high salt concentrations (up to 10% NaCl), and even survive in frozen foodstuff for a long period of time (Wemekamp-Kamphuis et al., 2002). Even though *L. monocytogenes* is a saprophytic microorganism, it can become an intracellular pathogen. This is achieved by expressing the internalins InlA and InlB, which are crucial for targeting epithelial cells, allowing *L. monocytogenes* to be taken up by eukaryotic cells (Cossart, 2011). However, as opposed to *Salmonella*, it escapes from the endosomal compartment where it is held at first, and stays in the cytosol of host cells. That is why it needs some key virulence factors such as lysteriolysin O and phospholipases C (Camejo et al., 2011). Once *L. monocytogenes* is divided in the cytosol, it can spread to neighboring cells by recruiting actin filaments, as was shown in a classic paper from Portnoy's lab (Portnoy, 2012; Tilney and Portnoy, 1989). Importantly, *L. monocytogenes* can escape from autophagy, which has awakened great attention in the past years (Lam et al., 2012; Yoshikawa et al., 2009). Moreover, it was recently found that *Listeria* deeply alters mitochondria by causing fragmentation of the mitochondrial network (Stavru et al., 2011). Two master regulators drive the expression of the genes that are needed for pathogenesis: Sigma B Factor and PrfA. The former factor acts mainly to help *Listeria* to deal with the gastrointestinal tract environment, which is characterized by high osmolarity, low oxygen pressure, low pH, and the presence of bile (Garner et al., 2006; Wemekamp-Kamphuis et al., 2004). On the other hand, PrfA is crucial for the expression of virulence factors once *L. monocytogenes* reaches the interior of host cells (Freitag et al., 2009). Interestingly, the listerial biofilms found in the environment are also regulated by PrfA (Luo et al., 2013; Zhou et al., 2011).

Although sporadic, listeriosis has a very high mortality rate (Temple and Nahata, 2000). Pregnant women, the elderly, infants, and immunocompromised patients are more susceptible to suffer from this disease. During uncontrolled infections in these susceptible populations, *L. monocytogenes* can cross not only the intestinal epithelium but also the blood–brain barrier and the placental barrier, which can lead to meningitis, encephalitis, and spontaneous abortion (Kaur et al., 2007; Lecuit, 2005).

Luckily, as *L. monocytogenes* is a gram-positive bacterium, a large number of bacteriocins from LAB prove to be very effective in killing this pathogen (Drider et al., 2006). Furthermore, there are several patents for the usage of bacteriocins as food preservatives, mainly because they are active against *L. monocytogenes* (Benmechernene et al., 2013; Berjeaud et al., 2001; Hugas et al., 1996; Vandenbergh et al., 1989). Even chimeric peptides have been recently designed that may help in getting rid of this potentially dangerous bacterium (Acuña et al., 2012).

23.3 Bacteriocins

23.3.1 Definition and General Considerations

Even though the word "bacteriocin" was employed for years to define antimicrobial compounds produced by bacteria and particularly those produced by LAB, currently it is used for small antimicrobial peptides synthesized by any bacterial strain. Bacteriocins are small ribosomally synthesized antimicrobial peptides produced by a bacterial strain (Cotter et al., 2005a). Bacteriocins produced by LAB microorganisms, like nisin and pediocin PA-1, are some of the most well-known members of the vast group of antimicrobial peptides produced by the members of all kingdoms of life (Brogden, 2005; Cotter et al., 2013; Hancock and Chapple, 1999). Although the bacteriocins form a heterogeneous group of their own (Chalón et al., 2012; Cotter et al., 2013), many of their members share the typical characteristics that constitute the hallmark of antimicrobial peptides: cationic, low–molecular weight peptides, usually amphiphilic or hydrophobic, and with the ability of interacting and even disrupting bacterial membranes.

23.3.2 Spectra of Action

The simplest explanation of the possible physiological role of bacteriocins is to mediate in a bacterial war for space and time in an ecological niche. A bacterial strain that produces a bacteriocin would have an edge over nonbacteriocinogenic strains. However, the key point is to eliminate only direct competitors without affecting other bacteria that might be useful. A simple example in our society would be as follows: Let us suppose that we are partners and owners of a restaurant. In principle, it would be desirable for us to eliminate other restaurants that offer similar products and take their place and clientele. However, it could have a negative impact if we affect other businesses as well, such as our raw material suppliers or our own customers. Therefore, we would need to do it in a very well-planned and directed manner. Bacteriocins are amazing products of evolution. Within a small molecule, they have the ability both to be lethal and to hit the right target alone. In general, bacteriocins are antimicrobial peptides with high specificity, able to kill only phylogenetically related bacteria, which would be direct competitors in the ecosystem. Producing strains were found in all bacterial genera explored. As the LAB are considered safe, for the moment the only bacteriocins currently employed in the food industry for food bioconservation are those produced by LAB and that is the reason why they are the most studied peptides.

23.3.3 Classification

Bacteriocins used to be classified into two groups: modified (class I) and unmodified (class II) bacteriocins (see Table 23.1). However, as some biochemical properties are shared with other peptides produced by the organisms of all kingdoms of life, a consensual classification and nomenclature was recently published (Arnison et al., 2013; Cotter et al., 2013). According to this proposal, bacteriocins can be included in the group of ribosomally synthesized peptides of low molecular weight, usually past or co-translationally

TABLE 23.1

Classification of Bacteriocins

Group	Structural/Biochemical Features	Examples/(Producer Organism)
Class I: posttranslationally modified bacteriocins		
Lanthipeptides	Modified amino acids lanthionine/methyl-lanthionine formed by Ser/Thr dehydration followed by addition of a Cys residue and formation of a thioether bond (Bierbaum and Sahl, 2009)	Nisin (*L. lactis*), Actagardine (*Actinoplanes liguriensis*), Duramycin (*Streptomyces* spp.)
Linaridins	Thioether cross-link but by means of different pathways than lanthipeptides (Sit et al., 2011)	Cypemycin (*Streptomyces* spp.)
Linear azol(in) e-containing peptides (LAPs)	Presence of heterocycles: thiazole, (methyl)oxazol, and, in reduced form, azoline, formed from residues Cys, Ser, and Thr of the precursor peptide (Melby et al., 2011)	Streptolisin S (*Streptococcus pyogenes*), microcin B17 (*E. coli*)
Cyanobactins	N- to C-terminals macrocyclized. Presence of heterocycles thiazol(in)e and oxazoline (Sivonen et al., 2010)	Ulicyclamide, ulithiacyclamide (cyanobacterium *Prochloron didemni*)
Thiopeptides	Nitrogenous six-membered macrocycle (piperidine, dehydropiperidine, pyridine) and heterocycles thiazole with dehydrated amino acids (Zhang and Liu, 2013)	Micrococcin P1 (*Actinobacteria, Bacillus, Staphylococcus*)
Bottromycins	Presence of a macrocyclic amidine, a decarboxylated thiazole at C-terminal, and C-methylated amino acids. Lack of N-terminal leader sequence (Huo et al., 2012)	Bottromycin A_2, B_2, C_2 (*Streptomyces* spp.)
Nucleopeptide	Heptapeptide–nucleotide conjugate (Severinov and Nair, 2012)	Microcin C7 (*Escherichia coli*)
Siderophore peptides	Linear peptide with a trimer of *N*-(2,3-dihydroxybenzoyl)-L-serine (DHBS) linked to C-terminal residue (Duquesne et al., 2007)	Microcins E492 (*Klebsiella pneumoniae*), H47, I47, and MccM (*E. coli*)
Lasso-peptides	Macrolactam at N-terminal by bonding of the N-terminus with the side chain of a Glu/Asp at position 8/9 (Severinov et al., 2007)	Siamycin I, II, Microcin J25, (*Actinobacteria, Rhodococcus, Proteobaceria, E. coli*)
Sactipeptides	Cross-link between a Cα and sulfur of a Cys. Biosynthesis involves single radical SAM pathway (Murphy et al., 2011)	Subtilosin A (*Bacillus*)
Glycocins	Glycosylated peptides in Cys residues. Disulfide bonds 158 and 159 (Oman et al., 2011)	Sublancin 162 (*Bacillus subtilis*), glycocin F (*Lactobacillus plantarum*)
Class II: unmodified linear or cyclic peptides		
Linear peptides	Pediocin-like (Drider et al., 2006), nonpediocin-like (Duquesne et al., 2007) and two peptides bacteriocins (Nissen-Meyer et al., 2010)	Pediocin PA-1 (*Pediococcus acidilactici*)—lactococcin G (*L. lactis*)
Head-to-tail cyclized peptides	Peptide bond between N and C terminals, 35–70 amino acids. Hydrophobic. Conserved tridimensional structure (4–5-helix bundle) (Sánchez-Hidalgo et al., 2011)	AS-48 (*Enterococcus faecalis*), gassericin A (*Lactobacillus gasseri*)

modified (RiPPs, ribosomally synthesized and posttranslationally modified peptides) along with many other bioactive peptides produced by ribosomes. This definition excludes the high–molecular weight membrane-lytic toxins and the peptidic metabolites produced by nonribosomal synthetic machinery, such as gramicidins.

An important feature of RiPPs members is that they are produced in a similar way by means of the processing of the precursor. This leads to the necessity of a unified nomenclature. In this way, the core region is usually the central part of the precursor peptide that will be processed in the final natural product. The core region is flanked by an N-terminal leader sequence, which is important for the recognition/guiding by the processing machinery and export enzymes, and by a C-terminal recognition sequence, which is important for the excision and cyclization typical of many of this vast group of peptides. In the classification presented in Table 23.1, linear bacteriocins produced by gram-negative bacteria having a siderophore bound to the C-terminal are considered to have suffered a posttranslational modification. In fact, siderophore synthesis requires several enzymes encoded by the genetic system of the bacteriocin (Thomas et al., 2004).

Moreover, linear bacteriocins that are not posttranslationally modified are grouped in class II. They are divided into two subgroups: linear and cyclic peptides. The linear-type bacteriocins include pediocin-like bacteriocins, two-peptide bacteriocins, and linear nonpediocin-like bacteriocins, including non-postranslationally modified microcins.

23.3.4 Mechanism of Action

Most bacteriocins act at the level of the bacterial membrane of the target cell. The presence of basic amino acids (Arg, Lys) in their primary structures would be important for the first step of binding of these peptides to the bacteria by means of electrostatic interaction between the positively charged peptides and the negatively charged molecules present in bacterial cell envelopes (Hancock and Chapple, 1999; Shai, 2002; Tossi et al., 2000), the anionic phospholipids phosphatidylglycerol and cardiolipin, the lipopolysaccharide of the outer membrane of gram-negative bacteria, or the lipoteicoic acids in gram-positive bacteria. Actually, the positive charge of antimicrobial peptides would also be determinant of the selectivity of bacteriocins toward microbial (negatively charged) instead of eukaryotic (zwiterionic) membranes (Shai, 2002).

The interaction of bacteriocins with microbial membranes involves partition equilibrium of the peptide between the polar aqueous media and the membrane, which has a lower dielectric constant (less polar environment) (White et al., 1998). However, partitioning of an extended peptide backbone into the hydrophobic core of the membrane is a thermodynamically unfavorable process (Almeida et al., 2012), because hydrogen bonding with other molecules is impaired. In this way, peptide–membrane interactions usually involve rearrangement of the secondary/tertiary structure of the peptide, from an extended conformation of the bacteriocin in the aqueous media to a folded conformation by means of intrachain hydrogen bonds formation (Fernández-Vidal et al., 2007). In addition, segregation of hydrophobic and polar regions along the peptide structure by means of the formation of amphipathic structures, like α-helix or sheets, is seen. Bilayer integrity assayed both in model membranes *in vitro* and in cell membranes *in vivo* is compromised and increase in permeability is observed for many bacteriocin-sensitive cell interactions (Breukink et al., 1999), leading to proton-motive force and electrochemical gradient dissipation (Chikindas et al., 1993; Maqueda et al., 2004; Minahk et al., 2000; Moll et al., 1996), drop of internal adenosine triphosphate (ATP) pool (Moll et al., 1996), and even efflux of metabolites like sugars and amino acids (Héchard and Sahl, 2002; van Belkum et al., 1991).

Different models of pore formation and bilayer disruption have been proposed mainly for explaining the permeabilization of model membranes (Brogden, 2005; Hancock and Chapple, 1999; Montville and Chen, 1998; Shai, 2002). In this way, peptides are thought to concentrate and aggregate at the surface of the membranes after the first step of electrostatic binding, and solubilize the lipids by a detergent-like action (carpet model). Alternatively, formation of pores upon important conformational changes was proposed. These pores could be solely composed of peptides (barrel stave model) or by peptides and lipids (toroidal pore model). However, direct structural data supporting these models are still scarce owing to the inherent difficulties of the high-resolution techniques for studying membrane-containing samples.

Moreover, bacteriocin concentrations needed for permeabilizing liposomes and model membranes are in the micromolar range, and thus actually up to one order higher than the minimal bactericidal concentration assayed in sensitive cells, which are in the nanomolar range. In addition, the relative narrow spectrum of bacteria that are killed by these peptides suggests the necessity of a mechanism of action mediated by a receptor or docking molecule participating in the binding and/or microbicidal activity of the bacteriocins with both high affinity and specificity. In this regard, lipid II has been demonstrated to be the target of nisin and other lantibiotics belonging to class I bacteriocins (Breukink et al., 1999), whereas the mannose phosphotransferase has been proposed as the receptor of several class II bacteriocins produced by LAB (Kjos et al., 2011) in gram-positive bacteria. On the other hand, the F_0 proton channel, ManYZ, and SdaC were identified as receptors for the membrane-permeabilizing microcins MccH47, MccE492, and MccV, respectively (Bieler et al., 2006; Gérard et al., 2005; Rodríguez and Laviña, 2003).

Although membrane interaction was demonstrated for most antimicrobial peptides, it cannot be said at the moment whether this is the sole or even the main mechanism of action of bacteriocins.

23.4 Future of Bacteriocins

23.4.1 Bacteriocins as Food Biopreservatives

The design and production of bacteriocins with greater antimicrobial activity, broad spectrum, and other desired physicochemical properties is the goal of many investigations in this field of science. A large number of researchers have devoted their studies to finding new bacteriocins, characterizing and studying the processing, secretion, and mechanism of action over the past three decades. Thanks to their effort, there is currently a variety of well-characterized antimicrobial peptides. Bacteriocins differ in the producing microorganism, posttranslational modifications, secretion mechanism, mechanism of action and spectrum, physicochemical properties, optimum working temperature, and resistance to proteolytic enzymes, among other aspects. The understanding of the mechanism of action of antimicrobial peptides is important not only from a biochemical point of view but also for the designing of new peptides with enhanced/broadened activities (Johnsen et al., 2005; Saavedra et al., 2004). Table 23.2 summarizes the desirable characteristics for bacteriocins that are useful as food preservatives. In the first place, bacteriocins should be totally innocuous to the consumers. In this regard, in vitro cytotoxicity assays were conducted in eukaryotic cell cultures for some bacteriocins. The ability to induce apoptosis, growth inhibition, and cell energy impairment was studied for that purpose. In addition, hemolytic activity was tested in some cases, which is a simple and reliable method for assessing potential cytotoxicity (Vaucher et al., 2010). However, different cytotoxicity values were obtained depending on the cell lines tested, the metabolic status, and the time of exposure to the antimicrobial agent (Murinda et al., 2003; Weyermann et al., 2005). Overall, a certain degree of toxicity was found for many tested bacteriocins. The concentration at which bacteriocins are cytotoxic is by far greater than the minimum inhibitory concentration (Lohans and Vederas, 2012). There are, however, exceptions; for example, the enterococcal cytolysin, a two-component lantibiotic, has widespread cytotoxic activity (Cox et al., 2005).

To date, there is no report about possible negative effects of bacteriocins in animal models (Lohans and Vederas, 2012). Furthermore, there are a few studies that were carried out to test bacteriocins as therapeutic agents in animal models. No toxicity was observed at the concentrations used (Bhunia et al., 1990; Cotter et al., 2013; Dabour et al., 2009; Rihakova et al., 2010; Salvucci et al., 2012). Indeed, there is a general acceptance that LAB bacteriocins are safe. Nisin has been used as a biopreservative since the 1950s. At that time, the controls were not as strict as they are today. Pediocin PA-1, a bacteriocin belonging to class IIa, is not used as an additive to add directly to food but as a preparation in milk called ALTA 2351® (Mills et al., 2011). So far, there is no report of side effects on consumers. For that reason, even if the use of a bacteriocin is not approved as a biopreservative, bacteriocinogenic strains that were not genetically modified could be used as starter culture for fermented food preparation.

Furthermore, bacteriocins should be safe also for intestinal microflora. Linear bacteriocins, which do not suffer posttranslational modifications belonging to class II, are very sensitive to intestinal proteases. Consequently, after being ingested with food, they are inactivated and do not have any effect on

TABLE 23.2

Properties of Bacteriocins for Food Preservation

Safety for consumer
Safety for the consumer's intestinal flora
Broad antimicrobial spectrum
Resistant to protease digestion present in the food
Resistant to the different processing steps (thermostable)
Good antimicrobial activity on a broad range of temperature
Good antimicrobial activity on a broad range of pH or salt concentration

intestinal microflora. Bacteriocins with posttranslational modifications, belonging to class I, are generally more resistant to proteases. Therefore, they are also more stable in foods that have natural proteases, which is good in such cases. On the other hand, it is very likely that a bacteriocin that is active against a determined food pathogen does not act on normal intestinal bacteria, owing to its high specificity.

Beyond these specific instances, it is generally desirable to have a wide spectrum of inhibition to eliminate pathogens and degrading contaminants of food. This is difficult because bacteriocins generally have a restricted antimicrobial spectrum and bacteriocins that inhibit gram-positive bacteria do not act on gram-negative bacteria and vice versa. There are a few peptides that display a broad spectrum of action (Chalón et al., 2012). A notable exception is enterocin AS-48, which has a wide spectrum of inhibition (Sánchez-Hidalgo et al., 2011). Interestingly, some bacteriocins can act against gram-negative bacteria if used in conjunction with chemical or physical hurdles that alter the permeability of the outer wall of gram-negative bacteria, as it was demonstrated for nisin (Chalón et al., 2012).

23.4.2 Physicochemical Properties of Bacteriocins

Bacteriocins are peptidic in nature, but due to their small size they are thermostable, that is, they can retain their activities and therefore their structures after being heated at high temperatures. In fact, bacteriocins only acquire their active conformation when bound to biological membranes. Thermostability is a useful property, as they can withstand heat treatments required for the cooking of many foods. However, beyond thermostability, bacteriocins may differ in other physicochemical properties. For instance, nisin Z, which is a natural variant of nisin that has the His27Asn mutation, has an increased diffusion rate but lower solubility at acidic pH compared to nisin A (Cotter et al., 2005b; Mulders et al., 1991).

Among the different physicochemical properties of bacteriocins, lipid affinity and solubility in aqueous solutions are key features (Rollema et al., 1995). In fact, an antimicrobial agent must have a homogeneous distribution in the food. As many bacteriocins are active at the level of biological membranes, they may have greater affinity to phases or regions of foods with higher proportions of lipids. Thus, they could be sequestered and concentrated in lipid phases.

23.4.3 Bacteriocins Genetically Engineered

If we envision an ideal bacteriocin for biotechnology use as a food bioprotector agent, that peptide should satisfy certain chemical, physical, and biological properties allowing its use in food. However, there is no perfect natural bacteriocin described so far.

Genes coding for bacteriocins can be manipulated to obtain other bacteriocins with desired properties that are best suited for determined biotechnological applications. Class II bacteriocins that are linear peptides not suffering large posttranslational modifications are easier to manipulate. In addition, they are also more easily heterologously expressed in other microorganisms. Furthermore, they do not require large enzymatic machinery for posttranslational modifications. As a matter of fact, genes coding for enzymes needed for bacteriocin class I maturation must be introduced together with the structural gene to achieve the correct heterologous expression (Acuña et al., 2011).

23.4.4 Genetically Engineered Class I Bacteriocins

Nisin, the star among bacteriocins, belongs to class I. It has been used as biopreservative for more than 60 years now and is considered safe by the Food and Drug Administration (FDA, 1988). For this reason, class I bacteriocins are the most promising peptides. The first mutants in this class of bacteriocins were obtained in order to study the mechanisms of posttranslational modifications and the relationship between structure and function. Cotter et al. (2005b) observed that different lantibiotic variants obtained by site-directed mutagenesis could be divided into different classes: (1) mutations affecting thiol bridge formation, (2) mutations in the unusual amino acids dehydroalanine and dehydrobutyrine, (3) mutations in the hinge region, and (4) mutations that alter the electrical charge.

Mutations affecting thiol bridge formation drastically diminished antimicrobial activity. However, different properties were observed in lantibiotics after suffering alterations in the primary structure, especially when the altered regions involved dehydroalanine (Dha), dehydrobutyrine (Dhb), or the hinge region. For example, gallidermin variants with increased susceptibility to proteases were obtained when conducted T14S mutation. Serine 14 undergoes dehydration to produce dehydrobutyrine. Both variants Dhb14S and Dhb14Dha were more sensitive to trypsin treatment (Ottenwälder et al., 1995). However, the variants Dhb14P and A12L were significantly more resistant to treatment with trypsin, but the antimicrobial activity was greatly reduced.

Furthermore, Rollema et al. obtained nisin variations with higher solubility and chemical stability. In fact, nisin Z mutants having a Lys residue in place of Asn27 or His31 retain antimicrobial activity and exhibit four- to sevenfold higher solubility than the natural peptide, respectively (Rollema et al., 1995). In addition, the antimicrobial activity of another nisin Z mutant (Dha5Dhb) dropped 2–10 times, but was more stable than the natural peptide at pH values below 5. A subtilisin variant with increased N-terminal hydrophobicity was also obtained, which was >3-fold more efficient than the natural bacteriocin (Liu and Hansen, 1992). Actually, it was the first lantibiotic genetically improved.

Moreover, the hinge region is a flexible joint between the thioether rings and is essential for the activity, for the maturation process and the exportation of the peptide (Chen et al., 1998). Some mutations in this region of nisin Z (N20K and M21K) did not alter the activity on gram-positive bacteria but surprisingly showed higher solubility that nisin Z and antimicrobial activity against the gram-negative bacteria *Shigella*, *Salmonella*, and *Pseudomonas*. In addition, other mutations in the same region showed higher solubility at pH 8 and higher thermal stability at neutral pH (Yuan et al., 2004). Subsequently, Field et al. generated a bank of nisin mutants by random mutagenesis. The mutant in the hinge region K22T showed higher activity against *Streptococcus agalactiae* than nisin. For this reason, saturation mutations were conducted in the region several mutants with increased activity against gram-positive pathogens *L. monocytogenes* and/or *S. aureus* were obtained (Field et al., 2008). In addition, M21V and K22T mutants named nisin V and T, respectively, exhibited potent activity against a broad range of gram-positive antibiotic-resistant pathogens, isolated veterinary, and food-borne pathogens (Field et al., 2010). Using a similar strategy, the same group of researchers identified new nisin mutants with increased antimicrobial activity against gram-positive bacteria (nisin A K12A) (Molloy et al., 2013). In addition, several mutants in the residue S29 are more active against gram-positive and gram-negative bacteria than nisin A. These are the first nisin A variants having improved activity against both types of bacteria, gram negative and gram positive (Field et al., 2012). It was also shown that the increased activity observed for different variants of nisin may be due in part to the greater capacity of the derivatives to diffuse into agar and not to increased specific activity (Rouse et al., 2012). Consequently, these derivatives could be useful for use as biopreservatives in foods containing similar polymers such as pectin, polysaccharides, gelatin, or guar gum.

Microcin J25, a lasso peptide, which is a class I bacteriocin, exhibits potent activity against gram-negative pathogens and resistance to intestinal proteases. It was modified to be susceptible to chymotrypsin. This derivative is degraded by intestinal enzymes in vivo and in vitro and its activity was not detectable in feces, ensuring that their use as food biopreservative does not affect the intestinal flora of consumers (Pomares et al., 2009).

23.4.5 Genetically Engineered Class II Bacteriocins

There are precedents for class II bacteriocins subject to genetic manipulation to either study basic biochemical aspects or to obtain other peptides that could fulfill the requirements of the food industry. Early works that focused on the effects of amino acid substitutions demonstrated that small changes in peptide primary structures can drastically diminish their activities (Fleury et al., 1996; Quadri et al., 1997; Rihakova et al., 2009). Moreover, fusing the N- and C-terminal halves of class IIa bacteriocins, it was concluded that the C-terminal half was responsible for most of the target specificity and is involved in the recognition by the immunity protein (Johnsen et al., 2005). As basic knowledge of the structure–function was deepening, there was remarkable progress in the development of peptides with promising biotechnology applications. Some linear bacteriocins have a disulfide bridge at the carboxyl terminus, which would be important to improve antimicrobial efficacy at elevated temperatures (Fimland et al., 2000). In fact, it was demonstrated that the introduction of extra cysteines in sakacin P led to a 10–20-fold increase in the antimicrobial activity and even broaden the antibacterial spectrum.

Generally, bacteriocins belonging to this class are chemically unstable and lose their activity following storage at room or low temperature after a few days. Johnsen et al. got a derivative of pediocin PA-1 by mutating a methionine residue by a hydrophobic amino acid. This pediocin PA-1 derivative did not oxidize easily and resulted in improved stability on storage for a long time, maintaining the original antimicrobial activity (Johnsen et al., 2000). Subsequently, other variants of pediocin PA-1 were obtained by combining its C-terminal half with the N-terminal half of enterocin A. The chimeric bacteriocin was more active against *Leuconostoc lactis* (Tominaga and Hatakeyama, 2007). In the same work but using DNA shuffling (Crameri et al., 1998), pediocin PA-1 N-terminus with the corresponding regions from 10 class IIa bacteriocins were exchanged. Bacteriocins with higher activity on different species of gram-positive bacteria were obtained (Tominaga and Hatakeyama, 2007).

Another important advance came when it was discovered that small fragments of enterocin CRL35 have antimicrobial activity. Moreover, these peptides enhanced the bactericidal activity when used in combination with full-length enterocin CRL35 (Saavedra et al., 2004; Salvucci et al., 2007).

All the aforementioned bacteriocins have shown activity on gram-positive bacteria. However, these bacteriocins can potentially be combined with microcins produced by (and active against) gram-negative bacteria (Acuña et al., 2011). The latter class of antimicrobial peptides is also formed by a heterogeneous group of peptides that display antimicrobial activity either through inhibition of critical enzymatic pathways or by disruption of the inner membrane permeability. An interesting example is given by the chimeric peptide ent35-colV. This is a successful combination of enterocin CRL35 originally isolated from *Enterococcus* and microcin V, produced by *E. coli*, which rendered a hybrid peptide that displayed the spectrum activity of both the antimicrobials (Acuña et al., 2012).

However, even though these fusions seem to be very promising because they allow tackling different types of pathogens at the same time, they probably encounter problems for approval at the national agencies such as the FDA. A number of concerns may arise not only for the modification of peptides but also because genetically modified microorganisms may be used for their production.

23.5 Conclusions

LAB were used in an unconscious way for the preparation of fermented foods for thousands of years. Indeed, they have very favorable properties for use as natural food biopreservatives because LAB produce inhibitory molecules to the growth of other microorganisms as a result of their metabolism. Furthermore, many BAL strains produce bacteriocins. Although nisin and pediocin PA-1 are already used as biopreservatives, the use of other bacteriocins has not yet been approved although they are great candidates for use in food biopreservation in the coming years. As a matter of fact, a number of patents involving the use of bacteriocins have been presented already. The systematic search for new bacteriocins in many species of bacteria and the understanding of their mechanisms of synthesis, secretion, and action by many laboratories worldwide have allowed the extending of the potential of most widely studied bacteriocins of LAB in terms of antimicrobial spectrum, potency, and physicochemical properties. On the other hand, the food preservation techniques are continuously evolving toward obtaining minimally processed foods, completely safe from a bacteriological point of view and with more shelf life.

In the future, bacteriocins will be designed with the required antibacterial spectrum of action and optimal physicochemical properties allowing their employment in the food industry for each particular foodstuff by adapting to the hurdles that are already in use.

REFERENCES

Aarts, H.J.M., Vos, P., Larsson, J.T., van Hoek, A.H.A.M., Huehn, S., Weijers, T., Grønlund, H.A., and Malorny, B. (2011). A multiplex ligation detection assay for the characterization of *Salmonella enterica* strains. *Int. J. Food Microbiol. 145*(Suppl. 1), S68–S78.

Abee, T., Groot, M.N., Tempelaars, M., Zwietering, M., Moezelaar, R., and van der Voort, M. (2011). Germination and outgrowth of spores of *Bacillus cereus* group members: Diversity and role of germinant receptors. *Food Microbiol. 28*, 199–208.

Acuña, L., Morero, R., and Bellomio, A. (2011). Development of wide-spectrum hybrid bacteriocins for food biopreservation. *Food Bioprocess Technol. 4*, 1029–1049.

Acuña, L., Picariello, G., Sesma, F., Morero, R.D., and Bellomio, A. (2012). A new hybrid bacteriocin, Ent35-MccV, displays antimicrobial activity against pathogenic Gram-positive and Gram-negative bacteria. *FEBS Open Bio 2*, 12–19.

Almeida, P.F., Ladokhin, A.S., and White, S.H. (2012). Hydrogen-bond energetics drive helix formation in membrane interfaces. *Biochim. Biophys. Acta 1818*, 178–182.

Altekruse, S. (1999). *Campylobacter jejuni*—An emerging foodborne pathogen. *Emerg. Infect. Dis. 5*, 28–35.

Ang, C.W., Jacobs, B.C., and Laman, J.D. (2004). The Guillain-Barré syndrome: A true case of molecular mimicry. *Trends Immunol. 25*, 61–66.

Arnison, P.G., Bibb, M.J., Bierbaum, G., Bowers, A.A., Bugni, T.S., Bulaj, G., Camarero, J.A. et al. (2013). Ribosomally synthesized and post-translationally modified peptide natural products: Overview and recommendations for a universal nomenclature. *Nat. Prod. Rep. 30*, 108–160.

Ashida, H., Ogawa, M., Mimuro, H., Kobayashi, T., Sanada, T., and Sasakawa, C. (2011). *Shigella* are versatile mucosal pathogens that circumvent the host innate immune system. *Curr. Opin. Immunol. 23*, 448–455.

Ashida, H., Ogawa, M., Mimuro, H., and Sasakawa, C. (2009). *Shigella* infection of intestinal epithelium and circumvention of the host innate defense system. *Curr. Top. Microbiol. Immunol. 337*, 231–255.

Belanger, A.E. and Shryock, T.R. (2007). Macrolide-resistant *Campylobacter*: The meat of the matter. *J. Antimicrob. Chemother. 60*, 715–723.

Benmechernene, Z., Fernandez-No, I., Kihal, M., Bohme, K., Calo-Mata, P., and Barros-Velazquez, J. (2013). Recent patents on bacteriocins: Food and biomedical applications. *Recent Patents DNA Gene Seq. 7*, 66–73.

Berjeaud, J.-M., Fremaux, C., Cenatiempo, Y., and Simon, L. (2001). Anti-Listeria bacteriocin. Application number: 01938357.9

Betancor, L., Pereira, M., Martinez, A., Giossa, G., Fookes, M., Flores, K., Barrios, P. et al. (2010). Prevalence of *Salmonella enterica* in poultry and eggs in Uruguay during an epidemic due to *Salmonella enterica* serovar Enteritidis. *J. Clin. Microbiol. 48*, 2413–2423.

Bharati, K. and Ganguly, N.K. (2011). Cholera toxin: A paradigm of a multifunctional protein. *Indian J. Med. Res. 133*, 179–187.

Bhunia, A.K., Johnson, M.C., Ray, B., and Belden, E.L. (1990). Antigenic property of pediocin AcH produced by *Pediococcus acidilactici* H. *J. Appl. Bacteriol. 69*, 211–215.

Bieler, S., Silva, F., Soto, C., and Belin, D. (2006). Bactericidal activity of both secreted and nonsecreted microcin E492 requires the mannose permease. *J. Bacteriol. 188*, 7049–7061.

Bierbaum, G. and Sahl, H.-G. (2009). Lantibiotics: Mode of action, biosynthesis and bioengineering. *Curr. Pharm. Biotechnol. 10*, 2–18.

Bishop, A.L. and Camilli, A. (2011). *Vibrio cholerae*: Lessons for mucosal vaccine design. *Expert Rev. Vaccines 10*, 79–94.

Borriello, S.P. (1995). Clostridial disease of the gut. *Clin. Infect. Dis. 20*(Suppl. 2), S242–S250.

Bottone, E.J. (2010). *Bacillus cereus*, a volatile human pathogen. *Clin. Microbiol. Rev. 23*, 382–398.

Breukink, E., Wiedemann, I., van Kraaij, C., Kuipers, O.P., Sahl, H., and de Kruijff, B. (1999). Use of the cell wall precursor lipid II by a pore-forming peptide antibiotic. *Science 286*, 2361–2364.

Brogden, K.A. (2005). Antimicrobial peptides: Pore formers or metabolic inhibitors in bacteria? *Nat. Rev. Microbiol. 3*, 238–250.

Brown, A.C. (2011). *Understanding Food: Principles and Preparation*, 4th Edition. Belmont: Wadsworth Cengage Learning.

Bücker, R., Krug, S.M., Rosenthal, R., Günzel, D., Fromm, A., Zeitz, M., Chakraborty, T., Fromm, M., Epple, H.-J., and Schulzke, J.-D. (2011). Aerolysin from *Aeromonas hydrophila* perturbs tight junction integrity and cell lesion repair in intestinal epithelial HT-29/B6 cells. *J. Infect. Dis. 204*, 1283–1292.

Camejo, A., Carvalho, F., Reis, O., Leitão, E., Sousa, S., and Cabanes, D. (2011). The arsenal of virulence factors deployed by *Listeria monocytogenes* to promote its cell infection cycle. *Virulence 2*, 379–394.

Centers for Disease Control and Prevention (CDC). (2013). Incidence and trends of infection with pathogens transmitted commonly through food: Foodborne diseases active surveillance network, 10 U.S. sites, 1996–2012. *MMWR Morb. Mortal. Wkly. Rep. 62*, 283–287.

Chalón, M.C., Acuña, L., Morero, R.D., Minahk, C.J., and Bellomio, A. (2012). Membrane-active bacteriocins to control *Salmonella* in foods: Are they the definite hurdle? *Food Res. Int. 45*, 735–744.

Chang, Z., Lu, S., Chen, L., Jin, Q., and Yang, J. (2012). Causative species and serotypes of shigellosis in mainland China: Systematic review and meta-analysis. *PLoS One 7*, e52515.

Chen, P., Novak, J., Kirk, M., Barnes, S., Qi, F., and Caufield, P.W. (1998). Structure-activity study of the lantibiotic mutacin II from *Streptococcus mutans* T8 by a gene replacement strategy. *Appl. Environ. Microbiol. 64*, 2335–2340.

Chikindas, M.L., García-Garcerá, M.J., Driessen, A.J., Ledeboer, A.M., Nissen-Meyer, J., Nes, I.F., Abee, T., Konings, W.N., and Venema, G. (1993). Pediocin PA-1, a bacteriocin from *Pediococcus acidilactici* PAC1.0, forms hydrophilic pores in the cytoplasmic membrane of target cells. *Appl. Environ. Microbiol. 59*, 3577–3584.

Cobo Molinos, A., Abriouel, H., López, R.L., Valdivia, E., Omar, N.B., and Gálvez, A. (2008). Combined physico-chemical treatments based on enterocin AS-48 for inactivation of Gram-negative bacteria in soybean sprouts. *Food Chem. Toxicol. 46*, 2912–2921.

Coker, A.O., Isokpehi, R.D., Thomas, B.N., Amisu, K.O., and Obi, C.L. (2002). Human campylobacteriosis in developing countries. *Emerg. Infect. Dis. 8*, 237–243.

Cooke, F.J., Wain, J., and Threlfall, E.J. (2006). Fluoroquinolone resistance in *Salmonella typhi*. *BMJ 333*, 353–354.

Cossart, P. (2011). Illuminating the landscape of host-pathogen interactions with the bacterium *Listeria monocytogenes*. *Proc. Natl. Acad. Sci. U.S.A. 108*, 19484–19491.

Cotter, P.D., Hill, C., and Ross, R.P. (2005a). Bacteriocins: Developing innate immunity for food. *Nat. Rev. Microbiol. 3*, 777–788.

Cotter, P.D., Hill, C., and Ross, R.P. (2005b). Bacterial lantibiotics: Strategies to improve therapeutic potential. *Curr. Protein Pept. Sci. 6*, 61–75.

Cotter, P.D., Ross, R.P., and Hill, C. (2013). Bacteriocins: A viable alternative to antibiotics? *Nat. Rev. Microbiol. 11*, 95–105.

Cox, C.R., Coburn, P.S., and Gilmore, M.S. (2005). Enterococcal cytolysin: A novel two component peptide system that serves as a bacterial defense against eukaryotic and prokaryotic cells. *Curr. Protein Pept. Sci. 6*, 77–84.

Crameri, A., Raillard, S.A., Bermudez, E., and Stemmer, W.P. (1998). DNA shuffling of a family of genes from diverse species accelerates directed evolution. *Nature 391*, 288–291.

Dabour, N., Zihler, A., Kheadr, E., Lacroix, C., and Fliss, I. (2009). In vivo study on the effectiveness of pediocin PA-1 and *Pediococcus acidilactici* UL5 at inhibiting *Listeria monocytogenes*. *Int. J. Food Microbiol. 133*, 225–233.

da Silva, E.P. and De Martinis, E.C.P. (2013). Current knowledge and perspectives on biofilm formation: The case of *Listeria monocytogenes*. *Appl. Microbiol. Biotechnol. 97*, 957–968.

de Jong, A., Stephan, B., and Silley, P. (2012a). Fluoroquinolone resistance of *Escherichia coli* and *Salmonella* from healthy livestock and poultry in the EU. *J. Appl. Microbiol. 112*, 239–245.

de Jong, B. and Ekdahl, K. (2006). The comparative burden of salmonellosis in the European Union member states, associated and candidate countries. *BMC Public Health 6*, 4.

de Jong, H.K., Parry, C.M., van der Poll, T., and Wiersinga, W.J. (2012b). Host-pathogen interaction in invasive Salmonellosis. *PLoS Pathog. 8*, e1002933.

de Paula, C.M.D., Geimba, M.P., do Amaral, P.H., and Tondo, E.C. (2010). Antimicrobial resistance and PCR-ribotyping of *Shigella* responsible for foodborne outbreaks occurred in southern Brazil. *Braz. J. Microbiol. 41*, 966–977.

Dieckmann, R., Helmuth, R., Erhard, M., and Malorny, B. (2008). Rapid classification and identification of Salmonellae at the species and subspecies levels by whole-cell matrix-assisted laser desorption ionization-time of flight mass spectrometry. *Appl. Environ. Microbiol. 74*, 7767–7778.

Dieye, Y., Ameiss, K., Mellata, M., and Curtiss, R., 3rd. (2009). The *Salmonella* pathogenicity island (SPI) 1 contributes more than SPI2 to the colonization of the chicken by *Salmonella enterica* serovar Typhimurium. *BMC Microbiol. 9*, 3.

Drider, D., Fimland, G., Héchard, Y., McMullen, L.M., and Prévost, H. (2006). The continuing story of class IIa bacteriocins. *Microbiol. Mol. Biol. Rev. 70*, 564–582.

Duquesne, S., Destoumieux-Garzón, D., Peduzzi, J., and Rebuffat, S. (2007). Microcins, gene-encoded antibacterial peptides from enterobacteria. *Nat. Prod. Rep. 24*, 708–734.

Elshafie, S.S., Asim, M., Ashour, A., Elhiday, A.H., Mohsen, T., and Doiphode, S. (2010). *Campylobacter* peritonitis complicating continuous ambulatory peritoneal dialysis: Report of three cases and review of the literature. *Perit. Dial. Int. 30*, 99–104.

Eurosurveillance Editorial Team. (2013). The European Union summary report on trends and sources of zoonoses, zoonotic agents and food-borne outbreaks in 2011 has been published. *Euro Surveill. 18*, 20449.

FDA Food and Drug Administration (1988). Nisin preparation: Affirmation of GRAS status as a direct human food ingredient. *CFR: Code of Federal Regulations 21*, 184.1538. URL: http://www.accessdata.fda.gov/scripts/cdrh/cfdocs/cfCFR/CFRSearch.cfm?CFRPart=184&showFR=1.

Fernández-Vidal, M., Jayasinghe, S., Ladokhin, A.S., and White, S.H. (2007). Folding amphipathic helices into membranes: Amphiphilicity trumps hydrophobicity. *J. Mol. Biol. 370*, 459–470.

Field, D., Begley, M., O'Connor, P.M., Daly, K.M., Hugenholtz, F., Cotter, P.D., Hill, C., and Ross, R.P. (2012). Bioengineered nisin A derivatives with enhanced activity against both Gram positive and Gram negative pathogens. *PLoS One 7*, e46884.

Field, D., Connor, P.M.O., Cotter, P.D., Hill, C., and Ross, R.P. (2008). The generation of nisin variants with enhanced activity against specific Gram-positive pathogens. *Mol. Microbiol. 69*, 218–230.

Field, D., Quigley, L., O'Connor, P.M., Rea, M.C., Daly, K., Cotter, P.D., Hill, C., and Ross, R.P. (2010). Studies with bioengineered nisin peptides highlight the broad-spectrum potency of nisin V. *Microb. Biotechnol. 3*, 473–486.

Fimland, G., Johnsen, L., Axelsson, L., Brurberg, M.B., Nes, I.F., Eijsink, V.G., and Nissen-Meyer, J. (2000). A C-terminal disulfide bridge in pediocin-like bacteriocins renders bacteriocin activity less temperature dependent and is a major determinant of the antimicrobial spectrum. *J. Bacteriol. 182*, 2643–2648.

Fitzhenry, R.J., Pickard, D.J., Hartland, E.L., Reece, S., Dougan, G., Phillips, A.D., and Frankel, G. (2002). Intimin type influences the site of human intestinal mucosal colonisation by enterohaemorrhagic *Escherichia coli* O157:H7. *Gut 50*, 180–185.

Fivaz, M., Abrami, L., Tsitrin, Y., and van der Goot, F.G. (2001). Aerolysin from *Aeromonas hydrophila* and related toxins. *Curr. Top. Microbiol. Immunol. 257*, 35–52.

Fleury, Y., Dayem, M.A., Montagne, J.J., Chaboisseau, E., Le Caer, J.P., Nicolas, P., and Delfour, A. (1996). Covalent structure, synthesis, and structure-function studies of mesentericin Y 105(37), a defensive peptide from Gram-positive bacteria *Leuconostoc mesenteroides*. *J. Biol. Chem. 271*, 14421–14429.

Fluit, A.C. (2005). Towards more virulent and antibiotic-resistant *Salmonella*? *FEMS Immunol. Med. Microbiol. 43*, 1–11.

Fluit, A.C. and Schmitz, F.-J. (2004). Resistance integrons and super-integrons. *Clin. Microbiol. Infect. 10*, 272–288.

Freitag, N.E., Port, G.C., and Miner, M.D. (2009). *Listeria monocytogenes*—From saprophyte to intracellular pathogen. *Nat. Rev. Microbiol. 7*, 623.

Gálvez, A., Abriouel, H., López, R.L., and Ben Omar, N. (2007). Bacteriocin-based strategies for food biopreservation. *Int. J. Food Microbiol. 120*, 51–70.

Garner, M.R., Njaa, B.L., Wiedmann, M., and Boor, K.J. (2006). Sigma B contributes to *Listeria monocytogenes* gastrointestinal infection but not to systemic spread in the guinea pig infection model. *Infect. Immun. 74*, 876–886.

Gérard, F., Pradel, N., and Wu, L.-F. (2005). Bactericidal activity of colicin V is mediated by an inner membrane protein, SdaC, of *Escherichia coli*. *J. Bacteriol. 187*, 1945–1950.

Ghenghesh, K.S., Ahmed, S.F., El-Khalek, R.A., Al-Gendy, A., and Klena, J. (2008). Aeromonas-associated infections in developing countries. *J. Infect. Dev. Ctries. 2*, 81–98.

Gordon, M.A., Kankwatira, A.M.K., Mwafulirwa, G., Walsh, A.L., Hopkins, M.J., Parry, C.M., Faragher, E.B., Zijlstra, E.E., Heyderman, R.S., and Molyneux, M.E. (2010). Invasive non-typhoid Salmonellae establish systemic intracellular infection in HIV-infected adults: An emerging disease pathogenesis. *Clin. Infect. Dis. 50*, 953–962.

Granum, P.E. (1994). *Bacillus cereus* and its toxins. *Soc. Appl. Bacteriol. Symp. Ser. 23*, 61S–66S.

Gray, M.L. and Killinger, A.H. (1966). *Listeria monocytogenes* and listeric infections. *Bacteriol. Rev. 30*, 309–382.

Hancock, R.E. and Chapple, D.S. (1999). Peptide antibiotics. *Antimicrob. Agents Chemother. 43*, 1317–1323.

Havelaar, A.H., Ivarsson, S., Löfdahl, M., and Nauta, M.J. (2012). Estimating the true incidence of campylobacteriosis and salmonellosis in the European Union, 2009. *Epidemiol. Infect. 1*, 1–10.

Héchard, Y. and Sahl, H.G. (2002). Mode of action of modified and unmodified bacteriocins from Gram-positive bacteria. *Biochimie 84*, 545–557.

Hugas, M.M., Garriga, T.M., Monfort, B.J.M., and Ylla, U.J. (1996). Bacteriocin from *Enterococcus faecium* active against *Listeria monocytogenes*. Application number: 94924890.0

Huo, L., Rachid, S., Stadler, M., Wenzel, S.C., and Müller, R. (2012). Synthetic biotechnology to study and engineer ribosomal bottromycin biosynthesis. *Chem. Biol. 19*, 1278–1287.

Hutkins, R.W. (2006). *Microbiology and Technology of Fermented Foods*. Chicago, IL: IFT Press.

Ibarra, J.A. and Steele-Mortimer, O. (2009). *Salmonella*—The ultimate insider. *Salmonella* virulence factors that modulate intracellular survival. *Cell. Microbiol. 11*, 1579–1586.

Inglis, G.D. and Kalischuk, L.D. (2004). Direct quantification of *Campylobacter jejuni* and *Campylobacter lanienae* in feces of cattle by real-time quantitative PCR. *Appl. Environ. Microbiol. 70*, 2296–2306.

Jensen, G.B., Hansen, B.M., Eilenberg, J., and Mahillon, J. (2003). The hidden lifestyles of *Bacillus cereus* and relatives. *Environ. Microbiol. 5*, 631–640.

Johnsen, L., Fimland, G., Eijsink, V., and Nissen-Meyer, J. (2000). Engineering increased stability in the anti-microbial peptide pediocin PA-1. *Appl. Environ. Microbiol. 66*, 4798–4802.

Johnsen, L., Fimland, G., and Nissen-Meyer, J. (2005). The C-terminal domain of pediocin-like antimi-crobial peptides (class IIa bacteriocins) is involved in specific recognition of the C-terminal part of cognate immunity proteins and in determining the antimicrobial spectrum. *J. Biol. Chem. 280*, 9243–9250.

Karmali, M.A., Gannon, V., and Sargeant, J.M. (2010). Verocytotoxin-producing *Escherichia coli* (VTEC). *Vet. Microbiol. 140*, 360–370.

Kaur, S., Malik, S.V.S., Vaidya, V.M., and Barbuddhe, S.B. (2007). *Listeria monocytogenes* in spontaneous abortions in humans and its detection by multiplex PCR. *J. Appl. Microbiol. 103*, 1889–1896.

Kjos, M., Borrero, J., Opsata, M., Birri, D.J., Holo, H., Cintas, L.M., Snipen, L., Hernández, P.E., Nes, I.F., and Diep, D.B. (2011). Target recognition, resistance, immunity and genome mining of class II bacteriocins from Gram-positive bacteria. *Microbiology (Reading, Engl.) 157*, 3256–3267.

Koluman, A. and Dikici, A. (2013). Antimicrobial resistance of emerging foodborne pathogens: Status quo and global trends. *Crit. Rev. Microbiol. 39*, 57–69.

Kweon, M.-N. (2008). Shigellosis: The current status of vaccine development. *Curr. Opin. Infect. Dis. 21*, 313–318.

Lahiri, A., Eswarappa, S.M., Das, P., and Chakravortty, D. (2010). Altering the balance between pathogen con-taining vacuoles and lysosomes: A lesson from *Salmonella*. *Virulence 1*, 325–329.

Lam, G.Y., Czuczman, M.A., Higgins, D.E., and Brumell, J.H. (2012). Interactions of *Listeria monocytogenes* with the autophagy system of host cells. *Adv. Immunol. 113*, 7–18.

Le Loir, Y., Baron, F., and Gautier, M. (2003). *Staphylococcus aureus* and food poisoning. *Genet. Mol. Res. 2*, 63–76.

Lecuit, M. (2005). Understanding how *Listeria monocytogenes* targets and crosses host barriers. *Clin. Microbiol. Infect. 11*, 430–436.

Leistner, L. (2000). Basic aspects of food preservation by hurdle technology. *Int. J. Food Microbiol. 55*, 181–186.

Lindsay, J.A. (1996). *Clostridium perfringens* type A enterotoxin (CPE): More than just explosive diarrhea. *Crit. Rev. Microbiol. 22*, 257–277.

Lingwood, C.A., Law, H., Richardson, S., Petric, M., Brunton, J.L., De Grandis, S., and Karmali, M. (1987). Glycolipid binding of purified and recombinant *Escherichia coli* produced verotoxin in vitro. *J. Biol. Chem. 262*, 8834–8839.

Liu, W. and Hansen, J.N. (1992). Enhancement of the chemical and antimicrobial properties of subtilin by site-directed mutagenesis. *J. Biol. Chem. 267*, 25078–25085.

Lloyd, S.J., Ritchie, J.M., and Torres, A.G. (2012). Fimbriation and curliation in *Escherichia coli* O157:H7: A paradigm of intestinal and environmental colonization. *Gut Microbes 3*, 272–276.

Lohans, C.T. and Vederas, J.C. (2012). Development of class IIa bacteriocins as therapeutic agents. *Int. J. Microbiol. 2012*, 386410.

Loś, J.M., Loś, M., Węgrzyn, A., and Węgrzyn, G. (2012). Altruism of Shiga toxin-producing *Escherichia coli*: Recent hypothesis versus experimental results. *Front. Cell Infect. Microbiol. 2*, 166.

Louwen, R., Horst-Kreft, D., de Boer, A.G., van der Graaf, L., de Knegt, G., Hamersma, M., Heikema, A.P. et al. (2013). A novel link between *Campylobacter jejuni* bacteriophage defence, virulence and Guillain-Barré syndrome. *Eur. J. Clin. Microbiol. Infect. Dis. 32*, 207–226.

Luo, Q., Shang, J., Feng, X., Guo, X., Zhang, L., and Zhou, Q. (2013). PrfA led to reduced biofilm formation and contributed to altered gene expression patterns in biofilm-forming *Listeria monocytogenes*. *Curr. Microbiol. 67*, 372–378.

Majowicz, S.E., Musto, J., Scallan, E., Angulo, F.J., Kirk, M., O'Brien, S.J., Jones, T.F., Fazil, A., Hoekstra, R.M., and International Collaboration on Enteric Disease "Burden of Illness" Studies (2010). The global burden of nontyphoidal *Salmonella* gastroenteritis. *Clin. Infect. Dis. 50*, 882–889.

Majtan, J., Majtanova, L., and Majtan, V. (2010). Increasing trend of resistance to nalidixic acid and emerging ceftriaxone and ciprofloxacin resistance in *Salmonella enterica* serovar Typhimurium in Slovakia, 2005 to 2009. *Diagn. Microbiol. Infect. Dis. 68*, 86–88.

Maqueda, M., Gálvez, A., Bueno, M.M., Sanchez-Barrena, M.J., González, C., Albert, A., Rico, M., and Valdivia, E. (2004). Peptide AS-48: Prototype of a new class of cyclic bacteriocins. *Curr. Protein Pept. Sci. 5*, 399–416.

Márquez-González, M., Cabrera-Díaz, E., Hardin, M.D., Harris, K.B., Lucia, L.M., and Castillo, A. (2012). Survival and germination of *Clostridium perfringens* spores during heating and cooling of ground pork. *J. Food Prot. 75*, 682–689.

Martin, L.B. (2012). Vaccines for typhoid fever and other salmonelloses. *Curr. Opin. Infect. Dis. 25*, 489–499.

Martinez-Becerra, F.J., Kissmann, J.M., Diaz-McNair, J., Choudhari, S.P., Quick, A.M., Mellado-Sanchez, G., Clements, J.D., Pasetti, M.F., and Picking, W.L. (2012). Broadly protective *Shigella* vaccine based on type III secretion apparatus proteins. *Infect. Immun. 80*, 1222–1231.

McMullan, R., Edwards, P.J., Kelly, M.J., Millar, B.C., Rooney, P.J., and Moore, J.E. (2007). Food-poisoning and commercial air travel. *Travel Med. Infect. Dis. 5*, 276–286.

Mead, P.S. and Griffin, P.M. (1998). *Escherichia coli* O157:H7. *Lancet 352*, 1207–1212.

Medalla, F., Sjölund-Karlsson, M., Shin, S., Harvey, E., Joyce, K., Theobald, L., Nygren, B.L. et al. (2011). Ciprofloxacin-Resistant *Salmonella enterica* serotype Typhi, United States, 1999–2008. *Emerg. Infect. Dis. 17*, 1095–1098.

Melby, J.O., Nard, N.J., and Mitchell, D.A. (2011). Thiazole/oxazole-modified microcins: Complex natural products from ribosomal templates. *Curr. Opin. Chem. Biol. 15*, 369–378.

Messaoudi, S., Kergourlay, G., Dalgalarrondo, M., Choiset, Y., Ferchichi, M., Prévost, H., Pilet, M.-F., Chobert, J.-M., Manai, M., and Dousset, X. (2012). Purification and characterization of a new bacteriocin active against *Campylobacter* produced by *Lactobacillus salivarius* SMXD51. *Food Microbiol. 32*, 129–134.

Mills, S., Stanton, C., Hill, C., and Ross, R.P. (2011). New developments and applications of bacteriocins and peptides in foods. *Annu. Rev. Food Sci. Technol. 2*, 299–329.

Minahk, C.J., Farías, M.E., Sesma, F., and Morero, R.D. (2000). Effect of enterocin CRL35 on *Listeria monocytogenes* cell membrane. *FEMS Microbiol. Lett. 192*, 79–83.

Miriagou, V., Tassios, P.T., Legakis, N.J., and Tzouvelekis, L.S. (2004). Expanded-spectrum cephalosporin resistance in non-typhoid *Salmonella*. *Int. J. Antimicrob. Agents 23*, 547–555.

Mitchell, L.A. and Koval, M. (2010). Specificity of interaction between *Clostridium perfringens* enterotoxin and claudin-family tight junction proteins. *Toxins (Basel) 2*, 1595–1611.

Moll, G., Ubbink-Kok, T., Hildeng-Hauge, H., Nissen-Meyer, J., Nes, I.F., Konings, W.N., and Driessen, A.J. (1996). Lactococcin G is a potassium ion-conducting, two-component bacteriocin. *J. Bacteriol. 178*, 600–605.

Molloy, E.M., Field, D., O'Connor, P.M., Cotter, P.D., Hill, C., and Ross, R.P. (2013). Saturation mutagenesis of lysine 12 leads to the identification of derivatives of nisin A with enhanced antimicrobial activity. *PLoS One 8*, e58530.

Montville, T.J. and Chen, Y. (1998). Mechanistic action of pediocin and nisin: Recent progress and unresolved questions. *Appl. Microbiol. Biotechnol. 50*, 511–519.

Mulders, J.W., Boerrigter, I.J., Rollema, H.S., Siezen, R.J., and de Vos, W.M. (1991). Identification and characterization of the lantibiotic nisin Z, a natural nisin variant. *Eur. J. Biochem. 201*, 581–584.

Mundy, R., Schüller, S., Girard, F., Fairbrother, J.M., Phillips, A.D., and Frankel, G. (2007). Functional studies of intimin in vivo and *ex vivo*: Implications for host specificity and tissue tropism. *Microbiology 153*, 959–967.

Murinda, S.E., Rashid, K.A., and Roberts, R.F. (2003). In vitro assessment of the cytotoxicity of nisin, pediocin, and selected colicins on simian virus 40-transfected human colon and Vero monkey kidney cells with trypan blue staining viability assays. *J. Food Prot. 66*, 847–853.

Murphy, K., O'Sullivan, O., Rea, M.C., Cotter, P.D., Ross, R.P., and Hill, C. (2011). Genome mining for radical SAM protein determinants reveals multiple sactibiotic-like gene clusters. *PLoS One 6*, e20852.

Nelson, J.M., Chiller, T.M., Powers, J.H., and Angulo, F.J. (2007). Fluoroquinolone-resistant *Campylobacter* species and the withdrawal of fluoroquinolones from use in poultry: A public health success story. *Clin. Infect. Dis. 44*, 977–980.

Nguyen, Y. and Sperandio, V. (2012). Enterohemorrhagic *E. coli* (EHEC) pathogenesis. *Front. Cell. Infect. Microbiol. 2*, 90.

Nissen-Meyer, J., Oppegård, C., Rogne, P., Haugen, H.S., and Kristiansen, P.E. (2010). Structure and mode-of-action of the two-peptide (class-IIb) bacteriocins. *Probiotics Antimicrob. Proteins 2*, 52–60.

Olsen, S.J., Miller, G., Breuer, T., Kennedy, M., Higgins, C., Walford, J., McKee, G., Fox, K., Bibb, W., and Mead, P. (2002). A waterborne outbreak of *Escherichia coli* O157:H7 infections and hemolytic uremic syndrome: Implications for rural water systems. *Emerg. Infect. Dis. 8*, 370–375.

Oman, T.J., Boettcher, J.M., Wang, H., Okalibe, X.N., and van der Donk, W.A. (2011). Sublancin is not a lantibiotic but an S-linked glycopeptide. *Nat. Chem. Biol. 7*, 78–80.

Ottenwälder, B., Kupke, T., Brecht, S., Gnau, V., Metzger, J., Jung, G., and Götz, F. (1995). Isolation and characterization of genetically engineered gallidermin and epidermin analogs. *Appl. Environ. Microbiol. 61*, 3894–3903.

Paton, J.C. and Paton, A.W. (1998). Pathogenesis and diagnosis of Shiga toxin-producing *Escherichia coli* infections. *Clin. Microbiol. Rev. 11*, 450–479.

Patton, B.S., Lonergan, S.M., Cutler, S.A., Stahl, C.H., and Dickson, J.S. (2008). Application of colicin E1 as a prefabrication intervention strategy. *J. Food Prot. 71*, 2519–2522.

Peck, M.W., Stringer, S.C., and Carter, A.T. (2011). *Clostridium botulinum* in the post-genomic era. *Food Microbiol. 28*, 183–191.

Penner, J.L. (1988). The genus *Campylobacter*: A decade of progress. *Clin. Microbiol. Rev. 1*, 157–172.

Peterson, K.M. (2002). Expression of *Vibrio cholerae* virulence genes in response to environmental signals. *Curr. Issues Intest. Microbiol. 3*, 29–38.

Peterson, M.C. (1994). Clinical aspects of *Campylobacter jejuni* infections in adults. *West. J. Med. 161*, 148–152.

Pomares, M.F., Salomón, R.A., Pavlova, O., Severinov, K., Farías, R., and Vincent, P.A. (2009). Potential applicability of chymotrypsin-susceptible microcin J25 derivatives to food preservation. *Appl. Environ. Microbiol. 75*, 5734–5738.

Portnoy, D.A. (2012). Yogi Berra, Forrest Gump, and the discovery of *Listeria* actin comet tails. *Mol. Biol. Cell 23*, 1141–1145.

Quadri, L.E., Yan, L.Z., Stiles, M.E., and Vederas, J.C. (1997). Effect of amino acid substitutions on the activity of carnobacteriocin B2. Overproduction of the antimicrobial peptide, its engineered variants, and its precursor in *Escherichia coli. J. Biol. Chem. 272*, 3384–3388.

Raffatellu, M., Chessa, D., Wilson, R.P., Tükel, Ç., Akçelik, M., and Bäumler, A.J. (2006). Capsule-mediated immune evasion: A new hypothesis explaining aspects of typhoid fever pathogenesis. *Infect. Immun. 74*, 19–27.

Ricke, S.C., Dunkley, C.S., and Durant, J.A. (2013). A review on development of novel strategies for controlling *Salmonella* Enteritidis colonization in laying hens: Fiber-based molt diets. *Poult. Sci. 92*, 502–525.

Rihakova, J., Cappelier, J.-M., Hue, I., Demnerova, K., Fédérighi, M., Prévost, H., and Drider, D. (2010). In vivo activities of recombinant divercin V41 and its structural variants against *Listeria monocytogenes*. *Antimicrob. Agents Chemother. 54*, 563–564.

Rihakova, J., Petit, V.W., Demnerova, K., Prévost, H., Rebuffat, S., and Drider, D. (2009). Insights into structure-activity relationships in the C-terminal region of divercin V41, a class IIa bacteriocin with high-level antilisterial activity. *Appl. Environ. Microbiol. 75*, 1811–1819.

Riley, L.W., Remis, R.S., Helgerson, S.D., McGee, H.B., Wells, J.G., Davis, B.R., Hebert, R.J. et al. (1983). Hemorrhagic colitis associated with a rare *Escherichia coli* serotype. *N. Engl. J. Med. 308*, 681–685.

Rodríguez, E. and Laviña, M. (2003). The proton channel is the minimal structure of ATP synthase necessary and sufficient for microcin H47 antibiotic action. *Antimicrob. Agents Chemother. 47*, 181–187.

Rollema, H.S., Kuipers, O.P., Both, P., de Vos, W.M., and Siezen, R.J. (1995). Improvement of solubility and stability of the antimicrobial peptide nisin by protein engineering. *Appl. Environ. Microbiol. 61*, 2873–2878.

Rouse, S., Field, D., Daly, K.M., O'Connor, P.M., Cotter, P.D., Hill, C., and Ross, R.P. (2012). Bioengineered nisin derivatives with enhanced activity in complex matrices. *Microb. Biotechnol. 5*, 501–508.

Roux, F., Sproston, E., Rotariu, O., MacRae, M., Sheppard, S.K., Bessell, P., Smith-Palmer, A. et al. (2013). Elucidating the aetiology of human *Campylobacter coli* infections. *PLoS One 8*, e64504.

Saavedra, L., Minahk, C., de Ruiz Holgado, A.P., and Sesma, F. (2004). Enhancement of the enterocin CRL35 activity by a synthetic peptide derived from the NH2-terminal sequence. *Antimicrob. Agents Chemother. 48*, 2778–2781.

Safa, A., Nair, G.B., and Kong, R.Y.C. (2010). Evolution of new variants of *Vibrio cholerae* O1. *Trends Microbiol. 18*, 46–54.

Salvucci, E., Saavedra, L., Hebert, E.M., Haro, C., and Sesma, F. (2012). Enterocin CRL35 inhibits *Listeria monocytogenes* in a murine model. *Foodborne Pathog. Dis. 9*, 68–74.

Salvucci, E., Saavedra, L., and Sesma, F. (2007). Short peptides derived from the NH2-terminus of subclass IIa bacteriocin enterocin CRL35 show antimicrobial activity. *J. Antimicrob. Chemother. 59*, 1102–1108.

Sánchez-Hidalgo, M., Montalbán-López, M., Cebrián, R., Valdivia, E., Martínez-Bueno, M., and Maqueda, M. (2011). AS-48 bacteriocin: Close to perfection. *Cell. Mol. Life Sci. 68*, 2845–2857.

Sánchez-Jiménez, M.M., Cardona-Castro, N., Canu, N., Uzzau, S., and Rubino, S. (2010). Distribution of pathogenicity islands among Colombian isolates of *Salmonella*. *J. Infect. Dev. Ctries. 4*, 555–559.

Sánchez-Vargas, F.M., Abu-El-Haija, M.A., and Gómez-Duarte, O.G. (2011). *Salmonella* infections: An update on epidemiology, management, and prevention. *Travel Med. Infect. Dis. 9*, 263–277.

Savini, V., Favaro, M., Fontana, C., Catavitello, C., Balbinot, A., Talia, M., Febbo, F., and D'Antonio, D. (2009). *Bacillus cereus* heteroresistance to carbapenems in a cancer patient. *J. Hosp. Infect. 71*, 288–290.

Scallan, E., Griffin, P.M., Angulo, F.J., Tauxe, R.V., and Hoekstra, R.M. (2011). Foodborne illness acquired in the United States—Unspecified agents. *Emerg. Infect. Dis. 17*, 16–22.

Scheiring, J., Andreoli, S.P., and Zimmerhackl, L.B. (2008). Treatment and outcome of Shiga-toxin-associated hemolytic uremic syndrome (HUS). *Pediatr. Nephrol. 23*, 1749–1760.

Severinov, K. and Nair, S.K. (2012). Microcin C: Biosynthesis and mechanisms of bacterial resistance. *Future Microbiol. 7*, 281–289.

Severinov, K., Semenova, E., Kazakov, A., Kazakov, T., and Gelfand, M.S. (2007). Low-molecular-weight post-translationally modified microcins. *Mol. Microbiol. 65*, 1380–1394.

Shai, Y. (2002). Mode of action of membrane active antimicrobial peptides. *Biopolymers 66*, 236–248.

Sit, C.S., Yoganathan, S., and Vederas, J.C. (2011). Biosynthesis of aminovinyl-cysteine-containing peptides and its application in the production of potential drug candidates. *Acc. Chem. Res. 44*, 261–268.

Sivonen, K., Leikoski, N., Fewer, D.P., and Jokela, J. (2010). Cyanobactins-ribosomal cyclic peptides produced by cyanobacteria. *Appl. Microbiol. Biotechnol. 86*, 1213–1225.

Stavru, F., Bouillaud, F., Sartori, A., Ricquier, D., and Cossart, P. (2011). *Listeria monocytogenes* transiently alters mitochondrial dynamics during infection. *Proc. Natl. Acad. Sci. U.S.A. 108*, 3612–3617.

Stevens, M.P., Humphrey, T.J., and Maskell, D.J. (2009). Molecular insights into farm animal and zoonotic *Salmonella* infections. *Philos. Trans. R. Soc. Lond., B: Biol. Sci. 364*, 2709–2723.

Svetoch, E.A. and Stern, N.J. (2010). Bacteriocins to control *Campylobacter* spp. in poultry—A review. *Poult. Sci. 89*, 1763–1768.

Swerdlow, D.L., Woodruff, B.A., Brady, R.C., Griffin, P.M., Tippen, S., Donnell, H.D., Jr., Geldreich, E., Payne, B.J., Meyer, A., Jr., and Wells, J.G. (1992). A waterborne outbreak in Missouri of *Escherichia coli* O157:H7 associated with bloody diarrhea and death. *Ann. Intern. Med. 117*, 812–819.

Taguchi, H., Yamaguchi, H., Osaki, T.Y., Yamamoto, T., Ogata, S., and Kamiya, S. (1997). Flow cytometric analysis for adhesion of *Vibrio cholerae* to human intestinal epithelial cell. *Eur. J. Epidemiol. 13*, 719–724.

Taylor, E.V., Holt, K.G., Mahon, B.E., Ayers, T., Norton, D., and Gould, L.H. (2012). Ground beef consumption patterns in the United States, FoodNet, 2006 through 2007. *J. Food Prot. 75*, 341–346.

Temple, M.E. and Nahata, M.C. (2000). Treatment of listeriosis. *Ann. Pharmacother. 34*, 656–661.

Thiele, I., Hyduke, D.R., Steeb, B., Fankam, G., Allen, D.K., Bazzani, S., Charusanti, P. et al. (2011). A community effort towards a knowledge-base and mathematical model of the human pathogen *Salmonella* Typhimurium LT2. *BMC Syst. Biol. 5*, 8.

Thomas, X., Destoumieux-Garzón, D., Peduzzi, J., Afonso, C., Blond, A., Birlirakis, N., Goulard, C. et al. (2004). Siderophore peptide, a new type of post-translationally modified antibacterial peptide with potent activity. *J. Biol. Chem. 279*, 28233–28242.

Tilney, L.G. and Portnoy, D.A. (1989). Actin filaments and the growth, movement, and spread of the intracellular bacterial parasite, *Listeria monocytogenes*. *J. Cell Biol. 109*, 1597–1608.

Toh, M., Moffitt, M.C., Henrichsen, L., Raftery, M., Barrow, K., Cox, J.M., Marquis, C.P., and Neilan, B.A. (2004). Cereulide, the emetic toxin of *Bacillus cereus*, is putatively a product of nonribosomal peptide synthesis. *J. Appl. Microbiol. 97*, 992–1000.

Tominaga, T. and Hatakeyama, Y. (2007). Development of innovative pediocin PA-1 by DNA shuffling among class IIa bacteriocins. *Appl. Environ. Microbiol. 73*, 5292–5299.

Tossi, A., Sandri, L., and Giangaspero, A. (2000). Amphipathic, alpha-helical antimicrobial peptides. *Biopolymers 55*, 4–30.

van Belkum, M.J., Kok, J., Venema, G., Holo, H., Nes, I.F., Konings, W.N., and Abee, T. (1991). The bacteriocin lactococcin A specifically increases permeability of lactococcal cytoplasmic membranes in a voltage-independent, protein-mediated manner. *J. Bacteriol. 173*, 7934–7941.

Vandenbergh, P.A., Pucci, M.J., Kunka, B.S., and Vedamuthu, E.R. (1989). Method for inhibiting *Listeria monocytogenes* using a bacteriocin. Application number: 89101125.6

Vaucher, R.A., da Motta, A., de S., and Brandelli, A. (2010). Evaluation of the in vitro cytotoxicity of the antimicrobial peptide P34. *Cell Biol. Int. 34*, 317–323.

Wemekamp-Kamphuis, H.H., Karatzas, A.K., Wouters, J.A., and Abee, T. (2002). Enhanced levels of cold shock proteins in *Listeria monocytogenes* LO28 upon exposure to low temperature and high hydrostatic pressure. *Appl. Environ. Microbiol. 68*, 456–463.

Wemekamp-Kamphuis, H.H., Wouters, J.A., de Leeuw, P.P.L.A., Hain, T., Chakraborty, T., and Abee, T. (2004). Identification of sigma factor sigma B-controlled genes and their impact on acid stress, high hydrostatic pressure, and freeze survival in *Listeria monocytogenes* EGD-e. *Appl. Environ. Microbiol. 70*, 3457–3466.

Weyermann, J., Lochmann, D., and Zimmer, A. (2005). A practical note on the use of cytotoxicity assays. *Int. J. Pharm. 288*, 369–376.

Wheeler, A. and Smith, H.S. (2013). Botulinum toxins: Mechanisms of action, antinociception and clinical applications. *Toxicology 306*, 124–146.

White, S.H., Wimley, W.C., Ladokhin, A.S., and Hristova, K. (1998). Protein folding in membranes: Determining energetics of peptide-bilayer interactions. *Methods Enzymol. 295*, 62–87.

Yoshikawa, Y., Ogawa, M., Hain, T., Yoshida, M., Fukumatsu, M., Kim, M., Mimuro, H. et al. (2009). *Listeria monocytogenes* ActA-mediated escape from autophagic recognition. *Nat. Cell Biol. 11*, 1233–1240.

Yuan, J., Zhang, Z.-Z., Chen, X.-Z., Yang, W., and Huan, L.-D. (2004). Site-directed mutagenesis of the hinge region of nisinZ and properties of nisinZ mutants. *Appl. Microbiol. Biotechnol. 64*, 806–815.

Zhang, Q. and Liu, W. (2013). Biosynthesis of thiopeptide antibiotics and their pathway engineering. *Nat. Prod. Rep. 30*, 218–226.

Zhou, Q., Feng, F., Wang, L., Feng, X., Yin, X., and Luo, Q. (2011). Virulence regulator PrfA is essential for biofilm formation in *Listeria monocytogenes* but not in *Listeria innocua*. *Curr. Microbiol. 63*, 186–192.

Zoja, C., Buelli, S., and Morigi, M. (2010). Shiga toxin-associated hemolytic uremic syndrome: Pathophysiology of endothelial dysfunction. *Pediatr. Nephrol. 25*, 2231–2240.

24

Biological Preservation of Foods

Osman Sagdic, Fatih Tornuk, İsmet Öztürk, Salih Karasu, and Mustafa T. Yilmaz

CONTENTS

24.1 Introduction

In recent years, with increasing consumer awareness on healthy consumption, tendency to consumption of natural/minimally processed foods that are free from chemical substances is increasing. Therefore, food manufacturers give a priority to the use of natural additives in food processing to meet customer acceptance and fulfill legal obligations. The success of modern food processing technology is dependent on the application of preservation technology without changing food quality throughout the stages until its consumption (Ross et al., 2002). Nowadays, biopreservation of foods is becoming very popular. This chapter describes the use of microorganisms and/or their metabolites, plant origin antimicrobial agents, and other biobased antimicrobials in food preservation.

24.2 Lactic Acid Bacteria and Their Metabolites

24.2.1 Role of Lactic Acid Bacteria in Biopreservation

Lactic acid bacteria (LAB) have an important role in the production of antimicrobial substances. LAB are common microorganisms in nature and play roles in the manufacture of a diversity of fermented foods such as cheese, pickle, yogurt, wine, and fermented sausage by becoming predominant culture in their microflora. As a result of LAB activity in fermented foods, some metabolites such as lactic acid, acetic acid, diacetyl, ethyl alcohol, acetone, acetaldehydes, carbon dioxide, and bacteriocins are produced. In addition to giving taste and flavor to the foods, these metabolites play an important role in preventing pathogen microorganisms' growth in foods (Toldra et al., 2001; Khalid, 2011).

LAB are generally abundant in the environments rich in nutritives and are naturally found in fermented dairy, meat, and vegetable foods (Carr et al., 2002). In addition to the contribution of LAB fermentation on textural and sensorial properties of foods, they also play an important role in the inactivation of saprophyte and pathogenic bacteria by producing antimicrobial substances (Hugas, 1998). Therefore, they are considered as strong biopreservation agents found in foods. Besides their strong biopreservative properties, LAB are safe microorganisms and are approved as *generally recognized as safe* (GRAS) (Ghanbari et al., 2013).

24.2.2 Bacteriocins and Other LAB Metabolites

24.2.2.1 Bacteriocins

LAB produce some antimicrobial peptides and proteins called bacteriocin. LAB bacteriocins have low molecular weight and are cationic, heat-stable, amphiphilic, and membrane-permeabilizing peptides (Zacharof and Lovitt, 2012). Bacteriocins are classified into three groups according to Klaenhammer (1993):

   I. *Group Bacteriocin (Nisin)*: This group of bacteriocins is defined as lantibiotics, which contains modified forms of some amino acids such as lanthionine and β-methyllanthionine and dehydrated amino acids (Hugas, 1998; Cleveland et al., 2001).

   II. *Group Bacteriocin (Pediocin, sakacin, carnobacteriocin, leucocin, and plantaricin)*: This group of bacteriocins includes peptides that are low molecular weight (<10 kDa) and heat stable and is composed of 30–60 amino acid chains except for lantonin. This group is classified into three subgroups: IIa: bacteriocin, which includes single peptide similar to pediocin, IIb: bacteriocin, which includes dipeptide, IIc: set-dependent secreted bacteriocin (Hugas, 1998; Cleveland et al., 2001).

   III. *Group Bacteriocin (Helveticin, acidophilucin, and lactacin)*: This group of bacteriocins is classified as heat-stable and high-molecular-weight proteins (Hugas, 1998; Cleveland et al., 2001).

In recent years, bacteriocins, especially nisin, have gained importance as biopreservatives in dairy, meat, vegetables, and egg products. Nisin is effective against gram-positive bacteria while it is ineffective against gram-negative ones (Zacharof and Lovitt, 2012). However, it was also claimed that nisin may possess antibacterial activity against some gram-negative bacteria in the case of destabilization of bacterial out membrane (Stevens et al., 1991).

Bacteriocins are incorporated into food matrix in three forms: (1) antimicrobial agents directly used in pure and semipure forms, (2) bacteriocin-based additives, and (3) use of protective starter cultures that have the ability to produce bacteriocins (Mills et al., 2011).

Bacteriocins show synergistic activity when used with some compounds (chelating agents such as EDTA, other bacteriocins as antimicrobial agents such as potassium sorbate, sodium lactate, sodium diacetate, nitrite, essential oils EOs, and plant extract), protein, and peptide agents. Some processing procedures such as thermal and nonthermal processes (heat, high hydrostatic pressure, and pulsed electric field) and food packaging systems (i.e., modified atmosphere packaging) also contribute to the efficiency of bacteriocins (Chen and Hoover, 2003; García et al., 2010; Mills et al., 2011).

Reuterin is an antimicrobial metabolite produced by *Lactobacillus reuteri* as a result of glycerol fermentation (Chen et al., 2001). It was also reported that reuterin can be produced by *Lactobacillus coryniformis* (Martín et al., 2005). *L. reuteri* is among heterofermentative LAB and is found in human/animal intestinal system and microflora of some fermented and probiotic foods (Langa et al., 2013). It has been demonstrated that reuterin can be used as an antimicrobial substance against spoilage and pathogenic bacteria, viruses, fungi, and protozoa (Chung et al., 1989; El-Ziney and Debevere, 1998; El-Ziney et al., 1999; Arques et al., 2004, 2008). Antimicrobial efficiency of reuterin depends on environmental and intrinsic factors such as temperature, pH, and salt content (Langa et al., 2013). However, it was noted that reuterin can dissolve in water and is effective in a wide range of pH and resistant to lipolytic and proteolytic enzymes (Axelsson et al., 1989). *L. reuteri* produces a high amount of reuterin when grown in an environment rich in glycerol (Schaefer et al., 2010).

24.2.2.2 Cyclic Dipeptides

Cyclic dipeptides (2,5 dioxopiperazines) are among the most prevalent peptide derivatives in nature (Crowley et al., 2013). These peptides are produced by some LAB and have antifungal effect on some fungi. It was also shown that they have antimicrobial activity on some gram-negative bacteria (Niku-Paavola et al., 1999; Dal Bello et al., 2007). However, although some *Lactobacillus* species (*Lactobacillus plantarum* and *Lactobacillus casei*) are capable of producing cyclic dipeptides, biochemical pathways of these substances have not been discovered completely (Niku-Paavola et al., 1999; Li et al., 2012; Crowley et al., 2013).

24.2.2.3 Organic Acids

LAB can produce organic acids such as lactic acid, acetic acid, propionic acid, and formic acid by carbohydrate metabolism (Crowley et al., 2013). Organic acids show its bactericidal or bacteriostatic activity by reducing the pH of foods. Lactic and propionic acids, which are known as weak acids, have common use in food and are approved as GRAS by the US Food and Drug Administration (FDA). Organic acids that have low pH values and strong acidifying potential cannot permeate through cell membrane and may show their inhibitory effect on membrane surface enzymes, whereas weak acidifying potential organic acids are able to permeate easily through cell membrane and, therefore, show their antimicrobial activity on cell metabolism by reducing cytoplasmic pH (Eklund, 1989).

Low pH value and high proton concentration affect permeability and intercellular functions of the microbial cells by causing protein denaturation, losing viability, and disturbing DNA (Erkmen and Bozoğlu, 2008a). It was observed that antimicrobial effect of organic acids is higher than strong inorganic acids (Sorrells et al., 1989). Acetic and propionic acids show synergistic effect with lactic acid (Schnurer and Magnusso, 2005). Lactic acid is one of the most common organic acids and can be used as a food preservative like acetic acid. Lactic acid, which is produced by starter cultures, also shows preservative effects on fermented foods such as butter, cheese, and bread dough, and it also used in surface washing solutions for decontamination of ewe and poultry carcasses (Hutton et al., 1991; Bostan et al., 1995). Other organic acids can be used as a preservative in the production of salted meat, fishery and vegetable products, infant formula, bread and other baked foods, jam, jelly, and confectionary products (Celikyurt and Arici, 2008).

24.2.2.4 Hydrogen Peroxide

Hydrogen peroxide (H_2O_2) has strong antimicrobial activity on both vegetative and spore forms of bacteria, molds, and yeasts (Askar et al., 1999; Erkmen and Bozoglu, 2008b). Most of the LAB possess protein oxidase enzyme, which breaks down the protein to H_2O_2 in aerobic conditions (Yang, 2000; Schnurer and Magnusson, 2005; Erkmen and Bozoglu, 2008b). H_2O_2 produced by LAB accumulates in the environment because of the inability of LAB cells to break down H_2O_2 using catalase enzyme (Yang, 2000). By means of oxidation of sulfhydryl groups, H_2O_2 causes inhibition of glucose-carrying system, hexokinase, and glyceraldehyde 3-phosphate dehydrogenase activity of microbial cell membrane (Yang, 2000; Juven et al., 1991). H_2O_2 leads to bacterial free radicals such as superoxide and hydroxyl to damage the DNA of bacterial cells (Yang, 2000).

H_2O_2 is used as a disinfection agent in aseptic food packaging and as a preservative in some foods especially dairy products (Andres, 1981). In recent years, H_2O_2 is used in the surface sterilization of some vegetables and fruits (Forney et al., 1991; Simmons et al., 1997; Sapers and Simmons, 1998; Sapers et al., 1999).

24.2.2.5 Carbon Dioxide

Carbon dioxide (CO_2) is produced by heterofermentative LAB by hexose fermentation. As a result of enzymatic decarboxylation, CO_2 shows its inhibitory effect by enabling anaerobic conditions. It was also stated that CO_2 accumulates in cell membrane's double lipid layer and damages the permeability properties of cell membrane (Lindgren and Dobrogosz, 1990; Yang, 2000). Antimicrobial activity of CO_2 is variable among microorganism groups. It has been shown that CO_2 is capable of reducing the numbers of bacteria and fungi by 10% and 20%–50%, respectively (Yang, 2000).

24.2.2.6 Diacetyl

Diacetyl, which gives the characteristic aroma and flavor of butter, is produced by LAB during citrate fermentation and shows antimicrobial activity at lower pH values (Schnurer and Magnusson, 2005; Erkmen and Bozoglu, 2008b). In addition to butter and other dairy products, it can also be formed in wine, brandy, roasted coffee, silage, and other fermented products (Jay, 1982). Diacetyl production is suppressed in hexose fermentation while it is produced at high amounts in citrate fermentation by means of conversion of citrate/pyruvate into diacetyl (Lindgren and Dobrogosz, 1990). Diacetyl shows bacteriostatic activity on gram-positive bacteria while it is bactericidal on gram-negative ones (Jay, 1982).

24.2.2.7 Ethyl Alcohol

Ethyl alcohol is produced as a result of sugar fermentation by heterofermentative LAB such as *Leuconostoc mesenteroides* (Schlegen, 1986). Ethyl alcohol shows bactericidal effect on vegetative cells of pathogenic bacteria while it is not effective on bacterial spores. Lethal effect of ethyl alcohol on microorganisms depends on its molecular weight, concentration, and treatment time. It was noted that the most effective concentration of ethyl alcohol is 70% (Volk and Wheeler, 1973). Antimicrobial activity of ethyl alcohol results from the denaturation of proteins present in the cell (Volk and Wheeler, 1973; Banwart, 1989; Pelczar et al., 1993). It also damages the lipid structure of cytoplasmic membrane (Pelczar et al., 1993). Drawbacks of ethyl alcohol are its weak diffusion and inactivation by organic acids. However, it increases the antimicrobial activity of some antimicrobial compounds (Shapero et al., 1978).

24.2.2.8 Fatty Acids

Some *Lactococcus* and *Lactobacillus* species can produce considerable amounts of fatty acid as a result of their lipolytic activity. It was demonstrated that fatty acids have antifungal activity depending on their chain length and the pH value of environment (Yang, 2000; Schnurer and Magnusson, 2005; Erkmen and Bozoglu, 2008b). It was reported that the mixture of 10 mN capric (C:10) and lauric (C:12) acids showed antimicrobial effect on *Candida albicans* (Bergsson et al., 2001). It was also shown that a fatty acid, which has 12C chain length, showed antifungal effect on a wide spectrum of fungi (Sjogren et al., 2003).

24.2.2.9 Phenyl Lactic Acid

The considerable antifungal activity of phenyl lactic acid increases as a result of synergistic effects with other metabolites of LAB (Yang, 2000). It was determined that phenyl lactic acid and 4-hydroxyl phenyl lactic acid, which were produced by *L. plantarum* 21B isolated from bread dough, delayed the formation of *A. niger* filaments for 7 days (Lavermicocca et al., 2000). It was also reported that phenyl lactic acid and acetic acid showed antifungal effects on *Aspergillus*, *Fusarium*, and *Penicillium*, which cause spoilage in bread dough (Gerez et al., 2010).

24.2.2.10 Natamycin

Natamycin ($C_{33}H_{47}NO_{13}$), which is also known as pimaricin, is polyene macrolide antibiotic and produced by *Streptomyces* sp. (*Streptomyces natalensis*, *Streptomyces chattanoogensis*, and *Streptomyces lydicus*) in aerobic conditions (Kallinteri et al., 2013). Natamycin is used as an antifungal agent since it is ineffective on bacteria, protozoa, and viruses. Natamycin was approved as GRAS fungal antibiotic by FDA and more than 40 countries to be used as an antifungal food additive. It is used the food processing especially in meat and dairy industry (EFSA, 2009a,b; Resaa et al., 2014).

24.3 Antagonistic Yeasts

24.3.1 Antagonistic Activity of Yeasts

Yeasts may influence the quality of fermented foods in both negative and positive directions. Besides their contribution to taste and aroma, they may cause spoilage of foods (Lowes et al., 2000). Yeasts show antagonistic effect on some yeasts and molds owing to their killer toxins (Liu and Tsao, 2010). Antagonistic effect of killer yeasts has gained importance since 1963 after the identification of killer phenotypes of *Saccharomyces cerevisiae* (Bevan and Makower, 1963). Killer yeasts produce *myosin*, which is called killer protein. It was reported that killer toxins are composed of protein and glycoprotein structures and their molecular weight varies from 1.8 to 1.87 kDa (Buzzini et al., 2007). Killing mechanism of K1 toxin is formed in two stages. At first, killer toxin attacks 1–6-β-D-glucan receptor of yeast cell wall, and this process is controlled by β-chain. At the second stage, α-chain affects yeast plasma membrane. This causes nonuniform flow of potassium in the cell and finally death of the cell (Sesti et al., 2001). First studies that were carried out on killer concept revealed that double-chain RNA (dsRNA) was combined with virus-like particles in yeast (Marquina et al., 2002). Since killer character depended on cytoplasmic factor and different chromosomal gene, it is a complex phenomenon. Numerous kinds of killer toxins have been reported, and it was noted that their genomes show three different characteristics: dsRNA virus (*S. cerevisiae*), linear double-chain DNA plasmid (*Kluyveromyces lactis* and *Pichia acaciae*), and chromosomal gene (*Williopsis californica* and *Pichia farinose*) (Schmitt and Breinig, 2002, 2006). In addition to the antifungal activity of killer yeasts, it was also stated that some killer yeasts (*Hansenula anomala*, *Williopsis mrakii*, *Kluyveromyces drosophylarum*, *K. lactis*, *Candida tropicalis*, etc.) are effective against some gram-positive pathogenic and nonpathogenic bacteria (Polonelli and Morace, 1986; Izgu and Altinbay, 1997; Psani and Kotzekidou, 2006; Waema et al., 2009; Meneghin et al., 2010).

24.3.2 Use of Antagonistic Yeasts in Food Preservation

Yeasts and molds can cause spoilage and loss of the quality of cheese, yogurt, bakery products, and fermented meat products. In addition to desired antifungal effects of killer yeasts in foods, sometimes it is not desired when yeast growth is required. *Debaryomyces hansenii* is used as a biocontrol agent in cheese and yogurt against the growth of some molds (*Aspergillus* sp., *Byssochlamys fulva*, *Byssochlamys nivea*, *Cladosporium* sp., *Eurotium chevalieri*, *Penicillium candidum*, and *Penicillium roqueforti*) (Liu and Tsao, 2009). It was also shown that *Williopsis saturnus* var. *saturnus* can be used as a biocontrol agent against some yeasts (*Candida kefir*, *Kluyveromyces marxianus*, *S. cerevisiae*, and *Saccharomyces bayanus*) and molds (*Byssochlamys*, *Eurotium*, and *Penicillium*) in flour and yogurt (Liu and Tsao, 2010). It was determined that as a result of inoculation to cereal grain, *Pichia anomala* increased its microbiological quality by inhibiting Enterobacteriaceae and mold population and improved its nutritional value by increasing the protein content and decreasing the phytate concentration (Olstorpe and Passoth, 2010). Numerous studies were carried out focusing on the use of killer yeast in the biopreservation of harvested fruits and vegetables. After harvesting the defensive mechanism of fruits and vegetables weakens and wounded and overripe products can be attacked easily by molds. Therefore, some chemical substances are used for fruit and vegetable preservation after harvesting. Studies showed that killer yeasts and their toxins can be used as an alternative for chemical substances (Helbig, 2002; Masih and Paul, 2002; Vero et al., 2002; Zhang et al., 2007; Rosa-Magri et al., 2011).

24.4 Use of Enzymes for Biopreservation of Foods

In biological systems, some naturally found enzymes serve as antagonistic agents against microorganisms. On the other hand, those enzymes constitute another group of natural antimicrobial agents used in the biopreservation of foods, which are susceptible to microbial attacks. Some microorganisms as well as eukaryotic organisms such as animals, plants, and humans produce them. Food systems which contain such antagonistic enzymes increase stability of the foods against microbial activity. On the other hand, they can be purified and used in the biopreservation of foods (Holzapfel et al., 1995). These enzymes include oxidases, lipases, proteases, lactoperoxidase (LP), myeloperoxidase, amylases, lysostaphin, conalbumin, and avidin endolysins. Some of them that are important in food industry are described in this section.

24.4.1 Glucose Oxidase

Oxidation of glucose can be catalyzed by glucose oxidase (GOX) enzyme yielding gluconic acid and hydrogen peroxide in the presence of molecular O_2 (Massa et al., 2001). GOX is produced by molds including *Penicillium* and *Aspergillus* species as well as *Botrytis cinerea* (Lium et al., 1998; Hafiz et al., 2003; El-Sherbeny et al., 2005). GOX is an industrially important food additive approved as GRAS. Besides several applications of GOX in foods such as stabilization of soft drinks and prevention of the Maillard reaction, it has been well demonstrated that the GOX/glucose system has antibacterial activity against both gram-negative and gram-positive bacteria as well as fungi (Holzapfel et al., 1995; Leiter et al., 2004; Smith and Hong-Shum, 2011). Inhibitory effect of GOX is dependent on enzyme concentration and glucose level as a substrate (Holzapfel et al., 1995). Cytotoxicity of H_2O_2 is the mainspring of the antimicrobial action. However, lowering of pH by the production of gluconic acid may also negatively affect the growth of microorganisms (Fuglsang et al., 1995).

Several studies have been conducted to elucidate the convenience of GOX as an antimicrobial agent in food systems. GOX enzyme treatment was shown to be inefficient for extending the shelf life of chicken (Jeong et al., 1992). Yoo and Rand (2006) reported that 0.5–2.0 U/mL GOX combined with 1.0, 2.0, and 4.0 mg/mL glucose inhibited *Pseudomonas fragi* growth in fish extract media while viable cell numbers in fish media showed clear growth inhibition with combinations of 1.0–2.0 U/mL GOX and 8.0–16.0 mg/mL glucose. Total psychrotrophs and *Pseudomonas* spp. counts of shrimp were effectively reduced by dipping them in a 4% glucose aqueous solution containing GOX and catalase during refrigerated storage (Dondero et al., 1993).

24.4.2 Lysozyme

Lysozyme, an enzyme that was first discovered by Fleming in nose secretions and tissues in 1922, is a peptidoglycan *N*-acetyl-muramoylhydrolase (Mir, 1977; Masschalck and Michiels, 2003). Lysozyme is present in animal tissues, organs, and serum as well as in tears, milk, and cervical mucus. Egg white is a rich and easy source of lysozyme, accounting for 3.5% of the total egg white proteins (Cegielska-Radziejewska et al., 2008). However, this enzyme is widely used in food formulations for antimicrobial purposes.

Lysozyme is involved in a group of enzymes that break down the cell walls of gram-positive bacteria by splitting β(1–4) binds between *N*-acetyl-muramic acid and *N*-acetyl-glucosamine of the peptidoglycan, which are the components making up bacterial cell walls (Bernkop-Schnurch et al., 1998). Antibacterial activity of lysozyme is limited to gram-positive bacteria; it is not active against gram negatives, including food-borne pathogens (Ibrahim et al., 1991). It is because of the fact that although peptidoglycan layer may be very thin in the cell walls of gram-negative bacteria, it is sandwiched between the inner and outer membranes and protected against enzyme attack.

24.4.3 Lactoperoxidase

LP is an enzyme that is included in the group of peroxidases and found in plants and animals. It forms some reactive products with antimicrobial activity by catalyzing the oxidation of a number of organic and inorganic substrates such as bromide and iodide (Kussendrager and van Hooijdonk, 2000). On the other hand, in the food preservation systems, thiocyanate, a commonly distributed compound present in animal tissue and secretions, is a natural substrate of LP. LP oxidizes thiocyanate to hypothiocyanate using hydrogen peroxide (Fulgsang et al., 1995). As a result of this process, less or more stable oxidation products, some of which are strong antimicrobial, arise (Popper and Knorr, 1997). The system that is called LP inactivation system has been known to be a major part of antimicrobial activity of raw milk. Although LP and thiocyanate naturally occur in milk adequately for antimicrobial activity, hydrogen peroxide, which is not a native component of milk, is produced by catalase-negative bacteria including LAB using glucose oxidase (Bjorck et al., 1975). Pasteurization of milk results in the complete inactivation of the enzyme LP; however, residual activity in normal pasteurized milk is in the region of 70% (Barrett et al., 1996). Degree of antimicrobial activity of LP varies depending on the group or species of microorganisms. Generally, gram positive bacteria such as streptococci and lactobacilli seem to be more sensitive to LP system than gram-negatives due to the differences in cell wall structure and their different barrier properties (Kussendrager and van Hooijdonk, 2000). Use of this system in the preservation of milk and dairy products has been investigated. Garcia-Graells et al. (2000) tested the efficiency of combined treatment with high hydrostatic pressure and the LP system in milk against *Escherichia coli* and *Listeria innocua*. The LP system alone exerted a bacteriostatic effect on both species for at least 24 h at room temperature with no inactivation of the strains while milk as a substrate provided a considerable protective effect on *E. coli* and *L. innocua* against pressure inactivation. In another study, two strains of *E. coli* and one strain each of *Salmonella* Typhimurium and *Pseudomonas aeruginosa* were killed by the bactericidal activity of the LP–thiocyanate–H_2O_2 system in milk and in a buffer solution (Reiter et al., 1976). Nakada et al. (1996) showed that the incorporation of LP to yoghurt made from pasteurized milk in the concentration of 5 ppm suppressed the bacterial acidification almost completely during 14 days of storage at 10°C without leading a remarkable reduction in the viable count of the culture bacteria.

24.5 Bacteriophages and Endolysins

24.5.1 Bacteriophages

Bacteriophages or phages, which were first discovered by an English bacteriologist Frederick Twort, are widespread viruses that infect and multiply in bacterial cells (Walker, 2006). Phages are completely host specific, infect only specific strains, and cannot locate in other cells including other bacterial strains, humans, and plants (Hagens and Loessner, 2007). High specificity of bacteriophages to the targeted host is determined by bacterial cell wall receptors leaving untouched the remaining microbiota (Sillankorva et al., 2012). Bacteriophages show their antibacterial activity by disrupting bacterial metabolism and causing the bacterium to lyse (Garcia et al., 2008). Lysis of the host bacterial cell occurs as a result of two mechanisms: (1) phage replication where genetic material is the only component that enters the cell and replicates rapidly, and (2) without phage replication, a sufficient number of phages adhere to the cell and lyse it through alteration of the membrane electric potential, and/or the activity of cell wall–degrading enzymes (EFSA, 2009a,b). However, for the survival of bacteriophages, they need a favorable medium where they come into contact with the host cell.

24.5.1.1 Use of Bacteriophages in Food Preservation

Bacteriophages can be naturally found in a variety of foods such as wine, ground beef, chicken, fish, and fermented products (Davis et al., 1985; Kennedy et al., 1986; Hsu et al., 2002; Sillankorva et al., 2012). This indicates that bacteriophages are widely distributed in food contact environment. In fact, they can be abundantly found in any locations populated by bacterial hosts. In terms of human health, the presence of bacteriophages in foods is not a concern. However, bacteriophages can cause significant economic losses in fermented food industry since phages contaminated to foods slow down or prevent the fermentation process by infecting starter cultures (Emond and Moineau, 2007). It is well known that bacteriophages are responsible for destroying lactic acid fermentation by the genera *Streptococcus, Lactobacillus,* and *Lactococcus* in dairy industry (Sturino and Klaenhammer, 2004). Phages are generally resistant to environmental conditions. Conventional pasteurization treatment for foods is inadequate for phage inactivation, and it requires more severe temperature and time conditions. For instance, International Dairy Federation recommends a temperature exposure of 90°C for 15 min for complete inactivation of bacteriophages in dairy industries (Svensson and Christiansson, 1991). Moreover, different techniques such as the use of biocides, high pressure, and photocatalysis have been tested for their efficiency for phage destruction, and instead of treating them separately, their combination has been recommended for better inactivation (Guglielmotti et al., 2012).

There is a considerable literature relating to the use of bacteriophages for food preservation as well as the prevention of animal and plant infections. Table 24.1 summarizes some studies conducted to provide the biocontrol of some food products. As seen in Table 24.1, bacteriophages have been used in different animal and plant origin foods, and efficacies of bacteriophages were variable. Survival of bacteria against phage applications in foods may be affected from a combination of different factors such as phage and bacteria concentration, food composition (fat, sugar, salt, protein content, etc.), temperature, and intrinsic factors of food including pH and water content.

24.5.2 Endolysins

Bacteriophages perform in two different ways to release from the infected bacterial cells: (1) filamentous phages are squeezed out from the cell without killing them, while (2) nonfilamentous ones destroy the bacterial cell wall by lysing it with phage-encoded lytic enzymes (Borysowsky et al., 2006). Rigid peptidoglycan layer in the cell wall provides the integrity and shape of the bacteria. To overcome this layer, double-stranded DNA phages produce a holin–endolysin-based enzyme system at the end of the replication cycle (Oliviera et al., 2012). Endolysins (lysins) are bacteriophage enzymes that hydrolyze peptidoglycan layers in conjunction with small hydrophobic proteins termed holins (Turner et al., 2007). In consequence of the cleavage of peptidoglycan, the cell loses its capacity to withstand the internal osmotic pressure and is ultimately exposed to hypotonic lysis (Oliviera et al., 2012).

24.5.2.1 Use of Endolysins in Food Preservation

Endolysins are capable of degrading the peptidoglycan of gram-positive bacteria that are much thicker than that of the gram negative when applied externally to the bacterial cell, thereby playing a role as potential antibacterial agents (Garcia et al., 2010a). Main advantages of endolysins as antimicrobial agents are (1) their host specificity, (2) rapid efficiency, (3) distinct mode of action, and (4) bactericidal

TABLE 24.1

Use of Bacteriophages in Food Preservation

Target Microorganism	Food Tested	Result	References
P. fragi	Raw milk	Significant reduction	Ellis et al. (1973)
S. aureus	Cheese milk (curd)	Complete elimination	Garcia et al. (2007)
S. aureus	Fresh-type cheese	Complete elimination	Bueno et al. (2012)
S. aureus	Hard-type cheese	Ongoing reduction during ripening	Bueno et al. (2012)
Salmonella Enteritidis	Cheddar cheese	1–2 log decrease during ripening	Modi et al. (2001)
Enterobacter sakazakii	Infant formula milk	Two phages exhibited host specificity for *E. sakazakii*	Kim et al. (2007)
L. monocytogenes	Raw milk cheese	Complete elimination during ripening	Carlton et al. (2005)
Campylobacter jejuni	Chicken skin	1.3 and 0.3 log reduction during storage	Atterbury et al. (2003)
S. Enteritidis	Chicken skin	Complete elimination depending on the phage concentration	Goode et al. (2003)
C. jejuni	Chicken skin	Significant reduction	Goode et al. (2003)
S. Enteritidis	Chicken carcass	Different reduction rates were achieved	Higgins et al. (2005)
E. coli O157:H7	Beef surface	Complete elimination	O'Flynn et al. (2004)
E. coli O157:H7	Tomato	>94% reduction	Abuladze et al. (2008)
E. coli O157:H7	Spinach	>99.5% reduction	Abuladze et al. (2008)
E. coli O157:H7	Broccoli	>97% reduction	Abuladze et al. (2008)
E. coli O157:H7	Ground beef	94.5% reduction	Abuladze et al. (2008)
L. monocytogenes	Chocolate milk	Complete elimination	Guenthner et al. (2009)
L. monocytogenes	Cheddar cheese brine	Complete elimination	Guenthner et al. (2009)
L. monocytogenes	Turkey breast	1.5 log reduction	Guenthner et al. (2009)
L. monocytogenes	Mixed seafood	2.5 log reduction	Guenthner et al. (2009)
S. Typhimurium	Fresh egg	0.9 log reduction	Spricigo et al. (2013)
S. Enteritidis	Fresh egg	0.9 log reduction	Spricigo et al. (2013)
S. Typhimurium	Chicken breast	2.2 log reduction	Spricigo et al. (2013)
S. Enteritidis	Chicken breast	0.9 log reduction	Spricigo et al. (2013)
S. Typhimurium	Lettuce	3.9 log reduction	Spricigo et al. (2013)
S. Enteritidis	Lettuce	2.2 log reduction	Spricigo et al. (2013)
S. Typhimurium	Turkey meat, chocolate milk	>5 log reduction	Guenthner et al. (2012)
S. Typhimurium	Hot dog, mixed seafood	>3 log reduction	Guenthner et al. (2012)
S. Typhimurium	Egg yolk	Effect was observed for 2 days	Guenthner et al. (2012)

activity independent from the antibiotic susceptibility pattern (Loessner, 2005; Borysowsky et al., 2006). Moreover, when considering the use of phage endolysins for controlling food-borne diseases, their stability on food conditions and food processing facilities, temperature and pH tolerances, costs, etc., should be considered (Oliviera et al., 2012). Consumer perception has also been taken into consideration since endolysin production is mainly performed by genetically modified organisms.

There is still limited research relating to the use of endolysins in food biopreservation. However, existing studies have mainly focused on dairy products. In pasteurized milk, a purified protein was able to kill *Staphylococcus aureus* rapidly, and the pathogen was not detected after 4 h of incubation at 37°C. Other bacteria belonging to different genera were not affected (Obeso et al., 2008). Garcia et al. (2010b) observed a synergistic interaction between purified endolysin and nisin against *S. aureus* in pasteurized milk, and cell numbers decreased to undetectable levels in 6 h. Nisin also enhanced the lytic activity of LysH5 on cell suspensions in in vitro conditions eightfold. *Listeria* bacteriophage endolysin LysZ5 was able to kill *Listeria monocytogenes* growing in soya milk, with the pathogen concentration reduced by more than 4 log after 3 h incubation at 4°C (Zhang et al., 2012).

24.6 Essential Oils and Their Constituents

24.6.1 Plant Extracts

Plants produce a high variety of secondary metabolites to protect themselves against predators and microbial pathogens due to their biocidal properties against microbes or herbivores. Some metabolites are also involved in defense mechanisms against abiotic stress (e.g., UV-B exposure) and have important role in the interaction of plants with other organisms (Schafer and Wink, 2009; Bassolé and Juliani, 2012). Most of the known secondary metabolites are involved in plant chemical defense systems, and they are formed as a response of interaction of plants with predators throughout the millions of years (Bassolé and Juliani, 2012). There are three major chemical groups of these secondary metabolites, including terpenes, phenyl propenoids, and N- and S-including compounds (Wink, 1999; Bassolé and Juliani, 2012). EOs have important place among these compounds because of commercial value and are used by the flavor and fragrance industries (van de Braak and Leijten, 1999; Bassolé and Juliani, 2012).

24.6.2 Essential Oils and Their Constituents

EOs are volatile and fragrant compounds with an oily consistency typically produced by plants. They differ from other fixed oils in terms of their rapid evaporation when exposed to heat. They can be liquid at room temperature although a few of them show solid like behavior and have different colors ranging from pale yellow to emerald green and from blue to dark brownish red (Balz, 1999). They can be extracted from all plant organs, that is, buds, flowers, leaves, stems, twigs, seeds, fruits, roots, wood, or bark, and are stored in secretory cells, cavities, canals, epidermal cells, or glandular trichomes (Bakkali et al., 2008; Bassolé and Juliani, 2012). The total EO content of plants is generally lower than 1% (Bowles, 2003; Djilani and Dicko, 2012); however, in some cases, for example, clove (*Syzygium aromaticum*) and nutmeg (*Myristica fragrans*), it exceeds 10%. EOs have hydrophobic character and are soluble in alcohol, nonpolar or weakly polar solvents, waxes, and oils. Their solubility in water is less, and they have lower density than water (Gupta et al., 2010; Martín et al., 2010; Djilani and Dicko, 2012). Due to their molecular structure, EOs are very sensitive to oxidation (Djilani and Dicko, 2012).

Various methods such as hydrodistillation, steam, and steam/water distillation are used for the extraction of EOs from plant material (Margaris et al., 1982; Bowles, 2003; Surburg and Panten, 2006; Djilani and Dicko, 2012). In addition to these common methods, some methods including solvent extraction, aqueous infusion, cold or hot pressing, effleurage, supercritical fluid extraction, and phytonic process, have been developed for the extraction of EOs (Lahlou, 2004; Surburg and Panten, 2006; Pourmortazavi and Hajimirsadeghi, 2007; Martínez, 2008; Da Porto et al., 2009; Hunter, 2009; Djilani and Dicko, 2012). Extraction method significantly affects the chemical composition and the product yield of EOs (Donelian et al., 2009).

EOs are composed of more than 60 individual components classified in different groups such as hydrocarbons (e.g., pinene, limonene, the bisabolene), alcohols (e.g., linalool and santalol), acids (e.g., benzoic acid and geranic acid), aldehydes (e.g., citral), cyclic aldehydes (e.g., cuminal), ketones (e.g., camphor), lactones (e.g., bergaptene), phenols (e.g., eugenol), phenolic ethers (e.g., anethole), oxides (e.g., 1,8 cineole), and esters (e.g., geranyl acetate) (Deans et al., 1992; Djilani and Dicko, 2012). Hydrocarbons, which contain only carbon and hydrogen, are major compounds of EOs and are included in terpenes (Svoboda et al., 1999; Djilani and Dicko, 2012). Esters give a pleasant smell to the oils and are commonly found in a large number of EOs. However, esters and ketones have lower antibacterial activity in comparison to other EO constituents (Carson and Riley, 1995; Inouye et al., 2001; Tajkarimi et al., 2010; Bassolé and Juliani, 2012).

Oxides or cyclic ethers have strong odors, and these molecules give characteristic odors of EOs (Djilani and Dicko, 2012). Lactones have high molecular weight, and pressed oils are the major source of lactones. Aldehydes are common EO components that are unstable and sensitive to oxidation. Many aldehydes are mucous membrane irritants and are skin sensitizers (Djilani and Dicko, 2012). Aldehydes and ketones have higher antibacterial activity than other compounds (Carson and Riley, 1995; Inouye et al., 2001; Tajkarimi et al., 2010; Bassolé and Juliani, 2012). Ketones are not very common in the majority of EOs; they are relatively stable molecules and are not particularly important as fragrances or flavor substances. Phenolic components form the major group of constituents of EOs with strong antimicrobial activity, and they are mainly responsible for their antimicrobial activity. On the other hand, some minor components have shown to increase the antimicrobial efficiency of EOs through their synergistic effect with major phenolic compounds (Burt, 2004).

24.6.3 Mode of Antimicrobial Action

EOs are composed of a diversity of constituents many of which are included in terpenoids while some non terpenoid compounds are also found. Numerous studies have been conducted to determine the mode of antimicrobial activity of EOs; however, it is not still well understood (Hyldgaard et al., 2012). Given the presence of a large number of constituents in EOs, it is acceptable to assume that EOs can have very different mechanisms to show their antimicrobial activity.

Antimicrobial action is mainly related to the chemical structure of the active component present in EO. Phenolic compounds such as carvacrol, thymol, and eugenol have been demonstrated to have similar antimicrobial mechanisms. Hydrophobicity of those compounds is critical, which enables them to be capable of breaking down lipids in bacterial cell membrane and mitochondria and making them more permeable, thereby causing leakage of intracellular components (Kalemba and Kunicka, 2003; Prabuseenivasan et al., 2006; Ait-Ouazzou et al., 2011). Phenolics have also been shown to disrupt the cellular membrane, inhibit ATPase activity, and release intracellular ATP and other constituents of microorganisms (Lambert et al., 2001; Burt, 2004; Gill and Holley, 2006; Negi, 2012).

Most of the works about the mode of antimicrobial action of EOs have focused on prokaryotes; little is known about their interaction with yeast and fungi. Coriander EOs show antifungal activity against *Candida* cells by inhibiting germ tube formation and with synergetic effect with amphotericin B (Silva et al., 2011). Carvacrol has been reported to disrupt ion homeostasis and cause dose-dependent Ca (2+) bursts in *S. cerevisiae* cell, thus showing antifungal activity (Rao et al., 2010). Dill EO causes morphological changes in the cells of *Aspergillus flavus* and a reduction in the ergosterol quantity. Exposure to dill oil also results in an increase in mitochondrial membrane potential and dose-dependent decrease in ATPase and dehydrogenase activities of the cells (Tian et al., 2012).

24.6.4 Use of Essential Oils in Food Preservation

EOs and their constituents have been broadly tested in vitro as well as biopreservation of a diversity of foods such as minced beef, cheese, vegetables, salads, fish products, and chicken (Shelef et al., 1984; Mendoza-Yepes et al., 1997; Menon et al., 2001; Skandamis et al., 2001; Mejlholm and Dalgaard, 2002; Sagdic et al., 2002; Singh et al., 2002; Baydar et al., 2004; Ozkan et al., 2004; Arici et al., 2005; Ozcan et al., 2006; Sagdic et al., 2009; Ozkan et al., 2010). In general, it has been found that a higher level of EO is required to achieve comparable effect in foods (Shelef, 1984; Burt, 2004). Antimicrobial efficiency of EOs in food systems is affected from intrinsic and extrinsic factors such as pH, nutrient composition, use of preservatives, antioxidants, temperature, and packaging type (e.g., vacuum and modified atmosphere). It is presumed that the occurrence of fat and protein with high levels in food decreases the antimicrobial efficiency of EOs (Aureli et al., 1992; Pandit and Shelef, 1994; Tassou et al., 1995; Gutierrez et al., 2008). Lower water content of food is also important since it may slow down the progress of antimicrobial agent to the target area in the cell (Smith-Palmer et al., 2001). Antimicrobial activity of EOs also increases with the decreasing temperature and oxygen level in the packaging (Tassou et al., 1995; Tsigarida et al., 2000; Skandamis et al., 2001). Decrease in pH improves the dissolubility of EOs in the lipids found in the cell membrane; therefore, it contributes to their antimicrobial activity (Juven et al., 1994).

24.7 Chitin and Chitosan

Chitin is a naturally abundant polysaccharide as structural components of insects, crustacean exoskeletons, and fungal cell walls, and is composed of N-acetyl glucosamine and N-glucosamine units linked to each other via β-1,4 glycosidic bonds. The molecule is termed chitin when the percent of

N-acetyl-glucosamine units is higher than 50%. Chitin is today widely produced from crab and shrimp shell wastes commercially. Chitosan is the most important derivative of chitin produced by partial deacetylation in alkali conditions or by hydrolysis in the presence of chitin deacetylase enzyme (Rinaudo, 2006). Chitin is insoluble in water and many organic solvents. Despite its wide availability, deficiency in the solubility of chitin restricts its utilization and commercialization. Chitosan is also insoluble in water; however, it can be readily solved in weak acidic conditions below pH 6 because of its alkali structure and primary amino groups (Pillai et al., 2009).

Today, applications of chitin and chitosan in food industry as well as in other fields are well documented. Apart from the solubility limitations, properties of chitin such as biodegradability, nontoxicity, physiological inertness, antibacterial properties, hydrophobicity, gel-forming properties, and affinity for proteins have gained it many industrial applications (Rinaudo, 2006). Chitosan has also many advantages including biocompatibility, nontoxicity, and biofunctionality. In 2005, shrimp-origin chitosan was approved as GRAS by FDA to be used in foods for multiple technological effects (No et al., 2007).

24.7.1 Use of Chitin and Chitosan in Biopreservation of Foods

Through the observation of antimicrobial activity of chitin, chitosan, and other derivatives against different groups of bacteria, yeast, and fungi, the use of those materials in food biopreservation has drawn considerable attention in the last two decades. Because of the solubility of chitosan at pH values lower than pH 6, its antimicrobial activity is much higher than chitin (Shadidi et al., 1999). Mode of antimicrobial action of chitin and chitosan is not known exactly. In many cases, the antimicrobial activity is attributed to the interaction between positively charged chitosan molecules and negatively charged microbial cell membranes, causing the leakage of proteinaceous and other intracellular constituents, thereby the death of microbial cells (Young et al., 1982; Papineau et al., 1991; Sudarshan et al., 1992; Fang et al., 1994; No et al., 2007).

With the increasing consumer demand for eliminating the use of chemical additives and plastic packaging materials in foods, chitosan has gained much more importance. Chitosan, as a biocontrol agent for extending shelf life of foods by inhibiting undesirable microorganisms, has been evaluated in food systems in two different approaches: an antimicrobial food additive and a packaging material. Chitosan-origin active packaging films allow to extend food preservation and to reduce the use of chemical preservatives. Incorporation of chitosan into plastic and polymeric matrices is also possible (Muzzarelli et al., 2012). Moreover, incorporation of antimicrobial compounds from different sources into chitosan films has also been tested to strengthen the antimicrobial activity of chitosan. In agriculture, chitosan has been used as a preventive agent for fruits and vegetables against postharvest fungal pathogens. Devlieghere et al. (2004) coated whole strawberry and mixed lettuce by dipping them in 0.5% chitosan (w/v) in 2% lactic acid/Na lactate solution and provided immediate decreases in LAB, total aerobic psychrotrophic bacteria, and yeast counts of the samples stored at 4°C. However, in this study, coated lettuce was considered as sensorially unacceptable by the panelists. Park et al. (2005) investigated antifungal efficacies of chitosan-based coatings applied on fresh strawberries against *Cladosporium* sp., *Rhizopus* sp., total coliforms, and total aerobic counts (TACs). Coating treatment reduced TAC and coliforms while the percent of *Rhizopus*- and *Cladosporium*-infected strawberries was significantly lowered by the coating on the fruits during the refrigerated storage for 20 days. Chien et al. (2007) found similar inhibition effects on the TAC of sliced mango fruits coated with chitosan.

Chitosan has also been used in processed foods as biocontrol agent in the form of both edible film and antimicrobial additive. Coma et al. (2003) used chitosan for the coating of Emmental cheese to inhibit the growth of *P. aeruginosa* inoculated to cheese surface. Edible chitosan coating was successful to increase the microbial lag phase of *P. aeruginosa* and decrease the maximum density of contaminating strains. Ha and Lee (2001) achieved 2–3 log decrease in *S. cerevisiae* and *Pseudomonas fluorescence* populations of artificially contaminated banana-flavored milk by an addition of 0.03% chitosan. Juneja et al. (2006) investigated the inhibition of *Clostridium perfringens* spore germination and outgrowth by chitosan incorporation during the chilling of cooked ground beef (25% fat) and turkey (7% fat). Addition of chitosan to beef or turkey resulted in concentration- and time-dependent inhibition of *C. perfringens* spore germination and outgrowth, and 3% chitosan reduced by 4–5 log cfu/g spore germination and outgrowth during cooling of the cooked beef or turkey. In another study, the use of chitosan glutamate as a natural food preservative in mayonnaise and mayonnaise-based shrimp salad against *Salmonella* Enteritidis, *Zygosaccharomyces bailii*, or *L. fructivorans* was investigated (Roller and Covill, 2000). In mayonnaise-containing chitosan, *L. fructivorans* was inactivated while *Z. bailii* counts were reduced by approximately 1–2 log cfu/g. TAC, yeasts, and LAB counts were also lower in shrimp salads incorporated with chitosan than those in control samples.

REFERENCES

Abuladze, T., M. Li, M.Y. Menetrez, T. Dean, A. Senecal, and A. Sulakvelidze. 2008. Bacteriophages reduce experimental contamination of hard surfaces, tomato, spinach, broccoli, and ground beef by *Escherichia coli* O157:H7. *Appl. Environ. Microbiol.* 7: 6230–6238.

Ait-Ouazzou, A., L. Cherrat, L. Espina, S. Lorán, C. Rota, and R. Pagán. 2011. The antimicrobial activity of hydrophobic essential oil constituents acting alone or in combined processes of food preservation. *Innov. Food Sci. Emerg. Technol.* 12: 320–329.

Andres, C. 1981. FDA approval opens way for aseptic packaging of shelf-stable milk, juices in U.S. *Food Process.* 35: 70–71.

Arici, M., O. Sagdic, and U. Gecgel. 2005. Antibacterial effect of Turkish black cumin (*Nigella sativa* L.) oils. *Grasas y Aceites* 56: 259–262.

Arques, J.L., J. Fernandez, P. Gaya, M. Nunez, E. Rodriguez, and M. Medina. 2004. Antimicrobial activity of reuterin in combination with nisin against food-borne pathogens. *Int. J. Food Microbiol.* 95: 225–229.

Arques, J.L., E. Rodriguez, M. Nunez, and M. Medina. 2008. Antimicrobial activity of nisin, reuterin, and the lactoperoxidase system on *Listeria monocytogenes* and *Staphylococcus aureus* in cuajada, a semisolid dairy product manufactured in Spain. *J. Dairy Sci.* 91: 70–75.

Askar, M., B. Aslim, and Y. Beyatlı. 1999. Et ürünlerinden izole edilen *Pediococcus acidilactici* suşlarının bazı metabolik ve antimikrobiyal aktivitelerinin incelenmesi. *Turk. J. Vet. Anim. Sci.* 23: 467–474.

Atterbury, R.J., P.L. Connerton, C.E.R. Dodd, C.E.D. Rees, and I.F. Connerton. 2003. Application of host-specific bacteriophages to the surface of chicken skin leads to a reduction in recovery of *Campylobacter jejuni*. *Appl. Environ. Microbiol.* 69: 6302–6306.

Aureli, P., A. Costantini, and S. Zolea. 1992. Antibacterial activity of some plant essential oils against *Listeria monocytogenes*. *J. Food Prot.* 55: 344–348.

Axelsson, L.T., T.C. Chung, W.J. Dobrogosz, and S.E. Lindgren. 1989. Production of a broad spectrum antimicrobial substance by *Lactobacillus reuteri*. *Microb. Ecol. Health Dis.* 2: 131–136.

Bakkali, F., S. Averbeck, D. Averbeck, and M. Idaomar. 2008. Biological effects of essential oils—A review. *Food Chem. Toxicol.* 46: 446–475.

Balz, R. 1999. *The Healing Power of Essential Oils*, 1st edn. Lotus Press: Twin Lakes, WI, pp. 27–80.

Banwart, G.J. 1989. *Basic Food Microbiology*. AVI Book/Von Nostrand Reinhold: New York, 773pp.

Barrett, N., A. Grandison, and M. Lewis. 1996. Contribution of the lactoperoxidase system to the keeping quality of pasteurised milk. *Process Optimisation and Minimal Processing of Foods, Proceedings of the Second Main Meeting*, eds. J.C. Oliveira and M.E. Hendrickx, Vol. 1, pp. 60–65.

Bassolé, I.H.N. and H.R. Juliani. 2012. Essential oils in combination and their antimicrobial properties. *Molecules* 17: 3989–4006.

Baydar, H., O. Sagdic, G. Ozkan, and T. Karadogan. 2004. Antibacterial activity and composition of essential oils from *Origanum*, *Thymbra* and *Satureja* species with commercial importance in Turkey. *Food Control* 10: 169–172.

Bergsson, G., J. Arnfinnsson, O. Steingrimsson, and H. Thormar. 2001. In vitro killing of *Candida albicans* by fatty acids and monoglycerides. *Antimicrob. Agents Chemother.* 45: 3209–3212.

Bernkop-Schnurch, A., S. Krist, M. Vehabovic, and C. Valenta. 1998. Synthesis and evaluation of lysozyme derivatives exhibiting an enhanced antimicrobial action. *Eur. J. Pharm. Sci.* 6: 301–306.

Bevan, E.A. and M. Makower. 1963. The physiological basis of the killer character in yeast. *Proceedings of the 11th International Congress on Gene*, ed. S.J. Goerts, Vol. 1, pp. 202–203.

Bjorck, L., C.G. Rosen, V. Marshall, and B. Reiter. 1975. Antibacterial activity of the lactoperoxidase system in milk against *Pseudomonas* and other gram-negative bacteria. *Appl. Microbiol.* 30: 199–204.

Borysowsky, J., B. Weber-Dabrowska, and A. Gorski. 2006. Bacteriophage endolysins as a novel class of anti-bacterial agents. *Exp. Biol. Med.* 231: 366–377.

Bostan, K., M. Ugur, O. Ozgen, and H. Aksu. 1995. Laktik asit solusyonlarına daldırmanın broiler karkaslarının mikrobiyolojik kalitesine etkisi. *J. Fac. Vet. Med. Univ. Istanbul* 21: 443–451.

Bowles, E.J. 2003. *The Chemistry of Aromatherapeutic Oils*, 3rd edn. Griffin Press: Sydney, New South Wales, Australia.

Bueno, E., P. Garcia, B. Martinez, and A. Rodriguez. 2012. Phage inactivation of *Staphylococcus aureus* in fresh and hard-type cheeses. *Int. J. Food Microbiol.* 158: 23–27.

Burt, S. 2004. Essential oils: Their antibacterial properties and potential applications in foods: A review. *Int. J. Food Microbiol.* 94: 223–253.

Buzzini, P., B. Turchetti, and A.E. Vaughan-Martini. 2007. The use of killer sensitivity patterns for biotyping yeast strains: The state of the art, potentialities and limitations. *FEMS Yeast Res.* 7: 749–760.

Carr, F.J., D. Chill, and N. Maida. 2002. The lactic acid bacteria: A literature survey. *Crit. Rev. Microbiol.* 28: 281–370.

Carlton, R.M., W.H. Noordman, B. Biswas, E.D. de Meester, and M.J. Loessner. 2005. Bacteriophage P100 for control of *Listeria monocytogenes* in foods: Genome sequence, bioinformatics analyses, oral toxicity study, and application. *Regul. Toxicol. Pharmacol.* 3: 301–312.

Carson, C.F. and T.V. Riley. 1995. Antimicrobial activity of the major components of the essential oil of *Melaleuca alternifolia*. *J. Appl. Bacteriol.* 78: 264–269.

Cegielska-Radziejewska, R., G. Leśnierowski, and J. Kijowski. 2008. Properties and application of egg white lysozyme and its modified preparations—A review. *Pol. J. Food Nutr. Sci.* 58: 5–10.

Celikyurt, G. and M. Arici. 2008. Gıda Koruyucusu Olarak Mikrobiyal Kaynaklı Organik Asitler ve Önemi, Türkiye 10. Gıda Kongresi, Bildiri kitabı, Erzurum, pp. 1023–1026.

Chen, C.-N., H.-F. Liang, M.-H. Lin, and H.-W. Sung. 2001. A natural sterilant (reuterin) fermented from glycerol using *Lactobacillus reuteri*: Fermentation conditions. *J. Med. Biol. Eng.* 21: 205–212.

Chen, H. and D.G. Hoover. 2003. Bacteriocins and their food applications. *Compr. Rev. Food Sci. Food Saf.* 2: 82–100.

Chien, P., F. Sheu, and F. Yang. 2007. Effects of edible chitosan coating on quality and shelf life of sliced mango fruit. *J. Food Eng.* 78: 225–229.

Chung, T.C., L. Axelsson, S.E. Lindgren, and W.J. Dobrogosz. 1989. In vitro studies on reuterin synthesis by *Lactobacillus reuteri*. *Microb. Ecol. Health Dis.* 2: 137–144.

Cleveland, J., T.J. Montville, I.F. Nes, and M.L. Chikindas. 2001. Bacteriocins: Safe, natural antimicrobials for food preservation. *Int. J. Food Microbiol.* 71: 1–20.

Crowley, S., J. Mahony, and D. Van Sinderen. 2013. Current perspectives on antifungal lactic acid bacteria as natural bio-preservatives. *Trends Food Sci. Technol.* 33: 93–109.

Coma, V., A. Deschamps, and A. Martial-Gros. 2003. Bioactive packaging materials from edible chitosan polymer—Antimicrobial activity assessment on dairy-related contaminants. *J. Food Sci.* 68: 2788–2792.

Da Porto, C., D. Decorti, and I. Kikic. 2009. Flavour compounds of *Lavandula angustifolia* L. to use in food manufacturing: Comparison of three different extraction methods. *Food Chem.* 112: 1072–1078.

Dal Bello, F., C.I. Clarke, L.A.M. Ryan, H. Ulmer, T.J. Schober, K. Strom, J. Sjogren, D. van Sinderen, J. Schnurer, and E.K. Arendt. 2007. Improvement of the quality and shelf-life of wheat bread by fermentation with the antifungal strain *Lactobacillus plantarum* FST 1.7. *J. Cereal Sci.* 45: 309–318.

Davis, C., N.F.A. Silveira, and G.H. Fleet. 1985. Occurrence and properties of bacteriophages of *Leuconostoc oenos* in Australian wines. *Appl. Environ. Microbiol.* 50: 872–876.

Deans, S.G., K.P. Svoboda, M. Gundidza, and E.Y. Brechany. 1992. Essential oil profiles of several temperate and tropical aromatic plants: Their antimicrobial and antioxidant activities. *Acta Hortic.* 306: 229–232.

Devlieghere, F., A. Vermeulen, and J. Debevere. 2004. Chitosan: Antimicrobial activity, interactions with food components and applicability as a coating on fruit and vegetables. *Food Microbiol.* 21: 703–714.

Djilani, A. and Dicko, A. 2012. The therapeutic benefits of essential oils. *Nutrition, Well Being and Health*, Bouayed, J. (ed.). InTech: Croatia, Europe, pp. 157–178.

Dondero, M., W. Egana, W. Tarky, A. Cifuentes, and J.A. Torres. 1993. Glucose oxidase/catalase improves preservation of shrimp (*Heterocarpus reedi*). *J. Food Sci.* 58: 774–779.

Donelian, A., L.H.C. Carlson, T.J. Lopes, and R.A.F. Machado 2009. Comparison of extraction of patchouli (*Pogostemon cablin*) essential oil with supercritical CO$_2$ and by steam distillation. *J. Supercrit. Fluids* 48: 15–20.

EFSA, European Food Safety Authority. 2009a. The use and mode of action of bacteriophages in food production. *EFSA J.* 1076: 1–26.

EFSA, 2009b. Scientific opinion of EFSA prepared by the panel on food additives and nutrient sources added to food (ANS) on the use of natamycin (E 235) as a food additive. *EFSA J.* 7: 1412–1437.

El-Ziney, M.G. and J.M. Debevere. 1998. The effect of reuterin on *Listeria monocytogenes* and *Escherichia coli* O157:H7 in milk and cottage cheese. *J. Food Prot.* 61: 1275–1280.

El-Ziney, M.G., T. van den Tempel, J. Debevere, and M. Jakobsen. 1999. Application of reuterin produced by *Lactobacillus reuteri* 12002 for meat decontamination and preservation. *J. Food Prot.* 62: 257–261.

Ellis, D.E., P.A. Whitman, and R.T. Marshall. 1973. Effects of homologous bacteriophage on growth of *Pseudomonas fragi* WY in milk. *Appl. Microbiol.* 25: 24–25.

El-Sherbeny, G.A., A.A. Shindia, and Y.M.M.M. Sheriff. 2005. Optimization of various factors affecting glucose oxidase activity produced by *Aspergillus niger*. *Int. J. Agric. Biol.* 7: 953–958.

Eklund, T. 1989. *Organic Acids and Esters, Mechanism of Action of Food Preservation Procedures*. Elsevier Applied Science: New York, p. 161.

Emond, E. and S. Moineau, 2007. Bacteriophages and food fermentation. *Bacteriophage: Genetics and Molecular Biology*, Mc Grath, S. and van Sinderen, D. (eds.). Caister Academic Press: Norfolk, U.K.

Erkmen, O. and T.F. Bozoglu. 2008a. *Chemical Preservatives and Natural Antimicrobial Compounds, Food Microbiol. 3 Food Preservation*. İlke Yayinevi, Ankara, Turkey, pp. 71–124.

Erkmen, O. and T.F. Bozoglu. 2008b. *Food Microbiol. 4 Beneficial Uses of Microorganisms for Food Preservation and Health*. İlke Yayinevi, Ankara, Turkey, pp. 39–49.

Fang, S.W., C.F. Li, and D.Y.C. Shih. 1994. Antifungal activity of chitosan and its preservative effect on low-sugar candied kumquat. *J. Food Prot.* 57: 136–140.

Forney, C.F., R.E. Rij, R. Denis-Arrue, and J.L. Smilanic. 1991. Vapor phase hydrogen peroxide inhibits postharvest decay of table grapes. *Hortscience* 26: 1512–1514.

Fuglsang, C.C., C. Johansen, S. Christgau, and J. Adler-Nissen. 1995. Antimicrobial enzymes: Applications and future potential in the food industry. *Trends Food Sci. Technol.* 6: 390–396.

Garcia, P., C. Madera, B. Martinez, and A. Rodriguez. 2007. Biocontrol of *Staphylococcus aureus* in curd manufacturing processes using bacteriophages. *Int. Dairy J.* 17: 1232–1239.

Garcia, P., B. Martinez, J.M. Obeso, and A. Rodriguez. 2008. Bacteriophages and their application in food safety. *Lett. Appl. Microbiol.* 47: 479–485.

Garcia, P., B. Martinez, L. Rodriguez, and A. Rodriguez. 2010b. Synergy between the phage endolysin LysH5 and nisin to kill *Staphylococcus aureus* in pasteurized milk. *Int. J. Food Microbiol.* 141: 151–155.

Garcia, P., L. Rodriguez, A. Rodriguez, and B. Martinez. 2010a. Food biopreservation: Promising strategies using bacteriocins, bacteriophages and endolysins. *Trends Food Sci. Technol.* 21: 373–382.

Garcia-Graells, C., C. Valckx, and C.W. Michiels. 2000. Inactivation of *Escherichia coli* and *Listeria innocua* in milk by combined treatment with high hydrostatic pressure and the lactoperoxidase system. *Appl. Environ. Microbiol.* 66: 4173–4179.

Gerez, C.L., M.I. Torino, G. Rollán, and G.F. de Valdez. 2010. Prevention of bread mould spoilage by using lactic acid bacteria with antifungal properties. *Food Control* 20: 144–148.

Ghanbari, M., M. Jami, K.J. Domig, and W. Kneifel. 2013. Seafood biopreservation by lactic acid bacteria—A review. *LWT—Food Sci. Technol.* 54: 315–324.

Gill, A.O. and R.A. Holley. 2006. Disruption of *Escherichia coli*, *Listeria monocytogenes* and *Lactobacillus sakei* cellular membranes by plant oil aromatics. *Int. J. Food Microbiol.* 108: 1–9.

Goode, D., V.M. Allen, and P.A. Barrow. 2003. Reduction of experimental *Salmonella* and *Campylobacter* contamination of chicken skin by application of lytic bacteriophages. *Appl. Environ. Microbiol.* 69: 5032–5036.

Guenthner, S., O. Herzig, L. Fieseler, J. Klumpp, and M.J. Loessner. 2012. Biocontrol of *Salmonella* Typhimurium in RTE foods with the virulent bacteriophage FO1-E2. *Int. J. Food Microbiol.* 154: 66–72.

Guenthner, S., D. Huwyler, S. Richard, and M.J. Loessner. 2009. Virulent bacteriophage for efficient biocontrol of *Listeria monocytogenes* in ready-to-eat foods. *Appl. Environ. Microbiol.* 75: 93–100.

Guglielmotti, D.M., D.J. Mercanti, J.A. Reinheimer, and A.L. Quiberoni. 2012. Review: Efficiency of physical and chemical treatments on the inactivation of dairy bacteriophages. *Front. Microbiol.* 2: 1–11.

Gupta, V., P. Mittal, P. Bansal, S.L. Khokra, and D. Kaushik. 2010. Pharmacological potential of *Matricaria recutita*. *Int. J. Pharm. Sci. Drug Res.* 2: 12–16.

Gutierrez, J., C. Barry-Ryan, and P. Bourke. 2008. The antimicrobial efficacy of plant essential oil combinations and interactions with food ingredients. *Int. J. Food Microbiol.* 124: 91–97.

Ha, T. and S. Lee. 2001. Utilization of chitosan to improve the quality of processed milk. *J. Korean Soc. Food Sci. Nutr.* 30: 630–634.

Hafiz, M., Hamid, M. Khalil–ur–Rehman, Anjum, Z., Asgher, M. 2003. Optimization of various parameters for the production of glucose oxidase from rice polishing using *Aspergillus niger. Biotechnology* 2: 1–7.

Hagens, S. and M.J. Loessner. 2007. Application of bacteriophages for detection and control of foodborne pathogens. *Appl. Microbiol. Biotechnol.* 76: 513–519.

Helbig, J. 2002. Ability of the antagonistic yeast *Cryptococcus albidus* to control *Botrytis cinerea* in strawberry. *Biocontrol* 47: 85–99.

Higgins, J.P., S.E. Higgins, K.L. Guenther, W. Huff, A.M. Donoghue, D.J. Donoghue, and B.M. Hargis. 2005. Use of a specific bacteriophage treatment to reduce *Salmonella* in poultry products. *Poultry Sci.* 84: 1141–1145.

Holzapfel, W.F., R. Geisen, and U. Schillinger. 1995. Biological preservation of foods with reference to protective cultures, bacteriocins and food-grade enzymes. *Int. J. Food Microbiol.* 24: 343–362.

Hsu, F.C., Y.S.C. Shieh, and M.D. Sobsey. 2002. Enteric bacteriophages as potential fecal indicators in ground beef and poultry meat. *J. Food Prot.* 65: 93–99.

Hugas, M. 1998. Bacteriocinogenic lactic acid bacteria for the biopreservation meat and meat products. *Meat Sci.* 49: 139–150.

Hunter, M. 2009. *Essential Oils: Art, Agriculture, Science, Industry and Entrepreneurship.* Nova Science Publishers, Inc.: New York.

Hutton, M.T., P.A. Chehakk, and J.H. Hanlin. 1991. Inhibition of botulinum toxin production by *Pediococcus acidilactici* in temperature abused refrigerated foods. *J. Food Saf.* 11: 255–267.

Hyldgaard, M., T. Mygind, and R.L. Meyer. 2012. Essential oils in food preservation: Mode of action, synergies, and interactions with food matrix components. *Front. Microbiol.* 3: 1–24.

Ibrahim, H.R., A. Kato, and K. Kobayashi. 1991. Antimicrobial effects of lysozyme against gram-negative bacteria due to covalent binding of palmitic acid. *J. Agric. Food Chem.* 39: 2077–2082.

Inouye, S., H. Yamaguchi, and T. Takizawa. 2001. Screening of the antibacterial effects of a variety of essential oils on respiratory tract pathogens, using a modified dilution assay method. *J. Infect. Chemother.* 7: 251–254.

Izgü, F. and D. Altinbay. 1997. Killer toxins of certain yeast strains have potential growth inhibitory activity on gram-positive pathogenic bacteria. *Microbios* 89: 15–22.

Jay, J.M. 1982. Antimicrobial properties of diacetyl. *Appl. Environ. Microbiol.* 44: 525–532.

Juneja, V.K., H. Thippareddi, L. Bari, Y. Inatsu, S. Kawamoto, and M. Friedman. 2006. Chitosan protects cooked ground beef and turkey against *Clostridium perfringens* spores during chilling. *J. Food Sci.* 71: 236–240.

Juven, B.J., J. Kanner, F. Schved, and H. Weisslowicz. 1994. Factors that interact with the antibacterial action of thyme essential oil and its active constituents. *J. Appl. Bacteriol.* 76: 626–631.

Juven, B.J., R.J. Meinersmann, and N.J. Stern. 1991. Antagonistic effects of lactobacilli and pediococci to control intestinal colonization by human enteropathogens in live poultry. *J. Appl. Bacteriol.* 70: 95–103.

Jeong, D.K., M.A. Harrison, J.F. Frank, and L. Wicker. 1992. Trials on the antibacterial effect of glucose oxidase on chicken breast skin and muscle. *J. Food Saf.* 13: 43–49.

Kalemba, D. and A. Kunicka. 2003. Antibacterial and antifungal properties of essential oils. *Curr. Med. Chem.* 10: 813–829.

Kallinteri, L.D., O.K. Kostoula, and I.N. Savvaidis. 2013. Efficacy of nisin and/or natamycin to improve the shelf-life of Galotyri cheese. *Food Microbiol.* 36: 176–181.

Kennedy, J.E., C.I. Wei, and J.L. Oblinger. 1986. Distribution of coliphages in various foods. *J. Food Prot.* 49: 944–951.

Khalid, K. 2011. An overview of lactic acid bacteria. *Int. J. Biosci.* 1: 1–13.

Kim, K.P., J. Klumpp, and M.J. Loessner. 2007. *Enterobacter sakazakii* bacteriophages can prevent bacterial growth in reconstituted infant formula. *Int. J. Food Microbiol.* 115: 195–203.

Klaenhammer, T.R. 1993. Genetics of bacteriocins produced by lactic acid bacteria. *FEMS Microbiol. Rev.* 12: 39–85.

Kussendrager, K.D. and A.C.M. van Hooijdonk. 2000. Lactoperoxidase: Physico-chemical properties, occurrence, mechanism of action and applications. *Br. J. Nutr.* 84: 19–25.

Lahlou, M. 2004. Methods to study the phytochemistry and bioactivity of essential oils. *Phytother. Res.* 18: 435–448.

Lambert, R.J.W., P.N. Skandamis, P.J. Coote, and G.J.E. Nychas. 2001. A study of the minimum inhibitory concentration and mode of action of oregano essential oil, thymol and carvacrol. *J. Appl. Microbiol.* 91: 453–462.

Langa, S., J.M. Landete, I. Martín-Cabrejas, E. Rodríguez, J.L. Arqués, and M. Medina. 2013. In situ reuterin production by *Lactobacillus reuteri* in dairy products. *Food Control* 33: 200–206.

Lavermicocca, P., F. Valerio, A. Evidente, S. Lazzaroni, A. Corsetti, and M. Gobbetti. 2000. Purification and characterization of novel antifungal compounds from the sourdough *Lactobacillus plantarum* strain 21B. *Appl. Environ. Microbiol.* 66: 4084–4090.

Leiter, E., F. Marx, H. Pusztahelyi, H. Haas, and I. Pocsi. 2004. *Penicillium chrysogenum* glucose oxidase—A study on its antifungal effects. *J. Appl. Microbiol.* 97: 1201–1209.

Li, H., L. Liu, S. Zhang, W. Cui, and J. Lv. 2012. Identification of antifungal compounds produced by *Lactobacillus casei* AST18. *Curr. Microbiol.* 65: 156–161.

Lindgren, S.W. and W.J. Dobrogosz. 1990. Antagonistic activities of lactic acid bacteria in food and feed fermentation. *FEMS Microbiol. Rev.* 87: 149–164.

Liu, S.Q. and M. Tsao. 2009. Biocontrol of dairy moulds by antagonistic dairy yeast *Debaryomyces hansenii* in yoghurt and cheese at elevated temperatures. *Food Control* 20: 852–855.

Liu, S.Q. and M. Tsao. 2010. Biocontrol of spoilage yeasts and moulds by *Williopsis saturnus* var. *saturnus* in yoghurt. *Nutr. Food Sci.* 40: 166–175.

Lium, S., S. Oelejeklaus, B. Gerhardh, and B. Tudzynki. 1998. Purification and characterization of glucose oxidase of *Botrytis cinerea*. *Physiol. Mol. Plant Pathol.* 53: 123–132.

Loessner, M.J. 2005. Bacteriophage endolysins—Current state of research and applications. *Curr. Opin. Biotechnol.* 8: 480–487.

Lowes, K.F., C.A. Shearman, J. Payne, D. MacKenzie, D.B. Archer, R.J. Merry, and M.J. Gasson. 2000. Prevention of yeast spoilage in feed and food by the yeast mycocin HMK. *Appl. Environ. Microbiol.* 66: 1066–1076.

Margaris, N., A. Koedam, and D. Vokou. 1982. *Aromatic Plants: Basic and Applied Aspects*. Martinus Nijhoff Publishers: The Hague, the Netherlands.

Marquina, D., A. Santos, and J.M. Peinado. 2002. Biology of killer yeasts. *Int. Microbiol.* 5: 65–71.

Martín, A., S. Varona, A. Navarrete, and M.J. Cocero. 2010. Encapsulation and co-precipitation processes with supercritical fluids: Applications with essential oils. *Open Chem. Eng. J.* 4: 31–41.

Martín, R., M. Olivares, M.L. Marín, J. Xaus, L. Fernández, and J.M. Rodríguez. 2005. Characterization of a reuterin-producing *Lactobacillus coryniformis* strain isolated from a goat's milk cheese. *Int. J. Food Microbiol.* 104: 267–277.

Martínez, J.L. 2008. *Supercritical Fluid Extraction of Nutraceuticals and Bioactive Compounds*. CRC Press: Boca Raton, FL.

Masih, E.I. and B. Paul. 2002. Secretion of beta-1,3-glucanases by the yeast *Pichia membranifaciens* and its possible role in the biocontrol of *Botrytis cinerea* causing grey mold disease of the grapevine. *Curr. Microbiol.* 44: 391–395.

Massa, S., G.F. Petruccioli, C. Altieri, M. Sinigaglia, and G. Spano. 2001. Growth inhibition by glucose oxidase system of enterotoxic *Escherichia coli* and *Salmonella derby*: In vitro studies. *World J. Microbiol. Biotechnol.* 17: 287–291.

Masschalck, B. and C.W. Michiels. 2003. Antimicrobial properties of lysozyme in relation to foodborne vegetative bacteria. *Crit. Rev. Microbiol.* 29: 191–214.

Mejlholm, O. and P. Dalgaard. 2002. Antimicrobial effect of essential oils on the seafood spoilage microorganism *Photobacterium phosphoreum* in liquid media and fish products. *Lett. Appl. Microbiol.* 34: 27–31.

Mendoza-Yepes, M.J., L.E. Sanchez-Hidalgo, G. Maertens, and F. Marin-Iniesta. 1997. Inhibition of *Listeria monocytogenes* and other bacteria by a plant essential oil (dmc) in spanish soft cheese. *J. Food Saf.* 17: 47–55.

Meneghin, M.C., V.R. Reis, and S.R. Ceccato-Antonini. 2010. Inhibition of bacteria contaminating alcoholic fermentations by killer yeasts. *Braz. Arch. Biol. Technol.* 53: 1043–1050.

Menon, V.K. and S.R. Garg. 2001. Inhibitory effect of clove oil on *Listeria monocytogenes* in meat and cheese. *Food Microbiol.* 18: 647–650.

Mills, S., C. Stanton, C. Hill, and R.P. Ross. 2011. New developments and applications of bacteriocins and peptides in foods. *Ann. Rev. Food Sci. Technol.* 2: 299–329.

Mir, M.A. 1977. Lysozyme: A brief review. *Postgrad. Med. J.* 53: 257–259.

Modi, R., Y. Hirvi, A. Hill, and M.W. Griffiths. 2001. Effect of phage on survival of *Salmonella* Enteritidis during manufacture and storage of cheddar cheese made from raw and pasteurized milk. *J. Food Prot.* 64: 927–933.

Muzzarelli, R.A.A., J. Boudrant, D. Meyer, N. Manno, M. DeMarchis, and M.G. Paoletti. 2012. Current views on fungal chitin/chitosan, human chitinases, food preservation, glucans, pectins and inulin: A tribute to Henri Braconnot, precursor of the carbohydrate polymers science, on the chitin bicentennial. *Carbohyd. Polym.* 87: 995–1012.

Nakada, M., S. Dosako, R. Hirano, M. Ooka, and I. Nakajima. 1996. Lactoperoxidase suppresses acid production in yoghurt during storage under refrigeration. *Int. Dairy J.* 6: 33–42.

Negi, P.S. 2012. Plant extracts for the control of bacterial growth: Efficacy, stability and safety issues for food application. *Int. J. Food Microbiol.* 156: 7–17.

Niku-Paavola, M.L., A. Laitila, T. Mattila-Sandholm, and A. Haikara. 1999. New types of antimicrobial compounds produced by *Lactobacillus plantarum*. *J. Appl. Microbiol.* 86: 29–35.

No, H.K., S.P. Meyers, W. Prinyawiwatkul, and X. Xu. 2007. Applications of chitosan for improvement of quality and shelf life of foods: A review. *J. Food Sci.* 72: 87–100.

Obeso, J.M., B. Martinez, A. Rodriguez, and P. Garcia. 2008. Lytic activity of the recombinant staphylococcal bacteriophage phiH5 endolysin active against *Staphylococcus aureus* in milk. *Int. J. Food Microbiol.* 128: 212–218.

O'Flynn, G., R.P. Ross, G.F. Fitzgerald, and A. Coffey. 2004. Evaluation of a cocktail of three bacteriophages for biocontrol of *Escherichia coli* O157:H7. *Appl. Environ. Microbiol.* 70: 3417–3424.

Oliviera, H., J. Azeredo, R. Lavigne, and L.D. Kluskens. 2012. Bacteriophage endolysins as a response to emerging foodborne pathogens. *Trends Food Sci. Technol.* 28: 103–115.

Olstorpe, M. and V. Passoth. 2010. *Pichia anomala* in grain biopreservation. *Antonie Van Leeuwenhoek* 99: 57–62.

Ozcan, M.M., O. Sagdic, and G. Ozkan. 2006. Inhibitory effects of spice essential oils on the growth of *Bacillus* species. *J. Med. Food* 9: 418–421.

Ozkan, G., O. Sagdic, N.G. Baydar, and H. Baydar. 2004. Antioxidant and antibacterial activities of *Rosa damascena* flower extracts. *Food Sci. Technol. Int.* 10: 277–281.

Ozkan, G., O. Sagdic, R.S. Gokturk, O. Unal, and S. Albayrak. 2010. Study on chemical composition and biological activities of essential oil and extract from *Salvia pisidica*. *LWT—Food Sci. Technol.* 43: 186–190.

Pandit, V.A. and L.A. Shelef. 1994. Sensitivity of *Listeria monocytogenes* to rosemary (*Rosmarinus officinalis* L.). *Food Microbiol.* 11: 57–63.

Papineau, A.M., D.G. Hoover, D. Knorr, and D.F. Farkas. 1991. Antimicrobial effect of water soluble chitosans with high hydrostatic pressure. *Food Biotechnol.* 5: 45–57.

Park, S., S.D. Stan, M.A. Daeschei, and Y. Zhao. 2005. Antifungal coatings on fresh strawberries (*Fragaria* × *ananassa*) to control mold growth during cold storage. *J. Food Sci.* 70: 202–207.

Pelczar, M.J., E.C.S. Chan, and N.R. Krieg. 1993. *Microbiology Concepts and Applications*. International Edition, McGraw-Hill, Inc.: New York, p. 896.

Pillai, C.K.S., W. Paul, and C.P. Sharma. 2009. Chitin and chitosan polymers: Chemistry, solubility and fiber formation. *Prog. Polym. Sci.* 34: 641–678.

Polonelli, L. and G. Morace. 1986. Reevaluation of the yeast killer phenomenon. *J. Clin. Microbiol.* 24: 866–869.

Popper, L. and D. Knorr. 1997. Inactivation of yeast and filamentous fungi by the lactoperoxidase-hydrogen peroxide-thiocyanate-system. *Nahrung* 41: 29–33.

Pourmortazavi, S.M. and S.S. Hajimirsadeghi. 2007. Supercritical fluid extraction in plant essential and volatile oil analysis. *J. Chromatogr. A* 1163: 2–24.

Prabuseenivasan, S., M. Jayakumar, and S. Ignacimuthu. 2006. In vitro antibacterial activity of some plant essential oils. *BMC Complement. Altern. Med.* 6: 39–46.

Psani, M. and P. Kotzekidou. 2006. Technological characteristics of yeast strains and their potential as starter adjuncts in Greek-style black olive fermentation. *World J. Microbiol. Biotechnol.* 22: 1329–1336.

Rao, A., Y. Zhang, S. Muend, and R. Rao. 2010. Mechanism of antifungal activity of terpenoid phenols resembles calcium stress and inhibition of the TOR pathway. *Antimicrob. Agents Chemother.* 54: 5062–5069.

Reiter, B., V.M. Marshall, L. Björck, and C.G. Rosen. 1976. Nonspecific bactericidal activity of the lactoperoxidases-thiocyanate-hydrogen peroxide system of milk against *Escherichia coli* and some gram-negative pathogens. *Infect. Immun.* 13: 800–807.

Resaa, O.C.P., R.J. Jagus, and L.N. Gerschenson. 2014. Natamycin efficiency for controlling yeast growth in models systems and on cheese surfaces. *Food Control* 35: 101–108.

Rinaudo, M. 2006. Chitin and chitosan: Properties and applications. *Prog. Polym. Sci.* 31: 603–632.

Roller, S. and N. Covill. 2000. The antimicrobial properties of chitosan in mayonnaise and mayonnaise-based shrimp salads. *J. Food Prot.* 63: 202–209.

Rosa-Magri, M.M., S.M. Tauk-Tornisielo, and S.R. Ceccato-Antonini. 2011. Bioprospection of yeasts as bio-control agents against phytopathogenic molds. *Braz. Archiv. Biol. Technol.* 54: 1–5.

Ross, R.P., S. Morgan, and C. Hill. 2002. Preservation and fermentation: Past, present and future. *Int. J. Food Microbiol.* 79: 3–16.

Sagdic, O., A. Kuscu, M. Ozcan, and S. Ozcelik. 2002. Effects of Turkish spice extracts at various concentrations on the growth of *Escherichia coli* O157:H7. *Food Microbiol.* 19: 473–480.

Sagdic, O., G. Ozkan, A. Aksoy, and H. Yetim. 2009. Bioactivities of essential oil and extract of *Thymus argaeus*, Turkish endemic wild thyme. *J. Sci. Food Agric.* 89: 791–795.

Sapers, G.M., R.L. Miller, S.W. Choi, and P.H. Cooke. 1999. Structure and composition of mushrooms as affected by hydrogen peroxide wash. *J. Food Sci.* 64: 889–892.

Sapers, G.M. and G.F. Simmons. 1998. Hydrogen peroxide disinfection of minimally processed fruits and vegetables. *Food Technol.* 52: 48–52.

Schafer, H. and M. Wink. 2009. Medicinally important secondary metabolites in recombinant microorganisms or plants: Progress in alkaloid biosynthesis. *Biotechnol. J.* 4: 1684–1703.

Schaefer, L., T.A. Auchtung, K.E. Hermans, D. Whitehead, B. Borhan, and R.A. Britton. 2010. The antimicrobial compound reuterin (3-hydroxypropionaldehyde) induces oxidative stress via interaction with thiol groups. *Microbiology* 156: 1589–1599.

Schlegen, H.G. 1986. *General Microbiology*. Cambridge University Press: Cambridge, U.K., pp. 587s.

Sesti, F., T.M. Shih, N. Nkolaeva, and S.A.N. Goldstein. 2001. Immunity to K1 killer toxin: Internal TOK1 blockade. *Cell* 105: 637–644.

Schmitt, M.J. and F. Breinig. 2002. The viral killer system in yeast: From molecular biology to application. *FEMS Microbiol. Rev.* 26: 257–276.

Schmitt, M.J. and F. Breinig. 2006. Yeast viral killer toxins: Lethality and self-protection. *Nat. Rev. Microbiol.* 4: 212–221.

Schnürer, J. and J. Magnusson. 2005. Antifungal lactic acid bacteria as biopreservatives. *Trends Food Sci. Technol.* 16: 70–78.

Shadidi, F., J.K.V. Arachchi, and Y. Jeon. 1999. Food applications of chitin and chitosan. *Trends Food Sci. Technol.* 10: 37–51.

Shapero, M., D.A. Nelson, and T.P. Labuza. 1978. Ethanol inhibition of *Staphylococcus aureus* at limited water activity. *J. Food Sci.* 43: 1467–1469.

Shelef, L.A. 1984. Antimicrobial effects of spices. *J. Food Saf.* 6: 29–44.

Shelef, L.A., E.K. Jyothi, and M.A. Bulgarellii. 1984. Growth of enteropathogenic and spoilage bacteria in sage-containing broth and foods. *J. Food Sci.* 49: 737–740.

Sillankorva, S.M., H. Oliveira, and J. Azeredo. 2012. Bacteriophages and their role in food safety. *Int. J. Microbiol.* 2012: 1–13.

Silva, F., S. Ferreira, A. Duarte, D.I. Mendonça, and F.C. Domingues. 2011. Antifungal activity of *Coriandrum sativum* essential oil, its mode of action against *Candida* species and potential synergism with amphotericin B. *Phytomedicine* 19: 42–47.

Simmons, G.F., J.L. Similanic, S. John, and D.A. Margosan. 1997. Reduction of microbial populations on prunes by vapor-phase hydrogen peroxide. *J. Food Prot.* 60: 188–191.

Singh, N., R.K. Singh, A.K. Bhunia, and R.L. Stroshine. 2002. Efficacy of chlorine dioxide, ozone, and thyme essential oil or a sequential washing in killing *Escherichia coli* O157:H7 on lettuce and baby carrots. *LWT—Food Sci. Technol.* 35: 720–729.

Sjögren, J., J. Magnusson, A. Broberg, J. Schnürer, and L. Kene. 2003. Antifungal 3-hydroxy fatty acids from *Lactobacillus plantarum* MiLAB 14. *Appl. Environ. Microbiol.* 69: 7554–7557.

Skandamis, P., K. Koutsoumanis, K. Fasseas, and G.J.E. Nychas. 2001. Inhibition of oregano essential oil and EDTA on *Escherichia coli* O157: H7. *Ital. J. Food Sci.* 13: 65–75.

Smith, J. and L. Hong-Shum. 2011. *Food Additives Data Book*, 2nd edn. Wiley-Blackwell Publishing: Oxford, U.K.

Smith-Palmer, A., J. Stewart, and L. Fyfe. 2001. The potential application of plant essential oils as natural food preservatives in soft cheese. *Food Microbiol.* 18: 463–470.

Spricigo, D.A., C. Bardina, P. Cortes, and M. Llagostera. 2013. Use of a bacteriophage cocktail to control *Salmonella* in food and the food industry. *Int. J. Food Microbiol.* 165: 169–174.

Sorrells, K.M., D.C. Enigl, and J.R. Hatfield. 1989. The effect of pH, acidulant, time and temperature on the growth and survival of *Listeria monocytogenes*. *J. Food Prot.* 52: 571–573.

Stevens, K.A., B.W. Sheldon, N.A. Klapes, and T.R. Klaenhammer. 1991. Nisin treatment for inactivation of *Salmonella* species and other gram-negative bacteria. *Appl. Environ. Microbiol.* 57: 3613–3615.

Sturino, J.M. and T.R. Klaenhammer. 2004. Bacteriophage defense systems and strategies for lactic acid bacteria. *Adv. Appl. Microbiol.* 56: 331–378.

Sudarshan, N.R., D.G. Hoover, and D. Knorr. 1992. Antibacterial action of chitosan. *Food Biotechnol.* 6: 257–272.

Surburg, H. and J. Panten. 2006. Common fragrance and flavor materials—preparation, properties and uses, 5th edn. Wiley-VCH, Weinheim.

Svensson, U. and A. Christiansson. 1991. Methods for phage monitoring. *Bull. Int. Dairy Fed.* 263: 29–39.

Svoboda, K., J. Hampson, and T. Hunter. 1999. Secretory tissues: Storage and chemical variation of essential oils in secretory tissues of higher plants and their bioactivity. *Int. J. Aromather.* 9: 124–131.

Tajkarimi, M.M., S.A. Ibrahima, and D.O. Cliver. 2010. Antimicrobial herb and spice compounds in food. *Food Control* 21: 1199–1218.

Tassou, C.C., E.H. Drosinos, and G.J.E. Nychas. 1995. Effects of essential oil from mint (*Mentha piperita*) on *Salmonella enteritidis* and *Listeria monocytogenes* in model food systems at 4°C and 10°C. *J. Appl. Bacteriol.* 78: 593–600.

Tian, J., X.Q. Ban, H. Zeng, J.S. He, Y.X. Chen, and Y.W. Wang. 2012. The mechanism of antifungal action of essential oil from dill (*Anethum graveolens* L.) on *Aspergillus flavus*. *PLoS One* 7(1): e30147.

Toldra, F., Y. Sanz, and M. Lores. 2001. *Meat Fermentation Technology*. Marcel Dekker Incorporated: New York.

Tsigarida, E., P. Skandamis, and G.J.E. Nychas. 2000. Behaviour of *Listeria monocytogenes* and autochthonous flora on meat stored under aerobic, vacuum and modified atmosphere packaging conditions with or without the presence of oregano essential oil at 5°C. *J. Appl. Microbiol.* 89: 901–909.

Turner, M.S., F. Waldherr, M.J. Loessner, and P.M. Giffard. 2007. Antimicrobial activity of lysostaphin and a *Listeria monocytogenes* bacteriophage endolysin produced and secreted by lactic acid bacteria. *Syst. Appl. Microbiol.* 30: 58–67.

van de Braak, S.A.A.J. and G.C.J.J. Leijten. 1999. *Essential Oils and Oleoresins: A Survey in the Netherlands and Other Major Markets in the European Union*. CBI, Centre for the Promotion of Imports from Developing Countries: Rotterdam, the Netherlands, p. 116.

Vero, S., P. Mondino, J. Burgueno, M. Soubes, and M. Wisniewski. 2002. Characterization of biocontrol activity of two yeast strains from Uruguay against blue mold of apple. *Postharvest Biol. Technol.* 26: 91–98.

Volk, W.A. and M.F. Wheeler. 1973. *Basic Microbiology*. Philadelphia: JB Lippincott.

Waema, S., S. Maneesri, and P. Masniyom. 2009. Isolation and identification of killer yeast from fermented vegetables. *Asian J. Food Agro-Indus.* 2: 126–134.

Walker, K. 2006. Use of bacteriophages as novel food additives. Available at: http://www.iflr.msu.edu/uploads/files/109/Student%20Papers/USE_OF_BACTERIOPHAGES_AS_NOVEL_FOOD_ADDITIVES.pdf. Accessed in May 2013.

Wink, M. 1999. *Functions of Plant Secondary Metabolites and Their Exploitation in Biotechnology*. Sheffield Academic Press: Sheffield, U.K., p. 362.

Yang, Z. 2000. *Antimicrobial Compounds and Extracellular Polysaccharides Produced by Lactic Acid Bacteria: Structures and Properties*. University of Helsinki Department of Food Technology: Helsinki, Finland, p. 61s.

Yoo, W. and A.G. Rand. 2006. Antibacterial effect of glucose oxidase on growth of *Pseudomonas fragi* as related to pH. *J. Food Sci.* 60: 868–871.

Young, D.H., H. Kohle, and H. Kauss. 1982. Effect of chitosan on membrane permeability of suspension cultured Glycinemax and *Phaseolus vulgaris* cells. *Plant Physiol.* 70: 1449–1454.

Zacharof, M.P. and R.W. Lovitt. 2012. Bacteriocins produced by lactic acid bacteria a review article. *APCBEE Procedia* 2: 50–56.

Zhang, H., H. Bao, C. Billington, J.A. Hudson, and R. Wang. 2012. Isolation and lytic activity of the *Listeria* bacteriophage endolysin LysZ5 against *Listeria monocytogenes* in soya milk. *Food Microbiol.* 31: 133–136.

Zhang, H.Y., L. Wang, Y. Dong, S. Jiang, J. Cao, and R.J. Meng. 2007. Postharvest biological control of gray mold decay of strawberry with *Rhodotorula glutinis*. *Biol. Control.* 40: 287–292.

Section IV

Modeling Microbial Growth in Food

25

Estimating the Shelf Life of Minimally Processed Foods: An Approach Integrating Process Engineering and Growth Kinetics Models

Paulo Ricardo Santos da Silva

CONTENTS

25.1 Introduction

In recent years, the behavior of the consumer has been modified in relation to food consumption. In search of better quality of life and healthy meals, the modern consumer prefers natural food, without additives and preservatives. Simultaneously, this consumer wishes quickness and convenience in the preparation of the food. Minimally processed and refrigerated fruits and vegetables are a foodstuff that meets the needs of this group of consumers.

In order to meet the consumer needs, the decrease in the quantity of preservatives or their elimination from food makes the food products more susceptible to deterioration caused by microorganisms. The fruits and related products have low pH, which inhibits the action of bacteria. Therefore, their deterioration is promoted by fungi (Jay 2005). There are many different species of filamentous fungi. Those that produce heat-resistant spores are especially important to the food industry because they have the ability to resist heat treatment typically applied by this industrial sector. In this group of microorganisms, the species deserving mention are *Byssochlamys fulva*, *Byssochlamys nivea*, *Neosartorya fischeri*, and *Talaromyces flavus* (Kotzekidou 1997).

The technological processes involved in the production of minimally processed fruits (peeling, slicing, and intensive handling) provoke the release of nutrients and intracellular enzymes, accelerating the degradation process. Therefore, the shelf life of these products is shorter than that of the raw materials. In this context, the greatest challenge of this sector of the food industry is to understand the deterioration process of minimally processed fruits, especially those promoted by microorganisms, controlling them, and thus extending their shelf life.

The action of the microorganisms in foods can be minimized by the proper adjustment of operating variables during the processing and the storage of the product. In this regard, predictive microbiology emerges as a useful tool to establish an equation to simulate different operational conditions and quickly evaluate their effects on microbial growth.

During food processing, environmental conditions can vary as a function of time or space in the product. The models that describe the heat and mass transfer phenomena are able to estimate the values of the environmental variables (temperature, pH, water activity) in the food. In turn, the equation used in predictive microbiology allows predicting the growth behavior of the microorganism in such conditions. This demonstrates the potential benefit of integrating the knowledge of predictive microbiology with process engineering models. A better evaluation of industrial operations in microbiological aspects is possible with this approach. In such case, it is even possible to establish the conditions that inhibit the growth of bacteria and molds. Thus, the shelf life of minimally processed fruits can be extended. However, this approach is incipient in the studies found in the literature, especially in evaluating the behavior of fungi in foods. The lack of integrating equations of predictive microbiology with heat and mass transfer models was cited in the works of Mafart (2005), Lebert and Lebert (2006), and Gougouli and Koutsoumanis (2010).

The goal of this chapter is therefore to demonstrate how the knowledge of integrating heat transfer models with predictive microbiology can be used to extend the shelf life of minimally processed fruits. Initially, the predictive microbiology models are reviewed. The fundamentals of heat transfer are also described. Then the major studies found in the literature in this field of knowledge are presented. At the end of the chapter, a case study is presented and discussed. In this case study, the effects of the operating variables of the cooling process and of product's size on the shelf life of minimally processed papaya pulp contaminated with spores of heat-resistant fungi are evaluated.

25.2 Predictive Microbiology and Growth Kinetics of Microorganisms

Predictive food microbiology involves the study of the microbial growth responses to environmental factors such as temperature, water activity, pH, and gaseous atmosphere in terms of kinetic and probabilistic models. The kinetic models concern the rate of growth of microorganisms. The probabilistic models predict the likelihood of some event, such as a spore germinating or a detectable amount of toxin being formed, within a given period of time (Ross and McMeekin 1994). This work is focused on the kinetic models.

From the point of view of predictive microbiology, the shelf life of a food can be evaluated as a reliable model describing satisfactorily the growth curve of the target microorganism is available. For filamentous fungus, this growth curve is obtained by plotting the colony radius versus time, as shown in Figure 25.1. The growth curve consists of three phases: the lag phase, the growth phase, and the stationary phase. Sometimes, the stationary phase is not observed in the growth curve depending on the

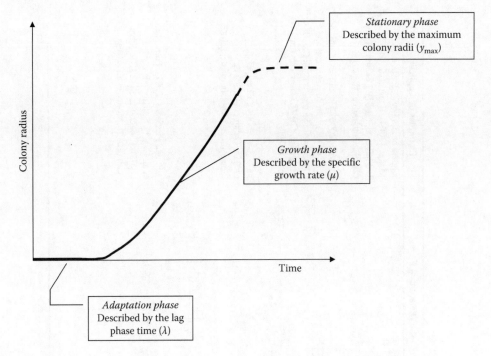

Colony radius

Time

Stationary phase
Described by the maximum
colony radii (y_{max})

Growth phase
Described by the specific
growth rate (μ)

Adaptation phase
Described by the lag
phase time (λ)

FIGURE 25.1 Typical growth curve for fungi in which the size of the colony is plotted over time.

growth conditions in which the fungus is submitted. For this reason, this phase of the growth curve is shown as a dashed line.

According to Whiting and Buchanan (1993), the kinetic microbial models can be divided into two categories: primary predictive models and secondary predictive models.

25.2.1 Primary Predictive Models

Primary predictive growth models describe the increase in the microbial population over time under fixed environmental conditions. They are characterized by two parameters: the specific growth rate (μ) and the lag phase duration (λ).

Table 25.1 shows the main primary predictive models found in the literature. Some studies in which each model was applied to describe the growth of molds are also listed in this table. In the Table, each model is presented in the integrated form and in the differential form. The former is used at isothermal conditions; the latter, at nonisothermal conditions.

25.2.1.1 Linear Model

The linear model (Equation 25.1) is the simplest model. It is very useful to describe the growth of filamentous fungus. This model establishes a linear relationship between the size of the colony's fungi and time after the lag phase. Thus, the instantaneous growth rate (dy/dt) is a constant over the entire period of the growth phase. The parameters of this model can be obtained by simple linear regression.

25.2.1.2 Logistic Model

The logistic model (Equation 25.2) assumes that the instantaneous radial growth rate (dy/dt) is proportional to the colony's radius in that moment. However, the growth of the colony is not unlimited. According to this model, the depletion of the available nutrients and the production of inhibiting metabolites by the microorganisms limit the size of the colony to a maximum value y_{max} (Corradini and Peleg 2007).

TABLE 25.1

Main Primary Predictive Models Found in the Literature to Describe the Growth Behavior of Fungi

Model	Equation in the Integrated Form	Equation in the Differential Form	Examples of Application
Linear model	$y = y_0 + \mu(t - \lambda)$	$\dfrac{dy}{dt} = \begin{cases} 0 & (t \le \lambda) \\ \mu & (t > \lambda) \end{cases}$ With $y(0) = y_0$ (25.1)	Cuppers et al. (1997), Parra and Magan (2004), Baert et al. (2007), Lahlali et al. (2007), Lee and Magan (2010), Gougouli et al. (2011)
Logistic model	$y = y_0 + \dfrac{y_{max}}{1 + \exp\left[\left(4\mu_{max}/y_{max}\right)(\lambda - t) + 2\right]}$	$\dfrac{dy}{dt} = \dfrac{4\mu}{y_{max}}\left[1 - \dfrac{y}{y_{max}}\right] y$ With $y(0) = y_0 + \dfrac{y_{max}}{1 + \exp\left[\left(4\mu/y_{max}\right)\lambda + 2\right]}$ (25.2)	Liu et al. (2003), Hamidi-Esfahani et al. (2004), van de Lagemaat and Pyle (2005), Sant'Ana et al. (2010a), Feng et al. (2010), Zimmermann et al. (2011)
Modified Gompertz model	$y = y_0 + y_{max} \exp\left\{-\exp\left[\dfrac{\mu_{max} \exp(1)}{y_{max}}(\lambda - t) + 1\right]\right\}$	$\dfrac{dy}{dt} = \mu \cdot \exp(1) \cdot y \cdot \ln\left(\dfrac{y_{max}}{y}\right)$ With $y(0) = y_0 + y_{max} \cdot \exp\left\{-\exp\left[-\dfrac{\mu \cdot \exp(1)}{y_{max}} \cdot \lambda + 1\right]\right\}$ (25.3)	Pardo et al. (2004), Cheng et al. (2010), Galati et al. (2011), Gougouli and Koutsoumanis, (2012)
Simplified Baranyi's model	$y = y_0 + \mu\left\{t + \dfrac{1}{\mu}\ln\left[\exp(-\mu t) + \exp(-\mu\lambda) - \exp(-\mu t - \mu\lambda)\right]\right\}$	$\dfrac{dy}{dt} = \mu\left[\dfrac{1}{1 + \exp(-Q)}\right]\left[1 - \exp(y - y_{max})\right]$ $\dfrac{dQ}{dt} = \mu$ With $y(0) = y_0$ $Q(0) = Q_0$ (25.4)	Valík and Piecková (2001), Patriarca et al. (2001), Samapundo et al. (2005), Panagou et al. (2010), Silva et al. (2010), Mousa et al. (2011)

Source: Longhi, D.A. et al., *J. Theor. Biol.*, 335, 88, 2013.

25.2.1.3 Modified Gompertz Model

The modified Gompertz model is based on the hypothesis that the instantaneous growth rate of the microorganism (dN/dt) is a logarithmic function of the population size (N) in that moment. In the integrated form, the equation consists of a double exponential function, which describes an asymmetric sigmoid curve (Baranyi and Roberts 1995, Nakashima et al. 2000, Lebert and Lebert 2006). The modified Gompertz model was adapted to describe the growth kinetics of fungi. So, the size of the population (N) in the original equation was replaced by the size of the colony (y), assuming the final form given by Equation 25.3.

25.2.1.4 Baranyi's Model

Baranyi's model was developed from a new approach in which the growth of the microorganisms is equated using biological principles. This model assumes that the instantaneous population growth rate (dN/dt) is proportional to the size of the population (N) in that moment. In addition, the instantaneous growth rate depends on the initial physiological state of the cells in the inoculation moment. On the other hand, the physiological state is a function of a critical substance existing inside the cells (enzyme, specific RNA, or DNA). The formation kinetics of this substance is described by the Michaelis–Menten model. The concentration of this substance inside the cells is considered a bottleneck in the beginning of the growth (Baranyi and Roberts 1995, Baranyi et al. 1995, Bellara et al. 2000, van Impe et al. 2005). In the differential form, Baranyi's model is constituted by two equations (Equation 25.4). The first differential equation describes the time evolution of the microbial population. The second differential equation takes the time evolution of the physiological state of the microorganism into account (Lebert and Lebert 2006).

According to Garcia et al. (2009), the original Baranyi's model can be simplified in order to describe the growth behavior of fungi. They pointed out that the mycelium is constantly growing in the peripheral region of the colony in which the culture media is always new and unexploited by the microorganism. Therefore, there is no limitation to the growth of the fungi in this region. So, in this case, the stationary phase is not achieved and the related term in the original Baranyi's equation can be eliminated. The final version of the integrated Baranyi's model for describing the fungi growth is given by Equation 25.4.

25.2.2 Secondary Predictive Models

Secondary predictive models describe the influence of environmental factors (temperature, pH, water activity) on the parameters of the primary predictive models, that is, specific growth rate (μ) and lag phase duration (λ). Several secondary models are found in the literature. Table 25.2 shows the most common secondary models.

25.2.2.1 Square Root Model

The Belehrádek model or square root model (Equation 25.5) was developed originally by Ohta and Hirahara. They observed that the square root of the rate of degradation of nucleotides of fish meat was proportional to the temperature. Ratkwosky and colleagues have introduced this model in studies of microbiology (Ross and McMeekin 1994). This equation is particularly useful in studies of the growth behavior of bacteria in refrigerated products (Lebert and Lebert 2006). According to Ross and McMeekin (1994), extended versions of the square root model (Equation 25.6) are available in the literature to describe the growth of microorganism at high temperatures.

25.2.2.2 Arrhenius Model

The Arrhenius model is generally used in the study of the kinetics of the chemical reaction that describes the effect of temperature on the specific reaction rate. In the field of predictive microbiology, this model is used in the linear form (Equation 25.7). Davey (1989) proposed an extended version

TABLE 25.2

Main Secondary Predictive Models Found in the Literature to Describe the Effect of the Environmental Variables on the Growth Rate (μ) and Lag Phase Time (λ)

Model	Equation	
Belehrádek model and extended	$\sqrt{\mu} = b\left(T - T_{min}\right)$	(25.5)
square root model	$\sqrt{\mu} = b\left(T - T_{min}\right)\left\{1 - \exp\left[c\left(T - T_{max}\right)\right]\right\}$	(25.6)
Arrhenius model	$\ln\left(\mu\right) = \ln\left(A_A\right) - \dfrac{E_A}{RT}$	(25.7)
Arrhenius–Davey model	$\ln\left(\mu\right) = C_0 + \dfrac{C_1}{T} + \dfrac{C_2}{T^2}$	(25.8)
Gibson's model	$\ln\left(\mu\right) = C_0 + C_1\sqrt{\left(1 - A_W\right)} + C_2\left(1 - A_W\right)$	(25.9)
Polynomial models	$z = a_0 + \displaystyle\sum_{i=1}^{n} a_i X_i + \sum_{i=1}^{n}\sum_{j=1}^{m} a_{i,j} X_i X_j$	(25.10)
Hyperbolic model	$\ln\left(\lambda\right) = \dfrac{p}{T - q}$	(25.11)

of this model—the Arrhenius–Davey model (Equation 25.8)—in order to evaluate the influence of the incubation temperature on the growth rate of bacteria (Samapundo et al. 2005). The Arrhenius–Davey model is also applied to evaluate the effect of others environmental factors (pH, water activity) other than temperature on the growth kinetics of microorganisms (Panagou et al. 2003).

25.2.2.3 Gibson Model

The Gibson model deserves special attention because it was the first equation developed specifically for fungi from the study of the growth behavior of these microorganisms (Garcia et al. 2009). Gibson et al. (1994) studied the effect of water activity on the growth rate of four species of *Aspergillus*: *Aspergillus flavus*, *Aspergillus nomius*, *Aspergillus oryzae*, and *Aspergillus parasiticus*. They noted a square root–type relation between the logarithm of the growth rate and the water activity. Therefore, they proposed this model (Equation 25.9).

25.2.2.4 Polynomial Models

Polynomial models (Equation 25.10) are purely empirical equations. These models are built from multiple linear regressions in which the influence of each environmental factor (temperature, pH, water activity) is correlated with the growth rate or lag phase time (Ross and McMeekin 1994, Garcia et al. 2009). Although these models are much criticized in the literature because of their lack of a biologic basis, a review conducted by Garcia et al. (2009) revealed that 62% of the analyzed studies have used this model.

25.2.2.5 Hyperbolic Model

The models described earlier are mainly applied to fit the effect of environmental factors on the growth rate, although they can also be used for modeling the lag phase time. The hyperbolic model (Equation 25.11) was developed by Zwietering and colleagues specifically to describe the influence of temperature on the lag phase time (van Impe et al. 1992).

As mentioned previously, in addition to the primary and secondary models listed earlier, there are many other equations in the literature. For more detailed analysis, readers can consult the works of Ross and McMeekin (1994), Rosso et al. (1995), van Impe et al. (2005), and Lebert and Lebert (2006) among other references.

25.3 Heat Transfer Models

The heat transfer phenomenon plays a central role in several operations in the food industry such as drying, evaporation, pasteurization, cooling, and freezing. The heating of the foods is performed with different purposes such as to reduce the microbial population in the food, to inactivate enzymes, to reduce product moisture, and to modify the functionality of certain components. The cooling of the foods aims to avoid or reduce deteriorative chemical and enzymatic reactions that take place in a food during its storage and to inhibit microbial growth (Welti-Chanes et al. 2005).

The heat is transferred in a material when there are gradients of temperature inside it or temperature differences between the material and the surroundings. Generally, heat transfer is governed by the three well-known mechanisms of transport—conduction, convection, and radiation. Radiation is important in high-temperature applications such as baking and grilling, which are not commonly integrated with predictive microbiology. For this reason, in this work, only the principles of the conduction and the convection mechanisms will be addressed, as shown in Figure 25.2. If the reader is interested in more details about heat transfer, the author suggests referring to the work of Incropera and De Witt (1998) and Çengel (1998).

25.3.1 Conduction Heat Transfer

The conduction of heat is a mechanism of heat transfer that takes place in solid and stagnant liquids. It is established as there are different energetic states among the atoms and molecules of the material. The energy of the vibration of the material particles in the hot regions is greater than in the cold regions. So, the energy is transferred from particle to particle, producing heat transfer. The steady-state conduction heat transfer is governed by Fourier's law (Equation 25.12), which states that the heat flux in the x direction (q_x'') is directly proportional to the temperature gradient in that direction in the material:

$$q_x'' = -k \frac{\partial T}{\partial x} \tag{25.12}$$

In this Equation, k is the thermal conductivity of the material.

Many processes in food technology are transient in nature. Therefore, the steady-state conduction heat transfer is limited. In these cases, the main goal of the heat transfer analysis is to determine the

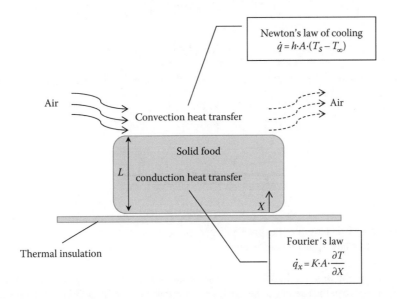

FIGURE 25.2 Heat transfer mechanisms within a solid food and between it and the air of the surroundings.

temperature distribution inside the food over time. The solution of heat diffusion equation allows achieving this goal. Assuming that (1) no heat is generated inside the food, (2) the product has constant thermophysical properties, and (3) the food is slab shaped and so the heat transfer takes place only in one direction (*x* direction, for example), the one-dimensional unsteady-state heat diffusion equation simplifies to the following equation:

$$\frac{\partial T}{\partial t} = \alpha \left(\frac{\partial^2 T}{\partial x^2} \right)$$

(25.13)

In Equation 25.13, α is the thermal diffusivity of the material.

The analytical solutions to Equation 25.13 are available for very limited boundary conditions. Considering the food mechanism shown in Figure 25.2 and assuming the initial condition and the boundary conditions applicable to the system, we have the following:

- At $t = 0$, $T = T_0$
- At $x = 0$, $\partial T/\partial x = 0$
- At $x = L$, $h(T_S - T_\infty) = -k \partial T/\partial x$

In this case, the transient temperature distribution within the food is expressed by Equation 25.14 (Incropera and De Witt 1998):

$$\frac{T - T_\infty}{T_0 - T_\infty} = \sum_{n=1}^{\infty} C_n \exp\left(-\xi_n^2 Fo\right) \cos\left(\xi_n x^*\right)$$

(25.14)

In this equation,

T_0 is the initial temperature of the product

T_∞ is the temperature of the surroundings

The variable x^* is the dimensionless position within the product (*x/L*)

The parameters C_n are given by Equation 25.15

ξ_n are the positive roots of the transcendental Equation 25.16

$$C_n = \frac{4 sen \xi_n}{2\xi_n + sen\left(2\xi_n\right)}$$

(25.15)

$$\xi_n tg\left(\xi_n\right) = Bi$$

(25.16)

In Equations 25.14 and 25.16, the parameters *Fo* and *Bi* are the Fourier number and the Biot number, respectively, defined as follows:

$$Fo \equiv \frac{\alpha t}{L_C^2}$$

(25.17)

$$Bi \equiv \frac{hL_C}{k}$$

(25.18)

where

L_C is the characteristic length of the product's geometry

h is the convective heat transfer coefficient

Note that to evaluate the Biot number, it is necessary to know aspects related to the convection heat transfer phenomenon between the surface of the food and the heating or cooling fluid. This issue is addressed in the following section.

25.3.2 Convection Heat Transfer

Convection is the mechanism of heat transfer that takes place between the surface of a solid (or a stagnant liquid) at a temperature T_S and a moving fluid at a temperature T_∞ for $T_S \neq T_\infty$. The convection heat transfer comprises the combined effect of conduction and advection. Conduction is the mechanism of heat transfer established between the solid surface and the layer of the fluid immediately adjacent to the surface. Advection is a result of the fluid's bulk motion allowing the contact between the fractions of hot fluid and cold fluid randomly, thus increasing the thermal exchange.

According to the driving force for moving fluid, there is the forced convection and the free convection. In the forced convection heat transfer, the fluid flow is promoted by external forces such as pumps or blowers. In the free convection, the motion of the fluid is induced by the density difference in the gas or liquid caused by temperature difference.

The convective heat flux from or to the surface of a material can be calculated using Newton's law of cooling (Equation 25.19):

$$q'' = h\left(T_S - T_\infty\right) \tag{25.19}$$

The fundamental parameter in Equation 25.19 is the convective heat transfer coefficient (h), which is a function of the geometry of the product, the flow regime (laminar or turbulent flow), the velocity of the fluid, and the transport properties of the fluid (thermal diffusivity, kinematic viscosity). The value of the forced convective heat transfer coefficient can be estimated experimentally or using empirical correlations of the following form:

$$Nu = \frac{hL_C}{k} = cRe^m Pr^n \tag{25.20}$$

where
  Nu is the dimensionless number of Nusselt
  Re is the dimensionless number of Reynolds
  Pr is the dimensionless number of Prandtl

The Reynolds and Prandtl numbers are defined as follows:

$$Re \equiv \frac{U_\infty L_C}{v} \tag{25.21}$$

$$Pr \equiv \frac{v}{\alpha} \tag{25.22}$$

For a fluid flowing parallel to the slab surface in a laminar regime of flow, the values of C, m, and n constants in Equation 25.20 are 0.664, 1/2, and 1/3, respectively. It is important to note that the aforementioned values of the constants are valid when $Re_L \leq 10^5$ and $Pr \geq 0.6$ (Incropera and De Witt 1998). For other geometries or turbulent regime of flow, the constants in Equation 25.20 are given in any textbook of heat transfer.

Due to the conduction and the convection heat transfer phenomena described earlier during food processing, there are spatial variations of temperature within the product over time. Temperature is the main environmental factor affecting the growth of microorganisms. Therefore, the integration between heat transfer knowledge and predictive microbiology seems to be essential to better understand the food operations in the industry. This issue will be addressed in the following section.

25.4 Integration between Predictive Microbiology and Heat Transfer Models

This section presents examples of studies in which predictive microbiology knowledge was integrated with heat transfer models for a better assessment of the operations in the food industry. This approach allows simulating several operational scenarios, identifying the conditions in which the microbial growth is inhibited, and thus extending the shelf life of food products, in microbiological terms. However, this subject is relatively new in the literature. Hence, there are few studies about this issue, mainly related to the growth of fungi in foods. The traditional approach comprises the integration between microbial kinetics and heat transfer models to study the inactivation process of bacteria and fungi spores, such as in the work of Gil et al. (2006), Mackey et al. (2006), and Valdramidis et al. (2006). This chapter will focus on modeling integration to study the cooling processes.

25.4.1 Growth of *Escherichia coli* in Cooling Process

Bellara et al. (2000) combined bacterial growth modeling with a heat transfer model describing the spatial temperature changes within a cylindrical solid object filled with all-purpose tween (APT) agar to check the feasibility of the integration between engineering and microbial models. The growth of *E. coli* W3110 over time was measured indirectly by optical density. Baranyi's model was the primary predictive model used to describe bacterial growth. The effect of the temperature on the specific growth rate was modeled by the square root model of Ratkowsky. The temperature profile in the vessel was evaluated by the numerical solution of the unsteady-state heat conduction equation in cylindrical coordinates. The integrated model was validated and an excellent agreement was found between experimental data and simulation results.

Using the integrated model, some simulations were performed to describe the growth of *E. coli* in the vessel in different cooling regimes varying the cooling time and temperature. A slow decrease in temperature was observed in the center of the vessel and so the bacterial growth at this point was faster than in the peripheral regions. Bellara et al. (2000) also studied the effect of reducing vessel radius on bacterial growth, keeping all other dimensions constant. The results showed that bacterial growth is extremely sensitive to vessel radius. The authors thus concluded that the model could play an important part in the design of food packaging.

Ben Yaghlene et al. (2009) conducted similar studies with this microorganism growing in an infinite slab. They included a convective boundary condition in the model. The effects of the convective heat transfer coefficient, the thickness of the product, and the chilling temperatures were investigated. They concluded that the bacterial growth kinetics were faster at the center of the product than at its surface. Moreover, the importance of the value of the convective heat transfer coefficient and of the product thickness on bacterial growth during cooling was also pointed out.

25.4.2 Growth of *Clostridium perfringens* during Failure in Cooling Process

Amézquita et al. (2005) combined the growth predictive models with transient heat transfer equations to analyze the effect of failures during the cooling of cooked boneless cured ham on the growth of *C. perfringens*. The growth predictive model was constituted by the primary model of Baranyi and the secondary model of the square root of Ratkowsky to describe bacterial growth in dynamic conditions. The heat transfer model assumed an ellipsoidal shape for the product, a variable initial temperature distribution, and the combination of convective, radiative, and evaporative boundary conditions. The model also considered that the boneless cooked ham is isotropic and heterogeneous with respect to the thermal properties, that is, temperature and spatial dependency.

The integrated model was solved using the finite element method. The mathematical model was validated with experimental data obtained in actual processing conditions considering three cooling scenarios. In the first case, the standard procedure established by the Food Safety and Inspection Service (FSIS) of the US Department of Agriculture with respect to the temperature of cooling air was applied. In the second case, a hypothetical deviation from FSIS cooling guidelines caused by a failure in the refrigeration equipment for a total downtime of 1 h was assumed in the simulation. In the third case, the total downtime was extended to 6 h.

The validation procedures have shown good agreement between predicted and observed microorganism growth in each cooling scenarios. For the first and second cases, a fail-safe condition was observed, that is, the model overestimated the actual value of bacterial population. In the third case, the model underestimated the observations and so a fail-dangerous side was observed. Amézquita et al. (2005) concluded that the integration between engineering and microbial modeling offers a useful quantitative tool to support several food safety management strategies such as hazard analysis and critical control points (HACCP) and microbiological risk assessment.

A similar study was conducted by Cepeda et al. (2013), focusing on the heat transfer phenomenon. However, in this case, the cooling process of a three-dimensional and irregular-shaped ready-to-eat product was analyzed. Again, the heat transfer model considered conduction as the governing equation, subject to convection, radiation, and moisture evaporation boundary conditions. The effects of the air temperature and the initial product temperature on the heat transfer were discussed in the work. The influence of heat loss due to evaporation on products with casings and without casings was also presented. The heat transfer model was validated in four commercial facilities. The results predicted by the heat model were in good agreement with the observed temperature values. The predictive model developed by Amézquita et al. (2005) was integrated with this heat transfer model in order to evaluate the safety of processed meat products in the meat industry. Using the integrated model, Cepeda et al. (2013) identified conditions in which the product would likely be unsuitable for distribution and consumption. The authors have pointed out that the integration between microbial and heat transfer models can help evaluate which cooling deviations have the potential to cause food safety risk.

25.4.3 Integration Modeling to Study the Growth Behavior of Fungi

The works described in the previous sections showed the integration between predictive microbiology and process engineering models to study the growth behavior of bacteria. The studies with fungi are recent and scarce. The works of Gougouli and Koutsoumanis (2010), Gougouli et al. (2011), Gougouli and Koutsoumanis (2012), and Silva et al. (2013) are examples of the application of this approach for the study of fungi's growth.

Gougouli and Koutsoumanis (2010) studied the growth of filamentous fungi *Aspergillus niger* and *Penicillium expansum* at isothermal and fluctuating temperature conditions. They used malt extract agar as their growth medium with pH adjusted to 4.2, simulating the pH of yogurt. The linear model was chosen as the primary predictive model to describe the increase in the diameter of the colony over time. The influence of temperature on the growth parameters of this primary model was expressed by the Cardinal Model with Inflection (CMI) developed by Rosso et al. (1993). The developed model was validated against the observed fungal growth at fluctuating temperature scenarios including single temperature shifts before and after the end of the lag phase time and at continuous periodic temperature fluctuations. The model provided accurate predictions at fluctuating temperature conditions when temperature shifts were within the growth phase. For storage conditions in which temperatures shift close to or outside the growth regions, the prediction errors increased. However, the predictions were on the fail-safe side. Gougouli and Koutsoumanis (2010) concluded that the model could be used to evaluate the shelf life of yogurt or in the assessment of fungal spoilage risk and optimization of food quality.

Similar studies were also conducted by Gougouli et al. (2011) and Gougouli and Koutsoumanis (2012). In the first work, Gougouli et al. (2011) developed predictive models to investigate the effect of the storage temperature and the inoculums size on the mycelium growth kinetics of 12 fungal species isolated from a yogurt production environment (*Penicillium expansum, Penicillium commune, Penicillium chrysogenum, Penicillium corylophilum, Penicillium spinulosum, Penicillium purpurogenum, Aspergillus niger, Aspergillus flavus, Fusarium oxysporum, Mucor circineloides, Rhizopus oryzae*, and *Cladosporium cladosporioides*). The linear model and the CMI were chosen as the primary and secondary models, respectively, to describe the growth of fungi. The predictive model was used to estimate the shelf life of yogurt at various storage temperatures. The authors proposed that the developed model could be used as a basis for the selection of appropriate conditions for challenge tests to detect the fungi's presence in yogurt production, which will lead to effective decision making. The model validation step has shown good performance of the model in predicting the shelf life of yogurt. The predictions were close to or slightly higher than the experimental data and therefore on the fail-safe side regarding the application of the model for the challenge test design.

In the second work, Gougouli and Koutsoumanis (2012) studied the germination process of the spores of *P. expansum* and *A. niger* at constant and dynamic temperature conditions. In this work, the modified Gompertz equation was selected as the primary model to describe the percentage of germinated spores over time for both fungi tested. The effects of the temperature on the parameters of the Gompertz model were modeled using the CMI. Faster germination for *P. expansum* was observed at temperatures between 20°C and 27.5°C. For *A. niger*, optimal germination was observed at 35°C. At dynamic temperature conditions, Gougouli and Koutsoumanis (2012) observed that the germination of spores was strongly affected by the type of temperature shift as with a greater shift (about 25°C), a significant overprediction of the model was observed after the shift. Moreover, the percentage of germinated spores at the time of the shift and the fungal species also influenced the germination process. They concluded that the model helped understand the germination behavior of the fungi and it could improve the control of fungi in foods.

As noted from the studies described earlier, the temperatures considered in the microbial models were not evaluated from heat transfer phenomena. To overcome this gap, Silva et al. (2013) combined predictive models with heat transfer equations to investigate the effect of the variations in refrigeration air velocity (from 0.5 to 5 m/s) and the initial product temperature (from 16°C to 28°C) on the time taken for stored papaya pulp to be spoiled by *B. fulva*. The primary predictive model chosen to evaluate the increase of colony radius over time was the modified Gompertz equation. The extended square root equation was used as the secondary model to describe the effect of temperature on the radial growth rate while the influence of temperature on the duration of the lag phase time was better described by the hyperbolic model. Using the integrated model, Silva et al. (2013) observed that the shelf life of papaya pulp was sensitive to changes in initial product temperature and refrigeration air stream velocity, mainly at high temperatures and low air velocity. The detection time of *B. fulva* growing in papaya pulp varied from 24 to 87 h. Depending on storage conditions, the simulations showed that reduction in shelf life reaches 76%, compared to the storage at the reference temperature (14°C).

The aforementioned works are examples of how the integration between predictive models and heat transfer equations can be used to better understand the effects of processing and storage conditions on food quality and safety. In the following section, a case study will be presented in which this issue is addressed.

25.5 Case Study: Growth of *B. fulva* in Papaya Pulp

In this case study, the shelf life of chilled papaya pulp contaminated with spores of *B. fulva* is evaluated using mathematical models. The analysis includes the evaluation of the effect of the product's size (length and thickness of the slice) and of the conditions of the air from the cooling system (temperature and velocity) on the time taken by the microorganism to spoil the food. It is an extension of the work of Silva et al. (2013) previously described. The results of this work are restricted to the growth behavior of the strain NRRL 1125 of *B. fulva* Olliver and Smith teleomorph used in the experiments for which the mathematical models were built. Different strains of the same microorganism can lead to distinct results.

25.5.1 Fruit and Microorganism

Papaya is one of the most widely cultivated fruits in the world, mainly in tropical and subtropical regions. According to the Food and Agriculture Organization (FAO 2013), India and Brazil are the largest producers of this fruit in the world. This fruit has a high nutritive value. It is low in calories (32 kcal/100 g) and rich in vitamins and minerals such as vitamin A, vitamin C, riboflavin, folate, calcium, thiamine, iron, niacin, potassium, and fibers (Santana et al. 2003, Rocha et al. 2005, Krishna et al. 2008). The pH of this fruit lies in the range of 4.5–6.0 due to its low content of organic acids. The low acidity of papaya gives it a nutritional advantage because it allows its consumption by people sensitive to acidic fruits (Santana et al. 2003, Lima et al. 2009). On the other hand, this poses a problem for papaya processors as low acidity promotes enzymatic activity and the growth of microorganisms such as fungi.

There are several species of fungi relevant in the fruit-processing industry. *Byssochlamys fulva* deserves special attention as (1) it is widely distributed in the environment, especially in the soil (Tournas 1994); (2) it is a typical spoilage microorganism for fruit products, even when the number of spores in the contaminated fruit is low (Sant'Ana et al. 2010a); (3) this microorganism is able to produce ascospores that can resist the heat treatments normally applied to hot-packed canned products in the food industry (Kotzekidou 1997, Ferreira et al. 2011); (4) this fungus is able to grow under low oxygen level or low pH environments (Chapman et al. 2007, Panagou et al. 2010, Taniwaki et al. 2010); (5) it is capable of producing high concentrations of mycotoxins in food even greater than the upper limits defined as free from health risks to the consumers (Sant'Ana et al. 2010b); and (6) its colonies have a high radial growth rate, which can reach 20 mm/day (Valík and Piecková 2001).

25.5.2 Physical System Studied and the Mathematical Model

Figure 25.3 illustrates the physical system that was modeled mathematically and simulated in this case study. It consists of a flat slab-shaped slice of papaya on a polystyrene tray covered by a sheet of polyethylene film in the upper surface along which a convective boundary condition is established. The initial temperature of the product is T_0. Its thickness and length are L_A and L_B, respectively. The temperature and velocity of the air in the convective boundary condition are T_∞ and U_∞, respectively.

25.5.2.1 Heat Transfer Model

The temperature within and on the surface of the product over time can be obtained by the solution of the diffusion heat transfer Equation 25.13, described in Section 25.3.1. The initial condition and the boundary conditions applicable to the system are the same as those listed in Section 25.3.1. Thus, the analytical solution proposed by Equations 25.14 through 25.16 can be used in this case. In order to estimate the convective heat transfer coefficient (h), the correlation presented by Equation 25.20 can also be applied.

As Equation 25.14 is constituted by a series of infinite terms, a previous convergence study was conducted to set the number of the terms used in the analytical solution. This analysis showed that using the first four terms of the series gave results with good accuracy without increasing the computational effort.

25.5.2.2 Growth Predictive Model

The predictive model for describing the growth of *B. fulva* in papaya pulp was developed and validated by Silva et al. (2013) under dynamic temperature conditions. According to this model, the modified Gompertz equation was chosen to evaluate the increase in the colony radius over time. Considering the nonisothermal nature of the phenomenon, the differential form of the modified Gompertz model (Equation 25.3) had to be used.

The influence of the temperature on the growth rate and on the lag phase time was estimated using the extended square root model (Equation 25.6) and the hyperbolic model (Equation 25.11), respectively. The values of the parameters of the growth predictive model and the heat transfer model are listed in Table 25.3.

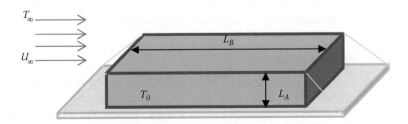

FIGURE 25.3 Illustration of the physical system that was modeled in the case study.

TABLE 25.3

Values of the Parameters of the Integrated Model Used in the Simulations Performed

	Parameter	Value	Source
Heat transfer model	k_{AIR}	0.026 W/(m K)[a]	ASHRAE (1981) and
	α_{AIR}	2.25×10^{-5} m²/s[a]	Incropera and DeWitt (1998)
	ν_{AIR}	1.59×10^{-5} m²/s[a]	
	Pr	0.707[a]	
	T_{AIR}	10°C–14°C	Assumed by the model
	U_{AIR}	1–5 m/s	
	T_0	28°C	
	k_{PAPAYA}	0.60 W/(m K)	Kurozawa et al. (2005)
	α_{PAPAYA}	1.11×10^{-7} m²/s	
	L_A	0.02–0.06 m	Assumed by the model
	L_B	0.10–0.20 m	
Microbial kinetics model	b	0.143	Silva et al. (2013)
	c	0.145	
	T_{min}	274.8 K	
	T_{max}	319.8 K	
	p	105.05	
	q	265.80	
	y_0	0	
	y_{max}	5.02 cm	
	$y_{CRITICAL}$	1.5 mm	Gibson et al. (1994)

[a] Thermal properties of air assessed at 300 K.

25.5.2.3 Integrated Model and Simulations

The heat transfer model was integrated with the growth predictive model as illustrated in Figure 25.4. The heat transfer model provided the evolution of the temperature at the product surface under given operating conditions over time. These temperatures were used to calculate the radial growth rate and the lag phase time, using the secondary model; and, finally, using the differential form of the primary model, the colony radius was calculated at each point in time during the simulation. The time unit used in both mathematical models was 1 min. The heat transfer simulation was run until the temperature of the product's surface achieved 14°C. According to Spoto and Gutierrez (2006), temperatures below this value can cause cooling damage to the papaya fruit. The integrated model was resolved using an algorithm implemented in MATLAB® version 5.3. As the microbial growth model was a differential equation, it was resolved using the fourth-order Runge–Kutta method.

The values of the parameters of the growth predictive model and the heat transfer model are listed in Table 25.3. In order to investigate the effect of product size on the product's shelf life, nine simulations were run combining different values of the thickness and length of the papaya slice, within the ranges given in Table 25.3. In these simulations, the temperature and velocity of the air of the cooling system were adjusted to 14°C and 5 m/s, respectively. At a second moment, nine simulations were performed to study the influence of the conditions of the air in the cooling system (temperature and velocity) on the product's shelf life. In this case, the thickness and the length of the papaya slice were set to 0.04 and 0.15 m, respectively.

25.5.3 Results of the Simulations

The goal of the case study is to evaluate the shelf life of chilled papaya pulp in different processing conditions. The product was defined as spoiled when the colony radius had reached 1.5 mm (critical radius). According to Gibson et al. (1994), fungi colonies of this size are visible to the naked eye and so the consumers reject the food. For all purposes, the shelf life is assumed to be the time needed for the colonies of *B. fulva* to reach the critical radius.

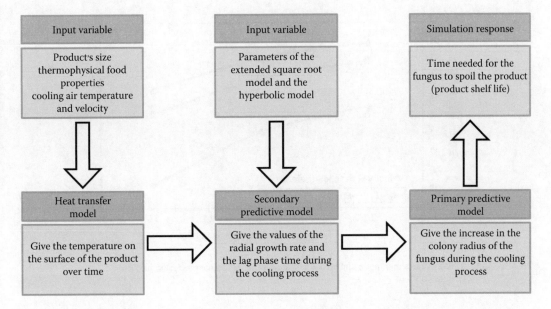

FIGURE 25.4 Structure of the integrated model implemented in MATLAB to simulate different cooling scenarios.

Figure 25.5 shows the effect of the length and thickness of the product on the product's shelf life. By keeping the air-cooling conditions constant (5 m/s for air stream velocity, 14°C for air temperature) and increasing the length of the slice of the papaya, the shelf life is reduced. Moreover, thicker papaya slices have shorter shelf life. In these conditions, the product volume is larger and so more energy needs to be removed from the food to reduce its temperature. As a consequence, the fungi spores are subjected to high temperatures for a long time, germinating rapidly.

Figure 25.6 illustrates the effect of temperature and velocity of the cooling air on the product's shelf life. As can be seen, for a constant product size (15 cm long and 4 cm in thickness), increasing the temperature of the cooling air decreases the shelf life of the papaya. The reduction in the shelf life is more intense for low velocity of the cooling air. In these conditions, the cooling rate is reduced as low velocity of the air corresponds to a weak convective heat transfer. Therefore, the time taken

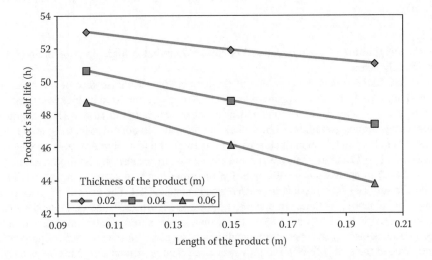

FIGURE 25.5 Effect of the length and thickness of a product on the product's shelf life.

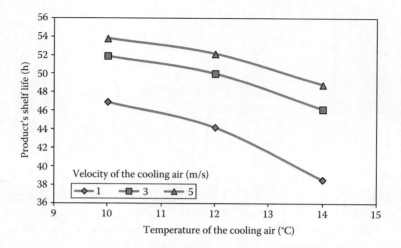

FIGURE 25.6   Effect of the temperature and velocity of the cooling air on the product's shelf life.

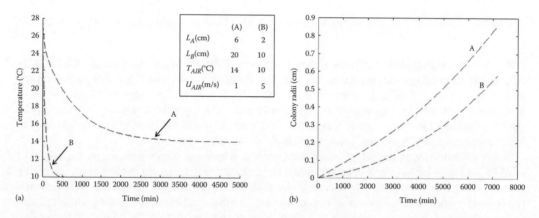

(a)   (b)

FIGURE 25.7   Comparison of two cooling scenarios (A and B) of the slice of papaya pulp. In (a), the evolution of the temperature on the surface of the product over time is shown. In (b), the growth curve of *B. fulva* is illustrated in each cooling scenario.

to reduce the temperature of the product will be longer than that at high air velocity and spores will germinate rapidly again.

Finally, Figure 25.7a depicts the evolution of the temperature on the surface of the product in two extreme conditions. In A, the slices of papaya that were 20 cm long and 6 cm in thickness were subjected to the cooling process using air at 14°C and with a velocity of 1 m/s. In this case, we have a large product undergoing a slow cooling condition. In B, we have an opposite situation in which the food has smaller dimensions (10 cm long and 2 cm in thickness) and it is subjected to a more favorable cooling condition (air temperature of 10°C and air velocity of 5 m/s). Note that in scenario B, the food reaches the target temperature (14°C) in approximately 100 min. On the other hand, in scenario A, over 3000 min are required for the surface of the product to reach a steady state (14°C). The growth behavior of *B. fulva* is meaningfully influenced by these temperature profiles, as Figure 25.7b shows. Thus, the shelf life of papaya pulp in scenario B (3320 min) is 86% greater than in scenario A (1780 min). These results reveal that the proper adjustment of the operational conditions associated with the analysis of product size can significantly extend the shelf life of minimally processed fruit as demonstrated for the papaya pulp in this case study.

REFERENCES

Amézquita, A., C. L. Weller, L. Wang, H. Thippareddi, and D. E. Burson. 2005. Development of an integrated model for heat transfer and dynamic growth of *Clostridium perfringens* during the cooling of cooked boneless ham. *International Journal of Food Microbiology* 101: 123–144.

Baert, K., A. Valero, B. De Meulenaer, S. Samapundo, M. M. Ahmed, J. Debevere, and F. Devlieghere. 2007. Modeling the effect of temperature on the growth rate and lag phase of *Penicillium expansum* in apples. *International Journal of Food Microbiology* 118: 139–150.

Baranyi, J. and T. A. Roberts. 1995. Mathematics of predictive food microbiology. *International Journal of Food Microbiology* 26: 199–218.

Baranyi, J., T. P. Robinson, A. Kaloti, and B. M. Mackey. 1995. Predicting growth of *Brochothrix thermosphacta* at changing temperature. *International Journal of Food Microbiology* 27: 61–75.

Bellara, S. R., C. M. McFarlane, C. R. Thomas, and P. J. Fryer. 2000. The growth of *Escherichia coli* in a food simulant during conduction cooling: Combining engineering and microbiological modeling. *Chemical Engineering Science* 55: 6085–6095.

Ben Yaghlene, H., I. Leguerinel, M. Hamdi, and P. Mafart. 2009. A new predictive dynamic model describing the effect of the ambient temperature and the convective heat transfer coefficient on bacterial growth. *International Journal of Food Microbiology* 133: 48–61.

Çengel, Y. 1998. *Heat Transfer: A Practical Approach*. São Paulo, Brazil: McGraw Hill.

Cepeda, J. F., C. L. Weller, H. Thippareddi, M. Negahban, and J. Subbiah. 2013. Modeling cooling of ready-to-eat meats by 3D finite element analysis: Validation in meat processing facilities. *Journal of Food Engineering* 116: 450–461.

Chapman, B., E. Winley, A. S. W. Fong, A. D. Hocking, C. M. Stewart, and K. A. Buckle. 2007. Ascospore inactivation and germination by high pressure processing is affected by ascospore age. *Innovative Food Science and Emerging Technology* 8: 531–534.

Cheng, K.-C., A. Demirci, J. M. Catchmark, and V. M. Puri. 2010. Modeling of pullulan fermentation by using a color variant strain of *Aureobasidium pullulans*. *Journal of Food Engineering* 98: 353–359.

Corradini, M. G. and M. Peleg. 2007. The non-linear kinetics of microbial inactivation and growth in foods. In: Brul, S., S. van Gerwen, and M. Zwietering (Eds.), *Modelling Microorganisms in Food*. Boca Raton, FL: CRC Press, pp. 129–160.

Cuppers, H. G. A. M., S. Oomes, and S. Brul. 1997. A model for combined effects of temperature and salt concentration on growth rate of food spoilage molds. *Applied and Environmental Microbiology* 63: 3764–3769.

Davey, K. R. 1989. A predictive model for combined temperature and water activity on microbial growth during the growth phase. *Journal of Applied Bacteriology* 67: 483–488.

FAO—Food and Agriculture Organization of the United Nations—FAOSTAT 2011. Available at: http://www.fao.org/corp/statistics. Accessed on February 19, 2013.

Feng, Y.-L., W.-Q. Li, X.-Q. Wu, J.-W. Cheng, and S.-Y. Ma. 2010. Statistical optimization of media for mycelial growth and exo-polysaccharide production by *Lentinus edodes* and a kinetic model study of two growth morphologies. *Biochemical Engineering Journal* 49: 104–112.

Ferreira, E. H. R., L. M. P. Masson, A. Rosenthal, M. L. Souza, and P. R. Massaguer. 2011. Termorresistência de fungos filamentosos isolados de néctares de frutas envasados assepticamente. *Brazilian Journal of Food Technology* 14: 164–171.

Galati, S., L. Giannuzzi, and S. A. Giner. 2011. Modelling the effect of temperature and water activity on the growth of *Aspergillus parasiticus* on irradiated Argentinian flint maize. *Journal of Stored Products Research* 47: 1–7.

Garcia, D., A. J. Ramos, V. Sanchis, and S. Marín. 2009. Predicting mycotoxins in foods: A review. *Food Microbiology* 26: 757–769.

Gibson, A. M., J. I. Baranyi, J. I. Pitt, M. J. Eyles, and T. A. Roberts. 1994. Predicting fungal growth: The effect of water activity on *Aspergillus flavus* and related species. *International Journal of Food Microbiology* 23: 419–431.

Gil, M. M., T. R. S. Brandão, and C. L. M. Silva. 2006. A modified Gompertz model to predict microbial inactivation under time-varying temperature conditions. *Journal of Food Engineering* 76: 89–94.

Gougouli, M., K. Kalantzi, E. Beletsiotis, and K. P. Koutsoumanis. 2011. Development and application of predictive models for fungal growth as tools to improve quality control in yogurt production. *Food Microbiology* 28: 1453–1462.

Gougouli, M. and K. P. Koutsoumanis. 2010. Modelling growth of *Penicillium expansum* and *Aspergillus niger* at constant and fluctuating temperature conditions. *International Journal of Food Microbiology* 140: 254–262.

Gougouli, M. and K. P. Koutsoumanis. 2012. Modeling germination of fungal spores at constant and fluctuating temperature conditions. *International Journal of Food Microbiology* 152: 153–161.

Hamidi-Esfahani, Z., S. A. Shojaosadati, and A. Rinzema. 2004. Modelling of simultaneous effect of moisture and temperature on *A. niger* growth in solid-state fermentation. *Biochemical Engineering Journal* 21: 265–272.

Incropera, F. P. and D. P. De Witt. 1998. *Fundamentos de transferência de calor e de massa*, 4ª edição. Rio de Janeiro, Brazil: LTC.

Jay, J. M. 2005. *Microbiologia de Alimentos*, 6ª edição. Porto Alegre, Brazil: Artmed.

Kotzekidou, P. 1997. Heat resistance of *Byssochlamys nivea*, *Byssochlamys fulva* and *Neosartorya fischeri* isolated from canned tomato paste. *Journal of Food Science* 62: 410–437.

Krishna, K. L., M. Paridhavi, and J. A. Patel. 2008. Review on nutritional, medicinal and pharmacological properties of Papaya (*Carica papaya* Linn.). *Natural Product Radiance* 74: 364–373.

Kurozawa, L. E., A. A. El-Aouar, M. R. Simões, P. M. Azoubel, and F. E. X. Murr. 2005. Determination of thermal conductivity and thermal diffusivity of papaya (*Carica papaya* L.) as a function of temperature. In: *2nd Mercosur Congress on Chemical Engineering—4th Mercosur Congress on Process Systems Engineering*, Rio de Janeiro, Brazil.

Lahlali, R., M. N. Serrhini, D. Friel, and M. H. Jijakli. 2007. Predictive modelling of temperature and water activity (solutes) on the in vitro radial growth of *Botrytis cinerea* Pers. *International Journal of Food Microbiology* 114: 1–9.

Lebert, I. and A. Lebert. 2006. Quantitative prediction of microbial behavior during food processing using an integrated modeling approach: A review. *International Journal of Refrigeration* 29: 968–984.

Lee, H. B. and N. Magan. 2010. The influence of environmental factors on growth and interactions between *Embellisia allii* and *Fusarium oxysporum* f. sp. *cepae* isolated from garlic. *International Journal of Food Microbiology* 138: 238–242.

Lima, L. M., P. L. D. Morais, E. V. Medeiros, V. Mendonça, I. F. Xavier, and G. A. Leite. 2009. Qualidade pós-colheita do mamão formosa 'Tainung 01' comercializado em diferentes estabelecimentos no município de Mossoró—RN. *Revista Brasileira de Fruticultura Jaboticabal* 31(3): 902–906.

Liu, J.-Z., L.-P. Weng, Q.-L. Zhang, H. Xu, and L.-N. Ji. 2003. A mathematical model for gluconic acid fermentation by *Aspergillus niger*. *Biochemical Engineering Journal* 14: 137–141.

Longhi, D. A., F. Dalcanton, G. M. F. Aragão, B. A. M. Carciofi, and J. B. Laurindo. 2013. Assessing the prediction ability of different mathematical models for the growth of *Lactobacillus plantarum* under non-isothermal conditions. *Journal of Theoretical Biology* 335: 88–96.

Mackey, B. M., A. F. Kelly, J. A. Colvin, P. T. Robbins, and P. J. Fryer. 2006. Predicting the thermal inactivation of bacteria in a solid matrix: Simulation studies on the relative effects of microbial thermal resistance parameters and process conditions. *International Journal of Food Microbiology* 107: 295–303.

Mafart, P. 2005. Food engineering and predictive microbiology: On the necessity to combine biological and physical kinetics. *International Journal of Food Microbiology* 100: 239–251.

Mousa, W., F. M. Ghazali, S. Jinap, H. M. Ghazali, and S. Radu. 2011. Modelling the effect of water activity and temperature on growth rate and aflatoxin production by two isolates of *Aspergillus flavus* on paddy. *Journal of Applied Microbiology* 111: 1262–1274.

Nakashima, S. M. K., C. D. André, and B. D. G. M. Franco. 2000. Aspectos básicos da Microbiologia Preditiva. *Brazilian Journal of Food Technology* 3: 41–51.

Panagou, E. Z., S. Chelonas, I. Chatzipavlidis, and G. E. Nychas. 2010. Modelling the effect of temperature and water activity on the growth rate and growth/no growth interface of *Byssochlamys fulva* and *Byssochlamys nivea*. *Food Microbiology* 27: 618–627.

Panagou, E. Z., P. N. Skandamis, G. J. E., and Nychas. 2003. Modelling the combined effect of temperature, pH and Aw on the growth rate of *Monascus ruber*, a heat-resistant fungus isolated from green table olives. *Journal of Applied Microbiology* 94: 146–156.

Pardo, E., S. Marín, A. Solsona, V. Sanchis, and A. J. Ramos. 2004. Modeling of germination and growth of ochratoxigenic isolates of *Aspergillus ochraceus* as affected by water activity and temperature on a barley-based medium. *Food Microbiology* 21: 267–274.

Parra, R. and N. Magan. 2004. Modelling the effect of temperature and water activity on growth of *Aspergillus niger* strains and applications for food spoilage moulds. *Journal of Applied Microbiology* 97: 429–438.

Patriarca, A., G. Vaamond, V. F. Pinto, and R. Comerio. 2001. Influence of water activity and temperature on growth of *Wallemia sebi*: Application of a predictive model. *International Journal of Food Microbiology* 68: 61–67.

Rocha, R. H. C., S. R. C. Nascimento, J. B. Menezes, G. H. S. Nunes, and E. O. Silva. 2005. Qualidade pós-colheita do mamão formosa armazenado sob refrigeração. *Revista Brasileira de Fruticultura Jaboticabal* 27(3): 386–389.

Ross, T. and T. A. McMeekin. 1994. Predictive microbiology. *International Journal of Food Microbiology* 23: 241–264.

Rosso, L., J. R. Lobry, S. Bajard, and J. P. Flandrois. 1995. Convenient model to describe the combined effects of temperature and pH on microbial growth. *Applied and Environmental Microbiology* 61(2): 610–616.

Rosso, L., J. R. Lobry, and J. P. Flandrois. 1993. An unexpected correlation between cardinal temperatures of microbial growth highlighted by a new model. *Journal of Theoretical Biology* 162: 447–463.

Samapundo, S., F. Devlieghere, B. De Meulenaer, A. H. Geeraerd, J. F. van Impe, and J. M. Debevere. 2005. Predictive modelling of the individual and combined effect of water activity and temperature on the radial growth of *Fusarium verticilliodes* and *F. proliferatum* on corn. *International Journal of Food Microbiology* 105: 35–52.

Sant'Ana, A. S., P. Dantigny, A. C. Tahara, A. Rosenthal, and P. R. Massaguer. 2010a. Use of a logistic model to assess spoilage by *Byssochlamys fulva* in clarified apple juice. *International Journal of Food Microbiology* 137: 299–302.

Sant'Ana, A. S., R. C. Simas, C. A. A. Almeida, E. C. Cabral, R. H. Rauber, C. A. Mallmann, M. N. Eberlin, A. Rosenthal, and P. R. Massaguer. 2010b. Influence of package, type of apple juice and temperature on the production of patulin by *Byssochlamys nivea* and *Byssochlamys fulva*. *International Journal of Food Microbiology* 142: 156–163.

Santana, L. R. R., F. C. A. U. Matsuura, and R. L. Cardoso. 2003. Genótipos melhorados de mamão (*Carica papaya* L.): Avaliação tecnológica dos frutos na forma de sorvete. *Ciência e Tecnologia dos Alimentos* 23(Suppl.): 151–155.

Silva, A. R., A. S. Sant'Ana, and P. R. Massaguer. 2010. Modelling the lag time and growth rate of *Aspergillus* section *Nigri* IOC 4573 in mango nectar as a function of temperature and pH. *Journal of Applied Microbiology* 109: 1105–1116.

Silva, P. R. S., I. C. Tessaro, and L. D. F. Marczak. 2013. Integrating a kinetic microbial model with a heat transfer model to predict *Byssochlamys fulva* growth in refrigerated papaya pulp. *Journal of Food Engineering* 118: 279–288.

Spoto, M.H.F. and A.S.D. Gutierrez. 2006. Qualidade pós-colheita de frutas e hortaliças. In: Oetterer, M., Regitano-D'Arce, M.A.B., and Spoto, M.H.F. (Eds.), *Fundamentos de Ciência e Tecnologia de Alimentos*. Barueri: Manole, pp. 403–452.

Taniwaki, M. H., A. D. Hocking, J. I. Pitt, and G. H. Fleet. 2010. Growth and mycotoxin production by fungi in atmospheres containing 80% carbon dioxide and 20% oxygen. *International Journal of Food Microbiology* 143: 218–225.

Tournas, V. 1994. Heat resistant fungi of importance to the food and beverage industry. *Critical Review in Microbiology* 20: 243–263.

Valdramidis, V. P., A. H. Geeraerd, J. E. Gaze, A. Kondjoyan, A. R. Boyd, H. L. Shaw, and J. F. van Impe. 2006. Quantitative description of *Listeria monocytogenes* inactivation kinetics with temperature and water activity as the influencing factors; model prediction and methodological validation on dynamic data. *Journal of Food Engineering* 76: 79–88.

Valík, L. and E. Piecková. 2001. Growth modelling of heat-resistant fungi: The effect of water activity. *International Journal of Food Microbiology* 63: 11–17.

van de Lagemaat, J. and D. L. Pyle. 2005. Modeling the uptake and growth kinetics of *Penicillium glabrum* in a tannic acid containing solid state fermentation for tannase production. *Process Biochemistry* 40: 1773–1782.

van Impe, J. F., B. M. Nicolai, T. Martens, J. D. Baerdemaeker, and J. Vandewalle. 1992. Dynamic mathematical model to predict microbial growth and inactivation during food processing. *Applied Environmental Microbiology* 58: 2901–2909.

van Impe, J. F., F. Poschet, A. H. Geeraerd, and K. M. Vereecken 2005. Towards a novel class of predictive microbial growth models. *International Journal of Food Microbiology* 100: 97–105.

Welti-Chanes, J., F. Vergara-Balderas, and D. Bermúdez-Aguirre. 2005. Transport phenomena in food engineering: Basic concepts and advances. *Journal of Food Engineering* 67: 113–128.

Whiting, R. C. and R. L. Buchanan. 1993. A classification of models in predictive microbiology—A reply to K. R. Davey. *Food Microbiology* 10: 175–177.

Zimmermann, M., S. Miorelli, P. R. Massaguer, and G. M. F. Aragão. 2011. Modeling the influence of water activity and ascospore age on the growth of *Neosartorya fischeri* in pineapple juice. *LWT—Food Science and Technology* 44: 239–243.

26

Strategies for Controlling the Growth of Spoilage Yeasts in Foods

Carmen A. Campos, María F. Gliemmo, and Marcela P. Castro

CONTENTS

26.1 Introduction

Yeasts are commonly present in foods and exert different roles. On one hand, they make possible the elaboration of fermented foods such as bread and beverages. On the other hand, they can cause the spoilage of foods, and in some cases, they can provoke allergic reactions. Nonetheless, it must be stressed that pathogenic yeasts are not transmitted by food (Fleet 1992, Deák 2008).

Yeast spoilage is a matter of concern in the cases where the growth of bacteria is restricted, such as the case of low pH, high sugar/salt concentration, and/or preservative-containing foods. However, in other foods such as certain meats, dairy products, fruits, and vegetables, spoilage yeast activity can be important depending on the product and conditions (Deák 2008).

In fresh meats, it is well known that bacteria grow better and faster than yeasts. However, in processed meats where stress factors applied suppressed bacterial growth, yeast can develop. The species more frequently found are *Candida*, *Rhodotorula*, and *Debaryomyces*. Yeast spoilage in processed meats depends on the type of product. In frankfurters and sausages, surface discoloration, gas, swelling, production of taints, and off-flavors can be found (Nielsen et al. 2008).

Significance of yeast in dairy products has been underestimated for many years. It can be mentioned that sweetened condensed milks allow the development of yeasts since its intermediate water activity (a_w) restricts bacterial growth (Fleet 1992). Surveys of retail yoghurt showed that yeasts were present in many samples, and their fermentation caused gas production and modified the flavor. Main species isolated were *Candida* and *Kluyveromyces* (Mayoral et al. 2005). In the case of cheese, the role of yeast depends on the type and, in some varieties, contributes positively to the development of flavor. Cheeses subjected to spoilage by yeasts are cottage, soft varieties, and mold ripened cheeses. The main species found are *Candida lipolytica*, *Kluyveromyces marxianus*, and *Debaryomyces hansenii* (Fleet 1992).

Molds are the main spoiler agent for fresh fruits, but yeasts can also be present on the fruit surface and probably its deleterious activity does not begin until the skin is damaged by different mechanisms (over-ripening, action of molds, and mechanical injury). In particular, some reports of spoilage of fresh fruits mentioned strawberry, tomatoes, and figs (Deák 2008). It must be stressed that the role as spoilers of yeast is particularly important for processed fruits such as fruit juices, pulps, ready-to-eat fruits, and beverages based on juice fruits. In these products, yeasts are present on the surface and in processing equipment, and they ferment the sugars into alcohol. Moreover, formation of haze, production of CO_2, off-flavors, and color changes are the result of spoilage yeasts (Tournass et al. 2006). Refrigerated storage and preservative addition can retard spoilage fermentation, but yeast-resistant species such as *Zygosaccharomyces bailii* can become the main flora. Thermal treatment and other stress factors must be applied to solve this problem.

In ready-to-eat vegetables, shelf life is limited by the growth of lactic acid bacteria and yeasts. Yeasts are also present during the fermentation of olives, pickles, and sauerkraut. They can produce off-flavors, metabolize the produced lactic acid, and, as a consequence, increase pH, allowing the growth of spoilage bacteria. Moreover, texture can be damaged by the action of pectolytic enzymes. In sauerkraut, a pink discoloration is induced by *Rhodotorula* spp.

Yeasts are also responsible for the spoilage of vegetable salads and salad dressings. In these foods, the presence of acids and preservatives inhibits bacterial growth favoring yeast development. The same trend is observed for foods of reduced a_w. In high-sugar foods such as syrups, fruit juice concentrates, jams, and dried fruits and also for high-salt foods such as soya sauce and miso, the osmotolerant *Z. rouxii*, *Z. bailii*, and *D. hansenii* are the main spoilers.

Alcoholic beverages are produced thanks to the action of selected yeasts, but the uncontrolled growth of wild yeasts can lead to spoilage. Proper conditions of processing decrease the risk of spoilage; however, outbreaks occur due to the activity of ethanol and sulfur dioxide resistant strains such as *Z. bailii* and *S. cerevisiae*. These yeasts produce gas, turbidity, and sediment by the fermentation of residual sugar. It must be stressed that a low number of cells such as 5 CFU/10 L can promote spoilage during storage. This fact highlights the importance of the detection of low levels of yeasts.

In fermented foods and beverages, the limit between beneficial and spoilage activities is not clear. For example, *Brettanomyces/Dekkera* produces 4-ethyl phenol as a secondary metabolite, and levels lower than 400 µ/L of this compound contribute to the flavor, but levels higher than 620 µ/L impart a deteriorative aroma described as horse sweat (Loureiro and Malfeito-Ferreira 2003, Couto et al. 2005).

In summary, many species of yeast are present in foods, but only a small group acts as spoilers in food products; these yeasts are *Brettanomyces bruxellensis*, *Issatchenkia orientalis*, *D. hansenii*, *Kloeckera apiculata*, *Pichia membranifaciens*, *Z. bailii*, *Z. bisporus*, *Z. rouxii*, *S. cerevisiae*, *Z. pombe*, and *C. holmii*. Spoilage activity of these yeasts induces gas production, discoloration, off-flavors, surface growth, film formation, haze, and changes in texture (Loureiro and Querol 1999). Mentioned changes are responsible for economic losses. However, reports about them are rarely informed probably due to reasons of commercial confidentiality.

Different strategies can be applied to control the growth of spoilage yeasts; first of all, it is essential to limit contaminants by selecting good-quality raw materials, monitoring the processes designed to control microorganisms, and preventing cross-contamination. Secondly, it is important to apply isolation and identification methods that allow the rapid detection of low levels of cells. Finally, the knowledge of the effect of food composition, stress factors applied, and the interaction among them on growth and inactivation of yeasts are key factors for selecting the conditions to avoid spoilage. Mentioned strategies will be discussed.

26.2 Methods for Isolation and Identification of Yeasts

Isolation of deteriorative yeasts has become a major issue among food microbiology. The ecological niche (food) from which these yeasts are to be isolated determines the specific growing parameters that isolation media must contain. Yet, differences in nutrient composition and concentration between the food system and the growth medium may lead to poor recovery of these microorganisms. Thus, selective and differential media most frequently used are formulated with extreme ingredients such as organic acids, high sugar and/or salt concentrations, glycerol, and antibiotics, among others. Herein, a brief overview of the most representative examples found in the recent literature will be described. Many are the exhaustive and accurate pieces of scientific literature dedicated to these topics; if the reader wishes to explore further, Deák (2008), Jespersen and Jakobsen (1996), and Loureiro and Malfeito-Ferreira (2003) are highly recommended, among others.

In an international collaborative study, Hocking (2000) found that Tryptone glucose yeast extract (TGY), a nonselective medium, gave the highest counts of *Z. bailii*, *Z. pombe*, and *P. membranaefaciens*, and TGY amended with 0.5% acetic acid (TGYA) was the best medium for recovery of all three preservative-resistant yeasts. *Z. bailii* medium (ZBM) was found to be selective for *Z. bailii*, but counts of this yeast on ZBM were significantly lower than on TGYA.

Several authors have reported the fastidious growth behavior of dressing microbiota on various microbiological media, the requirement for extended incubation (10 days), or the need for inclusion of additional ingredients, including fructose, to improve recovery (Smittle and Flowers 1982, Smittle and Cirigliano 1992); however, Waite et al. (2009) proposed a twist to the problem with the ranch dressing agar, a newly formulated food-based medium, that allowed for optimum recovery of the excessive gas-producing spoilage yeast, *Torulaspora delbrueckii*.

Yeast spoilage of five different processed meat products (bacon, ham, salami, and two different liver patés) was assessed by plating appropriate dilutions on MYGP-agar (yeast extract, malt extract, bacto-peptone, glucose, and agar supplied with chloramphenicol and chlortetracycline). After incubation at 20°C for 5 days, 12 different yeast species were isolated, with *Candida zeylanoides*, *Trichosporon lignicola*, and the recently described *C. alimentaria* (Knutsen et al. 2007) being the predominant species (Nielsen et al. 2008).

On the one hand, the acquisition of reliable data at the appropriate time is a must in the food industry where safety, quality, and shelf life of products have to be ensured. Consequently, several rapid and automated methods have been developed based on novel physical, chemical, immunological, and molecular principles, as an alternative to classical methods. These typing methods have been developed on the analysis of total proteins, long-chain fatty acids, and isoenzymes (Fleet 1992, Deák and Beuchat 1996, Loureiro and Querol 1999). On the other hand, the identification of yeasts occurring in foods is mandatory to form a sound ecological basis for the improvement of processing technologies and the assessment of product quality. Therefore, the food laboratory should have the ability to identify yeast isolates rapidly and accurately as a routine part of process control and quality assurance. Recent progress in molecular biology has contributed to the development of powerful typing techniques. Restriction fragment length polymorphism of mitochondrial DNA, chromosomal DNA electrophoresis, restriction enzyme analysis of polymerase chain reaction (PCR)-amplified ribosomal DNA, random amplified polymorphic DNA assay, and real-time PCR are nowadays routinely performed by food microbiologists. Find the description of techniques and critical reviews of the bases of spoilage yeasts' molecular taxonomy in Casey and Dobson (2004), Deák (2008), Renouf and Lonvaud-Funel (2007), Renard et al. (2008), and Serpaggi et al. (2012), among others.

26.3 Factors Affecting Growth and Inactivation of Yeasts

Yeasts are the most predominant spoilage microflora in sugar-containing low-pH foods such as fruit juices, beverages, wine, dressings, and sauces. Many factors affect yeast growth; among chemical and physical factors within the food, the presence of acids, salts and sugar content, lipids, and additives can be mentioned. Besides, the content of salts or sugars determine food a_w, which is also a very important factor affecting yeast growth. The latter will be discussed in the stress factors section.

26.3.1 Food Composition

Even though many yeast species can tolerate a wide range of pH, from pH 1.5 to 10.0, most yeasts prefer growth in a slightly acidic medium, between 3.5 and 6.0, the pH found in most fruit juices, beverages and soft drinks, and salad dressings (Martorell et al. 2007). The tolerance to low pH exhibited by species such as *Z. bailii* and *P. membranofaciens* is linked with the more efficient or stable activity of plasma membrane ATPase, which regulates intracellular pH by expelling protons (Praphailong and Fleet 1997).

Salad dressings are complex food systems since they are made from a dispersed oil phase, an emulsifier, and a high concentration of organic acids in an aqueous phase, which also contains salt, sugar, and hydrocolloid (Castro et al. 2003). *Z. bailii* is the main spoilage yeast being its growth affected by the concentration of oil and the additives included in the formulation (Smittle and Flowers 1982). An oil concentration within the range of 11%–23% (w/w) exerted an inhibitory action on *Z. bailii* growth in acidified emulsions that modeled salad dressings. The increase in oil to 46% promoted the inactivation of the yeast (Castro et al. 2003). These trends might be attributed to the reduction of the aqueous phase, which determines an increase in the additive concentration in that phase. Moreover, oil acts as a modifier of the food microstructure, conducting to the immobilization of microorganisms (Wilson et al. 2002). Furthermore, the use of Tween 20 as an emulsifier modified *Z. bailii* growth in different ways depending on the concentration of Tween 20, oil, and potassium sorbate (Castro et al. 2003).

Foods with low sugar content, such as juices, jellies, and jams, contain acids, sweeteners, and hydrocolloids. These additives are needed for sensory purposes. Among the most used sweeteners is aspartame, though the use of the natural sweetener, stevia, is increasing in the last years (Scott-Thomas 2013). The presence of mentioned additives affects the growth of *Z. bailii*. It was reported that aspartame addition increased the yeast population at the stationary phase in acidified Sabouraud broth containing enough glucose or xylitol to get an a_w value of 0.985. However, when 250 ppm of potassium sorbate was added to the system and it was heated at 50°C, aspartame addition increased the yeast thermal inactivation rate (Gliemmo et al. 2013). On the other hand, addition of steviosides to an acidified Sabouraud broth containing enough glucose or xylitol to get an a_w of 0.985 promoted an increase in the minimum concentration needed to inhibit the growth of *Z. bailii* (Hracek et al. 2010). Probably, depending on the composition of the system, the yeast could be able to metabolize aspartame or steviosides.

Trends commented highlight the importance of considering the effect of system composition and the interaction between the ingredients and/or stress factors when evaluating spoilage yeast growth.

26.3.2 Stress Factors Applied for Processing and Storage

The microbial stability and safety of most foods are based on the use of several stress factors (hurdles) that should be overcome by the microorganisms. This strategy has been applied for centuries; products such as dried fruits, jams, jellies, and sausages can be taken as examples. It is essential to know the mechanism of action of each individual factor and its interaction with the rest of factors applied in order to optimize the preservation (Leistner 1995).

In food products that can be spoiled by yeasts, the stress factors commonly applied are thermal treatment, depression of pH and/or a_w, the addition of preservatives, and storage under refrigeration. In the next sections, mentioned stress factors and the interaction between them will be discussed.

26.3.2.1 Depression of Water Activity and Its Interaction with Other Stress Factors

Depression of a_w can be obtained by drying as it has been made for centuries for fruits or by the addition of solutes, mainly sugars or sodium chloride. Bacteria are more sensitive to a_w depression than yeast and molds. It is well known that as a_w is decreased, microbial lag phase is extended; growth rate is decreased and also does the maximum population at the stationary phase. At a certain value of a_w that depends on the solute used, microbial cells are not able to grow (Gould 1985). Regarding yeasts, some species such as *D. hansenii*, *Hemimysis anomala*, and *Candida pseudotropicalis* are salt tolerant and may grow at NaCl up to 11%. On the other hand, other species are osmophilic (*Z. rouxii* and *Z. bisporus*) and are able to grow in high-sugar systems ($a_w = 0.65$). The differences in a_w limits show the differences in osmoregulatory ability. The main strategies to overcome a_w depression is by expelling the ions out of the cells and by the synthesis of compatible solutes that helps to equilibrate a_w within the cell and to stabilize the membrane (Leistner and Russell 1991).

The effect of a_w depression on yeast growth depends on the presence of other stress factors. As it was previously commented, a_w is commonly applied together with high or low temperature, low pH, and the presence of preservatives. All these stress factors can interact between them conducting to synergic or antagonic actions. Some results concerning the interaction between stress factors will be discussed.

For example, the combined use of a_w depression to 0.985 or to 0.971 by glucose addition and the presence of potassium sorbate (50 ppm) promoted a greater decrease in the growth rate of *Z. bailii* in Sabouraud broth with pH 3.0 than each factor by itself (a_w depression by glucose addition and potassium sorbate) (Gliemmo et al. 2006a).

Storage temperature modifies growth rate and metabolic processes. The maximum temperature for the growth of *D. hansenii* and *S. cerevisiae* in broth (pH 5.8) was 33.6°C and 5.6°C, respectively. Depression of a_w to 0.982 by the addition of NaCl or KCl produced different effects: the use of NaCl or KCl raised the maximum growth temperature by 1.6°C for *D. hansenii*, suggesting that these solutes exerted a protective effect. However, maximum growth temperature dropped by 1.8°C in the case of *S. cerevisiae* when NaCl was used as a solute. On the contrary, a slight increase was produced by KCl. Mentioned trends were related with the fact that, in general, sodium cation was toxic for *S. cerevisiae* but improved the growth of *D. hansenii*. On the contrary, potassium cation did not exert any effect on *S. cerevisiae* but improved the growth of *D. hansenii* (Almagro et al. 2000).

The minimum temperature for growth is modified by a_w of the media, for example, a decrease in a_w to 0.92 by glucose addition increased the growth minimum temperature for *Z. bailii* from 6.5°C to 10°C (Jermini and Schmidt-Lorenz 1987). Moreover, the growth rate of *D. hansenii* and *S. cerevisiae* at low temperature—close to the minimum temperature for growth—was decreased by a_w depression due to the presence of NaCl or KCl in the broth. The effect was more pronounced for *S. cerevisiae*, the yeast that does not exhibit resistance to mentioned salts (Almagro et al. 2000).

For most yeasts, tolerance to a_w is decreased at extreme pH values. Some yeasts such as *Z. bailii* and *P. membranofaciens* are more tolerant to acidic pH while others such as *D. hansenii* and *S. cerevisiae* are more tolerant to pH values within 5.0–7.0 (Praphailong and Fleet 1997). The interactive effect of a_w and pH has not been analyzed in detail yet, and there is a lack of information about the molecular basis of these trends.

The interaction of a_w and thermal treatment will be discussed in the next section.

26.3.2.2 Thermal Treatment and Its Interaction with Other Stress Factors

A thermal treatment is commonly applied to fruit juices, fruit purees, sugar syrups, and carbonated beverages in order to control microbial growth. In particular, yeasts and molds exhibit high sensitivity to heat. As a consequence, pasteurization is one of the most common and effective method of preventing spoilage in acidic foods.

The mechanism of heat inactivation of yeast cell is not totally elucidated, but it is known that damage of cell membrane, ribosomes, and mitochondria is involved (Deák 2008). Decimal reduction times (D) for yeasts at 55°C are on the order of 5–10 min. However, these values depend on yeast species, the strain, and the composition of the medium. Moreover, cells from the stationary phase are more resistant to heat (Truong-Meyer et al. 1997). A decrease in pH from 3.95 to 2.62 promoted a decrease in *P. anomala* D value from 6 to 2 min (Tchango et al. 1997). Furthermore, in fruit juices, the composition also exerted an effect, for example, the *Z. bailii* $D_{50°C}$ value was 4.48 min for pineapple juice (pH 3.4) and was 2.04 for cranberry juice (pH 3.5) (Raso et al. 1998). Inclusion of preservatives also produced the same effect. The addition of 500 or 1000 ppm of potassium sorbate or sodium benzoate in the heating medium promoted more rapid inactivation rates for 12 yeast strains compared to control systems, exerting potassium sorbate a higher effect. Even a level of 100 ppm of the latter decreased D values of *D. hansenii*, *P. membranaefaciens*, *S. cerevisiae*, *Candida krusei*, *K. apiculata*, and *R. rubra* in an acidic media (pH 2.5–4.5) (Beuchat 1981). Gliemmo et al. (2006b) also reported that the addition of 250 ppm of potassium sorbate to Sabouraud broth of pH 3.0 and containing different polyols increased the rate of heat inactivation of *Z. bailii* significantly. The cooperative effect of preservatives on heat inactivation was enhanced by the decrease in pH. Probably, preservatives promoted injury during heating.

Water activity depression and the solute used affect thermal inactivation. It is commonly reported that as a_w decreases, heat resistance of microorganisms is increased. Corry (1976a,b) found that the heat resistance of *Z. rouxii* and *Z. pombe* was enhanced in solutions of sugars and polyols containing 0.1 M phosphate buffer, pH 6.5 at an a_w value of 0.95. Golden and Beuchat (1992) reported that the use of glucose or sucrose to depress a_w to 0.93 in broth (pH 4.5) enhanced the heat resistance of cells of *Z. rouxii*. It was postulated that the increase in heat resistance is linked with the dehydration of the cell that took place when a_w was diminished and with the concomitant increased in stability of cell protein in the dry state (Gibson 1973). However, it must be stressed that sensitivity to heat at a reduced a_w depends on the type of microorganism, the solute used, and its concentration. For example, no change in *Z. rouxii* and *Torulopsis globosa* D_{55} values was observed when a_w was reduced from 0.995 to 0.980 by the addition of sucrose, but the use of enough quantity of this sugar to get an a_w of 0.90 promoted an increase in D_{55} value (Gibson 1973). In broth (pH 3.0), a_w depression to 0.985 by the addition of xylitol or sorbitol produced no change in the rate of thermal inactivation of *Z. bailii*. But when glucose was used as a solute to get an a_w value of 0.985 or 0.971, the inactivation rate constant was increased. However, the addition of glucose to depress a_w to 0.90 promoted a decrease in thermal inactivation rate constant (Gliemmo et al. 2006b). Furthermore, when potassium sorbate (250 ppm) was added to the broth of a_w 0.985, an increase in the rate of thermal inactivation was observed. These results suggest a cooperative action of a_w, heat treatment, and the preservative on yeast cells. This trend could allow decrease in the severity of thermal treatments with no detrimental effect on the sterility of the product.

26.3.2.3 Preservatives and Its Interaction with Other Stress Factors

Preservatives have been added to foods for centuries. They act as microbial inhibitors and are used as a stress factor in combination with other factors. Many chemical compounds have a long history of safety use and exhibit different mechanisms of action depending on their chemical structure (Devlieghere et al. 2004).

Salts of lipophilic weak acids such as sodium benzoate or potassium sorbate are the most common inhibitors for yeasts (Sofos 2000). Their effectiveness is markedly influenced by food composition and by environmental factors. Generally, most yeasts are inhibited by low concentrations of sorbates or benzoates. However, yeasts belonging to *Zygosaccharomyces* exhibit a high resistance to mentioned preservatives and also to other adverse environmental factors (Thomas and Davenport 1985). This resistance is related to specific mechanisms such as a dissimilar transport and the ability to degrade or catabolize acids (Deak 2008).

It is well known that in the case of weak acids, the undissociated species are the form with antimicrobial action since they are able to enter into the cell. As a consequence, a decrease in pH favors the action of the preservative. As an example, at pH 3.0, the maximum concentration of benzoic acid that allowed the growth of *Yarrowia lipolytica* and *Z. bailii* was 250 ppm, and this level increased to 1200 ppm as pH increased to 7.0 (Praphailong and Fleet 1997).

In general, a_w depression and the solute used modified the antimicrobial action of preservatives. Regarding potassium sorbate, the minimum concentration to inhibit the growth of *Z. bailii* was decreased from 250 to 200 ppm when a_w was depressed to 0.985 by the addition of sorbitol, xylitol, or mannitol (Gliemmo et al. 2004). When evaluating the growth of *Z. bailii* in liquid media containing low levels of potassium sorbate (50–100 ppm), the inclusion of glucose and/or polyols to depress a_w to 0.971 promoted a synergic action in relation to growth inhibition (Gliemmo et al. 2006a).

It was observed that benzoic acid (0.05 mM) at pH 4.0 in liquid media reduced the growth of *D. hansenii* by 43% and *S. cerevisiae* by 22%, and the addition of NaCl or KCl to depress a_w to 0.982 enhanced the inhibitory activity of the preservative on *S. cerevisiae* but it had no effect on *D. hansenii* (Almagro et al. 2000). The latter trend was linked to the high resistance to NaCl exhibited by *D. hansenii*.

The joint use of sodium benzoate and potassium sorbate exerted a higher inhibition than the one observed for each antimicrobial alone (Cole and Keenan 1986, Battey et al. 2002).

In conclusion, decreasing the pH and/or a_w and the combined use of preservatives are strategies that would allow to diminish the amount of preservatives used keeping microbial stability.

Nowadays, consumers demand products with high quality and free from traditional preservatives that are considered unsafe because they have no connection with the food matrix and/or they would promote

TABLE 26.1

Novel Preservatives Proposed to Inhibit Spoilage Yeasts

Preservative—Other Stress Factor	Yeast Media	Effect	Reference
Citron and citral essential oil, 2-hexenal—mild heat treatment	*S. cerevisiae*—noncarbonated beverages	Heat treatment and flavor agents showed a synergic action on yeast inhibition	Belletti et al. (2007)
β-Glucanasa	*S. cerevisiae, C. albidus, Dekkera bruxellensis, P. membranifaciens, Z. bailii,* and *Z. bisporus*—liquid media and white wine	β-Glucanasa inhibited the growth of spoilage yeasts	Enrique et al. (2010)
Synthetic peptides and lactoferricin B (LfcinB)-derived peptides	Wine spoilage yeasts—liquid media and white wine	Lactoferricin B (LfcinB) inhibited the growth of spoilage yeast in laboratory media	Enrique et al. (2007)
15 Grape phenolic compounds (phenolic acids, stilbenes, and flavonoids)	Wine spoilage yeasts *D. bruxellensis, Hanseniaspora uvarum, Z. bailii,* and *Z. rouxii*	Pterostilbene (MICs = 32–128 µg/mL), resveratrol, and luteolin possessed the strongest inhibitory activity	Pastorkova et al. (2013)
Cinnamon and clove essential oil	*Z. bailii* and *Z. rouxii*—liquid media and tomato jam	Cinnamon oil was more effective than clove oil in yeast inhibition in vitro	Gliemmo and Campos (2013)
Mentha oil	*S. cerevisiae, Z. bailii, Aureobasidium pullulans, Catocala diversa, Pichia fermentans,* and *P. kluyveri, P. anomala, H. polymorpha*—laboratory media and fruit juices	Mentha oil was effective to inhibit the yeast, and this action was synergic with a thermal treatment	Tyagia et al. (2013)
Chitosan	*K. apiculata, M. pulcherrima, S. cerevisiae,* and *Pichia* spp.—apple juice	*K. apiculata* and *M. pulcherrima* were inactivated, but *S. cerevisiae* and *Pichia* spp. multiplied slowly	Kiskó et al. (2005)
Vanillin	*S. cerevisiae, Z. bailii,* and *Z. rouxii*	Vanillin exerted a biostatic effect on studied yeasts	Fitzgerald et al. (2003)

allergic reactions such as the case of sulfur dioxide (Belletti et al. 2007, Suh et al. 2007). In this context, the search for new preservatives has become a goal. Many natural products have been proposed as preservatives; among them, plant extracts and essential oils have been evaluated. In Table 26.1, a revision of the novel preservatives proposed to inhibit yeasts can be seen. Several studies report the effectiveness of plant extracts and essential oils to inhibit yeast growth in vitro, but it can be stressed that food composition decreases the activity significantly and that the effective concentrations usually impart a strong flavor. For these reasons, it is a must to use them in combination with other stress factors. For example, antimicrobial activity of clove and cinnamon oil was enhanced when used in combination with an atmosphere with low oxygen and high carbon dioxide (Matan et al. 2006).

26.3.2.4 Novel Preservation Technologies

Over the last years, researchers are looking for new technologies that minimize the changes in quality parameters occurring in foods after applying traditional methods of preservation. In general, novel technologies involve physical methods, and they are combined with other stress factors to assay the possibility to increase its effectiveness. The main novel technologies that are being assayed are high hydrostatic pressure, UV-C-irradiation, pulsed electric fields, and ultrasound with temperature (thermosonication).

The effectiveness of high-pressure treatment is related to the disruption of the cell membrane, perforation, and release of the cell wall (Marx et al. 2011). It has the advantage to be effective at ambient and refrigeration temperatures (Reyns et al. 2000). *Z. bailii* and *Zygoascus hellenicus* showed high resistance to high-pressure treatment. The high-pressure treatment may affect the pattern of inactivation curve of yeasts. A biphasic curve was observed for *Z. rouxii* and *Z. bailii* with a first phase following a first-order kinetic and a second phase exhibiting a *tail*. It may be due to the presence of high-resistant cells (Reyns et al. 2000). The efficiency of high-pressure treatment against yeasts is weakly dependent on pH, and it was higher in fruit juices than in buffers at the same pH, suggesting that other factors are involved in the inactivation (Reyns et al. 2000, Daryaei et al. 2010).

The UV-C-irradiation has germicidal action since it breaks down the DNA of microorganisms causing cell death. Spoilage yeasts of clear apple juice were more UV-C resistant than *E. coli* O157:H7. It should be considered in the selection of a target microorganism to assure microbial safety of juices (Gabriel 2012). Since UV irradiation has low penetration effect, it is used for the surface sterilization of foods by the application of intense short pulses of near infrared region wavelengths (Barbosa-Canovas et al. 2000).

The application of pulsed electric fields to microorganisms mainly produces deformation of the cells, apparent fusion of cells, and the formation of pores (Marx et al. 2011). This technology had been successfully assayed as an alternative system of microbiological control in wineries (Puértolas et al. 2009). Its use does not cause sensory changes in food since it is a cold pasteurization (Qin et al. 1996).

In thermosonication, cellular death mainly occurs due to erosion and disruption of the cellular membrane, formation of orifices on the surface, and release of intracellular contents by lysis of cells (Marx et al. 2011). This technology may be a viable option for juice pasteurization, but it produced changes in the pH and color of pineapple, grape, and cranberry juices suggesting that thermosonication may promote chemical reactions; therefore, more research is necessary (Adekunte et al. 2010, Bermúdez-Aguirre and Barbosa-Cánovas 2012).

Of the mentioned technologies, most studied are high-pressure and pulse electric fields. However, it is noteworthy that to be applied in the industry level, it is necessary to estimate costs through the knowledge of the energy involved in each technology (Bermúdez-Aguirre and Barbosa-Cánovas 2012).

26.4 Mathematical Model of Growth/Inactivation of Yeasts

The development of mathematical models describing kinetic growth (growth and thermal inactivation) or the growth/no growth (G/NG) boundary is a tool that allows the assessment and prediction of the responses of microorganisms to environmental conditions to which the microorganisms are resistant. Several predictive models have been found for spoilage yeasts that are described in Table 26.2.

Growth parameters obtained from kinetic growth models, also called primary models, are used in secondary models to know environment influence on growth. The main difference with G/NG studies is that in the former, samples with no growth are ignored. Besides, the use of less laborious microbiological techniques than traditional ones (e.g., microplate vs. plate count) allows studies of G/NG to assay more environmental factors, to increase the number of determinations, and to extend the storage time (Evans et al. 2004, Dang et al. 2010). Slight changes in environmental conditions near the G/NG boundary may result in large changes in growth probability. In general, when an inhibitory factor is more stringent, growth boundaries become less stressful for the other factors and NG zones increase (Jenkins et al. 2000, Dang et al. 2010, Mertens et al. 2011). The G/NG studies allowed some authors to model the no-growth regions by the combination of different factors without using chemical preservatives (Jenkins et al. 2000, Belletti et al. 2007, Dang et al. 2010).

The use of G/NG models is helpful for industry since they allow knowing the influence of small changes in the formulation of a product on its microbial stability faster than through shelf life studies or challenge tests. For example, Arroyo-López et al. (2012) do not recommend the combined use of citric acid and natamycin for the preservation of table olives in packaging since progressive concentrations of acid decrease the antifungal activity of natamycin.

TABLE 26.2

Several Predictive Models Applied to Analyze Factors Affecting Inhibition/Inactivation of Spoilage Yeasts

Model/s	Yeast—Media	Factors	Relevant Results	Reference
Probabilistic	*Z. bailii*—Sabouraud broth supplemented with fructose (7.5% w/v) and glucose (5.5% w/v) mimicking high-sugar acidified sauces	Acetic acid (0.0%–0.25% v/v), pH (3.0–5.0), a_w adjusted with NaCl (0.93–0.97), and temperature (22°C and 30°C)	The growth abruptly decreased by the reduction of 0.01 units of a_w or 0.5 units of pH when the other factors were stressful for the yeast. The decrease in temperature did not reduce the growth.	Dang et al. (2010)
Probabilistic	*Z. bailii*—mango puree at pH 3.5	Storage temperature (5°C–15°C–25°C), storage time (5 weeks), a_w adjusted with sucrose or water (0.990–0.970), and antimicrobial type (1000 ppm of potassium sorbate or sodium benzoate)	Regardless of a_w, 1000 ppm of sorbate inhibited the growth for 30 storage days at 6.4°C, while for benzoate, 3.6°C is necessary.	López-Malo and Palou (2000)
Survival analysis and Classification and Regression Trees (CART)	Cocktail of *S. cerevisiae, Z. bailii, Z. rouxii, Pichia sp., C. stellata, C. pintolo, H. anomala* and *S. carlsbergensis*—Man Rogosa Sharpe broth mimicking a fruit-based or alcoholic food or drink	Alcohol (0%–20% v/v), pH (1.58–6.69), sucrose (0%–55% w/v), sorbate (0–1000 ppm), and temperature (2°C–30°C)	Both models could successfully predict the probability of growth within 150-day period or the time to growth. However, the survival analysis model was more reliable than CART model.	Evans et al. (2004)
Logistic/ probabilistic model	Cocktail of *S. cerevisiae, W. anomalus,* and *C. boidinii*—yeast-malt-peptone–glucose broth mimicking table olive	Natamycin (0–30 mg/L), citric acid (0.00%–0.45%), and sodium chloride (3%–6%)	Antifungal effect of natamycin decreased by progressive concentrations of citric acid. Sodium chloride levels above 6% favored natamycin action.	Arroyo-López et al. (2012)
Logistic/ probabilistic model	*Z. bailii*—modified Sabouraud broth (a_w =0.95) mimicking acidic sauces with high sugar content	pH (3.0–5.0), acetic acid (0%–0.35% w/v), and lactic acid (0%–3.0% w/v)	Lactic acid stimulated *Z. bailii* growth, then it would be risky to use it in replacement of acetic acid to minimize sensory impact.	Vermeulen et al. (2008)
Logistic/ probabilistic model	Cocktail of equal proportions of *S. cerevisiae, C. lipolytica,* and *Z. bailii*—cold-filled ready-to-drink beverage	pH (2.8–3.8), titratable acidity (0.20%–0.60%), sugar (8.0°Brix–16.0°Brix), sodium benzoate (100–350 ppm), potassium sorbate (100–350 ppm)	Benzoate, sorbate, and pH controlled the probability of growth while the other factors did not have significant effects.	Battey et al. (2002)
Survival primary model with three parameters	*S. cerevisiae* and *D. hansenii* after growth in malt agar or in liquid media—strawberry-based product (48°Brix, a_w = 0.92, pH = 3.65)	Temperature 55°C–70°C	Since *S. cerevisiae* was the most heat-resistant strain, its use is recommended to test thermal process of low-pH sugared product.	Truong-Meyer et al. (1997)

(Continued)

TABLE 26.2 (Continued)

Several Predictive Models Applied to Analyze Factors Affecting Inhibition/Inactivation of Spoilage Yeasts

Model/s	Yeast—Media	Factors	Relevant Results	Reference
Primary and probabilistic	*Z. bailii*—yeast nitrogen broth mimicking high-acidic foods	Fructose (7.0%–32.0% w/v), sodium chloride (2.6%–4.2% w/v), pH (3.5–4.0), and acetic acid (1.8%–2.8% v/v)	The models developed may be used in food formulation or to screen the stability of commercial products.	Jenkins et al. (2000)
Primary and secondary	*Lactobacillus plantarum* and *D. hansenii*—olive juice mimicking conditions of Greek-style black olive fermentation	Sodium chloride (256.6–855.0 mM), potassium chloride/sodium chloride 1:1 (128.3–427.5 mM), calcium acetate (19.1–190 mM), calcium lactate (9.6–95 mM)	Calcium acetate and lactate acted synergistically on the growth of single cultures.	Tsapatsaris and Kotzekidou (2004)
Probabilistic	*S. cerevisiae*—noncarbonated beverages	Citron essential oil (0–500 ppm), citral (0–120), (*E*)-2-hexenal (0–50 ppm), thermal treatment length at 55°C (0–20 min), and yeast inoculum (1–5 log CFU/bottle)	The growth was inhibited by flavoring agents, and the effect was enhanced by thermal treatment. Citron was the most effective with a short time heating.	Belletti et al. (2007)
Primary model of growth	*Z. bailii*—Sabouraud broth at pH 3.00 mimicking low-sugar acidic products	Potassium sorbate (0.005%–0.010% w/w) and a_w (1.000–0.971, adjusted with glucose, xylitol, mannitol, or sorbitol)	From the solutes assayed, xylitol was that promoted the highest decrease in growth. A synergism between solutes and sorbate was observed.	Gliemmo et al. (2006a)
Primary model of thermal inactivation	*Z. bailii*—Sabouraud broth at pH 3.00 with a thermal treatment of 50°C for 30–60 min, mimicking low-sugar acidic products	Potassium sorbate (0.000%–0.025% w/w) and a_w (1.000–0.900, adjusted with glucose, xylitol, mannitol, or sorbitol)	The use of glucose enhanced the rate of heat inactivation. A synergic effect was observed by the combined use of sorbate and sorbitol, xylitol, or glucose.	Gliemmo et al. (2006b)
Primary and secondary	*S. cerevisiae*—1% peptone solution	Sorbic acid (0–50 ppm), a_w (0.93–0.97 adjusted with glucose), and pH (4–6 adjusted with hydrochloric acid)	At a level of sorbic acid of 25 ppm, an a_w of 0.94, and at pH 4, the growth was inhibited.	An-Erl King (1993)

Several G/NG models have been developed to study the effect of system composition modeling acidic sauces on *Z. bailii* probability to growth (Jenkins et al. 2000, Dang et al. 2010, Mertens et al. 2011). Furthermore, the effect of medium structure was evaluated. Results obtained using xanthan gum and carbopol as gelling agents showed an unexpected increase in yeast growth probability as a function of medium structure (Mertens et al. 2011).

After selecting the appropriate levels of factors for food formulation, they should be validated by shelf life studies or challenge tests. In the validation of a G/NG model in ketchup, *Z. bailii* grew in conditions where low probability of growth had been predicted, though this situation was reverse expressing antimicrobial concentrations on water bases instead of broth (Dang et al. 2011). Jenkins et al. (2000) could successfully validate a predictive model of time to the growth of *Z. bailii* in retail acid condiments from the US market.

In some cases, the growth of yeasts is desirable in foods, and predictive models are developed to find optimum levels of additives that maximize the growth; such is the case of *D. hansenii* in the olive fermentation (Table 26.2) (Tsapatsaris and Kotzekidou 2004).

26.5 Conclusion and Perspectives

Yeast spoilage is a matter of concern in systems where the growth of bacteria is restricted, such as low pH, high sugar/salt concentration, and/or preservative-containing foods. However, in other food products, spoilage yeast activity can be important depending on the product and processing conditions, and more work is necessary to know the occurrence of yeasts in different food commodities.

It must be emphasized that spoilage yeasts are responsible for economic losses. However, reports about them are rarely informed probably due to reasons of commercial confidentiality. As it was mentioned, different strategies can be applied to control the growth of spoilage yeasts. In the last years, several rapid and automated methods have been developed based on novel physical, chemical, immunological, and molecular principles, as an alternative to classical methods for the rapid detection and identification of yeasts, and it is a must for the expansion of the use of these methods by the industry.

It was demonstrated that the effect of food composition, stress factors applied, and the interaction among them on the growth and inactivation are key factors for selecting the conditions to avoid spoilage. For the adequate selection of stress factors is a must to know the biochemical and/or molecular basis of the interaction between them; hence, comprehensive studies on this field should be increased. Furthermore, the search for natural preservatives and the application of novel preservation technologies such as high pressure and pulse electric fields merit further research together with cost estimation to be applied at an industrial level.

The development of mathematical models describing kinetic growth (growth and thermal inactivation) or G/NG boundary as a function of different environmental conditions to which the microorganisms are resistant is essential to select the conditions that assure food quality.

REFERENCES

Adekunte, A., B. K. Tiwari, A. Scannell, P. J. Cullen, and C. O'Donnell. 2010. Modelling of yeast inactivation in sonicated tomato juice. *International Journal of Food Microbiology* 137:116–120.

Almagro, A., C. Prista, S. Castro, C. Quintas, A. Madeira Lopes, and J. Ramos. 2000. Effects of salts on *Debaryomyces hansenii* and *Saccharomyces cerevisiae* under stress conditions. *International Journal of Food Microbiology* 56:191–197.

An-Erl King, V. 1993. Studies on the control of the growth of *Saccharomyces cerevisiae* by using response surface methodology to achieve effective preservation at high water activities. *International Journal of Food Science and Technology* 28:519–529.

Arroyo-López, F. N., J. Bautista-Gallego, V. Romero-Gil, F. Rodríguez-Gómez, and A. Garrido-Fernández. 2012. Growth/no growth interfaces of table olive related yeasts for natamycin, citric acid and sodium chloride. *International Journal of Food Microbiology* 155:257–262.

Barbosa-Canovas, G. V., D. W. Schaffner, M. D. Pierson, and Q. H. Zhang. 2000. Pulsed light technology. *Journal of Food Science* 65(Suppl):82–85.

Battey, A. S., S. Duffy, and D. W. Schaffner. 2002. Modeling yeast spoilage in cold-filled ready-to-drink beverages with *Saccharomyces cerevisiae*, *Zygosaccharomyces bailii*, and *Candida lipolytica*. *Applied and Environmental Microbiology* 68:1901–1906.

Belletti, N., S. S. Kamdem, F. Patrignani, R. Lanciotti, A. Covelli, and F. Gardini. 2007. Antimicrobial activity of aroma compounds against *Saccharomyces cerevisiae* and improvement of microbiological stability of soft drinks as assessed by logistic regression. *Applied and Environmental Microbiology* 73:5580–5586.

Bermúdez-Aguirre, D. and G. V. Barbosa-Cánovas. 2012. Inactivation of *Saccharomyces cerevisiae* in pineapple, grape and cranberry juices under pulsed and continuous thermo-sonication treatments. *Journal of Food Engineering* 108:383–392.

Beuchat, L. R. 1981. Effects of potassium sorbate and sodium benzoate on inactivating yeasts heated in broths containing sodium chloride and sucrose. *Journal of Food Protection* 44:765–769.

Casey, G. D. and A. D. W. Dobson. 2004. Potential of using real-time PCR-based detection of spoilage yeast in fruit juice—A preliminary study. *International Journal of Food Microbiology* 91:327–335.

Castro, M. P., O. Garro, L. N. Gerschenson, and C. A. Campos. 2003. Interaction between potassium sorbate, oil and Tween 20 on the growth and inhibition of *Z. bailii* in model salad dressings. *Journal of Food Safety* 23:47–59.

Cole, M. B. and M. H. J. Keenan. 1986. Synergistic effects of weak-acid preservatives and pH on the growth of *Zygosaccharomyces bailii*. *Yeast* 2:93–100.

Corry, J. 1976a. The effect of sugars and polyols on the heat resistance and morphology of osmophilic yeasts. *Journal of Applied Bacteriology* 40:269–276.

Corry, J. 1976b. Sugar and polyol permeability of *Salmonella* and osmophilic yeast cell membranes measured by turbidimetry, and its relation to heat resistance. *Journal of Applied Bacteriology* 40:277–284.

Couto, J. A., F. Neves, F. Campos, and T. Hogg. 2005. Thermal inactivation of the wine spoilage yeasts *Dekkera/Brettanomyces*. *International Journal of Food Microbiology* 104:337–344.

Dang, T. D. T., L. Mertens, A. Vermeulen, A. H. Geeraerd, J. F. Van Impe, J. Debevere, and F. Devlieghere. 2010. Modeling the growth/no growth boundary of *Zygosaccharomyces bailii* in acidic conditions: A contribution to the alternative method to preserve foods without using chemical preservatives. *International Journal of Food Microbiology* 137:1–12.

Dang, T. D. T., A. Vermeulena, L. Mertens, A. H. Geeraerdd, J. F. Van Impe, and F. Devlieghere. 2011. The importance of expressing antimicrobial agents on water basis in growth/no growth interface models: A case study for *Zygosaccharomyces bailii*. *International Journal of Food Microbiology* 145:258–266.

Daryaei, H., J. Coventry, C. Versteeg, and F. Sherkat. 2010. Combined pH and high hydrostatic pressure effects on *Lactococcus* starter cultures and *Candida* spoilage yeasts in a fermented milk test system during cold storage. *Food Microbiology* 27:1051–1056.

Deák, T. 2008. Preservation: Inhibition and inactivation of yeasts. In *Handbook of Food Spoilage Yeasts*, ed. T. Deák, pp. 87–115. Boca Raton, FL: CRC Press/Taylor & Francis Group.

Deak, T. and L. R. Beuchat. 1996. *Handbook of Food Spoilage Yeasts*. Boca Raton, FL: CRC Press/Taylor & Francis Group.

Devlieghere, F., L. Vermeiren, and J. Debevere. 2004. New preservation technologies: Possibilities and limitations. *International Dairy Journal* 14:273–285.

Enrique, M., A. Ibañez, J. F. Marcos, M. Yuste, M. Martínez, S. Vallés, and P. Manzanares. 2010. β-Glucanases as a tool for the control of wine spoilage yeasts. *Journal of Food Science* 75:M41–M45.

Enrique, M., J. F. Marcos, M. Yuste, M. Martínez, S. Vallés, and P. Manzanares. 2007. Antimicrobial action of synthetic peptides towards wine spoilage yeasts. *International Journal of Food Microbiology* 118:318–325.

Evans, D. G., L. K. Everis, and G. D. Betts. 2004. Use of survival analysis and classification and regression trees to model the growth/no growth boundary of spoilage yeasts as affected by alcohol, pH, sucrose, sorbate and temperature. *International Journal of Food Microbiology* 92:55–67.

Fitzgerald, D. J., M. Stratford, and A. Narbad. 2003. Analysis of the inhibition of food spoilage yeasts by vanillin. *International Journal of Food Microbiology* 86:113–122.

Fleet, G. 1992. Spoilage yeasts. *Critical Reviews in Biotechnology* 12(112):1–44.

Gabriel, A. A. 2012. Inactivation of *Escherichia coli* O157:H7 and spoilage yeasts in germicidal UV-C-irradiated and heat-treated clear apple juice. *Food Control* 25:425–432.

Gibson, B. 1973. The effect of high sugar concentrations on the heat resistance of vegetative microorganisms. *Journal of Applied Bacteriology* 36:365–376.

Gliemmo, M. F. and C. A. Campos. 2013. Effect of the combined use of cinnamon oil and stevia on the growth rate of *Zygosaccharomyces bailii*. In *Worldwide Research Efforts in the Fighting against Microbial Pathogens: From Basic Research to Technological Developments*, ed. A. Mendez-Vilas, pp. 145–149. Boca Raton, FL: BrownWalker Press.

Gliemmo, M. F., C. A. Campos, and L. N. Gerschenson. 2004. Effect of sweet humectants on stability and antimicrobial action of sorbates. *Journal of Food Science* 69(2):39–44.

Gliemmo, M. F., C. A. Campos, and L. N. Gerschenson. 2006a. Effect of several humectants and potassium sorbate on the growth of *Zygosaccharomyces bailii* in model aqueous systems resembling low sugar products. *Journal of Food Engineering* 77:761–770.

Gliemmo, M. F., C. A. Campos, and L. N. Gerschenson. 2006b. Effect of sweet solutes and potassium sorbate on the thermal inactivation of *Z. bailii* in model aqueous systems. *Food Research International* 39:480–485.

Gliemmo, M. F., L. I. Schelegueda, L. N. Gerschenson, and C. A. Campos. 2013. Effect of aspartame and other additives on the growth and thermal inactivation of *Zygosaccharomyces bailii* in acidified aqueous systems. *Food Research International* 53:209–217.

Golden, D. A. and L. R. Beuchat. 1992. Interactive effects of solutes, potassium sorbate and incubation temperature on growth, heat resistance and tolerance to freezing of *Zygosaccharomyces rouxii*. *Journal of Applied Bacteriology* 73:524–530.

Gould, G. W. 1985. Present state of knowledge of water activity effects on microorganisms. In *Properties of Water in Foods in Relation to Quality and Stability*, eds. D. Simatos and J. L. Multon, pp. 229–245. Dordrecht, the Netherlands: Martinus Nijhoff Publishers.

Hocking, A. D. 2000. Media for preservative resistant yeasts: A collaborative study. *International Journal of Food Microbiology* 29:167–175.

Hracek, V. M., M. F. Gliemmo, and C. A. Campos. 2010. Effect of steviosides and system composition on stability and antimicrobial action of sorbates in acidified model aqueous systems. *Food Research International* 43:2171–2175.

Jenkins, P., P. G. Poulos, M. B. Cole, M. H. Vandeven, and J. D. Legan. 2000. The boundary for growth of *Zygosaccharomyces bailii* in acidified products described by models for time to growth and probability of growth. *Journal of Food Protection* 63:222–230.

Jermini, M. F. G. and W. Schmidt-Lorenz. 1987. Cardinal temperatures for growth of osmotolerant yeasts in broth at different water activity values. *Journal of Food Protection* 50:473–478.

Jespersen, L. and M. Jakobsen. 1996. Specific spoilage organisms in breweries and laboratory media for their detection. *International Journal of Food Microbiology* 33:139–155.

Kiskó, G., R. Sharp, and S. Roller. 2005. Chitosan inactivates spoilage yeasts but enhances survival of *Escherichia coli* O157:H7 in apple juice. *Journal of Applied Microbiology* 98(4):872–880.

Knutsen, A. K., V. Robert, G. A. Poot, W. Epping, M. Figge, and A. Holst-Jensen. 2007. Polyphasic re-examination of *Yarrowia lipolytica* strains and the description of three novel *Candida* species: *Candida oslonensis* sp. nov., *Candida alimentaria* sp. nov. and *Candida hollandica* sp. nov. *International Journal of Systematic and Evolutionary Microbiology* 57:2426–2435.

Leistner, L. 1995. Use of hurdle technology in food processing: Recent advances. In *Food Preservation by Moisture Control*, eds. G. V. Barbosa-Cánovas and J. Welti-Chanes, pp. 377–410. Lancaster, U.K.: Technomic Publishing Co.

Leistner, L. and N. J. Russell. 1991. Solutes and low water activity. In *Food Preservatives*, eds. N. J. Russell and G. W. Gould, pp. 111–134. New York: Van Nostrand Reinhold.

Lopez-Malo, A. and E. Palou. 2000. Modeling the growth/no growth interface of *Zygosaccharomyces bailii* in mango puree. *Journal of Food Science* 65:516–520.

Loureiro, V. and M. Malfeito-Ferreira. 2003. Spoilage yeasts in the wine industry. *International Journal of Food Microbiology* 86:23–50.

Loureiro, V. and A. Querol. 1999. The prevalence and control of spoilage yeasts in foods and beverages. *Trends in Food Science and Technology* 10:356–365.

Martorell, P., M. Stratford, H. Steels, M. T. Fernández-Espinar, and A. Querol. 2007. Physiological characterization of spoilage strains of *Zygosaccharomyces bailii* and *Zygosaccharomyces rouxii* isolated from high sugar environments. *International Journal of Food Microbiology* 114:234–242.

Marx, G., A. Moody, and D. Bermúdez-Aguirre. 2011. A comparative study on the structure of *Saccharomyces cerevisiae* under nonthermal technologies: High hydrostatic pressure, pulsed electric fields and thermo-sonication. *International Journal of Food Microbiology* 151:327–337.

Matan, N., H. Rimkeeree, A. J. Manson, P. Chompreda, V. Haruthaithanasan, and M. Parker. 2006. Antimicrobial activity of cinnamon and clove oils under modified atmosphere conditions. *International Journal of Food Microbiology* 107:180–185.

Mayoral, M. B., R. Martin, A. Sanz, P. E. Hernández, I. González, and T. García. 2005. Detection of *Kluyveromyces marxianus* and other spoilage yeasts in yoghurt using a PCR-culture technique. *International Journal of Food Microbiology* 105(1):27–34.

Mertens, L., E. Van Derlinden, T. D. T. Dang, A. M. Cappuyns, A. Vermeulen, J. Debevere, P. Moldenaers, F. Devlieghere, A. H. Geeraerd, and J. F. Van Impe. 2011. On the critical evaluation of growth/no growth assessment of *Zygosaccharomyces bailii* with optical density measurements: Liquid versus structured media. *Food Microbiology* 28:736–745.

Nielsen, D. S., T. Jacobsen, L. Jespersen, A. Granly Koch, and N. Arneborg. 2008. Occurrence and growth of yeasts in processed meat products—Implications for potential spoilage. *Meat Science* 80:919–926.

Pastorkova, E., T. Zakova, P. Landa, J. Novakova, J. Vadlejch, and L. Kokoska. 2013. Growth inhibitory effect of grape phenolics against wine spoilage yeasts and acetic acid bacteria. *International Journal of Food Microbiology* 161:209–213.

Praphailong, W. and G. H. Fleet. 1997. The effect of pH, sodium chloride, sucrose, sorbate and benzoate on the growth of food spoilage yeasts. *Food Microbiology* 14:459–468.

Puértolas, E., N. López, S. Condón, J. Raso, and I. Álvarez. 2009. Pulsed electric fields inactivation of wine spoilage yeast and bacteria. *International Journal of Food Microbiology* 130:49–55.

Qin, B., U. R. Pothakamury, G. V. Barbosa-Canovas, and B. G. Swanson. 1996. Nonthermal pasteurization of liquid foods using high-intensity pulsed electric fields. *Critical Reviews of Food Science and Nutrition* 36:603–607.

Raso J., M. L. Calderón, M. Góngora, G. V. Barbosa-Cánovas, and B. G. Swanson. 1998. Inactivation of *Zygosaccharomyces bailii* in fruit juices by heat, high hydrostatic pressure and pulsed electric fields. *Journal of Food Science* 63:1042–1044.

Renard, A., P. Gómez di Marco, M. Egea-Cortines, and J. Weiss. 2008. Application of whole genome amplification and quantitative PCR for detection and quantification of spoilage yeasts in orange juice. *International Journal of Food Microbiology* 126:195–201.

Renouf, V. and A. Lonvaud-Funel. 2007. Development of an enrichment medium to detect *Dekkera/Brettanomyces bruxellensis*, a spoilage wine yeast, on the surface of grape berries. *Microbiological Research* 162:154–167.

Reyns, K. M. F. A., C. C. F. Soontjens, K. Cornelis, C. A. Weemaes, M. E. Hendrickx, and C. W. Michiels. 2000. Kinetic analysis and modelling of combined high-pressure-temperature inactivation of the yeast *Zygosaccharomyces bailii*. *International Journal of Food Microbiology* 56:199–210.

Scott-Thomas, C. 2013. Stevia sees exponential growth in Europe. http://www.foodnavigator.com/Financial-Industry/Stevia-sees-exponential-growth-in-Europe (accessed December 6, 2013).

Serpaggi, V., F. Remize, G. Recorbet, E. Gaudot-Dumas, A. Sequeira-Le Grand, and H. Alexandre. 2012. Characterization of the "viable but nonculturable" (VBNC) state in the wine spoilage yeast *Brettanomyces*. *Food Microbiology* 30:438–447.

Smittle, R. B. and M. C. Cirigliano. 1992. Salad dressings. In *Compendium of Methods for the Microbiological Examination of Foods*, eds. C. Vanderzant and D. F. Splittstoesser, pp. 975–983. Washington, DC: American Public Health Association.

Smittle, R. B. and R. S. Flowers. 1982. Acid tolerant microorganisms involved in the spoilage of salad dressings. *Journal of Food Protection* 45:977–983.

Sofos, J. N. 2000. Sorbic acid. In *Natural Food Antimicrobial Systems*, ed. A. S. Naidú, pp. 721–749. Boca Raton, FL: CRC Press.

Suh, H. J., Y. H. Cho, M. S. Chung, and B. K. Kim. 2007. Preliminary data on sulphite intake from the Korean diet. *Journal of Food Composition and Analysis* 20:212–219.

Tchango, J. T., R. Tailliez, P. Eb, T. Njine, and J. P. Hornez. 1997. Heat resistance of the spoilage yeasts *Candida pelliculosa* and *Kloeckera apis* and pasteurization values for some tropical fruit juices and nectars. *Food Microbiology* 14:93–99.

Thomas, D. S. and R. R. Davenport. 1985. *Zygosaccharomyces bailii*—A profile of characteristics and spoilage activities. *Food Microbiology* 2:157–169.

Tournass, V. H., J. Heeres, and L. Burgess. 2006. Moulds and yeasts in fruit salads and fruit juices. *Food Microbiology* 23:684–688.

Truong-Meyer, X. M., P. Strehaiano, and J. P. Riba. 1997. Thermal inactivation of two yeast strains heated in a strawberry product: Experimental and kinetic model. *Chemical Engineering Journal* 65:99–104.

Tsapatsaris, S. and P. Kotzekidou. 2004. Application of central composite design and response surface methodology to the fermentation of olive juice by *Lactobacillus plantarum* and *Debaryomyces hansenii*. *International Journal of Food Microbiology* 95:157–168.

Tyagia, A. K., D. Gottardi, A. Malika, and M. E. Guerzoni. 2013. Anti-yeast activity of mentha oil and vapours through in vitro and in vivo (real fruit juices) assays. *Food Chemistry* 137:108–114.

Vermeulen, A., T. D. T. Dang, A. H. Geeraerd, K. Bernaerts, J. Debevere, J. Van Impe, and F. Devlieghere. 2008. Modelling the unexpected effect of acetic and lactic acid in combination with pH and aw on the growth/no growth interface of *Zygosaccharomyces bailii*. *International Journal of Food Microbiology* 124:79–90.

Waite, J. G., J. M. Jones, and A. E. Yousef. 2009. Isolation and identification of spoilage microorganisms using food-based media combined with rDNA sequencing: Ranch dressing as a model food. *Food Microbiology* 26:235–239.

Wilson, P. D. G., T. F. Brocklehurst, S. Arino, D. Thualt, M. Jakobsen, M. Lange, J. Farkas, J. W. T. Wimpenny, and J. F. Van Impe. 2002. Modelling microbial growth in structured foods: Towards a unified approach. *International Journal of Food Microbiology* 73:275–289.

Index